S0-CFM-185

THE SCIENCE OF ECOLOGY

THE SCIENCE OF ECOLOGY

Paul R. Ehrlich
STANFORD UNIVERSITY

Jonathan Roughgarden
STANFORD UNIVERSITY

MACMILLAN PUBLISHING COMPANY
New York

COLLIER MACMILLAN PUBLISHERS
London

Printed in the United States of America

Macmillan Publishing Company
866 Third Avenue, New York, New York 10022

Collier Macmillan Canada, Inc.

Library of Congress Cataloging-in-Publication Data

Ehrlich, Paul R.
The science of ecology.

Bibliography: p.
Includes index.
1. Ecology. I. Roughgarden, Jonathan. II. Title.
QH541.E353 1987 574.5 87-1660
ISBN 0-02-331700-0

Printing: 3 4 5 6 7 8 Year: 8 9 0 1 2 3 4 5 6

ISBN 0-02-331700-0

To ANNE, LISA, TRUDY, and TIMOTHY

PREFACE

What Ecology Is About

Ecology is about the living beings of Earth and how they interact with one another and with their nonliving environments. It is about complexity—a complexity that seems overwhelming: billions of years of evolution have produced as many as 30 *million* different kinds of organisms living today, all responding to other life forms and to a bewildering array of geologic, climatic, and other physical environments. But ecology is also about finitude, which often is just as startling: Many ecological communities are small; a few hours' assault with heavy earth-moving equipment may destroy them forever.

We, too, depend on these complex and finite systems of organisms. Their destruction removes important benefits to humanity, such as stabilization of the Earth's climate and the potential, now barely tapped, to supply us with foods and medicines, as well as esthetic delights. In the tropics—where most of the world's species are concentrated and where the human population is growing most rapidly—habitat destruction is already fast eliminating many animal and plant communities.

The science of ecology—which is charged with understanding this complexity and finitude—is thus concerned with subjects not only of great scientific interest but also of immense practical importance. From ecology one gains an appreciation of how the world works and of humanity's participation in, and dependence upon, natural systems. The scope of ecology is enormous, but even though the task of ecologists seems daunting, substantial progress has been made, as we hope to show in this book, and each day brings more discoveries.

The Purpose of This Book

This text is intended for those taking a one-semester upper-division course in ecology. We assume that readers will have had an introductory college course in biology and are familiar with Mendelian genetics as well as basic botany and zoology. Some of the material is expressed in mathematical terms, and a solid knowledge of high school algebra should prove sufficient background, although a few cases use very elementary calculus. Also, elementary statistics are used in several places.*

Because of the scope of the discipline, *The Science of Ecology* approaches the material with four principles in mind:

1. *We have presented a broad overview useful to many teaching approaches.* The rapid progress in ecology during the past decade has produced a diversity of opinion about what the "essence" of ecology really *is*. Today, the content of ecology courses is quite variable. Most courses overlap in covering a few classical topics, such as exponential growth, basic demography, a survey of a few kinds of population interactions, succession, and energy flow. However, the courses differ in their emphasis on theory, focus on controversy, the presentation of an evolutionary perspective, and the selection of examples. Thus, we have presented what we think is a very broad overview of ecology in the hope that it will be useful in most courses, with instructors choosing which parts to assign and providing their own emphasis in lecture.

2. *The material presented is proportional to its representation in recent research literature.* We have attempted to present material more or less in proportion to its coverage in recent research papers. In this regard, we are reporters, not judges. For example, much of the research in community ecology has concerned competition, even though this is only one of several important processes that influence the composition of communities. Hence, our coverage of competition is more extensive than that of other processes, which in many circumstances are equally or more important.

3. *We offer a balanced view of ecological models.* In the past decade, most of the controversy in ecology has been about the use of simple mathematical models. Almost all of the first models to be introduced have met the same fate: They have been successful in limited contexts and disappointing when applied beyond such contexts. Our approach has been to emphasize the positive while mentioning the limitations of the early models. Our hope has been to strike a balance between optimism and caution.

4. *We emphasize the interplay of various ecological approaches.* Our book stresses the interaction among theory, laboratory investigation, and field observation and experiment, an interplay that we believe has been central to the explosive advances in ecology during the last two decades.

* If you are not familiar with statistics, just remember that the notations $P < 0.05$ or $P < 0.01$ mean simply that the probability (chance) of the result being accidental (rather than representing a real difference or being caused by a real process) is less than 1 in 20 or 1 in 100, respectively. To see how statistical tests are performed in biology, consult the classic text by Sokal and Rohlf (1981).

How This Book Is Organized

The Science of Ecology is organized into seven parts, whose subject matter progresses in scale from small to large—from the physiology of individuals, the growth of populations, and social interactions within populations to species interactions, the distribution of communities, and the functioning of ecosystems.

Part 1, "The Individual and the Environment," deals with the relationship of an individual microorganism, plant, or animal to its physical environment and resources. Organisms are not inert or static; they constantly change in response to variations in their surroundings, and they exchange energy and materials with those surroundings. This dynamic relationship of individuals to their environment is the province of the subdiscipline of ecology known as *physiological ecology*.

Part 2, "Populations," turns to population ecology, the study of one kind or species of organism, with the objective of explaining the distribution and abundance of that kind of organism. In addition, at this level, we may also see how the genetic makeup of populations changes through time—that is, how they evolve. Evolutionary thinking underlies much of modern ecology. Also the proportions of different genes within populations change in response to ecological factors. In this part, we also explore how the size of a population affects its evolution and how natural selection influences features of the life history such as life expectancy. Throughout the book, we repeatedly ask how the ecological situation under discussion evolved, a question that traces to Charles Darwin.

Part 3, "Social Interactions," is concerned with behavioral ecology, which deals with interactions among individuals in the same population—the need for mates, the occupation of territories, and the like. For most organisms, other individuals of the same species are a critical element in the environment, partly because of the ubiquity of sex, whose presence and persistence is one of the enduring mysteries of ecology. Here we consider some possible reasons why organisms find it advantageous to reproduce sexually instead of vegetatively, by budding, or by some other asexual method. We also discuss the formation of groups that facilitate feeding and group defense and describe how the environment shapes these behaviors and communication within the groups.

Part 4, "Population Interactions," moves on to a look at a higher degree of complexity, community ecology—how different species interact with each other. Here we are concerned with predation, cooperation, and competition and also with how these interactions shape the joint evolution (coevolution) of the actors. Our discussion of community ecology begins here with a look at basic interactions between species.

Part 5, "Organization of Communities," goes beyond Part 4 by considering the cumulative effects of all the interactions among many species. It looks mostly at pieces of communities, groups of populations called *guilds*, that have roughly similar activities within a community. It also looks at the entire ecological community when considering what determines the total number of species in a community and how they are connected to one other in a *food web*.

Part 6, "The Distribution of Communities," describes the distribution of communities in space and time. These include marine communities including coral reefs, kelp forests, the intertidal zone, and the open ocean itself. On the continents, such communities include deserts, Mediterranean grasslands, chaparral, temperate, montane and tropical forests. This part also considers the ecology of the past—how the changing physical world and biotic communities have interacted. It uses the ecology of dinosaurs to illustrate both the interest in and the difficulties of reconstructing the past.

Part 7, "Ecosystems," is concerned with the sum of all the interactions of species within a community with each other and with their physical environments. A major goal of ecosystem ecology is to describe the total flows of energy and materials through an ecosystem and to draw conclusions about how they might change through time. Recent research has shown how natural ecosystems provide humanity with many indispensible "public services"—regulating the quality of the atmosphere, ameliorating the weather, maintaining the soils and nutrients essential to agriculture, controlling pests and disease carriers, and maintaining a vast genetic "library" of possible new foods, medicines, and industrial products. In this final part of the book, we explore the biogeochemical cycles—the large-scale cycles of water, nutrients, and energy that are essential to life—and show how our understanding these cycles ultimately may help to determine whether we can continue to live in the world. We also look at some ways in which humanity is perturbing these cycles, and in and the process putting its own future at risk.

Acknowledgments

We wish to thank our Stanford colleagues David Dobkin, Anne Ehrlich, Christopher Field, Harold Mooney, Dennis Murphy, Peter Vitousek, and Bruce Wilcox for reading and criticizing parts of the manuscript. Field and Vitousek did especially detailed reviews of Parts 1 and 7, respectively, and we are very grateful to them. Other ecologists reviewed the text for Macmillan at our request: James Ehleringer, University of Utah; Frank Golley, University of Georgia; Nancy Knowlton, Yale University; Richard Michod, University of Arizona; Peter Morin, Rutgers University; Donald Strong, Florida State University; and C. Richard Tracy, Colorado State University. All of them made extremely helpful comments. Since we have not followed all of the advice given us, the responsibility for any errors of omission or commission rests solely with us. In the end, readers should recognize that this book expresses our personal vision of ecology. The book is not a "definition" of ecology, and other ecologists legitimately may see the subject in a quite different light.

Brian K. Williams of Montara, California, did a superb job of final editing and coordination on this extremely ambitious project. Without his herculean efforts, the book might never have been completed, and we are deeply appreciative. Darryl Wheye was extraordinarily helpful with the preparation of the manuscript, laboring both at entering corrections and at cleaning up infelicities. She also played an important role in the preparation of the bibliography.

Many people were helpful in supplying material for illustrations; they are acknowledged in the captions. Carl May of the Biological Photo Service found many photographs and in so doing made a major contribution to the illustration program.

The staff of Stanford University's superb Falconer Biology Library, under the direction of Beth Weil, has been extremely helpful to this enterprise. Beth, Claire Shoens, Judy Levitt, Zoe Chandik, and Joan Dietrich have struggled to run down obscure references and obtain crucial material on interlibrary loans. Steven Masley and Pat Browne handled a mass of photocopying with care and dispatch.

Gregory Payne and Bob Rogers, our editors at Macmillan, have given us moral support throughout the project. Cyndie Clark-Huegel, Wayne S. Clark, Carla J. Simmons, Karen J. English-Loeb, Marjorie C. Leggitt, D. J. Simison, and Vantage Art, Inc. worked patiently with the complicated art program. Our production supervisor, Dora Rizzuto, did an important job skillfully and patiently.

Above all, we are very much in the debt of our many fellow ecologists, without whose research contributions we would have had nothing to write about.

Our work in ecology has been funded by the National Science Foundation (NSF), the Department of Energy, and the Koret Foundation of San Francisco. Their support is deeply appreciated.

The order of the authors' names on the title page is alphabetical, and, for better or worse, the book is a product of true co-authorship. Although we divided up the task of writing first drafts of chapters, each of us has worked extensively on all of them, and the result represents our joint view of the field as it stands today. The task of putting together the book has been extraordinarily demanding. However, it has also been rewarding and exciting, and we hope that we have been able to convey this to you.

P.R.E. and J.R.
Stanford, California

BRIEF CONTENTS

CONTENTS

PART 1

THE INDIVIDUAL AND THE ENVIRONMENT

The world is endowed with great irregularity. Continents of varied shapes and textures are separated by oceans of different depths and salinities. Weather systems, driven by the sun and shaped by the spinning earth, sweep over land and sea. As days turn into centuries and into eons, mountain ranges wear away, lakes dry up, land masses re-form. From these billions of years of changes in the earth's geology and climate have evolved millions of varied microorganisms, plants, and animals—different organisms in different areas, different communities that support our human community. As we shall see in this book, whether human beings can continue to live in the world depends on how well we can learn to understand, and perhaps ultimately to manage, these communities of organisms in their ever-changing physical environments. Trying to achieve this understanding is the task of ecology.

Our formal exploration begins with what is, in ecology, the simplest level of organization, that of the individual, which provides us with an essential building block for understanding phenomena at more complex levels of organization—namely, the population, community, and ecosystem levels. In the four chapters of Part 1, we describe the relationship of the individual microorganism, plant, or animal to its physical environment.

CHAPTER 1

The Beginning: An Introduction to Physiological Ecology

Every type of organism is unique. How, then, with as many as 30 million *kinds* of organisms in the world—all the plants and animals above ground, below ground, in oceans, lakes, mud, and on rocks—can we make sense of such variety? The answer is that we must seek principles that apply to most or all organisms, changing the focus where necessary for particular cases.

Ecology—the study of the relationship between organisms and their physical and biological environments—is primarily concerned with groups of organisms, such as populations, species, or communities. It is concerned with how many organisms there are, where they are, and what they do. Yet ecology begins with the understanding of how a single individual lives—what it does to obtain food, to survive in the physical environment, and to reproduce. Thus, we start with *physiological ecology*, the subdiscipline of ecology that is concerned with the dynamic relationship of individuals to their physical environments and resources. This subdiscipline is the focus of this first part of the book.

THE EXCHANGE BETWEEN ORGANISMS AND THEIR ENVIRONMENT

Organisms are not inert or static, like plastic plants. They constantly change in response to variations in their surroundings, and they exchange energy and materials with those surroundings. Thus, the state of an organism is influenced by its surroundings and, conversely, an organism can change those surroundings.

This important concept, that organisms exchange energy and matter with their surroundings—and affect and are affected by them—is true for both plants and animals. Plants can be viewed as rather decentralized energy receivers and regulators of coupled flows of gases, water, and nutrients. Animals are more centralized and tend to be seekers of energy and nutrients; they also regulate coupled flows of gases and water and, in many cases, carefully adjust their temperature to maintain their mobility. Physiological ecologists study these relationships between individuals and their environments by making measurements under controlled conditions, sometimes even using mobile laboratories in the field.

WHAT IS AN INDIVIDUAL?

From an ecological standpoint, it is not easy to show what an individual actually is. In biology, an individual is usually thought to be either (1) an independently living single cell or (2) a group of cells physically attached to one another that have descended from a single cell.

However, this usual concept of an individual is too restrictive to apply to many organisms. For instance, as Figure 1–1 shows, some plants such as beach grass (*Amophila breviligulata*) send out underground runners from which stems grow, each of which forms what looks like, from above the ground, an individual plant. Indeed, some forest stands of birch or aspen can be a single plant with many parts, as illustrated in Figure 1–2. Invertebrates such as sea anemones, corals (see Figure 1–3), hydroids, tunicates, and sponges often form large colonies through budding. Because great variation exists among species in the extent to which members of a colony are functionally independent of other members of the same colony, the exact meaning of an individual depends on the situation.

For sexually reproducing vascular plants, it is useful to distinguish between two kinds of individuals: the ecologically distinct and the genetically distinct (Harper 1977). A *ramet* is an ecological unit; it is the entity that is noticed in the field as an individual, such as a shoot from an underground runner, as in the beach grass of Figure 1–1; sprouts, as in Figure 1–4; or a plant grown from a cutting, as in grape vines. A ramet is an

FIGURE 1–1 Beach grass growing from a runner *(Amophila breviligulata). (Photo by J. Roughgarden.)*

ecological individual because it is a unit that is largely autonomous in its use of resources. A *genet* is a genetic individual; it is all the tissue that grows from a single fertilized egg. A genet may encompass many ramets, including ramets that are no longer connected to one another. The whole aspen stand shown in Figure 1–2 is a genet.

A colony of paper wasps is, to some extent, a genetic individual because there is one female, called the *queen*, that lays the eggs, while other wasps, all of which are her daughters, provision the nest. The individual wasps, other than the queen, might be considered ramets because each is responsible for gathering its own resources.

In the next three chapters, we will concentrate on individuals that are autonomous in their use of resources. The genetic relationship of such

FIGURE 1–2 Example of a genet: stand of aspen, Snake Range, Nevada. *(Photo by C. W. May/BPS.)*

FIGURE 1–3 (A) Sea anemone. Rocks covered with the anemone *Anthopleura elegantissima,* McClure's Beach, Point Reyes, California. *(Photo © 1981 by David J. Cross/BPS.)* (B) Polys of a gorgonian from St. John (U.S. Virgin Islands). *(Photo by J. Roughgarden.)*

individuals to one another will become important in Parts 2 and 3 when we study the evolution of social interactions.

WHAT IS THE ENVIRONMENT?

The environment of an organism can be described as everything that is outside the organism, but this definition may encompass too much to be meaningful. Because it is usually described only in terms of those features that affect the organism, let us formally define the *environment* as all the elements in an organism's surroundings that can influence its behavior, reproduction or survival.

It is traditional to distinguish between abiotic and biotic features of the environment. An *abiotic* feature is a physical characteristic of the place

FIGURE 1–4 Example of a ramet. Each sprout from this log in a hemlock forest constitutes a ramet. *(Photo by B. J. Miller, Fairfax, Virginia/BPS.)*

in which the organism lives, such as the temperature of the air or water. A *biotic* feature is another organism, such as a predator, competitor, or mate. As you might imagine, the distinction loses its usefulness when organisms affect one another primarily by modifying the physical characteristics of the habitat.

A *resource* is an object or area in the environment that is consumed or used up by a living organism. Food and space are resources; temperature is not. (However, a place for an animal to sit in that has a certain temperature may be a resource.) Several kinds of resources may be limiting at the same time. A *renewable* resource is one that is replenished; a *nonrenewable* resource is used up through time.

An organism's *habitat* is the place, or neighborhood, in which it usually lives. Its *microhabitat* is the kind of spot where it is usually found within its habitat. For example, near Stanford University in northern California, the habitat of the caterpillar of the checkerspot butterfly, *Euphydryas chalcedona*, is the California chaparral; its microhabitat is the leaves of two kinds of shrubs, *Diplacus aurantiacus* and *Scrophularia californica*.

An organism's *niche* is the way in which it obtains resources—its "occupation" as contrasted to its "address" or habitat. For example, some birds, such as flycatchers and swallows, catch insects in the air. Other birds, such as many warblers, mostly glean bugs from leaves and branches, and still others, such as quail, spend much of their time scouring the ground. Each of these styles of obtaining food represents a niche.

Organisms that derive their energy directly from the physical environment, such as from sunlight or inorganic chemical reactions, are called *autotrophs*. The most common are organisms that photosynthesize. However, some bacteria obtain energy from inorganic reactions such as converting NH_4^+ to NO_3^- and H_2S to SO_4^{--}. Organisms that derive their energy from the breakdown of compounds that were previously manufactured by autotrophs are called *heterotrophs*. Fungi, many bacteria, and, of course, all animals are in this category.

STAYING ALIVE OUTDOORS

We now have the basic context with which to approach the study of how organisms stay alive outdoors. In the next three chapters, we will examine how organisms are influenced by their surroundings and in turn change those surroundings.

In Chapter 2, we explain how a plant works—how the flows of carbon dioxide and water are inextricably coupled and how those flows also may be coupled to the uptake of nutrients from the soil. As we will show, a plant may not be able to move much—except slightly through its own growth, as when roots reach toward water—but while it doesn't move much itself, it does cause or control the movement of many materials in its environment. In that chapter, we also discuss water stress, although water availability also affects animals as well as plants.

In Chapter 3, we describe how an animal works—and how the animal's ability to move affects critically both itself and the environment. Because the ability to move depends on body temperature, we also describe how animals regulate their body temperature, although temperature regulation also may be carried out by plants. In addition, we consider how the animal's requirement for certain nutrients, plus its ability to choose several food types, affects how the animal "decides" what and where to eat.

In the final chapter in this part, Chapter 4, we examine how the individual organism is affected by physical forces in the environment—wind, waves, and currents. We also discuss the kinds of materials organisms are made of and some of the principles of mechanical design, both of which determine how organisms function in the presence of these physical forces.

CHAPTER 2

How a Plant Works: The Coupling of Water, Carbon Dioxide, Sunlight, and Nutrients

Let us begin our investigation of the interaction between an organism and the environment by thinking of a plant with three compartments: a root system, a central stem, and a canopy in which photosynthesis occurs. The basis for how this plant works is an inextricable coupling between the flows of water and carbon dioxide (CO_2).

THE WATER–CO_2 CONNECTION

A plant is continually exchanging CO_2 with the environment through tiny holes called stomates, whose sizes can be adjusted. *Stomates* are openings, usually found on the underside of leaves, through which gases enter and leave. A plant is always respiring—burning some of its stored carbohydrates—and thereby releasing CO_2. Furthermore, if enough light exists, the plant is also *photosynthesizing*, taking up CO_2 and converting it to carbohydrates. The net flow of CO_2 through the stomates may be positive or negative, depending on the relative rates of respiration and photosynthesis. The light level at which photosynthesis exactly balances respiration is called the *light compensation point*. If the light level is below the compensation point, then the net inflow of CO_2 is negative (more CO_2 is leaving the leaf through the stomates than entering). If the light level is higher than the compensation point, then the net inflow of CO_2 is positive.

At the same time that CO_2 is being exchanged through the stomates, liquid water is taken up by the roots, moves up through the stem, and then, as water vapor, passes out into the air through the stomates. Now we see the connection between CO_2 and water: Because both water and CO_2 flow through the same pores, the control of both of these processes is automatically coupled. However, this connection creates a basic dilemma: During the day, the plant cannot reduce its water loss through the stomates without simultaneously cutting back on its supply of CO_2 for photosynthesis.

Water and Structural Rigidity: Turgor Pressure

If water is limited, why should a plant release it at all? Why not hold on to it in some sort of impermeable container, like a cistern, and then draw from it when water is needed for internal biochemical functions? A major reason is that plants use water to obtain structural rigidity. Whereas animals rely on bones and shells made of hard materials when they need such rigidity, a plant (we may say with only slight exaggeration) uses water as part of its skeleton.

A plant cell is surrounded by a cell wall composed largely of cellulose, lignin, and a variety of small polysaccharides that bind the cellulose fibers into a tough, flexible net. Though not very elastic, this cell wall is quite porous and freely permeable to water and salts. Within the cell wall, a cell membrane also surrounds the cell, and it too is permeable to water, though not to salts.

The contents of the cell may be viewed as a water solution containing a high concentration of dissolved molecules. The water surrounding the cell usually has a lower concentration of dissolved materials than does the interior of the cell. Hence, because of osmosis, water tends to move into the cell from the outside, causing the cell to expand and push against the cell wall. This pressure of the cell against its cell wall, called *turgor pressure*, is what gives the nonwoody parts of the plant most of their rigidity. If the plant loses water from the extracellular matrix, eventually the direction of flow reverses, water moves from inside the cell to the outside, the turgor pressure drops to zero, and the plant wilts.

Thus, to maintain its turgor pressure, a plant must keep water in the extracellular matrix. The stomates are the connections between the external environment and this extracellular matrix. Without the stomates, the cells cannot get the CO_2 for photosynthesis, yet the instant they are opened, water vapor begins to exit as the CO_2 enters. Hence, there is no escape from the dilemma of water loss pitted against CO_2 gain.

Water Flow and Nutrients

All this water loss is not completely disadvantageous to a plant, for the water flow transports dissolved nutrients from the plant's roots to its leaves, including the nitrogen used in amino acids and proteins. It appears, however, that only a low rate of *transpiration*—loss of water vapor, primarily through the stomates—is required to provide the needed nutrients (Schulze and Bloom 1984).

A large fraction of the nitrogen in a leaf is used in the enzymes of photosynthesis (von Caemmerer and Farquhar 1981). The overall photosynthesis rate per unit of leaf is directly proportional to the nitrogen content in the leaf. Figure 2–1 illustrates this important relation for plants

FIGURE 2–1 Net photosynthesis (in nanomoles of carbon dioxide fixed per gram of leaf per second) as a function of the nitrogen content of the leaf (in millimoles per gram of leaf). *(From Field and Mooney 1982; by permission of the authors.)*

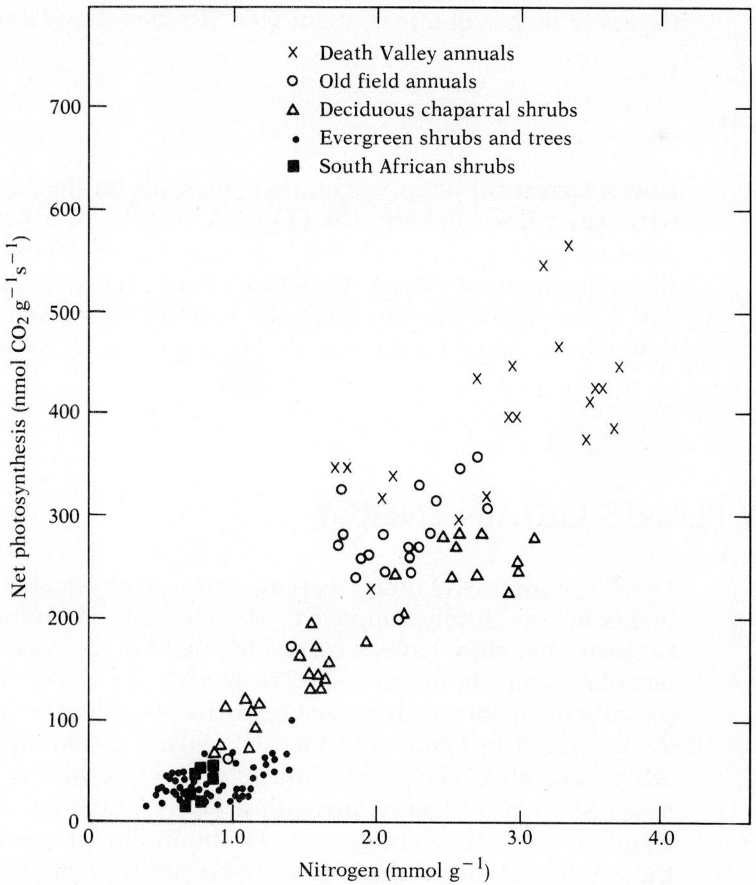

from several places: the desert at Death Valley, California; an old field in Illinois; and shrubs in South Africa (Field and Mooney 1985).

Nitrogen may be transported as NO_3^-, NH_4^+, or in amino acids such as glutamate. Similarly, phosphate is used in important compounds such as adenosine triphosphate (ATP), the phosphorylated sugars of the photosynthetic and glycolytic pathways, and DNA. It is transported as HPO_4^{--} or $H_2PO_4^-$, depending on the pH of the soil and sap. If a plant shuts down water flow, it also shuts down the flow of these nutrients.

Water and Cooling

The water flow also cools the plant as the water evaporates before leaving through the stomates. Desert plants, such as barrel cacti, which keep their stomates closed during the day to avoid water loss, have leaf temperatures 10–20°C above the air temperature. The highest nonlethal temperature for tissues of higher plants is about 60°C (Didden-Zopfy and Nobel 1982).

In Sum

How a terrestrial plant works, then, depends on the coupling of CO_2 flow with water flow. In turn, the CO_2 flow results from both photosynthesis and respiration. The evaporation of water from the extracellular matrix, where it maintains turgor pressure, causes the water flow. Both the CO_2 and water vapor pass through the stomates simultaneously. The water flow moves dissolved nutrients from the roots to the leaves. The evaporation of water also cools the leaves.

HOW PLANTS OBTAIN ENERGY

Plants are successful under a great variety of physical conditions. Because plants are so closely coupled to the physical environment, this success suggests that they have a correspondingly great variety of biochemical techniques for photosynthesis. The overall performance of a plant's photosynthetic apparatus is described by its photosynthetic action spectrum, as illustrated in Figure 2–2 for two seaweeds, one a green alga and the other a red alga. The *photosynthetic action spectrum* is simply a graph of how fast 1 cm^2 of leaf photosynthesizes as a function of the color (wavelength) of the light shining on it. To obtain the information illustrated by the graph, the color of the light is experimentally varied while its intensity is held constant.

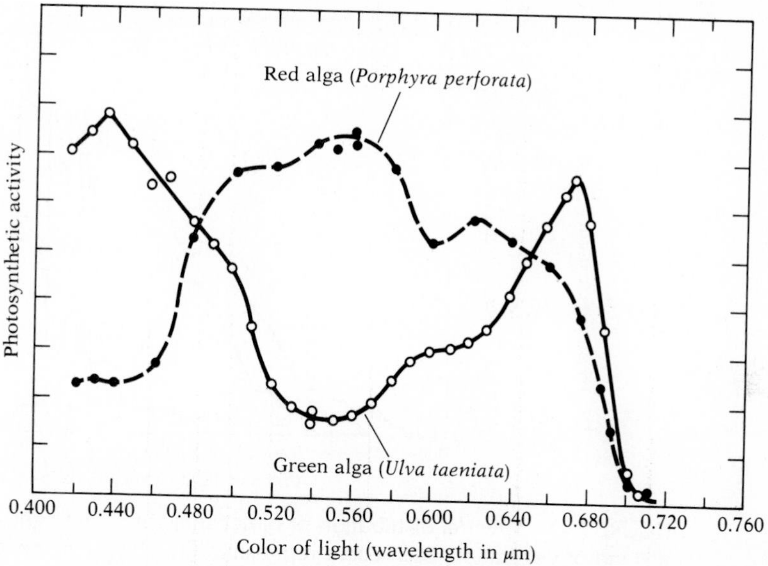

FIGURE 2–2 Action spectra of green and red algae. Wavelengths in micrometers (a micrometer equals one thousand nanometers); see Figure 2–3 for colors. *(After Haxo and Blinks 1950.)*

Most plants on land use light whose wavelength varies from 400 nm (blue) to 700 nm (red), colors visible to humans. (The unit, nm, is a nanometer, one-billionth of a meter (m); 1 nm = 10^{-9} m.) The majority of land plants are most efficient with red light, followed by blue light and then by green and yellow light. The colors at which the efficiency peaks coincide with the colors that the pigments, consisting of several kinds of chlorophyll and carotenoids, absorb best. A leaf's color is green because this color is reflected, not absorbed; this fact corresponds to a leaf's relatively low photosynthetic efficiency with green and yellow light.

Despite variations in efficiency with color, plants do photosynthesize with all the colors between about 400 and 700 nm. Nonetheless, small variations in efficiency may affect the success of plants in nature because often they are not exposed to direct sunlight, whose spectrum is illustrated in Figure 2–3. Instead, light is often filtered through the leaves of taller plants, making the light more yellow and green than direct sunlight. An example of the spectrum of light that has been filtered through the canopy also appears in Figure 2–3 and is illustrated in color plate A in the color section of this book.

Under water, the color and brightness of light vary even more dramatically than they do on land, as color plate B illustrates. Water filters out red light, leaving just blue and green as little as a meter or so below the surface. Algae show much more variation in their pigments than do land plants, and many of these pigments are active in the photosynthetic

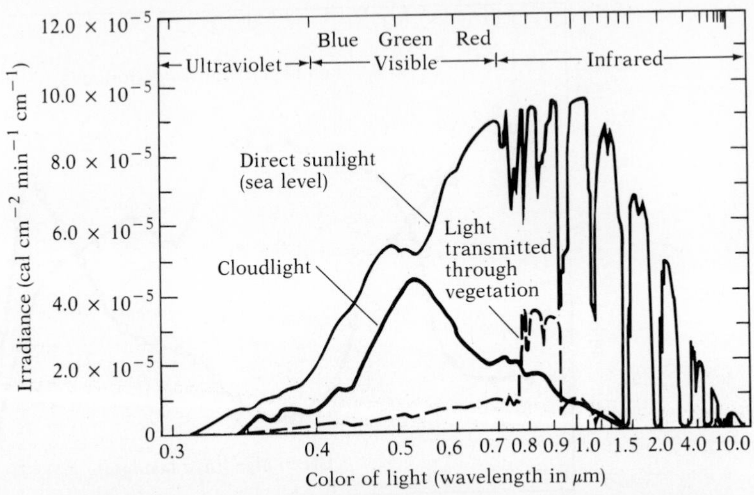

FIGURE 2–3 Spectral distribution of direct sunlight, cloud light, and light penetrating a stand of vegetation. Each curve represents the energy incident on a horizontal surface. Wavelengths in micrometers (a micrometer equals one thousand nanometers). *(After D. Gates 1965.)*

process, as previously shown in Figure 2–2. Indeed, some red algae actually photosynthesize most efficiently in green light (Haxo and Blinks 1950).

Photosynthesis can be divided into two parts. In the first part, light absorbed by chlorophyll leads to the synthesis of both ATP and NADPH (the reduced form of nicotinamide adenine dinucleotide phosphate), and to the release of O_2. ATP stores energy in phosphate bonds, and NADPH stores energy as the hydrogen atom, H, needed to reduce other compounds. This activity is called the *light reaction* of photosynthesis. In the second part, the ATP and NADPH are used to capture CO_2 and to incorporate it into a sugar. This second part is called the *dark reaction* of photosynthesis because it can occur in the dark, provided that ATP and NADPH are available. In the last decade, research has shown that the dark reaction of photosynthesis is carried out in a variety of ways among plants. Here we will review three biochemical pathways for the dark reaction. For more detail, see Salisbury and Ross (1978).

The C_3 Pathway

The capture of CO_2 and its incorporation into a sugar is called *carbon fixation*. The first known biochemical pathway for fixing carbon is called the C_3 pathway (Calvin and Bassham 1962). The defining feature of this pathway is that a phosphorylated five-carbon sugar (ribulose diphosphate)

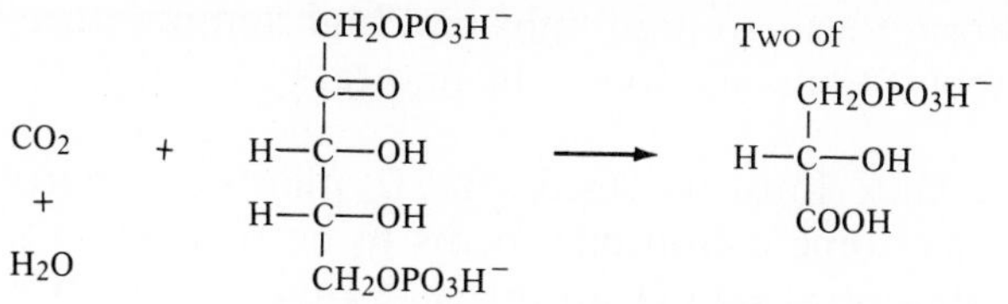

$$CO_2 + H_2O + \begin{array}{c} CH_2OPO_3H^- \\ | \\ C{=}O \\ | \\ H{-}C{-}OH \\ | \\ H{-}C{-}OH \\ | \\ CH_2OPO_3H^- \end{array} \longrightarrow \text{Two of } \begin{array}{c} CH_2OPO_3H^- \\ | \\ H{-}C{-}OH \\ | \\ COOH \end{array}$$

FIGURE 2–4 C_3 carbon-fixing reaction, with structures of molecules.

combines with one CO_2 molecule (and one H_2O molecule) to produce two three-carbon molecules (3-phosphoglycerate or 3-PGA), as illustrated in Figure 2–4. These three-carbon molecules then are routed into other pathways to make starch and sugars and to regenerate the original five-carbon sugar that united with the CO_2 molecule to begin with. During the regeneration of the five-carbon molecule, the ATP and NADPH made in the light reactions are used. This pathway is called the *C_3 Pathway* because the product of the carbon fixation is a molecule with three carbon atoms.

The C_4 Pathway

In the mid-1960s, another pathway for fixing carbon was discovered in two crop plants, sugar cane and corn (Hatch and Slack 1970). In this pathway, CO_2 (in the form of carbonic acid, HCO_3^-) combines with a phosphorylated three-carbon molecule (phosphoenolpyruvate or PEP) to yield a four-carbon molecule (oxaloacetate or OAA), as illustrated in Figure 2–5. This pathway is called the *C_4 pathway* because the product of carbon fixation is a molecule with four carbon atoms. Surprisingly, this four-carbon molecule is not routed immediately into other biochemical pathways. Instead, it moves into another type of cell. Here it releases its CO_2, which is then refixed by the C_3 pathway. The molecule that once again has three carbon atoms in it now shuttles back to the original cell, where it may participate again in fixing another CO_2 molecule.

The C_4 pathway is found primarily in plants from habitats that are tropical and have pronounced wet and dry seasons, such as tropical grasslands. It might seem that this pathway cannot be more efficient than the C_3 pathway because, in the end, it actually uses the C_3 pathway even though the CO_2 initially is fixed in a different way. Furthermore, the chemical reactions related to the transportation of the four-carbon molecule to another cell and then back again involve an energy expenditure that is absent in the C_3 pathway by itself. Yet per unit of leaf area, C_4 plants do

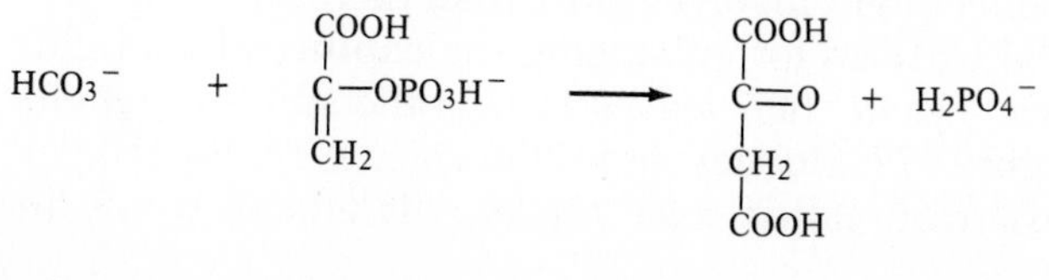

$$HCO_3^- + \begin{array}{c} COOH \\ | \\ C{-}OPO_3H^- \\ \| \\ CH_2 \end{array} \longrightarrow \begin{array}{c} COOH \\ | \\ C{=}O \\ | \\ CH_2 \\ | \\ COOH \end{array} + H_2PO_4^-$$

FIGURE 2–5 C_4 carbon-fixing reaction, with structures of molecules.

show a higher rate of photosynthesis in high light and high temperature than C_3 plants do. These conditions are found in habitats in which C_4 plants are common.

The explanation for the counterintuitive result that C_4 plants are more efficient than C_3 plants under some conditions seems to be this: The C_4 pathway acts as a pump, concentrating CO_2 in the cells that have the C_3 pathway. This pumping is important because O_2 and CO_2 compete for the enzyme that catalyzes the primary CO_2 fixation reaction in the C_3 pathway. The kinetics increasingly favor O_2 at high temperatures. With high temperature and light, and with plenty of oxygen, the enzyme initiates a reaction sequence that uses O_2 and releases CO_2 as a product, just the opposite of what the enzyme does when fixing CO_2. This curious reaction sequence, called *photorespiration*, lowers the *net* efficiency of the C_3 pathway. Sequestering the cells using the C_3 pathway in the center of the leaf (away from high O_2 concentrations) and then concentrating the CO_2 in them nearly eliminates the photorespiration. Under some conditions, this benefit outweighs the costs of transporting the CO_2 to the interior of the leaf (Ehleringer and Bjorkman 1977).

The CAM Pathway

As early as the seventeenth century, scientists observed unusual daily changes in the sourness of many succulent plants. These changes in taste, corresponding to changes in pH, have been interpreted more recently as resulting from a major variant on the C_4 pathway, a variant sufficiently different that it may be regarded as a third basic pathway. It is used by succulent plants, including many cacti, that are found in desert habitats.

For this pathway, the plants open their stomates at night, and CO_2 is fixed by the C_4 pathway. Then, during the day, the stomates close. Now the CO_2 is released within the plant and then refixed by the C_3 pathway. The advantage of this pathway is that it seems mainly to reduce water loss, because less water is lost through open stomates at night than during the day. This pathway is called the *CAM* (for crassulacean acid metabolism) *pathway*, a name that comes from a family of succulent plants, Crassulaceae, in which the pathway was intensively studied.

Light Collecting and Plant Structure

The ability of plants to collect light involves more than their biochemistry; it also involves structural features for enhancing the capture of sunlight. Some plants, for instance, orient their leaves to face the sun throughout the entire day (Wainwright 1977, Mooney and Ehleringer 1978, Ehleringer and Fortseth 1980). This trait, called *solar tracking*, is effective only in environments in which the plant can intercept direct solar rays. It is not effective for plants that grow in the diffuse light of the understory or in climates that are often cloudy.

The structural and biochemical features of a plant operate in concert; the biochemical traits do not, so to speak, come first and the structural features second. For instance, two desert annuals were compared, one using solar tracking, the other with a rosette of fixed horizontal leaves (Mooney and Ehleringer 1978). Yet their biochemical differences, combined with the presence or absence of the ability for solar tracking, resulted in both species having about the same total gain of carbon at the end of the day.

HOW WATER FLOWS THROUGH A PLANT

We have assumed that if water is available in the soil, it will flow through the plant and leave through the stomates. Why, however, does this activity occur? What force moves water inside the plant upward against the force of gravity? The answer is found in the physics of water flow. These principles also apply to the exchange of water between a reptilian egg and the soil in which it is buried, as we will discuss in Chapter 3.

Water flows from places where the pressure on it is high to places where the pressure on it is low. The speed of water flow is proportional to the *difference* in the pressure between the place it is flowing from and the place it is flowing to. That is,

$$\text{Flow rate} = g(P_{\text{source}} - P_{\text{destination}})$$

where P_{source} is the pressure at the source and $P_{\text{destination}}$ is the pressure at the destination. The constant of proportionality, g, is called the *conductivity* of the pathway along which the water is flowing. A narrow pipe has a lower conductivity than a wide pipe does because water flows more slowly through a narrow pipe than through a wide pipe for a given difference in the pressure at the ends of the pipe. The movement of liquid or gas caused by pressure differences is called *bulk flow*.

The Chemical Potential of Water

When the flow of water is accompanied by a change in phase, as when water evaporates from liquid to vapor, thermodynamics offers a generalization of the ideas that work for bulk flow. The *chemical potential* of water in some state—for example, as present in soil, in air, or in the leaf—is defined as the energy required to convert water from that state to a reference state. The reference state is often defined as pure liquid water under 1 atmosphere (1 atm) of pressure and at 20°C. The formula for the chemical potential takes into account temperature, pressure, and volume changes that occur during the procedure that converts water from the state it is in to the reference state. (The concept of energy considered here is called the *free energy* and has the symbol G in thermodynamics, where it is known as the *Gibbs free energy*.)

A change in phase proceeds spontaneously if it is *exothermic*—that is, if it releases energy to the environment. This indicates that the chemical potential of the condition before the phase change is higher than the chemical potential after the phase change. The speed of the phase change depends on the magnitude of the difference between the chemical potential of the initial and final states. This speed is analogous to the dependence of the speed of water flow on the pressure difference between the ends of the pipe.

The idea of the chemical potential of water is relevant to the biology of a plant because water tends to flow through a plant if the chemical potential of water in the soil is higher than that of water vapor in the air. Water flows down the gradient in its chemical potential, even though that may involve flowing up against the force of gravity, because the work involved in moving up the plant is included in the calculation of the chemical potential. When the potential is higher in the soil than in the air, the movement of water from the soil to the air is an exothermic process that occurs spontaneously by whatever pathways are available. The plant itself provides a pathway for the flow of water from the soil to the air. The stomates are the only elements of the path that the plant can control in the short term. Closing the stomates lowers the flow rate. The speed of flow for a given degree of stomatal opening depends on the magnitude of the difference between the potential in the soil and in the air.

Where the Water Is and Is Not

In nature, a high chemical potential for soil water combined with a low potential for water vapor in the air leads to the fastest rate of water flow through the plant. The chemical potential of water in the air is low if a gap exists between the amount of water actually in the air and the amount of water that will be in the air when allowed to come to equilibrium with standing water. The air is *saturated* if it holds as much water as it does when it has come to equilibrium with standing water. This state represents 100% humidity for a given temperature. If the air has more than 100% humidity, then condensation occurs, as when dew forms. Water flow through a plant is very slow in conditions of 100% humidity. However, some water may flow when the humidity in open air away from the plant is 100%, provided that the leaf temperature is higher than the open air temperature. In this circumstance, the temperature in the boundary layer next to the leaf is higher than that in open air, leading to a humidity in the boundary layer that is less than 100%.

The chemical potential of water in the ground is high if the soil pores are filled with water and low if the pores are nearly empty and the remaining water adheres tightly to the soil particles. The chemical potential of water in the soil is also low if it contains many dissolved salts and other molecules. When water is withdrawn from the soil, the potential of the remaining water continually drops as the remaining water is bound more and more tightly to the soil particles. Thus, as the water content of the soil drops, the tendency for the remaining water to evaporate into the air also decreases.

The Measure of Water Potential

The chemical potential of water is often simply called the *water potential.* That potential may be measured in units called *Pascals*. The chemical potential originally is defined with the dimensions of energy per unit mass. However, 1 gram (1 g) of water is 1 cm^3 in volume under standard conditions. Hence, the chemical potential of water also can be expressed in terms of energy per unit volume. Furthermore, energy per unit volume can be expressed as force per unit area, and this is pressure. Thus, units of pressure can be used for the chemical potential of water. A *Pascal* is a unit of pressure; 10^6 Pascals (or 1 Mega Pascal–1 MPa) nearly equal 10 atmospheres (9.87 atm, to be exact).

Potentials in the Air and Ground

The values for water potentials are usually negative numbers because they represent states that have a lower energy level than pure water has. For example, water vapor in air when the humidity is less than 100% has a lower water potential than liquid water does because liquid water spontaneously moves into the unsaturated air. However, since the water potential of the liquid water is typically near zero (by definition, it is exactly zero at the usual reference state of 1 atm pressure and 20°C), the water potential of vapor at less than 100% humidity must typically be less than zero. Here is the way these negative values might be used in practice. Because −100 is lower than −1, and because water flows from a high- to a low-water potential, it naturally moves from any place where the potential is −1 to a place where it is −100, provided that a path is available.

Table 2–1 shows the water potential of water vapor at various levels of relative humidity. The water potential of water vapor at 20°C and 50% relative humidity is about −100 MPa. The water potential of vapor at 92% relative humidity increases to −10 MPa, and it increases all the way to 0 at 100% relative humidity.

TABLE 2–1 Water potential of water vapor in the air

Water Potential (MPa)	Relative Humidity (%)
0	100.0
−0.67	99.5
−1.35	99.0
−2.72	98.0
−4.10	97.0
−6.91	95.0
−14.10	90.0
−30.10	80.0
−48.10	70.0
−68.70	60.0
−93.30	50.0
−Infinity	0

Source: After Larcher (1975).

Like air, soil can hold only a certain maximum amount of water, called its *field capacity*. Any water beyond this maximum leaves the soil as runoff. Sandy soils hold less water than do soils that are rich with clay. Figure 2–6 shows that the soil water potential approaches 0 MPa at field capacity. The soil is exceedingly dry at about −5 MPa. Crops usually wilt at soil water potentials from about −0.8 to −2 MPa, but some desert shrubs do not wilt until the water potential drops to −7 MPa. The water in the soil between the level of field capacity and that in which the plant wilts is considered to be the water that is available to the plant.

Water and Plant Stress

A high water flow through a plant occurs when the stomates are open and the humidity is relatively low. The high flow can continue only if the rate of water supply to the roots matches the rate of evaporation from the leaves. If it does not, the water potential in the leaves drops until the stomates close. Because the largest segment of the water potential gradient is typically that between the inside of the leaf and the air, this site is the most effective for flow-rate regulation. Hence, the stomates are ideally placed as regulatory structures.

The most stressful time for a plant is when both the air and soil are dry. Then the plant's water tends to leave whenever the stomates are opened. The advantage to a CAM plant in opening its stomates at night is that the humidity of the air is higher because the air temperature is lower,

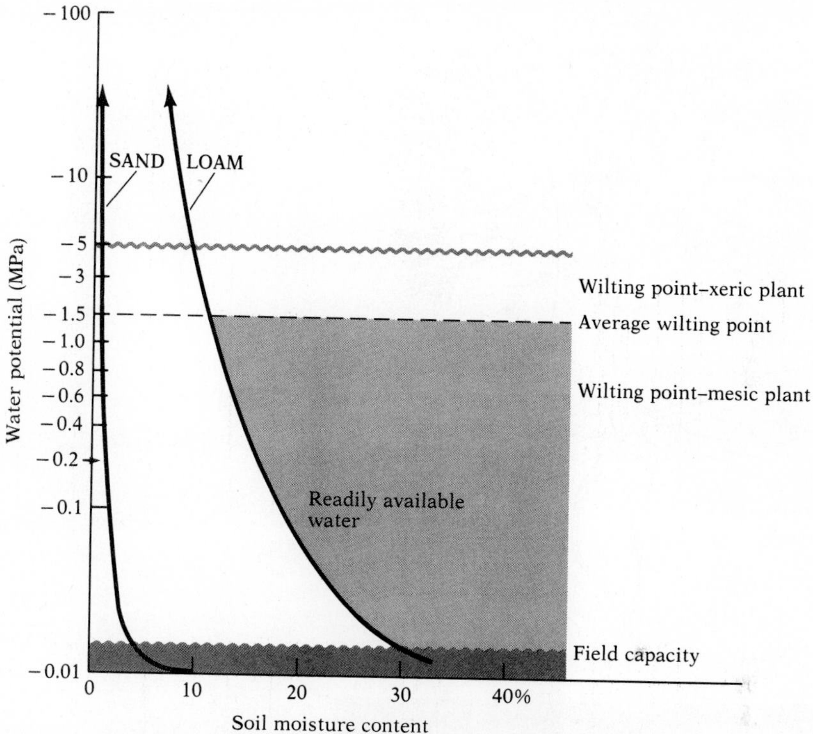

FIGURE 2–6 Soil water potentials in MPa. *(After Larcher 1975.)*

and hence the water potential in the air is higher at night. You may wish to consult Nobel (1983) for more information on the biophysics of water movement.

A mesquite tree, *Prosopis tamarugo*, which is native to the Atacama desert in northern Chile, offers an interesting example of how a plant regulates its water flow in an extremely drought-stressed habitat (Sudzuki 1969, Went 1975, Mooney *et al.* 1980b). This plant was studied in a region in which essentially no rainfall occurs. According to over 12 years of data, the maximum annual precipitation was less than 1 mm, a negligible amount. However, at the study location, underground water exists at depths of 5 m or more. Mesquite trees have a few long tap roots that extend to such depths and that have been known to supply at least some of their water. However, a dense mat of roots also exists, beginning at about 1 m and extending downward only about 0.5 m. With no rainfall, what could such a shallow root system do?

The mystery was heightened by the discovery that sap sometimes flows *down*, not up, in this plant. Furthermore, studies with radioactively labeled water showed that this mesquite actually absorbs water from the air, provided that the relative humidity is near 100% (Sudzuki 1969). In

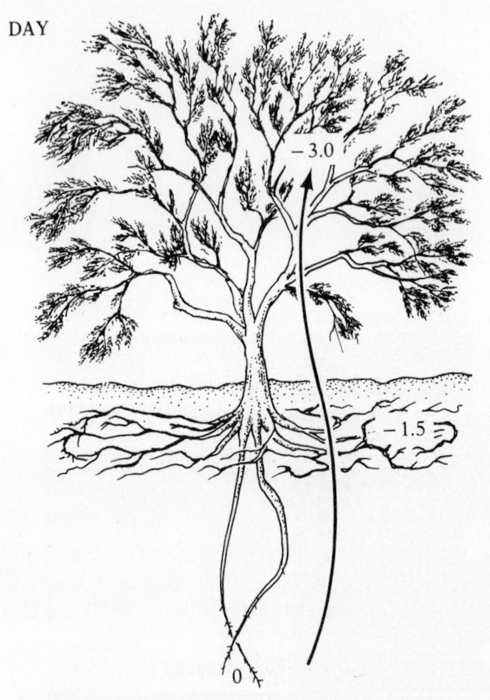

FIGURE 2–7 Water flow in a mesquite. *(After Sudzuki 1969.)*

sufficiently dry soil, and with saturated air, the water potential is higher in the air than it is in the soil; hence, the natural direction of flow is from the air into the ground. Indeed, the soil in the root mat is moist, and it initially appeared that this moisture was water that the plant captured from the air and that was transported down the stem into the soil. However, under field conditions, the relative humidity rarely reaches 100%, and then only during the night. Daytime relative humidities of less than 10% are common. Hence, it seems that the air generally is not humid enough for much water to be collected from it and transported to the ground, although this event sometimes may happen.

A possible solution to the puzzle of why mesquite has a shallow root system at all and why the soil around these roots is moist is that the root mat is a water storage organ acting as a buffer between a deep water-filled zone and the canopy (Mooney *et al.* 1980b). As illustrated in Figure 2–7, most of the water comes initially from the deep soil through a tap root. During the day in spring or summer, the water potential around the leaves is −3 MPa, in the root mat it is −1.5 MPa, and in the deep soil it is 0 MPa. Hence, the natural flow is from the deep soil to the root mat, and then on to the canopy. However, at night, the canopy water potential may increase to −1.5 MPa because the relative humidity increases as the air temperature drops. Now the natural flow is from the canopy *down* to the root mat and also from the deep water zone *up* to the root mat. Thus, the root mat is a buffer between the deep water zone and the canopy.

The reason this system works appears to be that no other plants are present in the vicinity to steal the water from the mesquite. A zone of moist soil in a less severe desert presumably would attract the roots from other individuals and thus could not function as a storage organ for the individual that obtained the water.

HOW PLANTS GET THEIR NUTRIENTS

Much of the drama of a plant's life takes place underground. Because plants often develop enormous root systems that collect water and nutrients, it might seem that a plant need only wait passively for water and nutrients to arrive. After all, as we have seen, soil water spontaneously moves through the plant on its way to the atmosphere when the water potential in the air is lower than that in the ground. This movement is indeed true for water in the immediate vicinity of the existing roots; however, the soil water near the roots must be replenished if transpiration is to continue. It seems that a plant abandons roots that are not producing a return in water or nutrients. Apparently, the way a root system gets to places of high water renewability is simply by growing most vigorously along the paths that are producing the highest yield.

Acquiring Nutrients

The uptake of nutrients is fundamentally different from the uptake of water because energy is required for the accumulation of nutrients; that is, the accumulation of nutrients is an active process. The concentration of nutrients in the roots is 100–1000 times the concentration in the soil, so the uptake of nutrients occurs *against* a concentration gradient. The root must be alive and metabolizing to produce the energy for this uptake. The root must have O_2 for its respiration. In loose soils, enough air may exist in the soil to supply the respiratory needs of the roots. In other cases, the morphology of the roots may provide air tunnels, as in the mangroves illustrated in Figure 2–8.

For a plant to continue growing, it must continue to find and to accumulate nutrients. Phosphate is a nutrient that diffuses slowly. It moves so slowly that it is almost like a seam of a precious ore in the gound. The plant's roots follow the seam and "mine" it. Phosphate is then recycled to the soil through the decay of fallen leaves and animal products.

Nitrate is the other nutrient needed in large quantities for plant growth. Nitrogen may be present in the soil as ammonium (NH_4^+), or as nitrate (NO_3^-). In aerated soils, bacteria acquire energy by converting ammonium to nitrate; hence, nitrogen most often is found as nitrate in aerated conditions. In soils without oxygen, some bacteria actually convert nitrate to ammonium. Although this process is an energetically unfavorable reaction, the nitrogen atom is needed by the bacteria to combine with the hydrogen atoms produced during respiration; hence, in nonaerated soils,

FIGURE 2–8 Structures in mangroves for aerating roots. Typical understory in a mangrove forest, and close up of roots are shown. *(Photos by J. Roughgarden.)*

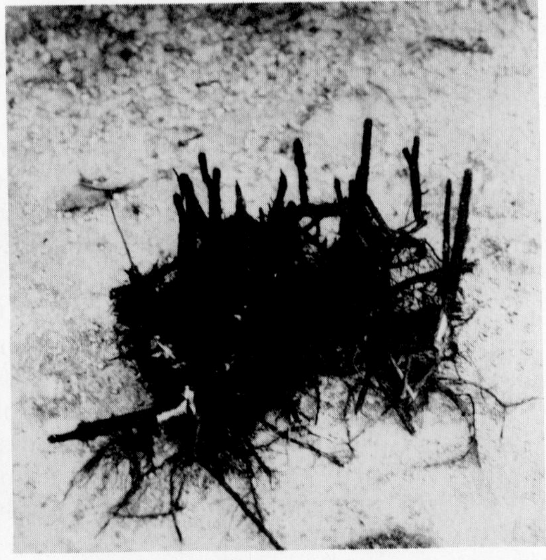

nitrogen is usually found as ammonium. The clay particles in soil typically carry negative charges. As a result, positively charged ammonium ions are bound to negatively charged soil particles and do not diffuse quickly through the soil. In contrast, nitrate quickly diffuses through the soil; hence, even though plants take up ammonium when grown in nutrient solutions, most plants in nature get the majority of their nitrogen as nitrate. It is unfortunate, however, that most soil nitrogen is in the form of nitrate, since nitrate is more easily lost from the soil than is ammonium, either by leaching or by conversion to N_2 (denitrification).

Nitrogen, like phosphorus, is returned to the soil with the decay of fallen leaf litter and animal products. In addition, atmospheric nitrogen is available to some plants because of a mutually beneficial relationship between these plants and bacteria that have the ability to fix atmospheric nitrogen. In the pea family (Leguminosae), about 10% of the approximately 12,000 species have been examined, and about 90% have an association with bacteria that can fix nitrogen (Salisbury and Ross 1978). Many woody plants in a variety of plant families also have recently been found to have associations with nitrogen-fixing bacteria, although the bacteria associated with these plants come from a different family than those associated with legumes. Generally, the plant supplies the bacteria with carbohydrates and the bacteria supply ammonium or other nitrogenous compounds in return. In the legumes, the nitrogen-fixing bacteria are housed in tiny bulbous structures on the roots called *root nodules*.

Fungus–Plant Associations

Roots also may have mutually beneficial associations with fungi called *mycorrhizae*—fungi that form a mutualistic relationship with the roots of plants, sometimes even penetrating cells of the roots, and aiding them in the uptake of nutrients in return for energy. These mycorrhizal fungi improve the rate of absorption of phosphate, and of other ions as well, and they receive sugars from the host plant.

These fungus–plant associations are important, especially in nutrient-poor soils and in plants without well-developed root hairs. Mycorrhizae are essential for some plants' survival. For example, European pines introduced into the United States failed to survive until they were infected with mycorrhizae from their native soils (Salisbury and Ross 1978).

HOW PLANTS "BEHAVE": GROWTH AS LOCOMOTION

Growth is a plant's locomotion. Through growth, a plant moves toward places where water, nutrients, and light are found. Also, plants produce leaf shapes that are appropriate to the microclimate that the leaf experiences. Plants can even show what might be called *aggression* through their growth. One plant can capture the light of another by overgrowing and shading it, and a plant can capture water from another by sending roots to the base of a neighboring plant to absorb its stem runoff.

A plant differs from a typical mobile animal in being comparatively decentralized. Different parts of a plant may experience greatly different microclimates and, as a result, acquire different physiological states. The plant as a whole can appear physiologically coordinated even though its various parts experience different environmental conditions (Field 1983). If the plant has only one stem leading to the root system, then the different parts of the canopy "see" the same picture of soil water availability. In this case, differences in the physiological state of different branches probably reflect differences in the microclimate within the canopy. However, if there are multiple stems or aerial roots, each part of the plant can see a different picture of conditions in both the soil and the canopy. In this case, the parts of the plant may become quite loosely coupled with one another. This decentralized style of life probably also occurs in colonial invertebrates such as corals.

CHAPTER 3

How Animals Work: Different Flows

Like plants, animals continually exchange matter and energy with the environment. The principles are much the same, but important differences exist. The relative importance of the various flows may be different for plants and animals. Animals do not photosynthesize, so the supply of oxygen rather than the supply of CO_2 is important to them. Furthermore, for animals much is known about the differences between the flows in terrestrial, freshwater, and marine environments.

Consider water relationships in animals. When an animal breathes, it releases water vapor—provided the air is not fully saturated. For this reason, the uptake of oxygen in animals is coupled with the release of water in a manner reminiscent of the way the uptake of CO_2 is coupled with the release of water in plants. Animals also lose water through the excretion of waste products and, in some mammals, through evaporation from the skin. Most animals obtain their water by drinking and, to a lesser extent, as a product of the breakdown of their food. In salt water, as on land, respiration is coupled with water loss. However, in fresh water, water is absorbed across the gills during respiration. Freshwater animals face a problem of eliminating the water they automatically take up during respiration. Water relationships also affect egg survivorship, as they do the survivorship of seeds in the soil.

ANIMALS AND BODY TEMPERATURE

Rather precise control of body temperature is important to many animals, especially mobile animals, probably because of the central role of locomotion in their daily lives. The speed of an animal's muscular activity depends on body temperature. A butterfly cannot fly until its temperature exceeds a certain threshold. A frog's breathing rate increases as a simple exponential function of its body temperature until the onset of thermal stress.

To describe the major flows of matter and energy between an animal and its surroundings, we will first consider how the flow of heat is related to an animal's body temperature in the same general way as we might look at the flow of heat in a leaf. We next consider how food "flows" into an animal by looking at the decisions an animal may make when foraging for food. Finally, we look at the flows of nitrogen, salt, and water. The flow of salt and water determines the osmotic state of the organism and also is relevant to marine and freshwater plants.

How Animals Attain a Working Temperature

The importance of temperature to an animal has been well illustrated in the land iguanas, *Conolophus pallidus* of the Galapagos (Christian and Tracy 1981). Figure 3–1 shows that a hatchling iguana can sprint faster as its body temperature increases. The lizard can barely run at all at 15°C and attains its maximum speed at 32°C. An individual's sprint speed, expressed as a percentage of its maximum speed, increases as a linear function of its body temperature for temperatures between 15 and 32°C. However, at about 40°C, the lizard begins panting, a sign of thermal stress.

Christian and Tracy (1981) also determined the susceptibility of hatchling iguanas to predation by hawks on Isla Santa Fe in the Galapagos. They showed that 67% of the observed attacks were successful on hatchlings whose body temperatures were computed to be less than 32°C, whereas only 19% of the attacks were successful on hatchlings whose body temperatures were computed to be equal to or greater than 32°C. Important effects of temperature similar to those on the mobility of hatchlings of the Galapagos iguana have been found in studies of other reptiles, including the velocity of the strike of the gopher snake, *Pituophis melanoleucus* (Greenwald 1974); survivorship in the side-blotched lizard, *Uta stansburiana* (Fox 1978); and the relation between the sprint speed and preferred body temperature of the Puerto Rican trunk-ground anole, *Anolis cristatellus* (Huey 1983).

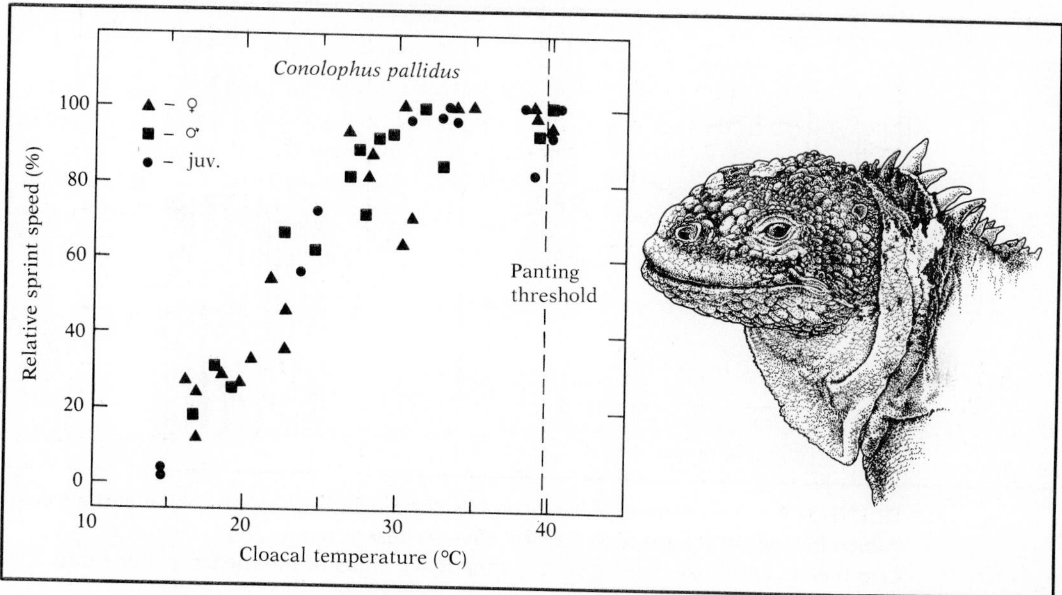

FIGURE 3–1 Sprint rates of individual Galapagos land iguanas, normalized to each individual's fastest sprint, as a function of body temperature. *(After Christian and Tracy 1981.)*

The "Thermal Personality"

Most animals have what might be called a "thermal personality," the manner in which an animal's body temperature relates to the temperature of the inanimate objects it encounters in its environment. To some extent, thermal personalities can be classified into three categories: nonregulators, heliothermic thermoregulators, and endothermic thermoregulators.

The *nonregulators* are animals that exert no physiological or behavioral control over their body temperature. Many animals are nonregulators: animals that burrow in soil, many nocturnal animals and those living in deep forest shade, and almost all very small invertebrates.

The *heliothermic thermoregulators* are animals whose body temperature equals, or nearly equals, the temperature of an inanimate object in the same position as the animal. However, heliothermic thermoregulators move around in the environment and maintain a fairly constant body temperature by regulating the amount of solar radiation they receive. Huey's (1983) study of the body temperature of the Puerto Rican lizard, *Anolis cristatellus*, offers a representative example. Figure 3–2 shows that these lizards, observed in an open park habitat, can maintain body temperatures that are practically constant at about 31°C throughout the day.

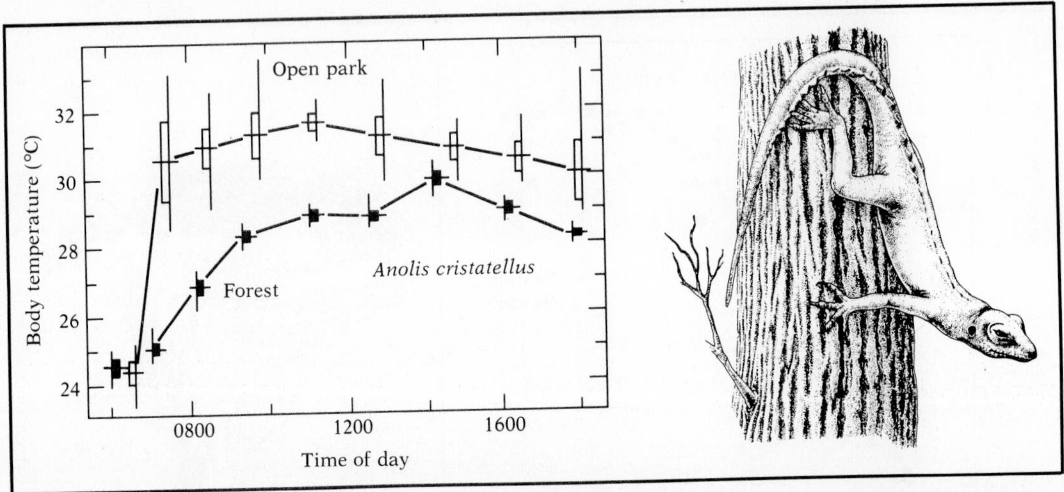

FIGURE 3–2 Body temperatures of *Anolis cristatellus* in an open habitat and a forest habitat in lowland Puerto Rico. Vertical lines represent ranges, horizontal lines indicate means, and boxes enclose 95% confidence limits of the means. *(After Huey 1983.)*

Further examples of heliothermic thermoregulators include some insects, such as butterflies and dragonflies, and many modern-day reptiles including lizards, snakes, and alligators.

The *endothermic thermoregulators* consist of birds and mammals that use their metabolism to produce body heat. They control heat loss physiologically by a variety of means, including regulating their blood flow to distal parts of the body, varying their breathing rate, and evaporating the water secreted from sweat glands. The endothermic thermoregulators also have insulating covers of fur or feathers. They use these physiological capabilities to achieve relatively constant body temperatures in spite of wide fluctuations in their thermal environment.

Even these three categories of thermal personalities do not encompass all the tactics by which animals can regulate their body temperature. For example:

- Limpets in the severe heat stress of the tropical intertidal zone can elevate their shell to let wind currents underneath. Then they attain a body temperature below the air temperature by evaporating water into the air currents.
- Large reptiles may develop body temperatures above the air temperature because their low surface-to-volume ratio limits cooling; these have been called *inertial thermoregulators*. They are discussed further in Chapter 22, in the section on the ecology of dinosaurs.
- Many animals indirectly control their thermal environment by the way they construct their shelters. Some termite colonies have brood

chambers in which the microclimate is remarkably constant. Some megapode birds do not sit on their eggs but incubate them in piles of decomposing vegetation, which the birds open and close to regulate the heat of decay.

Thus, the classification of thermal personalities should be taken as suggestive, not definitive. Moreover, organisms may adopt different thermal personalities at different ages and in different seasons. Finally, thermal classifications refer to the whole animal. However, recently the swordfish has been found to have local areas in which the temperature is metabolically regulated while the rest of the body remains at a temperature that is determined by the environment. The brain and eye of the swordfish are warmer than the surrounding water, and a tissue near the eye supplies the heat (Carey 1982).

Endothermic thermoregulators are sometimes called *warm-blooded* animals, and the rest are considered to be *cold-blooded.* This description is inaccurate. Anyone who has handled a lizard or snake after it has basked in the sun knows that these creatures can be very warm under natural conditions.

Endothermic thermoregulators live an energetically expensive life. Birds and mammals consume much more food than comparable-sized solar regulators because their body temperatures are maintained by the heat released during the respiratory breakdown of the food. As a result, one insectivorous bird consumes the food that could support many insectivorous lizards of the same size. This basic physiological fact underlies the relation between food supply and abundance for animals in nature.

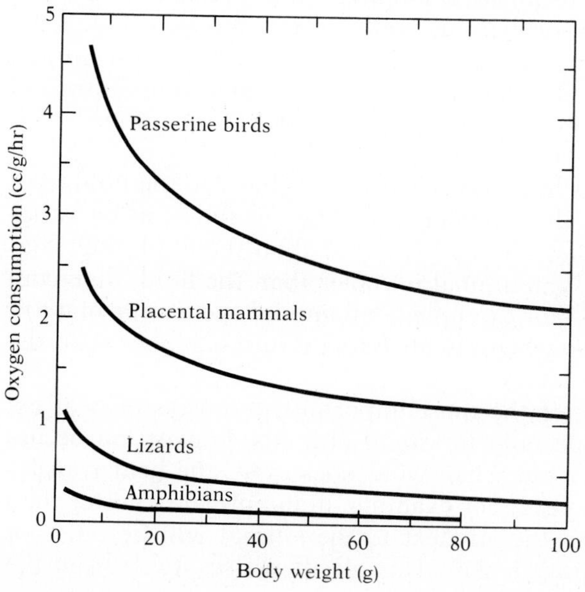

FIGURE 3–3 Metabolic rate versus body size for endotherms and non-endotherms. Body temperature for lizards is 37°C and for amphibians is 25°C. *(After Hill 1976.)*

An equally basic relation is that the energetic cost, per gram, of an animal within a thermoregulatory class decreases as the total body weight increases. For example, 1 g of field mouse is more energetically expensive than 1 g of brown rat, as demonstrated by Figure 3–3.

The Pathways of Heat Exchange

To find out how an animal's environment influences its body temperature, one must identify, and keep track of, all the major pathways by which heat is flowing into and out of the animal. The six main pathways of heat exchange between any terrestrial organism and its environment are solar absorption, infrared radiation, convection, conduction, evaporation, and the metabolic production of heat. Some of these pathways operate only in one direction and others are bidirectional. Let us examine them, one by one.

Solar Absorption. The absorption of sunlight may be a large source of heat for an animal. The amount of light that is absorbed depends on the distribution of colors in the light and on the color of the animal. Also, if scales, feathers, or fur are shiny, then some of the light will be reflected back to the environment without being absorbed. Some animals can change their color to exert a limited control over the amount of solar radiation they absorb (Norris 1967).

Infrared Radiation. Infrared radiation is invisible to humans. However, special films are sensitive to infrared radiation and enable one to use a 35-mm camera to take pictures with it, as illustrated in Figure 3–4. If the temperature of the ground and vegetation is higher than that of an animal, then the net flow of infrared radiation is into the animal. If an animal is hotter than the environment, then the net flow is out of the animal.

Convection. Convection is heat transfer from an object to the fluid (typically either air or water) that surrounds it. The net flow can be either into or out of an animal. If an animal is hotter than the fluid, then convection cools the animal. If an animal is cooler than the fluid, then convection warms the animal—for example, if an animal spends the night in a cool burrow and emerges when the air temperature is higher than the ground temperature.

Convection may be one of the most important pathways of heat exchange for an animal, especially for small animals. Convection occurs even when no wind exists, but when wind does exist, the heat transfer rate increases with wind speed. For example, if an animal is hotter than the air, even without wind the air next to the animal will become hot from contact with the animal's skin. This air then rises away from the

(A) (B)

FIGURE 3–4 Tree photographed (A) with panchromatic film with a light orange filter to tone the sky and (B) with infrared film. *(Photos © 1984 by David J. Cross/BPS.)*

animal, and cool air comes in to take its place. On the other hand, if wind is blowing, then much more air contacts the animal per unit time, and this air takes away much more heat than the small air currents generated by the animal's temperature alone.

The heat actually transferred through convection per unit time, for any given wind speed, depends on the difference between the animal's temperature and the air temperature. The larger the temperature difference, the larger the rate of heat flow. The effectiveness of convective heat transfer also depends on the shape of an animal, its surface texture, and how it is oriented toward the wind. Generally, if an animal is facing into the wind, it has a lower heat transfer rate than if it is broadside to the wind. Similarly, at most wind speeds, a rough surface lowers convective heat transfer by restricting air movement adjacent to the skin. Also, the convective transfer rate increases with the surface-to-volume ratio of the main trunk of the animal's body.

Conduction. Another pathway of energy exchange is the conduction of heat through physical contact between the animal and the ground, rocks, or vegetation. This pathway is bidirectional and may either cool or warm the animal, depending on whether the animal is warmer or cooler than the substrate with which it is in contact. This pathway is important for burrowing, crawling, and sessile animals and is usually negligible for others.

Evaporation. Now we come to the pathways associated with the animal's breathing and respiration. When the animal breathes, it exhales air that is saturated with water. Unless it inhales air that is saturated to begin with, the animal loses heat through evaporation of water into the air in the lungs. The amount of heat lost is determined by measuring the amount of air taken in a breath (which is much less than the lung's total capacity). Then, knowing the relative humidity of the inhaled air, one calculates the amount of water needed to raise the humidity to 100% at the temperature of the exhaled air. This amount is multiplied by the latent heat of vaporization (the heat that converts 1 g of liquid water into water vapor), and the result is the heat that is lost with each breath. By taking large breaths, and by taking them very quickly, as in panting, an animal can dissipate a great deal of body heat.

Some mammals release water on the surface of their skin through sweat glands. The effectiveness of sweating for cooling the animal depends on the evaporation rate. If the air is humid, the evaporation rate is low, and sweating loses effectiveness. Nonmammals also lose water through the skin. Templeton (1970) estimates the water loss through skin in lizards to be about 60% of the total water loss. The remaining 40% of the total water loss occurs through breathing. Furthermore, a large amount of water may be lost by evaporation from the surface of the eyes (Mautz 1980).

Metabolic Production of Heat. Last, and of major importance in birds and mammals, respiration provides a continual input of heat to the body. A minimum estimate of the heat from this source is found from the basal (resting) metabolic rate of the animal. Still more heat is generated by the animal's activity. The evaporative water loss of an endothermic regulator is less than its metabolic production of heat. The body temperature therefore increases above the air temperature, and the remaining heat leaves through convection and infrared radiation. A bird consumes about four to five times as much oxygen per hour as a similarly sized lizard, even when it is resting (Hill 1976). This higher basal metabolic rate is why a bird can attain the high body temperature necessary for its flight. Once it is flying, it produces additional heat from its muscular exertion, but it also loses more heat from convection.

Synthesizing Information on Energy Exchange: An Example

This list of six pathways of energy exchange between an organism and the environment can be used to predict what the body temperature of an animal will be under natural conditions. Morover, if the maximum and minimum tolerable temperatures for an animal are known, one can predict what places in the environment it can successfully occupy. Similarly, if an animal has a strongly preferred body temperature, one can predict where in the environment it will tend to be found throughout the day. To see how all this information can be synthesized, consider a simple example where energy exchange is mainly through solar absorption and convection. These are the main pathways for a small heliothermic regulator, such as a butterfly or small lizard weighing 5 g or less.

To account for all the pathways of energy flow between the organism and the environment, we simply add all the flows together. If the sum is negative, it means that the net flow of heat is out of the animal, and therefore the animal is getting colder. If the sum is positive, then the net flow is into the animal and it is getting hotter. If the sum is zero, it means that the animal is holding a constant temperature because the input of heat balances the output of heat. A constant temperature that results when the input rate balances the output rate is called a *steady-state temperature.* In the example considered here, the sum has only two terms, one to represent the heat gained from absorbing the sun's rays and the other the heat transferred from the animal to the air.

The *climate space* of an animal is defined as the set of environmental conditions in which the animal maintains a steady state temperature that is between its lethal limits (Porter and Gates 1969). For the most part, the animal must live within its climate space. It can move outside of the climate space for a short period only if it has a safe temperature to begin with and then returns to the climate space before the temperature goes beyond one of the lethal limits. The lethal limits may, however, change somewhat if the animal can acclimate to environmental extremes.

To calculate the limits of the climate space, we begin with the sum of the pathways of heat flow. This sum is set equal to zero to examine the steady state. The upper limit to the climate space occurs for the environmental conditions that produce a steady-state temperature equal to the maximum tolerable temperature. Similarly, the lower limit is determined by finding the environmental conditions that produce a steady-state temperature equal to the animal's minimum tolerable temperature.

The heat input from solar radiation can be labeled as Q_{sol}. The heat lost or gained through convection depends on the difference between the body temperature and the air temperature. The constant of proportionality is called the *convective heat transfer coefficient.* This coefficient depends on the wind speed and will be labeled as H_v, where the subscript indicates its dependence on wind velocity. The heat lost or gained to the air through convection is given by the following formulas:

$$\text{Convection when body temperature is } T_{max}\text{: } H_v(T_{max} - T_{air})$$

$$\text{Convection when body temperature is } T_{min}\text{: } H_v(T_{min} - T_{air})$$

where T_{air} is the air temperature, T_{max} is the maximum tolerable body

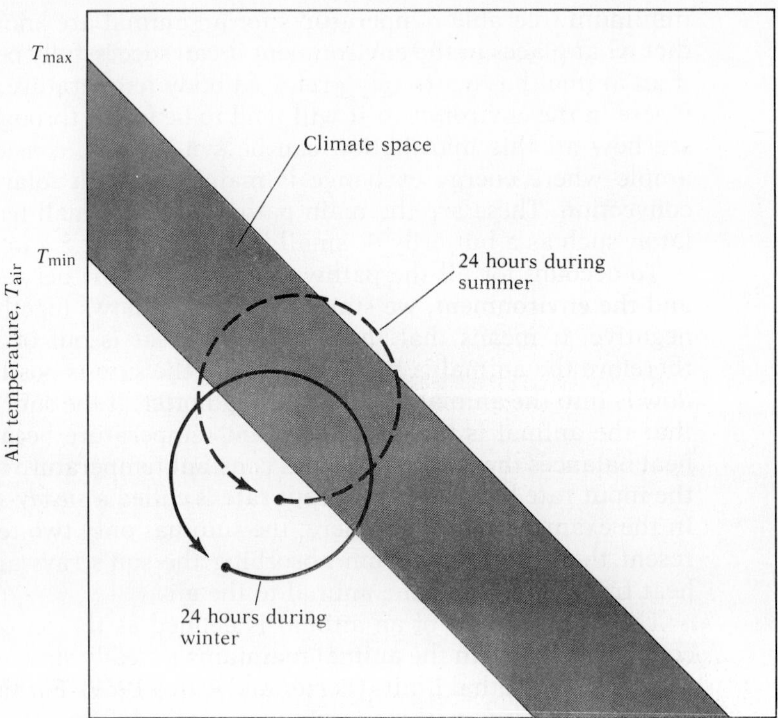

FIGURE 3–5 Climate space. The vertical axis is air temperature; the horizontal axis is solar radiation. See text for further explanation.

temperature, and T_{min} is the minimum tolerable body temperature. The convective heat transfer coefficient, H_v, refers to the entire animal at wind velocity, v. This coefficient is sometimes related to a coefficient that describes the heat loss *per unit area*, h_v. If so, the coefficient for the whole animal, H_v, equals h_v times the animal's surface area.

In this example, solar absorption and convection are the ingredients of the equation that sums up the inputs and outputs of heat to the animal. Hence, the body temperature has a steady-state value at T_{max} when

$$Q_{sol} - H_v(T_{max} - T_{air}) = 0$$

Any combination of Q_{sol} and T_{air} for which this equation is true is a point on the top boundary of the climate space. Similarly, the bottom boundary of the climate space is found from the equation

$$Q_{sol} - H_v(T_{min} - T_{air}) = 0$$

Any combination of Q_{sol} and T_{air} for which this equation is true is a point on the lower boundary of the climate space. The Q_{sol} has a positive sign because it is an input of heat. The convection term is an output of heat, provided that the body temperature is higher than the air temperature. If the body temperature is lower than the air temperature, the term changes sign to indicate that convection is now an input of heat.

The climate space can be graphed by plotting T_{air} on the vertical axis and Q_{sol} on the horizontal axis, as illustrated in Figure 3–5. Upon rearranging the equations so that T_{air} is on the left and Q_{sol}, together with everything else, is on the right, we obtain

$$T_{air} = -(1/H_v)Q_{sol} + T_{max}$$

and

$$T_{air} = -(1/H_v)Q_{sol} + T_{min}$$

Each of these equations graphs as a straight line with a slope of $-(1/H_v)$. The lines have different intercepts. The climate space is the set of all points between these two lines; these points represent combinations of air temperature and solar absorption that produce a steady-state body temperature for the animal that is within its lethal limits.

Other Situations, Other Pathways

For other situations, other pathways may have to be included. For example, conduction must be included for animals with extensive contact with the substrate, and infrared radiation must be included for large animals. Convective heat loss is less important than infrared radiation if an animal is large enough. For metabolic regulators, the heat from metabolism must be included, although for small invertebrates with a body temperature close to the air temperature, the heat generated from metabolism is low and approximately balances out the heat lost through evaporative water loss during breathing. Similarly, a climate space can be developed for a plant by taking into account solar absorption, convection, and the cooling effect of evaporative water loss from transpiration (Raschke 1956, Lange 1959, Gates 1965). Thus, each kind of organism and circumstance may have to be considered separately, but each may be approached in the same way by writing down all the important pathways of heat exchange, setting the sum equal to zero, and finding the environmental conditions that lead to the maximum and minimum tolerable temperatures.

Figure 3–6 illustrates examples of a climate space for an adult alligator (Spotila *et al.* 1973) and the garter snake, *Thamnophis e. elegans* (Scott *et al.* 1982), as presented in Tracy (1982). The slope of the top and bottom borders of the climate space is steep for large animals. This steepness indicates so great a sensitivity to overheating that adult alligators cannot be exposed to even moderate illumination without danger of overheating. This fact may help to account for why large crocodilians are found in association with bodies of water.

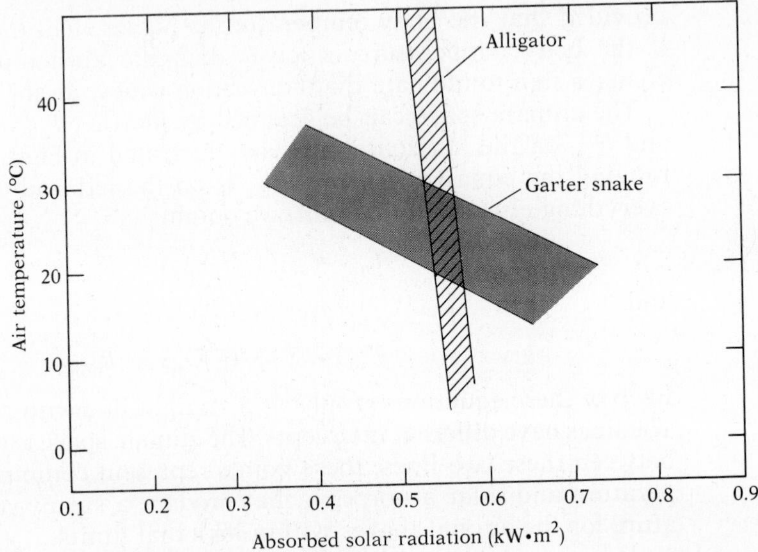

FIGURE 3–6 Climate space for an alligator and a garter snake. Wind speed is 0.1 m · s^{-1}. The alligator has a temperature range of 32–35°C and the garter snake a range of 26–32°C. *(After Tracy 1982.)*

The actual habitat typically presents only some of the conditions that are suitable for an animal. One can add to the climate–space graph curves that represent what is actually present in the environment. For example, T_{air} and Q_{sol} vary together during a 24-hour period, and one can plot curves representing these values for summer and winter days, as sketched in Figure 3–5. On a winter day, both the sunlight and air temperature are low as compared with that of a summer day, so the curve for a winter day is both lower and to the left of that for a summer day. In the figure, when a winter day begins, the air temperature at dawn is too cold, and the sun too low, for an acceptable temperature. Hence, the animal remains in its burrow. As the day proceeds, the curve cuts the lower boundary of the climate space, and the animal can begin its activity. The animal can remain active until some time before dusk, when the curve leaves the climate space. This is an *unimodal* activity pattern. In the summer, the animal can emerge at dawn, but the curve leaves the top border of the climate space some time before noon. The animal must then spend its midday in shelter, either in the shade or in a higher wind speed (e.g., high in the vegetation), but can be active again in the late afternoon until dusk. This is a *bimodal* activity pattern.

To use this theory in practice, one must use one of the most difficult parameters to measure, the convective heat transfer coefficient, H_v. The relationship between wind speed and H_v for an animal-shaped object was measured by Porter *et al.* (1973) with metal castings in the shape of a

FIGURE 3–7 Gold castings of *Dipsosaurus.* (*Photo by Warren Porter, University of Wisconsin–Madison.*)

lizard obtained from plaster-of-paris molds, as illustrated in Figure 3–7. Then, by placing a casting in a wind of known speed and by measuring the temperature changes in the casting, they determined the H_v for that wind speed. They repeated the measurement for many wind speeds and determined the relationship between H_v and the wind speed, v. They also showed that it is often an acceptable approximation to view a lizard as a cylinder. Mechanical engineers have developed tables of H_v for simple geometric objects such as cylinders. These tables can be used for animal shapes as well, although with some loss of accuracy. In practice, the activity schedule of a desert lizard, *Dipsosaurus dorsalis,* was predicted to within 30 minutes, using this theory based on heat exchange between the organism and the environment.

HOW ANIMALS OBTAIN ENERGY

The rate with which animals collect food depends on the way the animal behaves, as well as on the amount of food present. Capturing food is, for an animal, the counterpart of the uptake of light, CO_2, and nutrients for a plant; in both cases, the organism is taking energy and materials from the environment. The rate at which a plant gains materials can often be described in terms of how gases and liquids diffuse along concentration gradients. However, the rate at which an animal gains materials depends on how it behaves, and the principles of diffusion provide little help in describing animal foraging behavior.

Food-Seeking Strategies

During the last 10 years, animal ecologists have begun to understand the rates at which animals catch food in nature, and the choices they make of what to capture, in terms of *strategies.* The approach is to start with all the possible food types that are available to an animal and to consider the value that each food type offers to the animal, as well as the effort expended in getting it. From this list, one determines what the animal should take from the pool of available items to attain the highest net yield. This prediction of what the animal should do can then be compared with what it actually does. If prediction and data agree, and if other possibilities have been taken into account, then the animal can be said to forage *optimally.* If the animal does forage optimally, then this optimal foraging theory can be used to describe the rates and choices of its food uptake. Early reviews of this approach appear in Schoener (1971), Pyke *et al.* (1977), and Krebs and Davies (1978, chap. 2). Recent critical assessments include Heinrich (1983), who is generally negative, and Pyke (1984), who is generally positive. Here we offer a simple example of how this approach works and then discuss briefly its limitations and possible extensions.

Abundance and Quality in a Foraging Strategy

Suppose two prey types are available to an animal, which we can label 1 and 2. Two important features of a prey type are its abundance and its quality. Abundance is measured by the number of prey items encountered per day; to be encountered, a prey must come within catching distance of the predator. Quality, in this simple example, is measured by the length of time it takes the animal to consume the prey item once it has been captured.

Anyone who has watched a snake trying to swallow its catch, or even a house pet trying to devour a scrap, knows that the time taken to eat a prey item can be very long, from several minutes to over an hour. When an animal is occupied in eating prey, it generally cannot, at the same time, be searching for other prey, even though other prey might be even better than the one presently being devoured. Also, when the animal is handling the prey, it may be exposed to predation itself, or it may be leaving its nest or burrow unguarded. For these reasons, it may be appropriate to value a prey item in terms of the time it takes to consume it, with a good prey item being one that is quickly handled and a poor item one that requires a long time for handling. For other situations, other measures of quality are more appropriate, as we shall describe later.

If prey quality is measured in terms of handling time, one may hypothesize that the "objective" of the animal is to choose items in a way that minimizes the average total time spent both searching for and handling prey throughout the day. With this definition of the objective, it is a straightforward matter to determine the optimal foraging strategy of the animal.

Let us label the best prey as type 1 and the next best as type 2. In this example, the best prey is the one that is handled most quickly. The handling time for type 1 is t_1 and that for type 2 is t_2. By our labeling convention, $t_1 < t_2$. Next, let a_1 denote the abundance of type 1 and a_2 the abundance of type 2. The abundances, a_1 and a_2, are in units of number of encounters per unit searching time.

The best strategy is one of two possibilities. First, the animal may take only type 1 and ignore type 2. Second, the animal may take both. (A third strategy, taking only type 2, is never the best strategy because taking both type 1 and type 2 is always better than taking only type 2.) Let us, then, compare the average handling times of the two candidates for the optimal strategy.

The average time to find, catch, and handle a prey item for the strategy of taking only type 1, T_1, is

$$T_1 = \frac{1}{a_1} + t_1$$

The first term is the time between successful encounters (searching time). The second term is the time needed to process (processing time) an item that has been encountered and captured. Here is how this formula is developed. Because a_1 has the dimensions of number of prey encountered per unit searching time, its reciprocal must have the dimensions of time per encounter. This term is the average time spent searching for and catching prey. Once prey is caught, the handling time occurs. For the strategy of taking only type 1 to be the best strategy, it must have a lower average foraging time per prey item than the strategy of taking both prey types.

The average total foraging time for the strategy of taking both prey types, $T_{1\&2}$, is

$$T_{1\&2} = \frac{1}{(a_1 + a_2)} + \frac{a_1}{(a_1 + a_2)} t_1 + \frac{a_2}{(a_1 + a_2)} t_2$$

The first term, as before, is the time spent searching for and catching prey. Because both types are being taken, the total number of encounters per unit time is $a_1 + a_2$. The reciprocal of the total number of encounters per unit time is the time per encounter with either prey. The fraction of these encounters that involves prey of type 1 is just a_1 over the total number of encounters. When type 1 is encountered, the animal spends the t_1 in handling. Hence, the second term is the handling time spent after captures of type 1 prey, and the third term is the handling time spent after captures of type 2 prey.

Now the animal should take only type 1 and ignore type 2 if

$$T_1 < T_{1\&2}$$

It should take both type 1 and type 2 whenever they are encountered if

$$T_{1\&2} < T_1$$

When we substitute the expressions for T_1 and $T_{1\&2}$ into these inequalities, and rearrange, we obtain the following result. The animal should take only type 1 and ignore type 2 if

$$a_1 > \frac{1}{(t_2 - t_1)}$$

It should take both prey types when encountered if

$$a_1 < \frac{1}{(t_2 - t_1)}$$

This result is interesting because only a_1 is involved; the abundance of type 2 has no effect. If the abundance of type 1 prey is high enough, then type 2 should be ignored. If the abundance of type 1 is low, then type 2 should be added to the diet. Thus, whether type 2 is added to the diet is independent of the abundance of type 2. In particular, if type 1 is high enough so that it should be the only item in the diet, then increasing the amount of type 2 in the environment should not cause it to be added to the diet in the optimal strategy. Another prediction is that if the two items are fairly close in their handling times, so that t_2 is close to t_1, then the abundance of type 1 must be quite high for it to be the only item in the optimal diet. We should see finer discrimination when prey are abundant than when prey are rare.

Limitations of Simple Foraging Theory

Our example of how to calculate an optimal foraging strategy is a simple instance of the theory developed more fully in Schoener (1971), Pulliam (1974), and Charnov (1976a). Is this theory of foraging behavior too simple to be useful with real organisms? To answer this question, one should first consider that the theory itself can be made more realistic and that, to some degree, the same predictions emerge. For example, the model can easily be extended to more than two kinds of prey. The theory again predicts that whether a prey type is added to the diet does not depend on *its* abundance but does depend on the abundance of all the prey types with a higher rank.

Also, the prey do not have to be valued in terms of the handling time. Instead, the energy content of the prey types can be used, assuming that the handling times are approximately the same. If the prey are ranked by their energy content, then it is assumed that the organism's objective is to maximize the rate of energy capture. The theory once again predicts that whether a food type is added to the diet depends not on its own abundance but on the abundance of the food types with higher ranks.

Thus, the predictions are not *limited* to the assumptions of the simple foraging model just presented; they still follow in more general models. However, the predictions cannot be generalized indefinitely. Some of the assumptions are critical to the results. For example, the theory does not incorporate constraints on the ability of the animals to learn about the environment. After all, how can an animal find out how long it takes to handle a prey type if it does not catch it now and then? Even if the optimal diet excluded prey types whose ranks are lower than a cutoff, the animal would have to take such types once in a while just to see what they are.

The theory also does not incorporate changes in rank as a result of gaining experience in catching and handling prey. This notion is important because as a prey type becomes more abundant, the predator may acquire more experience with it and learn how to reduce its handling time or how to catch it more easily. As the animal acquires such experience, the rank of the item may increase. Hence, this prey may become incorporated into the diet even though it was initially predicted to be absent from the diet when it was rare with a low rank.

In sum, simple foraging theory has a chance of being approximately correct for actual organisms, but it also runs the risk of leaving out too many factors that could substantially change the predictions. Let us now turn to some case studies to see how well optimal foraging theory does at predicting the diets of actual organisms in their natural environment.

Foraging Theory and Prediction of Actual Diets

An immediate prediction of simple foraging theory is that animals should become more selective in their choice of prey as the overall abundance of prey increases and should become less selective as prey becomes scarce. This variation in selectivity has been observed in many organisms, ranging from barn owls (Herrera 1974) to starfish (Menge 1972).

In addition, experiments show that the degree of selectivity changes with food availability. Werner and Hall (1974) allowed a group of 10 bluegill sunfish to hunt for three size classes of *Daphnia* in an aquarium. The value of prey ranks with its size because the handling times are the same for the various sizes of *Daphnia*. When abundant food was placed in the aquarium, the fish took mostly the large prey and ignored the smaller size classes. In contrast, they took all size classes when prey were rare. Similarly, Pacala and Roughgarden (1982) varied the number of animals that were eating the food, rather than the quantity of food itself. The experiment used *Anolis* lizards in their natural habitat on the island of St. Maarten in the eastern Caribbean. These lizards prey primarily on the insects that occur on the forest floor, and, as with *Daphnia,* the value of a prey ranks with its size. Sixty animals from one species (*Anolis gingivinus*) were placed in each of two enclosures. Sixty animals from *A. gingivinus* plus 100 animals from another species (*A. wattsi*) were placed in each of two other enclosures. The individuals of *A. gingivinus* in the enclosures with 160 animals caught fewer prey, and took smaller prey, than did the individuals in enclosures with 60 animals.

These examples show how an animal's selectivity changes when the level of both high-ranked and low-ranked food is simultaneously varied. According to the theory, only the variation in the level of high-ranked food causes the addition to or removal of low-ranked types from the diet. Variation in the abundance of low-ranked types should be irrelevant. But is the diet really insensitive to the abundance of the low-ranked types? If the abundance of a low-ranked type is increased, while the abundance of each higher-ranked type is left unchanged, will the low-ranked type start to turn up in the forager's diet?

To answer this question, Krebs *et al.* (1977) used caged birds (great tits—*Parus major*) as predators and two sizes of meal worms as prey. When large and small prey were present in low density, the birds were not selective, in accordance with simple foraging theory. When the density of the large prey was increased sufficiently, the birds ignored the small prey, thus becoming highly selective, as predicted by foraging theory. Finally, keeping the density of large prey constant, the density of the small prey was increased so that they were twice as common as the large prey. The birds remained highly selective and ignored the small prey mostly, but not completely. In addition, the amount of small prey that was taken did depend somewhat on its abundance in the experimental environment. Thus, the experiment shows that the foraging behavior of these birds is approximately what would be expected by simple optimal foraging theory, but it is clearly not exactly what would be expected.

This failure of simple foraging theory to accord exactly with the actual foraging behavior of animals has two reasons. First, the theory is simple and could be made to fit nature more closely by adding more to it—by adding, for example, a description of how an animal detects the value of a potential food item and how its experience with an item may change its ranking of that item. Such additions to the theory might provide a nearly exact prediction of animal foraging in some circumstances. Second, there is more to an animal's life than catching food: courtship, territorial defense, and so forth. Possibly an animal's behavior has more than one objective; when a butterfly searches, for instance, it may be looking for both food (nectar) and mates. If so, even a complicated theory of foraging would not provide an appropriate explanation of the animal's behavior.

These qualifications notwithstanding, foraging theory may offer a useful beginning for understanding animal foraging in the field. For some applications, the difference between the predictions of foraging theory and what actually happens may not be important, even though such differences may be detectable.

Foraging Theory and Animal Movement: The Marginal Value Theorem

Foraging theory began with the notion of what animals choose to eat, but in recent years it has been extended to related questions such as when animals stop foraging in one area and move to another. For example, because flowers often are grouped in clusters, we could hypothesize that as a bee extracts nectar from the flowers in a cluster, the remaining nectar becomes harder to obtain. The question then arises of when the bee should leave one cluster of flowers in search of another that might offer a higher rate of return.

Foraging theory suggests a simple criterion that may be used by the bee in deciding when to leave a cluster of flowers. The animal is assumed to know what the average rate of food capture is throughout the entire habitat and what its rate of intake is in its current cluster. The average rate of food capture for the entire habitat is calculated over a long enough time for the transit time between many clusters to be included, together with the yield from those clusters. Presumably, this average rate for the habitat is known from the bee's accumulated experience there over several days or weeks. The criterion for when to leave the current cluster is to *leave when the rate of food intake from that cluster equals the average yield for the habitat*. The reason is simple: the yield from the current cluster will only decline, but if the animal leaves when the yield from the current cluster equals the average yield in the habitat, then it will, on the average, do just as well somewhere else as it is doing currently—and perhaps much better. Thus, the point at which the yield from the current cluster equals the habitat yield is the point at which the animal should leave and travel to another cluster. This criterion has come to be known as the *marginal value theorem* (Charnov 1976b).

Whitham (1977) tested the marginal value theorem with bumblebees in the southwestern United States. The bee, *Bombus sonorus*, pollinates the desert willow, *Chilopsis lincaris*. When the flowers are full of nectar, it may be collected both from a pool at the base of the corolla and from grooves in the flower that radiate out from the corolla base. A bee can take nectar from the pool much faster than from the floral grooves. The nectar in the pool is an overflow from the floral grooves. At the beginning of the day, when the pool at the base of each flower is filled, the bee should, according to the marginal value theorem, feed only from the pool. It should not feed from the floral grooves because it can get a higher feeding rate by moving to another flower than by exhausting the flower it is currently using. However, by late morning, the nectar pools have been used up, and now the best strategy should be to begin taking the nectar remaining in the floral grooves. This pattern of foraging, predicted by the theory, is what actually occurs.

The caloric gains and costs of nectar feeding for this bee were also estimated, and the time that the bee should remain in a flower at various points during the morning was predicted. The numerical predictions, however, were not very close to the actual behavior. Only the general pattern of nectar foraging by this bee accords with what is expected by optimal foraging theory.

Bumblebees have been well-studied as subjects of optimal foraging theory. Pyke (1978) and Zimmerman (1979) have also investigated movement patterns of bumblebees. In addition, recent extensions to foraging theory that explicitly include the role of environmental uncertainty were developed by Real (1981) and Real *et al.* (1982) and were illustrated using the foraging behavior of bees and wasps.

THE FLOW OF NITROGEN IN ANIMALS

Although the supply of nitrogen for plants is often limiting, for animals nitrogen is often present in excess. The reason for this difference is that, whereas plants obtain carbon from CO_2 from the air and nitrogen from the ground, animals generally obtain both carbon and nitrogen together in their food. Thus, in plants, the uptake of carbon and nitrogen is relatively decoupled, but in animals they are taken up simultaneously. For animals, the actual ratio of carbon to nitrogen that is taken depends on the relative amounts of protein and carbohydrate in the food.

Nitrogen is used for the synthesis of proteins and nucleic acids, but it is not itself a source of energy. When an animal's feeding rate yields more nitrogen than it is using for the synthesis of proteins and nucleic acids, the excess nitrogen is excreted in three categories of nitrogen-containing compounds—NH_4^+, urea, and uric acid. The ammonium ion, NH_4^+, is produced when the amino group is stripped off an amino acid; its production requires no energy. However, this ion is highly reactive and generally toxic to animals. It may be used as the main nitrogen excretory compound when the animal can afford to excrete a large amount of water, as in freshwater fish and other aquatic animals.

Urea is another important nitrogen excretory compound; it has one carbon and two nitrogen atoms. It is not nearly as toxic as ammonium, and high concentrations of this compound are found in some animals. In sharks, as we shall see, it is also used in the regulation of osmotic balance.

Urea is soluble in water. However, its excretion does not require as much water from the animal as does the excretion of ammonium because it can be excreted in high concentration. The synthesis of urea requires energy.

The most expensive and least water-requiring compound for excreting nitrogen is uric acid. This compound is not water soluble and is excreted with a small amount of water as a semisolid paste. Uric acid and some close chemical derivatives are excreted by birds (Dantzler 1982).

An important exception to the generalization that animals have excess nitrogen is that of nectar feeders, including adult butterflies and hummingbirds. Female adult butterflies in particular have a high demand for nitrogen because in this phase of their life cycle they are laying eggs. The paucity of amino acids and proteins in nectar, combined with the high demand for nitrogen, might make nitrogen a limiting resource for adult butterflies. Indeed, many butterflies accumulate nitrogen while feeding as caterpillars and manufacture eggs prior to emerging as adults. A few tropical *Heliconius* butterflies have the ability to extract amino acids from pollen (Gilbert 1972). Many hummingbirds also take insects as prey in addition to their nectar feeding, presumably to provide a supply of nitrogen.

THE FLOW OF WATER AND SALTS

The remaining major flows between animals and the environment involve water and salts. These flows vary in the way they operate across the animal kingdom, and among habitats, but the basics are as follows.

The Salinity of Cells and Body Fluids

Animal cells and body fluids may be classified as being more saline than, equal to, or less saline than the water in their environment. If the cell contents and body fluids are just as saline as the environment, then no net tendency exists for water to leave or enter the animal by osmosis across its membranes. Most marine invertebrates are in this category. They have no water balance problem unless they are somehow transported outside of their customary salinity range. When such marine invertebrates encounter brackish water, they may actually swell up as they take in water from the environment. For example, the starfish, *Asterias rubens*, gains up to 30% in volume when the sea water in its aquarium is diluted with fresh water to make a 60:40 ratio of sea water to fresh water. Conversely, it loses volume if placed in concentrated sea water (Prosser 1973).

Bony Fish versus Cartilaginous Fish

If the cell contents and body fluids are less saline than the water of the environment, then the water within the animal tends to flow out to the environment and salts tend to flow into the animal from the environment. Fish generally have tissues with fluids less salty than sea water. For fish, the sea is a desiccating environment. Bony fish have a different physiological technique for balancing this tendency to lose water than do cartilaginous fish such as sharks and rays.

Bony fish continually lose water from their gills. To replace it, a fish absorbs water in its gut; it "drinks" sea water. Apparently a fish actually obtains this water in an indirect manner. A fish moves enough salt into special tissues of the gut so that these tissues become more salty than the sea water. Then the direction of osmosis reverses, and water enters these tissues. The salt is then extracted, using metabolic energy, and apparently some of the salts (the divalent ions) are released in urine, while other salts (the monovalent ions) are released from the gills. In a sense, bony fish "pump" salts. By doing so, they obtain water that replaces their continual loss of water from the gills. Thus, a bony fish in the ocean is in a dilemma familiar to desert animals, which also continually lose water to the environment as they breathe. Since fish also exchange O_2 and CO_2 across their gills as they "breathe," they inevitably lose water just as a desert animal does.

However, cartilaginous fish, such as sharks and rays, do not lose water across their gill membranes. Their body fluids are as saline as sea water, perhaps slightly more so. What is novel here is that a shark's body fluids are not high in inorganic salts but instead are high in urea. Urea is often used as a waste product for the excretion of nitrogen, as we mentioned in the last section. Although the identity and concentration of the molecules in the body fluids of sharks and rays are not the same as those in sea water, the overall level of material dissolved in the body fluids is high enough that it is osmotically equivalent to sea water. As a result, water does not have a net tendency to flow out of the shark.

Both bony fish and sharks expend metabolic energy in the control of their water balance. Bony fish pump salt, and sharks synthesize urea. The costs of these two methods of controlling water balance are unknown, so their comparative efficiency cannot presently be evaluated.

Birds and Other Animals

Coastal and oceanic birds also have body fluids that are less saline than sea water. These birds regularly swallow sea water while catching their prey. Many sea birds, including petrels, gulls, pelicans, and penguins, have special salt glands that secrete a concentrated salt solution. These glands

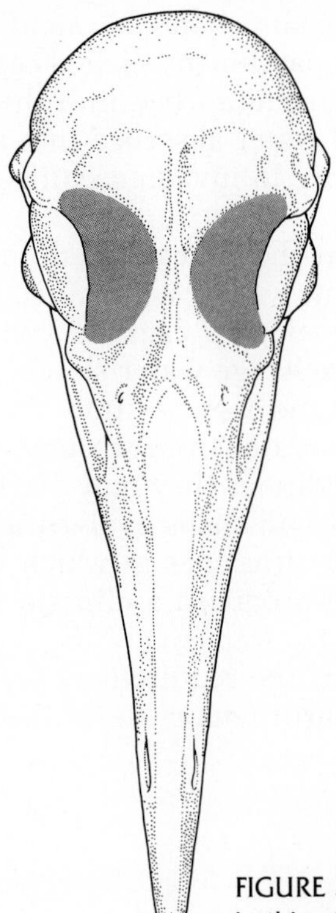

FIGURE 3–8 The position of the salt glands in a sea gull are shown in this picture of a skull by the shaded areas between the eyes. *(Figure courtesy of L. Argontatus.)*

are in the head and empty into nasal passages, as illustrated in Figure 3–8. Some marine reptiles, including the marine iguana of the Galapagos, and the loggerhead turtle also have salt glands. In a large group of marine iguanas on the Galapagos, some individuals can always be seen and heard "snorting" saline spray from their salt glands.

At the other extreme are the animals that live in fresh water. Because their body tissues are generally more saline than the water, these animals experience a continual influx of water from the environment. Freshwater mussels (*Anodonta cygnaea*), crayfish (*Astacus fluviatilis*), frogs (*Rana esculenta*), and freshwater fish simply excrete water in copious volumes of urine. Even freshwater protozoa, such as *Paramecium,* have the ability to pump water out of the cells. Furthermore, freshwater animals retain salt in their bodies and take up additional salt from the environment, using metabolic energy (Hill 1976).

The flow of water may be critical to the survivorship of eggs that are buried in the soil. Reptilian eggs have been classified into two main types: parchment-shelled and rigid-shelled (Packard *et al.* 1977). Parchment-shelled eggs have shells thin enough to be flexible, and these eggs rapidly exchange water and have the capacity to absorb much water (Tracy 1982). The survival and growth of embryos of parchment-shelled eggs appear to depend on the absorption of water; water absorbed into the developing embryo leads to larger hatchlings, and water absorbed into the albumin of the egg serves as an antibiotic barrier to invading pathogenic microorganisms.

It has been proposed that parchment-shelled eggs could have evolved in reptiles living in environments in which water is a limited resource for adult females but available to the developing eggs. In contrast, rigid-shelled eggs have a calcareous shell, exchange water slowly, and have a relatively large albumin content at the time of oviposition. The developing embryos appear not to need to absorb water from the environment, although if incubated under very dry conditions they may produce stunted hatchlings. It has been proposed that rigid shells evolved as protection against predators and pathogens under situations in which water is not limiting to adult females. Muth (1980) produced a climate space based jointly on the water and temperature requirements of the desert iguana, *Dipsosaurus dorsalis*. He concluded that the requirement for sufficiently moist soil along with a moderately warm temperature limited how far north the desert iguana lives.

CHAPTER 4

Substance, Form, and Function: The Stuff of Living Things

Animals and plants are often splendidly adaptable, yet the weight and strength of the basic materials that constitute living things also set limits. Bones and other structures act as mechanical levers, enabling organisms to run, swim, or fly, but by the same token the principles of engineering and of fluid mechanics also constrain what organisms can do. Finally, organisms are limited in their design because only a few basic "body plans" presently exist. Certainly we could design some interesting animals using existing biological materials and correct principles of mechanics, but unless genetic engineering allows genes to be artificially designed and synthesized, the design of animals is limited by the characteristics of their ancestors.

MOVING IN FLUIDS: FLOATING, SWIMMING, AND FLYING

The ecology of oceans and lakes shares principles with the ecology of terrestrial systems, but there are also profound differences. Whereas animals in oceans and lakes often spend all or nearly all their lives suspended in the water, no terrestrial organisms permanently live in the air. Even albatrosses, on the wing for most of their lives, eventually return to land to breed. Unlike animals of the air, animals in water are able to stay afloat with little energetic cost by using structures that produce buoyancy. Although a cubic centimeter of muscle or bone is heavier than a cubic centimeter of water and hence tends to sink, because an animal acquires the same *overall* density as water, the organism tends to remain at its depth without rising or sinking.

How Animals Obtain Buoyancy

Animals often obtain buoyancy by producing tissues, such as fat tissues, that are lighter than water and thereby counteract heavier tissues such as bone and muscle. Other methods of buoyancy involve structures with air pockets (Alexander 1968, Bligh *et al.* 1976). For instance, the *Nautilus* (Figure 4–1) has air chambers enclosed by a rigid wall and floats by maintaining gases in these chambers at a lower pressure than the pressure in the surrounding medium. The cuttlefish has a structure homologous to the air chambers of the *Nautilus* called the *cuttlebone.* This "bone" is a structure with chambers from which the animal withdraws water, leaving a partial vacuum.

Bony fish have structures with flexible walls called *swim bladders.* The volume in a swim bladder expands or shrinks when the fish changes depth. With a given quantity of gas in the bladder, only one depth exists at which the fish is exactly neutrally buoyant; at other depths, the fish must add or subtract gas to produce neutral buoyancy. For a fixed volume of gas, any slight ascent will lead to an expansion of the bladder, which in turn will make the fish rise still faster. Alternatively, any slight dive will compress the bladder, reducing its effective buoyancy, and lead to further loss of depth. Fish vary in the extent to which they can control the amount of

FIGURE 4–1 Nautilus. *(After Alexander 1968.)*

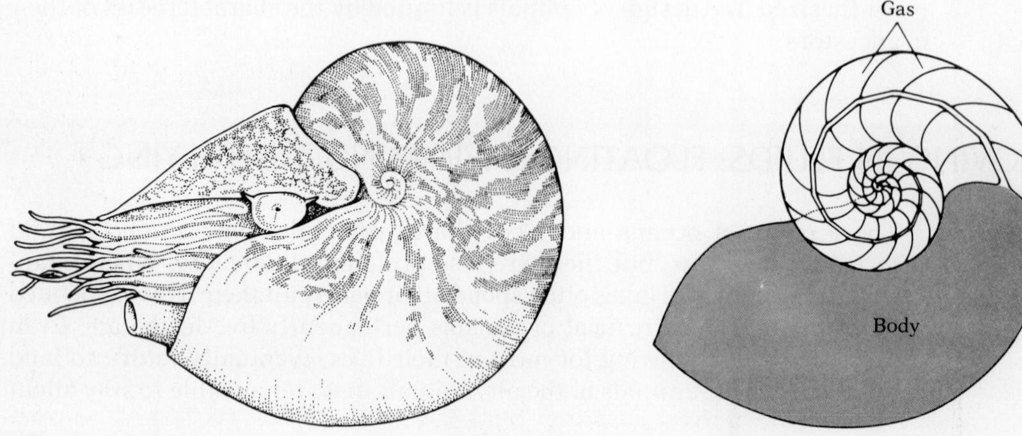

gas in their swim bladders, and those that cannot control it at all must be constrained to a narrow range of depths.

Similar considerations apply to diving mammals. As every scuba diver knows, using a "buoyancy compensator" type of life vest requires adding air at depth and letting air out while ascending to the surface. Also, a scuba diver who inhales at depth *must* exhale while doing "a free ascent" all the way up to the surface; otherwise the expanding air will rupture the lungs.

Many marine forms have air bladders of some kind. They are common in jellyfish, especially in large hydrozoan colonies such as the portuguese man-of-war. They are also common in marine algae, such as the sargasso weed that inhabits the Sargasso Sea, a gyre (area with a circular current) in the Atlantic Ocean south of Bermuda. Kelps also have large air bladders that allow them to reach the surface of the water, where the light is far more intense than at the bottom where they are attached.

Swimming, Drag, and the Reynolds Number

Many marine forms rely on swimming, rather than buoyancy mechanisms, to achieve their position above the sea float. However, to move forward in a fluid, an animal must overcome *drag*, a force resulting from the two components of pressure and friction. To see the first component, imagine that you have placed a small disk of plastic, 5 cm in diameter and 1 mm thick, in a stream of water. If you hold the disk with its full surface facing the direction of flow, you can feel the water push the disk downstream. This push is caused by the *momentum* of the flowing water. If you hold the disk with its edge to the direction of flow, you will again feel a force pushing the disk downstream, but it is produced largely by the water's *viscosity*—the resistance of a liquid to flowing over itself. Because the water very near the disk tends to stick to the disk and thus does not flow as fast as the water farther away from the disk, the water has to flow over itself to get past the disk. This imposes a force on the disk.

The momentum of the water, then, causes the *pressure component* of drag, and the viscosity of the water causes the *friction component* of drag. The total drag on an animal depends on its shape and speed. If it is big and bulky, then the drag is mostly pressure drag; the mark of a large pressure drag is a large wake behind the object. If the animal is streamlined, then the drag becomes increasingly restricted to friction drag.

Mechanical engineers have developed an index called the *Reynolds number* to describe the relative amount of the pressure and friction components in the drag. It is useful for understanding the forces operating on objects in any fluid, including water and air. The Reynolds number is defined as $(\rho ua)/n$, where ρ is the density of the fluid, n is the viscosity of the fluid, u is the speed of the fluid relative to the object, and a is the length of the object, usually measured along the cross section relative to

TABLE 4–1 Density and viscosity of fluids[a]

	Density, g/cm^3	Viscosity, (dyne·sec)/cm^2
Air	1.3×10^{-3}	1.8×10^{-4}
Water	1.00	0.010
Glycerin	1.26	15

Source: After Alexander (1968).

[a] The dyne is a unit of force; it is the force required to accelerate a free mass of 1 g at 1 cm/sec^2.

the direction of flow. The density and viscosity for three representative fluids are given in Table 4–1. The Reynolds number is a dimensionless quantity, and the units divide out, provided that the animal's size, a, is expressed in centimeters and the velocity of the water is expressed in centimeters per second.

It is possible to classify what sort of resistance an object encounters when moving in a fluid on the basis of the Reynolds number. If the Reynolds number is very low, less than 1, then motion is opposed mostly by viscous forces (that is, the drag is mostly friction drag). Because friction drag is proportional to surface area, small organisms tend not to be streamlined, but instead rather bulky, thereby minimizing the ratio of surface area to volume. If the Reynolds number is high, say, 1000 or more, then motion is opposed primarily by inertial forces (the drag is mostly pressure drag). Pressure drag is proportional to the area of the organism projected along the direction of movement. At high Reynolds numbers, organisms tend to be slender and streamlined. Between Reynolds numbers of 1 and 1000, both pressure and friction drag are important.

To calculate the Reynolds number, the animal's size and speed are multiplied together in the numerator of the formula. As a result, most biological situations involve large Reynolds numbers because the animal is big, is moving fast, or both. However, for tiny creatures suspended in the water column, the Reynolds number is very low. To them, water seems as viscous as molasses does to humans. Indeed, Koehl and Strickler (1981) have shown that the bristles on a copepod do not function as a comb to strain out food particles from the water, as had been believed. Instead, the combs actually function more as paddles that propel water in currents about the copepod.

The Principles of Flight

Swimming and flying are a natural consequence of moving in a fluid. The force against an object moving horizontally through a fluid can be divided into a component that acts against the front of the object and another component that acts either up or down on the object. The *drag,* as just discussed, is the component of the force that acts against the front of the object; the component that acts up or down is the *lift.* To swim or to fly, the animal must have wings or fins that produce upward lift and a minimum amount of drag. The cross-sectional shape of a wing or fin that works well depends on the Reynolds number. The cross-sectional shape of airplane wings works well for Reynolds numbers that are very high, on the order of 200,000. At lower Reynolds numbers, on the order of 50,000, the traditional airplane cross section is not as good as either a flat plate or a curved plate. At low Reynolds numbers, on the order of 100, the drag greatly exceeds the lift even for flat and curved plates, and horizontal swimming ceases to provide any vertical lift.

The tail flukes of a dolphin have the cross section of an airplane wing. They function at a Reynolds number of over a million, a value where this design is effective. In contrast, insect wings work at Reynolds numbers on the order of 5000, where the airplane type of cross section is much less effective in producing lift than is a flat plate. Birds work at Reynolds numbers between about 50,000 and 150,000. Thus, birds are more or less in the realm of Reynolds numbers where traditional airplanelike designs can be effective (Alexander 1968, Vogelmann 1982).

Birds, of course, cannot maintain equilibrium in the fluid they inhabit by changing the pressure of a gas in an internal bladder. They must move constantly relative to the air or fall to earth. Over the past 40 years, largely thanks to high-speed photography, the aerodynamics of birds have become rather well understood (e.g., Brown 1961, Pennycuick 1975). Like airplanes, birds generate lift by having their wings intercept the airflow (relative wind) at an angle known as the *angle of attack.* That angle and the shape of the wing mean that the air traveling over the top of the wing must travel a greater distance than that moving under the bottom. According to a theorem advanced in 1738 by the Austrian physicist Daniel Bernoulli, when the velocity is high, the pressure is low, and vice versa. For a bird's wing, greater pressure exists beneath the wing than above it, and the difference is the lift. You can easily demonstrate how lift works by blowing over the top of a strip of paper held lengthwise. If, however, the angle of attack becomes too large, the smooth flow of air over the upper surface of the wing separates from that surface and becomes turbulent; that is, the wing "stalls," and lift is lost.

When lift is equal to the weight of the airplane or bird, it will be able to remain at a given altitude. At that point, the lift of the flying object is balanced by air being thrust downward. In birds, most of the lift is generated by the inner portion of the wing (the so-called secondary feathers), the part that remains closed and together, presenting an airfoil much like an airplane's wing. Birds also have a structure, called the *alula,* composed of a few feathers on the "thumb" that form a leading-edge slot. This slot helps to maintain a smooth air flow at a higher angle of attack than would otherwise be possible. This device helps maintain flight at slow speeds and is similar to slots used on airplanes for the same purpose.

Something, however, must be capable of thrusting the airborne object through the air—to provide the *thrust* necessary to overcome more than drag and to provide the requisite lift. Interestingly, this problem is solved for birds in much the same manner as it is solved for propeller-driven airplanes. An aircraft's propeller blades are airfoils, just like airplane wings. The blades meet the relative wind at an angle of attack and generate lift according to Bernoulli's principle. They do so, however, at 90° to the direction in which the wing generates lift, and so they pull the airplane through the air.

Birds, of course, do not have propellers. On the downstroke of the wing, the primary feathers, toward the end of the wing, move *forward*. The upper surface of the primary feathers (indeed, the upper surface of much of the wings) is oriented partly forward and meets the relative wind at an angle of attack much like that of propeller blades. The downstroke provides most of the thrust for a bird. On the upstroke, the wings are often folded near the body as they move up and backward, and the primaries separate from one another to reduce the air resistance. Some slight thrust is produced on the upstroke by the primaries, and the secondaries still provide lift.

Differences in Flying Apparatus

The general principles of flight are the same for all birds, although many complexities exist, especially in the hovering flight of hummingbirds. A bird's niche is related to its flying apparatus. For example, albatrosses and gulls have long, slender, flat wings—ones with a high *aspect ratio* (the ratio of length to width) and a high wing loading (ratio of body weight to wing area). Their long, slender wings make them especially good at rapid gliding. At the tips of a bird's wings, vortices form that generate drag. The higher the aspect ratio, the smaller this drag is in proportion to the lift-producing portion of the wing. Albatrosses, for example, are able to spend the vast majority of their lives airborne, taking advantage of the difference in the speed of the wind adjacent to the waves (where it is slowed by friction) and the stronger wind 50 ft or so above them. They acquire momentum by descending with the wind and dissipate it by rising against the wind; their great weight (for a bird) and high speed keep them stable through minor perturbations in the airflow. Such "dynamic soaring" often permits the albatross to go many hours without flapping its wings, as long as it continually drifts downwind (McFarland *et al.* 1979). Albatrosses appear to be close to the structural limit of how long a wing can be and how high a wing loading is possible.

Marine birds have long, slender, soaring wings because their environment contains quite dependable winds; albatrosses are largely confined to oceanic regions with continuous winds. However, land birds that soar, such as vultures, which depend upon rising columns of heated air ("thermals") or updrafts caused when winds blow against ridges, must be more maneuverable. For example, vultures must be able to fly in tight spirals in thermals to gain altitude, something that cannot be done at high speed. They have lower aspect-ratio wings, and they reduce the drag caused by wing-tip vortices by spreading their primaries so that each individual feather acts as a small, high aspect-ratio wing. The slotting produced by the gaps in those feathers and the lower wing loading provided by broad wings allow vultures to take advantage of even small packets of rising air.

Other birds also have the shape of their wings and other features of their flight apparatus (such as the configuration of the tail, which aids in maneuvering) related to their niches. For example, swallows and falcons fly quickly as they hunt on the wing. Many songbirds and game birds have wings that permit twisting, dodging flight, so that they can escape predators and weave through brush. All birds have an intricate set of morphological adaptations for flight (see, e.g., Welty 1979), including an astonishing array of devices for reducing weight. Unfortunately, we have not the space to discuss further either those adaptations or the fascinating topic of the evolutionary origin of flight (see, e.g., Ostrom 1974, Martin 1983).

THE STUFF OF LIVING THINGS

The basic materials available for biological construction can be roughly classed in three categories (Wainwright *et al.* 1976). First are the tensile materials, materials that bend but do not break. Examples include cellulose, collagen, chitin, and silk. Next are the pliant materials, such as skin, which are flexible and often used as coverings. Last are the rigid materials, including wood, bone, arthropod cuticle, and stony exoskeletons such as shells and corals.

The Strength of Biological Materials

The strength of a material is described by noting its response when a force is applied to it. The force is given per unit area. The response of the material is given by its change in length as a result of the force. For example, consider the force that stretches a tendon as an animal runs. About 75% of the dry weight of a tendon is made up of collagen fibers. A tendon in a normally running vertebrate stretches and shows a 3% increase in length. This stretching is called a *strain* by mechanical engineers; it is the change in length caused by the force. The force itself, divided by the cross-sectional area of the tendon, is called the *stress*. The material is said to be elastic so long as its response (the strain) is proportional to the force on it (the stress).

The constant of proportionality is called *Young's modulus*. Unfortunately, the constant of proportionality is not very constant for many biological materials, and the Young's modulus may vary somewhat with the level of force applied and with the time over which it is applied. Nonetheless, the modulus is useful because it is a measure of how much deformation results when force is applied to the material. If a tendon is stretched by about 4% or more, it breaks. The force required to break the tendon is called its *tensile strength*.

Predicting the Strength of Organisms

In principle, one can use this kind of information to predict what kinds of environmental forces an animal or plant can withstand. For example, a kelp such as *Macrocystis* has a holdfast that anchors it to the bottom. It has a long stype ("stem") with blades ("leaves") and air bladders for flotation. By determining the drag on the kelp and the tensile strength of its stype, one can determine the water currents that it can survive before the stype breaks.

Kelps are often observed washed up on ocean shores after storms, where they become food for many organisms that feed on drift algae. Kelps regenerate from their holdfasts. By knowing the distribution of water speeds in areas where the kelps are growing, one can determine the expected time between breaks of the stype. Because the reproductive tissue of a kelp is in its blades, the suitability of the habitat to an individual kelp depends on whether it has time to regenerate its blades before the arrival of a sufficiently turbulent storm. Koehl (1984) uses kelp as an example to explore the general question of how a benthic organism withstands a force exerted on it by water currents.

A similar question is how heavy a tree's canopy may be before its trunk "buckles." If you push down on a straw, it will quickly buckle. If one were to push down on the trunk of a tree and the trunk were too thin, it too would buckle (King 1981). Also, an interesting problem is posed by trees such as coconut and palm trees, which cannot thicken their trunks. They appear to begin growth with a trunk that is thicker than that required to support their crown, and they may then grow past the point where their trunk provides adequate support. During storms, coconut trees often lose their crowns, either because the wind drag has exceeded the tensile strength of the trunk or because the trunk has buckled.

Perhaps the most dramatic situation in which water flow imparts stress to organisms is provided by the animals that occupy exposed, wave-swept locations in the rocky intertidal zone. The organisms that live in the intertidal zone typically are smaller than comparable subtidal organisms. To explain this general observation, Denny *et al.* (1985) explored the hypothesis that the small size reflects the inability of larger organisms to withstand wave action. Specifically, they point out that the force imparted by the acceleration of water is higher on a large organism than it is on a smaller organism, even after the increased strength of the larger organism is taken into account. Hence, the force imparted by a wave's acceleration potentially offers a mechanical limit to the size of an intertidal organism.

Denny *et al.* (1985) developed a model for how much force is imparted by wave action and measured the strength of wave action on the exposed coast of Washington State. They concluded that the size of several species of limpets, and of a sea urchin, could be limited by the mechanical stress caused by wave action but that the size of several species of thadid snails, and of a barnacle, was not limited by mechanical constraints, but presumably by biological factors. An important alternative hypothesis for the generalization that intertidal organisms are smaller than comparable subtidal organisms is that the feeding time is less in the intertidal zone, and so both growth rates may be slower and the optimal final size may be lower (Sebens 1982b).

RESPONSES TO LONG-TERM CHANGES IN THE ENVIRONMENT

The lives of organisms are coupled with the environment on a long time scale, in addition to the scale of minutes and days that we have considered so far. Almost all places in the world have seasons, as we discuss further in Part 6. In the temperate zones, one season is typically wet and warm and the other cold and relatively dry. Mediterranean climates have a dry, warm season followed by a wet, cold season. Tropical latitudes typically have a warm, dry season followed by a warm, wet season. All these seasonal patterns involve an alternation of very different physical conditions. In general, organisms have activities, rates of development, and possibly even life spans that are synchronized with the seasonal patterns in their habitats.

Typically, one period of the year may be considered favorable to growth and reproduction, and the other period may be considered unfavorable. Three kinds of strategies have evolved with respect to such alternation of good and bad times. In the first, organisms weather the bad times by going into a dormant state, as in the hibernation of mammals, the diapause of insects, and the desiccated state in which lungfish survive when their pools dry up between rainy seasons. In the second strategy, organisms leave the habitat altogether, as do the many birds that migrate from temperate latitudes to the tropics for the winter. In the third, organisms fit an entire generation or more within the favorable season, as do annual plants and many of the insects that pollinate them. The unfavorable season is then passed as a dormant life stage—seeds in the case of the plant, and usually a preadult stage in the insects.

In addition, the reproductive activities of organisms are almost always synchronized with the seasons, even if the organisms remain in the habitat throughout the year without entering a dormant state. The timing of a plant's flowering and fruiting or an animal's seasonal activities is called its *phenology* and is part of its *life history*. We will have more to say about how plants and animals organize their lives in Part 2, when we take up the evolution of an organism's life history in more detail.

PART 2

POPULATIONS

Population is one of the major concepts in biology. In its most common use, a population is a collection of organisms of the same species occupying a defined geographic area. A species, which is itself a population, may consist of local populations that are separated in distance and that exchange migrants with one another.

In Part 1, we were concerned with the ecology of an individual. However, for any *type* of any individual to exist, many representatives of that type are required; if only a few individuals representing some type are presently alive, it is highly likely, just on the basis of chance, that the type will become extinct.

Moreover, evolution occurs only in populations. An individual may develop, learn, and change during its life, but it cannot evolve; evolution occurs only when the composition of a population changes over time. Although we may think of evolutionary history as a progression of individual forms—as of fish with fleshy fins crawling from the ocean onto land, which then evolve legs, and so forth—these individuals are only samples, samples drawn from a population whose composition is changing.

Ecologists and evolutionists consider populations to be so central to their disciplines that they often speak of the entire science covered in this textbook as *population biology*. In this part, we examine the properties of single-species populations. We start with a chapter on population dynamics—how and why populations change in abundance, and especially the impressive properties of the phenomenon known as *exponential growth.* We examine the external constraints on such growth, and then consider the age composition of populations under various growth regimes. Finally, we examine constraints that appear when the growth of a population starts to outstrip the carrying capacity of the environment.

The next chapter presents both a Mendelian approach to evolution and a more approximate approach that does not require as much genetic information. We then consider more complex issues, especially the ecologically critical question, "What, if anything, does evolution optimize?"

The final chapter of Part 2 deals with issues of direct interest to working ecologists—such matters as the relationships between the evolution and dynamics of populations, how altruistic behavior can evolve, and whether characteristics that are beneficial for groups and deleterious for individuals can be preserved.

In a sense, this part is the heart of the book, for the concepts discussed in it are essential to a deep understanding of everything from why marine iguanas have salt glands to why nutrients are flushed from disturbed forest ecosystems.

CHAPTER 5

Population Dynamics: What Determines Abundance and Rarity

A population may seem like an abstraction. One can see, hold, and touch an individual and directly appreciate how it lives in its environment. In contrast, a *population*—a collection of organisms of the same species occupying a defined geographic area—usually contains too many individuals and occupies too large an area to be observed in its entirety. One rarely measures the size of a population by counting the individuals in it, because all the individuals cannot be located and their number may be too large. Instead, the size of a population is usually measured by counting the number in some subset of the population and then statistically estimating the total size. This use of statistics may heighten the perception that a population is an abstraction. Yet a population *is* indeed real and concrete.

A population is like a reservoir, the reservoir's volume of water analogous to a population's size and its area analogous to the population's range. To understand how much water is in the reservoir, one must understand the inputs and outputs—the streams leading into it, the streams leaving it, the amount of rainfall, the amount of evaporation. For a population, the inputs are births and immigration and the outputs are deaths and emigration, and the rates of these inputs and outputs determine the population's size. If output rates exceed input rates, the population cannot continue to exist in that place, even though a single individual might be able to survive there for a long time.

How does the concept of population relate to the ecology of individuals as discussed in Part 1? The key is that *the physiological and behavioral abilities of individuals* in an environment *determine the rates of birth and death* in that environment. From the standpoint of a population, the way a plant operates its stomates or a bird catches insects are not important by themselves; they are important because of how they influence the number of seeds a plant produces and the number of hatchlings a bird raises.

MEASURING A POPULATION'S SIZE

What causes some organisms to be plentiful or rare? What causes their abundance to change through time? Our investigation of populations begins with how to answer these questions, the subject called *population dynamics*—the study of constancy and change in population size.

For many animal populations, a common measuring technique is to mark certain animals and then recapture or resight the marked animals. An example is as follows.

Suppose we observe and mark 40 animals during a day in a particular area—say, in an acre of open woodland. The next day suppose we see 50 animals in the same area, 25 of which were marked on the previous day. To estimate the population size in the area, we use the following ratio:

$$\frac{\text{Number of marked animals}}{\text{Total number of animals}} = \frac{\text{Number of resighted animals}}{\text{Total seen on second visit}}$$

The idea here is that the fraction of resighted animals observed during the second visit to the site is the same as the fraction of marked animals within the total population. Because we know the number of animals that were marked, we can solve for the total population size. In our example, we have

$$\frac{40}{N} = \frac{25}{50}$$

When we solve for the total population size, N, we get an estimate of 80 animals. The estimate of a population's size obtained in this way is called the *Lincoln index* of the population.

Methods are available that extend this basic idea and include formulas for placing a statistical confidence interval around the estimated population size (Otis *et al.* 1978). This measurement technique rests on two assumptions. First, not much birth, death, immigration, or emigration must occur in the population during the 2 days of the census. Second, animals seen on the second day must be an unbiased sample of the animals that had been marked as well as those that were unmarked—that is, marked animals are no more or less likely to be resighted or recaptured than unmarked animals. Methods now exist for determining the correctness of these assumptions and for modifying the formula if they are not (e.g., Heckel and Roughgarden 1979). Also, we note that the estimates are accurate only when a high fraction of the population is marked. The statistical error in the estimate is inevitably large if the percentage of animals resighted on the second day is less than 50%.

EXPONENTIAL POPULATION GROWTH

All populations have the potential for rapid, enormous growth in numbers, a fact so dramatic that it can be compared to a bomb: the number of individuals in a population can increase by a chain reaction in a way that parallels the detonation of an atomic weapon. Let us show how this works.

How to Identify Exponential Growth

To begin, let us consider a closed population—that is, one for which immigration and emigration can be ignored and whose dynamics are determined only by birth and death rates. Suppose the number of births produced by each individual in a year is B and the fraction of individuals that die during the year is D. Suppose N_0 individuals are present at the beginning of year 0. What will be the number of individuals at the beginning of year 1? The answer is

$$N_1 = BN_0 + (1 - D)N_0$$

because BN_0 is the total number of births and $(1 - D)N_0$ is the total number that survive (did not die) during the year. Here it is helpful to simplify matters by defining a constant R such that

$$R = 1 + B - D$$

We can then rewrite the formula for N_1 as

$$N_1 = RN_0$$

The number of individuals in the population in year 2 can be found by using this formula once again; that is,

$$N_2 = RN_1$$

Because we know N_1 from N_0, we have

$$N_2 = R^2N_0$$

Indeed, if we know N_0 and R, we can project the population for all future times, say at year t, as

$$N_t = R^t N_0$$

This formula shows that the number of organisms in the population at each time is a factor times the number initially present. The factor is obtained by taking the constant, R, and raising it to the power, t. Thus, t is in the exponent. When the population size is described by this formula, the population is said to be in exponential growth. Specifically, *exponential growth* is growth in which the population increase in a period is a fixed percentage of the size of the population at the beginning of the period.

FIGURE 5–1 Population size through time for microorganisms plotted on both arithmetic and log scales. *(After MacArthur and Connell 1967.)*

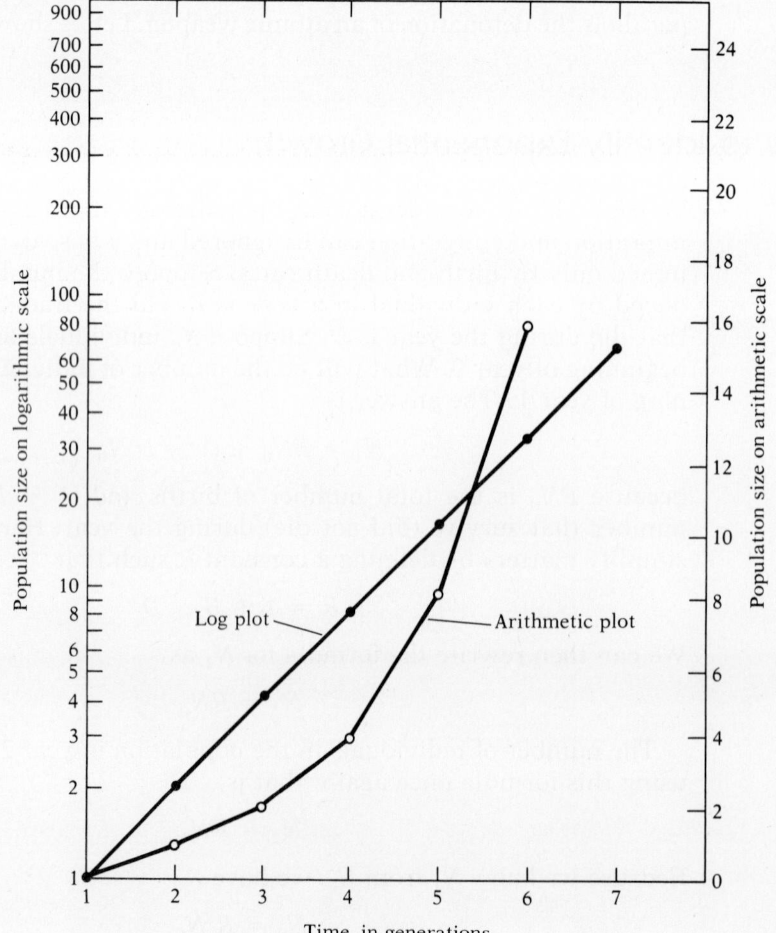

The way to tell whether a population is undergoing exponential growth is to plot the logarithm of N_t against t. Upon taking the log of both sides of the equation for exponential growth, we obtain

$$\begin{aligned}\log(N_t) &= \log(R^t N_0)\\ &= \log(R^t) + \log(N_0)\\ &= t\log(R) + \log(N_0)\end{aligned}$$

If, by plotting the log of N_t against t, one obtains a straight line where the slope is the log of R and the y-intercept is the log of N_0, then the population is growing exponentially, as illustrated in Figure 5–1. If, on the other hand, one plots the log of N_t against t and the graph is not a straight line, then the population is not growing exponentially.

A Population Bomb: Rabbits in Australia

Do populations grow exponentially in nature? Emphatically yes, though not very often or for very long. Perhaps the most dramatic instance that has been thoroughly studied was an explosion of the European rabbit, *Oryctolagus cuniculus*, in Australia. A map indicating the spread of the rabbit appears in Figure 5–2.

FIGURE 5–2 Contour lines for the explosion of a rabbit population in Australia, 1870–1960. Note multiple introductions after the original. *(After Myers 1971.)*

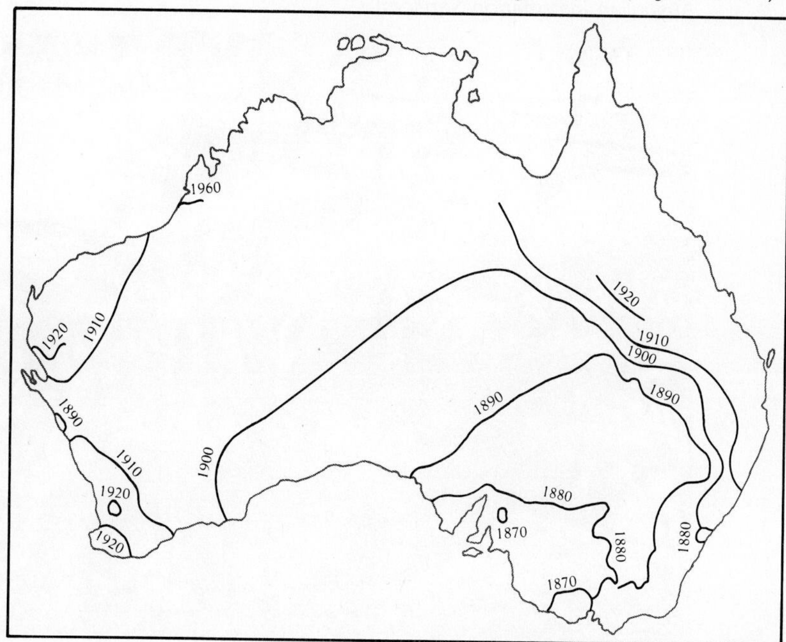

The rabbits were introduced by Thomas Austin, described as "an ardent acclimatizer who also introduced sparrows and who made several attempts to introduce rabbits before his efforts were crowned by undreamed success" (Kiddle 1961, quoted in Fenner and Ratcliffe 1965, p. 17). The introduction consisted of wild rabbits that had been collected in England and released on Austin's Australian estate in 1859. By 1865, 20,000 rabbits had been killed on his estate. As the population increase continued, the rabbits proceeded northward and westward with a speed which "we believe to be without parallel in the whole history of animal invasions," according to Fenner and Ratcliffe (1965, p. 23). The rabbits were first reported on the Queensland border in 1886, having crossed the state of New South Wales from south to north at a rate of 70 miles a year. Ultimately, rabbits appeared on the shores of the Indian Ocean—1100 miles away and only 16 years after their first sighting in southern Australia. Figure 5–3 shows an accumulation of rabbits around a water hole, depicting the high density that a rabbit population can attain. The following passage from Fenner and Ratcliffe (1965, pp. 23–24) indicates the severity of the Australian crisis:

> While the rabbit advanced, it was preceded by reports (growing ever more alarming) of its destructiveness in those areas where it had become established and had built up in numbers. With no experience and little biological understanding to guide them, the authorities tried to cope with the situation

FIGURE 5–3 European wild rabbits around a water hole at dusk. *(Photo by the Australian Information Service.)*

FIGURE 5–4 Rabbit fence in Australia. *(Photo by Richard Hobbs.)*

in various ways which proved ineffectual and sometimes very costly. The building of barrier fences deserves special mention, if only because of the vast scale of the effort involved [see Figure 5–4]. Many thousands of miles of rabbit-netted fencing were erected with the idea of protecting regions that were still rabbit-free. Perhaps the most famous was the 'No. 1 fence' in Western Australia that ran for over 1100 miles northward from a point on the south coast to the southern end of the Eighty Mile Beach. It took five years to build and was completed in 1907. The northern section of fence passed through virtually unexplored country at the edge of the Great Sandy Desert. Water had to be transported to the large number of men and draught animals that were employed, and much of the fencing material had to be packed on camels for the final stages on the journey. Although they may, in places, have brought temporary benefit, the barrier fences failed in their object of stemming the rabbit's spread, not only because they were often built too late, but because it proved impractical to maintain them at 100% efficiency, which was essential if they were to fulfill their expected role. Looking back, it is easy to condemn the effort as biologically unrealistic, and a waste of public funds; but at the time it seemed the only solution to a desperately serious problem.*

Range Expansion in an Exponentially Growing Population

As an exponentially growing population expands in range, the boundary of the population (its *wave front*) tends to move at a constant speed (Kendall 1948, Skellam 1951). Figure 5–5 shows the spread of the muskrat, *Ondatra zibethica*, from five individuals introduced into Bohemia in 1905, to 1927 (Elton 1958, from Ulbrich 1930). The square root of the area oc-

* Copyrighted by and reprinted with permission of Cambridge University Press from Frank Fenner and F. N. Ratcliffe, *Myxomatosis* (Cambridge: Cambridge University Press, 1965), pp. 23–24.

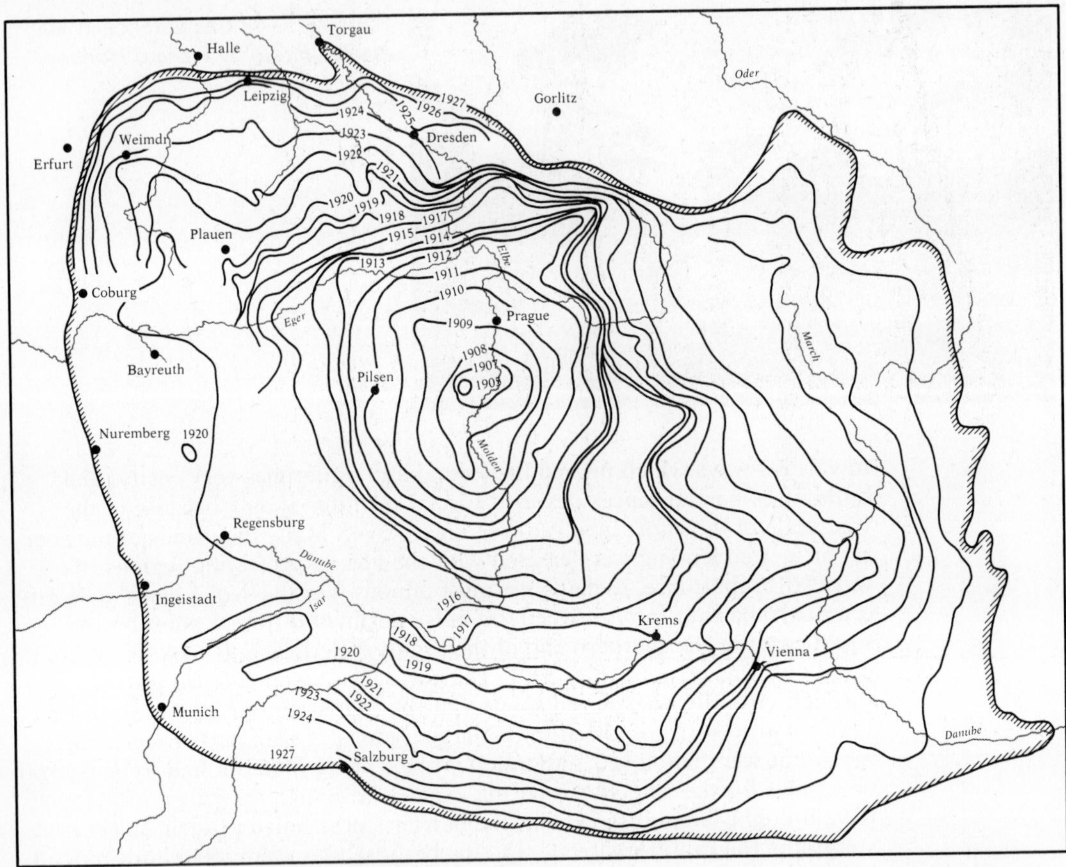

FIGURE 5–5 Contour lines for the spread of the muskrat *(Ondatra zibethica)*, 1905–1927. *(After Elton 1958.)*

cupied by the muskrat is plotted as a function of time in Figure 5–6, and a straight-line relation is evident (Skellam 1951). This observation is expected only when the habitat is more or less uniform throughout the region where the population is expanding and when the movement of each individual animal is more or less random in direction. In this case, the spread of the population can be approximated by equations that are related to those describing the diffusion of particles by Brownian motion.

Constraints on Exponential Growth

These two cases of exponential growth, the rabbit in Australia and the muskrat in central Europe, are examples involving introduced mammals, and as such are quite exceptional. In a review of the introductions of over

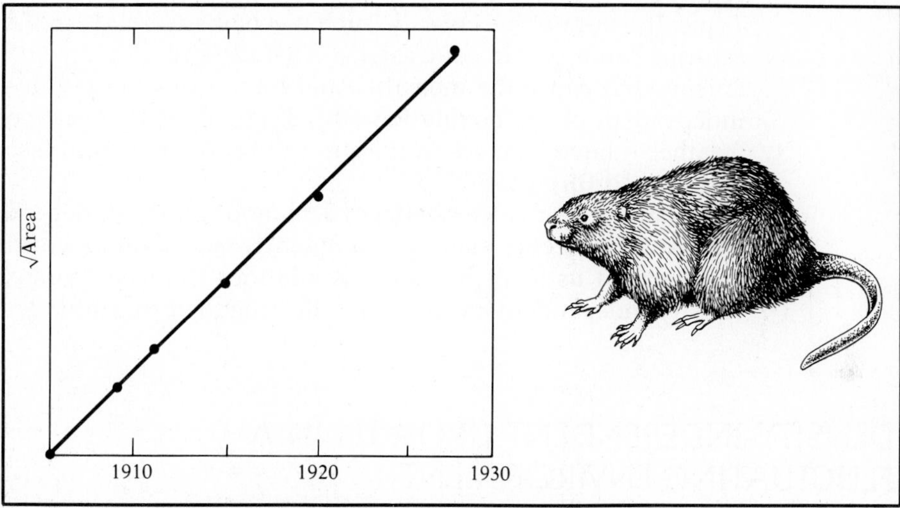

FIGURE 5–6 Straight-line relation between $\sqrt{\text{Area}}$ and time for the spread of muskrats in Europe. See Figure 5–5. *(After Skellam 1951.)*

200 species of mammals, De Vos *et al.* (1956) concluded that the great majority of the introductions failed or did not lead to explosive increase. Nonetheless, these two cases show that a population really can explode in number under natural conditions.

Why is such a population explosion rather unusual? One obvious possibility is that the population runs out of food, space, and other resources it consumes, so that the mortality rate increases, the birth rate decreases, or both occur before a gigantic increase in numbers is observed. However, mortality rate and birth rate are unlikely to be the same from year to year regardless of the amount of food, space, and other environmental resources. For example, even if a caterpillar has ample food, it still requires a suitable temperature for active feeding and digestion; if the temperature is cold, it will have to feed longer in order to reach the size at which it pupates (enters the resting stage, in which it is reorganized into a butterfly or moth), which means that the caterpillar will be susceptible to hazards (such as predation by birds) over a longer period of time. In short, even if ample food and space are present, mortality and birth rates may vary from year to year because of yearly variations in the physical conditions of the environment and exposure to hazard. This idea leads us to a less obvious reason why dramatic population explosions are rare.

Mechanisms that cause the mortality rate to rise or the birth rate to fall as the population size increases are called *density-dependent* factors.*

* *Density* is conventionally used rather than *size.* Population density is the number of individuals per unit area; if the area occupied by a population does not enlarge, an increase in density automatically means an increase in population size.

Typically, but not necessarily, such mechanisms relate to the depletion of limiting resources. In contrast, *density-independent* factors describe mechanisms that cause the mortality and birth rates to vary in a way that is independent of the population size. Typically, such mechanisms involve weather-related changes in the suitability of the habitat (see, e.g., Andrewartha and Birch 1954).

In nature, both density-dependent and density-independent types of mechanisms occur, and they frequently act simultaneously. To begin, however, let us focus on some populations that have almost exclusively density-independent mechanisms affecting their mortality and birth rates.

DENSITY-INDEPENDENT GROWTH IN A FLUCTUATING ENVIRONMENT

To understand a population's growth in a fluctuating environment, it helps to take a long-term view of the situation.

Averaging Over the Long Term

Suppose the growth factor, R, has a different value every year. Let R_t denote the growth factor in year t. Then the population size is determined by the R's in the following way. For N_1 we have

$$N_1 = R_0 N_0$$

Then for N_2 we have

$$\begin{aligned} N_2 &= R_1 N_1 \\ &= R_1 R_0 N_0 \end{aligned}$$

and so forth, so that for N_t we have

$$N_t = (R_{t-1} R_{t-2} \ldots R_1 R_0) N_0$$

Thus, the population size after a long time is determined by a product of the R's that were in effect during that time. Hence, after a long time, the population size is about the same as it was to begin with if the product of the R's throughout that time approximately equals 1.

A convenient way of thinking about the product of the R's is to use the *geometric average,* which is defined as the *nth root of a product of n numbers.* The geometric average for the R's in a fluctuating environment is

$$\text{Geometric average of } R = \sqrt[n]{R_0 R_1 \ldots R_n}$$

The geometric average of R is important because, after n time intervals, the population size in a fluctuating environment is the same as that of a population in a constant environment, provided that the R of the constant-environment population equals the geometric average of the R's for the fluctuating-environment population, and provided also that both populations had the same size to begin with. Hence, density-independent fluctuation in the growth rate, R, does not lead to a population explosion if the geometric average of the R's through time equals about 1. If the geometric average of the R's is much greater than 1, then the population eventually explodes; if the geometric average is much less than 1, it approaches extinction even if short-term population outbreaks occur.

It is significant that the geometric average, and not the usual arithmetic average, determines the fate of a population. The geometric average of a set of numbers is always less than the arithmetic average (unless all the numbers happen to be identical). For example, the arithmetic average of 0.5, 1.0, and 1.5 is 1.0, whereas the geometric average is the cube root of $(0.5 \times 1.0 \times 1.5)$, which is 0.909. Hence, if the population has an R of 0.5, 1.0, and 1.5 each roughly every third year, then it will approach extinction because the geometric average of its R is less than 1.

Density-Independent Dynamics of Butterflies

An illustration of natural extinction is provided by butterfly populations on Jasper Ridge, a biological preserve on the Stanford University campus. A checkerspot butterfly, *Euphydryas editha* (see color plate F), occurs in several local populations in the preserve. The adults fly in the spring. In this region (see color plates G–I), the caterpillars feed on leaves of two annual plants, a plantain (*Plantago erecta*) and Owl's clover (*Orthocarpus densiflorus*). Over 27 years of observation, the caterpillars have never become abundant enough to devour a significant fraction of the food supply. Instead, the feeding season for the caterpillars in the spring has been limited by the life cycle of the plants. That, in turn, has been limited by the amount of water held in the soil from the rains of the winter and earlier in the spring. With plentiful rain, the annual plants remain green until well into May, and many of the caterpillars reach the size at which they can successfully survive the dry summer in diapause, a special state of suspended animation. However, in drought years, the plants flower and senesce before most caterpillars can grow enough to reach the size for diapause. As a result, the prediapause mortality rate of caterpillars is exceptionally high in such years (Singer 1972, Singer and Ehrlich 1979). For these populations, only density-independent mechanisms have affected the mortality rate and birth rate of butterflies from year to year. It is significant that one of the local populations of butterflies on Jasper

Ridge became extinct in 1964. In 1966 the habitat was recolonized by butterflies from other local populations of the same species. Then in 1974 it became extinct again and has not subsequently been recolonized (Ehrlich *et al.* 1975, Ehrlich and Murphy 1981).

How Long to Extinction?

An entire species becomes extinct if all of its local populations become extinct simultaneously. The extinction of an entire species from the fluctuations in population size induced by density-independent mechanisms is a possibility. Yet some circumstances work against the extinction of an entire species, even though local populations within it regularly become extinct.

One circumstance that prolongs the life span of a species is that density-independent mechanisms may operate in different directions at different places within the species range. If one local population is on the decline, then chances are that some not too distant local population is on the rise. The population that has fared well can supply migrants to recolonize the place where a local population previously became extinct. This idea is known as "spreading the risk," to use a phrase popularized by den Boer (see, e.g., den Boer 1981).

Actually, spreading the risk of extinction occurs in two ways. First, the density-independent mechanisms that affect the population may function *independently* in each of the distinct local populations. For example, in drought-stressed habitats, small rain squalls may help some populations in one year and help others the next year. Whether one local population is touched by a squall is independent of whether another local population is touched. Such independence implies that only an improbable coincidence will lead to the simultaneous extinction of many local populations from density-independent mechanisms that operate on such a small scale.

Second, the biological situation of many of the local populations may be qualitatively different. Some local populations may be affected by certain density-independent mechanisms and other local populations by qualitatively different density-independent mechanisms. As Myers (1971) writes about populations of rabbits in Australia, "Each population is a case history in itself." *Euphydryas* butterfly populations also are affected by a unique set of density-independent mechanisms at each locale (Ehrlich and Murphy 1981). Populations of both *E. editha* and the related checkerspot *E. chalcedona* had varied responses to the California drought of the mid-1970s. Some, dependent on annual plants, were reduced in size or became extinct. However, an alpine population of *Euphydryas* feeding on perennial plants had its growing season extended and actually increased in size. Thus, the uniqueness of the biological situation of each local population can result in opposite consequences from the same large-scale environmental event. This effect makes it likely that in any year some local populations will prosper while others decline and reduces the probability of extinction for the entire species below that expected if the local populations are merely independent.

Demographic Stochasticity: Another Cause of Population Fluctuation

During the breeding season, organisms typically do not each have the same number of offspring; instead a *distribution* occurs in the number of offspring produced by a breeding pair. This distribution is particularly easy to see with nesting birds, because one can count the number of nestlings that are successfully fledged from each of many nests. For example, Lack (1966) notes that the number of birds raised per brood in different nests by the swift, *Apus apus,* at Oxford, England, varies between zero and three. If the number of offspring actually produced by each breeding pair can vary at random between, say, zero and three, even though the average is two, then the total population size will fluctuate somewhat as a result.

The fluctuation in population size caused by the random variation in offspring number that inevitably occurs even when the environment is practically uniform has been termed *demographic stochasticity* by May (1973). Demographic stochasticity has been studied in statistics using a model called a *birth-death process.* The assumptions of the model are that offspring are produced, one at a time, with a random time interval between births, and, similarly, that deaths are separated by random intervals of time. Imagine several replicates of a population, all started with the same population size. The model shows that the standard deviation in population size among these replicates comes to equal the average population size among the replicates.

Most ecologists have concentrated on environmental fluctuations as the cause of population fluctuation, because it is obvious that changing environmental conditions do affect the size of virtually all populations from year to year. However, it seems plausible that some population size fluctuation originates with demographic stochasticity; yet the relative importance of environmental and demographic stochasticity has not yet been determined in a natural population.

DEMOGRAPHY

As a result of birth and mortality, a population has a continual turnover. How fast is this turnover? When we visit a population in nature, how old are the individuals in that population? Here we answer these questions for a population that is growing exponentially. The study of the age structure in a population is called *demography*, a term once limited to studies of the age structure and dynamics of *Homo sapiens* (from the Greek *demos*, "common people") but now applied to any organism.

Predicting the Future

Consider a closed population (that is, with no immigration or emigration) that is censused every year. In the census we count the number of organisms of each age. We do not have to know their exact age; all that is needed is to place each organism in an *age class*. For example, suppose that each age class is 1 year long and that three age classes are sufficient; that is, suppose virtually no organism in this hypothetical population lives for more than 3 years.

The population is censused on, say, January 1 in the year t. We denote the number of organisms that were born during the preceding year as $n_{0,t}$; these organisms are the animals whose age is less than 1 year as of January 2. We also denote the number whose ages are between 1 and 2 years as $n_{1,t}$, and the number between 2 and 3 years as $n_{2,t}$. These three observations describe the age structure of the population at year t.

We can predict the age structure of this population through time based on knowing how well the organisms survive and reproduce. Let F_0 be the number of offspring produced, on the average, by each member of age class 0. Similarly, let F_1 be the number of offspring produced by each member of age class 1 and let F_2 be the same for age class 2. Then the number of new offspring for the census at year $t + 1$ is simply

$$n_{0,t+1} = F_0 n_{0,t} + F_1 n_{1,t} + F_2 n_{2,t}$$

The next step is to determine how many organisms will be in the older age classes at year $t + 1$. In a closed population, all of these organisms will have already been counted in year t, and those that live will simply be 1 year older when censused in year $t + 1$. Let P_0 be the fraction of organisms that survive through a year, given that they began the year in age class 0. Let P_1 be the fraction of organisms from age class 1 that survive the year. No P_2 exists because 3 years is assumed to be the maximum life span in this example. Hence, we have

$$n_{1,t+1} = P_0 n_{0,t}$$

because the number in age class 1 at time $t + 1$ is simply the fraction of those from age class 0 at time t that survive to the next census. Similarly,

$$n_{2,t+1} = P_1 n_{1,t}$$

is the number in age class 2 at year $t + 1$, and they all come from those in age class 1 at year t.

To measure the fertilities, F_0, F_1, and F_2, as well as the survival fractions, P_0 and P_1, one needs to follow the histories of a sample of individuals throughout their lives. A group of individuals of the same age is called a *cohort.* In practice, we must locate animals and mark each with a permanent tag; we must observe how long each individual lives and how many young each raises. Similarly, plants must be given permanent tags and records kept of their size and of the number of seeds they produce. It is possible, although often difficult, to get these data on natural populations.

An Illustration

Once the F's and P's are measured, we can predict what the census at year $t + 1$ will be, based on knowing what it was in year t, using the formulas just described. To illustrate, suppose F_0, F_1, and F_2 are 0.5, 1.0, and 0.75, respectively, and P_0 and P_1 are 0.666667 and 0.333333, respectively. Then, if we determine in the census at year t that 900, 300, and 100 animals are counted in age classes 0, 1, and 2, respectively, we predict that there will be the following number in the age classes at year $t + 1$:

$$n_{0,t+1} = 825 = (0.5)(900) + (1.0)(300) + (0.75)(100)$$

$$n_{1,t+1} = 600 = (0.666667)(900)$$

$$n_{2,t+1} = 100 = (0.333333)(300)$$

Discovering the Stable Age Distribution

Now what about year $t + 2$? If the environment is approximately the same for another year, it is tempting to use the equations once more—this time to predict the census at $t + 2$ from the prediction for $t + 1$, using the same F's and P's. Obviously, this procedure is a risky business. If the F's and P's were incorrectly measured, or if the census at year t was in error to begin with, then we are only compounding our errors. However, suppose that the measurement errors are not too large and that the environment is not varying much from year to year; let us see what happens. Upon using the formulas again, we obtain

$$n_{0,t+2} = 1088 = (0.5)(825) + (1.0)(600) + (0.75)(100)$$

$$n_{1,t+2} = 550 = (0.666667)(825)$$

$$n_{2,t+2} = 200 = (0.333333)(600)$$

Now, throwing caution to the wind, let us use the equations over and over again and see what the population will be every year for 10 years.

TABLE 5–1 An example of population projection

Year	n_0	n_1	n_2
t	900	300	100
$t + 1$	825	600	100
$t + 2$	1088	550	200
$t + 3$	1244	725	183
$t + 4$	1484	829	241
$t + 5$	1753	990	276
$t + 6$	2073	1168	330
$t + 7$	2452	1382	389
$t + 8$	2900	1634	461
$t + 9$	3431	1934	545
$t + 10$	4058	2287	645

We hope, of course, that the F's and P's do not change much during that time and that little measurement error occurred in the F's and P's and in the original census at year t. What we find is summarized in Table 5–1, with years t, $t + 1$ and $t + 2$ also included for completeness.

If one looks closely at this table, a regularity is apparent: The percentage of the population in each of the three age classes becomes constant, even though the population as a whole is growing. To show this more clearly, let us rewrite the table, with columns for the percentages in each age class and the population total, as shown in Table 5–2. Clearly, by about the fifth year, the age distribution in the population has settled into a constant pattern, even though the population is growing.

TABLE 5–2 Approach to the stable age distribution

Year	Percent in class 0	Percent in class 1	Percent in class 2	Total
t	69.2	23.1	7.7	1300
$t + 1$	54.1	39.3	6.6	1525
$t + 2$	59.2	29.9	10.9	1838
$t + 3$	57.8	33.7	8.5	2152
$t + 4$	58.1	32.4	9.5	2555
$t + 5$	58.1	32.7	9.2	3019
$t + 6$	58.0	32.7	9.2	3571
$t + 7$	58.1	32.7	9.2	4224
$t + 8$	58.1	32.7	9.2	4996
$t + 9$	58.1	32.7	9.2	5909
$t + 10$	58.1	32.7	9.2	6989

Another regularity is also evident: By the fifth year, the total population size increases by approximately the same factor each year. For example, in year $t + 6$ the population has increased by a factor of 1.18 from its value at $t + 5$. To see this change, note that $(3571/3019) = 1.18$. Similarly, the ratio of the size at $t + 7$ to that at $t + 6$ is $(4223/3571)$, which equals 1.18 again. Thus, by about the fifth year, the total population size, N_t, is growing according to the following rule:

$$N_{t+1} = RN_t$$

where $R = 1.18$. We have seen this rule before: It is the rule for exponential growth.

What Determines the Stable Age Distribution?

Thus we see that the population's age distribution settles into a constant pattern, and the total population size eventually grows exponentially, all after about 5 years. What determines the pattern of this ultimate age distribution, and what determines the ultimate exponential growth factor, R? Three possibilities exist: the F's and P's, the original census at year t, or both. To find out which of the three considerations affects the ultimate situation, let us consider a population with the same F's and P's as before but starting with a different initial census.

Suppose $n_{0,t}$, $n_{1,t}$, and $n_{2,t}$ are 100, 0, and 0, respectively; this census represents a population started, so to speak, from eggs. Then we calculate the population for year $t + 1$, then $t + 2$, and so forth, as before. The results of the calculation appear in Table 5–3, with the percentage in each age class as columns, together with another column for the total population size. Clearly, the ultimate pattern of the age distribution is exactly

TABLE 5–3 Another approach to the stable age distribution

Year	Percent in class 0	Percent in class 1	Percent in class 2	Total
t	100	0	0	100
$t + 1$	42.9	57.1	0	117
$t + 2$	62.2	22.6	15.1	147
$t + 3$	57.0	36.3	6.6	168
$t + 4$	58.2	31.7	10.1	202
$t + 5$	58.1	33.0	9.0	237
$t + 6$	58.0	32.7	9.3	281
$t + 7$	58.1	32.7	9.2	332
$t + 8$	58.1	32.7	9.2	393
$t + 9$	58.1	32.7	9.2	465
$t + 10$	58.1	32.7	9.2	550

what was obtained before, only here it has taken about 7 years to attain instead of the 5 years in the previous example. Furthermore, after these first 7 years, the ratio between the population sizes in consecutive years is always 1.18, exactly the value encountered before.

Thus, the initial census does not affect the results; the P's and F's are what determine both the ultimate pattern of the age distribution in the population and the ultimate factor by which the total population size increases each year. The ultimate age distribution is called the *stable age distribution.* The original census does, however, determine the time needed for the population to attain its ultimate condition. The initial census also sets the scale of the population size; that is, if two populations have the same initial percentages in each age class and also the same F_x's and P_x's, but one is 10 times larger than the other, then it will always be 10 times larger.

Predicting the Stable Age Distribution

Because the F's and P's determine the stable age distribution of a population and its ultimate exponential growth factor, R, it would be useful to know how to predict the stable age distribution and R without having to run a numerical example for every case. Fortunately, it is quite easy to find the equations that accomplish this task. Once again we will work out the equations, assuming three age classes, and then generalize to many age classes.

The main idea is to realize that when the stable age distribution is achieved, not only is the whole population increasing by a factor, R, each year but also the number *in* each age class is increasing by this same factor. That is, *after* the stable age distribution has been attained,

$$n_{0,t+1} = Rn_{0,t}$$

$$n_{1,t+1} = Rn_{1,t}$$

$$n_{2,t+1} = Rn_{2,t}$$

These formulas are important because the left-hand side of each equation for n at $t + 1$ can be replaced with R times n at t, thereby producing three equations in which the only variables are the n's at year t. Furthermore, the subscript, t, then becomes superfluous and can be omitted. Thus, after the stable age distribution has been attained, the numbers in each age class satisfy the following conditions:

$$Rn_0 = F_0n_0 + F_1n_1 + F_2n_2$$

$$Rn_1 = P_0n_0$$

$$Rn_2 = P_1n_1$$

To derive the equation for R, we start with

$$n_1 = \frac{P_0 n_0}{R}$$

$$n_2 = \frac{P_1 n_1}{R}$$

$$= \frac{P_1 P_0 n_0}{R^2}$$

Next, we substitute these expressions back into the first equation, multiply both sides by R^2, divide out the n_0, move everything to the left-hand side, and obtain a polynomial for R in terms of the P's and F's:

$$R^3 - F_0 R^2 - P_0 F_1 R - P_0 P_1 F_2 = 0$$

The R is a root of this polynomial. (A root is a value of R that makes the left-hand side of the equation equal to zero.) Here things could get complicated, for a polynomial has many roots, and which of these should we choose? We take the largest positive root. The easiest way to find this real root is to graph the polynomial and determine the highest place at which it crosses the R axis.

We also can find the stable age distribution. If we assume that the n's represent the fraction of the population in each age class, then the sum of the n's must be 1; or if the n's represent percentages, then their sum is 100. We may choose any convention we wish; thus, let us choose to have the sum of the n's equal 1. Therefore, we can say that

$$n_0 + \frac{P_0 n_0}{R} + \frac{P_1 P_0 n_0}{R^2} = 1$$

Hence, n_0 is

$$n_0 = \frac{1}{(1 + P_0/R + P_1 P_0/R^2)}$$

Therefore, the rest of the n's in the stable age distribution are

$$n_1 = \frac{P_0}{R(1 + P_0/R + P_1 P_0/R^2)}$$

$$n_2 = \frac{P_1 P_0}{R^2(1 + P_0/R + P_1 P_0/R^2)}$$

Thus, we first find the exponential growth factor, R, as the root of the polynomial, and then we calculate the stable age distribution. All this is determined solely from the data on fertility and survival.

An Illustration Continued

As an example, let us continue with the numerical values for the P's and F's that we used previously. Let us make a table of values of the polynomial for different values of R. Then, by inspection, we can see where the polynomial changes sign. Since a root of the polynomial is a value of R that makes the polynomial zero, a root must lie between values of R where the polynomial changes sign. Specifically, if we try evaluating the polynomial for R equal to 1, we find that

$$\begin{aligned}\text{Polynomial} &= (1)^3 - (1/2)(1)^2 - (2/3)(1)(1) - (2/3)(1/3)(3/4) \\ &= 1 - 1/2 - 2/3 - 1/6 \\ &= -1/3\end{aligned}$$

Similarly, if we try evaluating the polynomial for R equal to 2, we find that

$$\begin{aligned}\text{Polynomial} &= (2)^3 - (1/2)(2)^2 - (2/3)(1)(2) - (2/3)(1/3)(3/4) \\ &= 8 - 2 - 4/3 - 1/6 \\ &= 4.5\end{aligned}$$

Therefore, a root of the polynomial must lie somewhere between 1 and 2 because the polynomial changes sign between these two values. Indeed, we can be still more accurate. If we evaluate the polynomial for R between 1 and 2 in steps of 0.1, we obtain the results shown in Table 5–4. Now it is clear that a root lies between 1.1 and 1.2. In addition, if we evaluate the polynomial between 1.1 and 1.2 in steps of 0.01, we will see that a

TABLE 5–4 Search for R as the root of a polynomial

R	Polynomial
1.0	−0.333333
1.1	−0.174
1.2	0.0413335
1.3	0.318667
1.4	0.664001
1.5	1.08333
1.6	1.58267
1.7	2.168
1.8	2.84534
1.9	3.62067
2.0	4.5

root is present at about 1.18, as we know is true for this numerical example. Also, from further inspection of the polynomial, it is clear that this value is the largest positive root.

In general, the ultimate exponential growth rate of the population is found by determining the largest positive root of the polynomial in this way. Incidentally, the largest root may be less than 1, indicating a population that actually declines once the stable age distribution is attained, or 1, indicating that the size will stay constant once the stable age distribution has been achieved.

Now that we have found R as the root of the polynomial, we can also find the stable age distribution:

$$n_0 = \frac{1}{1 + (2/3)/1.18 + (1/3)(2/3)/(1.18)^2}$$
$$= 0.58$$

$$n_1 = \frac{(2/3)}{1.18[1 + (2/3)/1.18 + (1/3)(2/3)/(1.18)^2)]}$$
$$= 0.33$$

$$n_2 = \frac{(1/3)(2/3)}{(1.18)^2(1 + (2/3)/1.18 + (1/3)(2/3)/(1.18)^2)}$$
$$= 0.09$$

Notice that this is the age distribution we found earlier; it is approached from any initial condition.

Finally, to calculate the R and the stable age distribution with more age classes, say, n_0 to n_4 or more, the steps are all the same. To keep the notation as simple as possible, we can define a symbol that combines all the P's into one quantity. Also, this symbol has some interest in its own right.

What is the probability that an individual lives from birth to age x or more? To illustrate, suppose x is age 3. The individual must first live through the earliest age class, which has probability P_0; then it must live through the second age class, which has probability P_1. After the individual has lived through the second age class, it has reached age 3. Therefore, the probability of living to age 3 or more is simply the product, P_0P_1. In general, the probability of living to age x or more, which we denote as l_x, is the product of the P's from P_0 up through P_{x-1}; that is,

$$l_x = P_0P_1 \ldots P_{x-1}$$

If we plot l_x on the vertical axis and x on the horizontal axis, we obtain a curve called the *survivorship curve.* The survivorship curve always decreases as x increases; by definition, we take it as beginning at $l_0 = 1$.

The survivorship curve is useful in simplifying the formulas used to determine R and the stable age distribution. We can rewrite the formulas with three age classes as follows. The polynomial for R becomes

$$R^3 - l_0F_0R^2 - l_1F_1R - l_2F_2 = 0$$

and the stable age distribution is rewritten as

$$n_0 = \frac{l_0R^0}{l_0R^0 + l_1R^{-1} + l_2R^{-2}}$$

$$n_1 = \frac{l_1R^{-1}}{l_0R^0 + l_1R^{-1} + l_2R^{-2}}$$

$$n_2 = \frac{l_2R^{-2}}{l_0R^0 + l_1R^{-1} + l_2R^{-2}}$$

The same pattern can be extended to many age classes. In general, if the highest age class is w, the polynomial for R is

$$R^{w+1} - l_0F_0R^w - l_1F_1R^{w-1} - \cdots - l_{w-1}F_{w-1}R - l_wF_w = 0$$

and the formula for the stable age distribution is

$$n_x = \frac{l_xR^{-x}}{l_0R^0 + l_1R^{-1} + \cdots + l_wR^{-w}}$$

Demography and Deer Hunting in Scotland

Now let us consider a real example in which these kinds of data are useful. In 1957 the Nature Conservancy acquired the Isle of Rhum, in eastern Scotland, as a nature preserve. It was purchased with the aim of rehabilitating what was generally considered to be a degraded habitat, including the restoration of vegetation that had been lost to the grazing of sheep and deer (Lowe 1969). The sheep were removed, and the deer population was regularly hunted to remove one-sixth of the adults that had been counted in the spring census of that year. Here, then, is the question: Is the level of hunting on this isle high enough to prevent the deer population from increasing to a point where it will overgraze the vegetation, or is the level of hunting too high, so that the deer population will eventually die out?

Records of the survivorship and reproduction of the red deer (*Cervus elaphus*) from the Isle of Rhum allow us to answer the question (see Figure 5–7). The census is taken in the spring. The 0th age class consists of yearlings; these were born the previous summer and are about 10 months old by the time of the spring census. The next age class consists of the 2-year-

TABLE 5–5 Number of female red deer remaining from the 0th age class of 1957

Year	Number of deer remaining
1957	130 (initial number)
1958	130
1959	122
1960	98
1961	66
1962	40
1963	25
1964	18
1965	4

olds; they are about 1 year and 10 months old; and so forth. Most of the deer have died by the time 9 years have passed, so a fairly accurate picture can be developed using a total of nine age classes.

Let us focus on female deer. An estimated 130 female deer were in the 0th age class in 1957, and by the spring of 1966 only about 4 were left. The history of this group is presented in Table 5–5.

During this time, hunting was reducing the population by one-sixth each year. The mortality between 1957 and 1965 was caused by this hunting, as well as by falls from cliff edges, entanglement in abandoned fences, and malnutrition or starvation. These data allow us to determine the probability of survival for a female deer during these years:

FIGURE 5–7 A red deer *(Cervus elaphus)* from Scotland.

$$P_0 = \frac{130}{130} = 1.00$$

$$P_1 = \frac{122}{130} = 0.94$$

$$P_2 = \frac{98}{122} = 0.80$$

$$P_3 = \frac{66}{98} = 0.67$$

$$P_4 = \frac{40}{66} = 0.61$$

$$P_5 = \frac{25}{40} = 0.63$$

$$P_6 = \frac{18}{25} = 0.72$$

$$P_7 = \frac{4}{18} = 0.22$$

Next, we can calculate the survivorship curve, l_x, as follows:

$l_0 = 1$ (by definition)

$l_1 = P_0 = 1.00$

$l_2 = P_0P_1 = (1.00)(0.94) = 0.94$

$l_3 = P_0P_1P_2 = (1.00)(0.94)(0.80) = 0.75$

$l_4 = P_0P_1P_2P_3 = (1.00)(0.94)(0.80)(0.67) = .50$

Proceeding in this way, we also obtain

$l_5 = 0.31$

$l_6 = 0.19$

$l_7 = 0.14$

$l_8 = 0.03$

However, survival is only half of the story; the other half is reproduction. Although deer give birth singly, the fraction of the does that breed increases with age. Hence, the *average* number of offspring per doe increases with age, as shown in Table 5–6.

Still, the number of births itself is not exactly the information we need. First, we want the number of female births, because we are analyzing only the female component of the population. Second, we want the number of

TABLE 5–6 Average number of births to female red deer, including offspring of both sexes

Age class of mother	Average number of births
0	0
1	0
2	0.6
3	0.6
4	0.6
5	0.6
6	0.8
7	0.6
8	0.7

births that actually survive to the census in the spring, when they are counted as being in the 0th age class.

The young are born in the summer and must survive their first winter in order to be counted in the first census. The data suggest that the fraction of females among the newborn is approximately 0.5 (i.e., the sex ratio is 50:50). Also, the fecundity data suggest that the population of deer in 1957 should have produced about 175 newborn, whereas the number counted in the 0th age class the next spring was 154. Thus, it appears that the probability of surviving from birth to the first census is about 154/175, or 0.88. With this additional information, we can tabulate the fertilities as follows:

$$F_0 = (0.5)(0.88)(0) = 0$$

$$F_1 = (0.5)(0.88)(0) = 0$$

$$F_2 = (0.5)(0.88)(0.6) = 0.26$$

Similarly, for the remaining F's, we obtain

$$F_3 = 0.26$$

$$F_4 = 0.26$$

$$F_5 = 0.26$$

$$F_6 = 0.35$$

$$F_7 = 0.26$$

$$F_8 = 0.31$$

We can now write down the polynomial for the exponential growth rate, R, for this population of deer. If R is much greater than 1, then the hunting each year is not sufficient to prevent a population explosion of deer, but if R is much lower than 1, then the hunting will gradually drive the deer to extinction. If R is near 1, then the hunting program would seem to be on the right track, although small adjustments might be needed. The polynomial is

$$R^9 - (1)(0)R^8 - (1)(0)R^7 - (0.94)(0.26)R^6 - (0.75)(0.26)R^5$$
$$- (0.50)(0.26)R^4 - (0.31)(0.26)R^3 - (0.19)(0.35)R^2$$
$$- (0.14)(0.26)R - (0.03)(0.31)$$

When this polynomial is tabulated for values of R ranging from 0 to 1.5, we see that it has a root near $R = 0.9$, as Table 5–7 shows. Upon closer inspection of the interval between 0.9 and 1.0, we find that the root is at $R = 0.94$. This result suggests that the management policy of harvesting one-sixth of the deer population each year maintains an approximately constant population size, although a slight tendency exists toward population decline.

The age distribution in the population of 1957 can be compared with the stable age distribution that should result from the management policy of hunting one-sixth of the population each year. In 1957, the age structure was as shown in Table 5–8. In addition, 34 female deer older than the eighth age class were present that are not included in the calculations.

The fraction of individuals expected in the stable age distribution is shown in Table 5–9. Clearly, the population of 1957 is not far from the stable age distribution that is expected according to the new management

TABLE 5–7 The demographic polynomial for red deer

R	Polynomial
0	−0.0093
0.1	−0.0137008
0.2	−0.0201703
0.3	−0.0300665
0.4	−0.0457221
0.5	−0.0702844
0.6	−0.105826
0.7	−0.147397
0.8	−0.169243
0.9	−0.0975845
1.0	0.237801
1.1	1.18351
1.2	3.3872
1.3	7.98344
1.4	16.8609
1.5	33.0351

TABLE 5–8 Age distribution for red deer

Age class	Number of females in 1957	Fraction of total
0	130	0.18
1	114	0.16
2	113	0.16
3	81	0.11
4	78	0.11
5	59	0.08
6	65	0.09
7	55	0.08
8	25	0.03
Total	720	

policy of reducing the population by one-sixth each year. The stable age distribution peaks near age classes 1 and 2. In an increasing or constant population ($R \geq 1$), the stable age distribution continually decreases from the youngest class to the oldest class. However, in a declining population it is possible for an intermediate age class in the stable age distribution to have the highest fraction of the population.*

Extensions of Demographic Theory

During the last 40 years, ecologists have obtained information on the survival, reproduction, and age structure of many natural populations of both plants and animals. Some patterns seem to have emerged among the data

TABLE 5–9 Predicted stable age distribution for red deer

Age class	Fraction in stable age distribution
0	0.18
1	0.19
2	0.19
3	0.16
4	0.11
5	0.08
6	0.05
7	0.04
8	0.01

* For more information on the biology of red deer, see Darling (1937) and Clutton-Brock *et al.* (1982).

on survivorship for animals (Caughley 1977). Mammals show high mortality before puberty, followed by a plateau of low mortality, ending with high mortality. In birds, the mortality rate seems more constant throughout life than it does in mammals. Fish typically exhibit a huge mortality of juveniles before reproductive age is attained.

Of particular interest since the mid-1970s is the demography of plant populations, stimulated in large part by the review of Harper and White (1974). Examples of demographic studies with plants now include those by Law *et al.* (1977) with grasses, Sarukhan (1974, 1978) with buttercups and tropical trees, Bierzychudek (1982) with a forest herb, and others, many of which are reviewed in Solbrig (1980).

In addition, demography is being applied to the parts of an *individual plant*. For instance, the leaves on a tree can be considered a population, with each leaf having a survivorship curve. Similarly, grasses, including bamboo, have been considered as a population of stems from a common root system. Hence, the growth of an individual plant can be viewed in terms of a population of parts, or modules, such as its leaves and stems (White 1980).

Michod and Anderson (1980) present a useful discussion of some pitfalls one should avoid when estimating the survivorship curve from age structure data, as must be done when individuals have not been followed through time. Lenski and Service (1982) also have developed statistical methods for placing confidence limits on the growth rate of a population that has been calculated from data on survival and reproduction.

Demographic theory is both popular and useful for describing how a population size and its age structure change through time. Its principal limitation is that the equations may not include information about enough of the processes that influence a population's size and age structure. One of the key assumptions of the equations is that the population's survivorship curve, l_x, and its fertility curve, F_x, are constant through time. Another key assumption of the equations is that the population is closed and not subdivided. There is active research in theoretical ecology aimed at relaxing these restrictive assumptions. Cohen (1977) and Tuljapukar and Orzack (1980) have begun to extend demographic theory to include random fluctuation in the survivorship and fertility curves; Gurtin and MacCamy (1974) have made advances in bringing density dependence into demography; and Roughgarden *et al.* (1985) have developed demographic theory for an open population of space-limited sessile invertebrates.

DENSITY-DEPENDENT POPULATION GROWTH

A fascination of biology lies in discovering how organisms interact with each other and whether the interactions have interesting consequences. In this section, we begin with records of some bird populations that show a population dynamics quite different from what is expected of a closed population growing in a density-independent manner with a fluctuating exponential growth rate. The difference suggests the presence of density dependence, though it does not prove that it is present. We will then review some of the many mechanisms that can cause density dependence.

The Dynamics of Bird Populations in Ohio: Some Density Dependence

Birds have long been of interest to amateur naturalists, and their efforts have led to some of the best long-term data on the size of natural populations. In the United States, many Audubon Society chapters traditionally conduct a census, in the middle of December, of overwintering birds. Typically, the area surveyed is a circle 15 mi in diameter. Called the *Christmas Bird Count,* the results of this census are published each year in the journal *American Birds.*

Table 5–10 presents data on the counts of six kinds of birds for 1935–1981 from an area centered on Buckeye Lake Park in Ohio. The elevation is 880–1100 ft, and the area includes about 50% open farmland, 40% diciduous woodland, and 10% swamp and cattail marsh.

These bird populations have fluctuated in size from year to year, but none has exploded or become extinct. Both the population size and R are tabulated for each year. For example, in 1935 the R for the downy woodpecker was 1.15 because in 1936 it had increased from 124 to 143. Its R for that year is therefore (143/124), which equals 1.15. The geometric average of all the R's for the downy woodpecker, $(R_{1935}R_{1936} \cdots R_{1980})^{(1/46)}$, is 0.99. The average population size, $(N_{1935} + N_{1936} + \cdots + N_{1981})/47$, is 136.

This information suggests that the geometric average R happened to be about 1 in this location through these years, and hence that the population size has randomly bounced around a value that happens to equal about 136. However, closer inspection shows that the R's greater than 1 tend to be found in the years when the size is lower than 136 and that the R's less than 1 tend to occur when the size is higher than 136. Thus, it may not be just a coincidence that the geometric average of the R's is about 1 and that the population size has remained in the vicinity of 136 during the 47 years for which data are available.

We can statistically test whether the value of R is independent of the value of N or whether some relation exists between R and N. Table 5–11 presents a summary of the number of times that N was less than its average and R was less than 1. For the downy woodpecker, this figure is 10 times. Similarly for the other three possibilities. The table also shows the number of times we would expect both N to be less than its average and R to be less than 1 if they were each varying independently of the other.

TABLE 5–10 The Christmas census at Buckeye Lake Park, Ohio

	Downy woodpecker		Carolina chickadee		Tufted titmouse		White-breasted nuthatch		Cardinal		Flicker	
Year	*Size*	*R*	*Size*	*R*	*Size*	*R*	*Size*	*R*	*Size*	*R*	*Size*	*R*
1935	124	1.15	159	1.31	224	0.68	97	1.11	227	1.37	45	1.78
1936	143	0.58	208	0.67	152	1.26	108	0.64	310	0.61	80	0.19
1937	83	1.29	140	0.69	192	0.79	69	1.51	188	0.99	15	3.20
1938	107	1.20	96	2.56	151	1.02	104	0.74	186	1.17	48	0.29
1939	128	0.80	246	0.74	154	0.82	77	1.16	218	1.10	14	4.07
1940	102	1.23	181	1.36	126	1.48	89	1.40	239	0.79	57	0.53
1941	125	1.42	247	1.20	186	1.11	125	1.59	189	2.03	30	1.83
1942	177	0.87	297	0.91	206	1.03	199	0.72	383	0.99	55	0.73
1943	154	0.84	271	0.85	213	0.62	144	0.85	381	0.81	40	2.25
1944	130	0.79	231	0.37	132	0.81	123	0.29	309	0.46	90	0.22
1945	103	0.76	85	1.47	107	1.21	36	2.44	142	1.35	20	1.75
1946	78	0.64	125	0.95	129	0.67	88	0.59	192	1.08	35	1.11
1947	50	2.00	119	1.06	87	1.45	52	2.79	207	0.93	39	1.56
1948	100	1.54	126	1.59	126	1.22	145	0.97	192	2.09	61	0.69
1949	154	0.91	200	1.24	154	1.10	141	0.57	402	0.78	42	2.40
1950	140	1.02	248	0.72	170	0.61	81	1.20	313	0.72	101	0.71
1951	143	0.73	178	0.75	103	1.79	97	1.13	224	1.37	72	0.94
1952	105	2.12	133	1.68	184	1.38	110	1.53	307	1.14	68	1.96
1953	223	0.61	223	0.88	254	0.49	168	0.57	351	0.90	133	0.59
1954	137	0.84	196	1.09	125	1.04	95	1.28	316	0.70	78	0.87
1955	115	1.32	213	0.73	130	1.76	122	0.61	222	2.67	68	0.97
1956	152	1.61	155	1.63	229	1.31	75	2.12	593	0.94	66	2.80
1957	244	0.77	252	0.89	299	0.67	159	0.78	558	0.63	185	0.49
1958	189	1.06	225	0.93	199	0.90	124	0.77	354	0.77	90	0.78
1959	201	1.55	210	1.42	179	1.44	95	2.14	274	1.30	70	0.94
1960	312	0.65	299	0.63	257	0.75	203	0.67	355	0.74	66	1.45
1961	204	0.93	189	1.39	193	0.96	137	0.87	264	0.90	96	0.67
1962	190	1.04	262	0.71	185	0.85	119	0.93	238	1.50	64	2.22
1963	197	0.85	187	1.05	158	0.63	111	1.31	358	0.97	142	0.77
1964	167	1.20	196	0.63	99	1.64	145	0.85	348	0.90	110	1.00
1965	201	0.89	123	1.36	162	0.99	123	1.16	314	0.93	110	0.73
1966	178	0.70	167	0.92	160	0.89	143	0.62	293	0.91	80	0.34
1967	124	0.72	154	0.65	143	0.46	89	1.06	267	0.90	27	3.19
1968	89	1.62	100	1.78	66	3.79	94	1.93	239	1.47	86	1.62
1969	144	0.96	178	1.11	250	0.80	181	0.67	351	1.46	139	0.76
1970	138	0.81	197	0.40	200	0.49	122	0.36	514	0.83	106	0.50
1971	112	0.96	78	1.14	97	1.14	44	1.68	426	0.58	53	1.81
1972	108	1.17	89	1.52	111	1.33	74	1.32	248	1.14	96	0.67
1973	126	0.57	135	0.79	148	0.73	98	0.69	283	1.98	64	0.58
1974	72	1.49	106	1.28	108	1.54	68	0.84	559	0.65	37	1.51
1975	107	0.67	136	0.91	166	0.29	57	1.42	361	0.65	56	0.75

TABLE 5–10 (*continued*)

	Downy woodpecker		Carolina chickadee		Tufted titmouse		White-breasted nuthatch		Cardinal		Flicker	
Year	*Size*	*R*	*Size*	*R*	*Size*	*R*	*Size*	*R*	*Size*	*R*	*Size*	*R*
1976	72	1.47	124	0.92	48	2.31	81	0.90	236	1.12	42	1.12
1977	106	1.23	114	2.14	111	0.94	73	1.49	265	1.41	47	0.62
1978	130	0.48	244	0.27	104	0.34	109	0.36	373	0.55	29	1.62
1979	63	1.11	65	1.22	35	1.14	39	1.54	205	1.22	47	0.66
1980	70	0.97	79	1.10	40	1.53	60	1.20	251	0.66	31	0.84
1981	68		87		61		72		165		26	
Mean[a]	136	0.99	172	0.99	151	0.97	106	0.99	302	0.99	67	0.99

[a] Arithmetic for size, geometric for *R*.

For the downy woodpecker, this number is 14. The other possibilities are also tabulated. Thus, *N* less than 136 together with *R* less than 1 occur slightly more times than expected if *N* and *R* vary independently.

Is the association of a high *R* with a low population size statistically significant? Table 5–11 also provides the χ^2 statistic and the probability that a difference this large or more could result by chance if *N* and *R* vary independently. For the downy woodpecker, the probability is about 0.07, and this value is not small enough normally to be regarded as statistically

TABLE 5–11 A test of the independence of *N* and *R* for bird census data from Buckeye Lake Park, Ohio

	Downy woodpecker	Carolina chickadee	Tufted titmouse	White-breasted nuthatch	Cardinal	Flicker
Observed						
$N <$ ave., $R < 1$	10	7	6	5	7	9
$N <$ ave., $R >= 1$	15	15	16	19	17	17
$N >=$ ave., $R < 1$	15	16	17	18	20	17
$N >=$ ave., $R >= 1$	6	8	7	4	2	3
Expected (assuming no relation between *N* and *R*)						
$N <$ ave., $R < 1$	14	11	11	12	14	15
$N <$ ave., $R >= 1$	11	11	11	12	10	11
$N >=$ ave., $R < 1$	11	12	12	11	13	11
$N >=$ ave., $R >= 1$	10	12	12	11	9	9
χ^2 (with continuity correction)	3.37	4.27	7.06	14.72	15.59	9.72
P ($df = 1$)	0.067	0.039	0.0079	0.00012	0.000079	0.0018

significant. For some of the other species, however, the conclusion is inescapable that some negative relation exists between N and R. In the nuthatch and cardinal, for example, the evidence is excellent that R is greater than 1 when N is less than the average size and that R is less than 1 when N is greater than the average size.

The Alternative to Density Dependence

It is surprising to see how different a truly randomly growing, closed population is from the bird populations of Buckeye Lake Park. Let us construct a numerical example of a closed population with density-independent population dynamics and a growth rate that varies randomly from year to year. Suppose the chance is 50:50 that R is either $\frac{4}{5}$ or $\frac{5}{4}$ each year, and suppose that a coin is flipped each year to decide what R is. We use $\frac{4}{5}$ and $\frac{5}{4}$ because their geometric mean is 1, and these values are typical of the R's for the downy woodpecker in Table 5–11.

TABLE 5–12 Population size with a random R[a]

Time	Size	R	Time	Size	R	Time	Size	R	Time	Size	R
1	200	0.80	26	954	0.80	51	763	0.80	76	250	1.25
2	160	0.80	27	763	1.25	52	610	1.25	77	313	0.80
3	128	0.80	28	954	0.80	53	763	0.80	78	250	1.25
4	102	1.25	29	763	1.25	54	610	0.80	79	313	0.80
5	128	0.80	30	954	1.25	55	488	0.80	80	250	1.25
6	102	1.25	31	1192	0.80	56	391	0.80	81	313	0.80
7	128	0.80	32	954	1.25	57	313	1.25	82	250	1.25
8	102	0.80	33	1192	0.80	58	391	0.80	83	313	0.80
9	82	1.25	34	954	0.80	59	313	1.25	84	250	0.80
10	102	1.25	35	763	0.80	60	391	0.80	85	200	1.25
11	128	1.25	36	610	1.25	61	313	0.80	86	250	0.80
12	160	0.80	37	763	1.25	62	250	1.25	87	200	1.25
13	128	1.25	38	954	1.25	63	313	0.80	88	250	1.25
14	160	0.80	39	1192	0.80	64	250	0.80	89	313	1.25
15	128	1.25	40	954	0.80	65	200	1.25	90	391	1.25
16	160	0.80	41	763	1.25	66	250	0.80	91	488	0.80
17	128	1.25	42	954	1.25	67	200	1.25	92	391	1.25
18	160	1.25	43	1192	0.80	68	250	1.25	93	488	0.80
19	200	1.25	44	954	1.25	69	313	1.25	94	391	0.80
20	250	1.25	45	1192	0.80	70	391	0.80	95	313	0.80
21	313	1.25	46	954	0.80	71	313	0.80	96	250	1.25
22	391	1.25	47	763	1.25	72	250	0.80	97	313	0.80
23	488	1.25	48	954	0.80	73	200	0.80	98	250	0.80
24	610	1.25	49	763	1.25	74	160	1.25	99	200	0.80
25	763	1.25	50	954	0.80	75	200	1.25	100	160	1.25

[a] Average N is 450.

TABLE 5–13 A test of the independence of N and R for a numerical example of a population growing with a random R

Observed number of cases	
$N <$ ave., $R < 1$	31
$N <$ ave., $R >= 1$	34
$N >=$ ave., $R < 1$	19
$N >=$ ave., $R >= 1$	16
Expected (assuming no relation between N and R)	
$N <$ ave., $R < 1$	32.5
$N <$ ave., $R >= 1$	32.5
$N >=$ ave., $R < 1$	17.5
$N >=$ ave., $R >= 1$	17.5
$\chi^2 = 0.167$	
$P(df = 1) = 0.6756$	

Table 5–12 presents a simulation of what happens to a population that starts out with 200 individuals whose R is either 0.8 or 1.25 each year. In the first coin toss, R turned out to be 0.8. (This was tails.) So the next N is 160, because

$$N_{t+1} = RN_t = (0.8)(200) = 160$$

The coin was tossed again 99 times to obtain the values of R that are shown in the table. Notice the large swings in the population size, from low values to high ones, from 82 to over 1000 individuals. The average population size happened to be 450.

Table 5–13 presents the statistical analysis of a relation between R and the population size. The observed number of each case is, of course, close to what is theoretically expected in the absence of an association between R and N, because we constructed the example with a random R.

Mechanisms of Density Dependence

We conclude that many of the bird populations from Buckeye Lake Park whose census records we have examined do exhibit a negative association between R and the population size. However, what does this finding mean biologically? It is tempting to assert that density-dependent mechanisms are operating, perhaps involving the depletion of resources, such as food and suitable space to live in, but that conclusion would be premature. Something is going on, but what?

Our test has rejected the premise that we have a closed population whose R is independent of N. Therefore, either the population is not closed or the R in a local population decreases as N increases, or both. Some of the populations probably are mostly closed, with, say, no more than 10% migrants per year; if so, these populations must have mechanisms that cause R to decrease when N increases. Alternatively, some of the bird populations may be quite open. After bad years in Buckeye Lake Park, some neighboring local populations may exist that have fared much better, with the result that a net entry of birds has occurred. Conversely, in good years at Buckeye Lake Park, the neighboring local populations typically may have declines, with the result that birds from Buckeye Lake Park show a net migration out from the park to neighboring locations.

To decide which of these pictures is correct, we require more information. Information on bird movements is needed to determine how open or closed the population is. Information on how the birds use resources is needed in order to begin identifying the density-dependent mechanisms in detail.

Many mechanisms produce density dependence in some circumstances. No one of these has emerged as the most general and universally important, but the following nine mechanisms are commonplace.

Resource Depletion and Fecundity. Resource depletion may affect *fecundity* (the average number of eggs that a female produces) in two ways. First, the amount of food available may influence how fast animals grow in body size—as exemplified by fish, amphibians, reptiles, starfish, and corals, among other organisms—and body size, in turn, may determine the amount of reproductive tissue in an animal. Similarly, in plants, where the limiting resource is often light or water, high density can lead to smaller plants. The number of flowers may be proportional to the size of the plant, and thus, when the population size is high, the number of seeds produced per plant is lower than when the population size is low. Second, in organisms with a fixed schedule of body growth, such as most birds and mammals, the animal may not breed when the food supply is low, may resorb embryos, or may reduce its brood size by thinning.

Resource Depletion and Survival. In the absence of sufficient food, animals may starve and plants may wither, and more subtle effects may occur as well. Animals may have to spend more time looking for food and are often more susceptible to death from predation and exposure when doing so. Moreover, adults foraging during the breeding season often have young in nests or burrows, and increased foraging time implies less time to defend the young against nest predators.

Resource Depletion and Foraging Efficiency. Resource depletion leads to the use of lower-quality food, and the additional time spent searching for food lowers the net yield of even the desired food.

Space Depletion. Many mobile animals have *home ranges,* areas through which they regularly move in the course of their normal activities. Moreover, some animals *defend* all or part of their home range; a defended area is called a *territory*, as discussed in more detail in Part 3. As the population of a territorial animal increases, some individuals must occupy space that is less desirable because the risk of being captured by predators there is higher, the amount of food that can be gathered there is lower, the attractiveness or accessibility of the space to mates is lower, or the number of offspring that can be produced or raised in the space is lower. Furthermore, residents of marginal space are likely to be more exposed to environmental stress from, for example, storms or drought, which may produce higher mortality or lower fertility.

Increased Time in Social Interaction. When two animals meet, their interactions may vary from seemingly friendly greetings to aggressive disputes concerning territorial boundaries. The more encounters occur between an individual and others, the more time that individual may spend in social behavior. Such behavior consumes energy beyond that of a resting animal and may deprive it of time it would otherwise spend looking for food or watching out for predators or other threats to life.

Intraspecific Predation. Many organisms are generalized predators and do not hesitate to consume eggs or hatchlings of their own species. This behavior may act as a mechanism limiting population growth if the likelihood of an egg or juvenile being eaten by adults increases as the number of adults increases.

Search-Image Predation. Some insect-eating birds evidently develop an ability to recognize an otherwise cryptic insect. This ability, called a *search image,* is developed through repeated encounters with a particular type of prey. Such predator behavior can act as a density-dependent mechanism influencing the population growth of prey if the likelihood of a predator's capturing a certain type of prey increases as that type of prey becomes more common.

Migration. Individuals may tend to migrate more when the local population size is high than when it is low. This density-dependent migration can lead to a rather constant population size in particular areas within the population's overall range.

Reversed Density Dependence. The mechanisms we have just described, which lower fertility and increase mortality as density increases, generally operate when density is high. However, at the other extreme, mechanisms may exist that heighten fertility and lower mortality as density decreases. Mate location, group defense, and the care of young may be more effective in a population of intermediate size than in a population whose size is near zero. An advantage of higher density, where more is better, is sometimes referred to as an *Allee effect,* named after W. C. Allee, an ecologist who emphasized advantageous interindividual interactions.

The lengthy list of mechanisms that can cause density dependence in a population's dynamics might seem to suggest that the dynamics of all populations involve some density dependence. No doubt many populations do have density dependent mechanisms operating in typical years; however, for many other populations, this fact is not known. Density dependence, while far from rare, has yet to be demonstrated as common.

The size of the bird populations of Buckeye Lake Park in Ohio seems quite constant through time relative to the size of a hypothetical closed population with a random R. This observation suggests that somehow density dependence promotes a constancy of population size that a randomly growing population lacks. But is this idea really correct? As we will see, density dependence may cause a population's size to be rather constant, but it may also do just the opposite. That is, density dependence may cause population fluctuation and even population extinction.

Effects of Density Dependence: A Simple Model

With so many mechanisms available that can cause density dependence, how can we hope to discover if density dependence has a "typical" effect on a population's dynamics? Ultimately, we will have to investigate many of these mechanisms in detail, and, in particular cases, to see if a common theme emerges. Our search for generality may be facilitated, however, if we first investigate a simplified model of density dependence, a model that may provide clues to how density dependence may affect population dynamics.

We can develop a simple model for density dependence as follows: Let us suppose that the average number of births produced by each organism decreases linearly as the population size increases:

$$B(N) = B_0 - bN$$

In this formula, B_0 is the intercept with the vertical axis and b is the slope of the line. B_0 represents the density-independent component of the birth rate, and bN represents the density-dependent component.

Similarly, let us suppose that the fraction of organisms that die between census intervals increases linearly with population size:

$$D(N) = D_0 + dN$$

In this formula, D_0 is the intercept with the vertical axis and d is the slope. D_0 is the density-independent component of the death rate, and dN is the density-dependent component.

Using these formulas for the birth and death rates, we see that the population size at time $t + 1$ is, as before, the births plus the survivors from the preceding census:

$$N_{t+1} = (B_0 - bN_t)N_t + [1 - (D_0 + dN_t)]N_t$$

If no density dependence exists, then b and d are 0; in this case, we have exponential growth, with R equal to $1 + B_0 - D_0$, as before. However, if density dependence does exist, then the effective R varies as N varies. Indeed, by combining the birth and death processes, we get

$$N_{t+1} = R(N_t)N_t$$

where $R(N_t) = 1 + B_0 - D_0 - (b + d)N_t$.

This model is meaningful only if $R(N)$ is greater than or equal to 0; this statement places a requirement on N. If N becomes high enough, then $R(N)$ is negative and this model cannot be used. Instead, some other model would have to be devised.

Recall that a *steady-state* population size is one at which the birth rate equals the death rate. By definition, a steady-state population size does not change through time. A steady-state population size is said to be *stable* if the population's size eventually comes to equal that steady state even though the population size begins at some different value. Strictly speaking, the concept of a steady state is only concerned with whether a population size is constant through time; the concept does not apply to whether actual processes exist that can cause the population to achieve a steady-state size. The idea of stability goes a step further; it concerns whether processes exist that cause the population's size to attain a steady-state value even though the population size is initially at some other value.

Two steady states are present in this model. One is at the N where $R(N) = 1$, because at this N the birth rate equals the death rate, and hence $N_{t+1} = N_t$. This N is found by setting $R(N)$ equal to 1 and solving for N:

$$1 + B_0 - D_0 - (b + d)N = 1$$

so that

$$N = \frac{B_0 - D_0}{b + d}$$

Note that if little density dependence exists (that is, both b and d are small), then the steady-state N is large. If a great deal of density dependence exists (b and d are large), then the steady-state N is small.

The second steady state is $N = 0$. This steady state may seem trivial and uninteresting, but it is *very* important biologically because it represents population extinction.

Knowing *where* the steady states are is only half of the story. Are they, in fact, ever stable? The answer is that they can be, and a straightforward way to find out when is to analyze some numerical examples and then verify the results summarized in Table 5–14.

TABLE 5–14 Relation between R_0 and stability for the two possible equilibrium population sizes in the logistic equation

R_0	$N = 0$	$N = (B_0 - D_0)/(b + d)$
$0 < R_0 < 1$	Stable	Does not exist as a positive number
$1 < R_0 < 3$	Unstable	Stable
$3 < R_0 < 4$	Unstable	Unstable
$4 < R_0$	Invalid model—trajectories become negative	

In the table, R_0 is shorthand for $1 + B_0 - D_0$. If $R_0 < 1$, then the population becomes extinct. If $R_0 > 1$, a stable positive steady-state population size may exist. However, if $R_0 > 2$, then the population size tends to oscillate. Specifically, for R_0 between 2 and 3, these oscillations gradually disappear, but for R_0 between 3 and 4, the oscillations continue unabated. If $R_0 > 4$, the model should not be used; it should be replaced with another model, used with a shorter census interval, or some other modification made, because in this case the model predicts that the population size becomes negative. Figure 5–8 illustrates how this model predicts changes in population size through time for some selected values of R_0.

This simple model of density-dependent population dynamics suggests three conclusions:

- First, density dependence does not help a population avoid extinction. R_0 has to be greater than 1 to begin with for a density-dependent population to come to a positive steady state. Similarly, a density-independent population also has to have an R_0 greater than 1 to avoid extinction. However, density dependence does prevent a population from exploding.
- Second, in some circumstances, density dependence causes the population size actually to attain a steady-state value.
- Third, density dependence is not necessarily stabilizing, however, and whether it is or not depends on the population's R_0. A population with a high reproductive potential, as measured by R_0, tends to overshoot the steady-state size by producing too many offspring when the size is below the steady-state value. Conversely, when the population size is much above the steady-state value, the impact of the density dependence with a high R_0 on birth and survivorship is so large that a crash can occur.

We must caution, however, that these conclusions are derived from a simplified picture of how the mechanisms of density dependence operate. They do not automatically apply to all mechanisms.

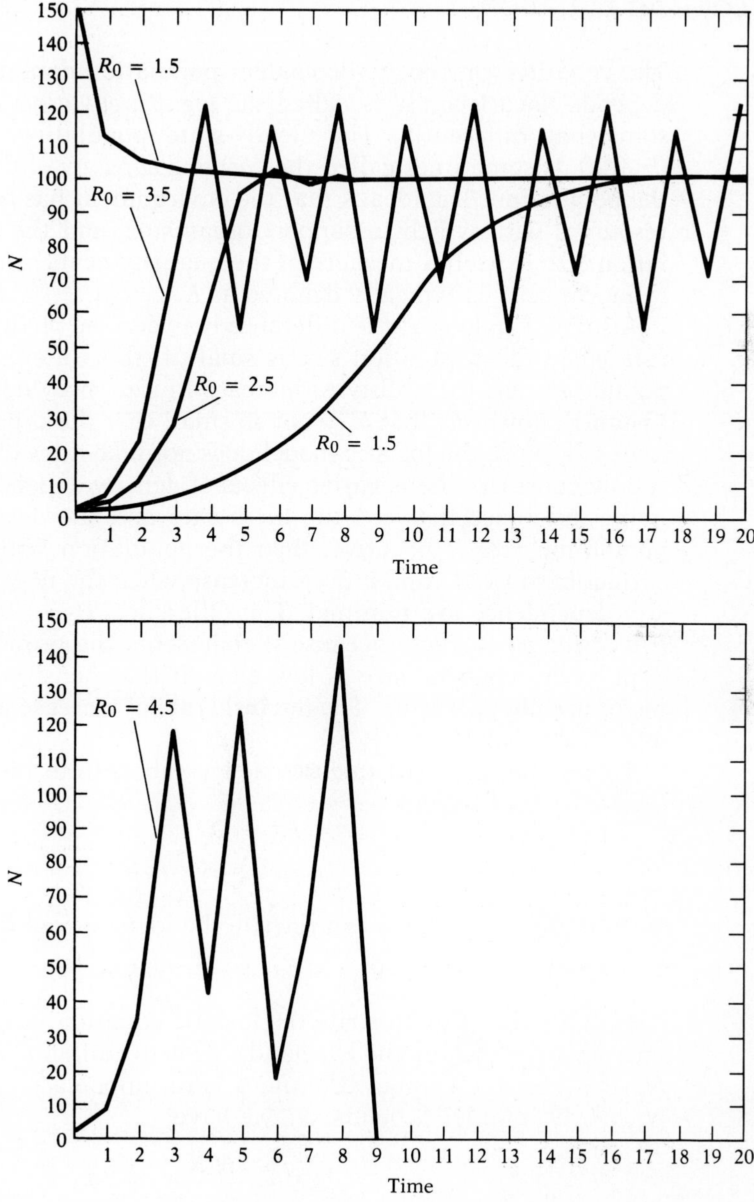

FIGURE 5–8 Population size through time as predicted by the logistic equation.

The Logistic Equation

The equation for density-dependent population dynamics in which $R(N)$ depends linearly on N is called the *logistic equation*. It is often written somewhat differently. The steady-state population size, $(B_0 - D_0)/(b + d)$, is sometimes called the *carrying capacity* of the environment for the population. The idea is that the environment has a certain amount of resources with which to support organisms, and the steady-state population size is itself a measure of the quantity of these resources. The carrying capacity is typically denoted as K.

Also, we can look at the difference between the birth rate and the death rate when the population size is small. If this difference is positive, the population has the ability to increase in size, provided N is small. (It is assumed, however, that N is not so small that the organisms cannot find mates, because the logistic model does not take Allee effects into account; it only considers the negative effects of density dependence.) In contrast, if the difference between the birth rate and the death rate at a small population size is negative, then the population will certainly become extinct because it cannot even increase when the negative effects of density dependence are minimal. The difference, $B_0 - D_0$, is called the *intrinsic rate of increase* because it represents the growth potential of the population when its size is low enough that density-dependent mechanisms are not operating. The intrinsic rate of increase is typically denoted as r. (Note that $R_0 = 1 + r$.)

To summarize, we define two new symbols in terms of the old ones:

$$r = B_0 - D_0$$

$$K = (B_0 - D_0)/(b + d)$$

With these symbols, we can rewrite the logistic equation as

$$N_{t+1} = [1 + r - (r/K)N_t]N_t$$

Still another way to write the logistic equation is as a so-called difference equation, where the left-hand side is the difference between N_{t+1} and N_t. If we let the symbol ΔN (the Δ is an uppercase Greek delta) denote $N_{t+1} - N_t$, and let N denote N_t, we have

$$\Delta N = \frac{rN(K - N)}{K}$$

which is perhaps the most familiar form of the logistic equation.

Using the parameters r and K in the logistic equation does save some space, but the saving is at the cost of obscuring the connection between a population's dynamics and the basic birth and death processes that cause those dynamics. If resources are added to an environment, K increases, but only because those resources affect the birth and death rates of individuals in that environment. Thus, the parameters r and K do not have an existence that is independent of the birth and death processes of population dynamics.

Measuring r and K

How do we measure the r and K of a population? In principle, if we have a closed population in a rather constant environment, we should do experiments such that the population is started at a size above K and at a size below K. These experiments should reveal whether a stable, steady-state population size is present and should illustrate how the population approaches the steady state, if it does. After these experiments are done, we may use the census data to provide points on a graph, in which $\Delta N/N$ is on the vertical axis and N is on the horizontal axis (Figure 5–9).

For example, suppose the population size in the experiment starts at 100, then moves to 150, and then to 175 organisms. In that case, the first ΔN is 50, and it occurred when N was 100. Thus, the first point to plot is at $\frac{1}{2}$ (this is $\frac{50}{100}$) on the vertical axis and 100 on the horizontal axis. The next ΔN is 25, which occurred when N was 150. Then the next point to plot is at $\frac{1}{6}$ (this is $\frac{25}{150}$) on the vertical axis and 150 on the horizontal axis, and so forth.

Some of the ΔN's will be negative; these values will come from the experiments in which N is started above K. For example, if the population size starts at 400, the next size may be 300 and then 250. The first ΔN from this experiment, then, will be -100, and it occurred when N was 400. Thus, a point should be plotted at $-\frac{1}{4}$ (this is $-\frac{100}{400}$) on the vertical axis and 400 on the horizontal axis. In this way, we can develop a collection of points of $\Delta N/N$ versus N.

The next step is to see whether the points $\Delta N/N$ versus N lie more or less on a straight line. If they do, then the intercept on the vertical axis can be identified with $1 + r$ and the slope with $-r/K$. For example, if the vertical-axis intercept equals 1.2, then r is 0.2; if the slope is -0.001, then K must be 200 because r was just found to be 0.2. If the points $\Delta N/N$ versus N do not lie on a straight line, then the logistic model is not appropriate, and some other formula that better fits the data must be used.

One reason the logistic equation frequently works, however, is simply that enough scatter usually is present in the data, and sufficiently few data are present, so that the deviation of the points from a straight line is not statistically significant. If so, the straight line is used by default because it is the simplest model that is consistent with the data. Of course, this reason is not a very satisfying one for using the logistic equation. It would be more desirable to know what the mechanisms of density de-

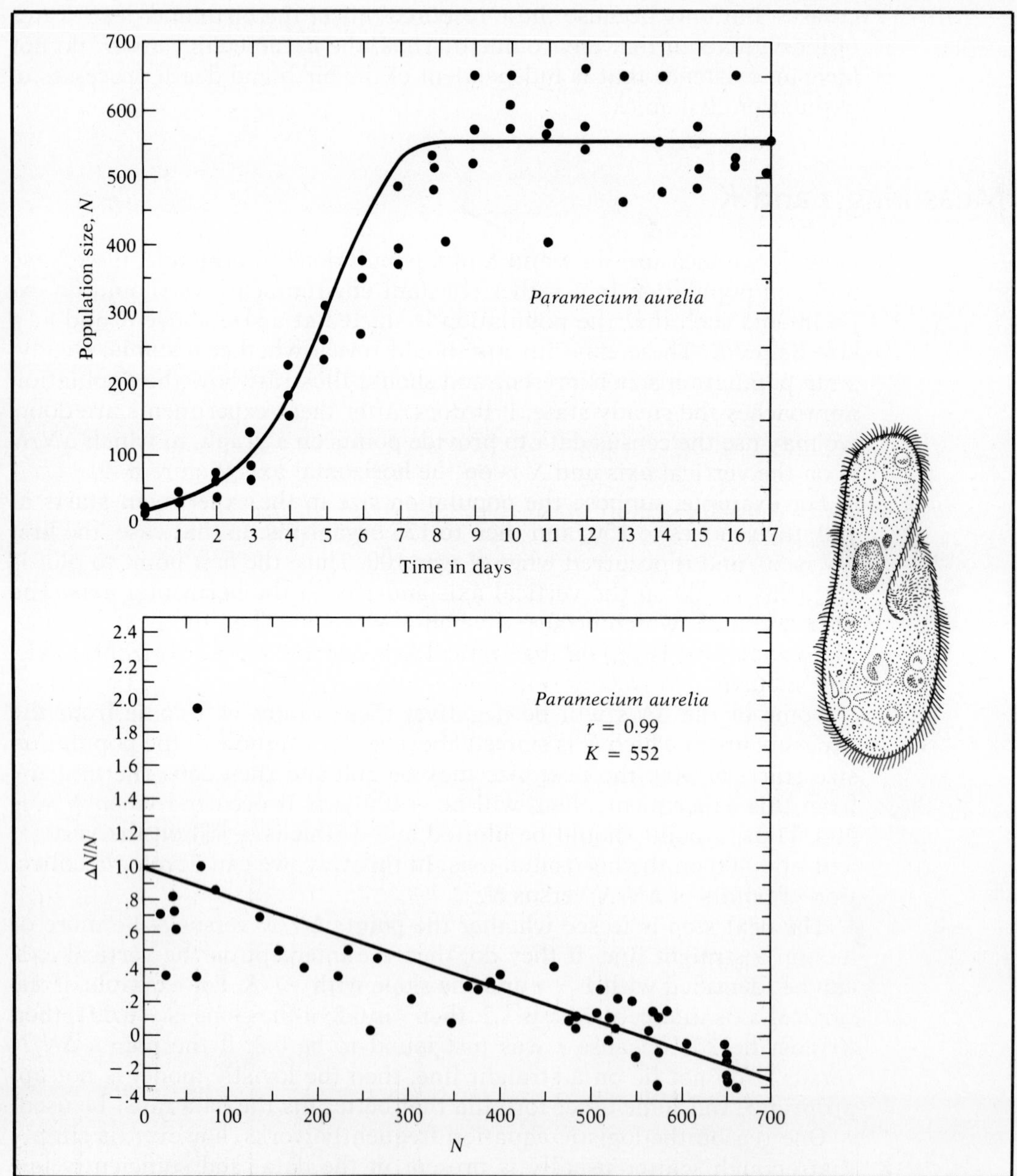

FIGURE 5–9 The fit of a logistic equation to data on the population size of *Paramecium* in a laboratory culture.

pendence are and, based on this knowledge, to show that the logistic equation offered an acceptable description of how those mechanisms affect a population's dynamics.*

Figure 5–9 illustrates the fit of a logistic equation to data on the population size of *Paramecium* in a laboratory culture. Notice that the data on $\Delta N/N$ versus N do more or less lie on a straight line and that the r and K obtained from the slope and intercept of this line do predict fairly well the population size of *Paramecium* through time.

Density Dependence and Plasticity in Plants

In plants, density dependence combines with the plasticity of plant growth to produce some regularities that are almost without parallel in animals, regularities known as the *rule of constant yield* and the *3/2 thinning rule.* Suppose many plots of ground are planted at different seed densities. Then the average weight per plant through time in each of the plots is measured. Next, we display the data on a graph, as illustrated in Figure 5–10, where the horizontal axis is the logarithm of the initial number of seeds per plot and the vertical axis is the logarithm of the average weight of a plant in each plot at different times (t).

Here is what typically happens: After one time period, the seedlings are small and do not interact with one another, as a result of which the average weight per seedling is the same across all plots. After another time period, the seedlings in the densest plots begin to interfere with each other's canopy or root system, but the seedlings in the sparse plots continue to grow without interacting. Because growth is occurring in all the plots, the average weight is higher in all plots at time 2 than at time 1. However, the sparse plots show more growth per seedling than the dense plots do because the individuals in the dense plots are interfering with each other's growth.

The first regularity is that in the dense plots, in which the individuals affect each other's growth, the total yield from the plot is constant. The *total yield* is defined as the sum of the weight of all of the plants in the plot; that is,

$$Y = WN$$

where W is the average weight per plant, N is the number of plants, and Y is the yield from the plot. Recalling that the graph has the logarithm of the weight per plant on the vertical axis and the logarithm of the num-

* If substantial error exists in the estimates of the population size, N, then analyzing ΔN itself, instead of $\Delta N/N$, is preferable. The ΔN should be fitted to a quadratic equation in N— a simple case of what is called a *polynomial regression.* The r and K are then derived from the coefficients in the quadratic equation.

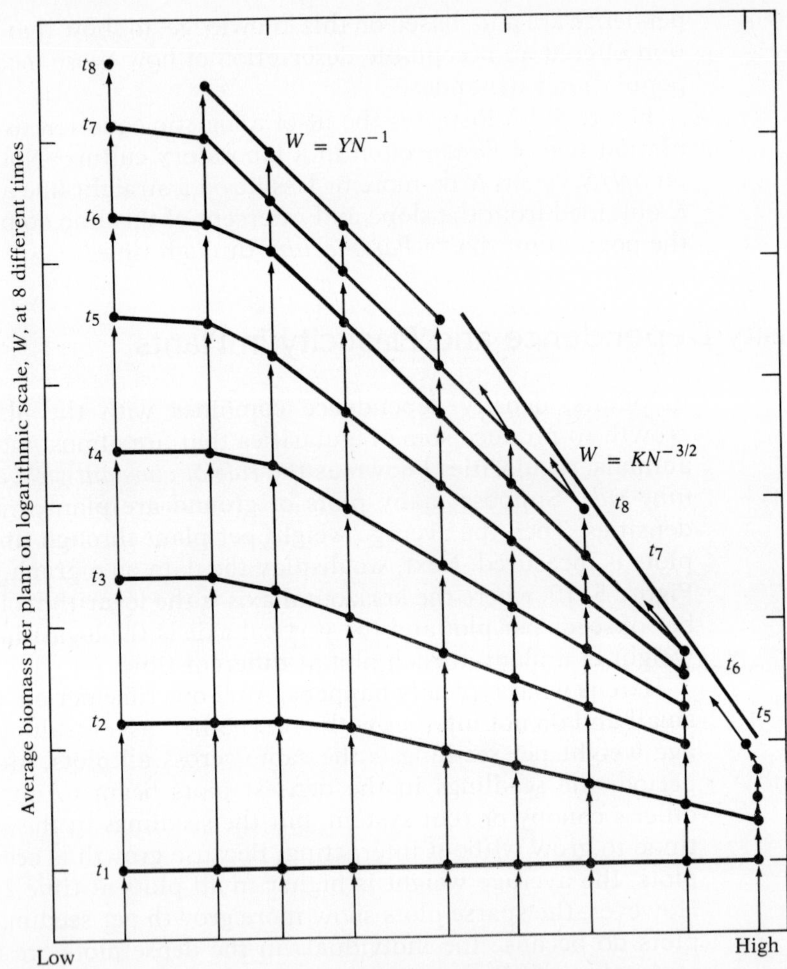

FIGURE 5–10 Graph showing the 3/2 thinning rule and the rule of constant yield. A dimensional interpretation of the 3/2 rule is that a plant's average mass, W is proportional to L^3, where L is a measure of the plant's length. Also, the number N, that can fit into a plot is inversely proportional to L^2. Solving for W as a function of N by eliminating L gives a relationship with $-3/2$ in the exponent. See text for further explanation of the figure. *(After White 1980.)*

ber of seedlings on the horizontal axis, we see among the dense plots the following relation:

$$W = \frac{Y}{N}$$

Hence,

$$\log W = (-1)\log N + \log Y$$

Thus, on the graph, among the dense plots in which the individuals are affecting each other's growth, a straight-line relation appears between log W and log N, with a slope of -1. In the graph, N remains at the initial seedling density because, so far, the interference among individuals has affected only growth, and not survival. This is called the *rule of constant yield.*

As more time passes, the interference among plants takes its toll in survivorship as well. The second regularity is that the average weight per plant, W, becomes related to the number of plants, N, as follows:

$$W = \frac{K}{N^{3/2}}$$

where k is a constant. Hence,

$$\log W = (-3/2)\log N + \log K$$

On the graph, once the number of plants per plot starts to decline as a result of crowding, the log of the weight of the survivors follows a straight-line relation with the log of the number of survivors, and the slope is $-3/2$. For a dimensional interpretation of this rule see the legend for Figure 5–10. As time passes, the point for a given plot shifts up and to the left along this line as the number of individuals in the plot continues to decline while the survivors continue to grow in weight. This finding is called the *3/2 thinning rule.*

This picture of how plant stands grow seems quite general, with examples from over 30 species of terrestrial herbs and trees known (White 1980).

CHAPTER 6

Evolution: Explaining the Properties of Organisms

Our modern understanding of evolution traces to Charles Darwin and Alfred Wallace, whose ideas were presented together at a meeting of the Linnean Society in 1858. Darwin's work on this theory subsequently appeared in his famous book *On the Origin of Species* (1859), and Wallace's work is summarized in two volumes entitled *The Geographical Distribution of Animals* (1876).

Since that time, the geographical distribution and abundance of organisms has tended to be the specialty of ecologists, whereas evolutionists have specialized in what determines an organism's traits. Yet these questions should be examined together: Mechanisms affecting the abundance and distribution of organisms may also affect their evolution, and vice versa. Therefore, let us begin by giving an introduction to pertinent parts of evolutionary theory.*

Since the early 1900s, two approaches to the study of evolution have developed, biometrical and genetical. These approaches differ in how they treat *inheritance,* the process that causes offspring to resemble their parents. The *biometrical* approach simply accepts the resemblance of offspring to parents and relies on a *description* of the closeness of that resemblance. The *genetical* approach requires specification of a precise genetic

* For a broader treatment, see the fine textbook by Futuyma (1979).

mechanism for the inheritance of the trait. The biometrical approach is not as fundamental as the genetical approach because it omits the mechanisms of inheritance; however, it is sometimes more useful because the genetic basis of many traits is unknown.† In this chapter, we will be principally concerned with the genetical approach.

THE GENE POOL

The genetical approach to evolution begins with the concept of a *gene pool,* the collection of all the genes from all the individuals in the population. Because organisms are not all alike, the gene pool at any time contains a variety of genes. The composition of the gene pool may change with the birth and death of organisms because genes enter the pool with births or are lost from it with deaths. According to the genetical approach, the study of evolution reduces to identifying the processes that affect the composition of the gene pool.

A small piece of the gene pool is all that is usually considered. It consists of all the genes found at a single locus. One, two, or several alleles may be present in the population at any particular locus. If all the individuals have the same allele at a particular locus, then the gene pool is said to be *fixed* at that locus. If two or more alleles are at the locus, the gene pool is said to be *polymorphic* at that locus.

In many species, each unspecialized cell of an individual has a pair of homologous chromosomes, one from each parent; this state is called *diploidy.* If an individual has different alleles at the same locus on each of its chromosomes, the individual is said to be *heterozygous.* If an individual has the same alleles at a particular locus on both of its matching chromosomes, then it is said to be *homozygous.*

Notice the difference between polymorphism and heterozygosity. Polymorphism refers to whether the *population* has two or more alleles at a particular locus, whereas heterozygosity refers to whether an *individual* has different alleles on its two chromosomes at a particular locus. If a locus is polymorphic, typically some of the individuals are homozygous for one of the alleles in the population, other individuals are homozygous for each of the other alleles, and the remaining individuals are heterozygous, with various combinations of alleles.

† An overview of the biometric approach can be found in Falconer (1981). For the genetic approach, see Hartl (1980), Roughgarden (1979), and Wallace (1981), all of whom also touch on the biometrical approach. For a sampling of articles tying ecology and population genetics together, see Shorrocks (1984).

GENOTYPIC RATIOS AND MATING RULES

A population's genetic variation is expressed in a regular pattern. To illustrate, consider a locus where the gene pool is polymorphic for two alleles. Suppose also that the individuals are diploid. Label the alleles as A_1 and A_2. Now, if the gene pool is 50% A_1 and 50% A_2, what will the individual genotypes be? Will the individuals be 50% A_1A_1 and 50% A_2A_2? Usually not. With a gene pool that is 50:50 for two alleles, the genotypes tend to be in the ratio 25% A_1A_1, 50% A_1A_2, and 25% A_2A_2. Furthermore, unless certain additional processes occur, these ratios remain constant through time. These ratios exemplify what are called the *Hardy–Weinberg ratios;* the claim that these ratios remain constant through time, unless certain processes occur, is called the *Hardy–Weinberg law.*

Why is the variation in the gene pool expressed in this regular way? The regularity reflects the mating rules in a population. The simplest mating rule is called *random union of gametes.* Imagine that gametes are released into water and that water currents cause the gametes containing A_1 and A_2 to be mixed together. Now focus on an unfertilized egg. What is the probability that it becomes fertilized with a sperm carrying A_1? If, say, one-fourth of the sperm carry A_1, then the probability is simply $\frac{1}{4}$. Similarly, the probability of the egg's being fertilized by a sperm with A_2 is $\frac{3}{4}$. To be more general, if the fraction of sperm with A_1 is p_m and the fraction with A_2 is q_m, then from the standpoint of a particular egg, the probability of being fertilized by a sperm with A_1 is p_m and with A_2 is q_m.

Next, consider what allele the egg itself is carrying. Let the fraction of eggs with A_1 be p_f and the fraction of eggs with A_2 be q_f. For example, suppose one-third of the eggs have A_1 and two-thirds have A_2. Therefore, after the eggs are fertilized, $\frac{1}{3} \times \frac{1}{4}$ zygotes contain A_1 alleles from both the sperm and the egg. Furthermore, $\frac{1}{3} \times \frac{3}{4}$ zygotes contain an A_2 allele from the sperm and an A_1 allele from the egg; these individuals are heterozygous at this locus. Similarly, we have $\frac{2}{3} \times \frac{1}{4}$ zygotes, with an A_1 allele from the sperm and an A_2 allele from the egg; these individuals are also heterozygous at this locus. Finally, we have $\frac{2}{3} \times \frac{3}{4}$ zygotes, with an A_2 allele from both the sperm and the egg. Thus, a random union of gametes leads to the following ratio of genotypes among the zygotes:

$$A_1A_1\colon\ p_m p_f = \frac{1}{3} \times \frac{1}{4}$$

$$A_1A_2\colon\ p_m q_f + q_m p_f = \frac{1}{3} \times \frac{3}{4} + \frac{2}{3} \times \frac{1}{4}$$

$$A_2A_2\colon\ q_m q_f = \frac{2}{3} \times \frac{3}{4}$$

This arrangement is the basis of the Hardy–Weinberg ratios. The sperm and eggs usually carry the same fraction of both alleles. If so, p_m and p_f are the same, as are q_m and q_f. Thus, typically, the ratios of genotypes among the zygotes are

$$A_1A_1: p^2$$

$$A_1A_2: 2pq$$

$$A_2A_2: q^2$$

In particular, if p and q both happen to be $\frac{1}{2}$, then the genotype ratios are 25%, 50%, and 25% for A_1A_1, A_1A_2, and A_2A_2, respectively, as discussed earlier.

In genetics, the proportions of the different alleles in the gene pool are called *gene frequencies,* and the proportions of the different genotypes are called the *genotype frequencies.* The gene frequencies, p and q, always sum to 1; the genotype frequencies do also. Notice, for example, that the Hardy–Weinberg ratios automatically sum to 1 (or 100%).

Random union of gametes is the simplest rule used to describe the mating in a population. Another rule is *random mating,* where *diploid adults* mate at random. (Contrast this rule with random union of gametes, where the *haploid* gametes fuse at random.) It can be shown that the Hardy–Weinberg ratios apply to random mating, just as they do to the random union of gametes.

The most important mating rule that does not lead to the Hardy–Weinberg ratios involves *inbreeding,* defined as the mating of organisms with relatives more often than expected by random encounter. *Outbreeding* is the opposite; it occurs when organisms mate with their relatives less often than would be expected by random encounter. Inbreeding is particularly important in populations in which the offspring do not disperse far from their place of birth. The most extreme form of inbreeding, *selfing,* often involves a stigma fertilized by pollen from an anther in the same flower.

To see how inbreeding differs from the random union of gametes, consider the special case of selfing. If a population initially contains 25% A_1A_1, 50% A_1A_2, and 25% A_2A_2, then what will happen after one generation of selfing? The A_1A_1 individuals produce A_1A_1 offspring, and similarly, the A_2A_2 individuals produce A_2A_2 offspring. When the heterozygotes self, each cross yields $\frac{1}{4}$ A_1A_1, $\frac{1}{2}$ A_1A_2, and $\frac{1}{4}$ A_2A_2 offspring, according to Mendel's laws. Hence, the heterozygotes are not replacing themselves. Only half of the offspring of heterozygotes are themselves heterozygotes; the rest are various homozygous genotypes.

Thus, selfing leads to a gradual loss of heterozygote individuals. Indeed, if the population is initially 25% A_1A_1, 50% A_1A_2, and 25% A_2A_2, then it eventually consists only of homozygotes, and the ultimate ratios are 50%

A_1A_1, 0% A_1A_2, and 50% A_2A_2. Less extreme systems of pure inbreeding, such as sib mating, also lead eventually to the total loss of heterozygosity, although more slowly. In contrast, inbreeding combined with random mating does not lead to the total loss of heterozygosity but to a reduction in the frequency of heterozygotes relative to the Hardy–Weinberg ratios.

MUTATION AND GENETIC DRIFT

The composition of the gene pool is continually subject to change as organisms are born and die. When an organism is born with a gene that is not a duplicate of a gene from either of its parents, it is said to carry a *mutation*. Mutations are always happening. The molecular machinery that duplicates the parental genes during the manufacture of gametes is not perfect. DNA is continually being rearranged, not only at the time of replication but also by viruslike particles that move pieces of the genome around. A *recurrent mutation* is a change from one allele to another allele that is already in the gene pool. In this situation, an offspring carries a gene that is not a copy of any gene in its parents but nevertheless is an allele that other individuals already possess. Most mutations are recurrent. In contrast, a *novel mutation* is a change to an allele that is not already in the population.

If a novel gene appears, one might ask why it is novel—why is the gene not already in the population? One reason might be that the gene is harmful to its carriers, possibly even lethal, and so disappears soon after its formation. Another possibility is that populations lose genes by a process called *genetic drift*, which occurs simply because every population is finite.

To illustrate how genetic drift works, consider a tiny population consisting of only two individuals that mate by random union of gametes. The gene pool at any locus contains only four genes. To make the next generation, in effect, four gametes are drawn from all those that these two individuals have produced, much as one might draw colored marbles from a basket. Suppose the adults have the genotypes A_1A_1 and A_2A_2, so that the pool initially consists of two A_1 and two A_2 alleles. The set of gametes produced by these adults resembles a large basket with many red and blue marbles in a 50:50 ratio. We might consider gametes carrying A_1 alleles as red marbles and gametes with A_2 alleles as blue marbles. To produce the next generation of adults, two eggs and two sperm must be "drawn" and fused to yield two diploid individuals.

Now here is the important point: Although the A_1 and A_2 alleles occur in a 50:50 ratio among the gametes in the pool, four gametes might be drawn *just by chance* that do *not* have a 50:50 ratio of A_1 to A_2 among them. For example, three A_1 gametes and one A_2 gamete might be drawn, or one A_1 gamete and three A_2 gametes. Of course, sometimes two of each allele will be drawn; if so, the gene pool in the next generation remains the same. However, much of the time, the gene pool changes from generation to generation simply because the gametes that are drawn to make up a particular generation are not, by chance, in the same ratio as they were in the previous generation. Such changes in gene frequencies are called *genetic drift.* Drift is the process by which the composition of the gene pool fluctuates erratically from generation to generation because the alleles drawn for a particular generation have slightly different frequencies than they did in the preceding generation.

At first glance, it might seem that genetic drift is a small complication that ought to be ignored—just some "noise" superimposed on the evolutionary process. However, genetic drift is important because it causes a *loss* of genes from the gene pool. As the gene pool's composition fluctuates from generation to generation, it eventually reaches a state in which it may actually lose an allele.

For example, suppose our population of two individuals comes to have three A_1 alleles and one A_2 allele. Then, when we come to draw four alleles again for the next generation, we might draw four A_1 alleles and no A_2 alleles. If this happens, then the gene pool will have lost the A_2 allele (and fixed the A_1 allele) until a new mutation arises that restores A_2 to the gene pool. Moreover, even when mutation does happen, the new mutant is itself susceptible to loss by genetic drift because the new mutant is rare when it starts out.

Thus, genetic variation is continually entering by mutation and leaving by genetic drift. Indeed, if other processes are sufficiently weak, the gene pool's composition is a dynamic steady state determined primarily by the balance of these two processes.

NATURAL SELECTION

Natural selection is the main process affecting the gene pool that involves ecology. *Natural selection* occurs if a difference exists in the fertility and/or survivorship of genetically different individuals. If individuals with different genotypes have the same fertility and survivorship, then no natural selection occurs, and the genotypes are said to be *selectively neutral.* Moreover, if individuals do differ in their fertility and survival, but these differences are not related to the genetic differences between them, then again no natural selection occurs. Thus, natural selection means that individuals differ in their fertility and/or survival *and* that the causes of the differences are genetically inherited by the offspring.

Natural selection affects the gene pool either by leading to a loss of variation that is rapid compared with the rate at which genetic drift causes the loss of variation or by conserving genetic variation and retarding its loss by genetic drift. To see how these possibilities arise, we begin with natural selection in an exponentially growing population. This type of natural selection is called *density-independent* selection because the strength of natural selection in this context is independent of the population size. Natural selection whose strength does depend on the population size is *density-dependent* selection, a topic covered in the next chapter.

How Natural Selection Works: A Model

To develop a useful model of natural selection, imagine a population with discrete generations and random union of gametes. This hypothetical population always consists of individuals of the same age, and the life cycle is as follows: The generation begins at time t when the organisms are zygotes, right after their fusion through a random union of gametes. The genotypes are in Hardy–Weinberg ratios at this point, because the zygotes are formed by random union of gametes. As the individuals age, they suffer mortality in the environment. Each genotype has a characteristic probability of surviving from the zygotic phase until the time of mating. Then the individuals release gametes, and each genotype has a characteristic fertility in the environment. Once the gametes have combined at random, the next generation is begun. Thus, the time period is exactly one generation long, and the beginning of each time period is placed at the zygotic phase.

To describe natural selection, we need to devise symbols for the survivorship of each type of individual. Let l_{11} be the probability that an A_1A_1 individual survives from its zygotic phase to the age at which mating occurs. The l_{11} is a number between 0 and 1. Similarly, let l_{12} and l_{22} describe the survival of individuals with A_1A_2 and A_2A_2 genotypes, respectively. To describe the fertility of the individuals, we use m_{11}, m_{12}, and m_{22} for genotypes A_1A_1, A_1A_2, and A_2A_2, respectively. However, the units for m may seem strange at first. An m for an individual is one-half of the number of gametes it makes that are actually incorporated into zygotes. For example, if an organism produces 1000 gametes, and 100 of them are incorporated into zygotes, then m is 50 for that individual. The reason one-half is used is that we are counting the zygotes at time $t + 1$ and it takes two gametes to make a zygote; thus, a gamete is only half of an offspring.

A count of the number of zygotes at the beginning of any time interval, say t, yields N_t. The fraction of the alleles contained in the zygotes at time t that are of type A_1 is denoted as p_t and the fraction of A_2 alleles among the zygotes at time t is q_t. With just two alleles, q_t equals $1 - p_t$.

Now we can predict how natural selection affects the gene pool. This prediction involves five steps:

First, because Hardy–Weinberg ratios are present at time t among the zygotes, each genotype has the following number:

- Number of A_1A_1 among zygotes: $p_t^2N_t$
- Number of A_1A_2 among zygotes: $2p_tq_tN_t$
- Number of A_2A_2 among zygotes: $q_t^2N_t$

Second, the number of adults of each genotype at the time of mating is the fraction of each type that survive times the number of zygotes of each type:

- Number of A_1A_1 among adults: $l_{11}(p_t^2N_t)$
- Number of A_1A_2 among adults: $l_{12}(2p_tq_tN_t)$
- Number of A_2A_2 among adults: $l_{22}(q_t^2N_t)$

Third, each of these adults is responsible for m_{11}, m_{12}, or m_{22} zygotes, depending on its genotype. Thus, the total number of zygotes, which is N_{t+1}, is m_{11} times the number of adults of type A_1A_1 plus m_{12} times the number of adults of type A_1A_2 plus m_{22} times the number of adults of type A_2A_2. This process is expressed symbolically as follows:

$$N_{t+1} = m_{11}l_{11}(p_t^2N_t) + m_{12}l_{12}(2p_tq_tN_t) + m_{22}l_{22}(q_t^2N_t)$$

This equation may be simplified by factoring out N_t, yielding

$$N_{t+1} = (p_t^2m_{11}l_{11} + 2p_tq_tm_{12}l_{12} + q_t^2m_{22}l_{22})N_t$$

Fourth, we may calculate the number of A_1 and A_2 genes among the zygotes at time $t + 1$. The A_1 genes among the zygotes come both from A_1A_1 homozygous adults and from the A_1A_2 heterozygous adults. Hence, the number of A_1 genes among the zygotes is:

$$\text{Number of } A_1 \text{ genes among zygotes} = (2)m_{11}l_{11}p_t^2N_t + (1)m_{12}l_{12}2p_tq_tN_t$$

Notice that we use a factor of (2) for the output from the homozygote adults because each of its offspring carries two A_2 genes. In contrast, we use a factor of (1) to emphasize that, for the output from the heterozygote adults, each offspring carries only one A_1 gene. Similarly, the number of A_2 genes among the zygotes is:

$$\text{Number of } A_2 \text{ genes among zygotes} = (2)m_{22}l_{22}q_t^2N_t + (1)m_{12}l_{12}2p_tq_tN_t$$

Fifth, and last, we can determine the fraction of the gene pool that is A_1 simply by writing out the number of A_1 genes divided by the total number of genes. The total number of genes in the gene pool is $2N_{t+1}$, because each individual carries two genes. Hence,

$$p_{t+1} = \frac{(2m_{11}l_{11}p_t^2N_t + m_{12}l_{12}2p_tq_tN_t)}{(2N_{t+1})}$$

Fortunately, we can simplify this by removing the (2) from both numerator and denominator. Furthermore, when we substitute the expression for N_{t+1} that we just developed, it is clear that N_t will also cancel out from the numerator and denominator. Finally, we should factor p_t from the numerator, leaving us with

$$p_{t+1} = \frac{(m_{11}l_{11}p_t + m_{12}l_{12}q_t)p_t}{(m_{11}l_{11}p_t^2 + m_{12}l_{12}2p_tq_t + m_{22}l_{22}q_t^2)}$$

In principle, the equations for p_{t+1} and N_{t+1} predict the population size and gene pool composition through time. If we know p and N at time 0, we can predict them at time 1. Then we can use the equations again and predict p and N at time 2, and so forth. Of course, if the m's and l's change in each generation, we would want to update their values each time the equations are used.

Fitness

These equations are famous in the study of evolution. They were originally developed by Sewall Wright, J. B. S. Haldane, and R. A. Fisher in the 1920s. To see what these equations predict, it is helpful to simplify them slightly before proceeding. Notice that the m's and l's always appear as a product. Hence, we can combine them into one set of symbols, W's. Let the W's be defined in terms of the m's and l's as follows:

$$W_{11} = m_{11}l_{11}$$

$$W_{12} = m_{12}l_{12}$$

$$W_{22} = m_{22}l_{22}$$

The symbol W is called the *fitness* of an organism; it combines both fertility and survivorship. The fitness of an organism is the number of offspring that it places in the next generation. As we will see, genes that produce organisms with a high fitness tend to increase in the gene pool through time. Because fitness includes both fertility and survival, natural selection is not correctly viewed as the "*survival* of the fittest," for the fertility of an individual is just as important as its survivorship. In fact, natural selection can occur if just differential fertility exists, without differential survival.

The concept of fitness is closely related to the exponential growth rate of a population. Indeed, if the gene pool happens to be fixed for, say A_1, then the population size at time $t + 1$ is just $W_{11} \times N_t$. This is the formula for exponential growth, where W_{11} is the exponential growth factor. The only reason for using the symbol W instead of the symbol R used in the last chapter is that the time interval must be one generation in this model, whereas R is used with any convenient time interval.

To summarize, a model for the way in which a population's size and gene pool change through time is

$$N_{t+1} = (p_t^2W_{11} + 2p_tq_tW_{12} + q_t^2W_{22})N_t$$

$$p_{t+1} = \frac{(p_tW_{11} + q_tW_{12})p_t}{p_t^2W_{11} + 2p_tq_tW_{12} + q_t^2W_{22}}$$

Laboratory Tests of the Model

Having derived these equations, let us ask: How well do they work? Figure 6–1 illustrates the accuracy with which these equations describe evolution in a laboratory population of *Drosophila pseudoobscura*. The genotypes are not single loci but are blocks of genes that segregate together and might be thought of as "supergenes." Two such blocks are *ST* and *CH*, and the genotypes are *ST-ST*, *ST-CH*, and *CH-CH*. Dobzhansky (1954) measured the fitness of these genotypes and predicted the fraction of *ST* through time for a laboratory population. Figure 6–1 shows four replicate experiments; the solid line is the predicted value, and the points are the experimental results.

Another example is presented in Figure 6–2 (Wallace 1963). Again, fruit flies are the organisms, but this time the species is *Drosophila melanogaster*. In this example, one of the alleles is a nearly recessive lethal. Natural selection gradually eliminates this lethal allele from the gene pool. The solid line presents the fraction of the lethal allele in the gene pool through time as predicted by the equation, assuming it is completely recessive. The points from the experiment are close to the predicted ones. However, the lethal gene is lost somewhat more rapidly than predicted because it is not completely recessive.

Predictions from the Model for Density-Independent Natural Selection

These laboratory tests offer some confidence in these equations. Let us now explore the model further; three interesting cases of predictions emerge.

Homozygote with Highest Fitness and Intermediate Heterozygote. Suppose that individuals homozygous for A_1 have the highest fitness, heterozygotes have an intermediate fitness, and homozygotes for A_2 have the lowest fitness. For example, suppose W_{11} is 2.0, W_{12} is 1.75, and W_{22} is 1.5. One can imagine that A_1 is a new mutation that is advantageous because its carriers have a higher fitness than do carriers of the "old" allele, A_2. If A_1 is a mutant, then its initial fraction in the gene pool is low, so p_0 is, say, 0.001. Suppose the population size, N_0, when the mutation arises is 1000, and we start the clock when the mutation arises. As Figure 6–3 shows, p_t gradually increases from its initially low value until it eventually takes over the entire gene pool. As A_1 spreads, the population growth factor increases. Initially, the population grows exponentially at a rate close to W_{22} because the gene pool consists almost entirely of A_2 at this time. However, as A_1 takes over the gene pool, the population growth factor approaches W_{11}.

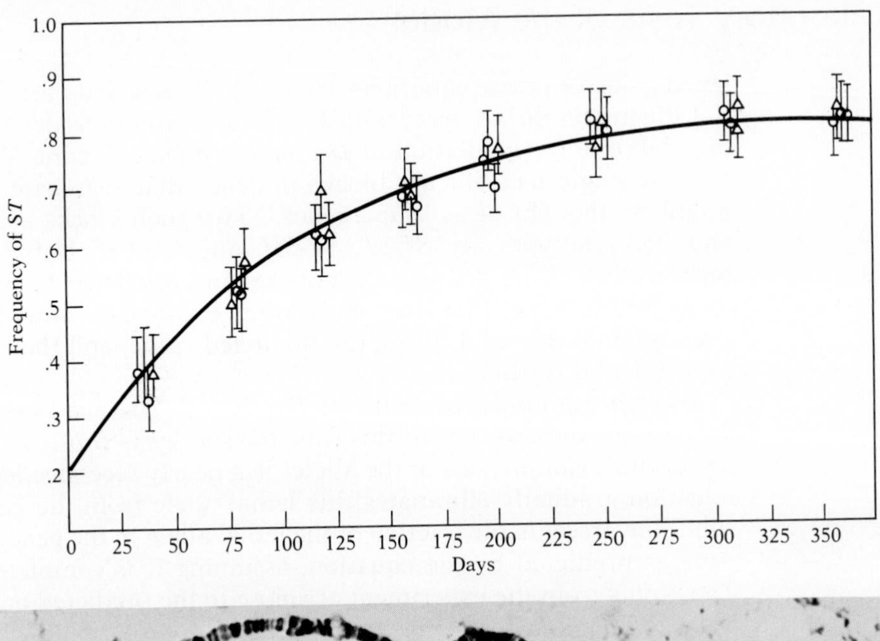

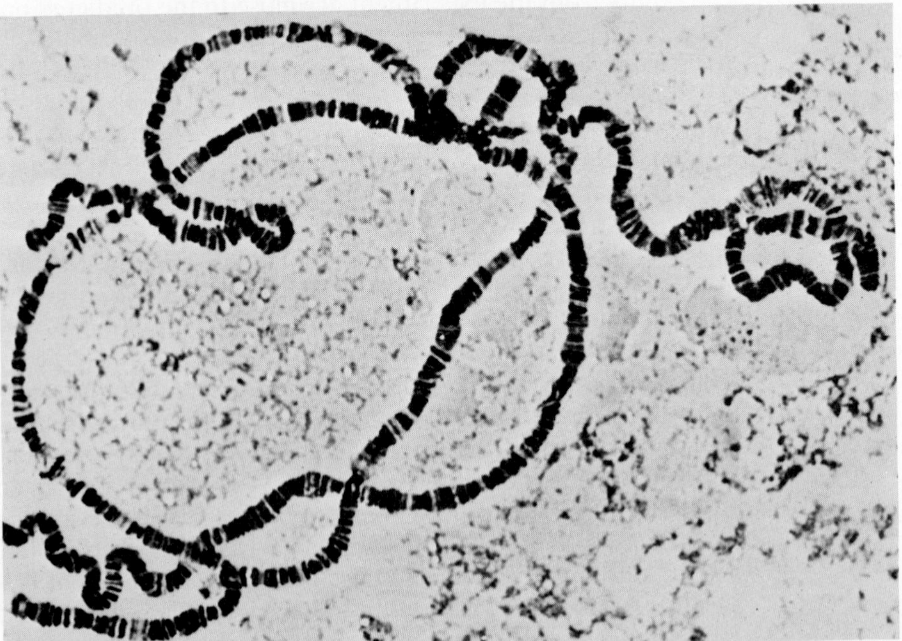

FIGURE 6–1 Evolution in a laboratory population of the fruit fly, *Drosophila pseudoobscura.* Heterozygotes have the highest fitness, and a stable polymorphism is attained. The genetic composition of an individual is determined with a microscope by examining chromosomes extracted from the salivary glands, as depicted below. *(From Dobzhansky 1954.)*

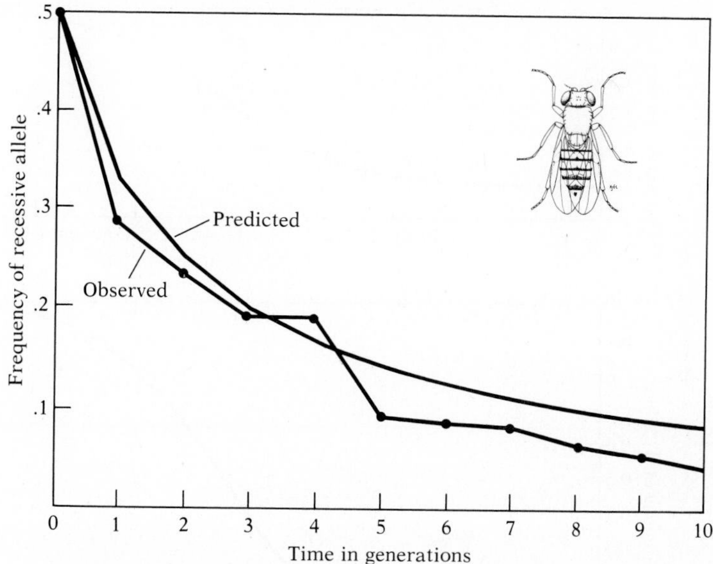

FIGURE 6–2 Evolution in a laboratory population of the fruit fly, *Drosophila melanogaster.* The frequency of a recessive lethal allele gradually declines through time. *(From Wallace 1963.)*

The previous example of the elimination of a nearly recessive lethal in *D. melanogaster* illustrates this case, except that the fitness of the heterozygote is not exactly intermediate between that of the two homozygotes.

Heterozygote with Highest Fitness. Now suppose the heterozygote has the highest fitness. This condition is called *heterosis.* It occurs when the heterozygote can function in ways that neither homozygote can; the heterozygote has, so to speak, the best of both worlds. Let W_{11} be 1.5, W_{12} be 2.0, and W_{22} be 1.75.

Two interesting initial conditions exist. One initial condition is, as before, a gene pool consisting mostly of A_2. Then, as soon as A_1 arises by mutation, it appears to be advantageous because its carriers have a higher fitness, W_{12}, than do the other members of the population, W_{11}, at that time. Hence, we expect p_t to increase if p is near 0 to begin with. However, this argument also applies to the other initial condition, where the gene pool consists mostly of A_1. As soon as A_2 arises by mutation, it too appears to be advantageous because its carriers have a higher fitness, W_{12}, than the other members of the population, W_{22} at that time. Hence, we expect q_t to increase (and p_t correspondingly to decrease) if q is near 0 to begin with.

Figure 6–4 illustrates these points: It shows that if p is near 0 to begin with, it increases and eventually approaches a specific value. Also, if p is near 1 to begin with (i.e., q is near zero), it decreases and eventually approaches the same specific value that it does from the other initial

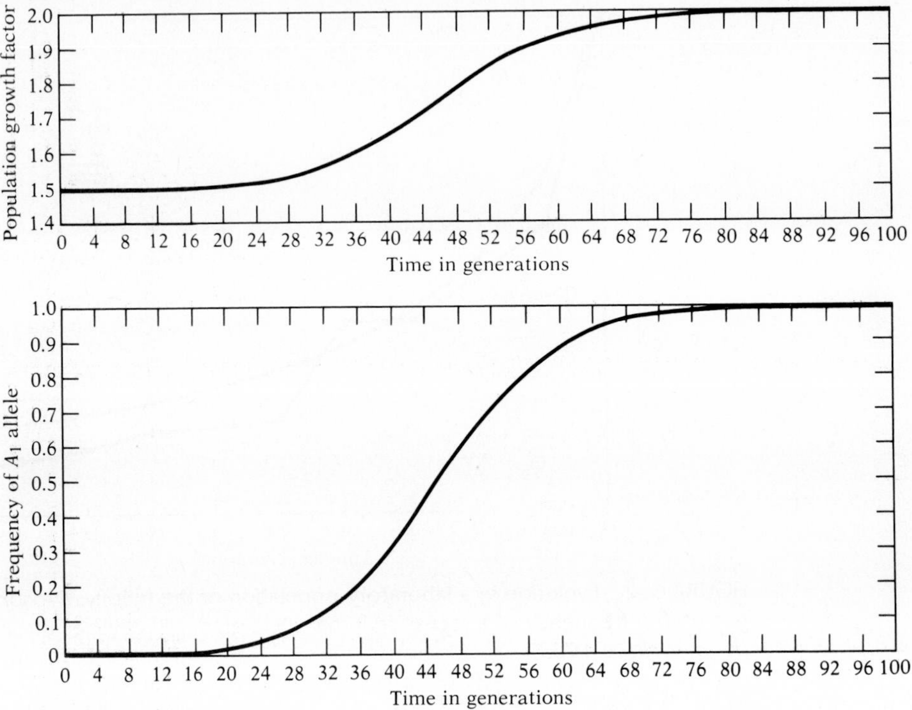

FIGURE 6–3 Plot of population growth factor and p_t as a function of t for directional selection.

condition. Thus, the gene pool becomes polymorphic, *and* a particular ratio of A_1 to A_2 is ultimately attained in the gene pool. Although we will not pursue this point further here, one can develop a formula that exactly predicts the ultimate ratio of A_1 and A_2.

Figure 6–4 also shows that the population growth rate is initially the same as the W of the prevalent genotype. That is, if p is near 0, the exponential growth factor of the population is near W_{22}; if p is near 1, the exponential growth factor is near W_{11}. In both cases, as evolution progresses, the exponential growth factor of the population increases in each generation as the final ratio of A_1 to A_2 is attained. As the population approaches this final condition, its exponential growth rate is higher than either W_{11} or W_{22} but is not as high as W_{12}. The final exponential growth rate cannot be as high as W_{12} because the population does not consist solely of heterozygotes.

The previous example involving the supergenes, *ST* and *CH*, in *D. pseudoobscura* illustrates this case, and we will meet it again in Part 4 in the context of mosquitos, malaria, and humans.

Heterozygote with Lowest Fitness. The case of the heterozygote with lowest fitness arises when two genes exist that are somehow incompatible.

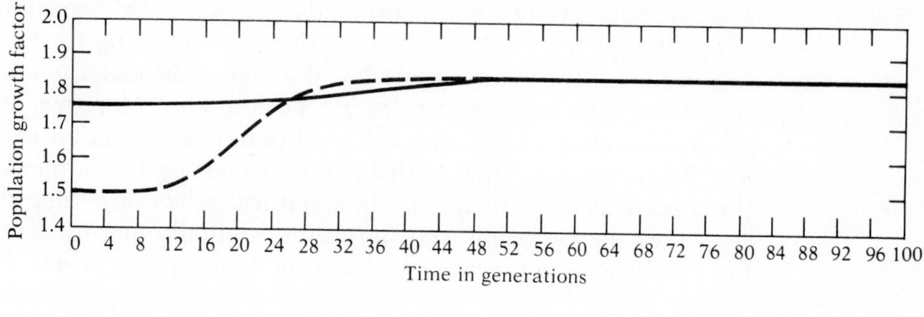

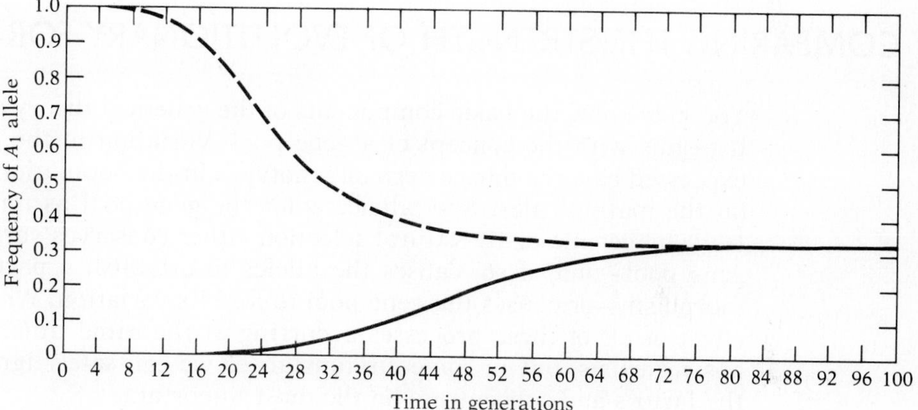

FIGURE 6–4 Plot of population growth factor and p_t as a function of t for heterosis. Dashed and solid lines refer to different initial conditions.

It is an interesting case because, from the standpoint of either gene, the other is disadvantageous. If the gene pool contains only A_2, then any mutation to A_1 is disadvantageous because W_{12} is lower than W_{22}, and the mutation does not spread. Similarly, if the gene pool consists of A_1, then a mutation to A_2 is disadvantageous and does not spread. This circumstance allows the gene pool to exist in either of two alternative states.

It does not matter whether W_{11} is higher or lower than W_{22}, so long as the heterozygote is smaller than either. For example, A_1A_1 homozygotes may have a higher fitness than A_2A_2 homozygotes. Nonetheless, if the gene pool initially consists only of A_2, any mutation to A_1 cannot spread because the heterozygotes have lower fitness than the A_2A_2 homozygotes do. Thus, the low fitness of the heterozygotes creates a barrier to the spread of a gene that confers the highest fitness once it is in a homozygous state.

The three cases of the model show that natural selection may have different effects on the gene pool. If homozygotes have the highest fitness, natural selection eliminates variation from the gene pool or prevents it from entering. If the heterozygote's fitness is lower than that of both homozygotes, natural selection again leads to a gene pool containing only one allele, though the allele may not be the one conferring the highest fitness. In contrast, if the heterozygote has the highest fitness, then natural selection causes the gene pool to retain both alleles. Moreover, in this case, natural selection causes the gene pool to have a particular ratio of alleles, and the numerical value of this ratio depends on the numerical values of the three fitness coefficients. Thus, natural selection either eliminates or conserves the variation that mutation originally provides, depending on the details of which genotypes have the highest and lowest fitnesses.

COMPARING THE STRENGTH OF EVOLUTIONARY FORCES

You now know the basic components of the genetical theory of evolution; it begins with the concept of a gene pool. Variation in the gene pool is expressed as a regular pattern of genotypes in the population, depending on the mating rules. New alleles enter the gene pool as mutations and leave by genetic drift. Natural selection either conserves variation in the gene pool—and, if so, causes the alleles to establish a particular polymorphism—or causes the gene pool to lose its variation. What is the net effect of all of these processes occurring at the same time? No general answer exists, but in some circumstances we can safely ignore some of the factors and concentrate on the most important.

The strength of natural selection is determined by the *difference* among the fitnesses. The greater the difference among the W's, the stronger the natural selection. The strength of natural selection is often expressed in a standard way. We take the largest W and divide all the W's by this value. We then obtain three numbers, the largest of which is 1. We label these new numbers w_{11}, w_{12}, and w_{22}; they are called the *relative fitnesses*. (Notice that we use lowercase w's for the relative fitnesses.) Next, we find the maximum difference among all possible pairs of the w's. For example, if the w's happen to be 1, 0.5, and 0.25, then the maximum difference between any two w's is 0.75. Let us call this maximum difference s.

The strength of genetic drift is determined by the population size. The larger the population size, the weaker the genetic drift. A rough criterion used to compare natural selection with genetic drift is this: If s is larger, by a factor of 10 or more, than the reciprocal of the population size, then natural selection is stronger than genetic drift (see, e.g., Crow and Kimura 1970). However, if s is smaller, by a factor of 0.1 or less, than the reciprocal of the population size, then genetic drift is stronger than natural selection. If s is close to the reciprocal of the population size, differing by a factor of, say, 0.1 to 10, then both forces should be considered.

To illustrate this idea, suppose the population size is 1000, so that its reciprocal is 0.001. Then if the relative fitnesses are 1.0000, 0.9999, and 0.9999, natural selection can probably be ignored relative to genetic drift.

However, if the relative fitnesses are 1.0000, 0.9900, and 0.9800, then genetic drift can probably be ignored relative to natural selection.

This comparison of the strength of forces provides a criterion for determining when two genotypes are selectively neutral. Recall that two genotypes are said to be selectively neutral if the fitness of individuals with these genotypes is the same. However, the fitness of these individuals need not be *exactly* the same. Instead, genotypes are selectively neutral, in practice, if genetic drift is stronger than whatever slight selection is present. Hence, two genotypes are effectively neutral if the difference between their fitnesses is less than about 0.1 times the reciprocal of the population size. The strength of recurrent mutation can be similarly compared to that of genetic drift. The rates at which mutations arise are on the order of 10^{-6} to 10^{-9}. Thus, for typical population sizes, genetic drift is either stronger than or about as strong as recurrent mutation.

THE NEUTRALITY CONTROVERSY

During the last 15 years, the main issues at the interface between population genetics and ecology have had to do with determining the strength of natural selection relative to genetic drift and determining the nature of natural selection itself. Research has focused on what has become known as the *neutrality controversy,* which is about whether the alleles at most loci are selectively neutral to each other. Specifically, when one observes that a natural population is polymorphic at a locus, it is not obvious whether the polymorphism represents a stable polymorphism caused by strong natural selection, as in heterozygote superiority, or is a snapshop of a gene pool containing selectively neutral alleles whose frequencies are varying randomly from generation to generation as a consequence of genetic drift. Let us explain the particular context in which this controversy arises.

The individuals of a population are biochemically different from one another. Human blood groups are a well-known example, and organisms generally differ from one another in other proteins as well. Since the amino acid sequence in a protein corresponds to some sequence of codons in the DNA, by investigating how individuals differ in their proteins, one obtains a glimpse into how the individuals differ genetically. Proteins are the signature of the genome.

Electrophoresis: A Tool for Detecting Genetic Variation

To determine whether two individuals have slightly different proteins that perform the same chemical reaction, a technique called *electrophoresis* is used. This is a procedure for separating different proteins on the basis of their electric charge. The basic idea is to put the proteins in a gelatin slab, and then to put the slab in an electric field. Typically, proteins that differ in one or more amino acids move through the gelatin at slightly different speeds. After some time—say, 6 hours or so—the different proteins are found at different locations in the gelatin slab, even though they started out from the same place. The actual speed with which a protein moves through the gelatin depends on its charge and shape. Two proteins that differ in only one amino acid typically differ enough in charge and shape to be distinguished by this technique.

The technique of electrophoresis was first applied in population genetics, with the aim of determining the amount of genetic variation within the gene pool at a *typical* locus. The first studies were of *Drosophila* by Lewontin and Hubby (1966) and of humans by Harris (1966). These studies focused on soluble enzymes, some with biochemical functions in the glycolytic pathway and others with functions that have not yet been related to known biochemical pathways. The hope was that these enzymes were a representative sample of the genome.

Let us briefly consider the data on *D. pseudoobscura* summarized by Lewontin (1974). From an analysis of 24 loci, 11 were found to be polymorphic, which is roughly half. The polymorphic loci had between 2 and 13 different alleles. At a locus for which the population is polymorphic, about one-fourth of the individuals are heterozygous. If the polymorphic and monomorphic loci are averaged together, we arrive at the summarizing statistic that about 15% of the individuals are heterozygous at a randomly chosen locus, and, similarly, that an individual is heterozygous at 15% of its loci.

As of 1984, over 1100 species have been investigated electrophoretically, each at over 13 loci, to determine their degree of genetic variation. A definitive survey of this information appears in Nevo *et al.* (1984). The original data are fairly representative of these many later studies. Moreover, intriguing additional patterns have become apparent. A trend has been noted toward more genetic variation in tropical species than in temperate ones. More genetic variation has been observed in species whose ranges span many habitat types as compared to those that occur only in one habitat type. Among vertebrates, reptiles have more variation than birds and mammals. Nevo *et al.* (1984) suggest that natural selection originating in the ecology of the species somehow causes these patterns. Whether or not it does is the heart of the neutrality controversy. What does cause the observed genetic variation?

Hypotheses Regarding Genetic Variation

To understand why this question is difficult to answer, consider two enzymes, which catalyze the same chemical reaction but which differ from each other in only one amino acid. Do organisms with these slightly different proteins have a large difference in fitness under natural conditions? Indeed, changing only one amino acid can make a large difference in the effectiveness of an enzyme. Some of the amino acids in an enzyme are close to the site at which the substrates bind, and an amino acid substitution there might destroy the protein's ability to function. However, other amino acids are located at positions farther from the active site of the enzyme. A substitution at these locations may not greatly affect enzyme function.

How much, however, is "not greatly"? Do organisms with real but slight differences in their enzymes have large enough differences in fitness for selection to be stronger than genetic drift? Furthermore, even if selection is stronger than genetic drift, does this selection account for the actual frequencies of the various alleles in populations? As we have seen, some forms of selection lead to polymorphism and others to fixation. If the variation of populations is to be explained, it is not enough to know only that selection is stronger than genetic drift; it is also necessary to know the form of the natural selection in order to determine how it affects genetic variation.

The hypotheses that might explain genetic variation can be classified into three main types:

Total Neutrality. By the total neutrality hypothesis, variation results solely from mutation and genetic drift, and no natural selection is involved. It refers primarily to polymorphism among DNA sequences that differ in the redundant third position of a codon (Kreitman 1983).

Sieve Selection. By the sieve selection hypothesis, variation results from mutation, genetic drift, and a special form of selection. The hypothesis assumes that most mutations are classified as either deleterious or neutral; advantageous mutations are considered rare. The deleterious mutations are those that seriously undermine the function of an enzyme; the neutral mutations are those that influence fitness to a degree less than 0.1 times the reciprocal of the population size. According to this hypothesis, natural selection eliminates the deleterious mutations but is indifferent to the neutral mutations. Natural selection is like a sieve, removing deleterious mutations while letting everything else pass through.

If this hypothesis is correct, the alleles that do occur at a polymorphic locus—that is, those that pass through the sieve—are selectively neutral to each other, although not to the alleles that have failed to pass through the sieve. If we focus on the neutral alleles, we should find that their frequencies fluctuate from generation to generation as a result of genetic drift combined with recurrent mutation; deleterious alleles, if present at all, should be rare and on their way out.

In this hypothesis, the way to explain the degree of polymorphism at any particular locus is to determine the rate of recurrent mutation to *neutral* alleles at that locus. If the enzyme produced by the locus happens to be very intolerant of amino acid substitution, then the mutation rate for neutral mutations is automatically low at that locus. At another locus, enzyme function may not be impaired by many kinds of substitutions; if so, the mutation rate to neutral mutations there is automatically high. The loci with high mutation rates for neutral mutations will accumulate more alleles in the face of genetic drift than will the loci with low mutation rates for neutral mutations.

According to this hypothesis, populations are typically polymorphic at roughly one-half of their loci because these loci produce enzymes that tolerate some amino acid substitutions without serious loss of enzyme function. Hence, these loci have a high enough rate of mutation to neutral alleles that several selectively neutral alleles have accumulated at each of them.

In the original view of the sieve selection hypothesis, the sieve originates in the internal biochemical environment of the organism. Whether an enzyme with an amino acid substitution can function is determined by whether that enzyme still has affinity for its substrates in the chemical environment of the cytoplasm, mitochondrion, nucleus, or wherever its activity normally occurs. In this view, natural selection is a result of conditions in an organism's internal environment. This version may be termed the *metabolic sieve.*

A variant of the original sieve hypothesis is that the sieve expresses performance in the external environment. It is postulated that, for any set of environmental conditions, certain alleles are selectively neutral but others are deleterious. Imagine, for example, an invertebrate in a high-temperature environment, and suppose that several alleles at a locus all perform effectively at high temperatures. Similarly, a different set of alleles may function effectively at low environmental temperatures. At locations in which the temperature remains high, a polymorphism will appear among the selectively neutral alleles that are favored at that location. The number of alleles in that polymorphism will be determined by genetic drift together with the mutation rate to that set of favored but selectively equivalent alleles. At locations in which the temperature remains low, the same picture would apply, only different alleles would be involved. This version may be termed the *ecological sieve.*

Balancing Selection. By the balancing selection hypothesis, variation results completely from natural selection. The form of selection involved is an extension of the idea of heterozygote superiority. In a one-locus, two-allele system with heterozygote superiority, the gene pool attains a stable polymorphism—that is, one in which the ratio of the number of alleles to each other in the gene pool remains constant through time. This form of natural selection is sometimes called *balancing selection* because selection against one of the homozygotes is countered, or balanced, by selection against the other, resulting in a stable polymorphism.

This idea can be extended to genetic systems that have more than two alleles at one locus, and potentially to genetic systems with multiple loci as well. Theoretically, a scheme of balancing selection exists that maintains a stable polymorphism among three, four, or more alleles. According to this hypothesis, the fitnesses of the various genotypes for an enzyme locus cause a stable polymorphism among the alleles that are observed, and in the ratios that are observed, while excluding the alleles that could be produced by mutation and yet are absent. The balancing selection might originate either internally or externally, as with the sieve selection hypothesis, but typically the polymorphism is viewed as reflecting an adaptive response to selection pressures from the external environment.

The Hypotheses and the Evidence. During the 15 years of dispute over these three hypotheses, the balancing selection hypothesis has incurred serious difficulties and, at this time, lacks supporting evidence. Lewontin and Hubby (1966) pointed out long ago that the sheer amount of variation discovered in natural populations is too large to be explained by any simple picture of heterozygote superiority cumulated over all the polymorphic loci. The problem is simple: Suppose the fitness difference at one locus is large enough to withstand genetic drift. Then extend this idea across the thousands of other loci that are also polymorphic. If the selection at each locus is strong, then the selection on the entire individual must be very strong. In fact, Lewontin and Hubby calculated that its strength would have to be too great to make any biological sense at all.

A further problem is that it is difficult to generate a stable polymorphism among many alleles. The strength of the balancing selection on each of the genotypes must be finely adjusted to have many alleles in a stable polymorphism; otherwise, one or two of the alleles will take over the gene pool, while the rest will be lost. The simplicity of heterozygote superiority with two alleles quickly disappears when the idea is extended to much more than two alleles. (This problem is theoretically similar to that of attaining coexistence among a large number of competing species in a community, as discussed in Part 5.) Finally, except for the relation of sickle cell anemia among human beings to selection favoring heterozygotes in places with endemic malaria (discussed in Part 4), virtually no examples are present in nature of balancing selection at the level of a single gene. (The example involving *D. pseudoobscura* presented earlier in Figure 6–1 involved selection on blocks of genes.)

Evidence is accumulating that sieve selection is involved at many loci. Much of the evidence, however, is indirect and concerns the compatibility of existing enzymatic polymorphisms with models of sieve selection; this evidence does not determine the cause of sieve selection (Nei *et al.* 1976). Other evidence of sieve selection relies on direct inspection of DNA sequences (Kreitman 1983). In the alcohol dehydrogenase locus of *D. melanogaster,* only 1 of the 43 polymorphisms that have been detected leads to an amino acid change, the rest being silent. The implication is that most amino acid changes in this enzyme are selectively deleterious.

The basis for sieve selection is not yet clear. The evidence amassed by Nevo *et al.* (1984) showing some relation between the degree of genetic variation and ecological features suggests that the sieve is at least partly ecological, but exactly what causes such a relation remains an unsolved problem. However, clues are provided by studies that relate the functional properties of particular enzymes to the ecological context.

Place and Powers (1979) discovered a gradation in the ratio of two alleles of an enzyme in the common killifish, *Fundulus heteroclitus,* found in brackish water along the Atlantic coast of North America (Figure 6–5). The enzyme, lactate dehydrogenase (LDH), catalyzes the reduction of pyruvate to lactate. Two alleles, B^a and B^b, are electrophoretically distinguishable. A steep gradient in the thermal environment along the Atlantic coast leads to about a 1°C change in annual mean water temperature per

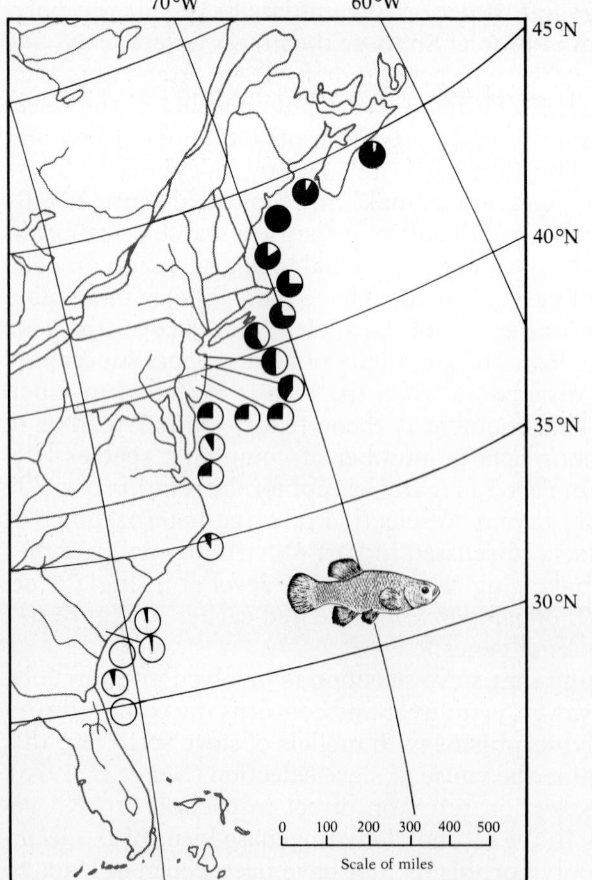

FIGURE 6–5 Cline of alleles at the LDH *B* locus in *Fundulus heteroclitus. (After Place and Powers 1979.)*

degree change in latitude. The most common allele in northern waters is B^b and in southern waters is B^a. The catalytic ability of B^b is highest at 20°C, and that of B^a is highest at about 30°C. Thus, a good correlation can be observed between the enzymatic capabilities of particular alleles and the environments each is found in. Nonetheless, the actual fitness in nature of the individuals carrying these alleles is not known. In particular, selection favoring the B^b allele in northern waters and the B^a allele in southern waters, combined with dispersal among locations, may cause the observed latitudinal gradient in allele frequency.*

THE BIOMETRICAL THEORY OF EVOLUTION

The biometrical theory offers another way of approaching the study of evolution. This theory is consistent with the genetical theory of evolution; that is, it is not a rival to it, nor does it offer what could be considered an alternative hypothesis as to how evolution occurs. Instead, the biometrical theory is a more approximate approach that sacrifices the long-term predictive power one has if the genetic basis for a trait is known. In return for this sacrifice, the biometrical theory is easier to use for short-term predictions of a population's response to natural selection. Moreover, its predictions more readily concern macroscopic traits because the theory deals with macroscopic aspects of the phenotype to begin with, and not with genes and their immediate biochemical products.

Describing Inheritance

The starting point for the biometrical theory of evolution is a description of how traits are inherited in phenotypic terms; that is, we must compare parental and offspring phenotypes. First, we calculate the average phenotype of two parents. If, say, the length of one parent is 100 mm and that of the other is 120 mm, their average is simply 110 mm. The average phenotype from the two parents is called the *midparent*. Next, we make a graph, plotting the midparent on the horizontal axis and plotting the phenotype of each offspring produced by these parents on the vertical axis. After points from many matings are plotted, we can obtain a graph describing the extent to which offspring resemble their parents. Finally, we fit a straight line through the points of this graph and obtain a formula

* Other studies of the biochemical function of the alleles at polymorphic loci and the environmental context in which these alleles occur include Day *et al.* (1974), McKechnie *et al.* (1975), Koehn (1978), Cavener and Clegg (1981), Burton and Feldman (1983), and Watt *et al.* (1983, 1985).

that can be used to project what the offspring phenotypes are for any pair of parents. The formula obtained in this way is a curve-fit to data on how a trait is inherited. It is not an explanation of the mechanism of inheritance but simply a description of the inheritance under the conditions of the experiment.

An example of this method of describing inheritance appears in Figure 6–6, taken from a study of the way several ecologically important traits are inherited in the great tit, *Parus major* (van Noordwijk *et al.* 1980). The horizontal axis is the average weight of the two parents, which ranges from about 15 to 19 g. The vertical axis is the weight of female offspring from these matings (another figure was made for the male offspring); this also varied from about 15 to 19 g. A line is fitted through the data; the slope of this line has a special name in genetics—the *heritability* of the trait. During the years 1955–1978 in a wooded site of the Institute for Ecological Research at Arnhem (The Netherlands), the heritability of body weight for *P. major* was 0.68 (±0.10) for female offspring and 0.50 (±0.09) for male offspring. The value of 0.68 is the slope of the line through the data illustrated in Figure 6–6.

FIGURE 6–6 Weight of offspring plotted against average weight of the parents in *Parus major.* The slope of the line is the heritability. *(After van Noordwijk* et al. *1980.)*

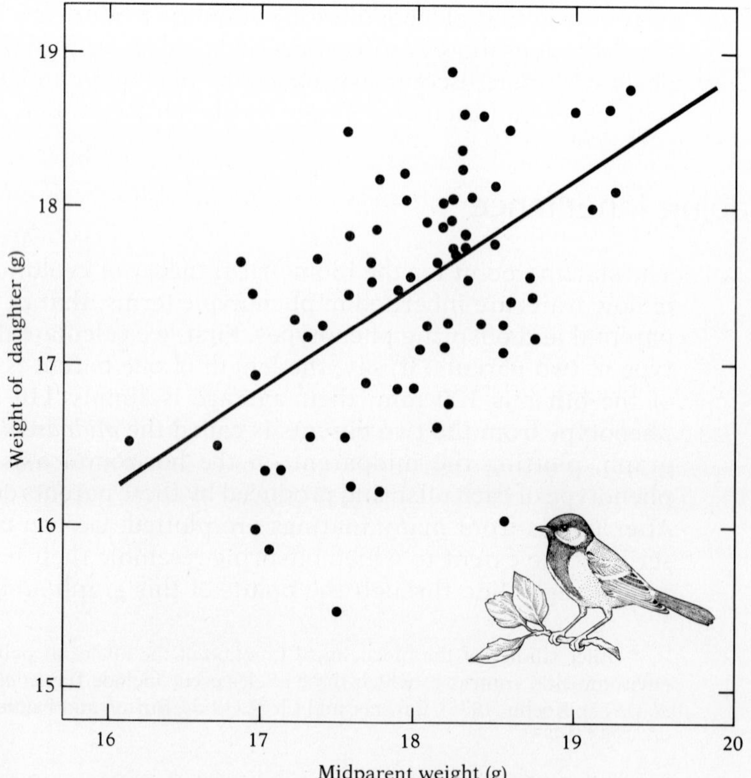

Natural Selection and the Average Value of a Trait

The data on natural selection and the average value of a trait allow one to predict how rapidly natural selection can cause the evolution of either a higher or lower body weight. Imagine differential survivorship of birds as a function of body weight. To illustrate, suppose that larger birds are more likely to survive than smaller birds (e.g., body fat may aid survivorship through the winter). Now denote the average weight of the breeding birds as X_{wt}. Also, denote as X_t what the average weight would be if all birds live to breed regardless of their weight; this is the average weight of breeding birds that would occur in the absence of natural selection. The next generation is found from the curve describing the inheritance. The slope to this curve, the heritability, is customarily denoted as h^2.* If the next generation's average weight is denoted as X_{t+1}, we have

$$X_{t+1} - X_t = h^2(X_{wt} - X_t)$$

The expression $(X_{wt} - X_t)$ on the right-hand side is called the *strength of selection.* It shows how far the selected birds differ from what the population mean would have been in the absence of selection. The expression $(X_{t+1} - X_t)$ is called the *response to selection.* It shows how much the population mean has shifted up in the next generation as a result of the natural selection in generation t. The heritability is always a number between 0 and 1. Hence, the response to selection is always less than the strength of selection.

In *P. major,* the heritability of male weight is about 0.5, and average weight is about 17.5 g. Suppose that, after recovering from a severe winter, the surviving birds weigh an average of 19 g. (Perhaps an ability to store fat provided insulation from the cold.) If so, the strength of selection is 1.5 g. The response, assuming a heritability of 0.5, will be 0.75 g. Hence, the average weight in the next generation will be 18.25 g. Moreover, if this process is continued for, say, three to four more generations, it is possible to imagine the average weight exceeding 20 g, a value that would represent a 12.5% increase. Indeed, after about twenty-five generations—perhaps 100 years—of this same strength of natural selection, the average weight could double, reaching 35 g, which would represent a bird so different from the present-day form to be noticed by even a casual observer. Before that happened, however, new selective factors (for example, more cumbersome flight) or perhaps exhaustion of genetic variability probably would have halted size increase.

The high heritability of ecologically important traits implies that these traits are potentially quite sensitive to natural selection. Hence, when we observe a trait in a natural population, we may be seeing it, as in a snapshot, while it is changing in a certain direction. If so, the natural selection

* Historically, the square root of the heritability was denoted as h, so we now live with a slightly cumbersome notation for the heritability itself.

causing the change is called *directional selection.* Alternatively, the current trait may itself be the one favored by selection, with differing traits having a lower fitness. If so, the natural selection on the trait is called *stabilizing selection.*

Reconciling the Biometrical Theory with the Genetical Theory

One of the triumphs of the great statistician and population geneticist R. A. Fisher was to demonstrate that the biometrical theory of evolution could be derived from the more fundamental genetical theory of evolution (Fisher 1930, 1958). The derivation began with the choice of an arbitrary trait determined by one locus and two alleles. With that genetic basis, Fisher calculated the regression line relating offspring traits to midparent traits in a random mating population. The slope of the line (the heritability) was then related to characteristics of the alleles at the locus determining the trait.

In his theoretical study, Fisher introduced a technical concept called *additive genetic variance** defined as the degree of phenotypic variation in the population caused by genetic variation, *provided* that the heterozygote has a phenotype intermediate between that of the two homozygotes. The additive genetic variance does not have a ready interpretation if the heterozygote phenotype is not intermediate between the homozygotes. Fisher went on to show that the heritability was equal to the additive genetic variance divided by the total phenotypic variance in the population. Thus, when the heterozygote is intermediate between the homozygotes, the heritability is interpreted in genetic terms as the *fraction* of the population's phenotypic variance caused by the population's genetic variation for the trait.

Natural Selection and the Variance of a Trait

The focus in both genetical and biometrical evolutionary theories is on the average trait and how natural selection can change the average trait, provided only that variation exists within the population for natural se-

* A measure of the amount of variation present among a set of numbers is called the *variance* of the numbers. The idea is intuitive. First, one calculates the average of a set of numbers. Then, for each number in the set, one determines how much it differs from the average. Next, these differences are squared, so that we have only positive quantities to consider. Finally, we take the average of these squared differences. The average of the squared differences between each of the numbers and the average itself is the variance. It measures how far the numbers typically are from their average. If the variance is low, then the numbers are all quite close to their average; if the variance is high, then at least some of the numbers are quite far from their average.

lection to act on. However, consider the degree of variation itself: Does natural selection mold not only the average trait but also variation around that average trait? The evidence on this point is scanty. In theoretical models of evolution, both at the phenotypic level and at the genetic level, the population variance for a trait emerges as a population property that is not nearly as malleable as the population average.

This point is easiest to illustrate with a simple one-locus, two-allele example. If the trait—say, body weight—for A_1A_1 is X_{11} and for the other genotypes is X_{12} and X_{22}, then the average body weight in the population is

$$X_{ave} = p^2X_{11} + 2pqX_{12} + q^2X_{22}$$

assuming that the Hardy–Weinberg relations approximately describe the genotype frequencies. Thus, if we know p (and of course q, which is $1 - p$), we have determined the average trait in the population. Natural selection can then directly affect the population average by determining the allele frequencies.

Now the problem is, as we will see, that p and q *also* directly determine the population variance, and so selection cannot directly influence both the mean and the variance independently; these population properties are coupled. The variance of body weight in the population is

$$\mathrm{Var}(X) = p^2(X_{11} - X_{ave})^2 + 2pq(X_{12} - X_{ave})^2 + q^2(X_{22} - X_{ave})^2 + V_e$$

Each of the first three terms (the ones with either p or q in them) represents the contribution to the population's variance provided by the genetic differences among the individuals. The final term, V_e, is the variance among individuals with the same genotype and is called the *environmental variance.*

Now notice that the total variance of the population is also set by p, so that if natural selection should cause a certain average trait to evolve, it will incidentally also produce a certain variance to the trait. Indeed, if natural selection is to mold the population variance, without affecting the population's average, it must act on additional loci, called *modifier loci,* that affect V_e without affecting the X's. All this points to a rather complicated scenario and suggests that genetic constraints exist that prevent any ready explanation of population variances in terms of natural selection, even though the average traits that occur may be strongly influenced by natural selection.

The possibility that genetic constraints impede the evolution of the population variance may be important in ecology. It may imply that no species can develop many kinds of individuals, that no single species can exhibit as many traits as several species combined (Roughgarden 1972). This proposition may be the reason why ecological communities containing many species exist rather than a "community" consisting of one highly polymorphic species.

Another constraint on the evolution of the phenotype is the existence of correlations among several seemingly distinct kinds of traits. In our discussion so far, we have focused on a single trait. However, natural selection acts on the whole organism. Plant and animal breeders have long known that selection for certain traits produces incidental effects on other traits as well (Lerner 1958). Hence, the response to selection on a specific trait may be influenced by the strength of selection on other traits with which it is correlated (as indicated previously in the discussion of size increase in *P. major*). To take account of correlations among traits, a multivariate phenotypic theory of evolution should be used, as developed in Karlin (1979).

OPTIMIZATION AND EVOLUTION

By thinking of an organism's traits in terms of how they affect survival and reproduction, we might hope to guess the selective pressure that caused the original evolution of the trait—or at least the selective pressure that maintains the trait in its current form. This way of thinking about traits quickly leads to the idea of a strategy, as discussed in Part 1 in the context of optimal foraging strategies. Are we justified in thinking about the evolution of traits in terms of strategies?

For example, suppose we discover that the function of fur is to insulate an animal to prevent heat loss in cold weather. We then determine the caloric cost of making fur, keeping it clean, and carrying it around, and we compare these costs with the calories conserved by the fur that would otherwise be dissipated to the environment through convection. With all these factors in mind, we could, in principle, compute the optimum fur thickness that strikes the best compromise between the costs and benefits of fur. Yet what are we entitled to conclude about this calculation? Can we presume that natural selection will cause the evolution of fur that is exactly as thick as this calculation predicts?

The answer to this question is not obvious, and it is important to understand why. One might say that we cannot hope to conclude that natural selection will cause the optimum fur thickness to evolve because we have not taken into account all the costs and benefits. Perhaps fur is used in mate recognition and courtship, and so its insulative value is only part of the picture. However, this is not the point. Suppose for this case that fur thickness has been found to be immaterial to mating success, and similarly for other possible implications of fur thickness. That is, suppose we have *correctly* determined what the optimum fur thickness is. What does *this* really have to do with the evolution of fur?

One way to answer this question is to relate the animal's fur to its fitness. We might hope to show that an animal with the optimum fur thickness has a higher fitness than animals with any other fur thickness, everything else being equal. If so, it might seem that evolution would produce animals with the optimum fur thickness. Is this approach correct? As we shall see, it is never exactly correct. Nonetheless, this approach may be a useful tool to help us understand how ecologically important traits evolve. You will have to make your own judgment on whether the risks of an optimization approach are worth taking relative to the insight that the optimization approach seems to provide.

Fisher's Fundamental Theorem of Natural Selection

The justification for thinking of evolution as leading to an increase in fitness begins with a theoretical result derived by R. A. Fisher, termed the *fundamental theorem of natural selection.* The model in which this result was first derived involves a trait determined by one locus with two alleles. The finding is that the average *relative* fitness, w, in the population increases in every generation (c.f., Figures 6–3, 6–4). It increases because the individuals with a low relative fitness comprise a progressively lower fraction of the population in each generation, and so the average fitness in the population progressively rises. The average fitness rises quickly at first, when individuals exist with many different fitness levels, and slows down as the population ultimately comes to contain only individuals with the highest fitness level. In the model that Fisher used, the relative fitness of each genotype was assumed to be constant through time. With this result, one could, in principle, determine what phenotype produces the highest fitness in the environment and then comfortably predict that the population will ultimately consist of individuals with that phenotype.

Yet Fisher's theorem is both delicate and easily misunderstood. It is delicate in that the theorem is highly sensitive to changes in the assumptions with which it was derived. Fisher's theorem pertains only to one component of the evolutionary process—namely, natural selection. In evolutionary models that also include recurrent mutation, recombination among two or more loci, genetic drift, and nonrandom mating, the overall outcome of the evolutionary process is generally *not* that of a population consisting only of individuals possessing the highest fitness level. A further limitation of Fisher's theorem is that even the kind of natural selection to which it pertains is very restricted. Essentially, the theorem refers to natural selection only in an exponentially growing population whose environment is constant through time. This assumption would seem to rule out all the ecologically interesting selection pressures that come from interactions among individuals and from a varying environment.

Yet these limitations are not necessarily fatal. First, in many instances, natural selection may be the strongest component of the evolutionary process. If so, the outcome of the overall evolutionary process is approximately predicted by considering only natural selection and by assuming that the other factors contribute only small details to a picture painted mostly by natural selection. Second, to some extent, Fisher's theorem can be replaced by other theorems that do apply to the forms of selection arising in an ecological context, as we will discuss in the next chapter. Yet even though ecologically relevant replacements for Fisher's theorem are becoming available, they are, *at best*, only approximate characterizations of the evolutionary process, because natural selection is only one component of the evolutionary process.

Examples of Optimization

Recall the concept of an optimal foraging strategy in Part 1. An animal was imagined to have a choice of several prey types, and we asked which types should be ignored and which should be taken to minimize the foraging time or to maximize the energetic return. The optimal diet was found to be insensitive to variations in the quantity of the nonincluded prey. For example, an insectivorous bird might accept caterpillars as prey, but if the value of ants is too low to be included in the diet, then ants should not be added to the diet even if they become very abundant. Experimental tests of this proposition revealed it to be approximately correct, but some low-ranked items did find their way into the diet if they were made extremely abundant. This discrepancy was explained as resulting from the birds' need to sample various prey in the environment in order to determine, and possibly update, their ranking of the available prey types.

Another example in which an optimization model has been tested concerns the optimum flowering time for an annual plant. Cohen (1966) hypothesized that a problem facing an annual plant during its limited growing season is to allocate the sugars it makes during photosynthesis either to the manufacture of more leaves or to the production of reproductive tissue, including flowers and seeds. He hypothesized that the plant's "objective" is to maximize the number of seeds it produces as of the end of the growing season. When a plant allocates its products of photosynthesis to the production of more leaves, those leaves, in turn, can produce still more photosynthetic products later in the season. However, if the plant waits until too near the end of the season, not enough time will be left in which to make seeds. In this context, the optimal strategy for the plant is an all-or-nothing strategy. It should begin by completely allocating its photosynthate to the production of leaves, and at a certain time that can be predicted from the optimization model, it should switch its allocation completely to the manufacture of seeds.

This model was tested using two annual wildflowers in the grasslands of California (King and Roughgarden 1983). The actual allocation schedule involves a rather rapid switch, but it is not as abrupt as the optimization model predicts, and the actual time of the switch is approximately when the model predicts. The discrepancy between the actual allocation schedule and the optimal strategy is explained by some mix of the following factors:

- The predicted optimal strategy is not actually optimal. Indeed, in a more sophisticated model, in which the growing season's length is a random variable (King and Roughgarden 1982), the optimal strategy is not an abrupt switch but a slightly more gradual change, much like that actually observed.
- The plants have not evolved the optimal strategy because they have not been in the habitat long enough or the habitat has changed in the recent past.
- The flowering-time trait is phenotypically correlated with other traits on which selection is also operating. This fact, together with the action of other components of the evolutionary process, prevents the evolution of an optimal strategy.

If one decides that an optimization method is acceptable, one must always find a suitable "objective function"—that is, a suitable criterion by which to evaluate the various strategies that will be considered. Simply choosing "fitness" as the criterion is not meaningful once the population's ecological context is taken into account; one must find the correct way to define and calculate fitness in an ecological context.

For example, suppose the environment changes from generation to generation and that the species population occupies several habitats, with migration back and forth. Because the fitness of an individual will depend on which generation and habitat it is in, which fitness should be used? If the fitness is to be some kind of average fitness, then what formula should be used for the average? In Chapter 7, we will see several examples of the way fitness can be defined in a complex ecological context.

An Evolutionarily Stable Strategy

With ecological selection pressures, it is often helpful to formulate a weaker concept of optimization than that used in Fisher's fundamental theorem of traditional population genetics. In Fisher's theorem, the average relative fitness continually rises in every generation as the population approaches the state in which all the members have the highest fitness possible with the available genetic variation. The point is that fitness is increasing *all the time,* and the end point is predictable as a state in which the average fitness is at a maximum.

Here is a weaker concept. Consider a set of phenotypes. Suppose a phenotype exists such that when the population consists completely of individuals with this phenotype, an individual with any *other* phenotype cannot enter the population and increase. That is, the population is resistant to invasion by any other phenotype from among the set of possible phenotypes. A phenotypic strategy with this property is called an *evolutionary stable strategy* (*ESS*) (Maynard Smith and Price 1973). This concept is especially useful for traits, such as cooperation and aggression, whose fitness depends on the ratio of phenotypes in the population.

The concept of an ESS is related to optimization because the initial increase of a rare phenotype in a population consisting of other phenotypes is governed by whether the fitness of the new phenotype is higher *at that time* than the fitness of the remaining phenotypes in the population. If so, the new phenotype will show a relative increase in its frequency by the next generation. On the other hand, if an established phenotype has a higher fitness than any other phenotype at that time, then that phenotype will remain established.

An ESS involves fitness maximization at only one point in time, the time when the new phenotype, so to speak, presents itself at the doorstep of the population. One generation later, the average relative fitness may go up or down because new ratios of the phenotypes to one another are now present, and so the relative fitnesses may have changed. Thus, the concept of an ESS leads to the idea that the optimal strategy is a strategy with the highest fitness, provided that it is already established and that any other strategies to which it is compared are introduced to the population in the form of rare individuals that carry out an alternative strategy. The concept of an ESS does not explain the evolutionary process that causes a strategy to become established. A further argument must be supplied to show how the ESS can become established to begin with because the ESS designation guarantees only that it will not lose its established state once that has somehow been attained.

The classical fundamental theorem is a stronger statement than an ESS because it indicates (1) that traits with the highest fitness increase when rare, (2) that the average fitness progressively increases in each generation, and (3) that the trait with the highest fitness eventually becomes established. Of course, the reason why such a strong statement is possible is that the kind of natural selection involved is so simple and all the rest of the components to the evolutionary process have been left out. The weaker concept of the ESS is necessary because selection in which the fitnesses depend on the ratio of phenotypes in the population can lead to many more kinds of evolutionary outcomes than classical natural selection based on constant relative fitnesses.

THE SPEED OF EVOLUTION IN THE FOSSIL RECORD

Fossil plants and animals provide the record of how evolution has actually occurred, and of special interest is how *fast* it has occurred. This question presently has two alternative answers.

Some recent paleontological research indicates that most fossil lineages show little sign of a continuing gradual process of evolutionary change. Instead they show *stasis*—species remaining essentially unmodified for millions of generations (Eldredge and Gould 1972, Stanley 1975, 1979, Gould 1980, 1983). These long periods of stasis are punctuated by relatively brief episodes of rapid evolutionary change coupled with speciation, the formation of new kinds of organisms. Such evolution, apparently progressing in fits and starts has been termed a *punctuated equilibrium.*

The punctuational view differs from the more widely held *gradualist* view (see, e.g., Charlesworth *et al.* 1982, Grant 1982). The punctuational view suggests the possibility that different mechanisms account for most *microevolution*—changes within populations through time—rather than for most *macroevolution*—speciation and creation of higher categories such as families and orders. In particular, species selection, the differential reproduction and survival of entire species, may be a source of major evolutionary trends (Stanley 1975). The views of the punctuationists have stimulated a healthy reexamination of the rates at which evolution occurs; what the final answer will be remains to be seen.

CHAPTER 7

Natural Selection in an Ecological Setting

Now we explore the major ecological selection pressures. These pressures result from variable environments, density-dependent interactions among organisms, population age structure, and social behavior.

SPATIALLY VARYING NATURAL SELECTION

Clines

Spatial variation in a selection pressure can lead to a gradient in gene frequencies or in phenotypes called a *cline*. Recall the gradient of gene frequencies at the LDH locus in killifish from Chapter 6 (Figure 6–5); this is an example of a cline. Another well-studied example involves color in the moth, *Amathes glareosa*, in the Shetland islands of Britain. In the northern Shetlands, a melanic form predominates that gives way to a light-colored form in the south (Kettlewell and Berry 1961). Similarly, the sulphur butterfly, *Colias philodice eriphyle,* in the Rocky Mountains of Colorado has a cline along altitudinal gradients in the amount of melanin in the wings (Watt 1968). Many clines in flowering time and physiological abilities are found among plants occurring from the shores of the Pacific Ocean to the top of the Sierra Nevada mountains (Clausen *et al.* 1941). On a smaller scale, where mine tailings occur next to normal soil, clines appear within 50 m in the ability of grasses and herbs to grow in the presence of zinc and other toxic heavy metals (Jain and Bradshaw 1966). Indeed, examples of geographical variation in the color and shape of organisms are myriad, and in many such instances a cline is involved.

Some clines pass through a polymorphic state in their center. Here the morphs from both extremes are found together. This is called a *morph-ratio cline*, and the Shetland islands moths are an example. Other clines simply have an intermediate phenotype in their center, and the phenotypic variation at the center may or may not be appreciably more than that at either extreme.

The central issue concerning clines is how accurately the steepness of the gradient in the gene frequency, or in the phenotype, parallels the steepness of the gradient in the selection pressure. The regular movement of organisms—the distance over which they disperse from where they were born—tends to smooth over sharp changes in a selection pressure.

Population geneticists have explored how closely the steepness of clines matches the spatial variation in the selection pressure in theoretical models. Simple formulas are available that predict how long the cline will be when the selection pressure changes abruptly in space from favoring one of the alleles to favoring the other (cf. Roughgarden 1979, pp. 240–255). The theoretical models concerning clines have not yet been tested in nature because of the difficulty in measuring spatial variation both for the selection pressure and for dispersal in the same system. However, a laboratory simulation of the evolution of a cline has been provided by Endler (1977) using *Drosophila*.

Causes of Clines

In some cases, the selection pressures responsible for geographical variation are known. Perhaps the most famous is that of *industrial melanism* in the moth, *Biston betularia*. Populations of these moths in polluted areas of Great Britain are made up mostly of dark (melanic) individuals, whereas those in unpolluted areas are made up primarily of light, variegated individuals (see Figure 7–1). That geographic pattern (Figure 7–2) developed in the last half of the nineteenth century, following industrialization.

The major factor in creating this geographic pattern was predation on the adult moths by birds (Kettlewell 1958, 1973). Where pollution was most intense, the melanic moths tended to be *cryptic*—that is, protectively colored—when they rested on sooty tree trunks. Birds ate the variegated moths and missed the melanics. Conversely, in unpolluted areas, the tree trunks were variegated and covered with lichens (which are killed by

FIGURE 7–1. Industrial melanism. Typical and melanic forms of the moth *Biston betularia* on different backgrounds. (A) Lichen-covered tree trunk in an unpolluted area. (B) Tree trunk in an industrial area where air pollution has covered the bark with grime and killed the lichens. The typical moth is camouflaged, and the melanic moth is prominent in (A); the reverse in (B). *(Drawing by Anne H. Ehrlich.)*

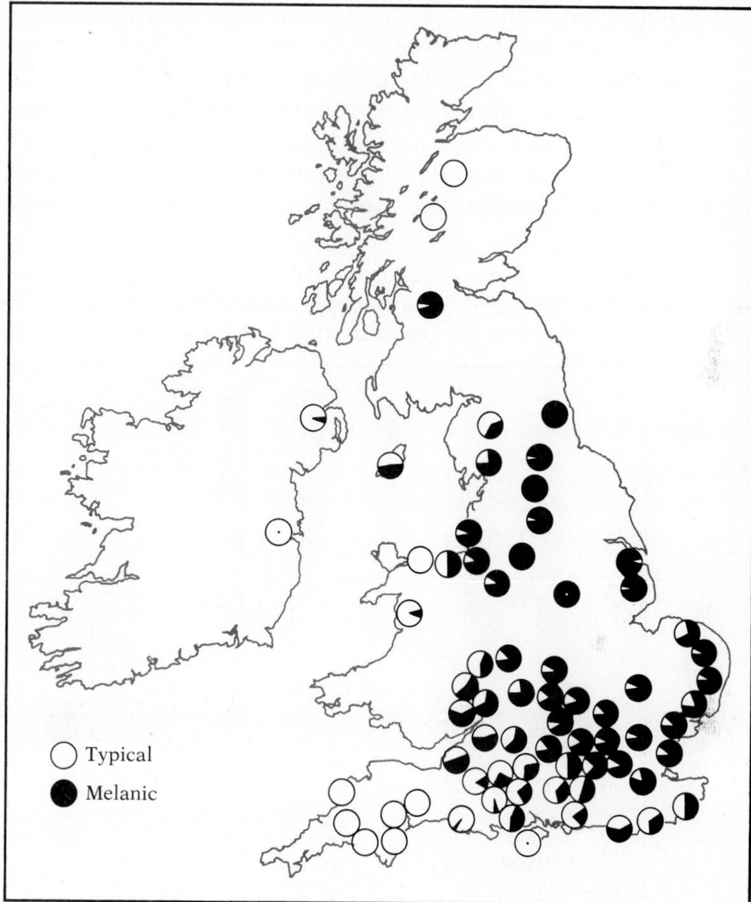

FIGURE 7–2. Distribution of *B. betularia* in the British Isles, showing the proportions of typical and melanic forms. *(After Kettlewell 1958.)*

pollution). In those areas, the variegated moths were cryptic, and the birds primarily ate the melanic moths.

Visual predation is also involved in maintaining geographic patterns in polymorphic land snails of the genus *Cepaea*. For instance, around Oxford, England, the percentage of yellow in the shells and the amount of banding are correlated with the snail's habitat. Predation by birds, especially thrushes, is largely responsible for the correlation (Cain and Sheppard 1950). The thrushes collected snails and then cracked them open on rocks—"thrush anvils." Frequencies of the morphs in snail populations at large are different from those in the predated portion of the population—that is, in the pile of shells around the anvil. For example, in a bog the difference between the predated and nonpredated portions of the population indicated that thrushes were selecting against the banded morph in the uniform herbage of the bog. Presumably the unbanded shells were better camouflaged against the relatively uniform background of the bog. In hedgerows the reverse was found—banded morphs were more favored in the rougher herbage of hedgerows.

Similarly, differential predation appears to be responsible for the presence of unbanded water snakes (*Nerodia sipedon*) on the islands of Lake Erie, whereas populations along the mainland shores are heavily banded (Figure 7–3). Banded young are differentially removed from the island

FIGURE 7–3. Map of islands in Lake Erie and the adjacent mainland, showing frequencies of banding types of populations of the water snake *Nerodia sipedon*. A = least banded; D = most banded. See Figure 7–4. *(After Camin and Ehrlich 1958.)*

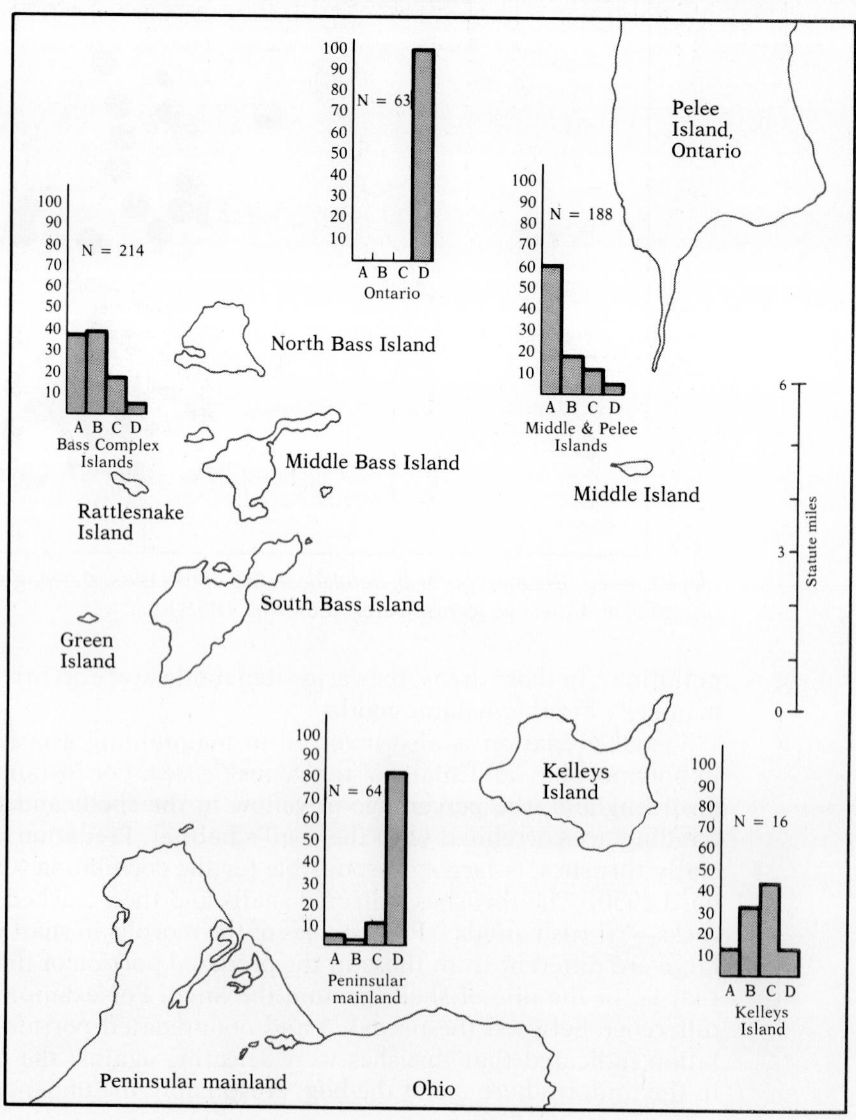

populations, presumably because they are more conspicuous against the flat limestone rocks of the island shorelines (Camin and Ehrlich 1958). In contrast, the banded patterns seem to be more cryptic in the more typical swamp habitats of the mainland. Gene flow from banded mainland populations seems partly to balance the selection pressure and prevent unbanded genes from becoming fixed on the islands (Figure 7–4).

Thus, one can see that environmental differences between locations—for example, polluted tree trunks versus unpolluted tree trunks, uniform herbage versus rough herbage, flat rocks versus swamps—can generate

FIGURE 7–4. Natural selection in water snakes (*N. sipedon*) on the islands of Lake Erie. See text for explanation. *(After Ehrlich* et al. *1974.)*

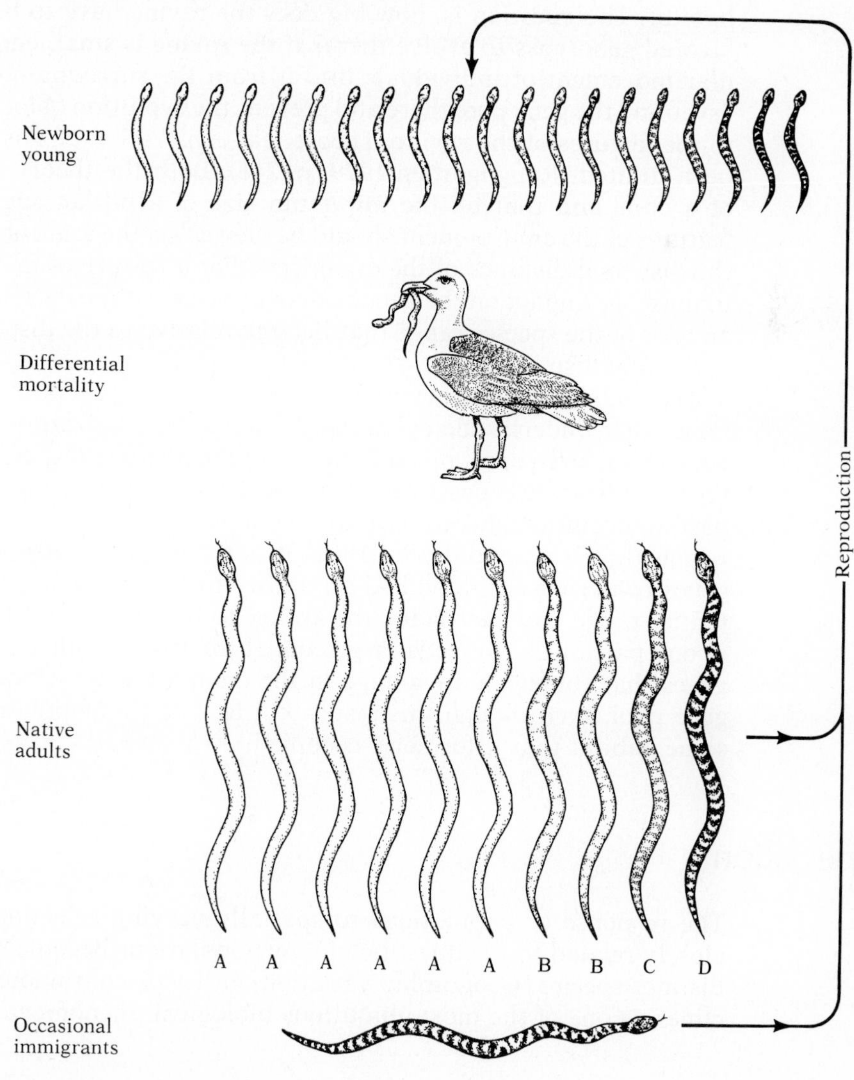

selection pressures that can cause the differentiation of populations, often generating clines in the process.

Pockets

An issue related to the evolution of clines is whether a population can respond to a *pocket* of natural selection. A pocket is a place in which selection favors one genotype, whereas selection favors another genotype everywhere else. Imagine, for example, a relatively moist ravine that slopes down the side of a dry mountain side. A different body color and other traits are possibly favored in the ravine relative to the surrounding habitat. The question is, how big does the ravine have to be in order for favored genotypes to evolve there? If the ravine is small enough, the regular movement of individuals into it from the surrounding habitat will "swamp" the gene pool there and prevent the evolution of local adaptation to the features of the ravine. Theoretical exploration of this question has been limited (Roughgarden 1979, p. 248). Both the theory on the length of a cline and that on the minimum size of a pocket suggest that the features of the environment should be viewed on the scale of several times the dispersal distance of the organisms. For a feature of the environment to make an impact on the gene pool of a species, it must occur throughout an area of the species range that is large relative to the dispersal abilities of the organisms.

The evolution of clines and the response to pockets of selection are issues that underlie the evolution of habitat specialization. By a *habitat*, we mean a large area, much larger than the average dispersal distance of the organisms. It is possible that the range occupied by a population is in part an evolutionary response by the population to the geography of the region that it potentially inhabits. It is theoretically possible that some genes confer on an individual an ability to live in some parts of the potential range, but sacrificing the ability to live in other parts. If the area of one part of the range is large enough relative to other parts, then the genes that confer an advantage in the one part will become fixed in the gene pool, even though this result will lead to the population's vacating some habitat that it formerly occupied.

Speciation

The response of populations to spatially varying selection pressures is closely related to the question of how populations become separated into distinct species. Geographic variation, either discontinuous or involving clines, is one of the most ubiquitous biological phenomena. Distinct spe-

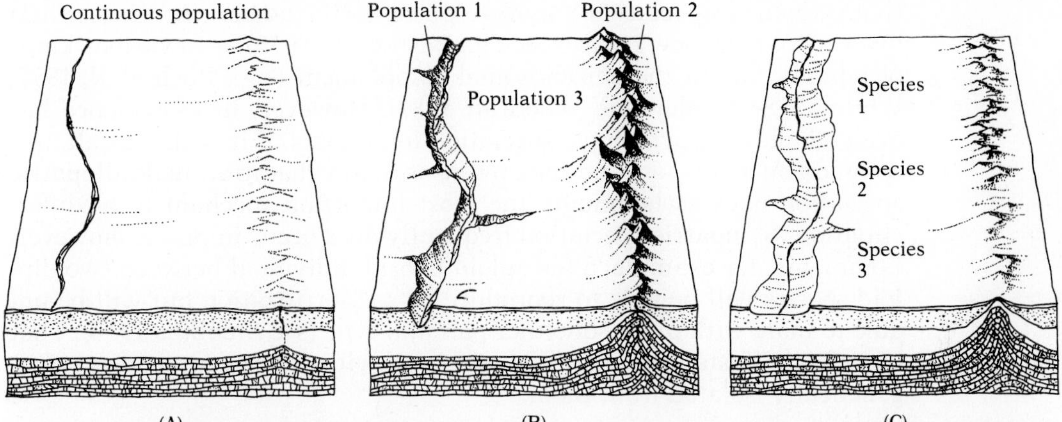

FIGURE 7–5. Classical model of allopatric speciation. (A) A single species occurs in an area with a meandering river and low hills. (B) Millions of years later, the river has eroded a deep canyon and a mountain range has developed, dividing the original population into three, which are genetically isolated from one another. (C) Millions of years later, the mountains have eroded away, the canyon is broad and shallow, and the three populations can come together; selection has led to differences among the populations, so that they look different and do not interbreed. *(After Ehrlich et al. 1976.)*

cies are usually defined as populations of organisms that are incapable of interbreeding.*

To some extent, *speciation*—that is, the process by which species are formed—is occurring gradually much of the time by the continuation of the processes described in the *Biston, Cepaea,* and *Nerodia* examples. Different selection pressures modify populations in different parts of the range of a species until they are so unlike one another that individuals from opposite ends of the range can no longer interbreed. This event occurs most readily if a barrier such as a mountain or a body of water forms in the middle of the range of the species, splitting it in two parts, so that each part evolves along a separate path. If the barrier subsequently disappears, when the two parts are subsequently reunited they may no longer be capable of interbreeding. If so, two species will have formed. This model of differentiation in isolation is diagrammed in Figure 7–5 and is called *geographic speciation* or *allopatric speciation* (allopatric populations are populations that occur in different places).

It is also possible that *sympatric speciation*—differentiation in the absence of geographic isolation—occurs in animals. Some observers claim

* This definition will suffice for our purposes, but the exact definition of species is a matter of some controversy. See, for example, Ehrlich (1961b), Mayr (1969), and Sokal and Crovello (1970).

that sympatric speciation may take place, for instance, when insect herbivores colonize new hosts (see, e.g., Bush 1969, 1975) or in various kinds of repatterning of the chromosomal complement (e.g., Bush *et al.* 1977, White 1978, Atchley and Woodruff 1981). However, little evidence has been offered that sympatric speciation in animals, if it occurs at all, is a common phenomenon, and the consensus is now that in animals allopatric speciation is overwhelmingly the most important mechanism of differentiation. Sympatric speciation frequently does occur in plants, however. Commonly, for example, a tetraploid hybrid individual between two diploid species will be able to reproduce by self-fertilization but will be unable to mate with plants of either parental type (see Stebbins 1950). That is, it is an "instant species," co-occurring with the parental species but genetically isolated from them.

TEMPORALLY VARYING NATURAL SELECTION

Selection pressures undoubtedly vary from generation to generation, and the *geometric mean* of the relative fitnesses plays the key role in determining the net fitness of a genotype over many generations. Recall our discussion in Chapter 5 on density-independent population growth in a variable environment. We found that if the geometric mean of a population's R was less than 1, its size would inevitably deline to 0—it would become extinct. If the geometric mean of R is greater than 1, its size tends to increase to very high values, even though it may pass through bottlenecks of low abundance.

A similar result is found for the spread of genes in the gene pool. Suppose two alleles are at one locus. Let us scale the relative fitnesses so that in each generation the relative fitness of the heterozygote equals 1. By this convention, the geometric mean of the heterozygote fitnesses necessarily equals 1. Haldane and Jayakar (1963) showed that if the geometric mean of the relative fitness of the A_1A_1 genotype is less than 1, then the *other* allele, A_2, cannot be eliminated. Similarly, if the geometric mean of the A_2A_2 genotype is also less than 1, then A_1 cannot be eliminated. Hence, if the geometric mean of both homozygotes is less than 1, then neither allele is eliminated, and a long-term polymorphic condition results.

The geometric mean is very sensitive to rare generations in which the fitness is low. Suppose, for example, that the relative fitness of the A_1A_1 genotype equals 2 every 9 years out of 10 but that in 1 year out of 10 it equals only 0.001. The arithmetic average is 1.8001 but the geometric mean is 0.9352, which is less than that of the heterozygote. If the arithmetic average were important, one would predict that the A_1 allele would eventually become fixed in the gene pool. However, the arithmetic mean is *not* the criterion to use. The rare year in which the A_1A_1 genotype is barely able to survive reduces its geometric mean fitness below 1 and hence ensures that the alternative allele, A_2, will be retained in the gene pool.

The evolutionary importance of rare events poses practical difficulties for field studies. As we will discuss in Part 5, in some environments interspecific competition for resources occurs in only rare years when resources are in unusually short supply. Yet it is precisely such rare years that may cause an evolutionary response to interspecific competition.

DENSITY-DEPENDENT NATURAL SELECTION

Many of the ecologically important selection pressures originate with density dependence. In this section, we discuss density-dependent natural selection within a single species; density-dependent selection involving two or more species will be introduced in Part 4, where we discuss population interactions.

Density-Dependent Selection Increases Population Size

Density-dependent selection tends to bring about a maximum population size (Anderson 1971, Roughgarden 1971, 1977). This result is the density-dependent analogue of Fisher's theorem for density-independent selection and is subject to the same limitations. That is, the population size is maximized only when density-dependent selection is quantitatively the most important force; if mutation, recombination, and correlation among characters are strongly involved, then the population size may be only approximately maximized.

The tendency of density-dependent selection to increase population abundance is illustrated in laboratory experiments with *Drosophila*. An early study (Buzzati-Traverso 1955) is summarized in Figure 7–6. The population size grows rapidly at first, and then it continues to increase gradually. The later period of gradual increase in population size is actually associated with genetic change (Ayala 1965, 1968, 1969) and reflects the evolution of flies better adapted to the experimental environment than were the flies initially introduced.

Density-Dependent Selection and Particular Traits: *r*- and *K*-Selection

The increase in population size just illustrated is an overall consequence of density-dependent evolution. It is also possible to pinpoint the effect of density-dependent evolution on particular traits. Consider a botanical example. An annual plant grows for some period of time, manufacturing

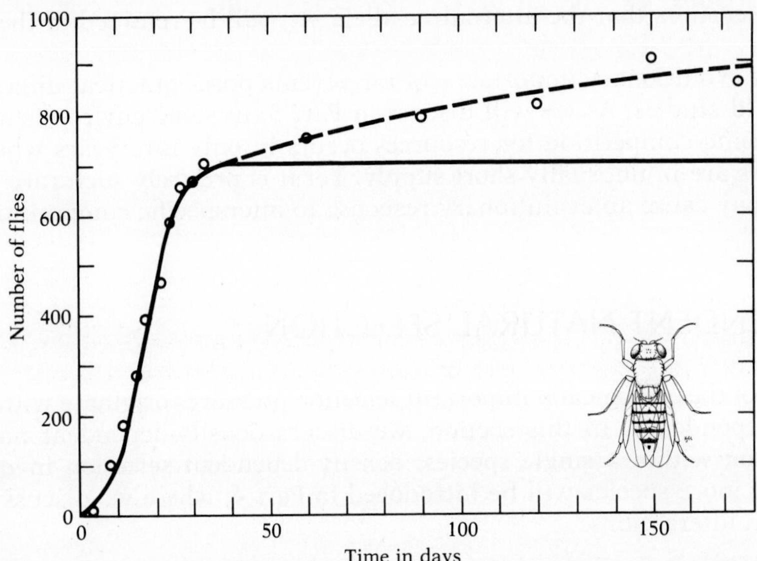

FIGURE 7–6. As evolution proceeds, the size of a laboratory population of *Drosophila melanogaster* increases. The solid line shows growth to the initial carrying capacity. The dotted line shows evolution for increased carrying capacity. *(After Buzzati-Traverso 1955.)*

leaves and other vegetative parts, and then switches, more or less abruptly, to the manufacture of flowers and seeds. The timing of the switch strongly influences the plant's eventual phenotype. An early switch leads to small plants, early flowering, and a short life span. A late switch leads generally to larger plants, late flowering, and a long life span. Let us focus on the switching time and see how density-dependent selection may influence this trait.

Suppose an early-flowering plant can produce seeds at a faster rate under uncrowded conditions than a late-flowering plant. Imagine that a late-flowering plant places much of its carbon in roots and supporting tissue in order to sustain its ultimate weight; in contrast, an early-flowering plant places proportionally more of its carbon in reproductive tissue. Next, suppose that a late-flowering plant is less susceptible to crowding than an early-flowering plant. Imagine that its roots extend deeper to find nutrients and that its tall stem allows it to receive some light in places where the early-flowering form would be shaded.

If these assumptions are true, then a population of early-flowering individuals has a higher intrinsic rate of increase, r, under uncrowded conditions than a population of late-flowering plants. However, a population of late-flowering individuals may attain a larger steady-state abundance than a population of early-flowering plants because the individuals may survive in crowded conditions in which the early-flowering plants would be shaded and at a loss for resources. In the terminology of the logistic equation, a population of late-flowering individuals has a higher K than does a population of early-flowering individuals.

If these assumptions are correct, we can predict that density-dependent selection will cause the evolution of late-flowering individuals in a steady state population because late flowering is the phenotype that produces the highest population size, or, in the terminology of the logistic equation, the highest K. In contrast, consider a population whose size is not approximately at a steady-state level. If the population is expanding in a density-independent manner, then the phenotypes with the highest r are those favored by natural selection. With such a population, we can predict that the early-flowering plants will evolve. Density-independent natural selection is often called *r-selection* and density-dependent selection is called *K-selection* (MacArthur 1962). This terminology emphasizes the distinction between the different aspects of the phenotype that are favored by natural selection in density-independent and density-dependent ecological contexts (Boyce 1984).

The common dandelion (*Taraxacum officinale*) was one of the first populations in which a phenotypic difference was explained by the difference between density-independent and density-dependent selection (Solbrig 1971). Density-dependent selection occurred at an undisturbed site and density-independent selection at a site kept heavily disturbed by weekly cutting with a lawn mower. The disturbed site had primarily small, early-flowering plants, and the undisturbed site had large, late-flowering plants.

Trade-offs in *r*- and *K*-Selection

The most rigorous test of the theory of r-selection and K-selection again comes from an experimental study with *Drosophila* (Mueller and Ayala 1981). Populations were maintained under conditions of low and high density to examine the results of r-selection and K-selection, respectively. After about 5 months (about 10 generations), the results presented in Figure 7–7 were obtained. The vertical axis is fitness (average per capita growth rate), and the horizontal axis is population density. In the K-selected populations, genotypes evolved that had a higher fitness at high density, and a lower fitness at low density, than did the genotypes from the r-selected populations.

The test by Mueller and Ayala is particularly surprising because it is not obvious that unspecified genes are present conferring a high fitness at high densities at the expense of a low fitness at low densities, and vice versa. Such genes imply that a trade-off exists between high-r and high-K traits such that an organism cannot possess both simultaneously. When discussing specific traits, such as the allocation of carbon fixed during photosynthesis to either the synthesis of vegetative or reproductive tissue,

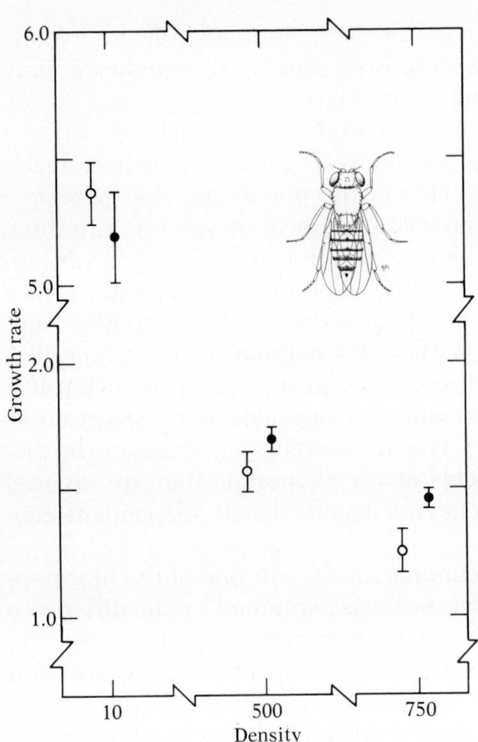

FIGURE 7–7. Fitness (per capita growth rate) as a function of population density for two laboratory strains of *Drosophila*. Open circles refer to a strain grown in an expanding environment, closed circles to a strain grown in an environment with density regulation. *(After Mueller and Ayala 1981.)*

the existence of a trade-off seems clear because the plant has a finite amount of photosynthate to use. It cannot escape the laws of conservation of mass and energy by allocating the same photosynthate to two places at the same time. The study by Mueller and Ayala did not focus on specific traits and provided no functional reason for suspecting the existence of a trade-off involving r and K. Yet evidently genes involved in such a trade-off do exist in the *Drosophila* populations studied.

EVOLUTION IN A POPULATION WITH AGE STRUCTURE

This section is about life, reproduction death, and immortality. It offers an evolutionary and ecological perspective on some of the basic questions that humankind has asked. The discussion is couched in the now-familiar vocabulary of age-specific mortality and fertility data, P_x and F_x.

First, *senescence* is defined as a decline in the probability of survival, P_x, as x increases. If P_x is the same for every x, then organisms can survive equally well regardless of their age. However, in a great many species—and *Homo sapiens* is no exception—the survival probability, P_x, declines after x has passed what might be considered a threshold age. With humans, P_x declines rapidly after 65 or 70 years of age, depending upon their sex and, to some extent, upon their nationality and other social variables. Why does senescence exist? Will modern medicine find a way to prevent it, perhaps with a "fountain of youth" pill?

The second issue concerns the timing of reproductive activity. Why, for example, do birds from some species lay only a few eggs each breeding season, whereas birds from other species lay many more? A bird's *clutch* is the set of eggs in its nest, and the *clutch size* is the number of eggs in the clutch. Figure 7–8 shows the relation between clutch size and latitude for several groups of birds (Cody 1971a). The clutch size generally increases as we pass from tropical latitudes to temperate latitudes. Moreover, as we pass from continental locations to locations on nearby oceanic islands, we see evidence of another relation, illustrated in Figure 7–9. In the temperate zone, the clutch size is smaller for birds on islands than it is for birds on the adjacent mainland. However, no such difference in clutch size exists between island and continental birds in the tropics.

We can roughly summarize these relations between clutch size and geographical location by saying that the clutch size increases in environments in which the density-independent mortality increases. The idea is that the degree of density-independent mortality is higher in temperate zone locations than it is in tropical locations, and also that oceanic islands in the temperate zone have less density-independent mortality because of a more moderate climate than do nearby continental locations. In the tropics, it is supposed that island and nearby continental locations share approximately the same climate and hence have about the same degree of density-independent mortality.

A more precise picture is seen in particular comparisons (Lack 1968). The black-headed gull (*Larus ridibundus*) feeds near shore, where it is susceptible to higher adult mortality (18% annually) than is the kittiwake (*Rissa tridactyla*), which feeds offshore (12% annual mortality). The clutch size in the near-shore gull is three eggs, the eggs incubate in 23 days, the time to fledgling is 30 days, and the age at first breeding is 2 years. The offshore-feeding kittiwake has a clutch of only two eggs, slower egg development (27 days), a longer time to fledgling (33 days), and an older age at first breeding (3 to 4 years). The picture that emerges is that birds from high-mortality environments tend to have higher clutch sizes, and faster egg development and juvenile maturation rates, than do comparable birds from low-mortality environments.

We should add, however, that these patterns are not automatically found in all groups of organisms. In lizards, for example, no simple latitudinal trend in clutch size is evident because clutch size is confounded with both body size and number of broods per season (Wilbur *et al.* 1974). Moreover, in the temperate zone, the distribution of most lizard species is centered in xeric habitats, making it difficult to find temperate forms

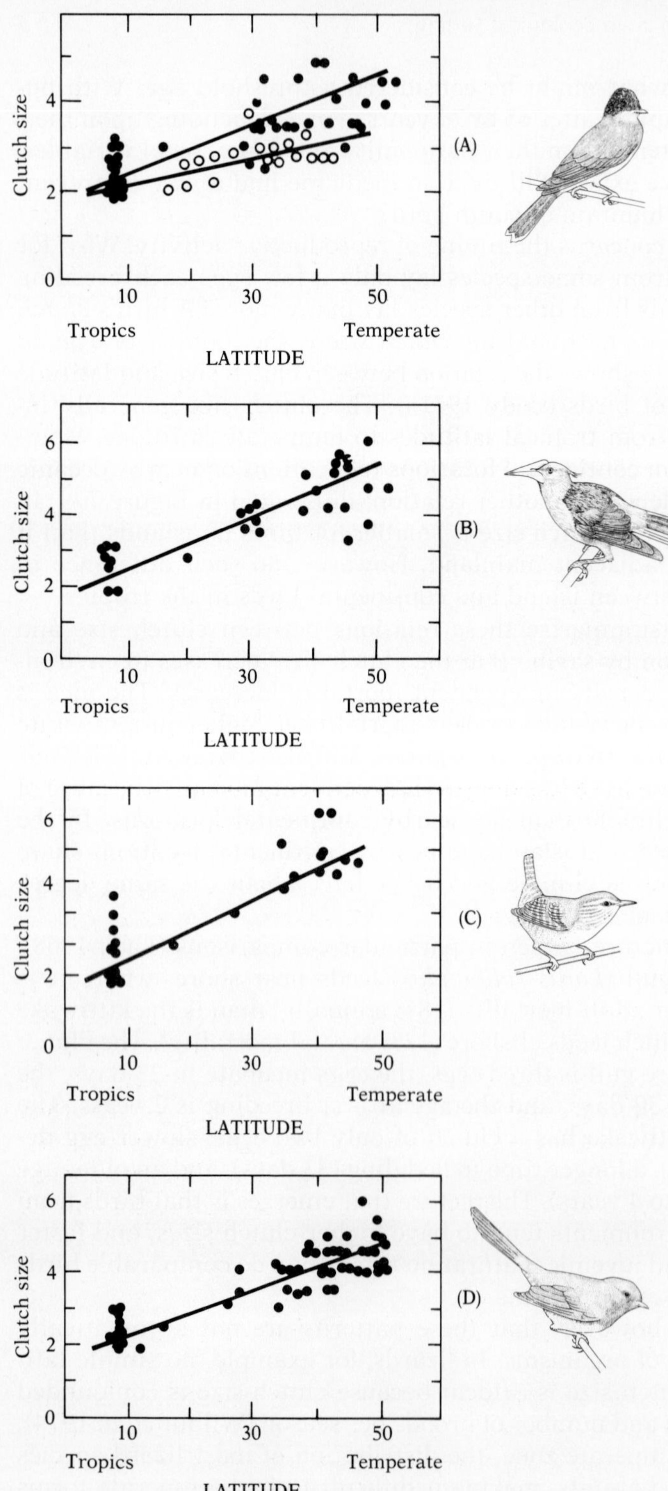

FIGURE 7–8. Latitudinal gradient in clutch size in four groups of birds. Open circles are South American; closed circles are North American. (A) Tyrant-flycatchers. (B) Icterids. (C) Wrens. (D) The slate-throated redstart (*Myioborus miniatus*). *(After Cody 1971.)*

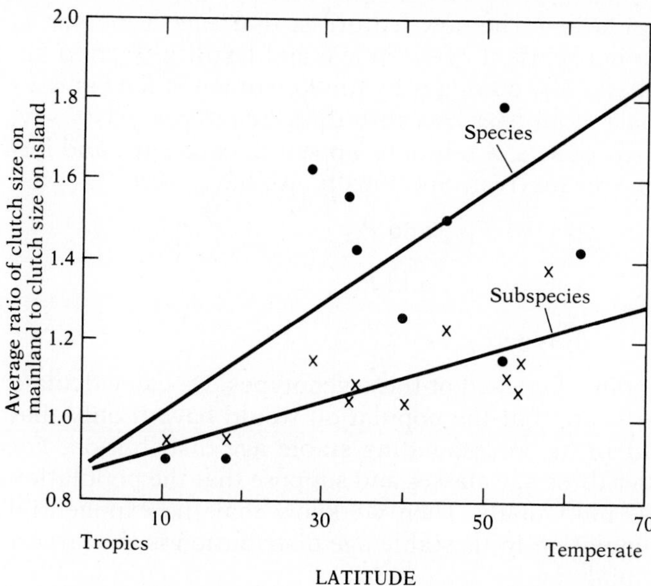

FIGURE 7–9. Average ratio of the clutch size of mainland birds to that of nearby island birds as a function of latitude. Dots indicate comparisons between species, and x's indicate comparisons between subspecies. *(After Cody 1971.)*

to compare with the tropical forms that live in forests, in mesic lowlands, and on mountain tops.

When the latitudinal and island-continent trends were discovered, they were originally interpreted in terms of the concepts of *r*-selection and *K*-selection that come from the theory of density-dependent selection. It was thought that in high density-independent mortality environments, the population is frequently in a state of exponential growth, resulting in selection for traits promoting a high *r* at the expense of a high *K*, and vice versa in environments of low density-independent mortality. However, adaptations in highly seasonal environments may be directed more to the rapid use of a "flush" of resources than to the degree of density-independent mortality in the environment (Boyce 1979). A further difficulty is that *r*-selection and *K*-selection theory does not include age structure; that theory is based on a model of a population with discrete generations. To discuss how the relationship between fertility and age evolves, it is more appropriate to present a theory describing how natural selection works in populations with age structure.

The Basic Theorems of Age-Specific Natural Selection

Norton (1928), in a paper presenting ideas nearly 50 years ahead of his time, established the main theoretical results that describe how evolution occurs in a density-independent, age-structured population. The model begins with the familiar one-locus, two-allele genetic system in an exponentially growing population. The new feature is that each genotype is allowed to have a distinct relation between age and fertility, F_x, and between age and survival, P_x. The mating is by random union of the gametes produced by individuals of all ages. Because three genotypes exist—say, A_1A_1, A_1A_2, and A_2A_2—three sets of relations appear between age and fertility and between age and survival. Specifically, we have

$$A_1A_1 \rightarrow F_{11,x} \text{ and } P_{11,x}$$

$$A_1A_2 \rightarrow F_{12,x} \text{ and } P_{12,x}$$

$$A_2A_2 \rightarrow F_{22,x} \text{ and } P_{22,x}$$

Now here is a key point: For each of these genotypes, we can calculate the exponential growth rate that the population would have if only that genotype were present in its corresponding stable age distribution. For example, let us consider three age classes and suppose that the population consisted only of A_1A_1 individuals. Then we know that the exponential growth rate of this population in its stable age distribution is the largest root, R, of the polynomial

$$R^3 - \ell_{11,0} F_{11,0} R^2 - \ell_{11,1} F_{11,1} R - \ell_{11,2} F_{11,2} = 0$$

where $\ell_{11,x}$ is the survivorship curve for genotype A_1A_1. We label this root R_{11}.

Similarly, we can calculate the ultimate exponential growth rate of the other genotypes. Thus, even though each genotype may differ from the others in many ways, as reflected in each genotype's having its own distinct relationship between age and fertility and between age and survival, we can assign to each genotype a single number that represents the exponential growth rate of a population with a stable age distribution that consisted only of that genotype. Of course, in reality a population could never consist only of heterozygotes because, when they mate, homozygotes are also produced. Nonetheless, R_{12} is, by definition, the exponential growth rate of a population consisting entirely of individuals with the phenotype of the heterozygote.

The main result of the theory of natural selection in a population with overlapping generations is that evolution leads to the maximization of the exponential growth rate, R. The genotype with the highest R is favored by natural selection, where R is computed as the exponential growth rate of a population consisting entirely of that genotype present in its stable age distribution.

The Evolution of Senescence

The first application of this result is to the evolution of senescence (Medawar 1946, Williams 1957, Hamilton 1966). The main idea is to consider genes that are expressed only at a certain age. Different genes are expressed in an organism throughout its development—some only in the embryo, others only after reproductive age is reached, and so forth. The question, then, is how strong the selection pressure is on a gene as a function of the age at which it is expressed.

For example, consider a hypothetical gene that increases survival ability by a standard amount, and suppose that this gene is expressed only at a particular age, a_1. Consider also another gene that increases survival ability by this same amount but that is expressed at a different age, say, a_2. Thus, these two genes confer an identical improvement in survival ability, but at different ages. Is the strength of selection favoring each of these alleles the same? The answer is no; the strength of selection depends on the *age* at which the trait is expressed. As makes sense intuitively, the later in life the improvement occurs, the smaller is its effect on R. Similarly, deleterious traits that act later in life are selected against less strongly than those that act early in life.

With this principle in mind, three mechanisms exist by which senescence can evolve:

1. Favorable mutations that affect older ages accumulate more slowly than comparable mutations that act early in life.
2. Deleterious mutations that act late in life are not as strongly opposed by natural selection as those that act early in life. Hence, late-acting deleterious mutations may be pushed to a higher frequency by recurrent mutation.
3. Genes may exist that affect two or more ages. Some genes may sacrifice an ability to survive at a late age while improving the ability to survive at an early age. Such genes will increase in frequency because they can confer a net increase in R. Indeed, formulas can be derived to assess how much improvement is needed at the early age to match the sacrifice at the later age and still result in a net increase in R (cf. Roughgarden 1979, chap. 19).

An important implication of this hypothesis is that senescence should not evolve by a single physiological mechanism, but by many. If so, no single cure can be found for senescence. This evolutionary hypothesis suggests, moreover, that there is no simple cure for human senescence because many kinds of medical procedures—surgery, drugs, transplants, and so forth—are needed to treat the multiple causes of senescence that have presumably evolved in our species.

Laboratory Studies of the Evolution of Senescence

An increasing number of studies now offer evidence on the evolution of senescence. Two strains of the flour beetle, *Tribolium castaneum*, were compared, one of which had been maintained by breeding only adults that were a few days old, the other by breeding adults that lived several weeks (Sokal 1974). The longevity of the first strain was significantly lower after 40 generations than that of the second, and Sokal suggested that the first strain accumulated late-acting deleterious mutants. A similar experiment did not detect a significant loss of longevity after about 11 generations in a strain bred only from young adults (Mertz 1975). However, strong evidence for the existence of genes for senescence has been obtained in *Drosophila melanogaster* (Rose and Charlesworth 1980). More recently, Luckinbill *et al.* (1985) and Rose (1984) have independently established, with further experiments using *D. melanogaster*, that selection for delayed reproduction leads to a longer life span accompanied by reduced early fecundity.

In addition, a comparison of species life histories offers indirect evidence that senescence is an evolved character. Senescence is predicted to evolve more readily in populations that are rapidly growing (high R) than in those that are slowly growing and in populations in which adult survival is low to begin with compared to populations in which adult survival is already high. This theoretical expectation accords nicely with general comparisons among species. Charlesworth (1980) notes that birds tend to have higher adult survival rates than do mammals of similar size, presumably because of their lower vulnerability to predation. Birds also tend to exhibit greater maximum life spans in captivity than do mammals of a similar size.

The Evolution of Latitudinal Gradients in Clutch Size

Similar ideas apply to the evolution of the relation between reproductive activity and age. The strength of selection favoring an increase in fertility at a particular age decreases with age. Furthermore, the effect on R decreases steeply with age in populations with high adult mortality and in populations that are rapidly growing. Hence, in populations with high adult mortality or rapid growth, or both, we may anticipate three possibilities similar to those discussed in the evolution of senescence:

1. Genes that promote fertility at early ages accumulate faster than genes that increase fertility to the same degree at later ages.
2. Genes that impede fertility at later ages are selected against more weakly than those that impede fertility at early ages. Hence they build up to higher frequencies because of recurrent mutation.
3. Alleles that sacrifice fertility at later ages to attain higher fertility at early ages may be favored.

These ideas seem to accord well with the latitudinal gradient in clutch size and with the differences in life history that were illustrated with data on gulls earlier in this section. Populations that come from habitats experiencing frequent catastrophes, that have high mortality from predation and other sources, or that typically colonize new habitats evolve higher fecundity at early ages than populations that do not have these characteristics.

These predictions for the evolution of life histories resemble those that arise with *r*-selection and *K*-selection discussed in the previous section. However, no mention of density dependence has been made. Our entire discussion has focused on evolution in an exponentially growing, age-structured population. In particular, we have no need to postulate that the offshore-feeding kittiwake, with a low annual mortality, a low clutch size, and a slow development time relative to the nearshore-feeding black-headed gull is closer to its carrying capacity than is the black-headed gull. This postulate would be required to apply *r*-selection and *K*-selection as a hypothesis to explain the difference in life history between these bird species. The postulate would also need independent supporting evidence before the involvement of *r*-selection and *K*-selection could be tentatively accepted. Nonetheless, if differences in the degree of density-dependent selection were found, then we would have to consider hypotheses related to *r*-selection and *K*-selection in addition to the age-specific selection discussed here.

An important additional class of traits are those in which a loss of survival is incurred during the course of reproductive activity. Reproductive activity often involves exposure to hazard or results in a depleted supply of fat reserves or other resources needed to survive through the winter or dry season. Also, in the laboratory, negative relations have been demonstrated between reproductive activity and survivorship in rotifers (Snell and King 1977) and *D. melanogaster* (Rose 1979), and further examples are cited in Stearns (1976). In this context, it is natural to ask how much loss in survivorship can be sustained in order to produce any particular increase in fertility. This question too has been studied theoretically, but less is known about the evolutionary coupling between F_x and P_x from laboratory studies than about the evolution of F_x and P_x curves considered by themselves.

The investigation of how evolution shapes the life histories of organisms is one of the most active and successful areas of evolutionary ecology. Yet it is well to reflect on what is still missing. Most glaring is the absence of an ecological context other than exponential growth. An early study by Murphy (1968) relating life histories to the pattern of temporal variation in the environment is only beginning to receive additional attention (Boyce and Daley 1980), and density dependence has yet to be considered in other than the most formal theoretical terms. More generally, evolutionary studies of life histories have yet to reach the stage in which they quantitatively account for clutch sizes and survivorship curves. They need to be coupled with more detailed studies, not of the life histories themselves—for the P_x and F_x curves are well known for some natural populations—but of what trade-offs are involved in molding life history traits.

KIN SELECTION

In this section we begin to probe how natural selection shapes the way organisms behave toward one another. We focus especially on *altruism*, which may be defined as actions performed by one individual at its expense that benefit another individual. Nature is not solely "red in tooth and claw," as is sometimes assumed. Animals of the same species frequently assist one another. Sometimes, when they help each other, all benefit, as when wolves or African wild dogs hunt together in packs, but this mode of behavior is known as *cooperation*, not altruism. Altruism occurs when animals aid one another and in doing so expose themselves to hazard they would otherwise not face or give up some resources they would otherwise use.

How does altruism evolve? One might think that such behavior would be eliminated by natural selection. We will see, however, that although natural selection favors traits that increase fitness, what counts as fitness is not always obvious, and an organism that in the short run acts in ways that decrease its own fitness may, in the long run, actually be raising its fitness.

Most altruism involves relatives. Parental care is such an obvious example of altruism that its evolution is usually not considered a problem. Clearly, an organism is increasing its own fitness by protecting and feeding its young, but how long should the parental care continue? At what point should a parent direct its efforts to raising more young and leave its elder offspring to fend for themselves? How gradually should the parental care taper off? Thus, although it is clear in principle that a parent is increasing its own fitness by providing care to its offspring, it is not obvious how much care it should provide or for how long.

The offspring themselves also may act to postpone the termination of parental care. The parent is not, in general, in total control of the situation. The fact that two or more parties are involved implies that the behavior finally observed is the result of several interacting strategies. In effect, a *game* is being played, and we cannot present an evolutionary analysis featuring only one party's point of view. We must consider the strategies of all participating parties and ascertain the result of all strategies operating simultaneously.

In addition, brothers and sisters frequently assist one another (or assist parents in raising younger siblings), though seemingly to a lesser extent than do parents with offspring. Is this an extension of the concept of parental care?

Finally, several groups show dramatic examples of altruism, especially the groups known as social insects, which are the wasps, bees, ants, and termites. *Social insects* often live in complex societies, with individuals performing specialized tasks, such as defending the nest, caring for the young, and collecting food. Of greatest importance, many of these specialized individuals are *sterile*; they themselves cannot reproduce, but they assist others to reproduce. How can sterility evolve as an extreme form of altruism? We will see again that kinship holds the key.

Kinship, however, is not the basis of all altruism, and in the last section of this chapter we will describe the notion of group selection, which suggests that *groups* of individuals may exhibit differential survival and reproduction and that focusing on the fitness of single individuals prevents our seeing that the basic idea of natural selection can be applied to aggregations.

Inclusive Fitness

W. D. Hamilton (1964) introduced the idea of *inclusive fitness*, which extends the classical concept of fitness to include the effects of an individual's behavior on the fitness of its relatives. The basic idea is straightforward. Suppose an individual animal, Y, is assisting a member of its own generation—a brother, sister, cousin, or even a more distant relative (which avoids, for the moment, having to compare benefits and costs across generations). Suppose Y has chosen a mate at random in a large population. If so, then any of Y's offspring, O, will carry one-half of Y's genome and one-half of the genome from its other parent. Because Y is mating at random in a large population, we assume that Y's mate does not have many genes in common with Y. Hence, Y must "presume" that its offspring, O, carries only one-half of Y's own set of genes.

The offspring from Y's brothers and sisters are Y's nieces and nephews. If Y's brothers and sisters have also mated at random in a large population, then each of Y's nieces and nephews will carry, on the average, one-fourth of Y's genome. This figure arises because Y's brothers and sisters have one-half of Y's genome to begin with. Each brother or sister contributes one-half of its own genome to the offspring, and this contribution is therefore one-quarter of Y's genome.

Now let us imagine that Y's "objective" is to place as many copies of its genome in the next generation as possible. Clearly, Y may raise one offspring of its own, or it may raise two nieces or nephews, and place the same total number of copies of its genome in the next generation. This calculation then offers a context in which to consider trade-offs. Suppose that Y is in a situation in which it may fail to produce offspring of its own. Perhaps it has a poor territory, or it has failed to attract a mate, or its mate has died. Let us assume also that its sib is in a position to raise a still larger family with Y's assistance. Then Y will place more copies of its own genome in the next generation if, for each of its own offspring that it does not produce, it causes more than two nieces and nephews to be produced.

This calculation is the basis of the *kin-selection* theory for the evolution of altruism. Individuals are viewed as giving up some fitness by reducing the production of their own offspring in order to obtain a much larger production of offspring from their relatives. In doing so, the altruistic individuals accomplish a net increase in the number of copies of their genome that they place in the next generation, and thus in their fitness.

Inclusive fitness is simply a formula that summarizes the costs and benefits of the behavioral trait. For example, suppose the trait consists of an individual's helping its brothers and sisters. Then a formula for the inclusive fitness is

$$\text{Inclusive fitness} = \text{fitness of nonaltruist} - \text{loss of altruist's offspring} + \tfrac{1}{2}\ \text{gain in offspring from brothers and sisters}$$

Notice the factor of $\frac{1}{2}$; it is the "value" of a niece or nephew relative to that of the altruist's own offspring. This factor depends on the degree of relationship between the donor and the recipient. If the altruistic act is toward a cousin, then the coefficient is $\frac{1}{4}$ because the cousin shares only one-fourth of the altruist's genome to begin with, rather than one-half, as with brothers and sisters.

Kin Selection and Hymenopteran Societies

The basic premise of the kin-selection theory for the evolution of altruism is that an altruistic trait evolves if the inclusive fitness of altruists is higher than that of individuals who do not perform the altruistic behavior. To use this theory, one must determine the costs and benefits of the altruistic behavior and also the genetic relationship between the donor and the potential recipients.

Hamilton introduced the idea of inclusive fitness while offering a hypothesis for the evolution of sterile casts in social insects from the order Hymenoptera (whose evolution is discussed further in Chapter 8). This order includes the bees, wasps, and ants; it does not include the termites, which are another group of social insects but which have a social structure and genetic system fundamentally different from those of hymenopteran societies.

The main facts that have to be explained about hymenopteran societies are illustrated by the biology of paper wasps from the genus *Polistes* (West Eberhard 1969). These wasps make a nest of a gray, paperlike material, which typically is suspended by a stalk from the underside of branches, roof overhangs, and the like. The wasp society is comparatively simple and consists of only three functional types of individuals, called *castes*: the female reproductives (queens); the workers, all of which are female; and the male reproductives (drones).

The life history of the paper wasp, *Polistes fuscatus* in Michigan, is typical (see Figure 7–10). The season begins in the spring with a female wasp that emerges from diapause. It carries a sperm packet that was acquired by mating at the end of the previous season. The wasp constructs a stalk and the first group of nest chambers. It lays an egg in each chamber. The eggs hatch and develop into larvae. Then the wasp feeds the larvae with captured prey. After the larvae have grown, their chambers are capped. The larvae then pupate and emerge as adult wasps.

All the newly emerged wasps are females. They do not leave to start nests of their own; instead, they assist the foundress, the queen, in raising more of her offspring. The queen at this time ceases her efforts to catch prey and to feed her newer offspring. Instead, she spends her time laying eggs, while her eldest offspring feed the younger offspring. In this way, the eldest offspring of the queen assist her in raising more offspring.

During the early summer, the offspring of the queen continue to consist of females. Indeed, the whole colony consists of females: the queen, the

FIGURE 7–10. Colony of a paper wasp.

workers, and the larvae. However, toward midsummer, some of the larvae emerge as males. At this point, the social organization of the colony disintegrates. The female workers relax their efforts at helping their mother to raise more offspring. Instead, they occasionally try to start nests of their own, and they may even cannibalize the larvae that their mother produced. Furthermore, the male wasps that have just emerged do not assist in feeding the young or in defending the nest. They remain in the vicinity and attempt to mate with wasps that are setting up new nests or that are preparing to enter diapause for the winter. Some of the female wasps do survive the winter, and the cycle is repeated.

Three main questions need to be answered:

1. Why do the eldest offspring of the queen assist her in raising more offspring instead of raising offspring of their own?
2. Why does the society disintegrate at the end of the season?
3. Why are no male workers in the nest?

The key to a hypothesis that answers these questions is that the genetics of sex in Hymenoptera rests on a sexual difference in ploidy. In Hymenoptera, females are *diploid* and are produced from fertilized eggs. Males are *haploid* and are produced from unfertilized eggs. Hamilton (1964) was the first to notice that this sexual difference in ploidy may be connected with the evolution of the special aspects of social behavior found among the Hymenoptera.

Because males are haploid, their sperm are produced through mitosis, not meiosis. Hence, *all the sperm from a male are genetically identical.* Now consider the implications of this fact for the eldest daughters of a queen. If the queen carries the sperm packet from one male, then all the sperm are genetically identical. Thus, sisters share three-fourths of their genome with one another. They share one-half of their genome because they were produced by identical sperm, and they share another one-fourth, on the average, because they were produced by eggs from the same mother. They receive only one-fourth of their genome in common from their mother because eggs are produced through meiosis, and so eggs are not genetically identical.

Thus, in a system of sex determination in which males are haploid and females diploid, full sisters share three-quarters of their genome. Notice that this amount is higher than the fraction shared between a mother and her own daughter (one-half). This point is critical to Hamilton's kin-selection theory for the evolution of social behavior in the Hymenoptera.

The answer to the first question can now be phrased in terms of the best strategy available to an elder daughter of the queen. If she can produce, say, n offspring of her own, she will place $\frac{n}{2}$ copies of her genome in the next generation. However, if she helps the queen produce n more female offspring than the queen would otherwise produce herself, then the elder daughter places $\frac{3n}{4}$ copies of her genome in the next generation.

Clearly, an elder daughter can *send* more copies of her own genome into the next generation by assisting the queen in making more daughters than by raising an equal number of her own offspring.

There is a catch, however. A daughter of the queen should help the queen raise only daughters; it should not help raise male offspring. A hymenopteran female shares only one-fourth of its genome with a brother. A brother is formed from an unfertilized egg and so has only one-fourth of its sister's genome.

Therefore, according to the kin-selection hypothesis, a queen should lay fertilized eggs in the early season. By doing so, it produces females, and they in turn assist in further reproductive activities. As the season draws to a close, however, the queen must begin to produce males as well, and at that time the workers should set forth on reproductive activities by themselves. The answer to the second question, then, is that the social system should disintegrate when the queen begins to produce males.

Meanwhile, males should not become sterile workers in the nest. A male shares one-half of its genome with its sisters, but it places a copy of its entire genome in each sperm. Hence, a male shares 100% of its genome with each of its own offspring. Clearly, by the kin-selection hypothesis, a male should produce its own offspring and not assist the queen in the production of more brothers or sisters. This answers the third question—why males do not participate in the hymenopteran society, according to the kin-selection hypothesis.

The Justification of Kin Selection

An important feature of the kin-selection hypothesis is that the strategy of every member of the society makes sense relative to the others. A social *system* results from individuals each pursuing its own strategy; it is the combination of all of these strategies. Thus, in trying to understand the evolution of social behavior, it is crucial not to restrict the focus to one kind of individual in the population.

To emphasize the importance of considering the strategies of all the organisms in a society simultaneously, Maynard Smith (1982) introduced the techniques of a subject called *game theory*, originally developed for economic theory. This approach has been used in investigating the evolution of a range of behaviors: aggression, territoriality, changes of sex with age (sequential hermaphroditism), altruism among unrelated individuals (reciprocal altruism), and the relationship between parent and offspring. As Maynard Smith points out, game theory ideas are implicit in many earlier writings on the evolution of behavior, yet this approach offers a unified theoretical framework for analyzing the simultaneous functioning of several strategies.

A major problem of researchers in evolutionary biology has been to see if, and how, the basic premise of Hamilton's kin-selection theory of altruism can be justified. Is it true that natural selection leads to the evolution of behavioral strategies that maximize inclusive fitness? Another issue is to determine the correct formula for the inclusive fitness in varying circumstances. In the preceding example, the population was assumed to be large enough so that matings occur between unrelated individuals. However, many societies are small from a genetic standpoint, consisting of, say, fewer than 100 individuals. Inbreeding necessarily occurs in such small populations. How is inclusive fitness defined in an inbred population? Important progress has been made in answering these questions; key references are Michod (1980) and Uyenoyama (1984).

Another possible explanation for the evolution of altruism is focused on how natural selection affects aggregations of individuals. We will discuss this possibility next.

GROUP SELECTION

Group selection leads to evolution by the differential extinction of genetically different groups within a species or the differential production of offspring from genetically different groups. This evolutionary mechanism requires that the species population somehow be divided into smaller units, at least for some time during the life cycle. Moreover, the gene pool within each of the smaller units cannot be identical; genetic variation between groups must exist for group selection to work, just as genetic variation among individuals must exist for traditional natural selection to work.

Early Ideas of Group Selection

The concept of group selection has been controversial. Early advocates of the idea (e.g., Wynne-Edwards 1962) thought group selection caused the evolution of traits promoting the "good of the species," in contrast to natural selection, which supposedly caused the evolution of selfish traits. The trait of territoriality, for instance, was viewed as a mechanism for population control that evolved through group selection in order to prevent overcrowding and exhaustion of environmental resources.

Early conceptions of group selection presented two difficulties:

1. They offered no clear theoretical picture of how group selection could actually operate. Such a theory is especially important because, for one to determine if group selection can cause the evolution of a trait opposed by natural selection, some way of assessing the relative strengths of these two forces must be available.

2. No traits could be exhibited whose existence could not be explained as a result of ordinary natural selection. Furthermore, opponents to the idea of group selection frequently took the position that if an evolutionary explanation based on traditional natural selection was consistent with the evidence, then one should not inquire whether group selection exists as a biological process. This latter position had a chilling effect on research into group selection, but fortunately that period has passed.

The Present Picture

Today two well-posed theoretical pictures have been presented on how group selection can work. Empirical examples also exist, as we will discuss.

In the first theoretical picture, the whole population is divided into many small subpopulations, and these subpopulations are assumed to become extinct at a rate that depends on the gene frequencies within them. If we suppose that one allele exists that leads to behavior benefiting the group and another allele exists that leads to selfish behavior, then we might assume that groups with a high frequency of the "selfish" alleles in them would become extinct more rapidly than groups with a high frequency of the "altruistic" alleles. However, it must be assumed also that ordinary natural selection increases the frequency of the selfish alleles within each group. Furthermore, we assume that when a group becomes extinct, it is replaced by a fair sample of individuals from the population as a whole. The between-group genetic variation is caused by genetic drift. Finally, we assume that a time exists within each generation when the groups exchange migrants.

Using this picture, Eshel (1972) established theoretically that if the migration among the groups is small enough, then group selection invariably leads to the altruistic allele being fixed, regardless of the strength of the opposing natural selection within each of the groups. Conversely, if the migration among the groups is high enough, natural selection invariably leads to fixation of the selfish allele, regardless of the group selection that is working in the other direction. Also, with an intermediate amount of migration, a polymorphism may occur between the altruistic and selfish alleles that is maintained by a balance between group selection and natural selection. (For a numerical illustration of these results, see Roughgarden 1979, pp. 283–292.) Finally, if an allele happens to be favored by both group selection and natural selection, it will evolve faster than it would if favored by natural selection alone. This point brings out the hazard of ignoring a possible role for group selection even in the evolution of traits favored by natural selection.

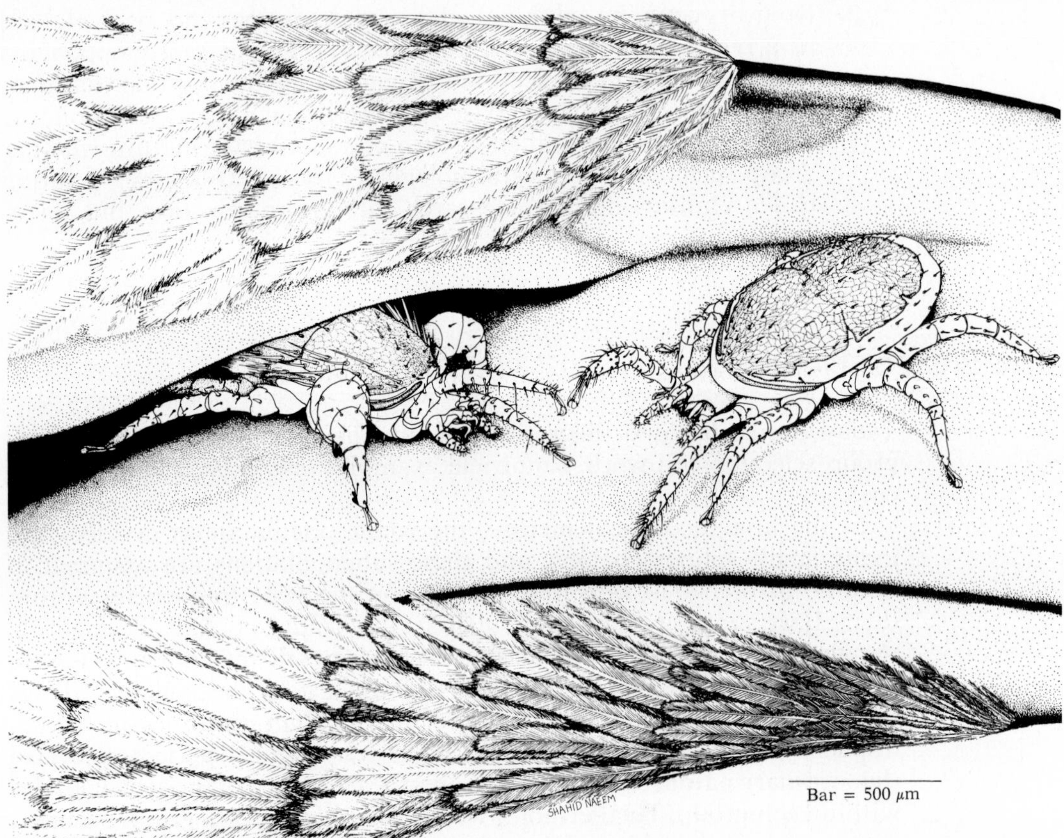

FIGURE 7–11. Hummingbird-dispersed mites that live in hummingbird-pollinated flowers. Shown here at the nare of a hummingbird. *(Drawing by Shahid Naeem, University of California, Berkeley.)*

This picture of how group selection works is important primarily because it suggested for the first time that group selection was possible and should not be dismissed. However, it is not easy to see how the assumptions of Eshel's model can apply to many populations.

A second theoretical picture consists of a population divided into small groups during most of the life cycle. All the groups merge at some point for mating and subsequently redistribute into small groups again. The groups contribute differentially to the total population, depending on their genetic composition. Using this picture, Wilson (1980) has shown that a gene that confers high group productivity can evolve even if it is neutral to, or opposed by, natural selection within the groups.

The first empirical studies demonstrating group selection used laboratory populations of the flour beetle, *Tribolium* (Wade 1976, 1977). More recent studies with this system allow a careful analysis of the interaction of the local population size with the amount of migration among the local populations on the progress of group selection (Wade and McCauley 1984). Colwell (1981, 1982) was the first to identify an example of group selection in a natural population. His example is close to the theoretical picture developed by D. S. Wilson, although some details are special. The example involes the sex ratio produced among the offspring of a female; it pertains to arthropods that occur in small groups. The groups are founded by several mated females. The progeny develop, and then mating occurs at random within the group of progeny from the founding females. Then mated females disperse to establish new groups. This life history can be found in wasps that mate within a single fig or host pupa and in mites that live in hummingbird-pollinated flowers (Figure 7–11).

The key observation is that the sex ratio among the offspring of a female in these species is typically biased toward females; that is, more females than males are present among the offspring of any particular female. As we describe in more detail in Part 3, traditional natural selection within a group usually favors females that produce a 50:50 ratio of females to males among their progeny. However, Colwell showed that groups having females whose progeny sex ratio is biased toward females contribute more to the total population than do groups whose females produce progeny in a 50:50 ratio. This fact evidently explains the evolution of a female-biased sex ratio in such organisms, a result that differs from what natural selection alone would produce in a large population without group structure.

At this point, you know the basics of what causes the abundance and distribution of organisms. You also know the basics of the main evolutionary processes and the extent to which they are founded in an ecological context. In Part 3, we will focus in more detail on behavioral traits within populations; in future chapters, we will explore how two or more populations interact as a basis for understanding ecological communities and ecosystems.

PART 3

SOCIAL INTERACTIONS

Animal behaviors seem at first glance to be incomprehensibly diverse—male elephant seals guarding harems, prairie chickens with complex courting rituals, frogs calling, sparrows flocking, wolves hunting in packs, ants foraging in colonies, and the schooling of fishes, just to name a few. Is this diversity of behaviors understandable? The answer is yes—if they are viewed within the context of evolutionary ecology. As with so many other areas of biology, a virtually meaningless jumble of observations begins to make sense once one asks how and why a phenomenon evolved.

In the following four chapters, we will often ask what ecological circumstances favor the evolution of a social system. In Chapter 8, we will introduce the problem by examining four different kinds of animal societies. In Chapter 9 we will address the ecology of sex, and in Chapter 10 that of mating systems. In the final chapter of this part, we will deal with a number of social interactions: sexual selection, territoriality, the sociology of attack and defense, and communication.

CHAPTER 8

The Evolution of Social Behavior: Some Examples of Animal Societies

This chapter deals with the ecology of animal *societies*—groups of conspecifics organized in a cooperative manner (Wilson 1975). Later on, we will examine some aspects of plant "sociality" as well. In an ecological context, each social behavior is presumed to contribute directly or indirectly to the production and survival of offspring in a specified environment. Most of these behaviors can be understood by considering individual selection, as we discussed in Part 2, although ideas from group selection may be needed in special cases.

To see why the ecological and evolutionary context of social behavior holds such attractive explanations—as well as several difficulties—let us look at some comparative studies performed in recent decades on four groups of quite different organisms: weaver finches, ground squirrels, bees and other social insects, and naked mole-rats.

WEAVER FINCHES

J. H. Crook, in a classic comparative study (1964) of the behavior of different species of weaver finches (Ploceidae—relatives of English sparrows), established a tradition of tying behavior to ecology. When differences in weaver finch social behavior are compared with differences in predation and the availability and distribution of food resources and nesting sites, patterns emerge. Weavers in Savanna habitats, for example, are more apt to have colonial nesting, feeding in flocks on seeds, and polygyny (one male mated to more than one female). In contrast, evergreen forests have weavers that are territorial, solitary feeders, and monogamous.

Weavers in the savanna, a semiarid grassland containing widely spaced trees, depend for food on the irregular mass production of seeds at scattered locations. Flocks of birds can find such concentrated but relatively unpredictable resources more efficiently than single individuals can (Ward and Zahavi 1973, Waser 1981). The birds form colonies (see Figure 8–1), weaving large, conspicuous nests, which probably insulate against the heat of the savanna, at the very tips of branches of trees. The colonial weavers presumably obtain protection from predators by the relative inaccessibility of their nests or by the selection of spiny acacias for their construction. Colony sites are generally near concentrations of resources.

In the forests, insect resources are much scarcer and more uniformly distributed than are the seeds of the savanna. This favors a wide spacing of nests and solitary foraging. Protection from predators is gained by camouflaging the isolated nests; the cooler forest presumably removes the need for the insulating structures of the savanna species.

The development of a polygynous mating system in the savanna and a monogamous one in the forest can also be explained on ecological grounds, as we shall discuss in Chapter 10.

FIGURE 8–1. Colony of white-browed sparrow-weavers at Lake Baringo, Kenya. (A) Nests of part of the colony. (B) Bird on a branch with the nest in the background. *(Photos by Paul R. Ehrlich.)*

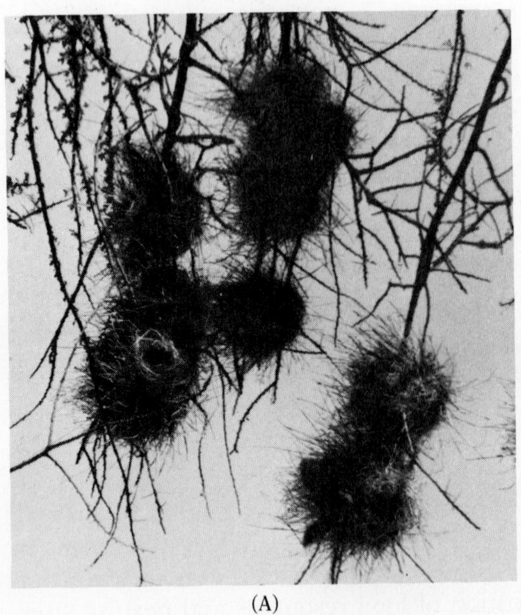

(A)

(B)

GROUND SQUIRRELS

The largest ground squirrels, marmots and woodchucks (*Marmota*), show a spectrum of sociality (Barash 1974, Armitage 1981). The woodchuck (*M. monax*) in eastern North America is basically asocial (Snyder *et al.* 1961). Adults are together only for mating, and young disperse in their birth year, during which time they reach about 33% of the weight of adults. Yearlings (young in their second season) reach about 80% of adult weight and sometimes become sexually mature.

The western yellow-bellied marmot, *M. flaviventris* (Figure 8–2), which lives largely in montane meadows in the Sierras, Rockies, and Cascades, is colonial, but it has individual home ranges and aggressive social encounters (Armitage 1962, 1975, 1984, Downhower 1968, Downhower and Armitage 1981, Armitage and Johns 1982). Yellow-bellied marmots disperse as yearlings when they are about 60% of adult weight. They do not breed until they are 2 years old.

The Olympic marmot, *M. olympus*, which occurs only on the upper slopes of the Olympic Mountains of coastal Washington, has the most highly developed social system of the three (Barash 1973). No individual home ranges are maintained, dominance relations are minimal, and social encounters are more frequent than they are in *M. flaviventris*. The social encounters consist mostly of "greeting behaviors," which are apparently involved in helping individuals recognize each other. Yearlings are only 30% of adult weight, and dispersal from crowded colonies is done by 2-year-olds. In small colonies, much less behavioral interaction occurs, and the young become permanent residents. Breeding does not take place until the young are 3 years old.

FIGURE 8–2. Yellow-bellied marmot at the mouth of its burrow in a subalpine meadow in Colorado. *(Photo by K. B. Armitage, University of Kansas/BPS.)*

This trend toward increasing sociality parallels a trend toward increasing severity of habitat. The asocial *M. monax* populations live in areas in which the growing season for the plants they depend on exceeds 5 months; the colonial *M. flaviventris*, in contrast, live in areas in which the season is 70–100 days long. The highly social *M. olympus* has the shortest season of the three—only 40–70 days.

Armitage (1981) considered this variation in sociality over several genera of ground squirrels in addition to *Marmota*. He concluded that, throughout the group, sociality was a life-history tactic that has evolved in relation to adult weight and age at first reproduction. Species with a large body size living in localities with a short growing season cannot produce young with a reasonable chance of survival if they disperse in their birth year or (in more severe surroundings) as yearlings. In these species, dispersal is delayed, and the parents make a larger reproductive investment by sharing resources with their offspring longer and helping to protect them from predators.

Why, though, should large size evolve? Why should not selection favor smaller individuals that can reproduce at a younger age? Four possible reasons are as follows:

1. Larger size provides protection against some predators (Waring 1965, Armitage 1981).
2. A wider variety of vegetation may be eaten by a larger animal, and the relative amount of protein required is less (Armitage 1982b).
3. More fat can be stored by a larger animal (Morisson 1960), and that fat will not be used as rapidly because of the proportionately lower metabolic weight of heavier individuals (Armitage 1981).
4. Larger individuals may be dominant and have disproportionate access to resources or mates.

Thus, a start has been made at sorting out the way environmental forces shape ground squirrel sociality.

BEES AND OTHER SOCIAL INSECTS

Understanding the ecological basis for the evolution of insect societies presents an even greater challenge than that of ground squirrels. What, for example, has led some groups of wasps to go down the evolutionary path exemplified by the paper wasps (discussed in Chapter 7), whereas others, such as the spider-hunting Pompilidae, do not form social groups? More is known about the evolution of societies in bees (basically just another group of wasps) than that of any other Hymenoptera. The key elements of true social behavior in insects consist of cooperative care of the

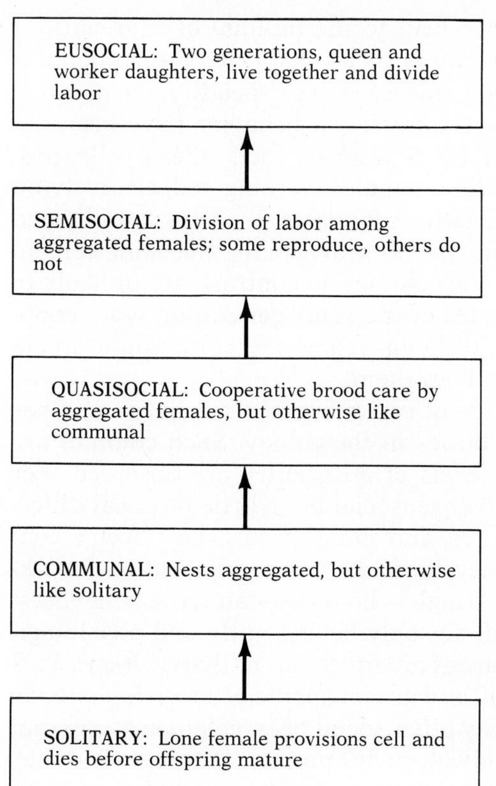

FIGURE 8–3. Possible sequence of evolution of social behavior in bees.

young, castes of more or less sterile workers that assist reproductives in the care of the young, and offspring remaining with parents to assist them during some period of their lives (Wilson 1975).

Michener (see, e.g., 1974) and co-workers have described a spectrum of behaviors in living bees that also represents the probable sequence in the evolution of true social behavior in most social bees (Figure 8–3). At one end of the spectrum are *solitary* bees, in which the females provision a cell with sufficient nectar and pollen to support the larva throughout its entire development. In most cases, the female dies before her offspring become adults; however, in some cases (as with many carpenter bees), the female may still maintain the nest when her offspring emerge from their pupae.

The next step in the sequence toward true sociality is the evolution of *communal* behavior. Here each female still builds and provisions her own nest, but the nests are aggregated. Not much is known about the ecological factors that favor communal nesting. Presumably, selection generally favors individuals staying in the area (and thus in the environment) in which they were successfully reared, unless the environment is subject to great unpredictability. This tendency can lead to the buildup of aggregations in the vicinity of previously successful nests as offspring "return home."

Quasisocial colonies add the first feature of true sociality, cooperative care of the young. Several levels of quasisocial behavior have been observed, from females doing a little bit of work on each other's cells (possibly by mistake) to as many as three females working cooperatively on the same cell. That some quasisociality originates in errors is indicated by the observation of two different species provisioning the same cells in a joint nest (Bohart 1955). *Semisocial* colonies, in contrast, are unlikely to be accidental in origin. Here females of the same generation work cooperatively and practice a division of labor—some lay eggs, while others with undeveloped ovaries work to help them.

The final step occurs when adults of two generations, a queen and her worker daughters, divide up the duties in the colony. Such colonies are truly social (*eusocial*). Different levels of eusociality are observed. For example, in the numerous primitively eusocial bees, little physical differentiation appears between the castes, and some species show even a continuum in behavior and ovarian development between queens and workers. In the highly eusocial tropical stingless honeybees and true honeybees, the castes are quite differentiated, not only behaviorally and physiologically but also morphologically. Queens are much larger than workers, lack the apparatus for gathering pollen, and are incapable of surviving outside of the colony. Unlike colonies of any other social bees, colonies of eusocial bees are perennial. As Michener (1974, p. 56) put it: "Barring accidents, predation, parasites, and disease, a colony of highly eusocial bees is, so far as is known, immortal in the sense that given suitable conditions, an *Amoeba* is immortal. It can go on dividing, growing, and dividing again, indefinitely. . . ."

Eusociality has evolved independently in 5 different groups of bees and some level of sociality in 14. What environmental pressures favor the development of social behavior? Field censuses (Michener 1964) show that solitary bees generally produce more offspring per female than social ones do. Why does this not prevent the evolution of sociality? Perhaps the chances of survival of entire colonies are higher than those of solitary nests (the field censuses do not count nests or colonies that have disappeared). One advantage of large colonies over individual nests is that they can control their internal environments. However, as Michener (1974) points out, control of physical environmental factors (such as temperature and humidity) by a small colony cannot be much better than that achieved by a single individual. On the other hand, protection against predators and parasites is definitely enhanced by the formation of colonies, for even a single female left in a nest while others are out foraging can reduce the chance of successful intrusion. Michener therefore argues that predator-parasite pressure has been a major force toward the evolution of social behavior in bees. Once sizable colonies are established, of course, they undoubtedly gain other advantages in environmental control and efficiency through division of labor.

Indeed, outside of human societies, the most thorough environmental control of all is exercised by the most social (eusocial) insects. These insects are bees and various other groups of the great orders Hymenoptera (all ants and certain wasps) and Isoptera (termites). All of these insects are characterized by nonreproductive individuals of two generations helping to promote the reproduction of others and cooperative care of the young (Wilson 1971, 1975, Michener 1974).

Social insects achieve control not only over mates, predators, food, and the like but often also, to a large degree, over their physical environments. This control is possible largely because all have sterile or juvenile worker castes that build nests, forage, feed reproductives, tend fungus farms, dry or ventilate nests, and so on. In termites and ants, a subclass of *soldiers* often exists—workers specialized for colony defense either with large mandibles, a gland that sprays a sticky defensive secretion, or both.

Probably the most extreme specialization in social insects is seen in the royal or reproductive castes, the kings and queens. They neither feed nor defend themselves—they are fed and reproduce. Termite queens in some species have evolved, in essence, into egg-producing machines. A tiny termite head and thorax is attached to a gigantic abdomen the size of a small sausage from which eggs constantly dribble. These eggs are carried away by workers, while other workers bring food to the eating end of the queen and still others groom her body. The queen lies enclosed forever in a special cell which she is too large to leave.

Social insects are among the most successful of all organisms. Leaf cutter ants (*Atta*) harvest a substantial portion of all of the vegetation in the New World tropics and use it to make a mulch in gardens in their nests. They feed exclusively on fungus, which they grow in these underground gardens. Ants in general, honeybees, bumblebees, yellowjackets, and other fully social hymenopterans are abundant almost everywhere. Termites are so common that in tropical areas they process about one-quarter of the Earth's net primary productivity and constitute a major pathway for the return of CO_2 to the atmosphere (Zimmerman *et al.* 1982). Many social insects build nests of great size and complexity, with climate controlled and with predators and competitors generally excluded.

Exactly how and why various insect societies evolved is still a subject of great debate, especially matters of haploidy and kin selection, as was discussed in Chapter 7.*

* For further information, see Alexander (1974), Trivers and Hare (1976), Alexander and Sherman (1977), and Brockmann (1984).

NAKED MOLE-RATS

Recently, investigators have discovered that eusociality is not restricted to the insects. Beneath the sun-baked sandy soils of Kenya and the Horn of Africa live colonies of perhaps the strangest of all mammals—naked mole-rats. As Figure 8–4 illustrates, they are a few inches long, nearly hairless, and have wrinkled pink skin and prominent incisors.

FIGURE 8–4. Naked mole-rats. (A) Adult holding a rock, showing digging incisors. (B) Adult feeding feces to a juvenile. It is possible that pheromones or symbiotic microorganisms that aid in digestion are passed around this way. *(Photos by J. Jarvis, University of Capetown.)* (C) Mole-rat bucket brigade. For explanation, see text. *(After Jarvis and Sale 1971.)*

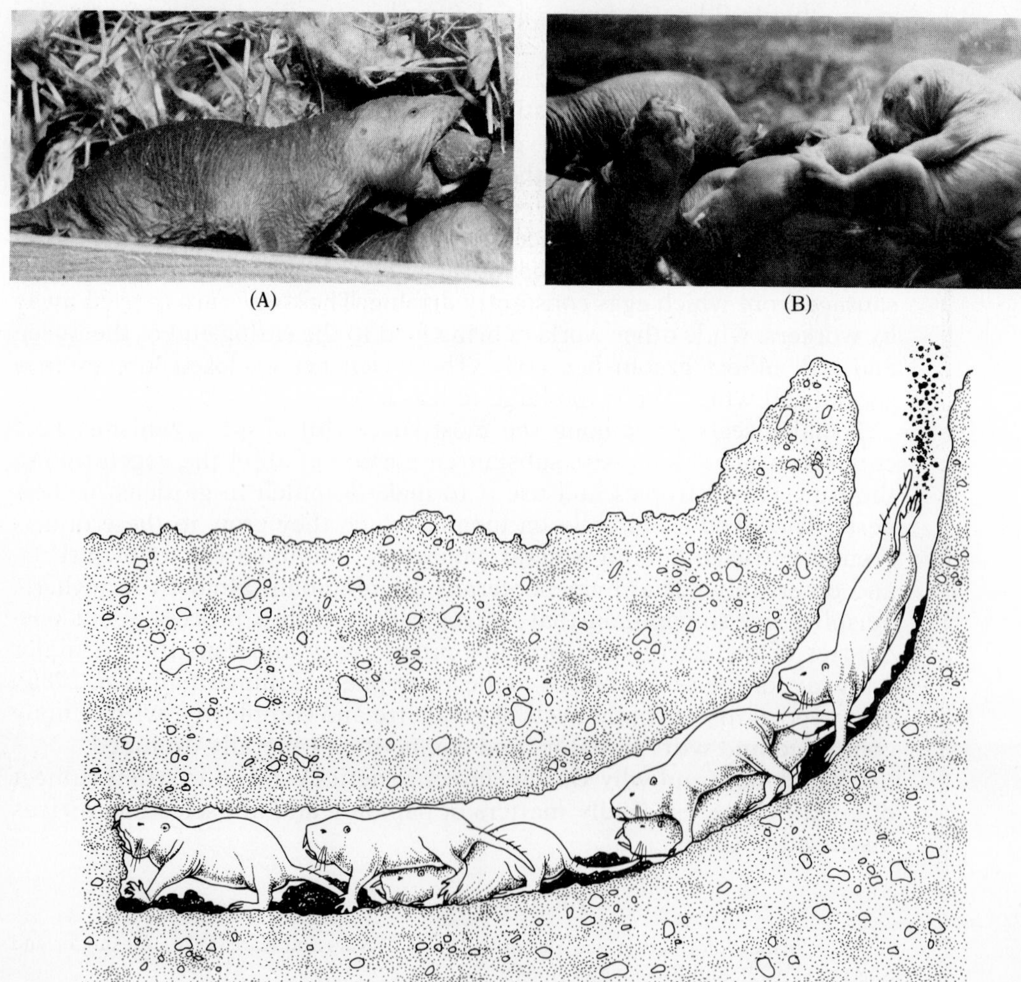

The naked mole-rats of a colony work in teams to excavate hundreds of yards of burrows. One rat works at the face of the burrow, chewing at the soil with its incisors. It then kicks the loosened earth to the animal just behind, which backs along the burrow with its load and eventually deposits it in a side branch leading to the surface. There another rat is responsible for kicking it out onto the surface. The loadless rat, meanwhile, returns to the face of the new burrow by straddling the chain of earth-laden rats backing away from the face (Figure 8–4*c*). In essence, then, the rats form a living bucket excavator (Jarvis and Sale 1971).

Not all of the rats participate in these efforts, however. Like the eusocial insects, mole-rats appear to exist as a series of castes (Jarvis 1981). One group, called *frequent workers,* does most of the nest construction and foraging for food. *Infrequent workers* are heavier individuals that do some of these tasks, but at less than half the performance level of the frequent workers. The mole-rat aristocracy is made up of even larger individuals that are *nonworkers.* They rarely do any digging or foraging. They may be a reproductive caste, because male nonworkers are most likely to mate with the single breeding female, the queen.

The queen produces up to four litters of as many as a dozen young annually. She also is the only female to suckle the young. All the other females, of all castes, appear to have quiescent ovaries and do not breed. However, members of all castes help in other ways to care for the young. The queen apparently suppresses reproduction in the other females, at least in part, by producing a chemical messenger in her urine; the urine is passed around the colony as the rats groom themselves during and after visiting communal toilet areas (J. Jarvis and B. Broll, unpublished observations).

The naked mole-rats have a division of labor based on castes, cooperative brood care, and only a single reproductive individual in a colony, and several generations of offspring assist the parent in the care of the young. Discovering why an insectlike eusociality has evolved in this one group of mammals is a challenge to ecologists. The rats are, of course, diploid; however, in spite of this, it seems likely that their evolutionary history was more like that of bees than of termites, for the cooperative care of the young was a crucial step in that history. As we will see, this kind of care is found in a variety of vertebrates, from lions (Bertram 1975) to birds (Emlen 1978), where young or siblings of the breeding birds aid the breeders to produce more broods but then themselves ordinarly become breeders later in life.

CHAPTER 9

The Ecology of Sex: Its Origin and the Ratio of Males to Females

Sex is the most important factor in determining the relationship of most organisms to a crucial component of their environment—other members of their own species. Two factors are prominent in the ecology of sex: the need for syngamy and parental investment.

Syngamy is the fusion of the sex cells. In animals, fusion usually implies that mating males and females must get together physically. In flowering plants (*angiosperms*), fusion requires a mechanism for transferring pollen from the anthers to a stigma; in nonflowering plants, fusion requires a mechanism for getting the sperm to the egg.

Parental investments are the resources individuals devote to each offspring. Often, in theory, a given investment could have been used to produce larger numbers of offspring, each receiving a smaller investment. Males normally make less of a parental investment in each offspring than females. Females, by definition, produce larger gametes and often a greater fraction of the parental care. Many of the observed features of social systems flow from this difference in parental investment.

WHY HAS SEX EVOLVED?

Why is reproduction in most organisms sexual? What environmental factors encourage reproduction of the sort that usually involves the participation of two parents and the mixing together of genetic characteristics from both in the offspring? These questions, fundamental ones in ecology, have not been answered, but progress is being made. To begin, let us consider the advantages of asexual reproduction; perhaps surprisingly, we will see that they are numerous, whereas those of sexual reproduction are comparatively few.

The Advantages of Asexual Reproduction

Many asexual populations exist. Among plants, asexual forms are often "weeds"—that is, forms adapted for colonizing temporary or newly opened habitats. These plants are good colonists in part because one individual is all that is needed to colonize an area. The common dandelion is one of the best-known asexual plants. Many sexual weeds also exist, however, and no clear relationship can be demonstrated between the stability of a habitat and the asexual or sexual nature of the plants found in it.

Other asexual plant populations occur locally in places in which a suitable pollinator is absent, as in the crucifer *Leavenworthia* (Rollins 1963). Such populations have been recently derived from sexual ancestors that were pollinated by an insect. Asexual reproduction is a response to the absence of the pollinator from places in which the plant can otherwise live successfully.

Asexuality is an advantage not only for populating habitats in which rapid reproduction is at a premium and in which mating is impractical, but also for a more basic reason. Consider what happens in sexual reproduction when a female is able to obtain the energy for two eggs. When she reproduces sexually, each egg carries a copy of only one-half of her genome, because the sperm from the male parent will contribute the remainder of each zygote's genetic material. However, if the female were to produce eggs asexually, then each egg would carry a copy of her entire genome. Thus, an asexually reproducing female would be able to place twice as many copies of her genome in the next generation as a sexually reproducing female. Maynard Smith has termed this fundamental disadvantage of sexual reproduction the *cost of meiosis.* The cost of meiosis seems high indeed—an effective loss of half of an individual's offspring.

The Advantages of Sexual Reproduction

The list of advantages of sexual reproduction is disappointingly short. The most frequent suggestion is that sexual reproduction is advantageous in a changing and unpredictable environment. Because variety is important here, an effective strategy for an organism is to produce many types of offspring in the hope that at least one will endure to survive and reproduce the next generation. Williams and Mitton (1973) have suggested a model based on a lottery for this situation.

The idea that sexual reproduction is advantageous in an unpredictably changing environment is consistent with the reproduction of many facultatively sexual forms such as protozoa (including amoebae), rotifers, and aphids. These forms reproduce asexually during summer months, and only at the end of the season do they reproduce sexually, producing offspring that will overwinter to start the next season. The life history of one of these forms, the black bean aphid, is shown in Figure 9–1.

FIGURE 9–1 The life history of an aphid *(Aphis fabae)*, which alternates between a primary host, the spindle tree, and an array of secondary hosts such as the broad bean. The dark oval under the spindle tree symbolizes a fertilized egg; all aphids are females except for the two with male symbols over them. Heavy arrows indicate the obligatory migration; dashed arrows indicate facultative migrations. *(After Baker 1978.)*

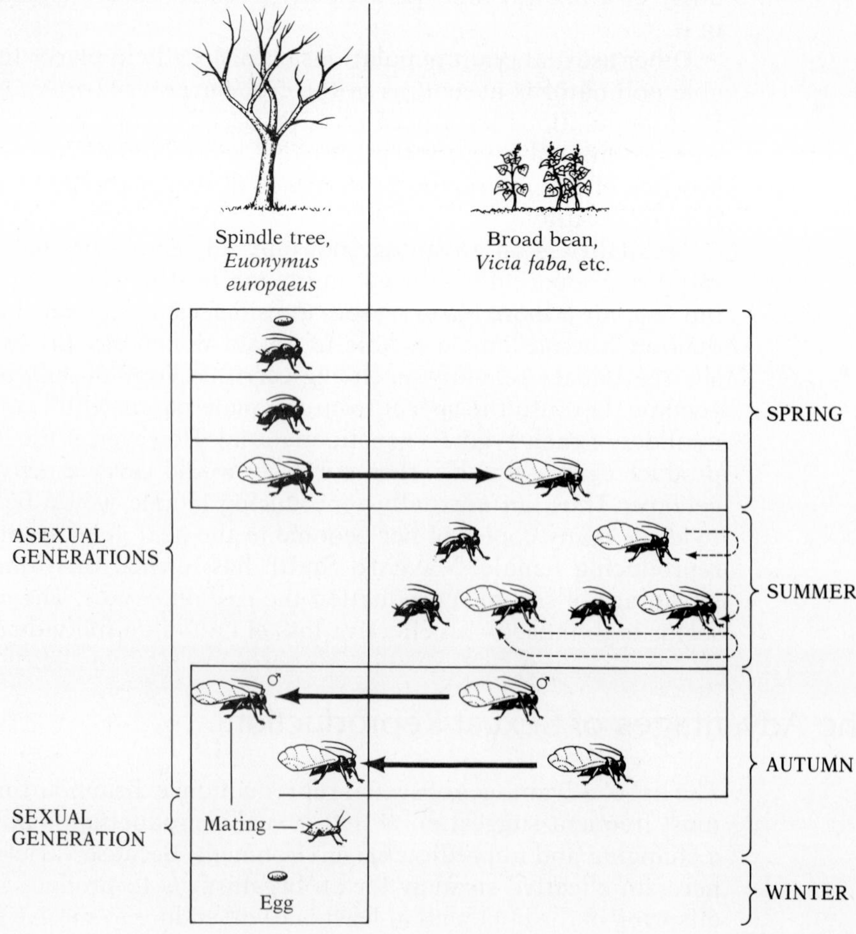

Nonetheless, the idea that sexual reproduction is advantageous in a fluctuating environment does not seem general enough to explain the enormous number of species that do reproduce sexually. Another conjecture (Bell 1982) is that sex has great advantages in heterogeneous *stable* environments, in which varied offspring outcompete asexual clones in all but those niches (which are relatively few) to which the clones are highly adapted.

Many investigators, in a long tradition, have sought ways in which sexual reproduction might be favored by group selection. Fisher (1930) originally suggested that the advantage to a species (that is, to a group) of sexually reproducing organisms is the faster production of individuals carrying *two or more* favorable mutations at different loci. For example, suppose A* and B* are both novel and favorable mutations that can arise at the A and B loci, respectively. In a sexually reproducing population, the mutation to A* can appear in one line and that to B* in another line; when individuals from these lines mate, they can produce offspring that carry both of these new genes.

This suggestion triggered a series of papers by population geneticists (Muller 1932, Crow and Kimura 1965, Maynard Smith 1968, and Karlin 1973), the main conclusion of which is that Fisher's original suggestion is often wrong. The reason Fisher's conjecture does not always apply is that mating brings the A* and B* genes together quickly, but it also moves them apart. If, for example, a parent happens to have both the A* and B* genes, its offspring may not, because the offspring's genotype will also depend on the other parent. In an asexual population, by contrast, once the A* and B* mutations have appeared in the *same* line, which will happen if one waits long enough, then mating will not dilute the line with this combination. Thus, the overall length of time for an asexual population to evolve completely to A* and B* may be about the same as that for a sexual population.

In sum, it is not clear that sexual reproduction really is an advantage to a species. Even if it were, the question remains as to how group selection could bring about the evolution of sexual reproduction in the face of the individual selective advantages of asexuality. Nevertheless, the idea that group selection is somehow involved in the evolution of sexual reproduction remains attractive because of evidence from systematics and taxonomy: No asexual lineage has ever produced a long evolutionary history (Stebbins 1950).

One suggestion on the ecological value of sex is a version of the "Red Queen hypothesis" (Van Valen 1973). Named for the character in *Through the Looking Glass* who had to run continually to stay in place, this hypothesis states that a species must continually evolve to survive in a world full of other evolving species. When this idea is extended to sex, the hypothesis is that competition and predator–prey relationships between sexual species keep sexuality going. For example, pathogens quickly "learn" to attack a genetically uniform host and wipe it out; only by continually changing its genotype can the host survive, forcing the pathogen to be sexual also if it is to be able to keep up the attack (Hamilton 1980). In fact, W. D. Hamilton has been quoted as believing "that disease will be found to be very important in the evolution of sex" (Maranto and Brownlee 1984, p. 27).

Whatever the answer to the dilemma of the evolution of sexual reproduction, it is important to remember that it is an evolved system—not a given. A major challenge to behavioral ecology is to understand the environmental circumstances that favor different reproductive strategies—not just asexual versus sexual, but such matters as hermaphroditism (male and female reproductive organs in the same individual) and differing ratios of the sexes.

SEX RATIO

The *sex ratio,* the proportion of males to females, is a fundamental feature of any sexually reproducing population. Like sex itself, it is an evolved characteristic of populations. A roughly 1:1 sex ratio generally is considered normal. One reason, of course, is that at meiosis the genes that determine sex segregate in a 1:1 ratio. However, because genes that can distort segregation are known (see, e.g., Sandler *et al.* 1959), the question arises as to why meiosis usually is not distorted. The answer was originally developed by Fisher (1930).

Fisher's Theory

Here is a simplified version of Fisher's argument. Suppose a population was made up of individuals genetically programmed to produce offspring in a ratio of two males to every female. In such a population, any mutant that produced, say, 1.5 males to 1.0 females would be favored. It would produce more of the rarer sex, which, in turn, would have a higher chance of mating than the commoner sex. Selection for sex ratio favors parents that produce more of the rarer sex, and this always pushes the ratio toward 1:1.

However, Fisher did not argue that the sex ratio should be even, only that the parental *investment* in males and females should itself be equal. This idea leads to some complex predictions, which seem well supported by the limited data available from actual populations. For example, in the European sparrowhawk, *Accipiter nisus,* females are twice as heavy as males at fledging, as shown in Figure 9–2. One might assume, therefore, that a female offspring requires twice the parental investment as would

FIGURE 9–2 Adult European sparrowhawks. (A) Male. (B) Female.

a male and that a parent fledging two males would make the same investment as one fledging a single female. Suppose that the sex ratio was 1:1 and that each pair fledged either two males or a single female. The pairs that fledged two males would have more grandchildren—for, on the average, the two male offspring would produce twice as many fledglings of their own as would the one female. Thus, the 1:1 sex ratio should not persist in the sparrowhawk, provided that differential parental investment occurred in the two sexes. Selection should favor matings that produce more of the "cheaper" sex (males) than of the more expensive (females), until the sex ratio shifted to 2:1. At that point, parents who had two male offspring would have, everything else being equal, the same number of grandchildren as those who had a single female offspring (for, on the average, only half of the males would find mates).

A sample of young sparrowhawks counted by Newton and Marquiss (1979) contained 1102 males and 1061 females. Does this ratio fail to conform to Fisher's theory? It does not, because it turns out that the weight differences are not indicative of differential investment. In fact, prior to fledging, the two sexes were fed the same amount. The females gain weight faster, but the males mature more rapidly (Maynard Smith 1978). Because the hawks feed their young for a time after fledging, we cannot be sure if differential investment exists; however, what is known supports a Fisherian prediction of a 1:1 ratio.

Work with grackles (a kind of blackbird), in which females are smaller at fledging, shows sex ratios with a bias toward females, conforming to Fisher's theory (Howe 1976, 1977b). However, in this case, one must assume that the weight difference accurately reflects differential investment, a fact that remains to be determined. More data should be gathered on a variety of organisms to test the differential investment theory (see, e.g., Charlesworth 1977, Charnov 1982). Still, it is difficult to see a reasonable alternative to Fisher's explanation of sex ratios.

Deviations from a 1:1 sex ratio that occur after the end of parental investment should not result in selection favoring parents that produce more of the rarer sex. For example, in the butterfly *Euphydryas editha*, adult males usually outnumber adult females by about 2:1, even though the genetic sex ratio is 1:1 and parents invest equally in male and female offspring (Ehrlich *et al.* 1984). The reason for the skewed ratio is primarily that the females suffer greater preadult mortality. However, the evolutionary advantage of producing a son is identical to that of producing a daughter.

To show this, let us suppose that butterflies mate only once, that no preadult mortality occurs in the males, and that 50% mortality occurs in the females (producing a 2:1 adult sex ratio). Thus, on the average, a son has a 50% chance of mating, because only half of the males will find a partner. A daughter will also have a 50% chance of mating, because although all surviving daughters are mated, only half of them survive to mate. The higher mortality of the females is thus directly balanced by their higher participation in mating because they are the rarer sex.

The analysis just outlined suggests plausible reasons why a 1:1 sex ratio should evolve if the parental investment in both sexes is equal. However, it does not explain why this state is most commonly achieved by two separate sexes combining gametes that have a 50% probability of producing each sex (Williams 1978). Why, for example, has a system not evolved that makes half of all females capable of producing only sons and the other half capable of producing only daughters? Why do such a variety of sex-determining mechanisms exist (see, e.g., Bull 1983)?

Biased Sex Ratios

Whatever the answer to these difficult questions, circumstances are known that lead to biased sex ratios, as when inbreeding is present (Hamilton 1967). In the extreme case, if a female knows that her daughters will be mated by her sons, she will maximize her reproductive contribution to future generations if she produces only enough sons to assure the fertilization of all of her daughters. One such extreme case has been described for the mite *Acarophenax tribolii* (Pyemotidae), which is *viviparous* (its eggs hatch internally and the young are born alive). A litter in *Acarophenax* consists of 1 male and 14 females. The male mates with all of its sisters while they are still inside the mother and then dies before it is born (Hamilton 1967). In another pyemotid, *Adactylidium,* six to nine offspring are produced viviparously, only one of which is male, and evidence exists that mating takes place before birth in this species also (Elbadry and Tawfik 1966). However, this situation may be accounted for by group selection, as in Colwell's study of hummingbird flower mites, which we described in Part 2.

Another circumstance in which the sex ratio is biased occurs when the probability of successful reproduction varies more from individual to individual in one sex than in the other (Trivers and Willard 1973). In many mammals, for example, males gather harems. In such situations, the chance of each female reproducing is much the same, whereas a relatively few dominant males do most or all of the reproducing—the less dominant males being out of luck.* Suppose that a higher variance in fitness is present for one sex than for the other and that a high parental investment produces a high fitness for the offspring. In that case, the sex ratio of the offspring of an individual should depend on that individual's ability to provide care. An individual able to provide excellent care should tend to produce offspring of the high-variance sex, on the theory that one of them will become a dominant reproducer. An individual unable to provide excellent care should tend to produce offspring of the low-variance sex, for they will almost certainly be mated wherever they end up on the fitness spectrum.

Data from polygynous mammal species—those in which one male normally mates with more than one female—including pigs, dogs, sheep, deer, and seals, fulfill this theoretical prediction. That is, sturdy females in good condition tend to produce sons; weaker females tend to produce daughters.

THE ENVIRONMENTAL DETERMINATION OF SEX

When the relative fitness of the two sexes is strongly influenced by environmental conditions and when an individual has little control over its environment, selection should favor environmental sex determination (Charnov and Bull 1977). For example, the "cleaner wrasse" (*Labroides dimidiatus*) of Indo-Pacific coral reefs normally occurs in social groups consisting of a single territorial male and a harem of several females. The wrasse feed on parasites living on other fishes. If the male is killed and a neighboring male does not take over the harem, the dominant female in

* Many societies are structured by a *dominance hierarchy* in which some individuals are physically dominant over others. The ranking is more or less permanent, and once it is established, by combat or other agonistic behavior, the society remains apparently at peace, because each individual knows its place. The first work on dominance hierarchies was done by Schjelderup-Ebbe (1922). For a fine overview, see Wilson (1975).

the harem rapidly changes into a male (Robertson 1972). Many other fishes of the wrasse (Labridae) and parrotfish (Scaridae) families show such *protogynous hermaphroditism*—functional females turning into males. In these fishes, fitness is related to size. Large males are the only males that are likely to breed, and they often can have access to several mates. Small males are out of luck. Small females, on the other hand, can breed. Thus, it seems that an optimal strategy is environmental sex determination. In this case, the best strategy is to remain female until conditions (such as the loss of a male or becoming large in size) favor a sex change.

A variety of nonhermaphroditic organisms show environmental sex determination. In some marine invertebrates, the male lives attached to the female. They produce planktonic (free-floating) larvae of indeterminate sex. Those that settle alone become females; those that settle on a female become males. In two genera of orchids, *Cycnoches* and *Catasetum*, plants that grow in bright sunlight become female and those that grow in shady spots become male. The reason that sunlight favors femaleness is unknown; probably it relates to the different costs of investing in seeds and pollen (Gregg 1975).

Why have not all organisms developed such a neat environmental sex determination system for maximizing fitness? Presumably, the reason is that costs are involved in retaining the ability to change. One cost might be that an early start at becoming one sex or the other produces a superior male or female, which places a premium on "deciding" early (Charnov and Bull 1977).

OPERATIONAL SEX RATIO

A final aspect of the sex ratio is critical to understanding the next chapter—how the environment shapes the evolution of mating systems. The sex ratio at the end of parental investment actually may bear little resemblance to the number of sexually active males and females that are accessible to each other at any time. As an extreme example, thousands of male insects or frogs may form assemblages in which females arrive one or a few at a time. In such a case, even though the sex ratio in the population is actually 1:1, the *operational sex ratio*—that of sexually ready males to fertilize females—will be strongly skewed in favor of males (Emlen and Oring 1977).

CHAPTER 10

Mating Systems: Who Mates with Whom, and Why

The female produces larger gametes—that is how femaleness is defined—and therefore usually has more at stake in each mating than the male does. Thus, the two sexes have related, but not identical, reproductive interests. This asymmetry is often enhanced when parental care is provided, because more often than not the female is the one who provides it.

Usually the male can mate more times than the female and has a smaller stake in each mating. Thus, being relatively indiscriminate is to its advantage. A mating with a less than optimum female may do little to reduce the chances of a better union shortly thereafter. In contrast, one mating may supply a female with the sperm it requires for *all* of its eggs. For the female, the wrong mate choice can be catastrophic. Hence, females tend to discriminate more than males (Bateman 1948, Trivers 1972).

This asymmetry has led Wilson to postulate that animals basically demonstrate *polygyny,* one male ordinarily mating with more than one female (Wilson 1975). Wilson suggests that *monogamy*—one male mating with one female—evolves from the more primitive polygyny when the environmental situation creates selection pressures that "operate to equalize total parental investment and literally force pairs to establish sexual bonds" (p. 327).

The most readily explicable situation, then, is polygyny. Males, as Bateman pointed out, show a higher variance in their breeding success than females. Because females normally can always get fertilized, they tend to be discriminating in their choice of mates. They are the *limiting sex*. Males, in contrast, cannot always get access to females. Usually they are the *limited sex*. Thus, males usually compete with each other for access to females—for example, by fighting over them (as do some frogs), or by gathering and defending harems (as do many mammals), or by defending attractive territories (many birds), or by entering into a communal sexual display (various insects, birds, and mammals).

WHAT DETERMINES THE KIND OF MATING SYSTEM?

One task of the behavioral ecologist is to discover the systems by which mates are chosen (Bateson 1983) and the environmental parameters that determine the evolutionary choices among such systems. Why, for example, do some male birds attempt to attract mates by defending individual territories containing resources, whereas others perform together in a sort of "beauty pageant" on *leks*, traditional display grounds? What ecological conditions lead to deviations from the primitive polygynous condition toward monogamy (the commonest condition in birds) or even *polyandry* (one female with several male consorts)?

The main ecological determinants of mating systems appear to be the distribution of resources in time and space (see, e.g., Emlen and Oring 1977). For example, suppose a bird species occupies an area in which the distribution of high-quality resources (say, insects to eat) is quite uniform. Suppose further that males occupy feeding territories throughout the area and that females remain in the male's territory. In such a situation, it would seem advantageous for females to seek out an unmated male rather than attempt to mate with an already mated male. That arrangement will tend to produce monogamous pairings, with the incidental consequence that the exploitation of resources will be rather evenly distributed. Furthermore, if the male aids in rearing the young, a female who mates monogamously will benefit from the undivided assistance of her consort in feeding and protecting the young. Still further, the lower density of individuals and nests in her territory will provide her with more food for her young and may reduce the chance of attracting a predator that might destroy the female or her brood.

Contrast this situation to one in which high-quality resources are patchily distributed. In that case, superior males will select and hold the territories richest in resources. Under those circumstances, once all of the males holding the richest territories are mated, an unmated female is presented with a more difficult choice. She can mate with a male on a less desirable site and still enjoy the advantages of monogamy, but at the price of exploiting a poorer base of resources. Alternatively, she can mate bigamously with a male on a rich territory, gaining the advantage of the

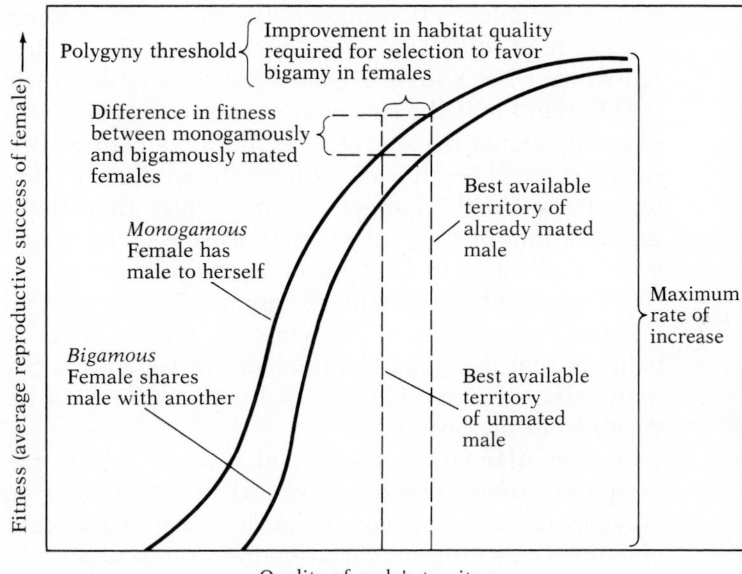

FIGURE 10–1 Graphic model of the conditions necessary for the evolution of polygyny. *After Orians 1969.)*

rich resources but accepting the disadvantages of sharing the territory with one or more other females.

At some point on a continuum, the difference between the high- and low-quality territories becomes sufficiently large that the balance of advantage and disadvantage shifts from monogamy to bigamy, as Figure 10–1 indicates. That point is known as the *polygyny threshold* (Verner and Willson 1966, Orians 1969; see also Weatherhead and Robertson 1977, Heisler 1981, Wittenberger 1981a, b).

If this set of suppositions is correct, it shows one way in which ecological factors can influence mating systems. Evidence of such effects appears in nature. Verner (1964) found that female marsh wrens (*Cistothorus palustris*) near Seattle, Washington, sometimes chose to mate with already mated males when bachelors were available. In addition, the number of females mated with each male was a function of the quality of the resources in the territory held by each male, the best of which contained something on the order of twice the resources of the worst—which normally were those of the bachelors. In contrast, European (winter) wrens (*Troglodytes troglodytes*), which are usually polygynous, are monogamous on islands of the northern British Isles, where the breeding densities and resource distribution apparently are such that no males hold marginal territories.

Other American passerine birds show a similar tendency toward polygyny when territories differ sharply in quality. Furthermore, in such situations, females that mate polygynously are successful as, or more successful, than those that mate monogamously. For instance, in a study of red-winged blackbirds (*Agelaius phoeniceus*) in Washington State, Holm (1973) found that the size of harems varied with the characteristics of the plant life in, not the size of, a territory. As Figure 10–2 illustrates, harems were larger in territories with scattered cattail (*Typha latifolia*) patches interspersed with channels of open water than they were in those with dense cattails or those in which bulrushes (*Scirpus* sp.) predominated. The number of young fledged per male and per female increased as the harem size increased from zero to six females. The range of productivity in marsh habitats, even those in the same vicinity, is large (Orians 1966). Holm believes that even small differences in vegetative cover in nesting areas may make a large difference in the availability of insect food and in security from predation.

In a similar study, Carey and Nolan (1975) correctly predicted that indigo buntings (*Passerina cyanea*) in an Indiana population would be polygynous on the basis of the diversity of habitats occupied (and the presumed concomitant variation in food resources). The lark bunting (*Calamospiza melanocorys*), another seed-eating, sparrowlike bird, also shows a high correlation between habitat quality and the occurrence of polygyny. In this case, the availability of shade for the nest site was the key variable, which predicted whether the male on a territory would be monogamous or bigamous (Pleszczynska 1978).

Finally, let us consider Crook's (1964, 1965) famous weaver finches. Most of the seed-eating savanna species, which have locally superabundant food resources, are polygnous. Here the males do not defend feeding territories, and the value of their help in feeding and defending the young may be small, because females may easily obtain the required food and the nesting situation provides protection against major predators. Thus, the disadvantages to the female of participating in a polygynous system (Verner 1964) are minimized, but she retains the possible advantage—being able to choose, by whatever means, to mate with the most desirable male regardless of his other commitments. On the other hand, among several insectivorous weaver species with dispersed, relatively scarce food resources and with the males defending feeding territories, 16 of the 19 species are monogamous.

We should note that polygyny is relatively rare in birds. Lack (1968) concludes that over 90% of bird species are monogamous. Verner and Willson (1966) similarly estimate that 95% of 291 North American passerine birds are monogamous, and of the remaining 14 polygynous species, 13 live in habitats (marshes, prairies, savannas) in which the density of resources, and their patchiness, are likely to be higher than in forests. In general, it may be that relatively few habitats have such variation in territory quality that a female is apt to increase her fitness by ignoring a bachelor in favor of the divided attentions of an already mated male on a superior territory.

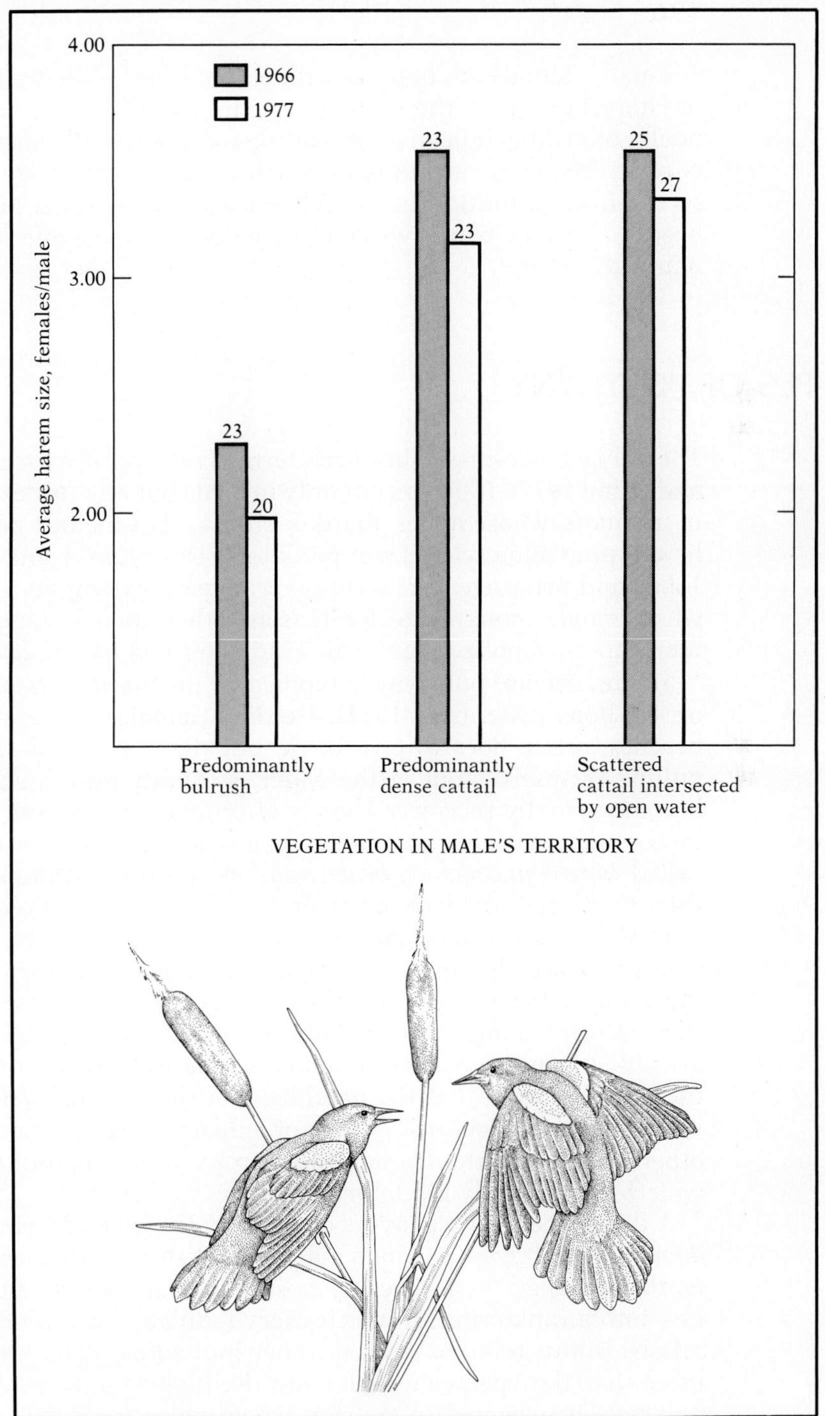

FIGURE 10–2 Relationship of average harem size in red-winged blackbirds *(Agelaius phoeniceus)* to the vegetation in the male's territory. Numbers over the bars are numbers of males. *(Data from Holm 1983.)*

Finally, let us note that the task of gathering evidence for the effects of territory quality on the mating systems of birds is hampered by the difficulty of making objective evaluations of quality. Measuring productivity is relatively easy, but measuring other factors is not—for example, how secure from predation birds feel in different habitats. Much more work needs to be done before we can be confident of the explanations just summarized.

TYPES OF POLYGYNY

The polygyny discussed has been termed *resource defense polygyny* (Emlen and Oring 1977). It occurs not only in birds but also, for example, possibly in marmots, whose males guard territories that include prime retreats for hibernating and escape from predators (Downhower and Armitage 1971, Johns and Armitage 1979). However, most polygyny in mammals occurs when females congregate for reasons other than mating and males attempt to monopolize them; this kind is termed *harem defense polygyny*.

Harem defense polygyny is typified by the mating system of eared seals or sea lions (Eisenberg 1981). Herds of females come ashore on rocks, beaches, or ice floes to give birth, but the number of sites suitable for pulling themselves out of the water is usually quite restricted. Females become sexually receptive shortly after giving birth, and they are mated by the males within whose breeding territory they occur. The males, often called *harem masters* or *beach masters,* patrol continuously, defending their territories and harems against other males (see Figure 10–3).

In some hooved animals, such as the African waterbuck, females and young stay together in small herds. During the mating season, these herds move into favored areas of habitat, where males have set up territories. Because the strongest males hold the most desirable areas, those males are able to control access to numerous mates (Jarman 1974). In this and the previous case, it is the patchiness of the environmental quality that permits males to control groups of females (harems) that aggregated for other reasons—in the sea lion the scarcity of hauling-out sites, and in the waterbuck probably predator defense.

A third type of polygyny recognized by Emlen and Oring (1977) is *male dominance polygyny,* in which males establish a dominance hierarchy; that is, they arrange themselves by hostile interactions—either overt or subtle—into a rank order in which every individual has a dominant or submissive status relative to every other individual. Females then select the mate that they perceive as having the highest status. Male dominance polygyny is expected to develop when males have little opportunity to control high-quality environments or herds of females. For example, resources might be either superabundant or so scattered in space or unpre-

FIGURE 10–3 Harem defense polygyny in elephant seals at Ano Nuevo, on the California coast south of San Francisco. Note the enlarged proboscis on the male in the foreground (which is showering itself with sand), which gives these seals their name. Males will often fight over females by bashing each other with their snouts. The male at the left-hand edge of the photo is watching his harem of four females, two with young, that are visible to his right. *(Photo by Paul R. Ehrlich.)*

dictable in time as to be difficult to defend. Alternatively, population density might be so high that energetic territorial defense becomes impossible.

The idea that defense of territories could be related to cost-benefit considerations is supported by investigations of natural populations. Birds have been especially well studied. For example, golden-winged sunbirds (*Nectarinia reichenowi*), in the mountains of Kenya, are nectar feeders—the ecological equivalent of New World hummingbirds. The birds defend feeding territories on stands of the mint *Leonotis nepetifolia* when nectar levels are below 2 μl per flower. When the levels are high, 6 μl (microliter) per flower, the birds are not territorial. At the higher levels, there is ample nectar for everyone, and so birds derive no compensatory gain for energetically defending a set of flowers to monopolize their output (Gill and Wolf 1975).

Iiwis (*Vestiaria coccinea*) Hawaiian honeycreepers, which extract nectar from the red flowers of the myrtaceous tree *Metrosideros collina*, similarly defend territories only when blooms are relatively scarce; however, they do so above a threshold density that makes it energetically possible for an individual to increase nectar availability by defending an area. Above that threshold, the efficiency of territorial defense decreases as the number of flowers per foraging area increases; when blooms become common, territorial behavior does not occur. Thus, at one end of the scale, resources are too scattered, and at the other too abundant, to make territoriality a viable strategy. When the resource distribution is unsuited for territoriality, the possibility of male dominance polygyny arises.

Ecological factors also control the degree of male dominance polygyny that occurs. If, as in certain frogs and butterflies or in *Scatophaga* flies, females must produce young within a brief time period in order to maximize the offspring's chances of survival, they tend to converge on males all at once, and breeding is explosive (Wells 1977, Odendaal *et al.* 1985). Male dominance is expressed in terms of a direct battle for access to the females. In this situation, the operational sex ratio (OSR) often is strongly skewed toward males, because receptive females at any given moment are much scarcer than males, and so the degree of polygyny may be low.

Many insects form male aggregations, often on hilltops, to which females putatively go to mate (see Thornhill and Alcock 1983 for a review). However, in the only case in which the mating success of males in and out of aggregations was actually measured, the males in a hilltop aggregation of the checkerspot *Euphydryas editha* were less successful than those patrolling the slopes below (Wheye and Ehrlich 1985, Ehrlich and Wheye 1986).

A fascinating case in which females do travel to locations where the males are found has been described by Parker (e.g., 1970b, 1974) in the dung fly, *Scatophaga stercoraria*. Females must lay their eggs in fresh cow pats, an ephemeral resource in which the maggots develop. Males remain in the vicinity of the pats, awaiting the arrival of females, which seek out the dung only when they are ready to lay eggs. Thus, the OSR is roughly 4:1. Males obtain mates by copulating with newly arrived females, by ousting a copulating male and taking his place, or by ousting a male that is guarding an ovipositing female and copulating with her. *Sperm precedence* occurs in *Scatophaga;* that is, sperm from later matings tend to displace those from earlier ones. About 80% of the eggs laid by a female after a second copulation will be fertilized by the second male.

The cow pat thus becomes a miniature arena in which males struggle over females. The behavior of the flies varies with their density. If it is low, copulation usually takes place on the pat. If it is high, the male carries a female off the pat to copulate, reducing the chance of a takeover by other males (Parker 1971). Copulation would be faster on the pat, where the temperature is higher, but this advantage is outweighed by the chance

FIGURE 10–4 Contact guarding by the damselfly, *Ischnura gemina.* After mating, the male remains in tandem with the female while she deposits eggs in aquatic vegetation, ensuring that his sperm are used to fertilize her eggs. *(Photo courtesy of John E. Hafernik, Jr., San Francisco State University.)*

of losing the female. Similarly, the male spends some 17 minutes guarding the ovipositing female—time that could be spent finding another mate. Parker (1970b) has shown, however, that this is time well spent in protecting some 80% of the male's genetic investment from cancellation by a subsequent male because of sperm precedence. Similar mate-guarding behavior is found in damselflies and dragonflies (Odonata), as shown in Figure 10–4, where males may remove sperm from previous matings from the female reproductive tract before inserting their own (see, e.g., Alcock 1979, 1982, Sherman 1983, Thornhill and Alcock 1983, Waage 1979a,b, 1984).

At a pond near Oxford, England, the breeding season of the European common toad, *Bufo bufo,* lasts for about 2 weeks, at the end of March and the beginning of April (Davies and Halliday 1979). During that time, males tend to stay at the pond, whereas the females arrive, spawn—by laying their eggs while in amplexus with a male that fertilizes the eggs as they enter the water—and then depart. The OSR, like that of *Scatophaga,* is about 4:1, and the result is the same kind of "scramble competition" among males for mates, as Figure 10–5 illustrates. Although only about one-fifth of the males studied bred successfully, almost two-fifths of those that did copulate managed it by forcibly taking over females already in amplexus with another male.

At the opposite end of the spectrum is *prolonged breeding,* which occurs when no strong selection pressure exists for females to reproduce at a specific, restricted time of year. Females then will not become receptive synchronously, and at a given time those ready to mate will become a

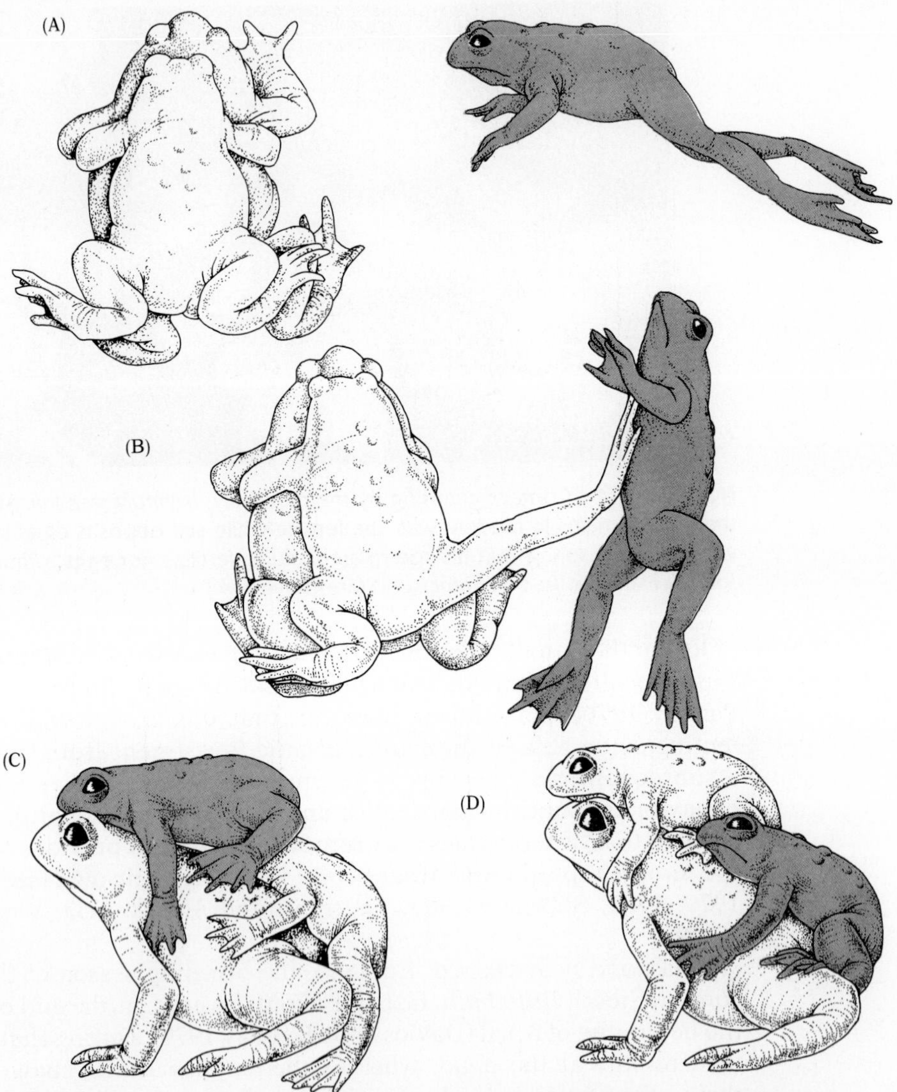

FIGURE 10–5 Scramble competition for mates in the European common toad. (A) An unpaired male launches an attack on a pair. (B) The paired male kicks him away. If the attacker succeeds in grasping the pair, he will attempt a takeover by either holding on to the front of the female and pushing the other male off (C) or squeezing in between the members of the mating pair (D). The attacking male is darker in all drawings. *(After Davies and Halliday 1979.)*

severely limiting resource for males. The OSR is then more strongly skewed toward males and gives rise to the opportunity for some males to monopolize a resource, females, that is patchy in time rather than space.

A close relative of the common toad is the prolonged breeding natterjack, *Bufo calamita*. In the shallow, warm ponds in which it spawns, breeding can be stretched over 4 months. Males conserve their energy, remaining in one place and attempting to lure ripe females with loud calls (A. Arak, in Vines 1982). A female seems to select her mate by his call, which is a measure of his size, age, and possibly his fitness, for the largest males are the oldest and thus have managed to survive longest.

The exact ecological conditions that constrain the breeding season of *B. bufo* but permit an extended season in *B. calamita* are not known. However, the strong skew of the OSR toward the males, and the long period during which female *B. calamita* ready to spawn are available, make herding or struggling over females an impractical strategy. However, males can compete with each other for the best calling stations and breeding territories. Thus, the natterjack shows elements of resource defense polygyny as well as male dominance polygyny.

The phenomenon of communal male display, in which females approach and then choose a mate from among competing male "advertisers," occurs not just in bullfrogs but also in several groups of insects (flies, dragonflies, beetles, cicadas), fishes, birds, and mammals. It is commonly, but not exclusively, found in prolonged breeders. The degree of polygyny is normally much higher than in explosive breeders, because in explosive breeders a few males cannot monopolize the numerous simultaneously receptive females.

Perhaps the most famous such communal displays are those put on by male birds of various species of the grouse family (Tetraonidae). Males of the sharp-tailed, sage, and black grouse, the prairie chicken, and the capercaillie compete against each other on leks—calling, displaying plumage, inflating air sacs on their necks, and going through repetitive ritualized dances. The dominant male on the display ground normally holds the central territory. He is also the most successful at mating females, and, as Figure 10–6 shows, the mating success of subordinate males is a function of their position in the dominance hierarchy (Kruijt and Hogan 1967, Wiley 1973, 1974). This situation leads to great variance in male reproductive contributions. For example, in 2 years of observation of a lek of the Eurasian black-grouse (*Lyrurus tetrix*), Kruijt and Hogan found that about one-third of the males were involved in some three-fourths of the matings. In another study, Koivisto (1965) reported that a single male performed 17 of 24 observed copulations.

Males of other promiscuously breeding tetraonids, the spruce, blue, and ruffed grouse, do not display on leks but on dispersed sites. In the blue grouse, these sites are defended territories; in the other two species, definite site boundaries are not defended, though male spruce grouse show agonistic behavior when other males approach. Whether males holding more desirable sites or emitting more seductive calls have greater reproductive success is not known. However, no durable bonds seem to be formed between males and females (Wiley 1974).

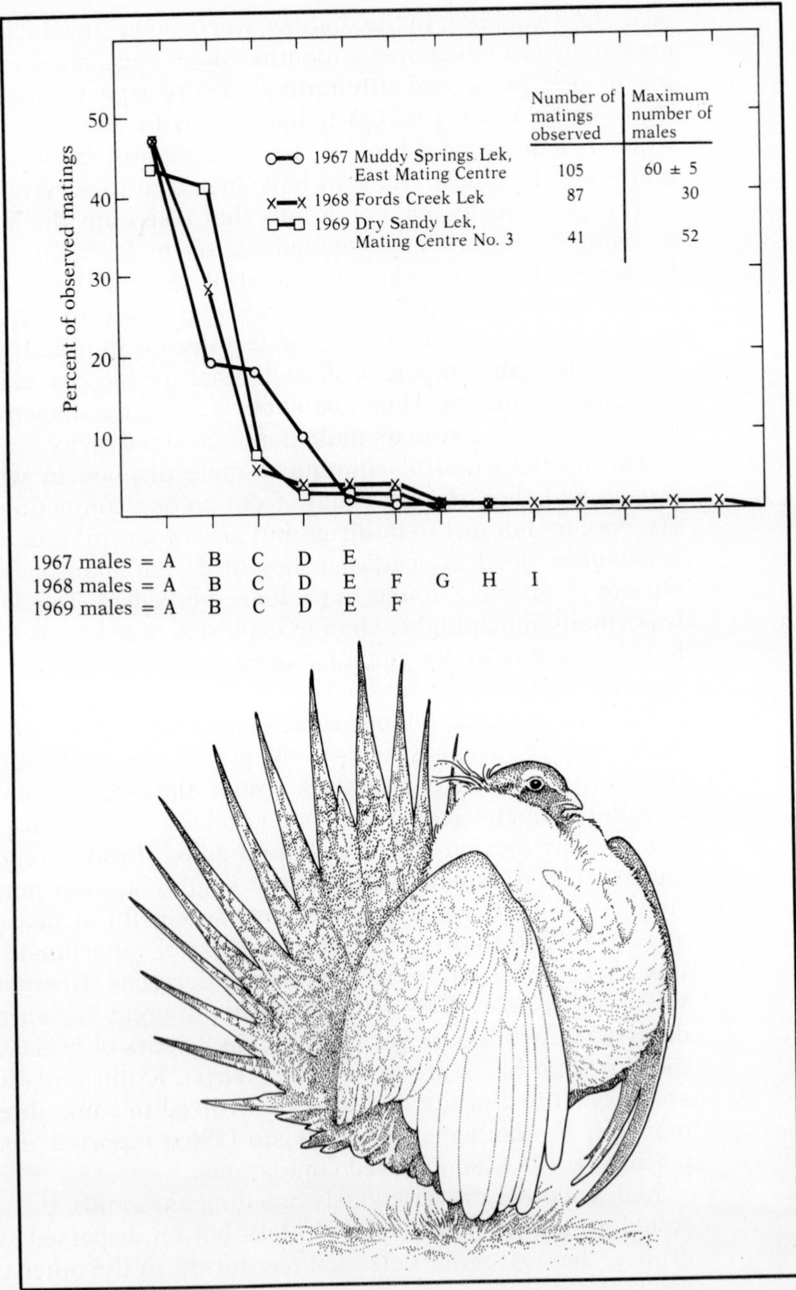

FIGURE 10–6 Graph of reproductive success versus position in the dominance hierarchy in the sage grouse. Letters indicate position of male in hierarchy: A = highest, B = next highest, etc. *(After Wiley 1973.)*

In still other tetraonids, males and females form associations. For example, in the red grouse, pairs stay together for about 6 months of the year, and males rarely have more than one mate. Males of this group sometimes aid in caring for the young, wheras those of the promiscuous species never do. The monogamous males may devote their energies to ensuring the survival of their young rather than to attempts at more copulations. They also can be certain of the paternity of a given female's clutch, whereas a promiscuous male probably cannot. In addition, in these monogamous species, males may breed in their first year; in the others, first-year breeding by males is rare or absent. Thus, as with the toads, in this group of birds we see closely related organisms with widely divergent mating systems. Such situations can be thought of as the results of natural experiments, challenging behavioral ecologists to determine the environmental variables responsible for the outcomes.

Wiley (1974) gives only one clear correlation of mating system with environment in these birds. Among the promiscuous species, males of those that live in open habitats display on leks, whereas most of those that live in forests tend to remain dispersed for their displays. Possibly this situation is due to the difficulty of detecting the approach of predators in the restricted visibility of the forest. The exception is the capercaillie, which lives in mature coniferous forests. However, it is the largest grouse, and thus presumably has fewer potential predators, and the males stay farther apart than those of the other species that also display on leks.

Other environmental factors, such as the pattern of food distribution, do not seem related to the mating system in the grouse. For example, hazel grouse and ruffed grouse live in similar habitats that presumably present individuals with similarly distributed food resources. However, the hazel grouse is monogamous, whereas the ruffed grouse is promiscuous with dispersed males—that is, males not on leks. Clearly, more investigation is needed.

Predation probably has played a role in the evolution of delayed reproduction of males in the species that form leks. Because a young, inexperienced male's chances of achieving dominance on the lek, and thus of successful copulation, would be very low, displaying itself the first year probably is not worth the risk of becoming conspicuous to a predator. Rather than enter the competition at all, it might find that a better strategy is to wait an additional year until the odds have improved of being successful on the lek. Female grouse, in contrast, have nothing to gain by delaying reproduction for a year, because all females produce one clutch of eggs each breeding season. This general pattern may hold for most species in which males display, whether or not leks are used.

Although many of the details of the ecological determinants of social structure in grouse remain to be worked out, the general outlines of an answer are emerging. Environmental influences operating directly, such as exposure to predation, and indirectly, such as physiological and other constraints on male and female size, interact to produce the observed diversity.

Some of the most interesting cases of communal displays have been described in fishes of the family Cichlidae. In Lake Malawi in Africa, different cichlid fish species all set up leks on the sandy lake bottom in a few meters of water near shore (McKaye 1983). Species that are morphologically identical but different in color build characteristic nests. For example, *Cyrtacara* ("*Haplochromis*") *eucinostomus* builds a mound of sand, as illustrated in Figure 10–7, whereas *C. argysoma* uses a flat area (McKaye 1983, McKaye and Louda, ms). Females school above the leks, which may be monospecific or include males of five or more different species. The females "sample" the available males and choose to spawn with one or more. Females may lay as little as a single egg with a given male. After fertilization, the eggs are scooped up by the female and carried in her mouth until they hatch, a form of behavior known as *mouthbrooding*.

This lek-centered social system is thus both polygynous and polyandrous. It contrasts to that of related cichlids in Lake Jilóa in Nicaragua, where monogamous pairs are formed before territories are established, spawning occurs on the substrate (mostly rock surfaces), and both adults guard the territory, the eggs (which are much smaller than those of the

FIGURE 10–7 A male cichlid fish *(Cyrtocara eucinostomus)* and its nest on a lek in Lake Malawi. The lek occupied by this species is some 4 km long at a depth of 3–9 m. Courting and mating occur on the lek in the morning, but females leave with the eggs, and males leave the lek in the evening to feed on zooplankton in deeper water. As many as 50,000 males may display simultaneously on this lek, making it the largest such assemblage ever reported. *(After McKaye 1983.)*

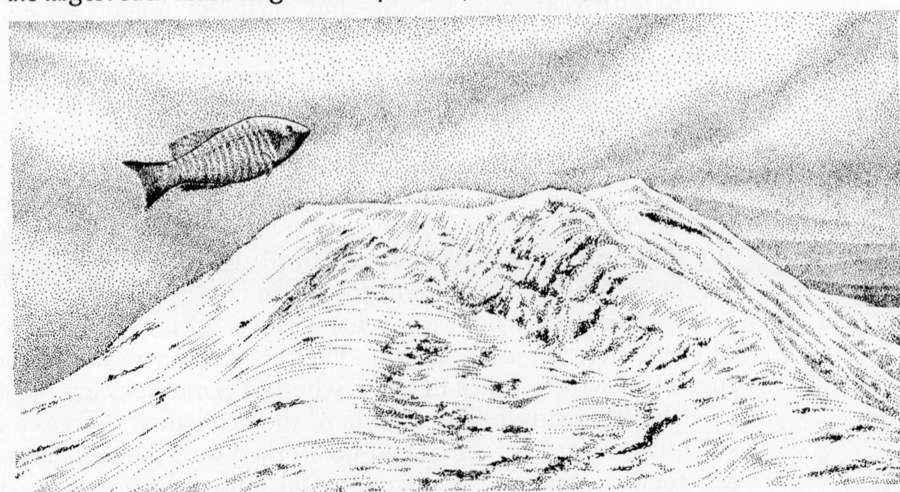

mouthbrooders), and the young (McKaye 1977). What environmental factors could account for such a difference? The conventional wisdom is that mouthbrooding is a response to predation pressure. Indeed, two fascinating groups of predators have evolved to prey on the large, energy-rich eggs of mouthbrooders.

However, the investigator responsible for the work on both systems thinks that the mouthbrooding could have evolved at least partly in response to the intense competition in Lake Malawi—where some 500 or more species of cichlids are present (McKaye 1983 and pers. comm.). Mouthbrooder territories, on the average, are smaller than those of the substrate spawners, and they are occupied for only 4–6 weeks. In contrast, the larger substrate territories must be defended for 12 weeks or more.

Some aspects of the mating systems are clearly a response to predator pressures. For example, cichlids in both lakes sometimes guard the young of other species along with their own young (McKaye and McKaye 1977, McKaye 1981b). The young that are adopted are selected carefully by size, and it appears that the act of thus enlarging the school containing their young enhances the survival probabilities of their own offspring.

Such behavior is not restricted to the family Cichlidae. McKaye and Oliver (1980) found predatory catfish caring for cichlid young along with their own. Why should interspecific brood care that is beneficial to both species occur in fishes and not in birds? A major reason is that in most birds the adults forage for the young, paying a high energetic price for provisioning each one. Adoption is expensive. In contrast, young fishes do not exact a high foraging cost from their adoptive parents; they feed on their own, often on plankton brought in by currents. Predation on a bird's nests also normally destroys the entire brood, whereas an attack by a predator usually results in the loss of only part of a fish brood. Thus, the only time the young of another species is found in a bird's nest is when it is a brood parasite such as a cuckoo (as we will describe in Part 4), which only gains, while the adoptive parent host only loses.

Perhaps the strangest feature yet found in any system of parental care was discovered by McKaye (1977) in Lake Jilóa, in which an herbivorous species, *Cichlasoma nicaraguense,* defends the young of the largest predator in the lake, *C. dovii,* as shown in Figure 10–8. This interaction, which will be examined in greater detail in Part 5, illustrates that community ecology, in this case complex predator–prey relationships, can dramatically modify patterns of parental care.

Some mammals show lek behavior quite similar to that of grouse and cichlids. For example, the Uganda kob, a medium-sized antelope, was once common in the nation for which it was named. Buechner and Schloeth (1965) found kob societies to be organized around leks several hundred meters across (see Figure 10–9). The males held small breeding territories about 15–30 m in diameter on the leks. The most successful males were

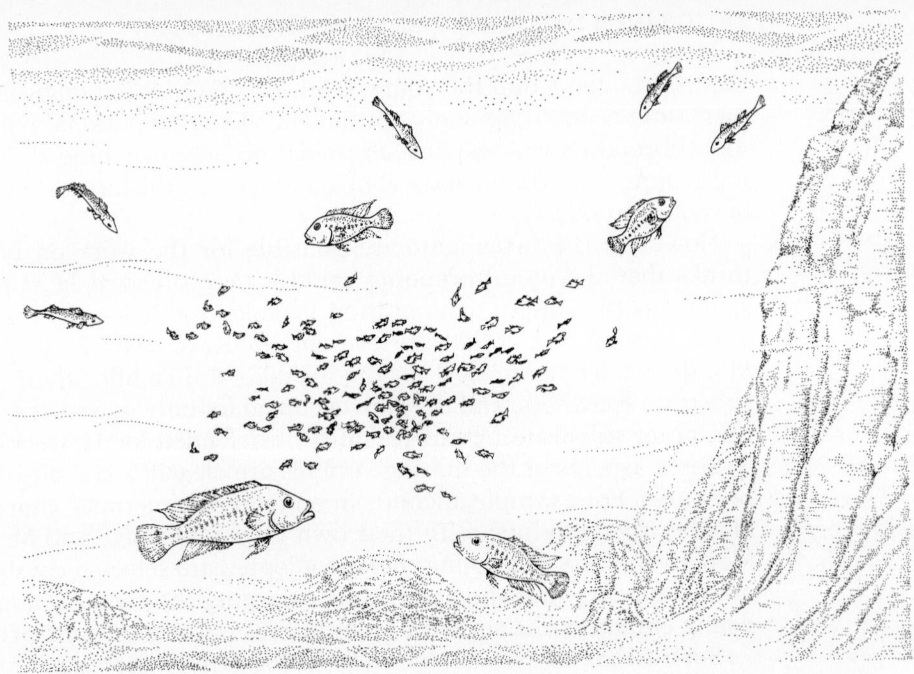

FIGURE 10–8 An herbivore's defense of the young of one of its predators. *Cichlasoma nicaraguense* guard *C. dovii* by remaining above or to the side of the school. The parental *C. dovii* stay below their brood. The torpedo-shaped fishes are one of the principal predators on older fry. *(After McKaye 1977.)*

FIGURE 10–9 Uganda kob on a lek. The male in the center is nuzzling the genital region of a female. *(After Buechner and Schloeth 1965.)*

those that occupied the more central territories of the lek, in which occur most of the successful copulations. For example, 5 out of 18 males, those on the most sexually active territories, accounted for 54% of the copulations (Floody and Arnold 1975). Only negligible sexual activity occurred on the peripheral territories of the lek, and males in the central territories defended them vigorously (Floody and Arnold 1975).

However, the distribution of matings over males in this example may be more uniform than the extreme differential mating on the lek would indicate. First, turnover of males on the most active territories is relatively rapid, and intruding males do sometimes manage to mate on the territories of others.

We should note a final way in which males in polygynous species promote their mating success. In addition to attempting to monopolize females, males may kill offspring of their mates that have been fathered by other males. Lion prides, for example, are based on females, which do most of the hunting. Coalitions of males, which often remain together for life, compete for ownership of prides (Packer and Pusey 1982, Pusey and Packer 1983). Sometimes the members of a male coalition are relatives; sometimes they are unrelated. Coalitions are much more likely to be able to monopolize the females of a pride than is a lone male (Bertram 1976, Bygott *et al.* 1979). Thus, a male benefits by belonging to a coalition, even though he would be more successful reproductively if he controlled a pride by himself (Koenig 1981).

On the average, a coalition of males controls a pride for only about 2 years, which puts a premium on reproducing rapidly (Bygott *et al.* 1979). When a new group takes over a pride, its members chase out large cubs and kill most of the small cubs, even though the lionesses may defend them fiercely. However, once the cubs are dead, the females come into estrus within days or weeks and mate again; if the cubs had lived, the lionesses might not have been sexually active again for as long as 18 months (Pusey and Packer 1983). Thus, infanticide increases the reproductive potential of the new males. Infanticide by newly dominant silverback males has been reported in mountain gorillas (Fossey 1983). Presumably it occurs for the same reason.

Infanticide is also apparently rather widespread in ground squirrels (Sherman 1981, McLean 1983), although its exact extent is not yet clear (G. R. Michener 1982, Sherman 1982). Here, as in lions, immigrant males are primarily responsible, but the way they benefit is less obvious, for female ground squirrels will not come into estrus and breed again in the same season. The primary advantage to the male that kills another male's offspring may be a mate in better condition the next season, because the mother will be able to spend more time feeding and less time on the lookout to protect her (now absent) offspring (McLean 1983). Other possible benefits include decreased competition for resources for the infanticidal male and its offspring.

POLYANDRY

Do ecological situations exist that would promote polyandry? A prerequisite would seem to be emancipation of the female from the duties of rearing young; therefore, polyandry obviously is unlikely to occur in mammals in which the females must both carry and nurse the young. In birds, however, males occasionally take over the brooding of eggs and rearing of young. Emlen and Oring (1977) believe that polyandry evolves when there is a long breeding season and destruction of clutches is frequent owing to high predation pressures. A premium is then placed on the physiological ability of females to lay replacement clutches, and the male takes over the energetically expensive chores of caring for the brood.

Polyandry is quite rare. In birds it has been documented in less than 1% of species in nature, the most thoroughly studied of which are in the order Charadriiformes (shorebirds) (Jenni 1974). For example, males of the jacana (*Jacana spinosa*)—in a marsh in tropical Costa Rica defend small territories in which they do most of the nest building, as shown in Figure 10–10. Females defend larger territories that encompass several

FIGURE 10–10 A male of the polyandrous northern jacana *(Jacana spinosa)* displayed next to its nest built on a raft of vegetation. The bird's large feet spread its weight so that it can stand and walk on aquatic vegetation, such as lily pads (leading to the common name *lily trotter*).

male territories and form simultaneous pair bonds with several males. They may copulate with all of their consorts in a single day, and they also help each male defend his territory against the others. The females refrain from sex with their mates only during the period of incubation and the first 6 weeks or so of the chick's life. If a predator destroys a clutch, copulation is resumed almost immediately, and a new clutch is laid within a few days (Jenni and Collier 1972).

Interestingly, infanticide by *females* has recently been suggested by observations of jacanas. Females ousting other females from their territories then apparently destroy the offspring in the nests, which the jacana males attempt to defend against the new female, just as lionesses attempt to defend their cubs against new males (Stephens 1982). The male presumably then breeds with the new female. However, the actual killing has not been observed, only the empty nest after a male defense against a female that had driven off his previous mate. If this behavior is confirmed, this incident will be the first time that infanticide has been reported as a female reproductive strategy (Hrdy 1979).

Another example of polyandry was discussed in the previous section; female mouthbrooding cichlids in Lake Malawi often spawn with several males.

COOPERATIVE BREEDING

In some birds, more than one pair cooperate in nest building, depositing a clutch of eggs in the nest, and/or helping to feed the brood (Lack 1968, Koenig 1981, Koenig and Pitelka 1981, Ligon and Ligon 1982). A wide variety of behaviors is included under this heading (Emlen 1978, Emlen and Vehrencamp 1983). At one extreme, the Florida scrub jay lives in groups consisting of a breeding pair and the young of that pair produced in the preceding year or two (Woolfenden 1973, 1975, Woolfenden and Fitzpatrick 1984). The young birds bring food to the hatchlings and help protect the nest from predators. At the other extreme, the groove-billed ani (*Crotophaga sulcirostris*) live in groups of up to four monogamous pairs, the females of which lay all their eggs in a communal nest. Members of the group share in the incubation and care of the young.

At one time, the situation of the ani was thought to be the ultimate in cooperation, but studies by Vehrencamp (1976, 1977) show that it is not. It turns out that females compete with each other to have the most eggs cared for. Females often visit the nest before they have laid any eggs and frequently roll out eggs laid previously by other females. After a female has laid, this behavior ceases, as females cannot identify their own eggs.

There is a linear dominance hierarchy in the female ani and the dominant females lay their eggs last, biasing their proportional representation in the communal clutch. Subordinate females have strategies for count-

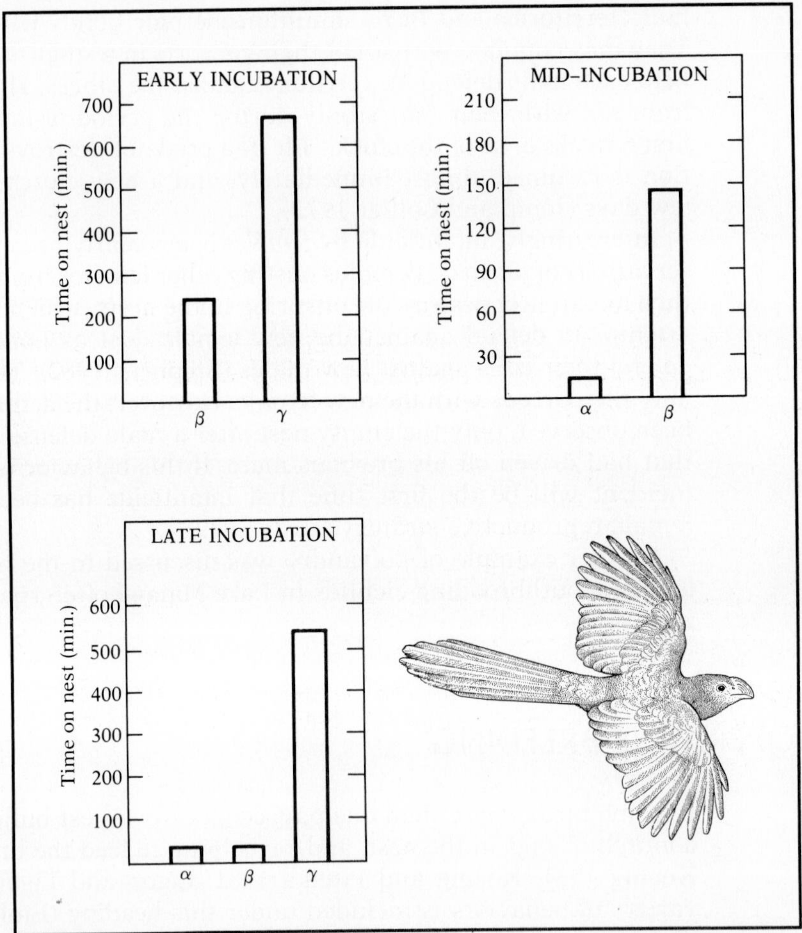

FIGURE 10–11 Graphs of time contributed by α, β, and γ (α dominant) female groove-billed ani at different points in the incubation period. *(After Vehrencamp, 1977.)*

ering this bias, but they are only partially successful, and the dominant females maintain a higher ratio of eggs actually incubated to eggs laid even though they contribute less to the incubation process (Figure 10–11).

The environmental conditions conducive to the evolution of cooperative breeding are thought to be diverse. Most emphasis has been placed on saturation of the habitat in the primarily benign environments in which cooperative breeding (and related group territoriality) is generally found. When survivorship is high, all areas available for territories are soon occupied, and openings for young birds are scarce or nonexistent (Ricklefs 1975). The opportunity is then present for nonbreeders unable to maintain territories to add to their inclusive fitness by aiding their breeding relatives, but the cost-benefit calculations are complicated (Emlen 1978, Gaston 1978).

Limitations on suitable nesting areas seem to be a prerequisite for the evolution of cooperative breeding in species that nest colonially (Brown 1974). This situation occurs, for example, in bee-eaters—birds of the genus *Merops*—of African savannas (Fry 1972, 1977, Emlen 1978). The species most intensively studied nest in unlined cavities at the end of meter-long tunnels bored in cliffs, normally near streams. Fry (1972) used observation trenches, which were dug down from the cliff top and gradually enlarged toward the cliff face until all the nest cavities were revealed. He was able to band all birds in a colony and follow all the activities at their nests.

The social system in bee-eaters is very complex. For example, three levels of relationships between individuals have been found in the white-fronted bee-eater, *Merops bulockoides* (Emlen 1981). It lives in large colonies, with between 20 and 120 in a nest. Primary breeding groups of two to seven individuals form "friendships" with specific members of other breeding groups and more tenuous bonds with all individuals in the colony. A satisfactory explanation of the ecological conditions leading to the evolution of all the features of this complex system remains to be developed.

Cooperative breeding also occurs in mammals. For example, in black-backed and golden jackals (*Canis mesomelas* and *C. aureus*), some offspring serve as "helpers at the den." These are offspring that, instead of breeding themselves, aid their parents in the provisioning and guarding of the young (Moehlman 1979, 1983, Montgomerie 1983).

MONOGAMY

Against the background of the preceding sections, monogamy—which many nonbiologists tend to regard as the natural state of affairs for higher organisms—might seem a rather rare situation, forming an unstable intermediate condition between polygyny and polyandry. However, as we have already noted, monogamy is the most common mating system in birds (Lack 1968, Orians 1969). Though relatively rare, it also occurs among a diverse set of mammals, and it is known in most other vertebrate groups.

No single explanation can account for all cases of monogamy. For birds, for instance, the consensus is that some 90% of the species are monogamous primarily because their young require the care of both parents and the quality of male territories does not vary sufficiently to cross the polygyny threshold. Not surprisingly, then, monogamy predominates in *altricial* birds—those in which the young are helpless when they hatch. Polygyny and polyandry are largely restricted to *precocial* birds—in which the young are able to move about and feed on hatching. In cases in which precocial birds are monogamous, care by the male parent—provisioning or predator protection—seems to be important to the survival of the young.

However, migratory ducks may be monogamous for another reason. Females apparently are vulnerable to predators when rearing young, so that the sex ratio tends to be biased toward males (Aldridge 1973). Male migratory ducks do not defend territories but defend females. In the process, the female is freed of harassment by unmated males, and some evidence suggests that, among other benefits, this situation may increase the feeding rate of females during the period of egg laying (Ashcroft 1976).

In mammals, monogamy seems to occur both when male territories are relatively uniform and when resources are sparse. In this situation, the costs of polygyny to the female may include food shortage, and presumably those costs outweigh any benefits derived from consorting with an already mated male (Clutton-Brock and Harvey 1978). In other words, the situation is below the polygyny threshold. In some gibbons, mated females aggressively stop unmated females from copulating with their mates (Tenaza 1975, 1976). In this case, the ecological basis of the behavior would seem to be that either resources are too scarce or predation too prevalent for the penalties of mate sharing to be borne by the female.

THE ECOLOGY OF HUMAN SEXUALITY

Now that we have sampled a variety of animal mating systems and their ecological correlates, it is appropriate to consider briefly the ecology of human sexuality.

The human mating system is unique among those of mammals. Females of other mammals, including all other primates, have a sharply defined *estrus*—a period of sexual receptivity or "heat"—during which they advertise their readiness for fertilization and availability for copulation. In contrast, human females conceal their fertilizability—not only from males but, to a large degree, from themselves. They are more or less continuously receptive, and as a result *Homo sapiens* show year-round, nearly constant sexuality, the results of which pervade all human societies.

The reasons for the evolutionary loss of estrus is the central mystery of human sexuality, and some serious barriers stand in the way of solving the mystery. One is the enormous plasticity of human behavior under the influence of *culture*—the accumulated nongenetic information that each society passes on from generation to generation. This plasticity not only makes the identification of genetic variation in sexual behavior, if any, extremely difficult; it also influences the judgments of investigators—who are always either males or females typically raised in male-dominated societies.

Another barrier is lack of information on the ecological situations in which the human mating system evolved. That evolution occurred over several million years after humanity diverged from the estrus-retaining great apes. Not enough time has gone by—only about 400 generations—since the invention of agriculture for selection to have made substantial genetic changes in the mating system. Therefore, we can be sure that the environment in which that system is now operating in most societies is not the one that controlled the genetic portion of its evolution.

It seems clear that loss of estrus is of no advantage to the male; presumably it evolved because it was advantageous to females to conceal their fertilizability. Symons (1980) gives two very speculative scenarios for loss of estrus. The first is based on the observation that female chimps in heat were more successful in begging meat from males. Females in ancestral human groups might have been better nourished if they had been continuously receptive to males. The second scenario is that marriage arrangements are mostly made by males, but a female's fitness may be increased if she is fertilized by a male other than the one designated by the social system. By concealing her time of ovulation, a female makes it more difficult for her husband to monitor her activities and increases the chance that she can be inseminated by other males. Symons himself admits that both scenarios are probably wrong, but they are certainly interesting. Note that the first of his scenarios could be part of the cause of marriage as a custom, whereas the second could be the result of marriage as a custom.

An enduring controversy exists over the degree to which male and female attitudes toward sex are the result of biological or cultural evolution. Symons reviews the evidence and concludes that human males and females have important innate differences in sexual behavior that can be traced to the differential investment in offspring that is typical of mammals. For example, males often tend to seek sex with more partners than females do and to seek younger partners. Seeking more partners makes evolutionary sense because males can increase their genetic contribution to subsequent generations by mating with many females, whereas females cannot increase theirs by mating with many males. Seeking younger partners can be accounted for by the male's attempting to father children with females in prime childbearing condition, whereas females find other criteria in a mate—for example, the ability to command resources—more important to the successful rearing of their young.

That an explanation "makes evolutionary sense," however, does not mean that evolution actually occurred that way. Nor should any particular moral value be placed on doing "what comes naturally."

THE ECOLOGY OF PLANT SEXUALITY

Most terrestrial plants and some marine algae show a variety of sexual systems partly analogous to animal mating systems. For example, in some species, each individual is hermaphroditic and bears perfect (bisexual) flowers. Other species are also hermaphroditic because flowers of both sexes—male flowers that produce pollen and female flowers that bear ovules—are found on the same plant. The latter kind of plants are termed *monoecious. Dioecious* plants, on the other hand, have male and female flowers on different plants (in which case the entire plant may be called male or female). To make things more complicated, some species have two kinds of individuals, either male or hermaphroditic, and other species have individuals that are either female or hermaphroditic. Further, in the jack-in-the-pulpit *Arisaema triphyllum,* a perennial herb (see Figure 10–12), an individual does not flower as a small plant; rather, it reproduces as a male when it becomes larger and then as a female when it

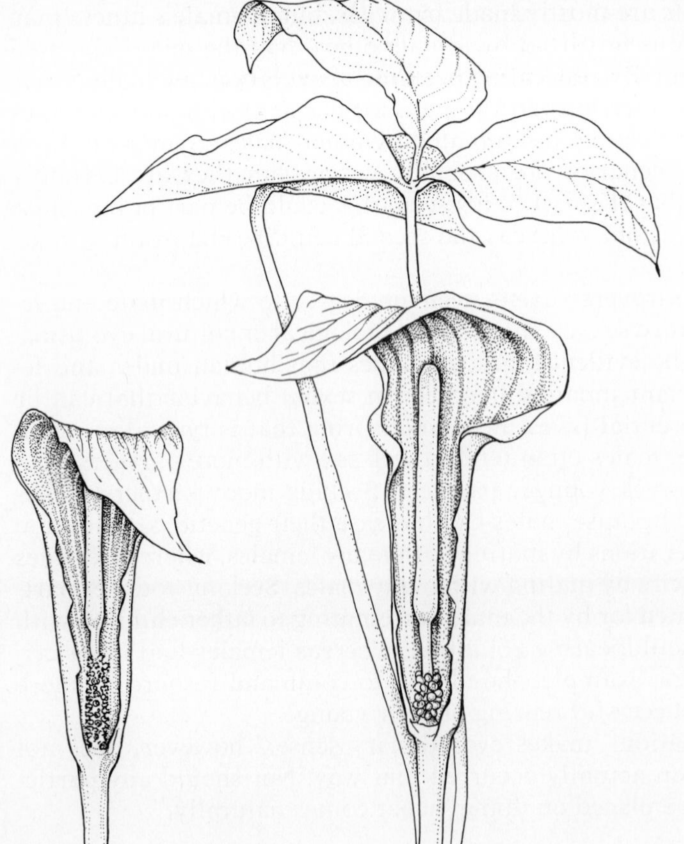

FIGURE 10–12 Jack-in-the-pulpit, showing different-sized male and female plants. *(After Bierzychudek 1982.)*

grows larger still. Individuals declining in size may go through the sequence in reverse (Bierzychudek 1982).

Recall that flowers do not produce eggs or sperm. Rather, flowers are structures of the sporophytic generation that produce male and/or female gametophytes, which in turn produce eggs or sperm. Thus, the usage of *male* and *female* to describe flowers is not exactly parallel with that in animals, but for the purpose of this discussion it is close enough.

The ecological questions about plant mating systems are quite similar to those asked about animals. The various systems do not occur uniformly throughout the plant kingdom. For example, dioecy is relatively rare, being found in something on the order of 5% of flowering plants (Bawa 1980). It seems to occur disproportionately in tropical plants, those that are perennial, and those that bear fleshy fruits. What environmental pressures have caused some plants to evolve dioecy?

This question traces all the way back to Charles Darwin, who raised it in 1877 in a book entitled *The Different Forms of Flowers on Plants of the Same Species.* Most biologists have focused on the advantages of outcrossing (reduction of inbreeding).*

Another pressure toward dioecy might be the evolution of pollination by large vertebrates that do not visit the flowers primarily for nectar but get their reward by consuming floral parts or pollen (Cox 1982). For example, the Samoan vine *Freycinetia reineckei* is largely dioecious, producing spikes of tiny flowers clustered in *inflorescences* that are themselves flowerlike, just as the "flower" in a daisy is actually an inflorescence composed of many tiny flowers. Most plants produce inflorescences containing flowers of only one sex, but occasionally plants are monoecious and produce inflorescences containing both male and female flowers, as shown in Figure 10–13.

F. reinecki is pollinated primarily by a large bat, the "flying fox" *Pteropus samoensis,* and a starling, *Aplonis atrifuscus,* that are attracted to the sweet, fleshy bracts (petal-like structures associated with the inflorescences) and to the pollen and the waxy material in which the pollen is embedded. Cox found that male and bisexual flowers suffered much greater damage than did female flowers. This situation appears to be a result of the feeding behavior of the bat, which devours the bracts of all inflorescences it visits and also eats the flower spikes of male and hermaphroditic inflorescences. In the latter, the female structures tend to be badly damaged, because they are closely associated with the pollen-bearing parts. For example, in one survey, 96% of the male inflorescences and 69% of the hermaphroditic ones were destroyed. In contrast, only 6% of the female inflorescences were destroyed.

* However, other possibilities exist, including more efficient resource allocation (Freeman *et al.* 1976, Bawa 1980, 1982, Givnish 1980, Herrera 1982), ability to utilize different resources and avoid competition between the sexes, and reduction of seed predator impact (Janzen 1971) if male and female plants are separate.

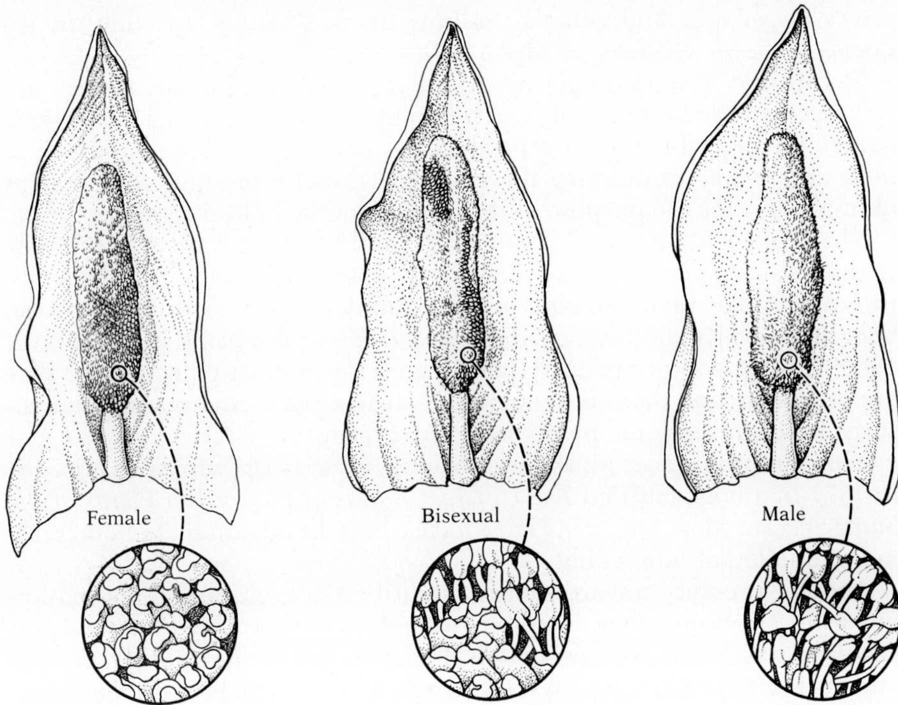

FIGURE 10–13 Diagram of the *Freycinetia* flower structure. *(After Cox, 1982.)*

Thus, one can see a possible evolutionary reason for the predominance of dioecy in *Freycinetia*. The plants with male inflorescences do not suffer fitness loss from the devouring of their flowers by the bat, for in the process the bat's face becomes coated with pollen, which it may then carry to a female inflorescence. If it visits a female, it will fertilize the flowers there while, in all probability, doing minimum damage to them. In contrast, while picking up pollen at a bisexual inflorescence, the bat will often destroy most or all of the female flowers at that inflorescence. If survival of female flowers in that inflorescence is not high enough to compensate for the smaller load of pollen picked up by the bat, the plant producing bisexual inflorescences will be at a selective disadvantage, and dioecy will be maintained.

However, why has *Freycinetia* not simply evolved the capacity to produce both male and female inflorescences on the same plant? The answer in this case appears to lie primarily in the architecture of the plant: Each vine can produce only a single inflorescence each year.

The example of *Freycinetia* only hints at the complexity of the problem of understanding how the environment has shaped the evolution of plant sexual systems. Plants, in contrast to animals, have developed an incredible array of sexual systems that can vary within every taxonomic level from the species (as we have seen in *Freycinetia*) on up. The same plant genus or family, for example, will often contain both self-fertilizing and outcrossing species, and the outcrossers may have evolved different systems for preventing self-fertilization. Few generalities seem widely accepted, in spite of extensive investigations and speculations (see, e.g., Vuilleumier 1967, Stebbins 1974, Lloyd 1980). One reasonably accepted notion is that outcrossing makes adaptation to changing environments easier, but that it must often be foregone by pioneer species or those that colonize by long-distance dispersal. Understanding the environmental factors shaping plant mating systems remains an important challenge to ecologists.

CHAPTER 11

Other Social Interactions: From Sexual Selection to Cultural Evolution

This chapter begins with a final topic related to sex: the ways natural selection operates within and between sexes. It then deals with a variety of nonsexual social interactions, all of which may be important to understanding the ecology of different groups of organisms.

SEXUAL SELECTION

Sexual selection has two basic forms—intrasexual and intersexual. In *intrasexual selection,* some members of one sex have more offspring than others because they compete with one another for opportunities to mate with the other sex. In *intersexual selection,* members of one sex create a reproductive differential in the other by preferring some individuals as mates. Usually, but not always, competition among males produces intrasexual selection, whereas females choosing among males produce intersexual selection.

The evolution of mating systems leads to situations in which sexual selection is bound to occur. Therefore, the environment, by influencing the evolution of mating systems, also indirectly influences the evolution of, for example, the enormous tails in male widowbirds (Anderson 1982), the bizarre plumes (see Figure 11–1 and color Plate K) and behavior of

FIGURE 11–1 This magnificent bird of paradise *(Diphyllodes magnificus)* is a common species in New Guinea. The male has a brown crown and nape. The back is a glossy deep red and the rump orange. The tail is black, and the wings are marked with glossy yellow-orange. The breast shield is deep glossy green, with a central band of glossy blue-green. The female is olive-brown on top and grayish below. *(Redrawn from a Gould painting.)*

male birds of paradise (Diamond 1981), the large size of female jacanas (which, let us recall, are polyandrous and in which the female presumably does the selecting), the calls of male frogs, the antlers of bull elk, the large size of male elephant seals, and other features that are involved in determining who mates with whom. Of course, ecological conditions help determine the strength of sexual selection, which in turn influences the evolution of mating systems.

Recently, Hamilton and Zuk (1982) proposed that, at least in birds, sexual selection may allow the choosing of mates that are more fit in characters other than those directly involved in mate attraction itself. Bird populations can carry high blood parasite loads. In these populations, one important aspect of the fitness of individuals—their resistance to the current strain of the parasite—is reflected in the vigor of their sexual display. Hamilton and Zuk found that a highly significant correlation appeared in North American passerine birds between the incidence of chronic blood parasites and the development of secondary sexual characters such as plumage brightness and variety and complexity of male song.

Sexual selection is a fascinating and controversial subject at present; understanding and documenting it thoroughly is a major challenge for evolutionary ecologists. The kinds of investigations that must be done on a wide variety of organisms are typified by the already classic work of T. H. Clutton-Brock, F. E. Guinness, S. D. Albon, and their colleagues in the Large Animal Research Group at Cambridge University. For a dozen years, they have been intensively studying a population of red deer (*Cervus elaphus*) on Rhum, a 13- by 14-km island just south of Skye, off the west coast of Scotland (Clutton-Brock *et al.* 1982; see also the demography of these deer discussed in Chapter 5). For the first time in a large animal species, the lifetime reproductive success of individual stags (males) and hinds (females) was recorded.

These data permitted determination of a crucial parameter, the *variance* in reproductive success. As would be expected in a polygynous species, the variance in lifetime reproductive success of stags was greater than that of hinds, as Figure 11–2 makes clear. It was not as great, however, as might have been expected from variance *within* seasons. In a given year, a few stags were quite successful, but success rates changed with age, and a few years later these same stags were replaced as reproductive leaders by others, thus reducing the lifetime variance among stags. Some hinds, in contrast, were consistently successful at reproduction throughout their lives; others consistently failed. This disparity increased the variance among hinds. Because both patterns—age-related reproductive success in males and consistent year-to-year reproductive differences among individual females—are common in polygynous mammals, Clutton-Brock *et al.* (1982) believe that short-term studies tend to overestimate the strength of sexual selection.

Another important finding of this research on the red deer is that the environmental factors affecting reproductive success are different for stags and hinds. For example, body size is crucial for stags, which battle over hinds—battles that the heavier males tend to win. Maternal investment during the first year and growth rate are the ultimate factors influencing stag reproduction. In contrast, the ultimate factor in hind success is the quality of the resources available, determined primarily by the hind's home range, which she inherits from her mother because the hinds tend to remain in groups with their mothers and sisters.

This difference, Clutton-Brock *et al.* (1982) point out, explains why hinds invest more in male calves during the first year, as evidenced by their suckling them more frequently. However, after the first year, hind investment in female offspring is higher, because daughters continue to use their mothers' home ranges and thus compete with them for resources. This behavior apparently roughly balances the maternal investment in the two sexes of offspring, preventing the evolution of the biased sex ratio that would follow from asymmetrical investment (as discussed in Chapter 8).

The red deer study highlights the relationship between sexual selection and natural selection. Differential *variation* in the reproductive success of the stags and hinds, the result of intersexual selection, interacts with dif-

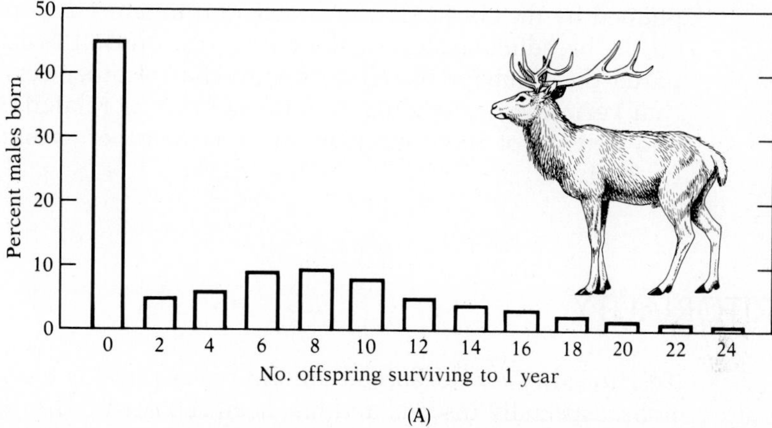

(A)

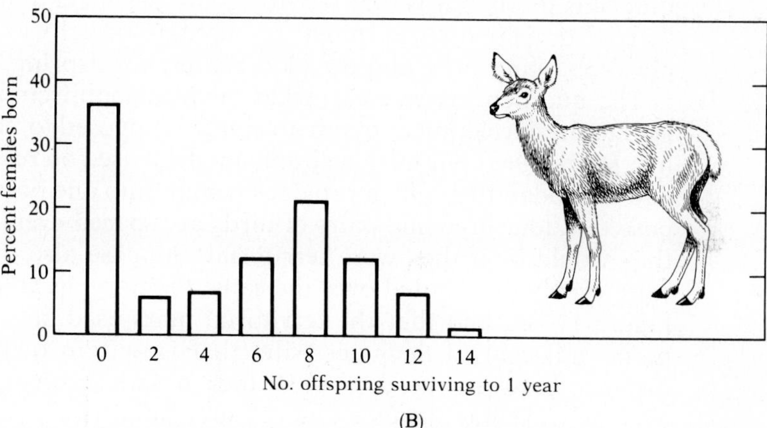

(B)

FIGURE 11–2 Variance in reproductive success in red deer stags (A) and hinds (B). Note that almost one-half of the stags and one-third of the hinds never successfully reproduced. *(After Clutton-Brock* et al. *1982.)*

ferent environmental influences on stag and hind reproduction (natural selection) to produce, among other things, the observed sexual dimorphism—stags are about 1.7 times heavier than hinds. Clutton-Brock points out (in Lewin 1982) that, in the red deer, sexual dimorphism would be produced even if both sexes had equal variance in reproductive success, because selection would still favor large males more than large females. It would not evolve, regardless of stag–hind differences in the variance of reproductive success, if body size itself were not selected for.

Clutton-Brock *et al.* (1982) believe further that, in species in which striking sexual dimorphism does not occur, the environmental factors operating on breeding success in the two sexes will prove to be similar. The general positive correlation between dimorphism and polygyny is explained by the common occurrence of different environmental influences on the breeding success of the two sexes. Clutton-Brock *et al.* have compared the behavior of red deer with that of other cervids and conclude that variation in breeding systems of deer "is related to the distribution and density of food supplies, as it is in other groups of vertebrates" (p. 304).

TERRITORIALITY

Territoriality is widespread among vertebrates, is found in some arthropods, especially insects, and has been reported in limpets. The literature on the adaptive significance of territoriality is substantial.* The theme that runs through it is that territories are defended to monopolize or protect one or more of three things: resources (especially food, but also hiding places, basking sites, and the like), mates, or offspring.

The question arises as to what environmental circumstances should lead to the development of territoriality as opposed to, say, colonial breeding. Horn (1968) put forth a simple model, based on resource distribution and dependability, which provides insight into one possible answer. Suppose that four breeding pairs of birds are spaced evenly over an area, as they would be if they were territorial. Suppose also that food resources were evenly distributed over the same territory, as shown by the dots in Figure 11–3A, and that the territorial pairs used only the four resource points adjacent to their nest site (that is, within their territory). If the distance between the dots is D, then a little geometry shows that the average traveling distance to a resource point is

$$\sqrt{\left(\frac{D}{2}\right)^2 + \left(\frac{D}{2}\right)^2} = 0.71D$$

On the other hand, if the birds nest colonially, as diagrammed in Figure 11–3B, and share all the resource points, the average distance traveled to food is

$$\frac{4\sqrt{(D/2)^2 + (D/2)^2} + 8\sqrt{(3D/2)^2 + (D/2)^2} + 4\sqrt{(3D/2)^2 + (3D/2)^2}}{(4 + 8 + 4)} = 1.50D$$

Thus, if the distribution of the food is uniform and predictable, and the energy budget for travel is a critical variable, one might expect territoriality to evolve. On the other hand, if the food resource is clumped and

* See, for example, Noble (1939), Allee *et al.* (1947), Lack (1954), Tinbergen (1957), Schoener (1968), Brown and Orians (1970), Wilson (1975), and Itô (1980).

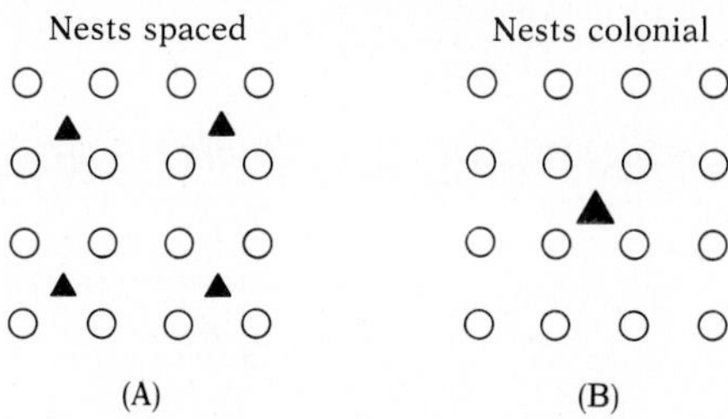

FIGURE 11–3 Two diagrams of cost-benefit territoriality. Nests shown as triangles; food as circles. See text for explanation. *(After Horn 1968.)*

predictable, one might expect a colony to be established near it. If the resources are clumped but unpredictable (so that all of the resource points must be continually investigated), the colonial species will travel an average of 1.50*D*. However, the territorial species would have to go an average of 1.93*D*. In this case, coloniality should be favored.

Horn's simple but elegant cost-benefit model considers only the cost of moving to the food and the benefit of obtaining it in determining the spacing pattern of the breeding birds, and yet it has proven surprisingly powerful. For instance, most seabirds forage for moving schools of fishes—notoriously clumped but unpredictable resource points—and most seabirds are, as predicted, colonial nesters.

The way in which variations in territorial behavior fit into an environmental context is shown nicely by the varied social organization of arctic sandpipers (Pitelka *et al.* 1974). As Figure 11–4 demonstrates, they have evolved two kinds of strategies for dealing with the variable and unpredictable insect food supplies of their arctic and subarctic breeding habitats: conservative and opportunistic.

One conservative strategy involves establishing territories that are of relatively constant size from year to year. For example, throughout Alaska, *Calidris alpina* defends territories that are large enough to provide a pair with sufficient food even in lean times. Low population density alone does not account for the large size of the territories on the tundra near Point Barrow on the north coast. The territories are strongly defended, and the removal of territorial males results in their prompt replacment by others that presumably are either nonterritorial or hold inferior territories (Holmes 1970). Apparently, under the rigorous conditions of the north, exclusive use of large areas is required if the territorial birds are to get enough food. Furthermore, in more southerly localities in the state, with a longer growing season and a richer and more dependable food supply, territories are smaller and breeding density is higher.

In contrast, opportunistic breeders seek out high-productivity, high-risk locations such as lowland marshes for their territories. They trade the chance, for example, of being flooded out by summer rainstorms—runoff in the tundra is rapid and flooding is common because the frozen ground has almost no water-absorbing capacity—for the dense insect fauna of the

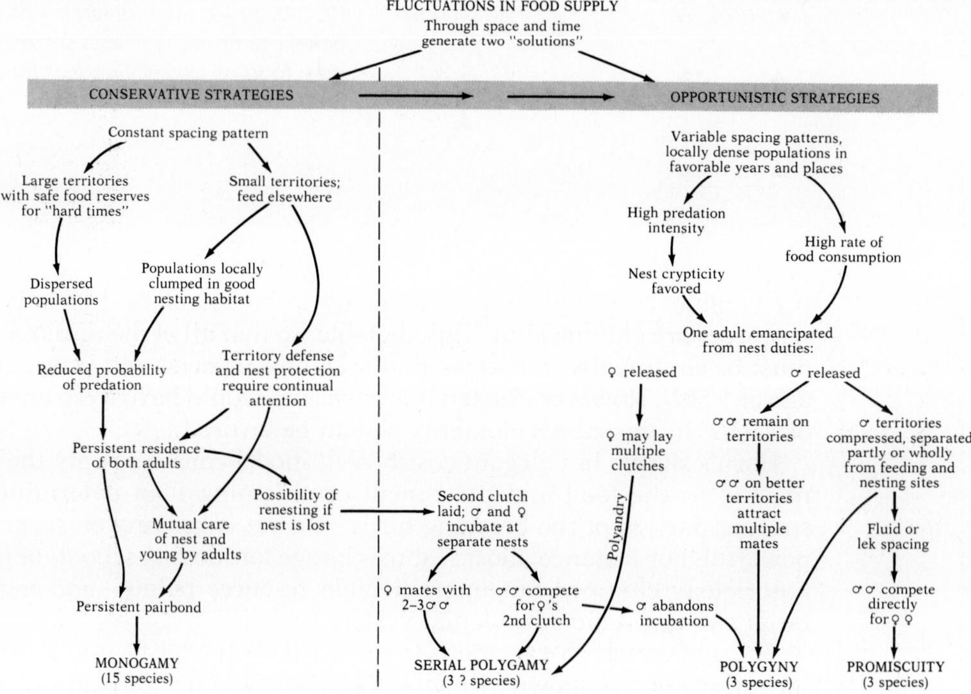

FIGURE 11–4 Model for the evolution of sandpiper social systems. Different environmental conditions (top) and different strategies lead to four different kinds of social systems (see text). *(After Pitelka* et al. *1974.)*

marshes. As shown in Figure 11–4, this behavior tends to clump the nesting territories, to produce high selection pressure for cryptic behavior (the high density of nests tending to attract predators), and to emancipate one adult, presumably to lessen activity around the nest and to lower pressure on the concentrated resources. This situation can lead to lek behavior, with male territories not encompassing nesting or feeding sites, or to polygyny, with the males holding superior territories doing most of the breeding.

For the most part, territories are defended only against conspecifics. This situation, of course, is eminently sensible in the case in which territories are used to protect access to mates. It is also usually a reasonable strategy with respect to resources. Defense of territory requires expenditures of energy and taking of risks. Individuals most likely to consume the same array of resources as the territory holder are conspecifics, so they are the most important to exclude from the territory. For example, as Figure 11–5 shows, hyena clans defend large territories against hyenas of other clans in the Ngorongoro Crater in Tanzania (Kruuk 1972). Not only do the different clans hunt the same prey species, but members of one clan attempt to join members of another clan feeding on their kill. When

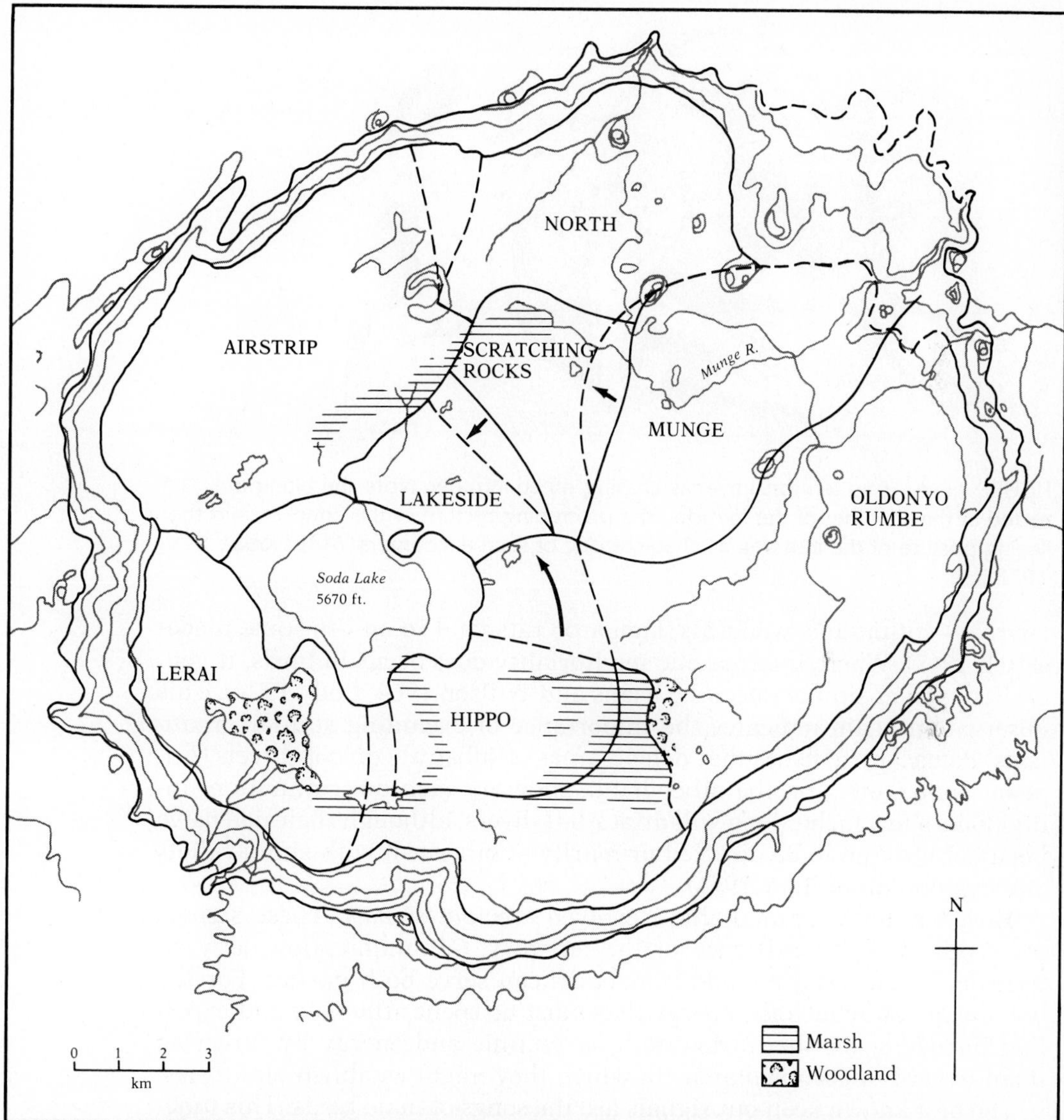

FIGURE 11–5 Hyena clan territories in Ngorogoro Crater. Dotted lines show uncertain boundaries; arrows show long-term boundary shifts. *(After Kruuk 1972.)*

such intruders are discovered, they are viciously attacked and driven off (see Figure 11–6).

When resource use of different species does overlap strongly, the species may be so different that they cannot exclude one another. Such is the case, for example, in the foothills of the Californian Sierra, where owls, hawks, bobcats, coyotes, and rattlesnakes share many of the same food re-

FIGURE 11–6 Two territorial hyenas chasing off an intruder. Note the biting directed at the shoulder of the intruder, the protruding rectum of the attacker, and the fleeing posture of the intruder, all characteristic of such encounters. *(After Kruuk 1972.)*

sources—cottontails, woodrats, kangaroo rats, and so on—in some places (Fitch 1947). Where interspecific territoriality does occur in birds, it generally involves similar species (Orians and Willson 1964, Cody 1974). This observation again indicates the importance of excluding similar organisms. Pomacentrid fishes that defend areas of substrate on coral reefs (*Eupomacentrus*, etc.), on the other hand, show no such discrimination. Individuals a few inches long will attack 6-ft divers, although their defensive behavior appears to be directed primarily at other fish with similar diets and egg predators (Low 1971).

Most territorial animals have evolved *keep-out signals.* These signals warn others of the existence of the territory. Communication between territory holders and potential intruders can serve both parties. For the holders, it can reduce the energy that must be spent attacking and expelling intruders; for the intruders, it saves time and energy by directing them toward unoccupied areas in which they might establish territories.

The best known keep-out signals are the songs of male birds. This function of bird songs has been elegantly demonstrated by J. R. Krebs (1971, 1977). He removed territorial male great tits (*Parus major*) from their territories and found that they were rapidly replaced by others that had not previously had territories. However, when he set up a loudspeaker that played great tit songs, the replacement did not occur; the song alone was sufficient to defend the territory. Moreover, Krebs showed that the song was important, not just the noise; a loudspeaker simply whistling did not keep other males out.

Individual great tits sing a variety of songs as they move about their territories, and loudspeakers that played a varied repertoire were more effective "guards" than those that just played one song. Krebs hypothesized that the variety persuaded territory seekers that an area was already heavily occupied and caused them to stay out. This suggestion is supported by data showing the lower reproductive success of great tit males holding territories in densely populated areas, although the idea is very controversial (see, e.g., Searcy 1983). Krebs christened his idea the *Beau Geste hypothesis* after the famous story of that name, in which attacking tribesmen were persuaded that a Foreign Legion fort was fully manned because the few surviving defenders had propped the corpses of their comrades in the embrasures and ran from one to the next firing their rifles.

THE SOCIOLOGY OF ATTACK AND DEFENSE

Obviously, one of the main reasons for social relationships among organisms is sexual reproduction, and we have seen how sexual behavior and environmental factors (especially resource distribution) seem to interact to shape mating systems. Additional environmental influences that favor social behavior involve obtaining food and avoiding being obtained as food. These influences may, or may not, have a relationship to the mating system.

Group Foraging

Many organisms, from lions, wolves, and wild dogs to surgeon fishes and army ants, seek their food in groups. Many others—weasels, most tigers, most fishes, ladybird beetles—are solitary hunters. What environmental factors or phyletic constraints have led to this difference?

Carnivores that live in groups also tend to be more successful at hunting if they hunt cooperatively. As Figure 11–7 makes clear, two cooperating hyenas, for example, are much more successful at obtaining wildebeest calves than are single hyenas, primarily because the wildebeest mother is ineffectual at protecting the calf against more than one hyena. Lions also are more successful when they hunt together than when they are alone (Schaller 1972, Caraco and Wolf 1975). Two lions seem to be the optimal hunting group. Although more lions may have a higher chance of getting the prey, each must also share the rewards of the hunt with the others. Therefore, as Figure 11–8 shows, the amount of meat obtained per hunter drops with hunting group sizes greater than two. Groups consisting of more than four lions in the wet season do less well than lone hunters.

These relationships between group size, hunting success, and food intake, which presumably hold in some form for most predators, provide

FIGURE 11–7 Group size and success of hyenas hunting wildebeest calves. *(After Kruuk 1972.)*

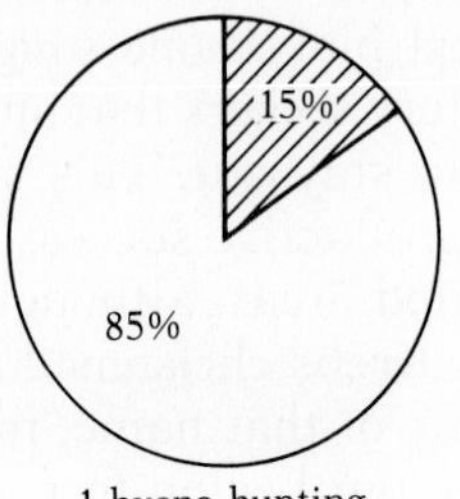

1 hyena hunting
(74 observations)

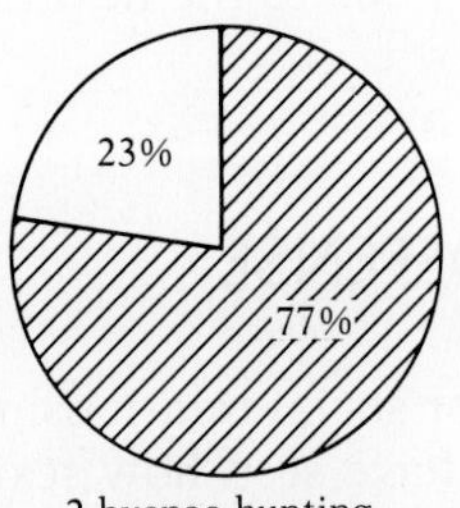

2 hyenas hunting
(18 observations)

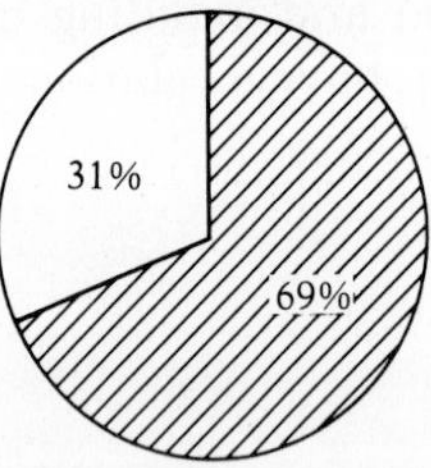

3 or more hyenas hunting
(16 observations)

Percent successful hunts

Percent unsuccessful hunts

the basic answer to the question, why hunt in groups? The answer is, to get more to eat. Carnivores can expand the size range of their prey if they cooperate to bring it down. Hyenas tend to stalk the small Thomson's gazelle alone, the much larger wildebeest in groups (Kruuk 1972). Groups also can often defend carcasses in situations in which single animals cannot. A group of hyenas sometimes can prevent a single lion from taking their kill and can manage to share a carcass with a group of lions.

These advantages of group hunting tend to be greatest in relatively open habitats. There opportunities for springing upon prey from ambush are minimal and those for running them down in packs great. The chances of concealing successful kills in the open are also smaller than they are in thick bush or dense forest. Therefore, it is not surprising that group hunt-

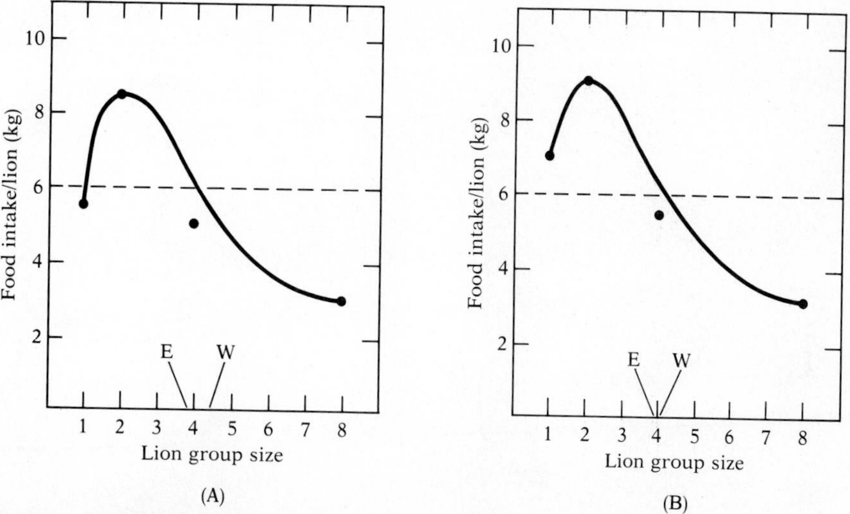

FIGURE 11–8 Group size and hunting success of lions in the eastern plains and western woodlands of the Serengeti in the wet season. (A) Preying on wildebeest. (B) Preying on zebra. Observed average group sizes in the east (*E*) and west (*W*) are indicated by arrows. The dashed line indicates minimum daily requirements of one lion. *(After Caraco and Wolf 1975.)*

ing in carnivores is largely limited to terrain such as prairies, savannas, and tundras, all with limited cover. Wolf packs on Isle Royale, however, hunt moose in the forest (Mech 1966). Here numbers add greatly to the chances of bringing down the prey, which are proportionately much larger and more powerful (see Figure 11–9) than, say, a wildebeest is to a hyena. The group-hunting advantage itself is considered the major force leading to social groups in these mammalian predators (Kleiman and Eisenberg 1973).

Raptorial birds sometimes hunt in pairs or larger groups. Eleanora's falcon forms living nets ("falcon walls") over Mediterranean islands, waiting for passerine birds migrating from Europe to Africa (Walter 1979). Frequently, those small, agile birds that are missed by one falcon will be caught by another. However, no sharing of the catch takes place. Other raptors—falcons, hawks, and eagles—frequently hunt in pairs or groups, especially when after hard-to-catch prey (Glutz von Blotzheim *et al.* 1971, Cade 1982). Again, as in the carnivores, success in groups tends to be higher than among solitary hunters.

Similar considerations explain group foraging in army ants, which often attack prey many times their own size. However, most ants forage singly outside of their nests, returning with food items that they are capable of handling themselves. When large or concentrated food sources are encountered, many of the ants will return home, leaving a scent trail, and will recruit other workers for the task of acquiring the food and de-

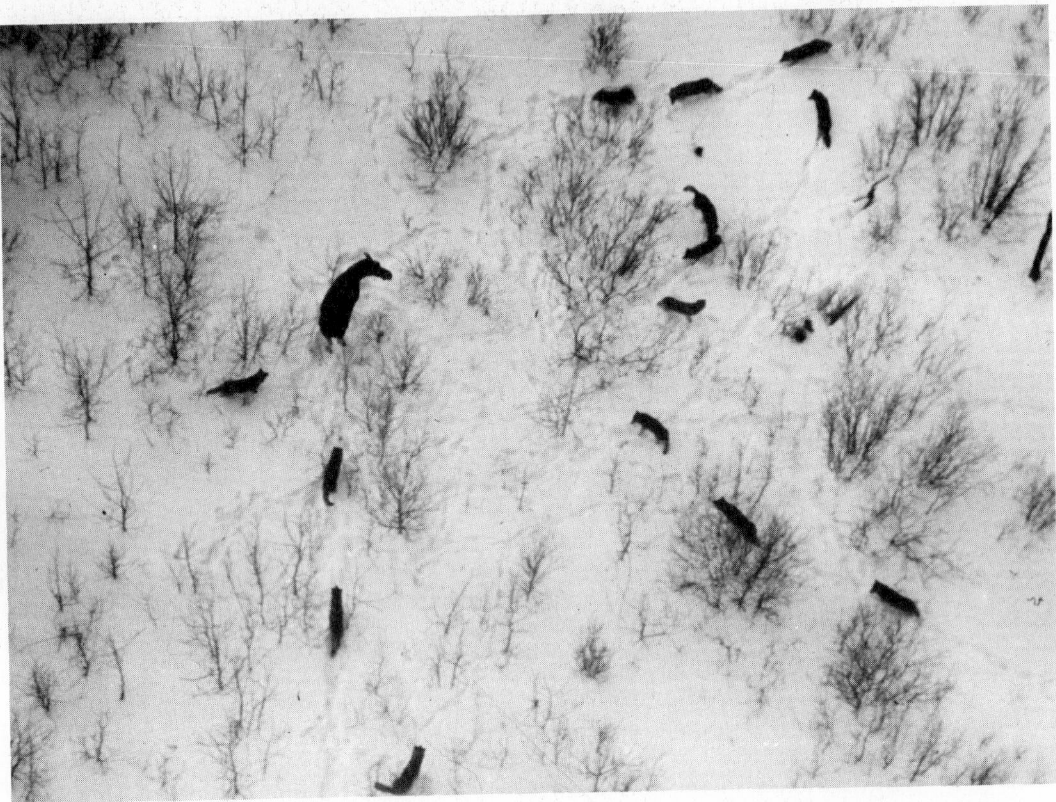

FIGURE 11–9 Wolves holding a moose at bay. Note that the only close individuals are behind the moose. The pack harassed the animal for 5 minutes and then left. *(© L. David Mech.)*

fending it from other ant species (Chadab and Rettenmeyer 1975, Wilson 1975). In this way, the ants manage to have the best of both worlds, hunting singly when it is efficient and cooperatively when it is not.

Feeding in groups may also make more food available. For example, flocks of birds may flush from cover insects that are then captured by other flock members. It may also avoid duplication of effort. Flocks of finches in the Mohave desert seem to move in a way that prevents rapid return to areas already gleaned (Cody 1971b). It has also been suggested that communal feeding makes it easier for individuals to find food, because they can imitate the successful moves of others or join them at dining on "finds" (Krebs 1974, Krebs *et al.* 1974).

Some surgeonfishes and parrotfishes (for example, *Acanthurus triostegus, Scarus croicensis*) increase their access to food by schooling. These fishes are grazers on tropical reefs, where much of the substrate is defended by territorial damselfishes or surgeonfishes (Barlow 1974, Ehrlich 1975, Robertson *et al.* 1976). By moving to the reef surface in schools, these fishes manage to overwhelm the defenders and feed successfully. In surgeonfishes, however, the foraging advantages of schooling may be much smaller than its protective advantages—a topic to which we will now turn.

Group Defense

In the open, animals find safety in numbers—such might appear to be almost a law of nature. Single individuals with no place to hide—on a prairie or savanna, in midair or midwater—tend to be more at risk for predation than those in groups (Williams 1964). Hamilton (1971) presented a model of the "geometry of the selfish herd," based on the idea that predators tend to attack the nearest prey individual. Therefore, prey lacking suitable cover will be impelled by self-interest to assemble in groups as each attempts to reduce its "domain of danger"—the area around it in which a predator might appear closer to it than to some other individual. Hamilton cites many instances in which individuals cluster closely together, each attempting to place others between it and a predator. However, his model does not state that the average individual is necessarily safer in a herd than alone—only that those strong, fast, or smart enough to get into the center of a herd are safer than those on the periphery.

Indeed, this behavior, which may well be the evolutionary source of the social behavior of many animals, in theory could work to the disadvantage of the group as a whole. Milling groups of prey sometimes appear to make a banquet for predators. If the major predators of a species learn to benefit from the herding/schooling behavior, advantages may begin to accrue to solitary individuals. Thus, a tension may exist between selection for aggregating and disaggregating behavior (Hamilton 1971). This tension may lead to varying degrees of grouping and, when seasonal predation patterns differ, to temporal variation in the urge to associate with conspecifics.

Does assembling in groups also sometimes increase average security? Can flocking protect the flock as a whole? Three of the usual reasons given for assuming that the answer is yes are as follows:

1. The sensory apparatus of the entire group is available to detect the approach of a predator.
2. Predators will be confused by a tightly packed group and have difficulty catching any individual.
3. The group may be able to defend against, or discourage, a predator when a single individual cannot.

A growing body of literature supports these points, in spite of the difficulty of doing the necessary investigations. Evidence has been found that in some circumstances solitary birds are more readily captured by raptors than are those in flocks (see, e.g., Page and Whitacre 1975). Starlings fly in a loose formation when above a hawk, but they cluster tightly when below one. This behavior discourages the predator, apparently because a stooping (downward-attacking) hawk attacks talons first; if it plunges into a flock, it can injure more delicate parts of its anatomy (Mohr 1960). In addition, field experiments indicate that, within a certain range of group sizes, large flocks of wild laughing doves (*Streptopelia*) have a greater probability than small flocks of detecting an approaching predator (Siegfried and Underhill 1975).

These experiments also illustrate some of the difficulties of trying to find simple relationships in behavioral ecology. A careful experimental design was used in which a hawk model was flown in a stereotyped manner over doves attracted to a bait (see Figure 11–10); the results were filmed so that the speed with which the birds took flight could be measured. The natural flocks tested ranged in size from 2 to 21 doves. Care was taken to eliminate statistically the possibility that the birds became accustomed to the model and reacted faster in later trials.

When the correlation coefficient was calculated, using all the model flights, no significant relationship was seen between flock size and reaction time (see Figure 11–11). The cause of the lack of correlation seemed to be the behavior of both very small and very large flocks. The former were exceptionally nervous, did not feed well, and experienced many false alarms. These flocks sometimes reacted faster than larger groups. At the other end of the spectrum, the largest flocks had high levels of agonistic

FIGURE 11–10 Hawk and dove experiments. *(After Siegfried and Underhill 1975.)*

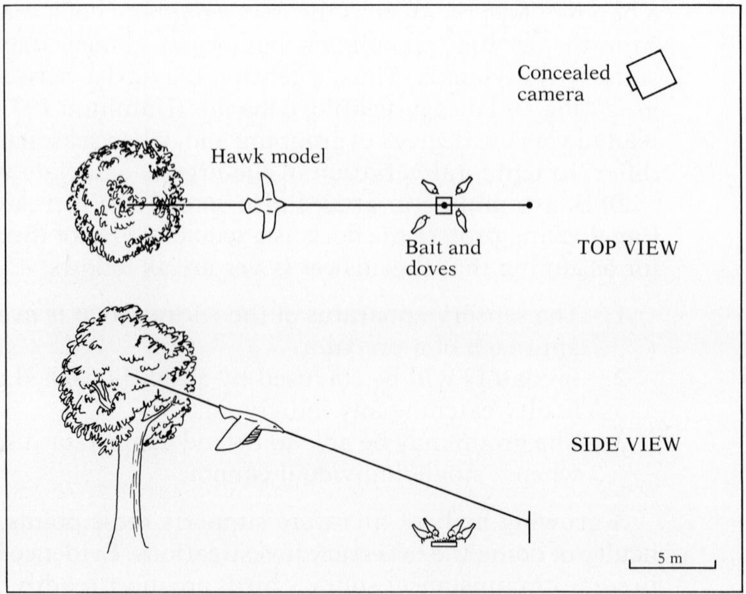

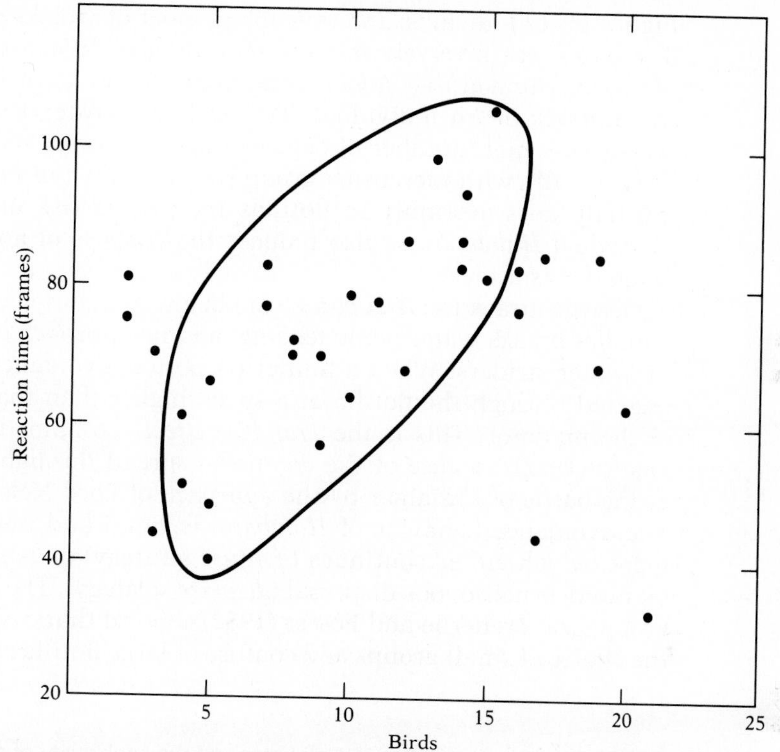

FIGURE 11–11 Correlation between flock size and reaction time when a model hawk is passed over a flock of feeding doves. Time is measured in the number of frames of motion picture film between the flight of the first dove and the time the model passed directly over the feeding point (the longer the time, the earlier the alarm). For the significance of circled points, see text. *(After Siegfried and Underhill 1975.)*

interactions between birds competing for food, the birds were generally less skittish, and the flocks had slow reaction times. Statistical analysis determined that flock sizes of between 4 and 15 showed an essentially linear relationship between size and reaction time (see the points surrounded by the line in Figure 11–11). No relationship was found between speed of reaction and proximity to the model.

The confusion of predators by flocks has recently been demonstrated nicely by studies of one of the few marine insects, the water strider *Halobates robustus* (Hemiptera). These creatures occur in assemblages (flotillas) that live on the surface of the ocean in the Galapagos Islands. The flotillas are densely packed, with 200–500 individuals per square meter, and may include up to 700. One way in which forming flotillas protects individual *Halobates* is a simple dilution effect. The probability of attack by a herring (*Sardinops*) on a flotilla was independent of group size, as

Figure 11–12 indicates, and thus the number of attacks per individual per 5 minutes was inversely related to group size. In essence, the formation of groups thinned the *Halobates* population and greatly reduced the risk of an attack on an individual (Treherne and Foster 1982). Furthermore, the success rate (number of captures per attack per individual) declined dramatically with increasing group size, as shown in Figure 11–13. Thus, not only does assembly in flotillas reduce the risk of an attack on an individual *Halobates,* it also reduces the chances of an individual being eaten if it is attacked.

Herring attack the *Halobates* from below, without warning. In contrast, a mullet breaks water while feeding, and its approach can be detected by the water striders. When a mullet is spotted, avoidance behavior is propagated through the flotilla at a speed higher than that of the approach of the predator. This is the *Trafalgar effect*—so named by Treherne and Foster (1982) because of the confusion spread through the Spanish fleet at the battle of Trafalgar by the approach of Lord Nelson's British ships. The avoidance behavior of *Halobates* is rapid and unpredictable movement, which either continues (*confuse* strategy) or is followed promptly by rapid, synchronous dispersal (*disperse* strategy). The choice depends on group size. Treherne and Foster (1982) showed that disperse tended to be the choice of small groups and confuse of large flotillas (see Figure 11–14).

FIGURE 11–12 Group size in *Halobates* and the probability of a herring attack. Vertical lines at top of each bar indicate the 95 percent confidence interval. *(After Treherne and Foster 1982.)*

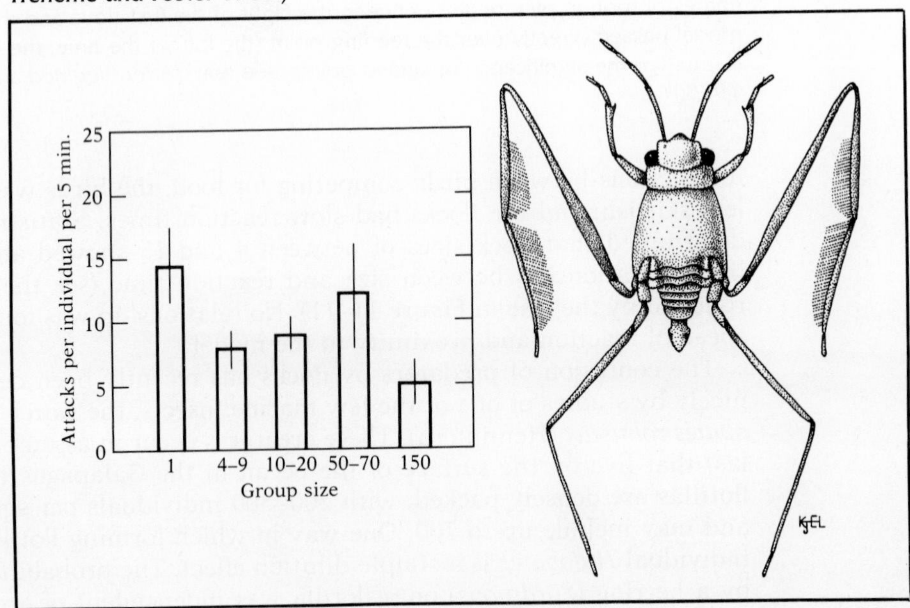

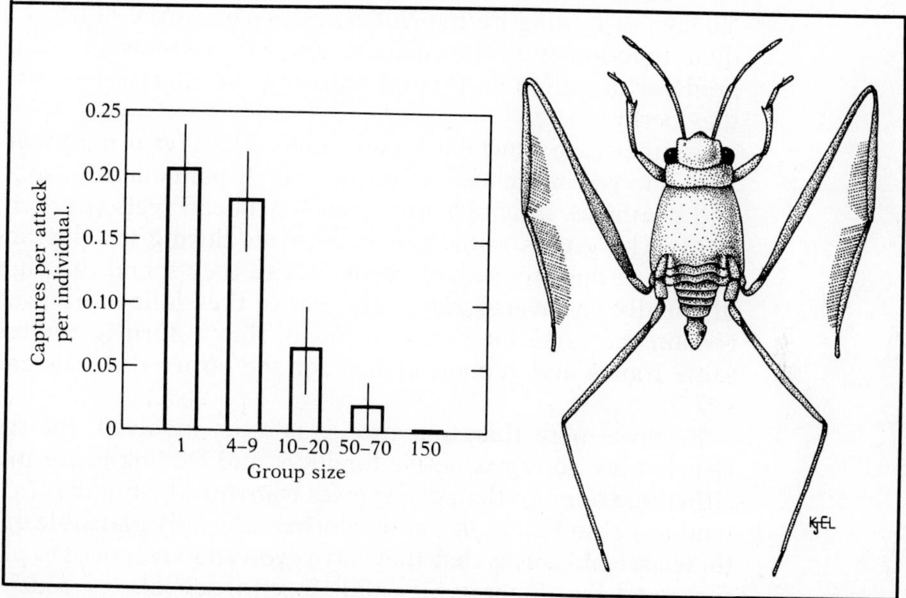

FIGURE 11–13 Success rate of attacks by herrings on *Halobates* as a function of group size. Vertical lines at top of each bar indicate 95% confidence interval. *(After Treherne and Foster 1982.)*

FIGURE 11–14 Choice of defense strategies by *Halobates* groups of different sizes when approached by herrings or predator model (see text). *(After Treherne and Foster 1982.)*

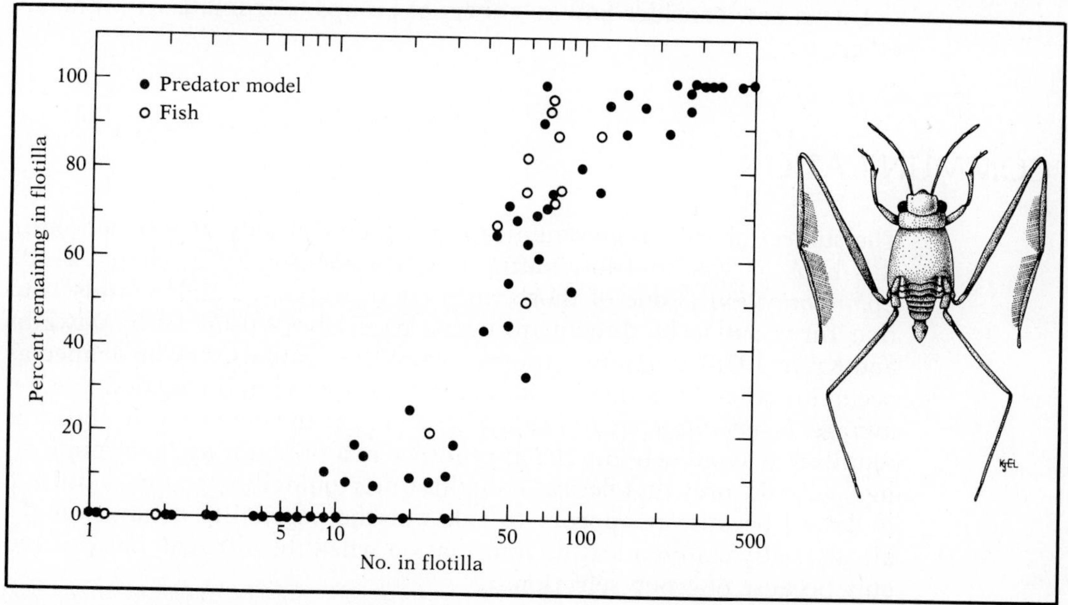

No sign of feeding by the *Halobates* is seen in the flotillas. It seems that their function is simply defense against predators by dilution, making predator detection easier and reducing the effectiveness of attacks when they occur.

Another case in which groups appear to have a purely defensive function is in resting schools of grunts, small pomadasyid fishes (*Haemulon*), and goatfishes (*Mulloidichthys*) on Caribbean reefs (Ehrlich and Ehrlich 1973). The grunts remain over the reefs during the day and migrate in schools at dusk to seagrass beds, where they spend the night feeding individually on invertebrates. The routes they follow are precise and may remain the same for years. At dawn, they return to the reefs along the same routes and remain at rest for the entire day (Ogden and Ehrlich 1977).

Because more than one species is often involved, the resting schools clearly play no reproductive function, and feeding is not involved. When either one or more than one species is involved, all individuals in a school tend to be similar in size and coloring. The only plausible explanation for these assemblages is that they have evolved in response to predation pressures and that heterotypic (multispecies) schools are a device for developing larger schools than would be possible if only one species were involved.

Of course, groups form for reasons other than mating, attack, and defense. These reasons are as diverse as protection from wind or cold (for example, in tight herds of musk oxen), passing around of gut symbiotes needed to digest cellulose (termites), and improved locomotion (possibly porpoises).

Let us now consider how members of groups coordinate with one another.

COMMUNICATION

The subject of animal communication, which can only be touched on in this book, is vast and fascinating (see, e.g., Wilson 1975, chaps. 8–10). *Communication* is one of those words that everyone "understands" but that turns out to be difficult to define precisely (Wilson 1975, Dawkins and Krebs 1978). *Communication* between two animals may be defined as occurring when an action by one creates a response in the second *that, on average, is advantageous to the first.* The part of the definition in italics is required to avoid defining the appearance of a predator as "communicating" with the prey that flees at its sight. Unfortunately, even this definition is flawed because it appears to rule out the possibility of an animal's altruistically communicating information when the altruism has evolved only because of group selection.

FIGURE 11–15 Light signals of eight different species of fireflies as they would appear in a time-lapse photograph. Arrows show direction of flight. Diagrammatic—not all eight species occur in the same locality. *(After Lloyd 1966.)*

Animals, of course, communicate with each other for a wide variety of purposes. Signaling between the sexes of readiness to mate is one of the most common forms of animal communication. Others are communications between parents and offspring, warnings that an animal is intruding upon another's territory or that a predator approaches, and announcements about the location of food. It is possible that animal signals have evolved to transmit the necessary information with a minimum of both energetic investment and risk on the part of the signaler. If so, signals, like other features of organisms, have evolved characteristics appropriate to the environments in which they are used.

One group of beetles, fireflies, appears to avoid predation by mating at night (Lloyd 1966). This situation presents a problem in mate finding that has been solved by the evolution of light signals, as demonstrated in Figure 11–15—a system that has led to some unusual risks, as we shall see in Part 4 in our discussion of light mimicry. Perhaps the most famous of all animal communication systems was evolved by an insect for signaling in the dark. Returning honeybee foragers transmit the distance and direction of a desirable food source through the nature of the dance they do in the dark among the other bees on the vertical face of a comb in the hive (von Frisch 1950, 1967; for a summary, see Michener 1974).

However, usually more than one way exists to signal in a given situation. Night mating has been solved by various moths, including giant silkworm moths (Saturniidae), by the perfecting of chemical communication. A virgin female moth releases a chemical signal (a *pheromone*), a *single molecule* of which can be detected by receptor organs on the antennae of a male moth. Thus, males are able to detect the female at a long distance, perhaps for more than a mile, and to fly upwind along the concentration gradient formed in the pheromonal plume downwind of the female (Marsh *et al.* 1978). Lepidopterists wishing to add to their collections of silkworm moths often place freshly emerged females in cages outdoors overnight and then collect the hordes of males than have been attracted.

In one arctiid moth, *Utetheisa ornatrix*, it has been shown by researchers in Thomas Eisner's laboratory that pheromone release creates not only a pattern in space but one in time as well (Conner *et al.* 1980). The moth releases its chemical signal in the form of short pulses about 1.5 seconds apart, which remain discrete for at least 60 cm downwind of the moth, as shown in Figure 11–16. Conner *et al.* (1980) believe that the temporal pattern may serve as a short-range orientation device, informing the male that he is close to his target and that simple upwind flight may no longer be adequate for orientation. The pulsing may also serve to reduce desensitization of the male's chemoreceptors, which otherwise would be receiving a strong, steady pheromonal bombardment.

FIGURE 11–16 Simulation of the pulsed pheromone signal from *Utetheisa ornatrix*. Air is being pulsed, visually "stained" with titanium chloride; note how the signal remains discrete. *(Photo by William Conner and T. Eisner, Cornell University.)*

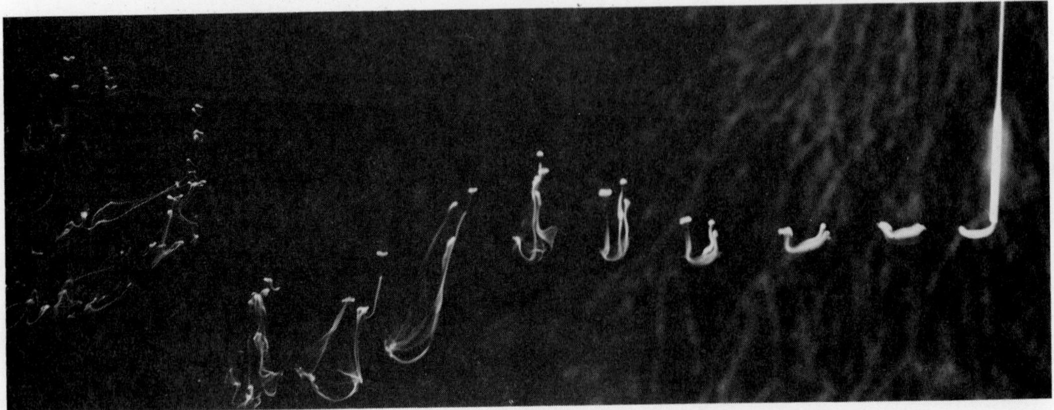

Chemical communication can be efficient when it is not important to transmit information *rapidly* over a great distance. For example, it allows a female moth to "advertise" for mates over an area far beyond that which would be possible by much faster light or sound signals. The females also appear to be surprisingly safe. For example, no predator is known that has learned to detect the giant silkworm moth's pheromone plume and follow it to the female. Chemical communication also is used effectively by social insects, which are normally in close contact with each other. In fact, evidence is growing that genetically based chemical differences are used by these insects to identify not only members of their own colonies but also kin (see, e.g., Greenberg 1979, Hölldobler and Michener 1980, C. D. Michener 1982).

However, in a crunch—as when a lion is stalking or a hawk swooping—releasing a smell to alert the group will not do. The speed with which the signal can be carried by diffusion or a breeze cannot compare with that of sound and light, both of which are essentially instantaneous in a local situation. Sound appears to be the medium of choice, all else being equal, when a sudden warning of a predator's approach is to be transmitted. The reasons that sound rather than light has been favored in this case seem evident. First, ears are multidirectional receivers; one does not have to be listening in a certain direction to receive the message. Second, squeeking or shrieking is less likely to call attention to an individual in a flock, or give away the position of an isolated individual, than jumping up and down.

Students of *alarm calls* have been interested in the question of why they are given at all (see e.g., Charnov and Krebs 1975, Staton 1978). Why should an organism be altruistic and take a chance on drawing attention to itself? Why not, after all, just flee on spotting a predator, leaving one's fellows to fall prey? Selection would seem to favor such behavior, but many other factors may be operating. As in the *Halobates,* an individual may have a much higher chance of surviving a predator's attack in an alerted group than on its own. Alternatively, the members of the group may be kin, and therefore the altruistic alerter may be increasing its inclusive fitness. The call may be of a quality that makes the caller difficult to locate (Marler 1955). The call may even be ventriloquial, misleading the predator about the caller's location (Perrins 1968). Charnov and Krebs (1975) claim that calling operates to the benefit of the caller—that it is manipulating the rest of the flock. For example, having seen a hawk, a bird has two pieces of information: a danger exists, and the danger is coming from a certain direction. It then passes on only the first piece of information to flock members, which must react without the second piece of information. The caller is then able to choose a position in the reacting flock that minimizes its own exposure or otherwise benefit from the general confusion created.

Sherman (1977) has tested some of these ideas in studies of Belding's ground squirrels (*Spermophilus beldingi*). Females were more frequently associated with relatives than males were, and Sherman found that they also gave many more alarm calls than did males upon spotting a predator. Females with young were more likely to call than those without young. Reproductive females that also had female relatives in the area did more calling than did those that lacked them. Furthermore, predators more frequently stalked calling individuals than mute ones. Overall, at least in these squirrels, the evidence supports the theory that alarm calling evolved because it increased the inclusive *fitness* of the caller—that is, it evolved by kin selection.

Of course, we have no reason to believe that all alarm calls evolved in a similar manner. For example, it has been claimed that alarm calls evolved in crocodilians because they are overwhelmingly valuable to the caller in the early years of life, when mortality is highest (Staton 1978). When a young crocodile's life is threatened, its distress calls quickly summon its mother. Whereas a sparrow's mother may be unimpressive to a hawk, rare is the predator that will take on a mother crocodile.

The communication by sound that people most often hear has nothing to do with warnings against predation—namely, the territorial songs of birds. Bird songs have been the subject of intensive study by those interested in the ontogeny, evolution, and ecology of communication (see, e.g., Konishi and Nottebohm 1969, Marler 1970, Marler and Mundinger 1971). From an ecological perspective, one of the most interesting discoveries has been that of local song *dialects* within species.* For example, young male white-crowned sparrows (*Zonotrichia leucophrys*) preferentially learn native over alien dialects when both are presented to them (Cunningham and Baker 1983). Furthermore, juvenile males remain within a few hundred meters of their birthplace during the period in which they are learning their songs (Baker *et al.* 1982a), and males show a stronger territorial reaction to the songs of males from adjacent dialect areas than to those from more distant dialect areas (Baker *et al.* 1981, Baker 1982a). Females, in turn, mate preferentially with males that sing in their native dialect, and they show a secondary preference for songs of males from a contiguously adjacent dialect area (Baker 1982b).

All of these data indicate that dialect boundaries should form substantial barriers to gene flow and should be an important factor in the structuring of populations of white-crowned sparrows (but see Petrinovich *et al.* 1981). This assumption is supported by studies of allozyme frequencies that show significant heterogeneity among samples from populations with different dialects and a greater genetic similarity between individuals from adjacent dialect areas than between those from more distant ones (Baker *et al.* 1982b).

* See, for example, Marler (1952), Marler and Tamura (1962), Nottebohm (1969), Conrads and Conrads (1971), King (1972), Kroodsma (1974), Baker (1975), Baptista (1975), and Payne (1981).

In addition to sounds, visual signals are important in the social systems of vertebrates. Visual signals were a focus of classical study by *ethologists*—those who investigate animal behavior in natural situations and interpret the results in an evolutionary framework. In 1914, Julian Huxley described the courtship behavior of great crested grebes (*Podiceps cristatus*), waterfowl that go through a complex series of gestures and dances before a pair mates (see Figure 11–17). His pioneering work began attempts to comprehend how simpler and more easily understandable motions evolved into complicated and seemingly exaggerated communicatory behavior—that is, became *ritualized*. The answer is still not entirely clear, and older explanations based on benefits to the signal sender and signal recipient, such as those by Tinbergen (1964) and Marler (1968), are being challenged by those who view communication principally as at-

FIGURE 11–17 Different courtship displays of the great crested grebes. *(After Huxley 1914.)*

tempts by one individual to manipulate another (Dawkins and Krebs 1978).

In the latter view, mates are not individuals with which to cooperate in reproduction but environmental objects to be subverted into helping one reproduce. To paraphrase Dawkins and Krebs, the swollen belly and upright posture of a female stickleback fish that elicits the male's mating behavior, as observed by Tinbergen (1951), is a manipulation of the male, a sign that the female has subverted its nervous system. However, this fascinating subject is more in the province of animal behavior than ecology. Suffice it to say that the debate continues.

The way in which the physical environment can shape the evolution of signals is nicely demonstrated by the feeding behavior of gulls. Feeding by adult herring gulls—regurgitation of half-digested food in front of the chicks—is elicited by the pecking of the baby gulls at a red spot on the parent's bill. The red spot on a long, thin, downward-pointed, moving object serves as a signal to the chicks (Tinbergen and Perdeck 1950). Herring gulls nest on flat land. Kittiwakes, in contrast, are a cliff-nesting gull, and their mating and chick-rearing behavior has adjusted evolutionarily to this setting (Cullen 1957). One of the adjustments has been in the begging signal. The red spot on the outside of the adult beak is gone, replaced by a red inner lining of the beak. This lining directs the chicks to obtain their food directly from the adult's mouth and lessens the chances of the food being dropped and lost forever down the cliff face.

The ecological significance of visual signals may be complex and must be interpreted with care. One of the best-known sets of such signals are the so-called poster colors of certain coral reef fishes, as exemplified by butterfly fishes and angelfishes (Chaetodontidae) (see color Plate D). The original supposition by Lorenz (1962) that they were territorial signals was incorrect, as many of the most spectacularly hued chaetodontid species are not territorial (Ehrlich *et al.* 1977a). The poster colors are now thought to serve various functions, which, in butterfly fishes, may include predator avoidance (Ehrlich *et al.* 1977a). The colors may also permit individuals to remain in contact with one another (Kelly and Hourigan 1983).

LEARNING AND CULTURAL EVOLUTION

The ability of many organisms to manipulate their environments is greatly enhanced by the passage of information from generation to generation extragenetically—that is, through learning and the creation of a *culture*. Obviously, social grouping enhances the possibilities for the development of culture, and significant cultures are restricted to social organisms. Cultural transmission is of most significance, of course, in *Homo sapiens* but is by no means restricted to it. For example, cultural trans-

FIGURE 11–18 A Japanese monkey *(Macaca fuscata)* washing a sweet potato. *(After Kawai 1965.)*

mission of such traits as sweet potato washing (see Figure 11–18) and separating wheat grains from sand by flotation has been documented in Japanese macaque monkeys (*Macaca fuscata*) (Kawamura 1963, Tsumori 1967). Indeed, the classic example of cultural transmission in nonhuman animals was observed in great tit and other bird populations in England. The birds learned to remove the foil caps on bottles of nonhomogenized milk and drink the cream. The habit spread rapidly through their populations—much too rapidly for genetic transmission to have been involved (Fisher and Hinde 1949, Hinde and Fisher 1952).

ECOLOGY AND BEHAVIOR

This brief introduction to behavioral ecology should, we hope, give you some insight into the speed with which progress is being made in this subdiscipline of ecology. The field has moved from an era of describing behavior to one in which successful attempts are beginning to be made to explain the ecological and evolutionary significance of behavioral patterns.

PART 4

POPULATION INTERACTIONS

Among the most important features of the environment of any organism are other organisms, including organisms of different species. Populations of one species change in size and evolve in response to the dynamics and evolution of populations of other species. Each population ordinarily interacts with numerous others—some intensely, others less so. Population interactions are, in a sense, the building blocks of community ecology.

In this part, we examine the zone of interface between population and community ecology—an area concerned with *strong interactions* between pairs of populations. Chapter 12 is concerned with competition. Chapter 13 describes the interaction between predator and prey, and Chapter 14 describes mutualistic interactions and mimicry.

CHAPTER 12

Competition: Two Organisms Using the Same Limited Resource

Any population interaction may have two aspects: codynamic and coevolutionary. When the population interaction involves reciprocal influences on population size, we call it *codynamic*. To the degree that two species affect each other's evolution, the interaction is *coevolutionary* as well.

In addition, when a limited resource is used by two or more individuals, they are competing. This kind of competition is called *exploitative*. When competition takes the form of direct conflict, it is known as *interference competition* (Park 1954, 1962, Miller 1967, Schoener 1973, 1974). Competition often involves both exploitative and interference components, as happens, for example, in interspecific territoriality.

EARLY STUDIES OF COMPETITION

Interest in competition traces at least as far back as Charles Darwin, who pointed out:

> As species of the same genus have usually, though by no means invariably, some similarity in habits and constitution, and always in structure, the struggle [for existence] will generally be more severe between species of the same genus, when they come into competition with one another, than between species of distinct genera [1859, p. 76].

Before the middle of the twentieth century, studies of competition were an abiding interest, and the almost exclusive domain, of plant ecologists. For instance, the distinguished British ecologist A. G. Tansley, in his 1914 presidential address to the First Annual General Meeting of the British Ecological Society, emphasized the importance of competition in plant community dynamics and urged his colleagues to carry out experiments to determine its precise effects. His call was heeded, resulting in a wide variety of investigations of competition between plant species.

Tansley himself (1917) reported on studies of two species of bedstraws, plants of the family Rubiaceae—*Galium saxatile,* which occurs abundantly on silica-rich soils, and its relative, *G. sylvestre,* which grows on limestone soils. Mixed and pure cultures of both were grown on different kinds of soil. The soil type influenced the rates of germination, (Figure 12-1), seedling survival, and ultimately the outcome of competition. On limestone soil, *G. sylvestre* germinates and grows vigorously, whereas *G. saxatile* shows a lower germination rate, and many of its seedlings become chlorotic—that is, lose their chlorophyll—and die. Those *G. saxatile* seedlings

FIGURE 12–1 Results of one set of germination experiments. *Galium sylvestre* has a higher germination percentage on both limestone and lime-poor (peat) soils, but its relative advantage depends on soil type. *(Data from Tansley 1917.)*

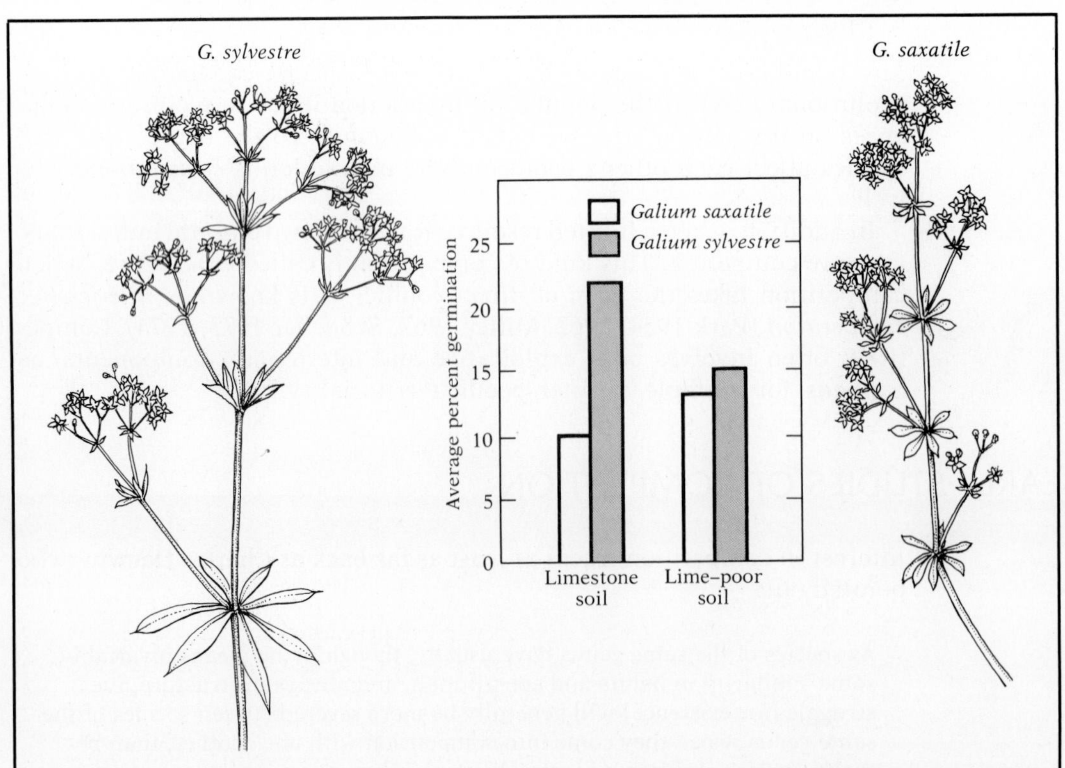

that survive eventually lose out in competition to *G. sylvestre*. But on peat, while *G. sylvestre* still has an edge in germination, it eventually covers less ground than *G. saxatile*.

Other experiments showed the importance of shading for the development of beech seedlings and loblolly pine seedlings and the role of competition from tree roots on forest herbs (Harley 1939, Chapman 1945). The latter researchers used trenching to cut the roots, a technique widely used in studies of forest ecology. In general, understory vegetation develops relatively thickly in trenched plots in which the roots of forest trees have been killed (Toumey and Keinholz 1931, Lutz 1945, Ellison and Houston 1958). The relative growth of roots and shoots under different conditions of competition was also explored (Nedrow 1937). In areas in which plants were allowed to grow freely, so that competition was primarily for light, the ratio of shoot to root was much higher than in areas in which the plants were separated and tied apart so that they were fully illuminated and needed to compete only for water.

Most of these early plant ecologists did not focus on the evolutionary reasons why some plants were better able to compete under certain conditions, whereas others were superior in other circumstances. The characteristics of the competitors were taken as givens. An exception was E. J. Salisbury, who in 1929, 15 years after Tansley, provided the British Ecological Society with another remarkable presidential address. In it he addressed many aspects of the evolution of the life history strategies of plants in relation to their abilities as competitors, among them the merits of sexual versus asexual reproduction, overground laterals versus underground stolons, simultaneous versus discontinuous germination of seeds, and few large versus many small seeds. He was, however, ahead of his time. Systematic attempts to give an evolutionary perspective to the role of competition in determining the distribution and abundance of organisms would come only after the development of some mathematical models and the rise of the "modern synthesis" of evolution (Jackson 1981).

The result of this early botanical activity was the rendering of a quite sophisticated picture of the importance of competition in determining the distribution and abundance of plants, a picture that was to be a great influence on ecological theory in this area.

THE ORIGINS OF MATHEMATICAL COMPETITION THEORY

Modern competition theory can be traced to experiments carried out by the microbiologist G. F. Gause (1934) and to Gause's interpretation of them in terms of a mathematical theory developed by Vito Volterra (1926) and elaborated by A. J. Lotka (1932). In his classic work *The Struggle for Existence,* Gause clearly shows his debt to plant ecologists. In the first chapter he states, "botanists have already recognized the necessity of having recourse to experiment in the investigation of competition phenomena" (p. 4). In the second chapter, "The Struggle for Existence in Natural Conditions," he then deals extensively with plant competition experiments. Here also, before discussing his own experiments, he expresses the idea that is sometimes associated with his name as *Gause's principle:* "It is admitted that as a result of competition two similar species scarcely ever occupy similar niches, but displace each other in such a manner that each takes possession of certain peculiar kinds of food and modes of life in which it has an advantage over its competitor" (p. 19).

We will examine this notion of *competitive exclusion* and the related idea of *limiting similarity* in Chapter 16 when niche theory is considered. For the moment, however, let us look at how Gause's own experiments supported this idea. As a background to those experiments, Gause quotes (on p. 90) the great physicist Boltzmann (1905):

> In order to utilize in the best manner [the passage of energy from the hot sun to the cold earth] the plants spread under the rays of the sun the immense surface of their leaves, and cause the solar energy before reaching the temperature level of the earth to make syntheses of which as yet we have no idea in our laboratories. The products of this chemical kitchen are the object of the struggle in the animal world.

Gause was determined to investigate Boltzmann's notion in a microcosm with a limited supply of energy in the form of nutrients into which populations of two species of protozoa would be introduced. He wished to determine whether one species would drive the other out of existence, by appropriating all the available energy, or whether the two would coexist. He thus pioneered the laboratory study of competition, working with yeasts and protozoa. Perhaps his best-known experiments involved pure and mixed cultures of two species of *Paramecium,* the ciliated protozia known as "slipper animalcules" that are familiar to anyone who has examined pond water under a microscope. The results of one experiment, in which, in Gause's words, "*Paramecium caudatum* gradually diminishes as a result of its being driven out by *P. aurelia,*" are shown in Figure 12–2. He concluded: "the process of competition under our conditions has always resulted in one species being entirely displaced by another, in complete agreement with the predictions of the mathematical theory" (p. 103).

Early mathematical work on competition focused on two species and used an extension of the logistic equation to cover interspecific competition as well as the intraspecific competition (density dependence) that the logistic equation already represents. Recall that the logistic equation, in discrete time, is

$$\Delta N = \frac{rN(K - N)}{K}$$

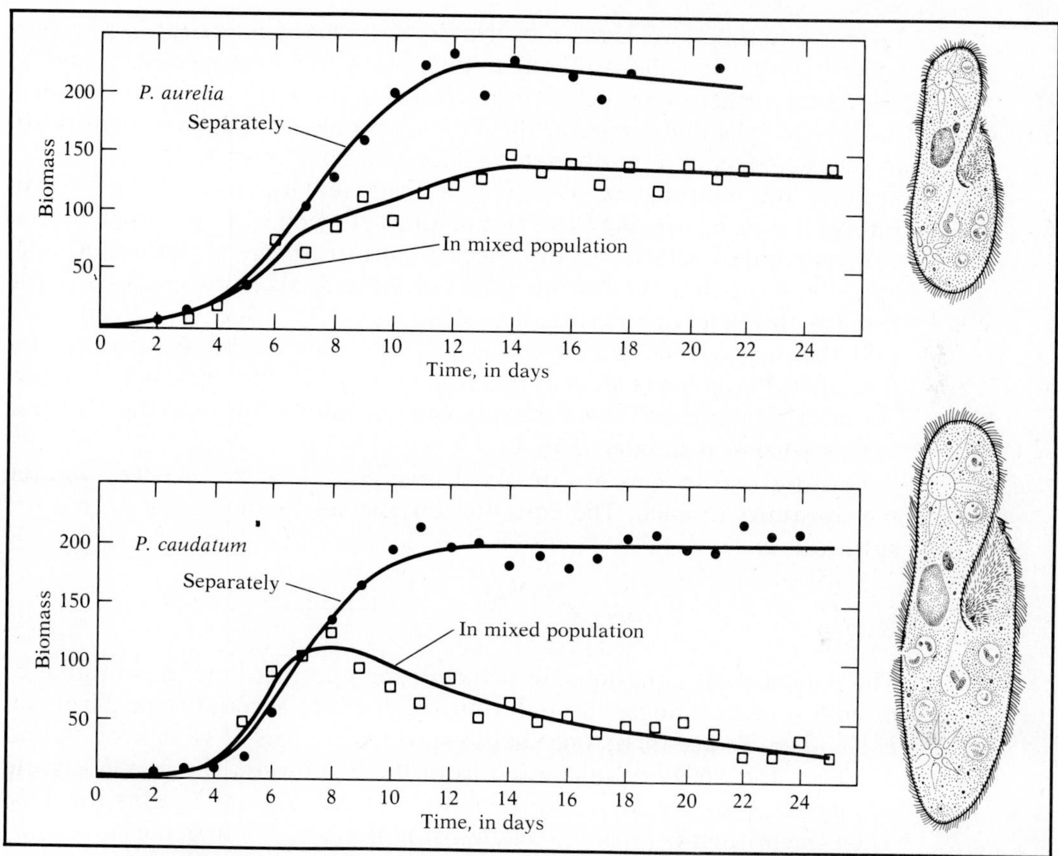

FIGURE 12–2 Growth of *P. aurelia* and *P. caudatum* populations separately and together. *(After Gause 1934.)*

where r is the intrinsic rate of increase and K is the carrying capacity. This equation can be extended to encompass competition from another species by including the other species' abundance in the expression in parentheses. However, this other species may not compete in the same way; that is, the effect on an individual of the first species caused by an individual of the second species may not be the same as the effect caused by another member of the first species. To take this possibility into account, we will use a coefficient to measure the effect of an individual of the second species *relative* to that of the first species. The coefficient is usually written as a Greek α, with subscripts to indicate the identity of the species involved. Specifically, $\alpha_{1,2}$ is the coefficient for the effect of an individual of species 2 against an individual of species 1. Now the equation for the population dynamics of species 1 becomes

$$\Delta N_1 = \frac{r_1 N_1 (K_1 - N_1 - \alpha_{1,2} N_2)}{K_1}$$

In this equation, we have subscripted everything referring to the first species with a 1. Thus, N_1 is the abundance of the first species; r_1 and K_1 are its intrinsic rate of increase and carrying capacity, respectively. Similarly, N_2 is the abundance of the second species, and, as just mentioned, $\alpha_{1,2}$ is the effect of an individual of species 2 on species 1.

From the way that α is used as a coefficient in front of N_2, we can interpret it as measuring the effect of an interspecific interaction between two individuals relative to the effect of an intraspecific interaction. Specifically, if α equals 1, then the effect of individuals from species 2 is the same as the effect of individuals from species 1. In this case, we could say that the interspecific interaction is just as strong as the intraspecific interaction. If α is lower than 1, then the strength of the intraspecific interaction is greater than that of the interspecific interaction, and the converse is true when α is greater than 1.

Because two species are involved, two equations are needed, one for the dynamics of each. The equation for species 2, analogous to that for species 1, is

$$\Delta N_2 = \frac{r_2 N_2 (K_2 - N_2 - \alpha_{2,1} N_1)}{K_2}$$

The pair of these equations, in principle, can predict how the abundance of both species changes through time, taking into account their effects on each other. To use them, one should start with a pair of values for N_1 and N_2. Then the ΔN's are calculated from the equations. The new N_1 is the original N_1 plus the ΔN_1, and the new N_2 is the original N_2 plus the ΔN_2. Then the process is repeated again, using the new N's to generate a third set of N's, and so forth. If you have access to a microcomputer, you can easily program this operation, using a language such as BASIC or Pascal.

The values of N_1 and N_2 through time that are predicted by these equations can be placed on a graph, where N_1 is on the horizontal axis and N_2 is on the vertical axis. The starting values of N_1 and N_2 provide the first point. The next values of N_1 and N_2 provide the second point, and so forth. The line beginning at the first point and ending with the last point to be calculated from the equations is called a *trajectory*. It shows graphically how the pair of populations changes through time. The graph usually has enough space to illustrate several trajectories, each representing a pair of populations that is started from a different place on the graph. The pattern that is apparent when many trajectories are shown on the graph is called the *flow*.

When the r's are small enough so that oscillations do not occur,* the trajectories take three basic patterns, as illustrated in Figure 12–3. In one pattern, all starting points at which both species are initially present produce trajectories that lead to the elimination of a certain species. In Figure

* Or if the model is written using differential equations to begin with.

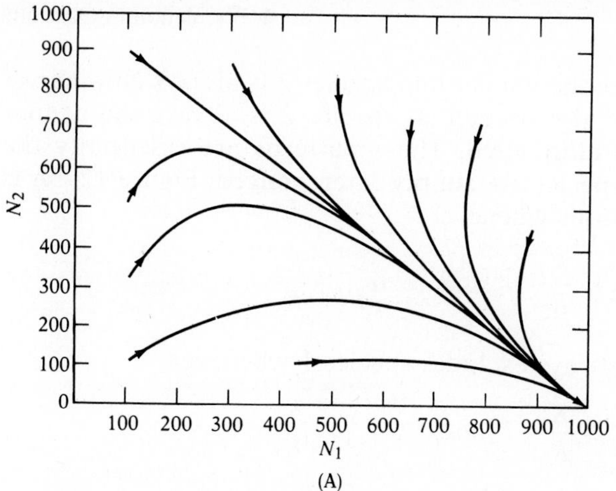

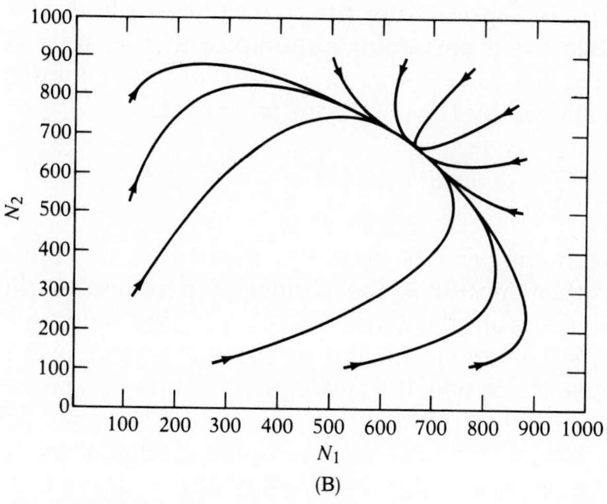

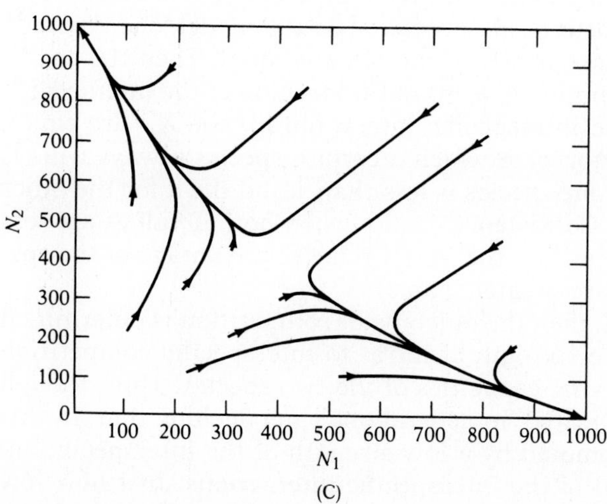

FIGURE 12–3 Examples of trajectories produced by the competition equations. (A) Species 1 excludes species 2 from any initial condition. (B) Both species coexist. (C) Either species can exclude the other, depending on the initial condition.

12–3A, species 1 is always the winner and species 2 is always eliminated. In the same vein, it may also occur that species 2 is always the winner and species 1 is always eliminated. These patterns are essentially the same; the names of the species are simply interchanged. Figure 12–3A is produced from the equations whenever

$$\alpha_{1,2} < \frac{K_1}{K_2} \quad \text{and} \quad \alpha_{2,1} > \frac{K_2}{K_1}$$

Alternatively, species 2 always excludes species 1 whenever

$$\alpha_{1,2} > \frac{K_1}{K_2} \quad \text{and} \quad \alpha_{2,1} < \frac{K_2}{K_1}$$

Another pattern that is possible is illustrated in Figure 12–3B. Here all the trajectories lead to a point representing the coexistence of both species. Moreover, each species has a particular abundance at this state of coexistence, which is called the *steady state* or *equilibrium state* for the system. This pattern is produced by the equations whenever

$$\alpha_{1,2} < \frac{K_1}{K_2} \quad \text{and} \quad \alpha_{2,1} < \frac{K_2}{K_1}$$

Finally, the third pattern that can be found is illustrated in Figure 12–3C. Here one or the other species is the winner, depending on the starting abundances. This case is similar to the first case in that long-term coexistence is impossible, but it differs in that no certain winner is evident. Here history, in the sense of which species had the advantage initially, is important in determining the identity of the eventual winner. All these mathematical conditions may be derived by simple graphical methods (for illustration see, e.g., Roughgarden 1979, chap. 22).

To find a biological interpretation for the mathematical conditions that determine which of the patterns are predicted by the model, one typically begins with the special case in which the K's are equal. Then the conditions can be interpreted simply in terms of the strength of the interspecific interaction relative to the intraspecific interaction. If the K's are equal, then the pattern of trajectories in which a certain species always wins is produced when the α for one species is less than 1 and the α for the other species is greater than 1. Coexistence occurs when both α values are less than 1. The outcome depends on the initial relative advantage of the species when both α values are greater than 1.

If the K's are not equal, then the outcome of competition is determined by comparing the relative strength of intra- to interspecific competition with the ratio of the carrying capacities of the two species. Thus, though the details may be complicated, in general one can conclude that, in this model, coexistence is promoted by a low strength of the interspecific interactions relative to that of the intraspecific interactions. Just how low

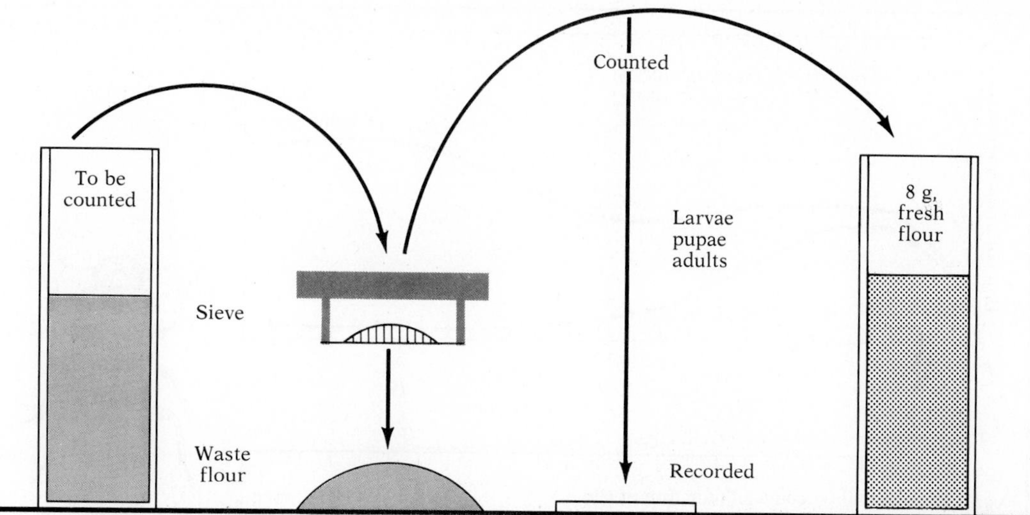

FIGURE 12–4 Technique used in *Tribolium* competition experiments. *(After Park 1962.)*

the strength of the interspecific interaction relative to the intraspecific interaction must be to allow coexistence depends on the resources available to each of the species, as measured by their carrying capacities. This model tends to predict the dynamics of competition in laboratory cultures of protozoa rather well (Vandermeer 1969).

However, the theory is less successful in describing competition between flour beetles of the genus *Tribolium*, which are major pests of stored grain. *Tribolium* are among the easiest insects to culture in the laboratory. A quarter ounce of whole wheat flour in a glass vial is a universe for these tiny insects. As illustrated in Figure 12–4, a vial may be seeded with a pair of the beetles; once a month the beetles may be sieved out of the flour, using a silk screen, and the various stages counted. They may then be replaced in the vial with fresh flour. Many replicates of each experiment can be done. The food supply is standardized, the shape of the environments is identical (a cylinder), and all the vials can be stored in a chamber with constant temperature and humidity.

Thus, competition experiments between two species, *T. confusum* (CF) and *T. castaneum* (CS), can be conducted in identical microcosms. Park and his students (e.g., 1962) found that one species or the other would always "win"; eventually either CF or CS would be alone in the culture. Furthermore, if the conditions of temperature and humidity were changed, the outcome of the competition could be altered in predictable ways. For example, if the vials were kept in hot-moist chambers, CS was always the victor; in a cool-arid environment, CF always won. In other environments, however, the results were not so straightforward. In a cool-moist environment, for instance, CF won in 71% and CS in just 29% of

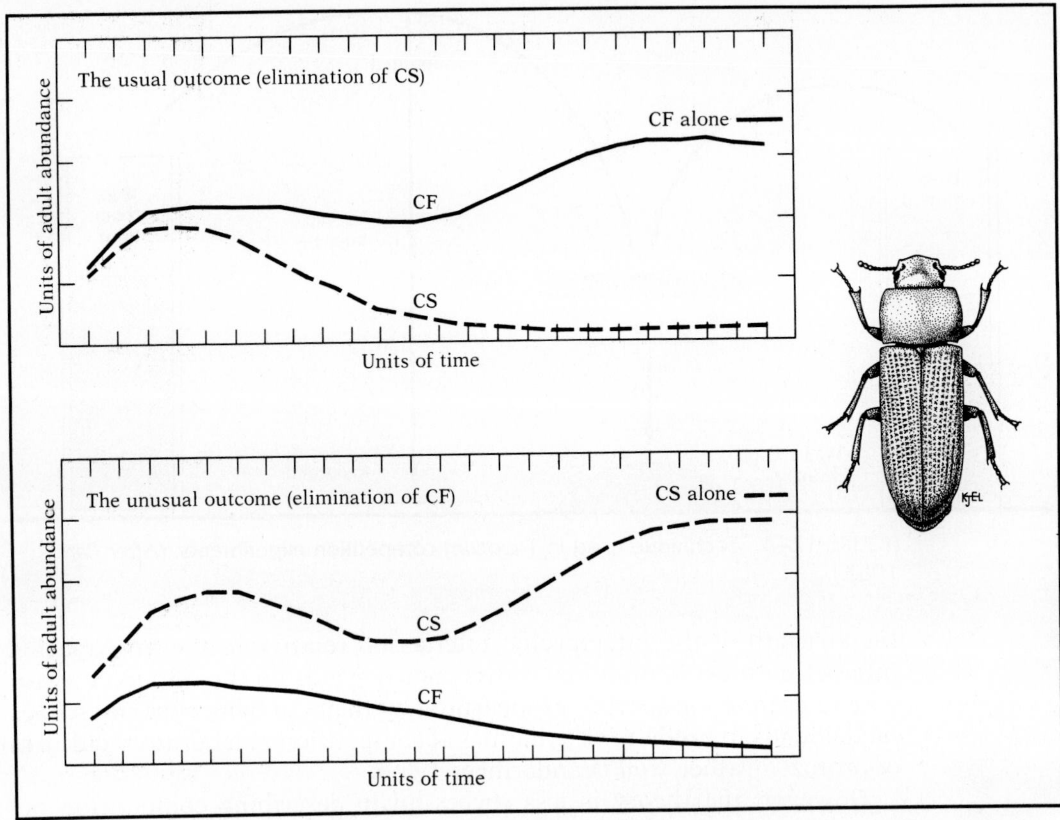

FIGURE 12–5 Smooth curves showing alternative outcomes for *Tribolium confusum* (CF) and *T. castaneum* (CS) when in competition in the cool-moist climate. The beetles are just a few millimeters long. *(After Park 1962.)*

the vials (see Figure 12–5). And, interestingly, the performance of each species when it was alone in these environments did not accurately indicate how it would do in the competition. In a cool-moist environment, CS alone was able to maintain larger populations than CF (see Figure 12–5).

The indeterminacy of the outcome in some environments clearly needed explaining. One hypothesis was advanced by population geneticists, who suggested that genetic variability was present for competitive ability in both CF and CS (Lerner and Ho 1961, King and Dawson 1971). In the hot-arid environment, presumably most CF could outcompete CS, but some CS were genetically better competitors than CF. The outcome in any particular vial would thus be determined by the genetic attributes of the individuals chosen to start the experiment. Experiments demonstrated that this genetic *founder effect* could influence the outcome. Strains could be selected that had changed competitive ability (thereby showing that genetic variability existed for that trait), and more consistent results appeared when inbred beetles (which have less genetic variability) were used as founders.

That report is not the entire story, however. Experiments by Mertz *et al.* (1976) indicated that the outcome of competition in environments that produce indeterminacy can be influenced by random changes in population sizes during the experiments. One species or the other becomes more common simply by chance, and then the more common species has a competitive advantage—an example of demographic stochasticity.

COMPETITION BETWEEN ANIMALS IN NATURE

Many observations have been made of natural situations in which competition seems to be at work—especially situations in which two similar species with overlapping ranges are more restricted in their use of resources or microhabitats where they occur together than where they are separate. In recent years, however, the existence of competition in nature has been documented experimentally with increasing frequency.

In a review of over 150 field experiments, Schoener (1983a) reported that competition was found in more than three-quarters of the species studied, with exploitative and interference competition being about equally common. Another survey, dealing with 527 experiments and 215 species, uncovered competition in about two-fifths of the experiments and over half the species (Connell 1983). Some of the results of these surveys are summarized in Tables 12–1 and 12–2. In general, leaf-eating herbivores and filter-feeders (such as clams) showed less competition than did plants, predators, scavengers, or nectar- or grain-eating herbivores. Marine organisms tended to compete more than terrestrial ones and large organisms more than small ones.

TABLE 12–1 Percentage of all experiments in each category that showed interspecific competition

	Terrestrial		**Marine**		**Freshwater**		**Total**	
	No. exp.	*%*	*No. exp.*	*%*	*No. exp.*	*%*	*No. exp.*	*%*
Plants	205	30	31	68	2	50	238	35
Herbivores	45	20	13	69	0	. . .	58	31
Phytophages*	(22)	(23)						
Carnivores	36	11	5	60	3	67	44	20
Total	286	26	49	67	5	60	340	32

Source: After Connell (1983).
*Phytophages are a subset of herbivores.

TABLE 12–2 Number of experiments in which competition was detected in terrestrial systems, broken down by whether or not enclosures were used to control the presence or absence of potential competitors

	Plants		**Herbivores (P, G, N)[a]**		**Carnivores**	
Competition	*Encl.*	*None*	*Encl.*	*None*	*Encl.*	*None*
Always present	15	59	(3, 8, 3)	(2, 8, 2)	1	6
Always absent	0	22	(1, 0, 0)	(5, 3, 0)	0	0
Sometimes present	3	42	(0, 0, 0)	(1, 1, 1)	0	0

Source: After Schoener (1983a).

[a] P = phytophages (eaters of plant tissues except for grains); G = grain eaters; N = nectarivores.

We will return to competition in Chapters 16 and 17, when the issue of how competition influences community structure is discussed. As you will see, the variability in the occurrence of competition between and within broad groups, as shown in the tables, adds an important complication to assessing its role.

CHAPTER 13

The Predator–Prey Interaction: The Ecology of Attack and Defense

When one organism attacks and injures or kills another in order to eat it, the attacking organism is called a *predator* and the organism that is injured or killed is called the *prey*. Talk of predators and prey usually conjures up visions of lions pouncing on zebras or wolves chasing caribou. Here we are using the terms in a broader sense to include both parasites and their hosts and herbivores and the plants they feed on.

That predators can have strong impacts on prey populations has been frequently demonstrated. For example, transplanting a prey species to an area in which its predators do not occur frequently has led to population explosions of the prey. *Opuntia* cactus was introduced into Australia from South America in the early twentieth century and its insect predators were left behind. It quickly spread to occupy millions of hectares of Queensland and northern New South Wales, effectively preventing the area from being used for farming or grazing.

In 1925, a small predator of *Opuntia,* the moth *Cactoblastis cactorum,* was transplanted from South America to Queensland. It made short work of the *Opuntia* populations, reducing them to scattered patches and keeping them at a low level to this day (see Figure 13–1). Only small patches of the cactus can now be found, and they are soon covered by the moths and destroyed. The moth also persists as small populations and dispersants. Neither species goes extinct and neither species can build a large population in a single area.

FIGURE 13–1 (A) Queensland forest plot choked with *Opuntia* cactus. (B) Same forest plot after introduction of South American cactus-eating moths, *Cactoblastis cactorum.*

THE ORIGINS OF PREDATION THEORY

Gause exposed test-tube populations of *Paramecium caudatum* to the attacks of the predacious protozoan *Didinium nasutum*. *Didinium* is a barrel-shaped, blundering predator that (in what may be an exaggeration) "attacks all objects coming into contact with its seizing organ . . . the collision with suitable food being due to chance" (Gause 1934, p. 114). When the two species were placed together in a test tube in water containing food for *Paramecium*, the results never varied. The *Didinium* found and ate all the *Paramecium*. Then the *Didinium* themselves starved to death, as represented in Figure 13–2A.

Gause next tried creating a refuge for the *Paramecium* by putting some glass wool in the bottom of the test tubes. Some of the *Paramecium* then were able to escape from the *Didinium*, which starved after the accessible *Paramecium* were devoured. The hidden *Paramecium* then repopulated the tube. Thus, the predator population went extinct and the prey population recovered (see Figure 13–2B).

Gause caused the two protozoa to coexist only by simulating migration from an outside source—occasionally introducing more *Didinium* to the system. Few totally isolated systems exist in nature—indeed, few exist in which a predator has only one prey and vice versa. Attempts to get a predator and prey to coexist in microcosms have been successful for only limited periods of time, when the microcosm was large enough to contain a great deal of structural heterogeneity—sites favorable to the organisms separated by barriers. Then migration among the sites prevented the extinction of either species in the entire microcosm, and their population sizes tended to oscillate asynchronously (see Figure 13–2C).

The basic mathematical models for the predator–prey interaction reveal an inherent tendency for predator and prey populations to oscillate with one another. Because other factors may act to stabilize the abundance of predator and prey, these populations do not necessarily oscillate, but they do have a tendency to oscillate. If the stabilizing factors are removed, then the tendency to oscillate becomes expressed.

The first demonstration of a tendency for predator and prey systems to oscillate comes from a simple model called the *Volterra equations*. The model has four assumptions:

1. The prey are assumed to grow exponentially in the absence of predators.
2. The rate at which prey are caught by predators is proportional to the product of the predator and prey population sizes. This assumption can be visualized as predation resulting from random collisions between predator and prey individuals.
3. The rate of birth of new predators is proportional to the rate of prey capture; for example, with every capture of, say, 1000 prey, a new predator is produced.

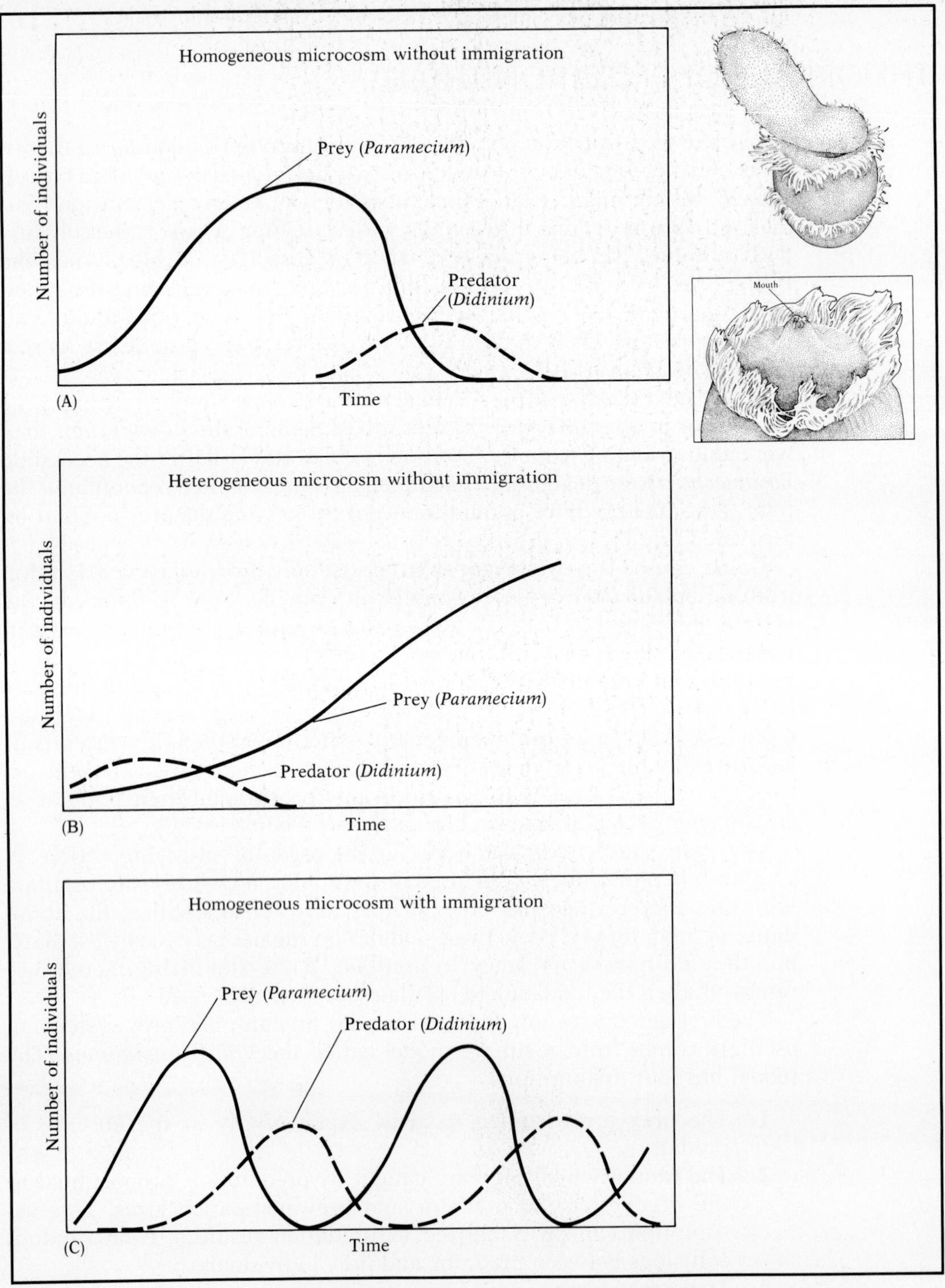

FIGURE 13–2 Predator–prey interactions of *Didinium* attacking a *Paramecium*. *(After Gause 1934.)*

4. The death rate in the predator population occurs at a density-independent rate.

Figure 13–3 is a graph in which the prey population size, V, is plotted along the horizontal axis and the predator population size, P, is plotted along the vertical axis. The graph shows several representative loops, each of which indicates an oscillation. Suppose the system is started with, say, 500 prey and 15 predators; then the number of prey increases quickly and the number of predators increases slowly. On the graph, the system moves to the right along a loop. By the time the prey population has increased to about 2700, the predator population itself starts increasing quickly because a lot of prey are being captured. On the graph, the system starts to move up along a loop. Now as the predators increase, the prey population starts to decline. Thus, on the graph, the system moves toward the upper left along a loop. Finally, the loss of prey causes the predator population to decline again, and, in this model, the starting values of 500 prey and 15 predators are exactly repeated. The cycle then starts all over again.

A special and peculiar property of the original Volterra predator–prey model is that every starting point lies on a closed loop. More specifically, in this model, the amplitude of the oscillations depends *only* on the starting values, whereas the period of the oscillation depends on the numerical values of the parameters in the model. Hence, the trajectories for this model consist of a family of closed loops, each nested within another and surrounding a single point that represents a steady state.

FIGURE 13–3 Family of oscillations produced with the Volterra predator-prey model.

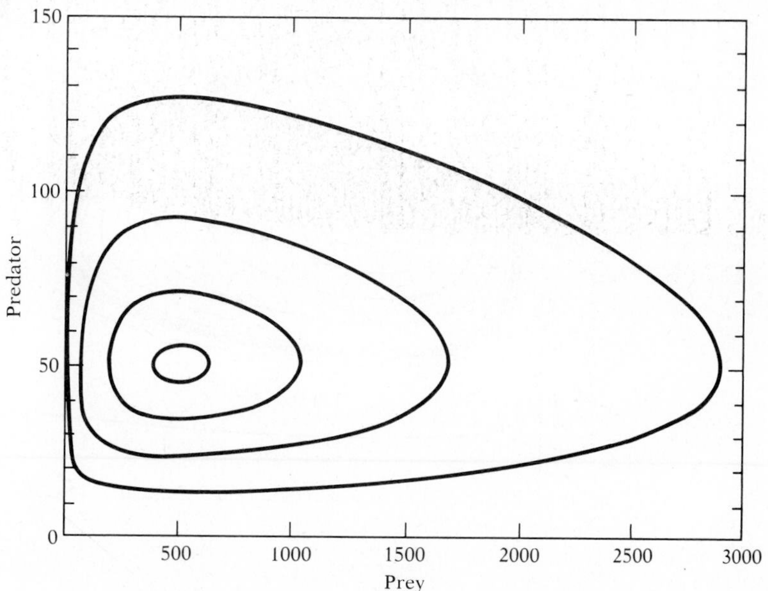

The Volterra model is only of historical interest, because it was the first model that suggested the tendency for predator and prey populations to oscillate. A somewhat more realistic model assumes that the prey grow logistically in the absence of the predator, not exponentially, as in the original Volterra model. In addition, it is more realistic to assume that predators can become satiated when prey are sufficiently abundant; in the original Volterra model, each predator continues merrily capturing prey at the same rate whenever it collides with a prey item, no matter how many prey items are present.

When the assumptions of logistic prey growth and predator satiation are added to the original Volterra model, then, provided both predator and prey can coexist, the trajectories may exhibit three patterns, depending on the strength of the density dependence in the prey population:

1. Density dependence in the prey population tends to stabilize the system; if it is strong, then the predator and prey eventually coexist at steady-state abundances, and they do not even oscillate during their approach to the steady state.
2. If the density dependence is only moderately strong, then the predator and prey eventually coexist without oscillating, but they show a damped oscillatory approach to their ultimate steady-state abundances.

FIGURE 13–4 A limit cycle oscillation produced with a predator-prey model that includes predator saturation and density dependence in the prey.

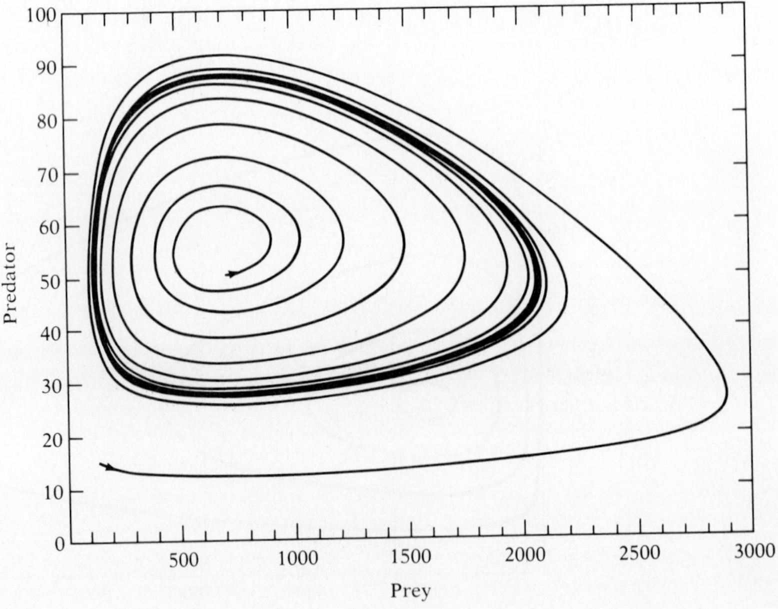

3. Finally, if the density dependence in the prey is weak, then the predator and prey may oscillate indefinitely (limit cycle).

Figure 13–4 illustrates a closed loop that is approached from any starting point on the graph. This loop is called a *limit cycle.* Notice the fundamental difference between a limit-cycle oscillation and the oscillation that occurred in the original Volterra model. The original model exhibited no single loop; rather, an infinity of loops were present, corresponding to all possible starting points. In contrast, the limit-cycle model shows only one loop, and trajectories from all starting points approach it. Thus, in a limit cycle, both the amplitude and period of the ultimate oscillation are determined by the numerical values of the parameters in the model.

Figure 13–4 illustrates the basis for an important prediction from models of predator–prey codynamics that Rosenzweig (1971) termed the *paradox of enrichment.* If the food supply of the prey is enriched, the consequence may be the eventual extinction of the prey population. Enriching the food supply of the prey effectively weakens the density dependence in the prey population and destablizes the system. If the predator and prey populations begin to oscillate as a result, the prey population may become extinct because it is especially vulnerable when its abundance is near the low mark during a cycle of the oscillation.

PREDATOR–PREY CODYNAMICS

We now have many models for the codynamics of predator and prey. The main ingredients in modeling a predator–prey interaction are called the functional response and the numerical response (Holling 1965). The *functional response* describes the rate at which an individual predator consumes prey as a function of the number of prey available to it. The *numerical response* describes the rate at which that predator produces young as a function of the number of prey available. The functional response describes the reaction of an *individual* predator to the availability of prey, and the numerical response describes the reaction of the predator *population* to the availability of prey.

Often interference competition is present among the predators. Interference is not restricted to the territorial behavior of large predators, such as the hyenas, but is found even in *parasitoid* wasps, whose interactions with their prey have been the focus of theoretical interest and laboratory experimentation (Hassell 1978). Parasitoids are predators that kill and consume only one prey individual during their lives. They are an important class of predatory insects, made up mostly of various wasps and flies; they may include 10% or more of all organisms (Askew 1971, Erwin 1982).

Parasitoids are smaller than the prey they attack, and mate finding is done by the female alone. She locates the prey and lays eggs in it, or she finds the prey's food plant and lays eggs on it (so that the prey eats them along with the plant). In either case, the parasitoid larvae feed on the prey, usually from within. When they are small, they behave much like parasites and do not cause gross damage. However, as they mature, they devour the prey, vital organs and all, usually leaving it virtually an empty sack. Basically they change from what would be thought of as a parasite into a predator.

The largest British ichneumonid wasp, *Rhyssa persuasoria,* is about 5 cm long, with an ovipositor of about the same length, as shown in Figure 13–5. It uses the ovipositor to penetrate the trunks of coniferous trees, finding and ovipositing in the wood-boring larvae of the wood wasp, a member of a different suborder of Hymenoptera, the sawflies. The sawfly larvae is located by the smell of a substance in a fungus that characteristically grows in the wood wasp feces.

Female *Rhyssa* interfere with each other by maintaining territories on tree trunks from which they expel other females of the same species (Spradbery 1969, 1970). Interference between parasitoids is not always so dramatic, however. Sometimes the tendency of parasitoids to leave an area increases only when their density increases. Models of parasitoid

FIGURE 13–5 *Rhyssa* ovipositing in a sawfly larva inside a tree trunk (the larva cannot be seen). Near the end of the wasp's abdomen, the ovipositor sheath can be seen looping to the right as the ovipositor itself is driven into the trunk of the tree.

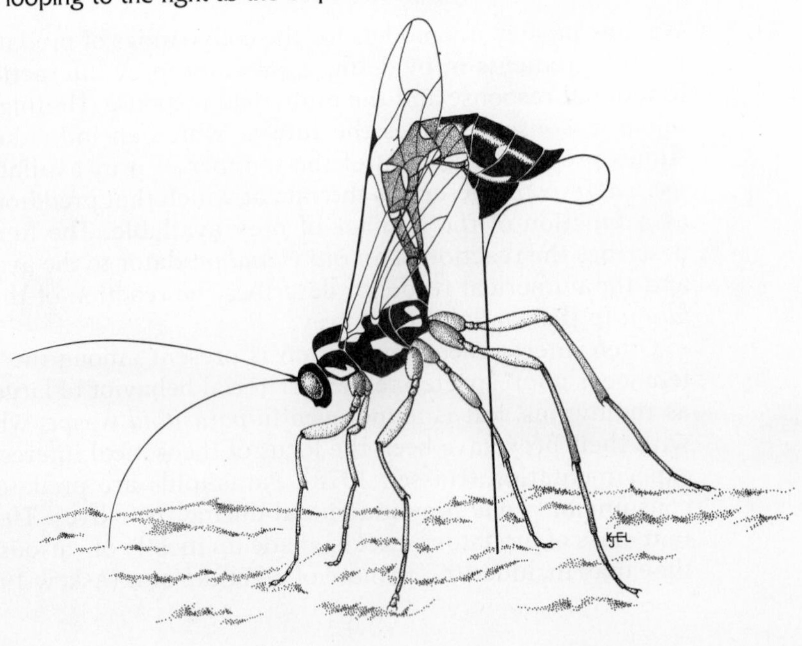

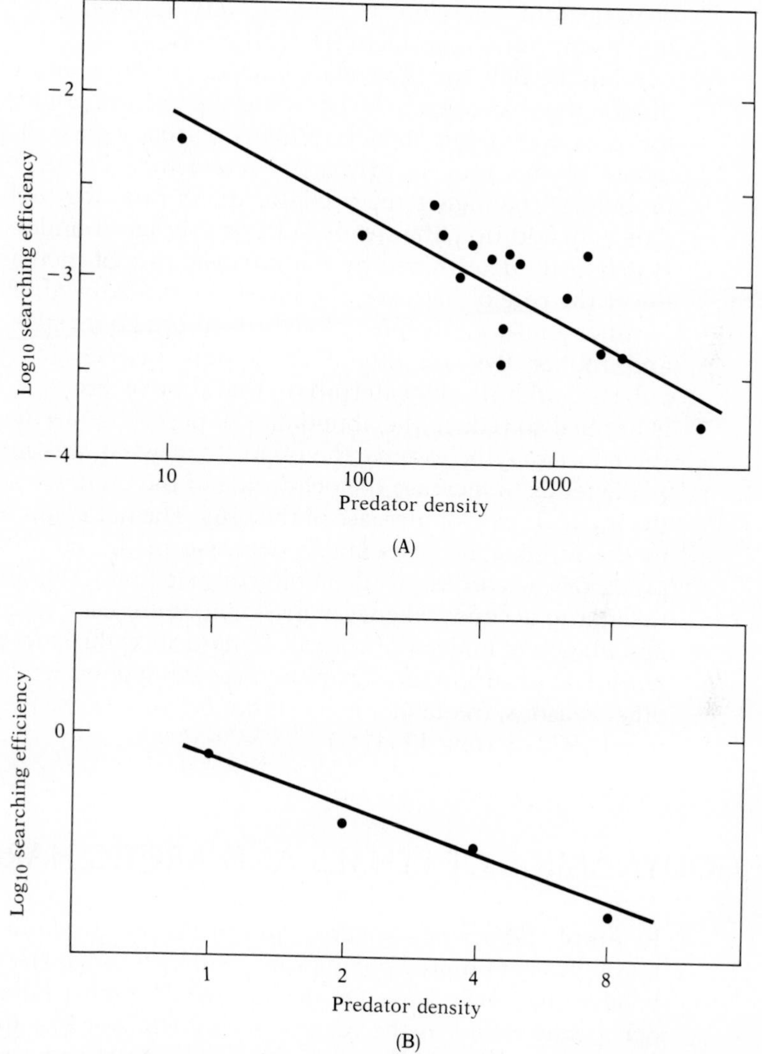

FIGURE 13–6 Graph of the relationship of parasitoid searching efficiency to density. (A) parasite on *Drosophila* larvae; (B) lady bug feeding on aphids. *(After Hassell 1978.)*

behavior quantify this effect with the use of a coefficient of interference (m), the slope of the function relating the searching efficiency of the parasitoid to its density, as shown in Figure 13–6.

Questions about foraging strategies, aggregation, and interference in parasitoids are closely related and are important to the design of programs of *biological control*—the use of predators to control organisms that attack humanity or its crops and domestic animals. The ecological models of parasitoid behavior are thus directly applicable to the solution of an important human problem (Hassell 1978).

Many models for the codynamics of predator and prey lead to a prediction that has come to be known as the *Volterra principle.* In many predator–prey codynamic models, either the steady state or the average abundance of the prey is primarily determined by the death rate of the predators; the higher the predator death rate, the higher the prey abundance. In addition, the steady-state or average abundance of the predators is primarily determined by the intrinsic rate of increase of the prey; the lower the rate of increase, the lower the predator abundance.

Among insects, the prey of arthropod predators such as wasps, spiders, and robber flys are often herbivorous insects—for example, weevils, aphids, and butterfly caterpillars that destroy crops. When an insecticide is applied to reduce the abundance of pests that are herbivores on crops, the effect may be exactly the opposite of what was intended. The insecticide tends to increase the death rate of the predator as well as to reduce the intrinsic rate of increase of the prey. The net result may be an increase in the number of pests and a decrease in the number of their natural predators, according to the Volterra principle. This principle should be taken into account whenever insecticides are used to control pests whose abundance is under the control of natural arthropod predators to begin with. The results of indiscriminant spraying of insecticides is, for this and other reasons, frequently an eventual *increase* in the pest population (Barducci 1972, Strong 1984).

THE CODYNAMICS OF LYNXES AND ARCTIC HARES

In simple laboratory systems, and in theory, predators may have a dramatic impact on prey populations, and vice versa. Have we any reason to believe that they can do so in nature? The most famous purported evidence was found in the records of the Hudson Bay Company's trade in furs. The number of furs of lynxes and snowshoe hares sold to the company annually is an index of the size of the populations of these animals. The population sizes of the two species fluctuate with peaks about every 10 years (Elton and Nicholson 1942). The lynxes, which eat the hares, generally have large populations at the same time or just after the hares do, as Figure 13–7 makes plain.

The codynamics of these species seem easily explained. When hares are abundant, the well-fed lynxes easily raise their kittens, and the lynx populations grow. The increased number of lynxes decimate the hare populations, leading in turn to food shortage and population decline for the lynxes. Relieved of the predation pressure of the lynxes, the hare populations rebound, and the oscillations continue.

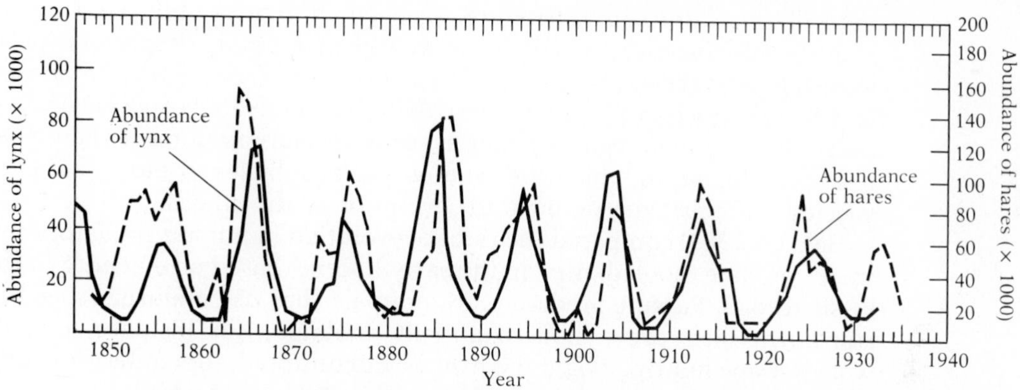

FIGURE 13–7 Abundance of lynx and hares in northern Canada as indicated by records of furs purchased by Hudson Bay Company stores (and other sources).

Unfortunately, this simple idea cannot be the entire explanation; indeed, it may not even be part of it. One difficulty is that, on an island from which the lynxes are absent, the hare populations fluctuate just the same (Keith 1963). The answer to the overall puzzle probably lies largely in the interactions of other factors with the lynx–hare predation system, especially the relationship of the hares to the plants they eat. Perhaps a decline in available food or a change in its quality helps to slow the hare population explosions. One theory is that lack of food ends hare outbreaks, and lynxes and other predators then reduce the hare populations to low densities. In addition, the suggestion has been made that the hares may assault the lynx, not vice versa, by being carriers of a lynx disease.

Whatever drives the hare–lynx cycle, laboratory systems show that, with refuges and migration, interactions of predators and prey can cause coupled cycles. Simple mathematical models, in which predators and prey interact only with each other, show the same pattern. Whether or not regular oscillations in natural populations are ever caused by predator–prey interactions remains to be seen.

PREDATOR–PREY COEVOLUTION

How much does the lynx influence the evolution of the hares and vice versa? Are antelopes always evolving more speed and deception for escaping from lions, and are the lions, in turn, continually evolving into more skillful hunters? Lynx populations certainly can place selective pressures on hare populations, and the ability to catch hares certainly can affect the reproductive success of a lynx. However, owls, wolves, and bobcats, among other animals, prey on hares as well. Lynxes eat a wide variety of small animals. Therefore, each population is only one element placing selection pressures on the other. When one suite of species evolves in response to pressures generated by another suite of species, the process is called *diffuse coevolution* (Gilbert 1975, Janzen 1980, Fox 1981). Diffuse coevolution may be as important as, or even more important than, *pairwise coevolution,* in which the relationship is sufficiently close that each species is the primary actor in the evolution of the other.

Bakker (1983) compared the rates of evolution in pursuit predators and their ungulate (hoofed) prey in the early Eocene, based on evidence in the fossil record. Pursuit predators are animals that use enduring speed to run down their prey; modern pursuit predators, for instance, include wolves, Cape hunting dogs, and hyenas. Pursuit predators differ from ambush predators, such as lions and leopards, which lie in wait for, or slowly stalk, their prey and must catch them in one short sprint. Bakker's morphological data show that both now-extinct mesonychid pursuit predators and their prey became faster and less dependent on the use of cover between 40 million and 60 million years ago. However, the ungulates grew faster more rapidly than the mesonychids, creating an *adaptive gap.*

A good fossil record exists for the genus *Pachyaena,* the only group of distant pursuit predators for the first 3 million years of the Eocene. The record shows, for example, a large adaptive gap between *P. gracilis* and its prey at the start of the record. However, that gap does not close for a period of roughly half a million generations. Only when new kinds of mesonychids replaced *Pachyaena* did the predators start to catch up. Bakker concludes that the *Pachyaena* were unable to evolve in response to the selection pressure presumed to exist because of the adaptive gap.

THE IMPACT OF PREDATION

The importance of predation in the ecology of communities is not often obvious, especially when the prey consists of very early life-history stages. Many marine organisms, fishes, and invertebrates produce enormous numbers of eggs. These eggs hatch and become planktonic larvae that may drift for weeks or months before transforming into the more familiar adult forms. During the egg and larval stages, these populations obviously are subject to intense predation pressure from plankton-feeding predators (planktivores). This interaction is difficult to observe and virtually impossible to measure. Yet its importance is clear from a comparison of the number of eggs produced with the number of transforming larvae.

Plant populations often are subject to intense predation on their early stages. Figure 13–8 shows a stand of California live oaks (*Quercus agrifolia*)

FIGURE 13–8 Live oak *(Quercus agrifolia)* in an area grazed by cattle in Santa Clara County, central California. Note the absence of seedlings. *(Photo by Paul R. Ehrlich.)*

in an area grazed by cattle. Note the total absence of seedlings. Cattle and squirrels are gradually exterminating the oak populations by devouring oak seedlings. The process is not obvious to the casual observer because of the long life of the adult oaks, which live for hundreds of years.

Elegant studies of the impact of predators on early stages have been done by Janzen (1969) on the attack of pea weevils (beetles of the family Bruchidae) on the seeds of legumes. Bruchids are a group of beetles related to leaf beetles that have coevolved with the legumes. Bruchids attack the majority of the legume species that have medium-sized seeds. Those species often have more than 80% of their seeds killed by bruchids.

Thus, one of the main problems in evaluating the impact of predation is to measure it over the entire life cycle of the prey organisms. In general, predation is probably a more important factor in codynamics than is usually thought because of the tendency to focus on examples of predation on adults—the lionesses of the pride pulling down a zebra or the pack of wolves harrying an aging bull moose. The steady attrition of zebra foals and moose calves is less dramatic but doubtless more important. And, after all, who can get excited by the image of a weevil pouncing on a pea?

MODES OF PREDATION

There are virtually as many ways for a predator to catch prey as there are predators. Nevertheless, it is often useful to categorize modes of attack, as in the pursuit–ambush dichotomy. Peckarsky (1982) employed a two-way attack classification in a review of the predator–prey interactions among aquatic insects, dividing them into searchers versus ambushers and engulfers versus piercers. Stoneflies (*Plecoptera*), for example, are searcher-engulfers. They seek out their prey and then chop them up or swallow them whole. Some true bugs, on the other hand, are ambusher-

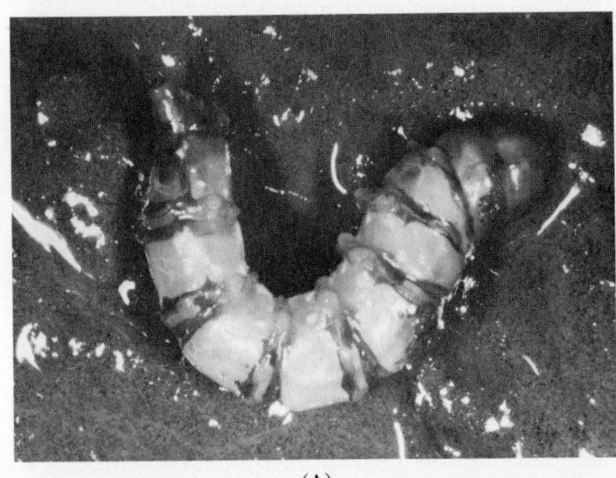

(A)

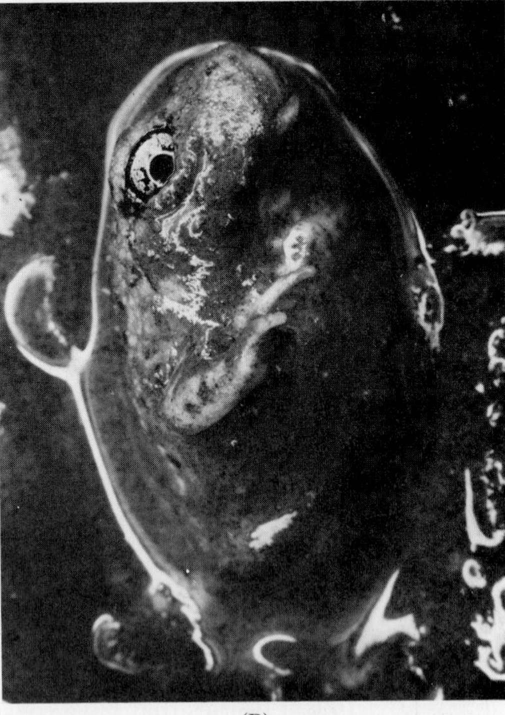

(B)

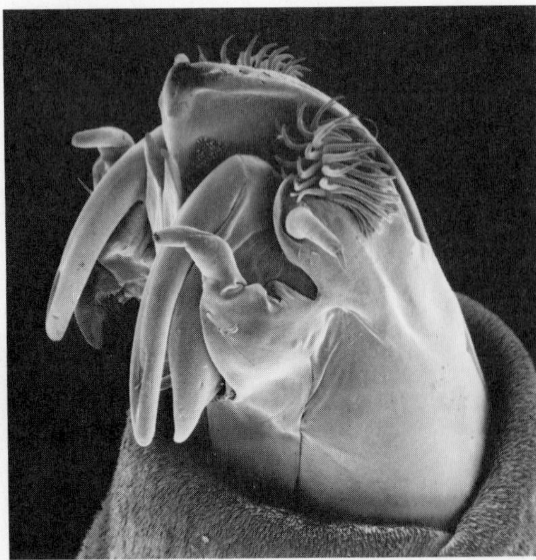

(C)

FIGURE 13–9 An insect acting as a predator on a toad. (A) A maggot of a large horsefly *(Tabanus punctifer)*, which lies submerged in mud. (B) A spadefoot toad *(Scaphiopus multiplicatus)* being pulled under mud by an attacking maggot. (C) An electron micrograph of the head of a maggot showing hooked fangs with which the toads are caught. The elongated depressions in the front of the fangs are thought to be openings of ducts from venom glands. *(Photos by T. Eisner, Cornell University.)*

piercers. They lie in wait for their prey, stab them with their mouth parts, inject toxins or proteolytic or paralytic enzymes, and then suck out the prey's body contents at leisure.

Many insects, in turn, fall victim to frogs and toads that operate as ambush predators. The anurans have visual systems especially adapted to detecting the movement of insects, and sticky tongues hinged at the front of their mouths that can snap out and capture flying prey. Occasionally, however, roles can be reversed. It was recently discovered that larvae of a horsefly (*Tabanus punctifer*) feed on toads (*Scaphiopus multiplicatus*) in Arizona. The large larvae bury themselves in mud, use hooked mandibles to pull the small toads partly into the mud (Figure 13–9), and kill them by sucking out blood and other body fluids (Jackman *et al.* 1983).

One of the most unusual attack modes has evolved under water in a group of cichlid fishes in Lake Malawi, Africa. Many cichlid species in the lake are mouthbrooders. After the eggs are fertilized, they are picked up by the females and carried in their mouths until they have hatched. The young are then transported in the females' mouths. Analysis of the stomach contents of certain other cichlids showed that they feed primarily on fish eggs and young (e.g., Greenwood 1974). Recently, field observations have tied the two phenomena together, showing that some cichlids have learned to exploit the resource of mouth-held eggs and young. The predators stalk brooding females and then ram them in the head (McKaye and Kocher 1983). This ordinarily causes the brooder involuntarily to eject the eggs or young, which are then snatched up by the predator.

MODES OF DEFENSE

Just as a vast array of attack strategies have evolved, a wide variety of defenses have coevolved along with them. Peckarsky (1980) found, for example, that some of the mayfly prey of stonefly predators could detect the presence of the stoneflies in a stream by noncontact chemical stimuli. Observation boxes, one of which is illustrated in Figure 13–10, containing the prey were placed in a stream, and then food coloring was introduced in screen tubes at the start of each experiment. The dispersion of the color permitted Peckarsky to map the area of highest chemical stimulus downstream of the tube. A screen tube containing a predator was substituted for the food coloring, and changes were mapped in the distribution of the prey. Some prey species moved out of the area of highest chemical stimulus when the predator was present in the tube, and moved back into the area when the predator was removed.

Peckarsky also found that that some mayfly species, on encountering stonefly predators, froze and assumed a "scorpion posture," as shown in Figure 13–11. This posture may increase the mayfly's apparent size and

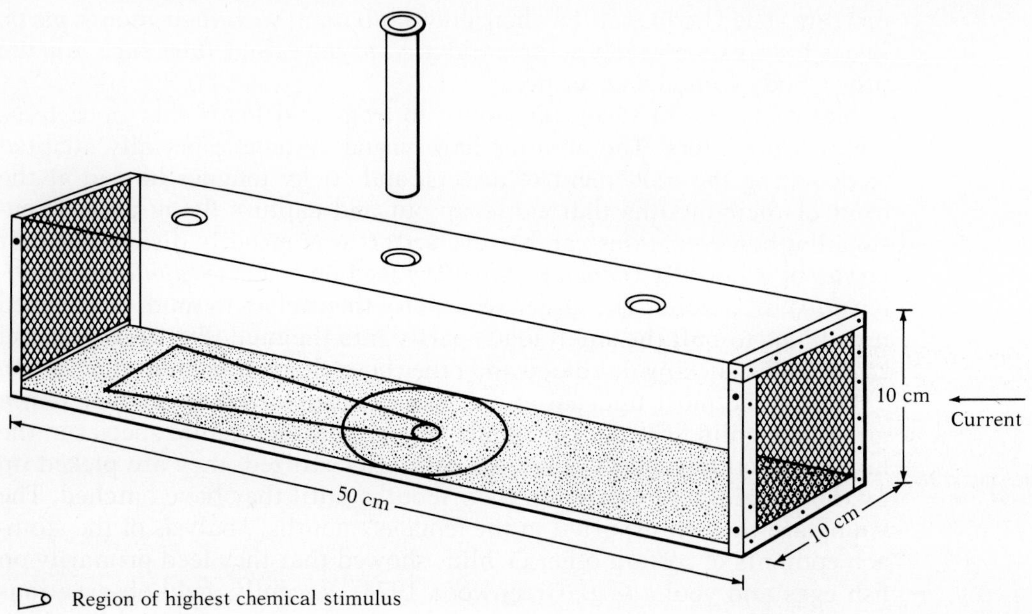

FIGURE 13–10 Observation box with a screen tube inserted. *(After Peckarsky 1980.)*

FIGURE 13–11 Mayfly nymph in a (A) normal resting posture, (B) low-intensity "scorpion" posture, (C) moderate-intensity scorpion posture, (D) high-intensity scorpion posture. *(After Peckarsky 1980.)*

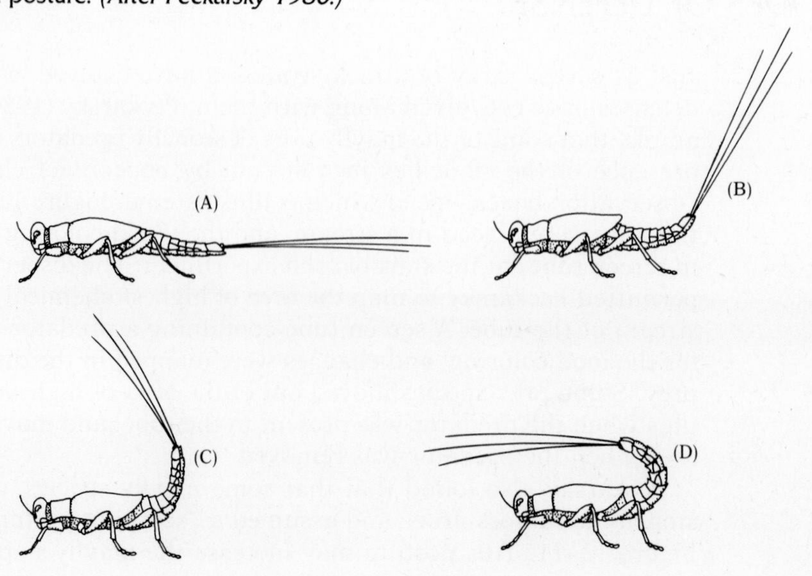

cause it to be rejected by the predator. If true, this act parallels the behavior of many vertebrates that erect hair, feathers, or skin flaps so that they appear larger when threatened.

Chemicals are not only employed by prey for detecting the presence of a predator; as anyone who has had an encounter with a skunk can testify, they may be used quite directly and effectively in defense. The ubiquity of chemical defense mechanisms in arthropods has been shown by Eisner and co-workers (see, e.g., Eisner 1958, 1960, 1970, Eisner and Meinwald 1966, Roach *et al.* 1980, Eisner *et al.* 1981).

One of the most detailed investigations has been of the mechanism by which bombardier beetles ward off the attacks of ants. When these beetles are bitten by ants (or by forceps simulating ants), the flexible end of the beetle's abdomen pivots toward the attacker. Muscular contractions push secretions from a pair of glands into a chamber, where they are mixed with crystalline enzymes. The mixture produces a chemical explosion and the projection, with a "pop," of a boiling hot, noxious spray that repels the ant (Aneshansley *et al.* 1969). As Figure 13–12 demonstrates, Eisner has discovered using ultra-high-speed photography that the spray is not

FIGURE 13–12 Bombardier beetle *(Stenaptinus insignis)* releasing its hot spray toward a forceps that is being used to imitate the attack of an ant. The beetle is secured by a wire attached to its back. *(Photo by T. Eisner and D. Aneshansley, Cornell University.)*

a continuous stream but rather a series of pulses. The chamber has evolved a structure that produces this pulsing and thus keeps the reaction going at the proper rate without internal overheating. Furthermore, the beetles have developed an intricate series of morphological and behavioral traits to ensure proper aiming and distribution of the spray (Eisner, pers. comm.; Eisner and Aneshansley 1982).

To find truly unusual predator defense strategies, let us descend once again into the waters of Lake Malawi. There one can find the young of mouthbrooding cichlids schooling with the young of large predatory catfish and being protected by the catfish (McKaye and Oliver 1980). The cichlid young are kept at the periphery of the school, where they apparently buffer the catfish young from lateral attack (see Figure 13–13). When the catfish parents are chased away, the cichlid young are devoured first (S. Louda and K. McKaye, pers. comm.). The cichlids are thought to benefit too, for they are able to feed more and grow more rapidly outside of the mother's mouth.

Across the world in Laguna de Jilóa, Nicaragua, defense strategies are even stranger. There adults of the cichlid *Cichlasoma nicaraguense* have been observed by McKaye (1977) guarding the young of the largest cichlid in the lake, *C. dovii*—a predator on *C. nicaraguense.* This peculiar behavior of prey altruistically helping the predator has been explained by the idea of kin selection (McKaye 1977, 1979). The offspring of the *C. nicaraguense* that help defend the predator's young may benefit because the impact of *C. dovii* is much more severe on the competitors of *C. nicaraguense* than on *C. nicaraguense* itself.

FIGURE 13–13 A predatory catfish *(Bagrus meridionalis)* in Lake Malawi, Africa, defending young cichlid fishes along with its own young. *(Drawn from a photograph by Kenneth R. McKaye.)*

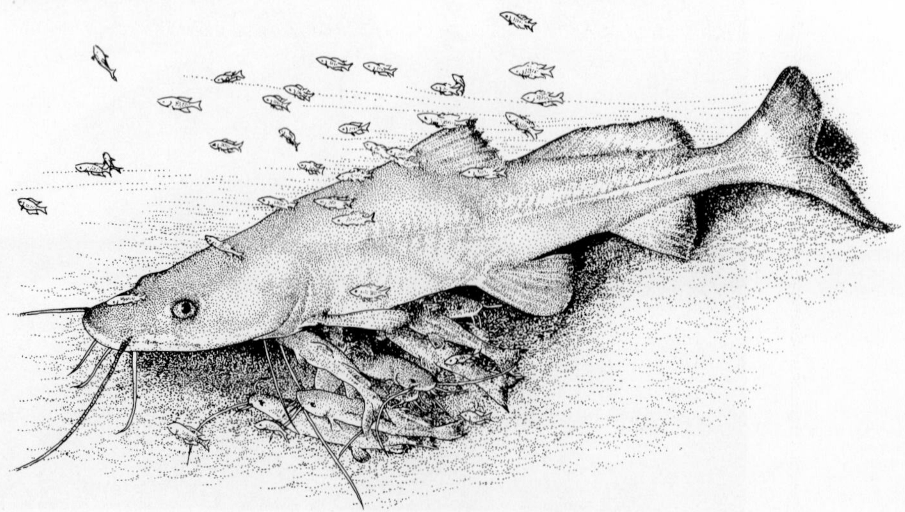

One of the most ingenious hunting systems used by predators is the echolocating "sonar" of bats. The bats generate ultrasonic pulses and then pinpoint their prey by the bearings of returning echoes (Griffin *et al.* 1960). Moths have countered this attack by developing paired "ears" at the base of their abdomens and elsewhere with which they can detect the bat sonar and attempt to take evasive action (Roeder and Treat 1960, Roeder 1962, 1964, Michelsen 1979). Moths capable of hearing the bats are 40% less likely to be captured than those that cannot. Some arctiid moths go further and produce sequences of ultrasonic clicks to "jam" the bat sonar (Fullard 1977, Fullard *et al.* 1979, Fullard and Barclay 1980).

In turn, the bats may attempt to counter the moths' defenses by shifting their sonar frequencies away from those to which the moths are most sensitive, or they may stop echolocating and use sounds produced by the moths to find them (Fenton and Fullard 1981). A bizarre concomitant to this story is the infestation of the ears of many moths by ear mites (Treat 1954). Colonies of the mites occupying an ear destroy its ability to pick up the sonar of the bats. However, the mites have developed behavioral patterns for colonizing the moths that normally ensure that only one ear is infested and the moth can still hear—an obvious advantage to the mite colony, which perishes if its host is devoured by a bat (Treat 1957, 1960).

Other insects also have evolved clever defenses against the threat of vertebrate attack. (One whole class is dealt with under mimicry in Chapter 14.) Birds have learned to detect the presence of caterpillars by the damage they do to leaves (Heinrich and Collins 1983). In response, palatable caterpillars have evolved feeding behaviors that tend to conceal their feeding damage from the predators. For example, rather than leaving the ragged remains of a leaf on the plant, they will often sever the unfinished portions (Heinrich and Collins 1983).

A variety of defenses can itself be a defense (Rand 1967, Clarke 1969, Milstead *et al.* 1974, Ricklefs and O'Rourke 1975, Sargent 1978). The basic idea is simple. If prey have a diversity of behaviors or appearances, predators have more difficulty predicting the behavior of the prey or forming search images. For example, whiptail lizard (*Cnemidophorus tigris*) seems to suffer less predation—as indicated by broken tails—when it lives with several congeners that have diverse escape behaviors than when it occurs alone (Schall and Pianka 1980).

PARASITE–HOST INTERACTIONS

When one organism attacks another but normally does not kill it outright, the predatory organism is usually called a *parasite* and its prey a *host.* Thus, a malarial organism, a protozoan of the genus *Plasmodium,* living in the red blood cell of a human being is a parasite. So is an oak moth

FIGURE 13–14 A lamprey attacking a fish. *(Tom Stack/Tom Stack and Associates/BPS.)*

caterpillar (*Phryganidia californica*) munching on the leaf of a *Quercus agrifolia*. For convenience—that is, because this division is traditional—we will consider parasites with animal hosts in this section and parasites (herbivores) with plant hosts in the next.

Some animal parasites, such as lampreys (shown in Figure 13–14), leaches, ticks, and mosquitos, are relatively large, obvious, and straightforward. They attach to the outside of their prey, suck their blood or other body fluids, and then detach themselves and go about their business. The host may be weakened by a massive attack but only rarely is killed outright.

Other parasites are less obvious. For example, sabre-toothed blenny fishes (*Plagiotremus*) of Australia's Great Barrier Reef are about the size and shape of a human forefinger. They live in holes in the reefs, from which they watch for the larger fishes that are their hosts. When a suitable host swims by, they dash out at high speed, bite off a small piece, and then dash back into their shelters—all in the space of a second or so. The closest terrestrial equivalents, horseflies (Tabanidae), are sluggish by comparison.

The most interesting parasites, however, are those such as various protozoa and worms, which have evolved complex life histories that involve multiple hosts. Malarial parasites (*Plasmodium*) are an example of the protozoa. Each species spends its life cycle alternately in two hosts—an

Anopheles mosquito and a vertebrate. Four species have as their vertebrate host *Homo sapiens,* and as a result they have caused one of the most widespread and serious diseases of humanity. The life cycle of the most lethal species of human malarial parasite, *Plasmodium falciparum,* is shown in Figure 13–15.

Thorough understanding of the complex ecology of a parasite such as *Plasmodium* is essential to its control. The environment must be satisfactory for *both* hosts, and the numbers and condition of the hosts must be suitable for maintenance of a *Plasmodium* population. Humanity attempts to control malaria primarily by altering the environment of the mosquito at the larval stage (draining swamps, introducing predators to aquatic environments), by changing the environment of the mosquito adults (by spraying pesticides), by preventing adults from reaching people (screens), and by making *H. sapiens* a less suitable host (improved nutrition, antimalarial drugs).

Unfortunately, *H. sapiens* also often alters the environment in ways that encourage malaria—most commonly by providing new aquatic habitats such as rain barrels, water-filled hoofprints of domestic animals, fish ponds, and reservoirs, which are sites for anopheline mosquito larvae. For example, around 1960 in the Demerara River estuary of Guyana, people began to convert a subsistence agroecosystem based on maize and cassava to a cash system based on cash cropping of rice (Desowitz 1981). The mosquito, *Anopheles aquasalis,* was the prime potential vector of malaria in the area. Whereas the subsistence system had provided few breeding habitats and its population density was low, the new rice fields became ideal nurseries.

Even so, malaria would not have become a problem because *A. aquasalis* prefer to suck blood from domestic animals rather than from *H. sapiens.* Unfortunately, however, the countryside was cleared for rice fields, and livestock that formerly were abundant were reduced because of mechanization and a lack of pasture. Lacking their preferred vertebrate hosts, the *A. aquasalis* turned to people for sustenance, resulting in a malaria epidemic with heavy mortality among children. The codynamic interaction of *A. aquasalis* and *H. sapiens* was underlined later when a successful malaria control program succeeded in reducing infant mortality to one-third of its previous level and produced a human population explosion (Desowitz 1981).

Similar codynamic situations have occurred in many parts of the tropics and subtropics (Desowitz 1981). For instance, the Kano River rice development scheme in Kenya resulted in habitat alterations that shifted the mosquito community from 99% *Mansonia* (a genus that does not carry malaria) to 65% *Anopheles gambiae* (an excellent vector). In general, clearing of tropical moist forest tends to promote malaria, because *Anopheles* tend to prefer sunlight for mating and need standing water for their larvae—two requisites that are in relatively short supply in the forest.

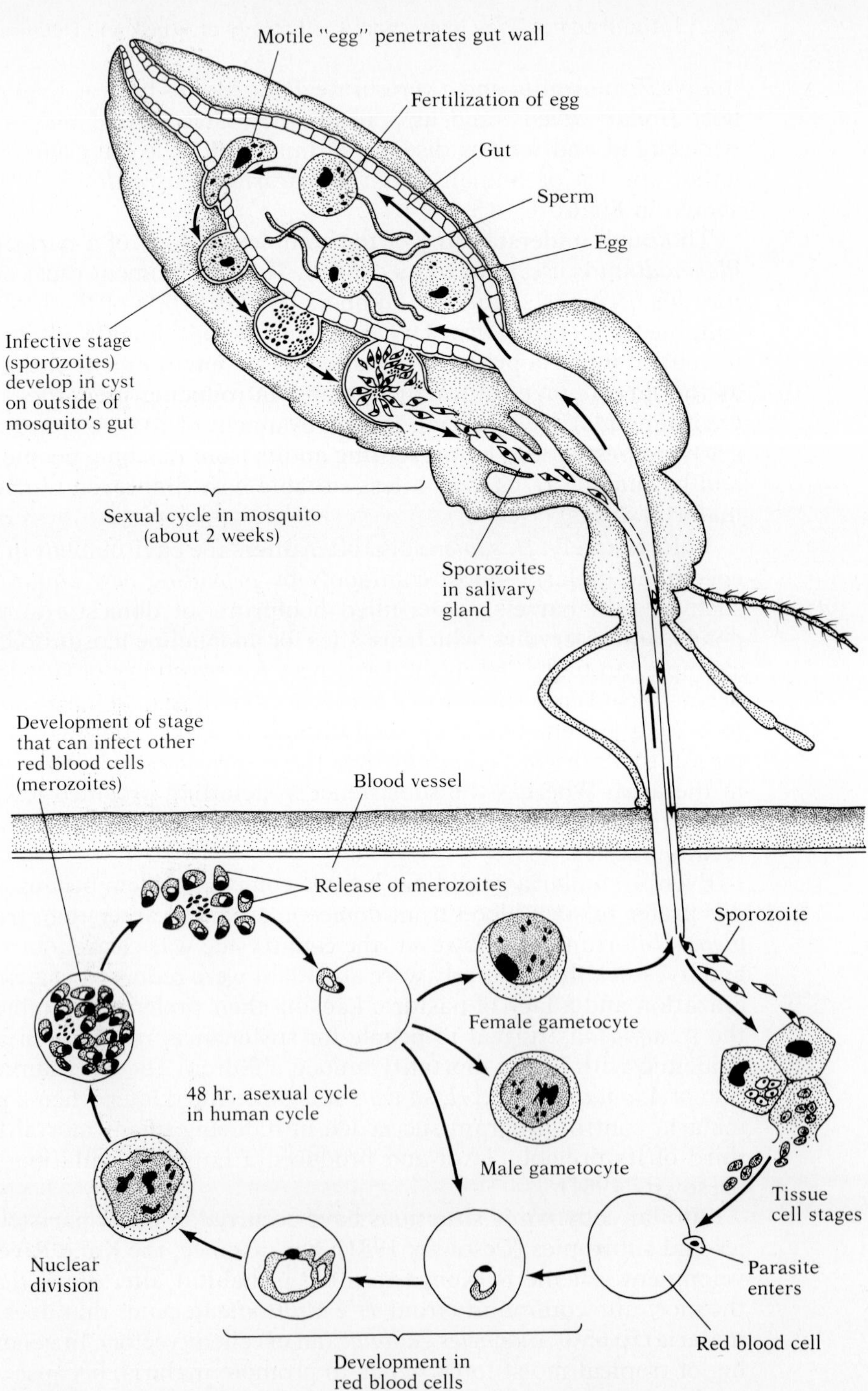

FIGURE 13–15 Diagram of the malaria (*Plasmodium falciparum*) life cycle.

That *H. sapiens* has coevolved with *Plasmodium* is indicated by varying degrees of malaria resistance shown by human populations. The most thoroughly studied case is that of resistance to *P. falciparum* shown in parts of Africa by children that are heterozygous (Hb^AHb^S) at the locus that controls the type of hemoglobin produced in the blood. Individuals with nothing but "normal" hemoglobin (Hb^AHb^A) are more susceptible to malaria; the other homozygotes (Hb^SHb^S) suffer a severe *sickle-cell anemia* and normally die before they reach reproductive age. However, the heterozygotes normally do not suffer from anemia and enjoy relative protection from malaria. In a malarial environment, they usually are more fit than either homozygote. The result, as explained in Chapter 6, is the maintenance of a stable polymorphism.

Malaria is not alone among protozoan diseases that alternate between insect and vertebrate hosts. The dreaded Gambian and Rhodesian *sleeping sicknesses* of Africa are caused by two hemoflagellates (protozoa that live in the bloodstream), *Trypanosoma gambiense* and *T. rhodesiense,* respectively. For *T. rhodesiense,* human beings are the only important vertebrate host, and the insect hosts are tsetse flies of the *Glossina palpalis* group. That tsetse fly stays in the vicinity of water and pupates in the shade. Those habits, and the inability of *T. gambiense* to colonize game herds, is probably what restricts Gambian sleeping sickness to forest belts (Hunter *et al.* 1966).

In contrast, *T. rhodesiense* can colonize wild animals, and its vectors of the *Glossina morsitans* group are open country organisms; that disease occurs in savanna areas. Similarly, *T. brucei,* which is also vectored by *G. morsitans,* appears in savanna areas and causes the disease known as *nagana* in domestic animals. The presence of the two sleeping sicknesses and nagana has had profound effects on the distribution and abundance of *H. sapiens* in Africa, long keeping human population densities low in some areas.

Certain blood flukes of the genus *Schistosoma* use snails and people (and other primates) as alternate hosts. In people, their attack causes a disease known as *schistosomiasis* or, in Africa, *bilharzia.* The life cycle of the *Schistosoma* that causes bilharzia, *S. haematobium,* is shown in Figure 13–16. As with malaria, development activities have influenced the size of schistosome populations. For example, the completion of the Aswan High Dam in Egypt permitted the conversion of large portions of the Nile floodplain from a one-crop irrigation system to a four-crop rotation system. This provided ideal conditions in irrigation canals for the snail hosts of *S. haematobium.* This event led to a fivefold increase in bilharzia along the Nile following completion of the dam (Ehrlich *et al.* 1977b). Various behavioral, immunological, and genetic studies indicate that schistosomes and other worms coevolve extensively with their hosts (for a fine review, see Holmes 1983).

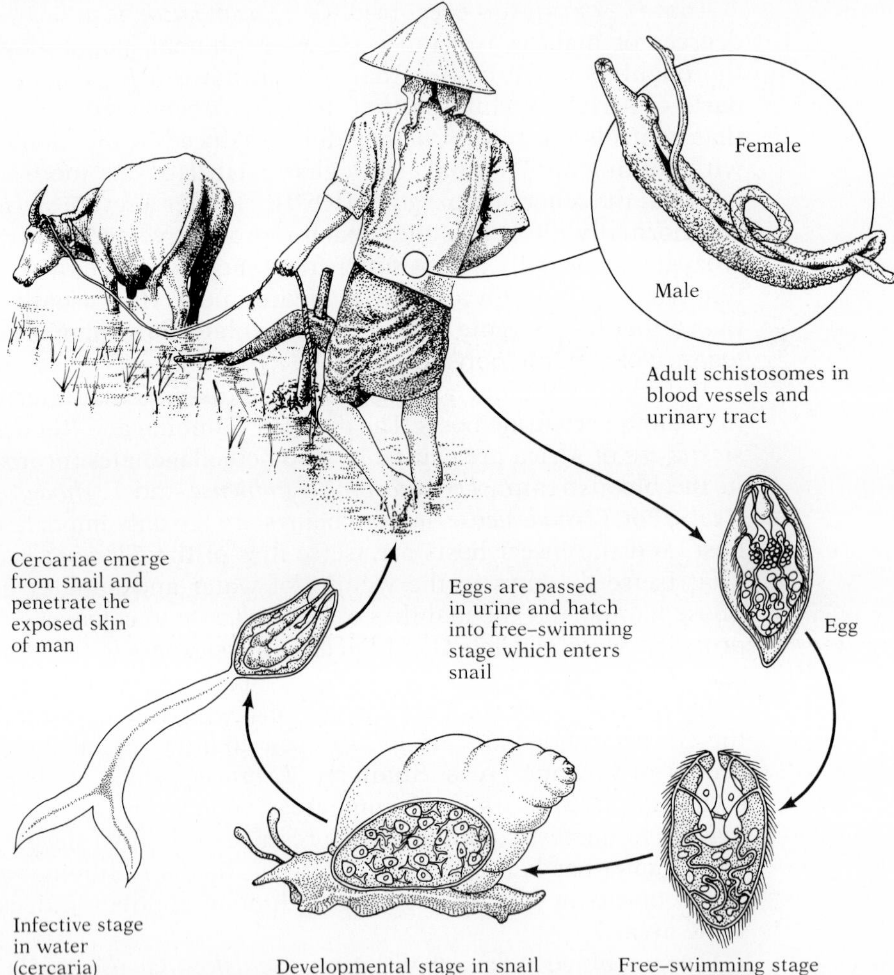

FIGURE 13–16 Diagram of the life cycle of *Schistosoma haematobium.*

The diversity of parasitic attack tactics is truly extraordinary. Consider, for example, avian brood parasites, the best known of which are cuckoos. They lay their eggs in the nests of other birds, eggs that usually mimic the host's eggs in appearance. Cuckoo young, when they hatch, push the host's young over the side of the nest and monopolize the food-transporting capabilities of their foster parents. The complexities of parasite–host codynamics and coevolution in these systems of brood parasites are just beginning to be unraveled (see, e.g., Smith 1968, Rothstein 1971, 1975). In some situations, the parasites may actually increase the fitness of their hosts and become, in effect, mutualists, as discussed further in Chapter

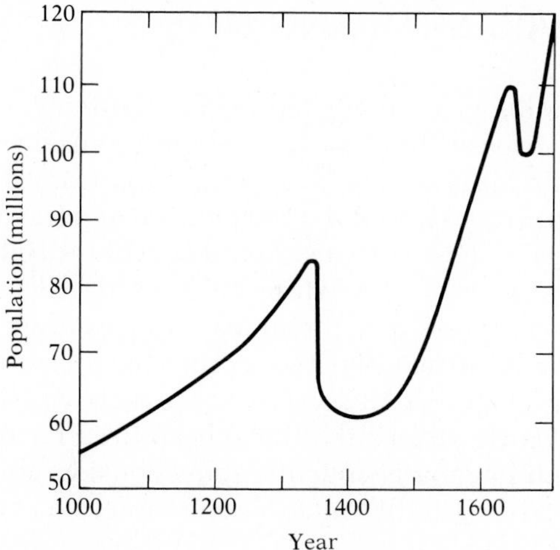

FIGURE 13–17 The effects of the bubonic plague on the size of the human population of Europe in the fourteenth and seventeenth centuries. The curve is estimated from historical accounts. *(After Ehrlich et al. 1977a.)*

14. For example, in Panama, the giant cowbird parasitizes the nests of oropendulas, large tropical orioles that build conspicuous pendulous nests in colonies. In some situations, the cowbird chick actually saves the lives of its oropendula nestmates by eating the larvae of botflies and blood-sucking mites that might otherwise kill them (Smith 1968).

Alternatively, near the other end of the parasite size scale from cuckoos, Bradleigh Vinson and colleagues have found that parasitoid wasps of the family Ichneumonidae have a mutualistic relationship with a virus (see, e.g., Edson *et al.* 1981). The virus can replicate only within the ovaries of the insect. In return, the virus suppresses the host caterpillar's immune system, preventing the defensive encapsulation of the wasp egg.

That many parasites can play an important role in the dynamics of host populations is in little doubt. People concerned with problems of biological control of organisms classically have used the introduction of parasites as a method of controlling insect and plant pests. For example, the chalcid wasp parasite *Metaphycus helvolus* has been responsible for the effective suppression of black-scale insects on citrus in many areas (DeBach 1974). In addition, much anecdotal evidence suggests that high death rates may result from parasite attacks; the famous black plague (see Figure 13–17) is a well-documented example, but many others exist as well. A pandemic (widespread epidemic) of the rinderpest virus at the end of the nineteenth century appears to be largely responsible for the distribution patterns of antelopes and other artiodactyls in East Africa today (May 1983a). Parasites are important contributors to mortality in populations of whales and dolphins (Delyamure 1955, Perrin and Powers 1980). Fungus infection ends outbreaks of the tropical moth *Zunacetha annulata* (Wolda and Foster 1978).

THE COEVOLUTION OF PARASITES AND HOSTS

Unfortunately, epidemiologists interested in the maintenance and transmission of infectious diseases have not concerned themselves much with the impact of those diseases on the host population (May 1983). One of the most general assumptions about the coevolution of parasites and hosts is that successful parasite–host systems should coevolve until the parasite is relatively harmless (Burnet and White 1972). After all, does not the parasite reduce its own fitness if it threatens its host—that is, threatens to destroy its habitat? Evidence that this occurs has been found, for example, with native antelopes in Africa that suffer little when they harbor *Trypanosoma brucei*. Cattle strains that have been bred for many generations in areas in which nagana is endemic show considerable morbidity, and newly imported cattle usually suffer a severe infection and die unless treated (Allison 1982).

However, some investigators have shown that the evolution of reduced virulence is not inevitable (see, e.g., Roughgarden 1975, May and Anderson 1983). What happens in a parasite–host coevolutionary sequence will depend on the connection between the production of transmission stages of the parasite and its virulence. For example, many viruses of insects must kill their hosts in order to be transmitted. In such a system, coevolution toward low virulence obviously is unlikely.

The evolutionary ecology of parasites is a fascinating and underexplored area. Parasites are extreme specialists in resource exploitation and are adapted to small, discontinuous, temporary environments (Price 1977, 1980).*

HERBIVORE–PLANT COEVOLUTION

In recent years, more attention has probably been focused on the relationship between herbivores and the plants they eat than on any other set of population interactions. In part this interest can be traced to the widespread acceptance of the notion that plant biochemicals play key roles in mediating the relationship between herbivores and plants (Ehrlich and Raven 1964). Throughout the butterflies, related species tend to feed on plants with the same kinds of "secondary" chemicals—chemicals with no obvious function in the plant's metabolism. Swallowtails of the *Papilio machaon* group (Papilionidae) feed on plants containing similar essential oils, cabbage butterflies (Pierinae) on plants containing mustard oil glu-

* For further reading on these and related topics, we recommend recent papers by Barrett (1983), Ehrman (1983), and May and Anderson (1983).

cosides. The cabbage butterflies do not eat plants such as anise, a swallowtail favorite; swallowtails refuse cabbage and nasturtium, favorites of pierines.

Ehrlich and Raven used this and other evidence to conclude that butterflies and plants had been running a "selectional race" in which the plants produced poisons in attempts to ward off herbivores and the butterflies countered by evolving the ability to detoxify certain compounds. Most butterfly species can feed on only one or a few plants of a single plant family. No butterfly can feed on plants of more than a few families, and no butterflies have evolved the ability to feed on some plants, such as ferns. These observations are signs of plant success. In contrast, the vulnerability of many plant groups to the attacks of restricted groups of butterflies—such as the ease with which monarch butterflies (Danainae) gobble milkweeds (Asclepiadaceae)—show instances in which plants have at least partly lost the selectional race.

These conclusions were controversial at first. Many scientists had thought that plant secondary compounds were excretory products, although in retrospect it seems strange to assume that plants would bother to make metabolically expensive, energy-rich excretory products to deposit in their leaves and flowers. It was also argued that herbivores did not put enough selection pressure on plants to cause them to evolve chemical responses. After all, the world was green and herbivore damage not very prominent.

In the late 1960s, however, evidence began to build that small herbivores could influence greatly the reproductive output of plants. For example, the impact of one small butterfly on the seed set of lupine plants in Colorado was shown often to be extremely high. The blue butterfly *Glaucopsyche lygdamus*, with a wingspread of about 25 mm, lays its eggs on the unopened inflorescences of lupines, as shown in Figure 13–18. The larvae of the butterfly feed on the buds and flowers. By carefully removing the *Glaucopsyche* eggs from every other inflorescence in lupine populations, one may compare seed sets in plants subject to attack by the herbivore with those free of attack. In some lupine populations, 85% or more of the flowers are destroyed by the larvae, with a corresponding loss of seed set (Breedlove and Ehrlich 1968, 1972).

One would expect, therefore, that plants with successful chemical defenses against *Glaucopsyche* would be at a considerable selective advantage. Because some populations of lupines are more heavily attacked than others (see the data in Table 13–1), the system seemed an ideal one in which to test the notion that plant biochemicals serve as defense mechanisms. Lupines were well known to contain poisonous alkaloids. A careful analysis of the alkaloid content of the plants in the various populations was undertaken. The working hypothesis was that the populations *least* subject to attack would have the *highest* alkaloid content.

FIGURE 13–18 Female blue butterfly, *Glaucopsyche lygdamus,* ovipositing on an unopened lupine inflorescence. *(Photo by Paul R. Ehrlich.)*

TABLE 13–1 Susceptibility of Colorado lupine populations to predation by *G. lygdamus*

Population		Use by *G. lygdamus*	
Locality	*Species of Lupinus*	*Predation*[a]	*Availability*[b]
Jacks Cabin	*L. bakeri*	1	8
	L. caudatus	19	1104
	L. floribundus	94	1762
Gold Basin	*L. bakeri*	3	0
	L. floribundus	86	2632
Upper Gold Basin	*L. bakeri*	3	104
	L. caudatus	30	880
	L. floribundus	195	1737

Source: After Dolinger *et al.* (1973).

[a] Numbers of *G. lygdamus* ovipositions on 100 lupine inflorescences (see text).

[b] Numbers of young flower buds per 100 lupine inflorescences (see text).

That hypothesis proved false: The *amount* of alkaloid was not closely related to the attack rate. However, a clear relationship *was* found between alkaloids and herbivory. The lupine populations that suffered the highest predation were the ones with a set of alkaloids that varied little from plant to plant. In contrast, those populations that were least damaged by *Glaucopsyche* showed great variation from plant to plant in the kinds of alkaloids they contained and in their total alkaloid content (Dolinger *et al.* 1973). These results are shown in Figure 13–19.

How can these differences be accounted for? The most likely explanation seems to be that by presenting the butterflies with a variable defense, the plants are able to retard the evolution of resistance to alkaloid poisoning in the butterfly population. Butterflies that developed as larvae on plants containing one mixture of alkaloids would lay their eggs on plants with different mixtures. Each generation of *Gluacopsyche* would not be exposed to identical selection pressures.

How is the variation maintained in the plant population? Most likely *frequency-dependent selection* is involved. Whenever an alkaloidal type starts to become common, successive generations of *G. lygdamus* tend to be exposed to the same array of poisons, and the butterflies start to become resistant to that array. This situation increases the pressure on that type, and it becomes less common. Therefore, the alkaloidal type in highest frequency is always selected against, and thus the population remains variable (Dolinger *et al.* 1973).

Why all lupine populations do not maintain such variability is not understood. *Glaucopsyche,* however, is not the only predator on lupines. Populations more exposed to the attacks of unspecialized large herbivores, such as deer and cattle, which consume parts of several plants in a single mouthful, may be better deterred by a high dose of a constant set of alkaloids. Certainly the variable strategy would have no effect on their ability to become resistant.

Early support for the importance of herbivory to plants and the defensive role of secondary compounds came from the now-classic work of Janzen (1966) on the mutualistic relationship between ants and certain acacia trees in Central America. Swollen thorn acacias (e.g., *Acacia cornigera*) have an obligatory relationship with ants that live in their swollen thorns. The ants get shelter and feed on special structures produced for them by the acacias, as shown in Figure 13–20. In return, the ants are active day and night and defend the acacia against herbivores; they also prune away competitive plants.

The value of the ants to the acacias was clearly demonstrated by experiments in which the growth and survivorship of trees was tested with and without ants and by the comparative occurrence of plant-eating insects on ant-guarded and ant-free *Acacia* (see Table 13–2). Furthermore, acacias not defended by ants had bitter-tasting leaves, indicating the pres-

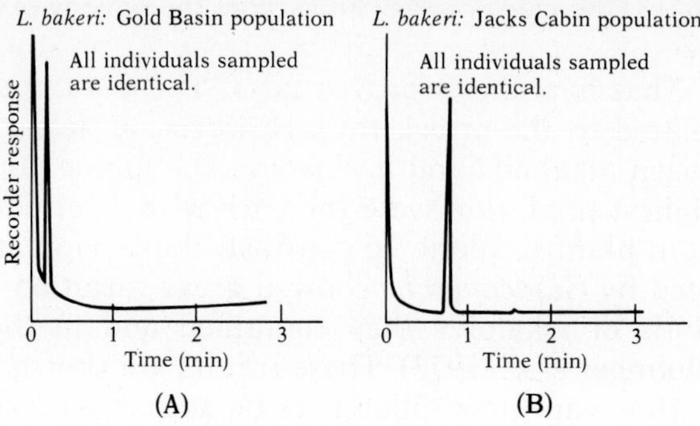

L. bakeri: Gold Basin population
All individuals sampled are identical.
Recorder response
Time (min)
0 1 2 3
L. bakeri: Jacks Cabin population
All individuals sampled are identical.
Time (min)
0 1 2 3

(A) (B)

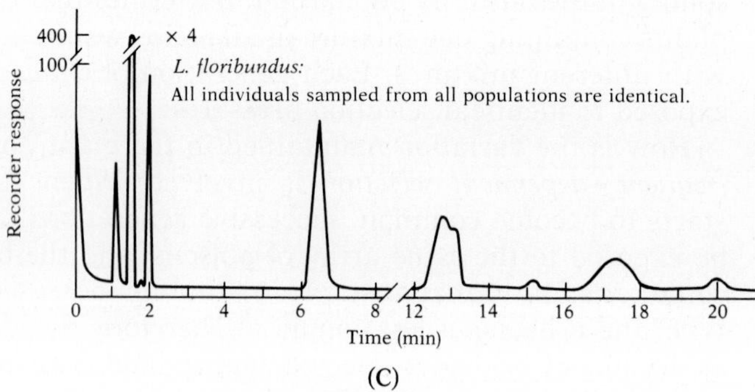

400
100
× 4
L. floribundus:
All individuals sampled from all populations are identical.
Recorder response
0 2 4 6 8 12 14 16 18 20
Time (min)

(C)

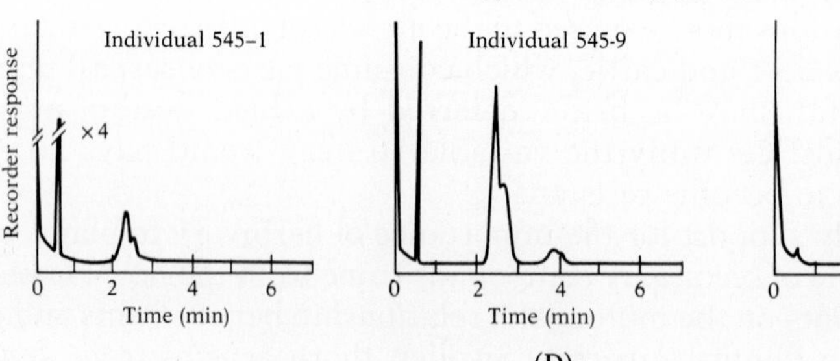

L. bakeri: Upper Gold Basin population
Individual 545–1
×4
Recorder response
0 2 4 6
Time (min)
Individual 545-9
0 2 4 6
Time (min)

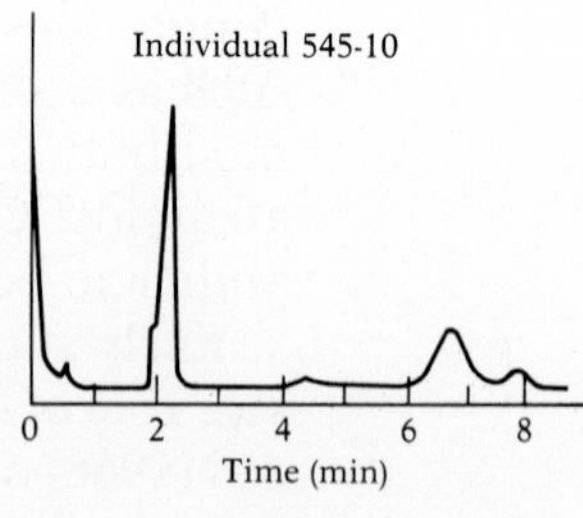

Individual 545-10
0 2 4 6 8
Time (min)

(D)

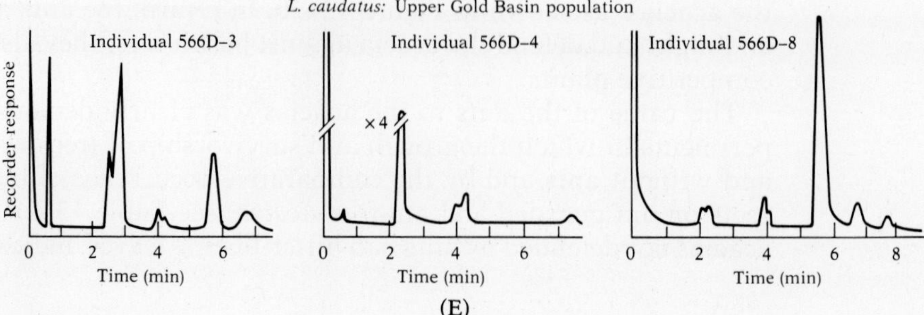

L. caudatus: Upper Gold Basin population
Individual 566D–3
Recorder response
0 2 4 6
Time (min)
Individual 566D–4
×4
0 2 4 6
Time (min)
Individual 566D–8
0 2 4 6 8
Time (min)

(E)

FIGURE 13–19 (*Facing page*) Gas chromatographic traces showing variations in lupine alkaloid contents in Colorado lupine *(Lupinus)* populations. Peaks indicate different alkaloids. Graphs (A–C) show results in uniform populations. (D–E) Three individuals each from two variable populations. *(After Dolinger* et al. *1973.)*

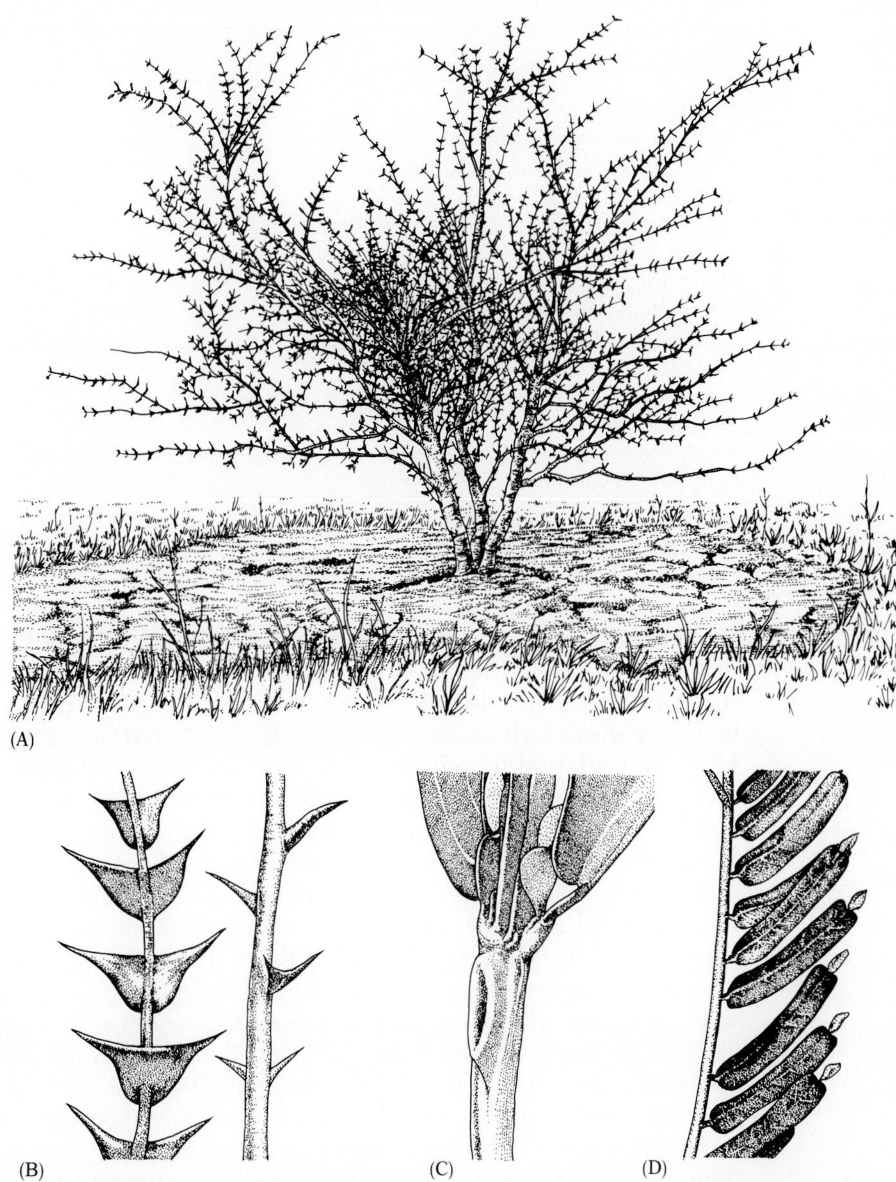

FIGURE 13–20 Ant acacias. (A) *Acacia collinsii* in a Nicaraguan pasture. Note the bare area at the base cleared by acacia ants, *Pseudomyrmex nigrocincta.* (B) Swollen thorns of *Acacia hindsii* in which the acacia ants live. (C) Nectary on a petiole of *Acacia cornigera* from which ants get nectar. (D) "Beltian bodies" (special protein-rich structures that ant-acacias produce to feed the ants) on the tips of *Acacia cornigera* leaflets. *(After Janzen 1966.)*

TABLE 13–2 Incidence of phytophagous insects on shoots of *Acacia cornigera* occupied and unoccupied by *Acacia ants,* between June 13 and July 29, 1964 (Temascal, Oaxaca, Mexico, first part of the rainy season)

	Occupied	**Unoccupied**
Daylight		
No. of shoots examined	1,241	1,109
Percentage of shoots with insects	2.7	38.5
Mean no. of insects per shoot	0.039	0.881
Mean no. of insects per shoot known to feed on *A. cornigera*	0.036	0.806
Nighttime		
No. of shoots examined	847	793
Percentage of shoots with insects	12.9	58.8
Mean no. of insects per shoot	0.226	2.707
Mean no. of insects per shoot known to feed on *A. cornigera*	0.220	2.665

From Janzen (1966).

ence of defensive chemicals, whereas swollen-thorn acacias did not. Ants are now known to protect other plants from herbivores in return for shelter or nectar (see, e.g., Inouye and Taylor 1979). However, ants may sometimes aid some herbivores while discouraging others, as when they tend lycaenid butterfly caterpillars in return for a sugary secretion, resulting in quite complex plant–ant–herbivore interactions (Horvitz and Schemske 1984). Indeed, plants influence a wide variety of interactions between insect herbivores and their natural enemies (Price *et al.* 1980).

An enormous literature now exists on patterns of chemical defense in plants (for an excellent review, see Futuyma 1983). Pioneering attempts to make generalizations about those patterns were made by Feeny (1975, 1976) and by Rhoades and Cates (1976). They focused on a particular contrast of plants and plant tissues: those that are predictable and easily available to herbivores, which Feeny christened *apparent* plants and tissues, and those that are unpredictable or ephemeral and thus relatively difficult for herbivores to locate, which are *unapparent* plants and tissues.

Forest trees and old leaves are apparent; short-lived annual and young leaves are unapparent. The investigators' basic hypothesis was that the tissues of apparent plants, tissues virtually certain to be discovered by their enemies, are expected to be defended by "quantitative" defenses that reduce their digestibility and are difficult for herbivores to evolve resistance against, yet also expensive metabolically for the plants to produce. In contrast, unapparent plants and tissues are expected to be defended by metabolically "cheap" qualitative defenses—toxins that are easier to produce and store but also easier for herbivores to circumvent. Rhoades and Cates (1976) also hypothesized that large, long-lived woody

FIGURE 13–21 The checkerspot butterfly, *Euphydryas chalcedona,* on Stanford University's Jasper Ridge Biological Preserve. (A) Adult male sunning on a sticky monkey-flower, *Diplacus aurantiaucs.* (B) Nearly mature larva feeding on *Diplacus.* (C) Younger larvae feeding on the California bee plant, *Scrophularia californica.* *(Photos by Paul R. Ehrlich.)*

species having many years to reproduce could afford to allocate proportionately more resources to the production of expensive defensive chemicals.

The apparent–unapparent dichotomy has proven a useful starting point for analysis, and data indicate that the predictability of occurrence of a plant is related to its mode of chemical defense. For example, Futuyma (1976) found that, in the northeastern United States, trees display a lower diversity of defensive chemicals than do herbaceous plants. He postulated that the higher diversity of the chemicals among herbaceous species made a given species less likely to be successfully attacked by the herbivores that were eating a neighbor.

The defensive chemicals of some apparent plants have been shown to slow the attack of even highly adapted herbivores. For example, the checkerspot butterfly, *Euphydryas chalcedona,* shown in Figure 13–21A, feeds on two principal food plants in the Outer Coast Range of central California. One is the sticky monkey flower, *Diplacus aurantiacus,* and the other the California bee plant, *Scrophularia californica;* both of these plants are shown in Figure 13–21 (B and C). The *Diplacus* is a common perennial shrub that loses all but its terminal leaves during the summer drought. The life cycle of the butterfly is closely timed to that of the plant (Mooney *et al.* 1980a, 1981c), as may be seen in Figure 13–22. The plant's leaves vary seasonally in their nitrogen content and are protected by large quantities of phenolic resin on their surfaces (Lincoln 1980).

E. chalcedona larvae feed preferentially on the *Diplacus* leaves with the highest nitrogen-to-resin ratio, and their ability to digest food plant is inhibited by the resin (Lincoln *et al.* 1982, Williams *et al.* 1983a). Larvae uniformly prefer feeding on *Scrophularia* to feeding on *Diplacus,* and grow better on it, regardless of the food plants of their parents or their own previous nutritional experience (Williams *et al.* 1983a,b). This case is an example of an herbivore persisting in the use of a plant that has rather effective defenses against it in an environment that contains a nutritionally superior food plant. The herbivore appears to persist because the *Scrophularia* is a less abundant and dependable resource than the *Diplacus* and grows in a thermally less suitable environment. The several other potential food plants in the area are ecologically unsuitable. Either they senesce too early for the larvae to be able to reach the proper developmental stage to survive the summer drought, or they grow in deep shade in which thermal conditions are not hospitable for the growth of butterfly larvae.

Other cases seem similar. Potential hosts may be determined by coevolution operating on the ability to deal with defensive chemicals, but the pattern of utilization within that range may be determined by ecological factors. For example, Holdren and Ehrlich (1982) found that in Gunnison County, Colorado, *Euphydryas editha* females oviposited exclusively on *Castilleja linariifolia,* even though laboratory experiments indicated that two related plants that grow intermixed with *C. linariifolia, Castilleja chromosa* and *Penstemon strictus,* are equally nutritious. The primary reasons for the restriction, as represented in Figure 13–23, seem to be that the phenology of *C. linariifolia* is more suited to the life cycle of *E. editha,* the abundance of *C. linariifolia* is high, and it is relatively drought resistent.

Such cases of *ecological monophagy* seemingly should lead eventually to the evolution of a superior ability to digest the ecologically more suitable host (Smiley 1978). Evidence to support this *digestive efficiency hypothesis* may be found, for example, in some *E. editha* populations in California, which are better able to digest their own food plants than each other's (Rausher 1982). Why similar host specificity has not yet evolved in Colorado is unknown.

Many such minor mysteries remain to be solved in the complexities of herbivore–plant interactions. Moreover, some problems exist with the ap-

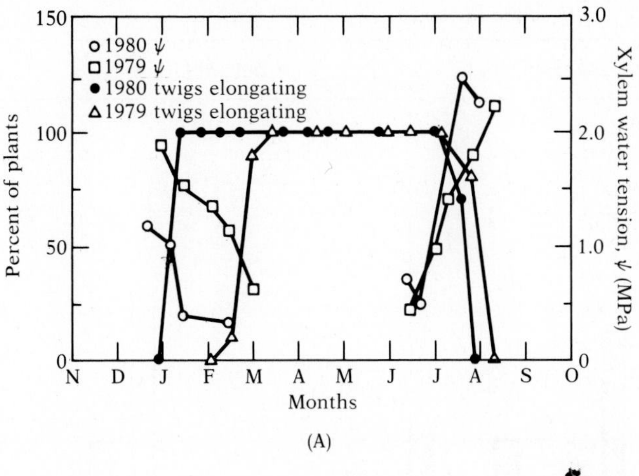

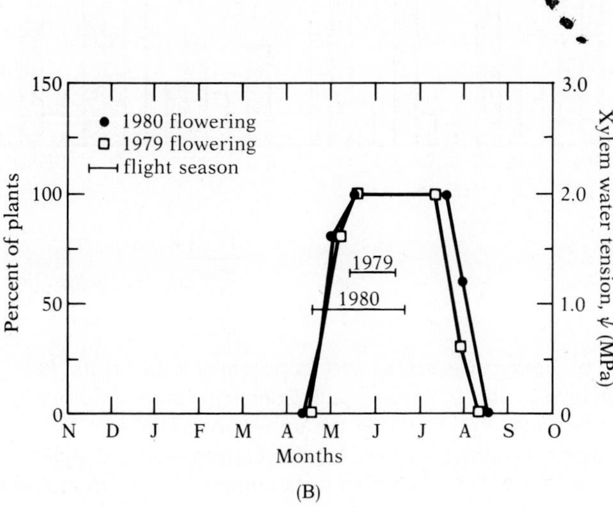

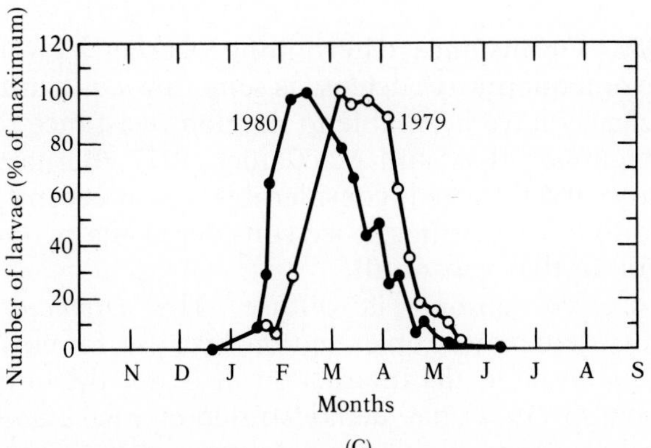

FIGURE 13–22 Phenology of *Diplacus* and *Euphydryas chalcedona* in 1979 and 1980. (A) Course of twig elongation and dawn water potential of *D. aurantiacus.* (B) Flowering period of *Diplacus* and flight season of *E. chalcedona.* (C) *E. chalcedona* larval activity as a percentage of the maximum observed on *Diplacus.* *(After Mooney* et al. *1981.)*

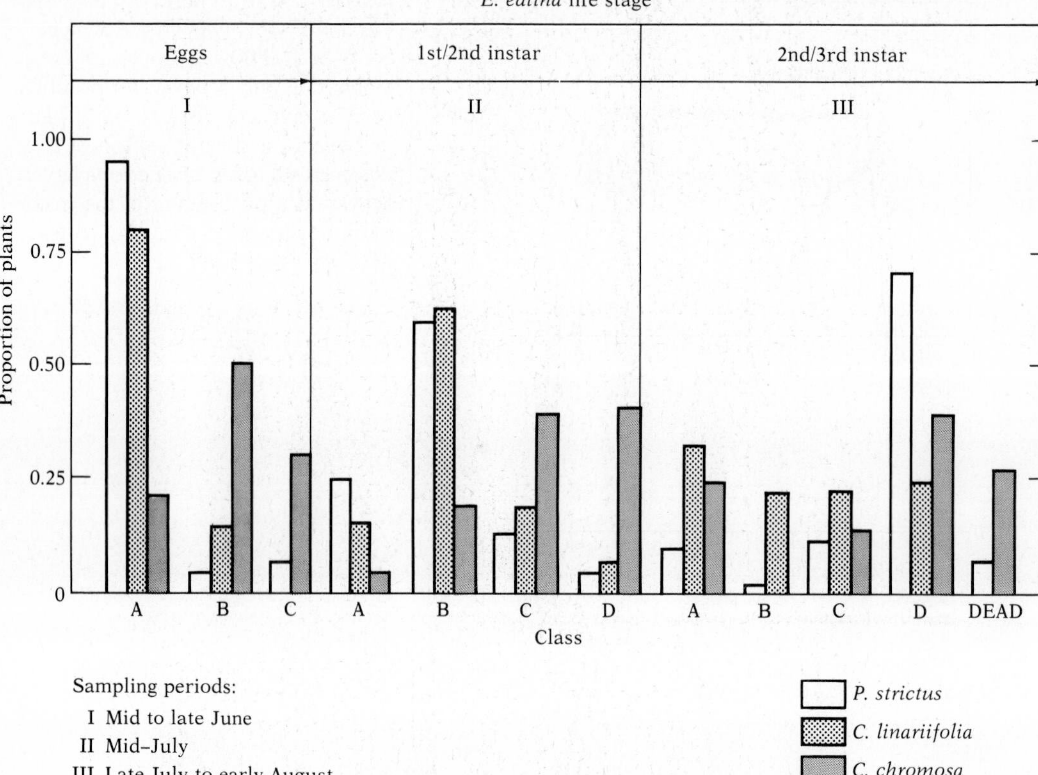

FIGURE 13–23 Phenology of *Euphydryas editha* and its potential food plants, *Penstemon strictus, Castilleja linariifolia,* and *C. chromosa* in Gunnison County, Colorado. Egg and larval stages of the butterfly are shown on the upper horizontal axis. Plant classes: A—flowers not yet open; B—flowers open; C—flowers senescing; D—plants senesced; DEAD—aboveground parts that have died back completely. *(From Holdren and Ehrlich 1982.)*

parancy theory (see Fox 1981). For instance, determining whether a given compound is a quantitative or a qualitative defense is sometimes difficult. In addition, some insects clearly have been able to develop resistance to digestibility-inhibiting compounds (Fox and MacCauley 1977, Bernays 1978, Berenbaum and Feeny 1981). Indeed, considerable doubt recently has been thrown on the notion that tannins, once considered the prime digestion inhibitors, function in that way at all.

Much of plant–herbivore coevolution is diffuse. The *Diplacus–Scrophularia–E. chalcedona* system may be an exception, because, for now, no other major actors appear to be in the interaction. It is not the only exception, however. Berenbaum (1981) has demonstrated stepwise coevolution in a study of furanocoumarins, highly toxic derivatives of another

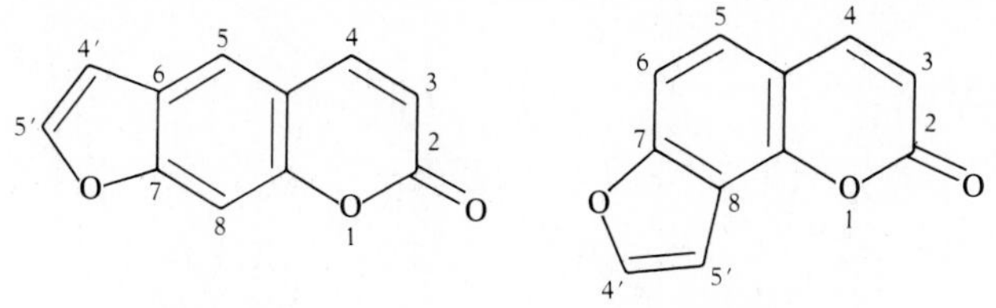

FIGURE 13–24 Furanocoumarin structures. (Molecular skeleton only, side chains not shown.) *(After Berenbaum 1981.)*

group of compounds, hydroxycoumarins, whose chemical structure appears in Figure 13–24. Berenbaum points out that by far the most diverse group of the parsley family, Umbelliferae, the Apoideae, is the only group in the family containing genera that produce furanocoumarins. She suggests that this success is due to diversification behind the furanocoumarin shield, a view supported by the species richness and wide distribution of those genera within the Apoideae that produce the compounds.

Furanocoumarins are highly toxic to general feeders such as the southern army worm (*Spodoptera eridania*). However, insect specialists on the Umbelliferae thrive on furanocoumarins. For example, the eastern black swallowtail butterfly, *Papilio polyxenes,* actually grows more rapidly on artificial diets containing a furanocoumarin concentration lethal to *Spodoptera* than it does on a diet free of it. Adaptive radiation by the insects is indicated by the richness of genera feeding on the furanocoumarin-containing plants in contrast to closely related genera feeding on plants lacking the furanocoumarins.

Berenbaum and Feeny (1981) have gone into even greater detail, relating patterns of herbivore diversity to the distribution of two different classes of furanocoumarins, linear and angular, as represented in Figure 13–25. They hypothesize that this situation indicates an escalation of the "arms race" between the coumarin-containing plants and herbivores.

Not all plant defenses against herbivores consist of being either unapparent or poisonous, as anyone who has had close contact with a cactus can testify. Indeed, some plants have developed quite sophisticated mechanical defense systems, as exemplified by the sensitive plants *Schrankia microphylla* and *Mimosa pudica* (Leguminosae). As Figure 13–26 shows, these plants have a mechanism that folds their leaflets out of the way and exposes their defensive thorns when disturbed (Eisner 1981).

Gilbert (1971, 1975, 1982) has illuminated an array of nonchemical defenses that *Passiflora* vines have evolved to use against *Heliconius* butterflies that have learned to deal with *Passiflora* chemical defenses. These include the following:

- Sharply pointed hairs that can pierce *Heliconius* caterpillars and fatally immobilize them.
- Extrafloral nectaries that attract predacious ants and other enemies of the caterpillars.

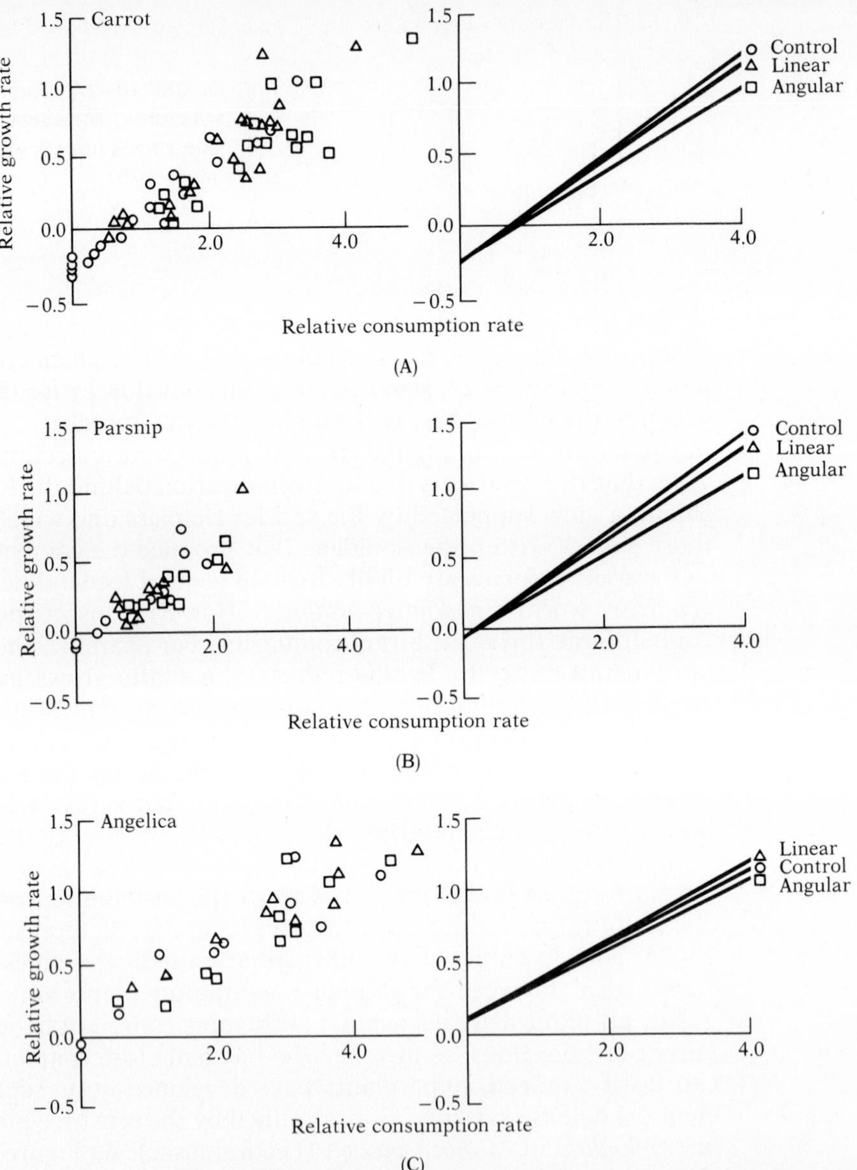

FIGURE 13–25 Growth efficiency (relative growth rate/relative consumption rate) of swallowtail butterfly *(Papilio polyxenes)* larvae that normally feed on umbelliferous plants containing linear furanocoumarins (LFs) when exposed to angular furanocoumarins (AFs). Larvae were raised on carrots (which contain no furanocoumarins), parsnip (LFs) and angelica (LFs and AFs). Each kind of larva then had its growth efficiency tested while feeding on the plant unmodified (control), with added LF (xantholoxin), and with added AF (angelica). As can be seen from the slopes of the regression lines on the right, the larvae feed less well on the angelica with its natural AFs than on the other plants, and added AFs reduce feeding efficiency on the others. Apparently herbivores specializing on plants with LFs (potent defenses against generalist herbivores) have encouraged some umbellifer species to evolve AFs—an escalation of a "coevolutionary arms race." *(After Berenbaum and Feeny 1981.)*

(A)

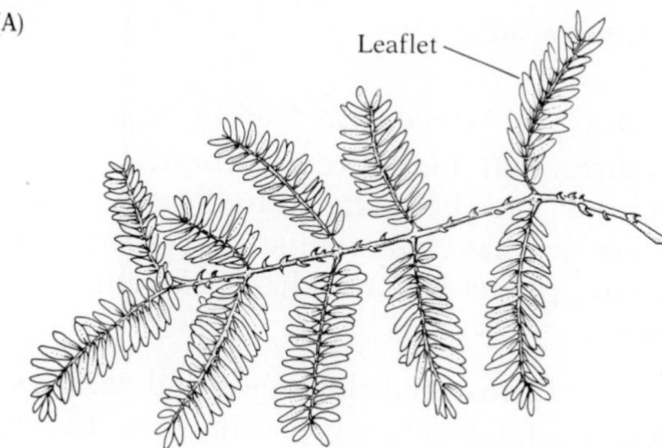

(B)

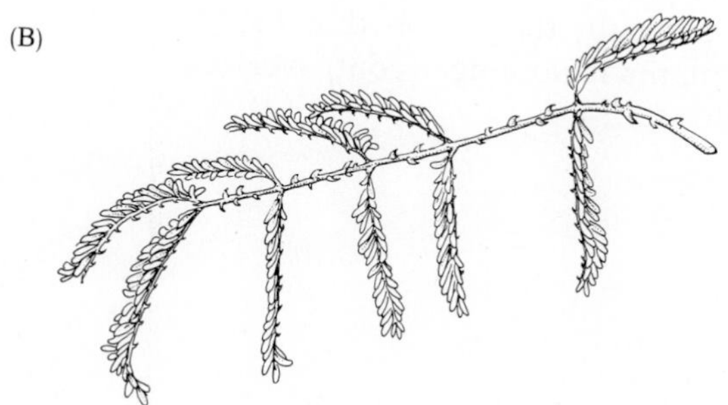

FIGURE 13–26 Portion of a compound leaf of a sensitive plant (A) before and (B) after a sensitive reaction. Note how folding of leaflets exposes spines. *(L. E. Gilbert, University of Texas at Austin/BPS.)*

FIGURE 13–27 *Heliconius* egg mimicry. Actual egg is on curling process at lower right. Ends of slender stipules (projecting from bracts) that are above it are egg mimics. *(Photo by Larry Gilbert/BPS.)*

- Leaf shapes that resemble non-*Passiflora* plants and do not resemble other species of *Passiflora*.
- Rapid shedding of the structures on which eggs tend to be laid.
- Egg mimics, yellow structures that look like *Heliconius* eggs, as shown in Figure 13–27, and that have been shown to discourage oviposition by *Heliconius* females that have cannibalistic larvae and that carefully search for egg-laying plants free of other *Heliconius* eggs (Williams and Gilbert 1981).

It is now widely accepted that, in spite of the cleverness and variety of plant defenses, herbivores can have significant impacts on growth, seed production, and recruitment of plants in nature* and thus on their distribution and abundance. As a result, the notion that herbivores place potent selection pressures on plants is no longer controversial.

* See, for example, Breedlove and Ehrlich (1968, 1972), Janzen (1971), Rockwood (1973), Manley *et al.* (1975), Waloff and Richards (1977), Rausher and Feeny (1980), Parker and Root (1981), Louda (1982a,b), and Dirzo (1984).

CHAPTER 14

Mutualism and Mimicry: "Beneficial" and "Deceitful" Relationships Between Species

Not all interspecific interactions result in harm to one of the organisms. In some interactions, called *commensalistic,* one species benefits and the other apparently suffers no harm. In other interactions, called *mutualistic,* both species benefit. One example of mutualism has already been discussed in Part 1, that of higher plants and mycorrhizal fungi.

POLLINATION

Examples of mutualistic plant–animal relationships are most abundant in the area of pollination. The lack of mobility of plants has caused them to evolve dependency on many agents in order to bring their gametes together. When those agents are animals, both codynamic and coevolutionary interactions often occur. In most cases, the animals also benefit from the relationship, ordinarily obtaining food from the plant. In some cases, such as that of pseudocopulation (to be described in the final section of this chapter), the pollinating animal is manipulated by the plant so that it gains no benefit or is even harmed (Wiens 1978, Little 1983).

The study of pollination by animals is one of the most active areas of ecology today.* It is of economic interest because of the large number of

* For reviews, see Feinsinger (1983), Jones and Little (1983), and Real (1983).

crops dependent upon pollination (in the United States alone, more than 90 are pollinated by insects). Pollination is also of theoretical interest because of the great diversity of systems that have evolved, the roles pollinators play in establishing community patterns, and often the great intimacy and intricacy of the interactions.

Pollination systems provide some of the best examples of tight coevolution (Janzen 1979, Wiebes 1979, Feinsinger 1983). A classic system is that of figs and fig wasps. Some 900 species of *Ficus* exist, and virtually every species is pollinated by its own species of tiny wasp of the family Agaonidae. In general, related figs are associated with related wasps. The part of the fig that is eaten by people is actually an enclosed inflorescence containing many flowers. The female enters a fig through a small opening, pollinates the flowers, lays eggs in their ovaries, and dies. The young wasps mature within the flowers, in which they cause tiny galls. Wingless males emerge from their galls, find female-containing galls, and mate with the females while they are still inside, using an extensible abdomen, as shown in Figure 14–1. The males then die without ever leaving the fig. The females collect pollen and then depart the fig to search for another fig in the proper stage of development.

Although they do not have an entirely mutualistic relationship, yucca plants and some yucca moths have a rather similar interdependency (Powell and Mackie 1966, Feinsinger 1983, John Addicott, pers. comm.). In fig–wasp and yucca–moth mutualisms, the distribution and abundance of one partner clearly are controlled by the other, and vice versa. For example, when Smyrna figs were first introduced into California late in the nineteenth century, they did not produce fruit; only when the proper wasps were introduced was their cultivation successful.

Perhaps stranger, but less tightly coevolved, is the relationship between orchids and male euglossine bees that pollinate them in the American tropics (van der Pijl and Dodson 1966, Dodson 1975, Williams 1983). In return for pollination, the bees do not obtain nectar or pollen for nourishment but instead collect a variety of floral fragrances. The use to which the bees put the scents is not certain, but the evidence is that they may

FIGURE 14–1 Fig wasp copulation. Female wasp inside a gall, and the male wasp is copulating with her from the outside through a hole he has bitten in the gall.

modify them into compounds to attract female bees. The bees collect scents from a wide variety of orchids (and a few other plants), and they are thought to be scantly affected by evolutionary events within the orchid community (Feinsinger 1983).

Diffuse coevolution between plants and pollinators is common. However, even in areas in which a variety of pollinators service each plant and pollinators visit a variety of plants, mutual adaptations are usually quite evident. *Buzz pollination* provides a glimpse of the complexities of these relationships. Flowers with anthers opening through pores at the top are visited by certain bees (not honeybees), which, by rapidly contracting their flight muscles, emit a loud buzz (Michener 1962, Wille 1963). These vibrations are transmitted to the anther, causing it to eject a cloud of pollen directed at the underside of the bee. Because many of the flowers with this sort of anther open downward when containing a bee, one might expect much of the pollen to simply miss the bee or fall off it "as if a salt shaker had been turned upside down" (Erickson and Buchmann 1983). It does not, however, because plants have negatively charged surface electrostatic fields, especially near terminal points such as flowers. Foraging bees, in contrast, tend to have positive charges, presumably because the electrons have been stripped away during flight. The difference in charge probably helps both in releasing the pollen and in attracting it to the bees so that little is lost to vagrant air currents (Buchmann and Hurley 1978).

Many other aspects of pollination biology are of interest to ecologists. For instance, competition between pollinators for nectar resources seems to be common (see, e.g., Inouye 1978). In addition, competition among plants for pollinators may be important in maintaining the diversity of colors, shapes, and flowering times in floral communities, but this suggestion needs further experimental study (Waser 1983). Ecologists are also investigating the conflicts between the needs of the plants and those of the pollinator (Baker and Hurd 1968, Heinrich and Raven 1972, Feinsinger 1983, Waser and Price 1983). From the plant's perspective, an ideal pollinator would move rapidly among plant individuals but remain constant to the species visited. Selection presumably favors the plant species producing just enough nectar per flower to attract a pollinator's visit. However, from the pollinator's perspective, it would be best to obtain a maximum of nourishment with a minimum of movement.

Finally, ecologists are beginning to understand why ants—which, in many places, are among the commonest insects—are so rarely involved in pollination systems. One reason may be that the nesting and brood-raising behavior of ants exposes them to a wide range of pathogenic fungi and other dangerous microorganisms. In response, the ants secrete large amounts of antibiotics, which secondarily inhibit proper pollen function (Beattie *et al.* 1984, Beattie 1985).

The vast majority of flowering plants are utterly dependent on animals for pollination. Biologists have little doubt that coevolution was involved in the diversification of flowering plants and pollinating insects such as bees, butterflies, flies, and beetles at the end of the Cretaceous period and the beginning of the Tertiary period (Baker and Hurd 1968, Proctor and Yeo 1972). The complex set of population interdependences that these pollination relationships now represent are important parts of the functional "glue" that ties together trophic levels within communities and ecosystems.

DISPERSAL AND ESTABLISHMENT OF SEEDS BY ANIMALS

Another broad area in which plants and animals have mutualistic relationships is the dispersal of propagules. Plants have evolved mechanisms for enlisting the aid of a variety of animals in this task (see Howe and Smallwood 1982, and Janzen 1983a, for access to the literature). Besides the well-known method of dispersal by passage through the digestive tracts of fruit-eating birds, bats, primates, and the like, other systems exist that are more obscure. For example, fruit- and seed-eating fishes in the Amazon basin contribute to the dispersal of trees in the vast floodplain forests along the river (Goulding 1980). Many of these interactions are mutualistic, with the animal receiving nourishment in return for transport.

However, some of the interactions are commensalistic or even parasitic, as any hiker knows who has been plagued by "burrs" or "sticktights"—fruits that, like those shown in Figure 14–2, have evolved barbs and hooks that can attach themselves to passing mammals or birds. For example, fruits of burdock (*Arctium*—Compositae) retain a ring of bracts that transform after they flower into extremely sharp hooks, which can cause great discomfort for the animal in whose fur they become entangled. Fruits of the genus *Pisonia* (cabbage tree and its relatives—Nyctaginaceae) in the

FIGURE 14–2 Fruits that have evolved devices for being dispersed by mammals. (A) Beggar-tick. (B) Unicorn. (C) Cocklebur.

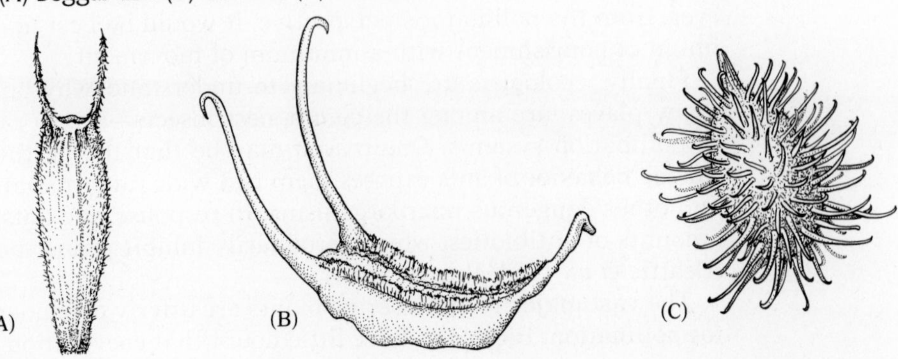

Pacific are sticky so that they can cling to the feathers of birds; they are dispersed by large birds such as herons and frigatebirds (Proctor and Proctor 1978). On some islands, birds and even reptiles become so entangled in masses of *Pisonia* fruits that they die.

Seeds that are transported are conveyed more or less at random by animals, wind, water, or explosive mechanisms of the plant. The odds of a seed's landing in a suitable patch of habitat for successful germination may be small indeed, and the odds of its being eaten if lying exposed may be great. Clearly, seeds that are buried have the advantage, and many species of plants have recruited ants for this task. Ant dispersal and burial of seeds are encouraged by the evolution of special nutritious tissues in the seed coat (Beattie 1985). As Figure 14–3 illustrates, ants carry the seeds to their nests, which are higher in essential nutrients, such as nitrogen, phosphorus, and potassium, than are surrounding soils (Beattie and Culver 1982). The germination and survival of the seeds of ant-dispersed species of violets taken into ant nests have been compared with those of seeds planted at random (Culver and Beattie 1980, Beattie 1983). Those in the ant nests were nearly three times as likely to emerge, were larger, and were more than four times as likely to survive for 2 years.

Animals as seed dispersers have had strong selective impacts on plants, as is apparent from the many special structures—from tasty fruits and eliasomes to nasty barbs—that plants have evolved to get the job done. Fruits dispersed by birds and diurnal mammals, for example, often have red, yellow, black, or blue fruits that attract the dispersers; moreover, because birds lack a sense of smell, the fruits they disperse are usually odorless. The pattern of color, the timing, and the quantity of presentation are related in complex ways to the available dispersers and the potential predators on the fruit and on the disperser.* Naturally, color is not important in bat-dispersed fruits that are consumed at night; instead, such fruits tend to give off a musty odor that attracts the bats.

Some tropical fruits are dispersed by birds that are strictly frugivorous, and those fruits, unlike others, provide a balanced diet of protein, fats, and vitamins in addition to the usual carbohydrates and water (Proctor and Proctor 1978). Plants may "choose" different strategies for dealing with dispersers. For example, Howe (1982) compared two Panamanian trees, one of which produces abundant mediocre fruit that attracts opportunistic, inefficient dispersers, whereas the other offers small quantities of high-quality, nutritious fruits that are sought by an efficient specialist disperser. That dispersers also play a key role in determining the spatial patterns of plant communities is also evident, but what about the reverse? What impact do the plants have on the animals?

* See, for example, Snow (1965, 1971), van der Pijl (1972), Morton (1973), McKey (1975), Howe and Estabrook (1977), Thompson and Willson (1978, 1979), Howe (1979b), Thompson (1981), and Willson and Thompson (1982).

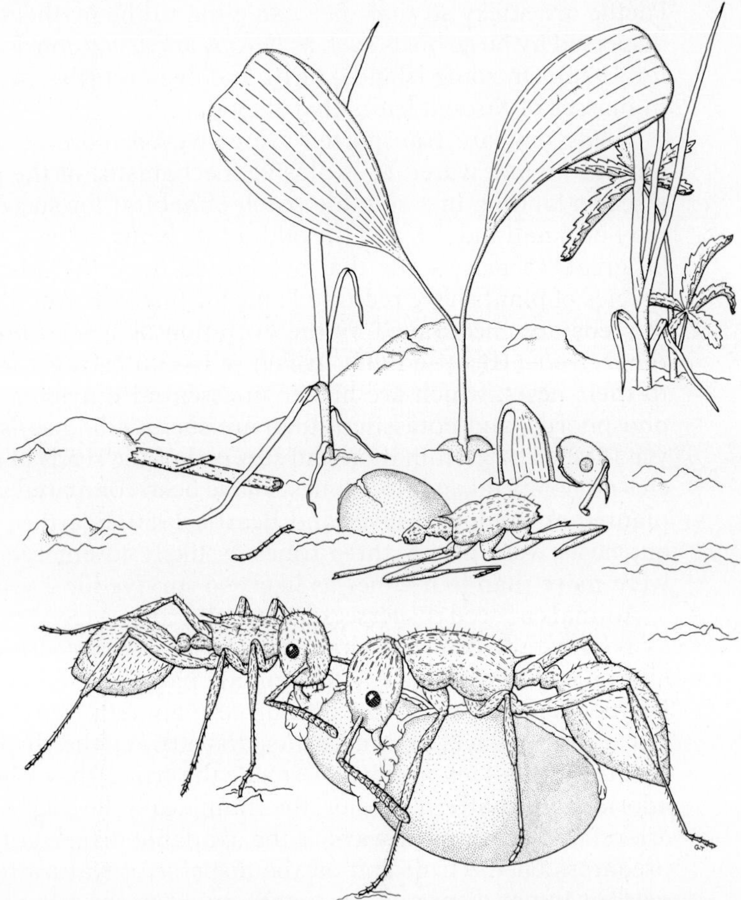

FIGURE 14–3 Ants carrying seeds back to their nest. In the foreground, two foragers of *Myrmica discontinua* are carrying violet *(Viola nuttallii)* seeds, holding them by their special nutritious tissues or *eliasomes.* Once the eliasomes have been removed, the seeds are abandoned, together with other nest waste such as dead workers and parts of insect prey, in garbage dumps close to the nest. In the background, a violet seed from the previous year remains half buried in the soil and nest waste, but a second seed has germinated and produced two healthy cotyledons. *(Drawing by Christine Turnbull.)*

Adaptations for fruit eating in animals are perhaps most clearly seen in the bills of birds, as shown in Figure 14–4. In general, however, the coevolutionary impact of the fruit–disperser system is less obvious in the animals than it is in the plants; most animal-dispersed fruits are eaten by a considerable variety of frugivores (Howe and Smallwood 1982). Overall, then, plant–disperser systems seem to be less tightly coevolved than plant–pollinator systems are (Feinsinger 1983). Nevertheless, the production of fruits by plant species plays a major role in the ecology of frugi-

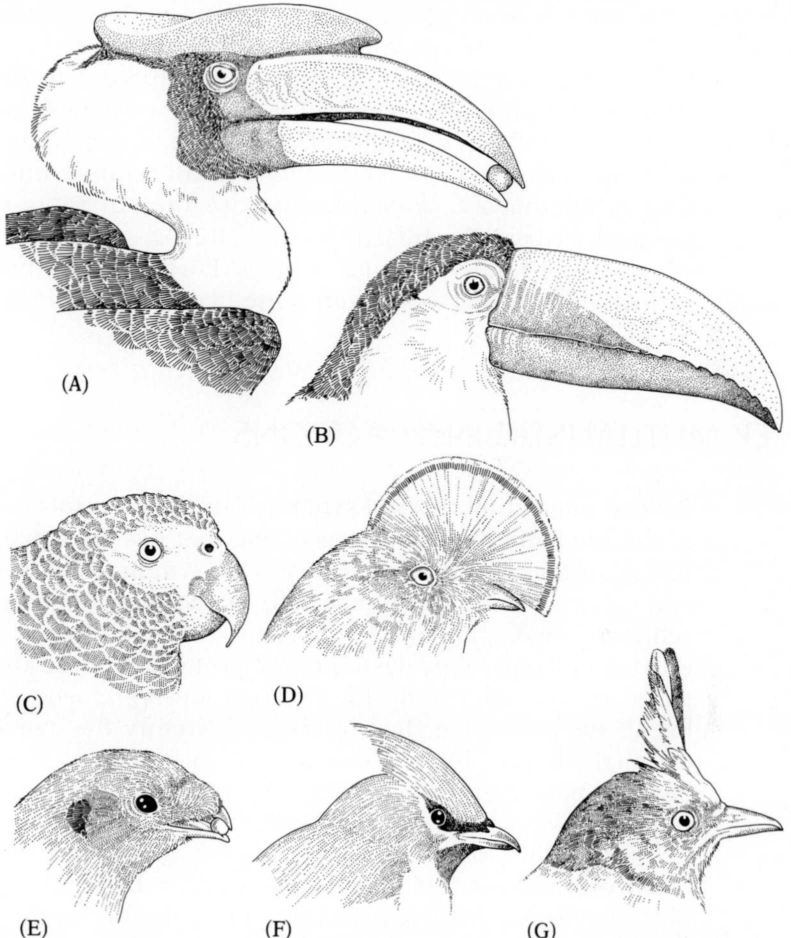

FIGURE 14–4 Bills of fruit-eating birds. (A) great hornbill, (B) black-mandibled toucan, (C) Grey parrot. (A–C) have sharped edge bills that can cut or peel big fruits. (D) Cock-of-the-rock and (E) green broadbill have small bills that open wide to swallow large fruits. (F) The Bohemian waxwing and (G) the black-crested bulbul have bills that allow them to eat both insects and small berries. *(After Proctor and Proctor, 1978.)*

vorous animals. Some plants serve as *keystone mutualists*—organisms whose presence is critical to the persistence of entire groups of species (Gilbert 1980). A plant that may play this role in some wet forest communities in the American tropics is the tree *Casearia corymbosa* (Flacourtiaceae). This species fruits during annual periods of scarcity; for weeks it is virtually the sole support of three species of obligate avian frugivores—a cotingid, *Tityra semifasciata,* and two species of *Ramphastos* toucans—that would probably go extinct in its absence (Howe 1977).

Relationships of plants with their dispersers can change dramatically from community to community. At a wet forest site at La Selva in Costa Rica, the *Casearia*–disperser dependency appears to be reciprocal. *Casearia* would probably go extinct if the *Tityra* disappeared. The *Ramphastos* toucans and less dependent frugivores that use the tree are inadequate seed dispersers. In contrast, a taxonomically similar but ecologically distinct population of *C. corymbosa* in a dry forest at Santa Rosa, Costa Rica, has seeds dispersed primarily by the yellow-green vireo (*Vireo flavoviridis*), which is not totally dependent on it. Furthermore, should the vireo go extinct, other dispersers would quite likely maintain the tree (Howe and Kerckhove 1979).

OTHER MUTUALISTIC INTERACTIONS

Several interesting types of symbiosis (intimate association) have evolved in marine habitats. For instance, one that has attracted the attention of biologists for more than a century is the symbiosis between sea anemones and certain damselfishes, especially of the genus *Amphiprion*.* The damselfishes feed on plankton in the vicinity of an anemone, and at the slightest sign of danger they dash into the protection of the anemone's stinging tentacles (see color Plate E). The relationship is asymmetrical because, for all the protection they receive, apparently the damselfishes only occasionally benefit the host anemone by bringing it bits of food. However, because both members benefit, this relationship is an instance of mutualism.

The anemone–damselfish interaction was used in a theoretical analysis of the evolution of symbioses (Roughgarden 1975). The evolution of symbiosis appears to involve two issues. First, the formation of the association requires a guest individual that makes an effort to find its future host. This search presumably involves a cost in fitness that must be more than offset by a gain in fitness if it succeeds in finding a suitable host to which to attach. Similarly, the host may make an effort to attract the guest if it is beneficial but repel the guest if it is harmful. Both of these considerations pertain to the "getting together" aspect of symbiosis. Second, the guest may exploit its host, but lowering the survivorship of the host involves forfeiting future benefits from staying with it.

The evolution of symbiosis was explored theoretically from the point of view of the evolutionary strategy of the guests; the investigation assumed that the properties of the host population were relatively constant (Roughgarden 1975). According to this analysis, mutualism can evolve only in restricted circumstances for two reasons.

* See, for example, Crespigny (1869), Verwey (1930), Davenport and Norris (1958), Mariscal (1972), Allen (1975), and Ehrlich (1975).

First, the potential host individuals must have a sufficiently high survivorship for the association to develop evolutionarily in the first place; otherwise, a potential guest individual might have a higher fitness by not searching for, and thus remaining unassociated with, a short-lived host. From this argument, it follows that more species of obligate mutualists should have evolved to live with long-lived hosts than with short-lived hosts, everything else being equal.

Second, the extent to which a guest forgoes exploiting its host should depend on how much that sacrifice actually improves the host's survival. If the host already survives sufficiently well, or if the guest can have little impact on the host's survival, then selection should favor little if any sacrifice to the guest in aiding its host. The evolution of mutualism seems to require that the host have an intermediate survivorship. It cannot be too low, for otherwise the association will not form; it cannot be too high, for otherwise the guest receives little benefit by improving it still further.

If the guest evolves some moderation in its exploitation of the host, it might be called a "gentle parasite," provided that the host is actually harmed by the guest nonetheless. For the guest to be more than a gentle parasite, that is, to be a mutualist, it must increase the survivorship of its host above that which exists in the absence of the guest. The guest is evolutionarily "unaware" of how well the host survives when it is not attached to it. Hence, the evolution of mutualism seems to involve a lucky coincidence. The guest must evolve in such a way as to cause the host to survive better with the guest than without it. However, if this should happen, then the host should solicit the guest actively, and a fully reciprocal mutualistic interaction can evolve.

Thus, we would expect mutualistic relationships such as the anemone–damselfish association to be relatively rare. Indeed, compared to parasitism and commensalism, they certainly are. Those mutualistic cases that have been identified seem to hold a special fascination for biologists. One of the most widely publicized mutualistic interactions are cleaning symbioses in the oceans, the best known occurring on coral reefs (Feder 1966, Hobson 1969, Ehrlich 1975). Scuba divers are familiar with the small, brightly colored, *cleaner* fishes that remain at specific spots (cleaning stations) on reefs and service a parade of "customers." The Pacific cleaner wrasses (*Labroides* spp.) do a little "dance" that advertises that they are in business, as illustrated in Figure 14–5. Diverse large fishes approach the station and adopt a cleaning posture that is characteristic of the species. The cleaner then goes over the cleanee, removing parasites and loose scales and often entering both gills and mouthes to clean.

Species that are obligate cleaners throughout their lives belong to different families in the Pacific (wrasses—Labridae) and Atlantic (gobies—Gobiidae). However, they are similar in color—black, with an electric blue or yellow stripe. This pattern apparently elicits the cleanee behavior, because predacious Pacific fishes placed in aquaria with Atlantic cleaners

FIGURE 14–5 Hawaiian cleaner fish *(Labroides phthiriophagus)* attending an injured butterfly fish *(Chaetodon miliaris).* The cleaning station is located in about 10 m of water on a volcanic intrusion about 7 miles off the coast of Lanai. *(Photo by Paul R. Ehrlich.)*

adopt the characteristic cleanee posture and will not attack the small cleaners.

The position of cleaning stations affects the distribution of fishes on the reef (Slobodkin and Fishelson 1974). However, some territorial damselfishes never go to the stations; rather, they are serviced by individual cleaners that appear to trapline—visit individual cleanees in a regular manner, much as orchid bees repeatedly visit a series of orchids. The removal of cleaners from a reef has been reported to lead to a decline in the populations of other fishes. However, the codynamic aspects of cleaner–cleanee relationships have not been worked out in detail.

Several stages in the development of cleaning behavior in reef fishes can be observed. Not all cleaners are obliged to clean. Some do it part time or, in the case of certain wrasses and angelfishes (Pomacanthidae), only when they are young (see, e.g., Randall and Helfman 1972). Fishes that originally picked small prey and bits of organic matter from the reef substrate are thought to have gradually extended their gleaning to the surface of larger fishes. The behavior benefited both, and the reciprocal signals evolved.

Mutualism has coevolved in terrestrial systems as well. For instance, blue butterflies (Lycaenidae), such as *Glaucopsyche lygdamus,* have a mutualistic relationship with ants. The ants tend the butterfly larvae and provide protection for them from predators and parasites, as shown in Figure 14–6 (Pierce and Mead 1981, Pierce and Easteal 1986, Pierce and Young 1986). In return, the ants receive a sugar solution from specialized glands in the grublike caterpillars.

Sugar-secreting caterpillars of the European large blue (*Maculinea arion*) are tended by a species of ants as the caterpillars feed on wild thyme plants. The caterpillars are cannibalistic; big caterpillars eat smaller caterpillars as well as the thyme (Ford 1945, Emmel 1975). When the caterpillars reach the last stage of growth, the ants remove the caterpillars from the thyme and carry them to their own nests, where the caterpillars live as social parasites. At that stage, the caterpillars are much like ant larvae in size, color, and skin texture. They beg food from the adult ants much as ant larvae do, and the ants feed the caterpillars. The caterpillars also dine on their hosts' young. The caterpillars pupate in the ant nest, and the adults emerge there, crawl out when their wings are dry, and fly away.

The large blue became extinct in England when grazing of sheep became uneconomical in the last areas occupied by the butterflies. The thyme thrived and grew thick, thereby changing the conditions needed by the ant species that was host to the blue. The ant populations declined and, as a result, the blue butterfly populations died out (Ratcliffe 1979).

FIGURE 14–6 (A) Ant tending a larva of the blue butterfly, *Glaucopsyche lygdamus.* (B) Ant seizing an attacking braconid wasp parasitoid in its mandibles and swinging its abdomen forward to spray with defensive chemicals. *(Photos courtesy of Naomi Pierce; photo (B) previously published in Pierce and Mead 1981.)*

(A)

(B)

MIMICRY

When, as in the case of the blue larvae resembling the ant larvae, one organism evolves a resemblance to another, the phenomenon is known as *mimicry* (Cott 1940, Wickler 1968, Owen 1980, Gilbert 1983). The organism imitated is the *model;* the imitator is the *mimic.* The case of the blue caterpillar is one of *aggressive mimicry;* the mimic benefits by being able to be fed by, and to feed on, the ants. Many examples of aggressive mimicry are in evidence. For instance, as Figure 14–7 demonstrates, certain mantids resemble flowers and thus are able to capture insects that at-

FIGURE 14–7 The African "devil's flower" mantid lying in wait for prey; its resemblance to a bloom is an example of aggressive mimicry. *(After Wickler 1968.)*

tempt to pollinate them. In a sense, mimics adopt a "deceitful" survival strategy.

One of the most curious examples of aggressive mimicry occurs on coral reefs. Sabre-toothed blennies of the genus *Aspidontus* bear a striking resemblance to the common cleaner wrasse, *Labroides dimidiatus.* The mimic does a little dance, just like *Labroides,* and elicits cleaning behavior from other fishes. It then proceeds to bite chunks out of them. The blennie does not even bother to flee after the dirty deed is done. Apparently, the blennie's cleaner appearance is sufficient to inhibit attack by its larger victims. Experienced fish soon learn to avoid painful "cleaning" stations established by *Aspidontus*—just as people would soon learn to avoid a barber shop if bits of ear frequently were cut off.

Equally curious is the aggressive mimicry practiced by females of fireflies of the genus *Photuris* (Lloyd 1975). The flashing lights of fireflies (which are actually misnamed beetles) are sexual signals. Males fly around and emit a species-specific pattern of flashes. Females perching on the ground or in low vegetation flash back species-specific replies. The number of flashes, their duration, and the interval before the female responds are important in the communication. Normally a male flash–female response dialogue is repeated about 5 to 10 times in the process of the male's joining the female.

The *Photuris* females mimic the mating signal of females of fireflies of other species, as indicated in Figure 14–8. Males of the other species are lured by the false advertising, only to be eaten by the *Photuris* (Lloyd 1975). The false signals could be derived either from the predator's own mating responses or from so-called locomotory flashes of unknown function that are commonly emitted by females of the predatory species when they alight or walk.

Less spectacular, but equally effective, are the tactics of the beetle *Myrmecaphodius excavaticollis,* which lives in the nests of fire ants and acquires cuticular hydrocarbons from its hosts. This chemical mimicry permits the beetles to roam among their ant hosts, obtain food from them, and eat the ant larvae with impunity (Vander Meer and Wojcik 1982).

A rather different form of aggressive mimicry is carried out by some orchids in order to obtain sexual services rather than nourishment. Orchids of the genera *Ophrys* and *Cryptostylis* take advantage of the naïveté of newly emerged male Hymenoptera. The orchids present various male bees and wasps with structures that visually, tactilely, and olfactorily simulate those of the females of the species, which normally emerge after the males (Kullenberg 1961, Little 1983, N. H. Williams 1983). The males attempt to copulate with the flowers, as illustrated in Figure 14–9, and in the process pollinate them. This *pseudocopulation* system works for a plant only until the females themselves emerge, at which point the males become smarter and shift their affections.

Are the male insects harmed by the orchids' hoax? No studies have

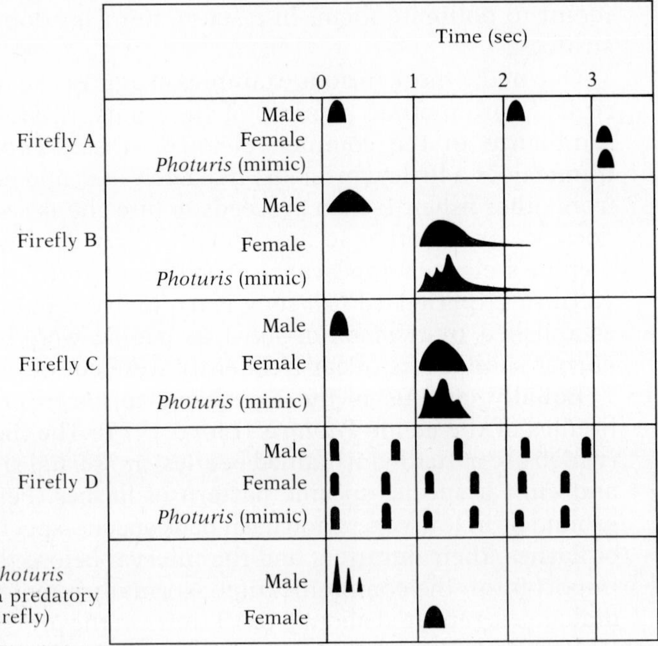

FIGURE 14–8 Light signal mimicry in *Photuris.* Male fireflies signal to the females, and predatory fireflies *(Photuris)* mimic the female's response. The male signal is shown on top, the female response immediately below, and the mimic's response below that. The bottom panel shows the mating signal of the *Photuris* firefly itself. *(After Lloyd 1975.)*

been done to answer this question. We could guess that they suffer some harm from greater exposure to predators or from loss of time for searching out early-emerging females. However, severe damage seems unlikely; otherwise, selection presumably would have sharpened the discrimination of the victims so that they were no longer lured by the mimic. Possibly the males even gain experience with the orchids that improves their later ability to copulate with females of their own species.

This example of mimicry in flowers is just one of many (N. H. Williams 1983). In some cases, flowers of one kind of plant that offers little or no nectar mimic those of another that offers abundant nectar (see, e.g., Brown and Kodric-Brown 1979, 1981). In others, a number of flowers present the same attractive ultraviolet patterns, a situation analogized by Watt and his collaborators (1974) to Müllerian mimicry (discussed below).

Protective mimicry, in which tasty or harmless organisms mimic distasteful or dangerous ones, is a common phenomenon. Toxic animals often advertise their poisonousness; that is, they are *aposematic*. It is hardly surprising, therefore, that less noxious animals often have taken advantage of this situation by evolving false advertising. Thus, many stingless

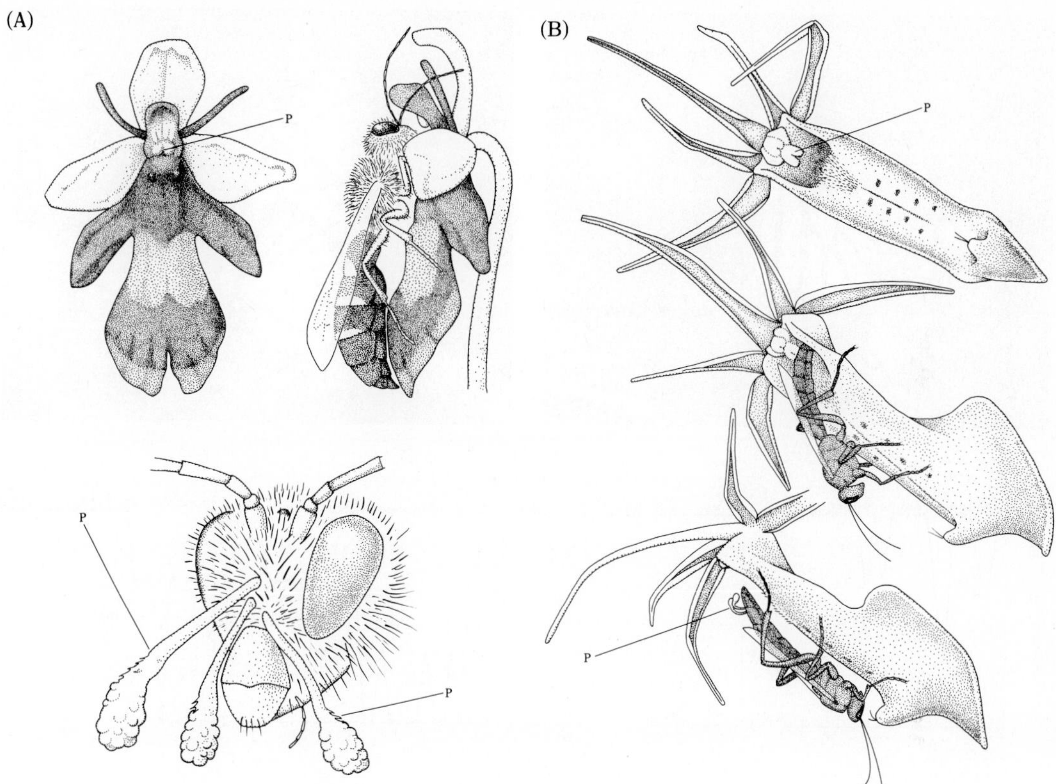

FIGURE 14–9 Example of pseudocopulation. Male insects attempt to copulate with flowers that resemble the females of their species. (A) *Upper left:* A flower of the European fly orchid, *Ophrys insectifera. P* represents pollinia, a coherent mass of pollen grains bearing a stick disk for attachment. *Upper right:* A male long-horned bee attempts to copulate with a flower and picks up pollinia. *Bottom:* A bee's head carrying pollinia. (B) *Upper right:* A flower of the Australian orchid, *Cryptostylis leptochila. Middle:* A male of the wasp *Lissopimpla semipunctata* attempts to copulate with flower and picks up pollinia on its abdomen. *Lower left:* A wasp with pollinia. *(Redrawn from Wickler 1968.)*

flies resemble bees and wasps (Wickler 1968). Certain harmless snakes look like venomous coral snakes (Greene and McDiarmid 1981). Small birds may converge in pattern and behavior to larger competitors in order to lessen attacks on themselves (Diamond 1982b). Edible butterflies mimic poisonous ones. When mimicry involves such false advertising, it is called *Batesian mimicry,* after the English naturalist Henry W. Bates, who first described such mimicry in 1862.

Batesian mimicry has been most thoroughly studied in butterflies. In a classic series of experiments, J. Brower (1958a–c) demonstrated the efficacy of mimicry in several model–mimic pairs of North American but-

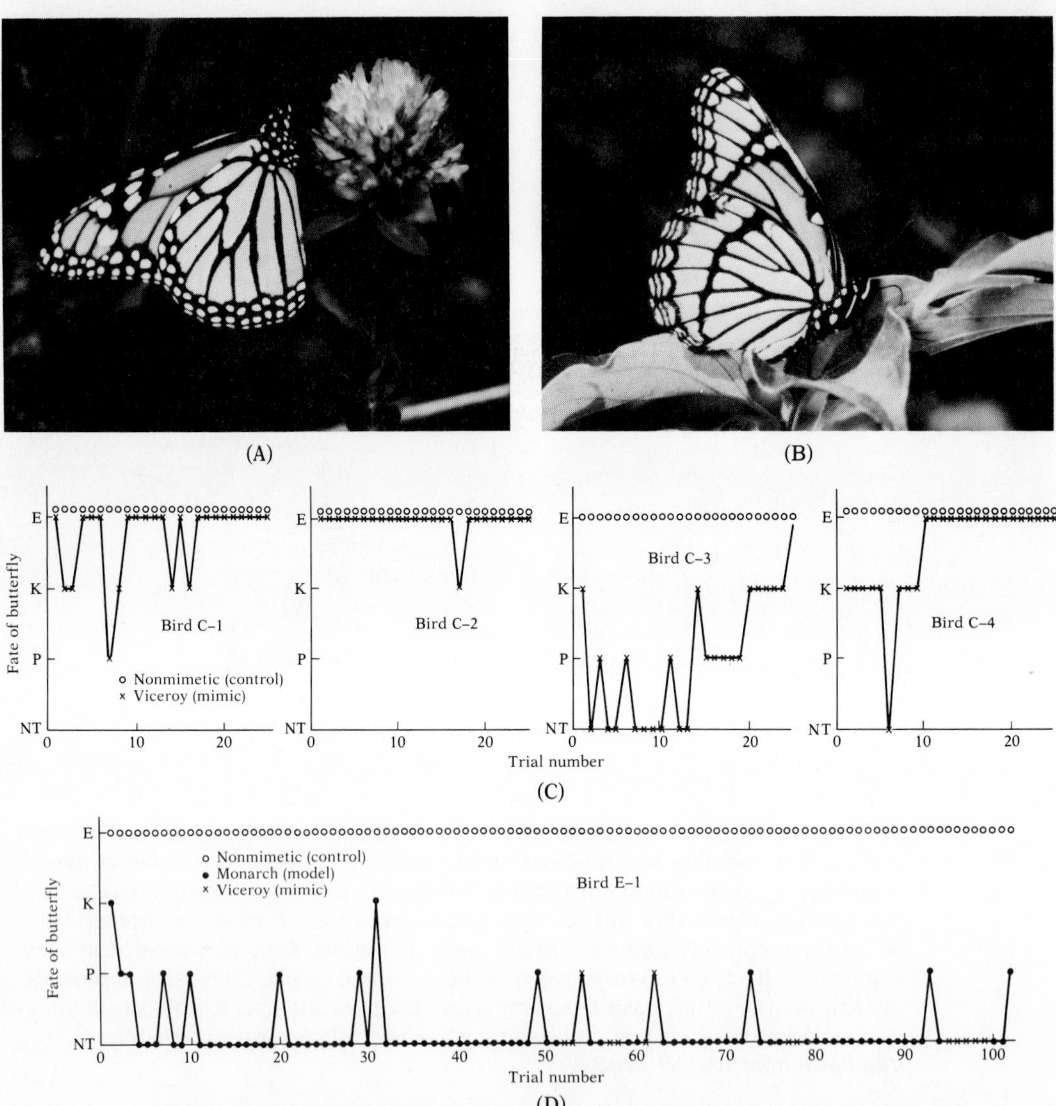

FIGURE 14–10 Monarch viceroy mimicry. (A) Monarch butterfly model. *(R. Humbert. Stanford University/BPS.)* (B) Viceroy butterfly mimic. *(R. Humbert, Stanford University/BPS.)* (C) Experiments showing that viceroys (mimics) are eaten by many birds, but not as much as the control, a non-mimetic butterfly that is always eaten. (D) Experiments showing that exposure to the monarch model leads to a refusal to eat the viceroy mimic. NT = not touched; P = pecked; K = killed; E = eaten. *(After J. Brower 1958a.)*

terflies. Birds that had never seen the butterflies in question were exposed to sequences of models and mimics. After learning that the models were distasteful, the birds were fooled by the mimics and shunned them. The results of a series of experiments using the distasteful monarch (*Danaus*

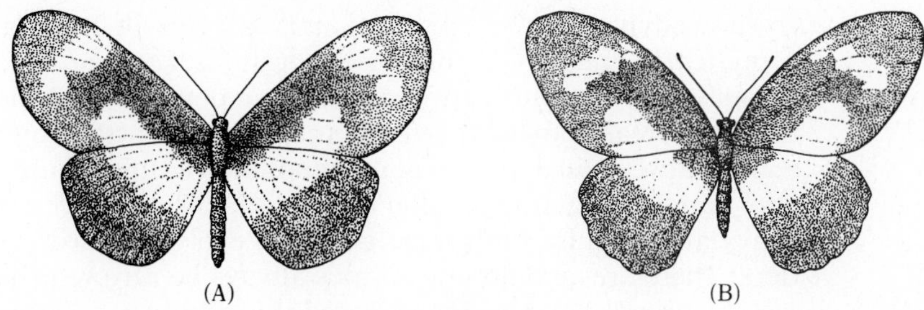

FIGURE 14–11 Batesian mimicry in tropical African butterflies. (A) Model from the subfamily Acraeinae of the family Nymphalidae (brush-footed butterflies). (B) Mimic from the family Papilionidae (swallowtail butterflies).

plexippus) as the model and the palatable viceroy (*Limenitis archippus*) as mimics are shown in Figure 14–10. These two species are in different subfamilies; their resemblance is purely superficial. Moreover, their resemblance is relatively inexact when compared to that of some model–mimic pairs in the tropics, as evident from the pair shown in Figure 14–11.

Observations of the geographic distribution and abundance of the other mimicry complex studied experimentally by Brower lent credence to the laboratory results (Brower and Brower 1962). In areas in which the toxic swallowtail *Battus philenor* was relatively abundant, so was the dark female form of the tiger swallowtail (*Papilio glaucus*) that mimics *Battus*. Another swallowtail mimicking *Battus, Papilio troilus,* was a closer mimic in areas in which *Battus* is common. Two nymphalid butterfly mimics of *Battus, Liminitis astyanax* and females of *Speyeria diana,* are confined in their distributions to locations in which *Battus* is common.

More recently, field experiments also have confirmed the efficacy of Batesian mimicry. For example, Sternberg *et al.* (1977) released day-flying moths (*Callosamia promethea*), some painted to resemble the yellow and black form of the palatable tiger swallowtail and others painted to resemble the toxic *B. philenor*. Those mimicking the toxic swallowtail were recaptured more frequently, as Table 14–1 indicates, survived longer, and were less often attacked by birds than those resembling the palatable tiger swallowtail.

TABLE 14–1 Recaptures of painted moths

	Days after release							
	0	*1*	*2*	*3*	*4*	*5*	*6*	**Total**[a]
Toxic-painted	54	26	12	4	2	2	2	102
Tiger-painted	54	7	3	5	3	3	0	75

Source: After Sternburg *et al.* (1977).

[a] The difference between the tiger- and toxic-painted recaptures is significant ($P < 0.025$).

The codynamic and coevolutionary aspects of Batesian mimicry are complicated and raise a number of questions. For example, how large may a population of mimics grow relative to a population of models before the mimicry breaks down? If most of the predators attacking butterflies with a given advertising pattern get a palatable victim, should not both model and mimic populations suffer heavy attack? Moreover, should not this heavy attack cause both populations to evolve toward protective coloration? Therefore, should not mimics always be rare relative to models?

J. Brower (1960) also investigated this question, using an experimental system in which mealworms painted in different patterns, some dipped in quinine (to make them unpalatable) and some not, served as models and mimics; starlings were the experimental predators. The results may be summarized as follows. If the model is distasteful enough, the mimics can be more common than the models, and the system still works to the mimic's advantage. This case is sometimes called the *poisoned beer principle,* which may be stated as follows: Even if only every tenth can of brand X beer contained poison and the other nine were delicious, people would shun brand X.

Perhaps the ultimate in Batesian mimicry has been shown in work by L. P. Brower (1970) and co-workers (1968) on monarch butterflies. These model butterflies owe their toxicity to cardiac glycosides obtained from their milkweed food plants. Shortly after eating a monarch loaded with these glycosides, a blue jay vomits, as depicted in Figure 14–12. Some monarch populations feed on milkweed species that do not contain the glycosides. However, they tend to escape predation because they are identical in appearance to poisonous monarchs—a phenomenon Brower *et al.* (1970) named automimicry, in which a set of individuals mimics another set of individuals of the same species.

In the tropics, many butterflies, as well as other organisms, are in mimicry complexes that do not involve false advertising because *all* the members of the complex are distasteful. This situation, known as *Müllerian mimicry* and shown in Figure 14–13, presumably evolves because it spreads the costs of predator training over several species. Suppose that in a given area a fledgling bird must eat a butterfly with a certain pattern four times (birds not being very intelligent) before it learns that the butterfly tastes awful. If a single aposematic butterfly population is present in the area, it therefore must sacrifice four individuals in order to train each new bird. If a Müllerian complex of eight species exists, each must sacrifice on the average only one-half of an individual in order to train each bird.

The degree to which coevolution in the strict sense has been involved in the development of Müllerian complexes is the subject of an ancient controversy.* The basic question was whether two aposematic species

* See Dixey (1897, 1909), Marshall (1908), Fisher (1927, 1930), J. R. G. Turner (1977, 1981), and Gilbert (1983).

FIGURE 14–12 The reaction of a blue jay *(Cyanocitta cristata)* to a monarch butterfly. The bird eats part of the butterfly (A) and then shows distress (B). Shortly thereafter it vomits (C) and subsequently will not eat monarchs. Birds familiar with monarchs also will ordinarily not eat viceroy butterflies, a species that mimics monarchs. *(Based on photos by Lincoln Brower.)*

would evolve toward each other when one was rare and the other common. Marshall (1908) took the position that evolution of the rare species toward the phenotype of the common one was all that would occur. Mutants of the common species toward the phenotype of the rare one would be at a disadvantage, because the experience of most predators would be with the common phenotype of the common species. Fisher (1927, 1930) concluded that convergence would occur only after a mutation in the rare species led to a major phenotypic shift that produced enough resemblance to the common species to confuse predators. Only then could the two species coevolve with each other. Gilbert (1983) points out that commonness and rarity are not species phenotypes but ecological variables. He suggests that Müllerian coevolution would occur most readily when two aposematic species shift numerical rank frequently in time and over their joint geographic distributions.

Mimicry complexes illustrate nicely the intricacies of codynamics and coevolution. The network of immediate interactions in butterfly mimicry involves not just the models and mimics but also the food plants of both and their predators. All may be important factors in the dynamics of populations of the others and in their evolution. If we consider just the coevolution of a Batesian mimic and its model, presumably it is to the

FIGURE 14–13 Butterflies from Latin American Müllerian mimicry complexes. Each pair of butterflies (left and right) are distasteful members of different species that mimic each other. *(L. E. Gilbert, University of Texas at Austin/BPS.)*

mimic's advantage to resemble the model as closely as possible. However, that situation is disadvantageous to the model, for the bigger the model–mimic difference, the more likely it is that a predator will learn to discriminate and attack the model less frequently after a pleasant experience with the mimic.

In theory, then, we might expect the model and the mimic to engage in a selectional race—the model evolving away from the mimic and the mimic trying to catch it. Examples of imperfect Batesian mimicry may represent a model's success in staying away from its mimic (Ehrlich 1970). Note that both model and mimic are coevolving with their food plants. Also note that both are coevolving with the predators involved in maintaining the mimicry, the predators presumably being selected for their better ability both to catch the butterflies and to discriminate between model and mimic.

This web of relationships can easily be stretched in the imagination to competitors of all of the organisms involved, and other predators upon them, and *their* predators and competitors—and so on, ad infinitum. At some point, of course, the influences become so slight as to be lost in the "noise" of the system.

PART 5

ORGANIZATION OF COMMUNITIES

We now approach one of the most complex groupings of organisms—the ecological community, defined as the collection of all organisms that live in a specific region. Most of the following five chapters deal with pieces of entire communities, systems involving from 2 to as many as 50 populations of different species living together. Although no ecological community is fully understood, pieces of some of the simpler ones are. In Part 4 we examined the main types of interactions that can occur between the populations in a community. Here we see how these interactions affect the properties of a community. This description will lay the basis for a worldwide survey of communities (Part 6) and our discussion of the most inclusive grouping, the ecosystem (Part 7).

CHAPTER 15

The Concept of a Community: From the Whole to the Parts

The concept of an ecological community originated with early studies of plant ecology, when vegetation throughout much of the world was classified and named. You are already familiar with communities recognized by their vegetation. These include deserts, savannas, chaparral, rain forests, deciduous forests, coniferous forests, tundra, and so forth. Many of these may be seen along a transect from a desert floor to a mountain top. Maps of characteristic plant communities throughout the world are now available in atlases published commercially and also by government agencies.

THE PARTS OF A COMMUNITY

In some contexts, the word *community* may not refer to the entire ecological community but to a large set of populations within it: all the plant populations of the eastern deciduous forest of northeastern America or the redwood forest of California's Sierra Nevada; all the fish of the Great Barrier Reef of Australia, the rodents of the Sonoran Desert, the birds of the chapparal, or the arthropods of the leaf litter of an oak woodland.

In general, however, the word *community* is reserved for the entire ecological community. The question then arises of how to define the parts of a community. A *guild* is a set of species that make a living in the same way (Root 1967). For example, the bird species feeding on insects by gleaning insects from leaves and branches comprise a guild, those that catch insects in the air another, and seed-eating birds still another. As with the term *community,* usage of the word *guild* is not rigid and can be applied to the populations comprising almost any functionally defined subset of a community (Jaksic 1981, MacMahon *et al.* 1981).

Another way to subdivide a community is to consider its *trophic levels,* which are typically larger than guilds, and may be thought of as feeding levels. Every community has pathways along which energy travels (as will be discussed in more detail in Part 7). In a terrestrial community, energy is originally captured by plants, is passed to herbivores and to decomposing organisms through leaf fall, and some is also passed from herbivores to carnivores and then to decomposers through waste products. A trophic level is a step along this energy path. Examples of trophic levels are as follows:*

- *Primary producer*—Nearly always a green plant, but includes any organism that binds energy from inorganic sources into the chemical bonds of organic compounds.
- *Primary consumer*—An herbivore.
- *Secondary consumer*—Predator on the herbivore.
- *Tertiary consumer*—Predator on the predator on the herbivore.
- *Decomposer*—At the end of the chain, an organism that lives by feeding on the carrion or wastes of other organisms.

Today no single standard exists for dividing a community into its parts. For instance, the smallest parts are the populations, but communities may have thousands of populations, each more or less open to migration—too many populations even to be listed, much less individually investigated. As a result, ecologists are continually seeking to identify functionally useful components of a community that are larger than single populations but that can still be considered as building blocks from which the whole community is assembled. Although the guild and the trophic level are helpful concepts, they are not perfect. The most common problem is that some populations cannot be assigned to only one guild or trophic level. For example, some birds both glean insects from leaves and catch them on the wing, and in the winter may eat fruits or seeds as well. Among most mosquito larvae, adult males feed on nectar and other plant juices, but adult females often feed on blood. Despite these difficulties, we will use these distinctions because nothing better is available.

* In communities with other paths of energy flow, these definitions may be supplemented with definitions of other compartments as needed.

IDENTIFYING COMMUNITIES

The practice of classifying communities leads us to wonder how discrete they are. Is a community a "package" of populations that are always found together? Is a community a natural unit in biology, a unit with a status analogous to that of other fundamental units such as the cell, organism, and population?

The question of whether a community is really a fundamental unit of organization has a long history of debate in ecology. One view is exemplified by the statements of British animal ecologist Charles Elton, who wrote in 1933 that "in any fairly limited area only a fraction of the forms that could theoretically do so actually form a community at any one time. . . . The animal community really is an organized community in that it apparently has 'limited membership' " (Elton 1933, p. 22 in the 1966 edition).

An alternative view is that a community is a haphazard and unstructured collection of individuals. In this view, the only properties shared by the individuals of a community are (1) that accidents of dispersal happened to bring them to the same place at the same time and (2) that accidents of environmental variation happened to allow them to survive at the same time. This school is identified with American plant ecologist Henry Gleason (1926), who wrote that "the vegetation of an area is merely the resultant of two factors, the fluctuating and fortuitous immigration of plants and an equally fluctuating and variable environment. As a result, there is no . . . reason for adhering to our old ideas of the definiteness and distinctness of plant associations [communities]" (Gleason 1955).

Although this debate continues in modified form today, as we shall see, we feel it is time to bury the hatchet. It is increasingly clear that no one view contains all the "truth." In some pieces of communities, membership is limited primarily by population interactions, particularly interspecific competition. In other pieces, interspecific competition is effectively absent, and predation is paramount. In still others, both competition and predation are equally important, and in some, mutualistic interactions are important. In other pieces of communities, interactions of any kind between species may be ignored. Finally, with certain communities it may be impossible to decompose the whole into pieces. Thus, in community ecology one must keep an open mind and deal effectively with all of these possibilities.

One approach to investigating whether a community is an organized collection of interacting populations or a haphazard collection of individuals is statistical. Originally used by plant community ecologists to identify and map plant communities, statistical methods are of two kinds: association analysis and ordination.

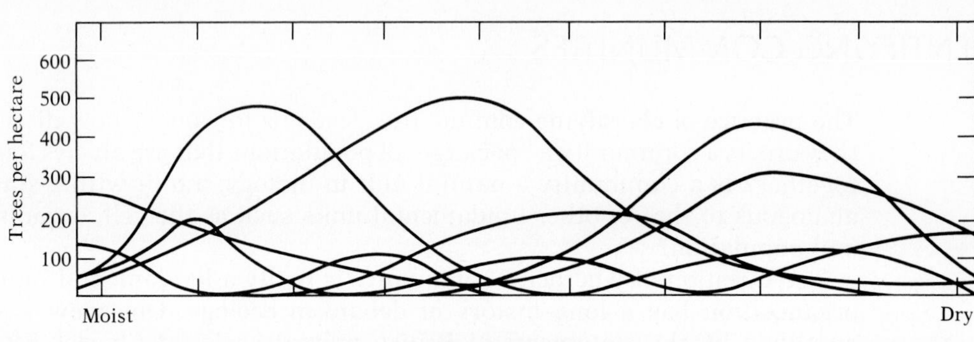

FIGURE 15–1 Abundance of several species of trees along a moisture gradient in the Santa Catalina Mountains, Arizona. *(Whittaker 1967.)*

Association analysis investigates the extent to which species co-occur. The quadrats are placed throughout an area, and species lists are made for each quadrat. The simplest test for co-occurrence is a contingency table, which detects an association between variables. This test for association is what we used to detect the presence of an association between R_t and N_{ave} in Chapter 5. However, more powerful and often more informative tests are available.* The groups that are found through an analysis of association may or may not have biological meaning. Also, some species are likely to be found in all quadrats and so will not be diagnostic of any group.

Ordination investigates the distribution of species along environmental gradients. Quadrats are placed at regular intervals along a transect that traverses an environmental gradient such as elevation, soil, or moisture, and the abundance, or biomass, of each species is measured for each quadrat.

If communities are packages of populations that are generally found together, one would expect to see the simultaneous rise and decline of sets of species along the transect, representing various community types distributed along the environmental gradient. Instead, as illustrated in Figure 15–1, what is actually found is that, for trees, each species seems to rise and fall in abundance along the transect independently of the others (Whittaker 1956, 1967, Bray and Curtis 1957, Loucks 1962).

Advocates of the idea that communities are discrete groups of species initially disputed the findings from ordination methods on the grounds that communities of different ages were being combined. They claimed that ordination studies merged communities of different successional stages together; the word *succession* refers to the sequential appearance of species during the colonization of a vacant habitat (as we discuss in

* Examples may be found in Williams and Lambert (1959, 1961) and Fager and McGowan (1963), together with a review in Pielou (1974, chap. 11).

more detail in Part 7). The rejoinder, however, is that if young and old communities are adjacent to one another as a result of recurring natural disturbances in the environment, such as tree falls from storms or damage from lightning-started fires, then perhaps it is just as legitimate to include communities of different ages in the ordination study as it is to include communities from sites with different slopes, elevations, or latitudes. In any case, it is clear that ordination studies of forest trees unanimously agree that communities of trees are not discrete bundles of species whose abundances rise and fall together.

More advanced ordination methods are also available in which the axis used for the ordination is a composite of several environmental axes. These techniques are useful if the important environmental axis is not known in advance of the study and if several environmental parameters are measured at each quadrat. Then statistical techniques are used to combine all the environmental measurements into the single axis with the most explanatory power. This method has been used, for example, by James (1971), Cody (1974, 1979), and Rice *et al.* (1983) when ordinating bird communities.

Both association analysis and ordination remain important techniques for the description of communities. Moreover, the results of these studies have made it clear that communities are generally not groups of species that are closely bound together. Because of this finding, we use *community* as a term of convenience; it refers to whatever organisms occur in some defined region. To explain the composition of a community, we must dissect it into pieces in a way that somehow reflects the major processes affecting the component populations. It is not clear how to do this, or even whether it can be done in all cases, but sometimes it can, as we shall see.

In the next four chapters, we will explore a representative variety of community pieces, charting a path that runs through many types of communities. In Chapter 16, we describe guilds composed of species that are competing for food; the examples are primarily from terrestrial habitats. In Chapter 17, we move to guilds composed of species whose members compete primarily for space; the examples here are primarily from marine habitats. In Chapter 18, we continue with pieces of marine and aquatic communities, passing through communities in temporary ponds and near streams and then back to land, where we explore community pieces that primarily consist of terrestrial plants and insects. Finally, in Chapter 19, we conclude with a discussion of patterns and approaches pertaining to the entire community, not just some of its components.

CHAPTER 16

Food-Limited Guilds: Coexistence and Niche Differences

The pieces of a community that have received the most attention in community ecology are collections of species that compete with one another for food. We refer to such a piece of a community as a *food-limited guild.*

Historically, phenomena from these guilds were described before investigation actually established that food was limiting and that competition occurred. Indeed, investigators during the 1960s and early 1970s often assumed that the description of certain patterns was itself sufficient evidence of the existence of competition. Reacting to such studies, researchers in the latter half of the 1970s made a determined effort to find out experimentally whether competition actually occurs and to improve the rigor and style of the inquiry on competition in communities. In addition, they explored the use of mathematical models to see if, even theoretically, competition could cause the patterns in food-limited guilds that had been attributed to competition.

LIMITED MEMBERSHIP: THE CLASSIC PATTERNS

The first pattern investigated in food-limited guilds was whether coexisting species differ from one another in the resources they use. The background to this search is the competitive exclusion principle discussed in Part 4. According to this principle, species must use different resources to

PLATE A Tree ferns in a forest of *Dicksonia antarctica* and *Eucalyptus regmans*. Notice the green light in the forest.

PLATE B Underwater photographer approaching a deep fore reef, Grand Cayman. Notice the blue light under water. *(See Chapter 2 for a discussion of the effect of light on plants.)*

PLATE C Caribbean coral. An example of a colonial life form. Note the distinct polys of each individual in the colony. *(See Chapter 1 for a discussion of the concept of the individual.)*

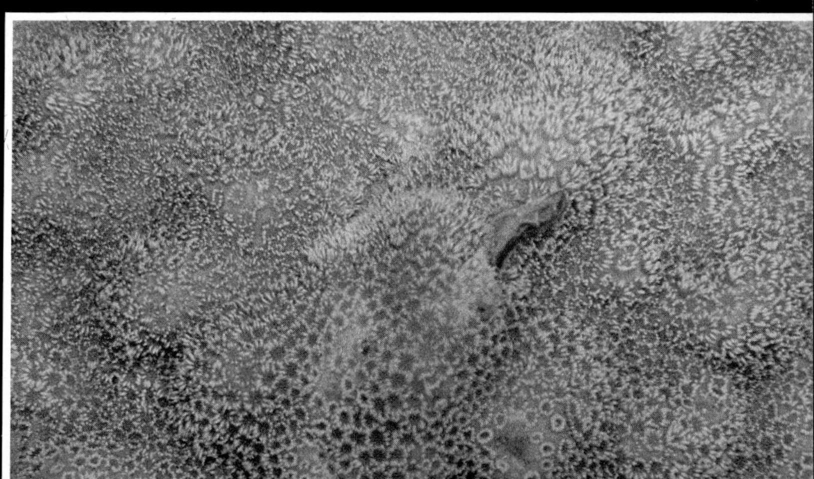

PLATE D A poster-colored butterfly fish, *Chaetodon lunula*, in midwater at Bora Bora, Society Islands. The striking colors of this and similar coral reef fishes are not territorial signals, as was previously believed. *(See Chapter 11 for discussion.)*

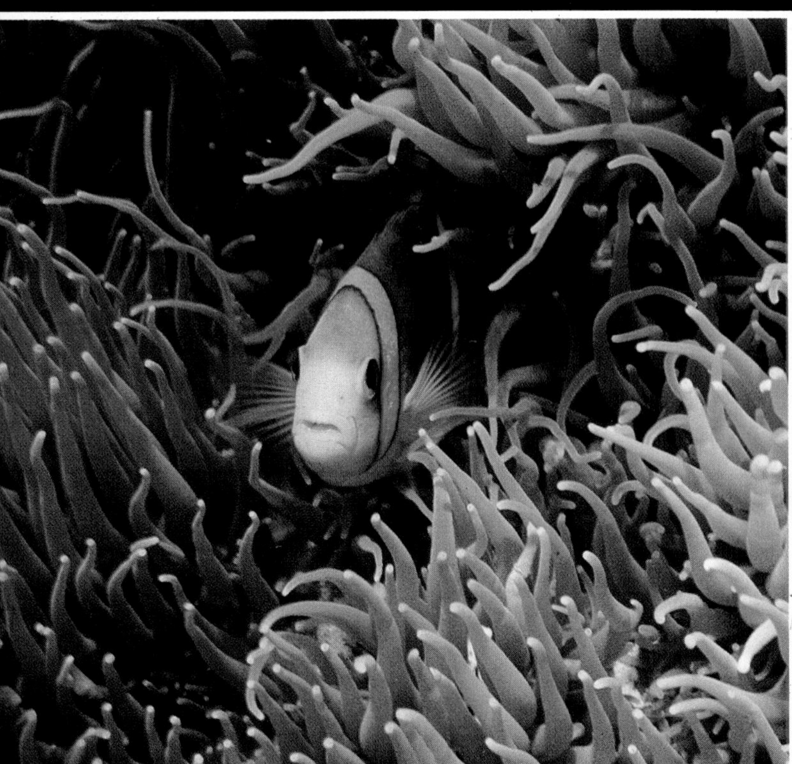

PLATE E Anemone fish, *Amphiprion bicinctus*, in the tentacles of its host anemone, Raiatea, Society Islands. This picture demonstrates an asymmetrical mutualistic relationship: The fish gets shelter from its enemies; the anemone is only occasionally fed by the fish. *(See Chapter 14 for a discussion of mutualism.)*

PLATE F A male Bay checkerspot butterfly, *Euphydryas editha bayensis*, taking nectar. The dynamics of populations of this northern California butterfly have been under intensive study since 1959 at Stanford University's Jasper Ridge Biological Preserve.

PLATE G A caterpillar of the Bay checkerspot feeding in the spring on a superficially grasslike annual plantain, *Plantago erecta*. The butterflies lay their eggs on or around this plant.

PLATE H *Plantago erecta* (small white flowers) growing intermixed with owl's clover (larger purple flowers, actually about 2 cm across), *Orthocarpus densiflorus*, on Jasper Ridge. Young *E. editha* larvae eat the plantain, but it generally senesces before they become large enough to pass the dry California summer in diapause. The larvae that survive are primarily those that manage to transfer to the owl's clover.

PLATE I Aerial photo of part of Jasper Ridge. Yellow "goldfield" flowers outline an area of serpentine soil. The flowerless grass on near the edge of the flowers is on sandstone soil, as is the chaparral at the upper right. The Bay checkerspot butterflies occur only in the serpentine area. *(See Chapter 5 for a discussion of population dynamics.)*

PLATE J *Anolis marmoratus* from Dominica. *Anolis* lizards are the principal ground-feeding insectivorous vertebrates of the West Indies, replacing the birds that carry out similar activities in the nearby continents. On some islands, such as Dominica, the lizards vary in color and size from place to place. This photo illustrates the lizards characteristic of various places in Dominica. *(See Chapter 15 for a discussion of the community of* Anolis *lizards in the Caribbean.)*

PLATE K Bird of paradise. The bright red and gold colors and long tail feathers of this bird of paradise are thought to be the result of sexual selection. *(See Chapter 11 for a discussion of sexual selection.)*

PHOTO CREDITS: *Plate A* by J. N. A. Lott, McMaster Univ./BPS. *Plate B* by S. K. Webster, Monterey Bay Aquarium/BPS. *Plate C* by J. Roughgarden. *Plates D–I* some extent in order to coexist. The use of different, or mostly different, types of limiting resources by two or more species of organisms is called *resource partitioning.*

The investigation of resource partitioning usually has been indirect, relying on a correlation between a morphological trait and what resources are used. This approach originates with Lack's (1947) classic study of the finches on the Galapagos islands. Lack noticed that the bill sizes of these birds depended on whether they co-occurred with a presumed competitor. On islands on which only one species of finch was feeding on the ground, the bill depth was approximately 10 mm. On islands where two or more ground-feeding species existed, however, the average bill depth for the smallest was approximately 8 mm, and the next larger species had bills whose depth was approximately 12 mm; no 10-mm form was found. Bill depth was presumed to be related to the size and thickness of the seeds that a bird eats. If so, a difference in bill size between two species would indicate that, to some extent, they eat different species of seeds or seeds from different places, thus indicating some degree of resource partitioning. Lack further suggested that the species on an island with more than one species *evolved* the differences in bill size in response to competition.

Almost 10 years later, Brown and Wilson (1956) brought to general attention Vaurie's (1951) similar observations with two insectivorous nuthatches whose ranges overlap in central Asia. These species have a bill size and coloration that differ from each other in areas in which their ranges overlap but have the same bill size and coloration at locations in which only one species is present. Brown and Wilson termed this phenomenon *character displacement*—that is, the characters of bill size and coloration are "displaced" away from each other at places where the species ranges overlap. Brown and Wilson suggested that character displacement can result from two different selection pressures. The first is interspecific competition that, as Lack postulated, causes the species to evolve differences in their resource use in the area sympatry. They considered the character displacement in bill size to have evolved under this selective pressure. The second is hybridization. If hybrids between the species are relatively infertile or inviable, then differences in traits that promote species recognition will be favored by selection; that is, selection will be against individuals that tend to hybridize. Brown and Wilson interpreted the character displacement in coloration as evolving by this selective pressure.

These studies set the stage for a more widespread use of size characters as indicators of resource partitioning. In 1959, G. Evelyn Hutchinson, in a famous paper entitled "Homage to Santa Rosalia, or Why Are There So Many Kinds of Animals?", suggested that coexisting species generally differ from one another by a standard amount. After reviewing data on the sizes of the feeding structures of several pairs of bird and mammal species, he suggested that the ratio of the sizes (in units of length) between the large and small members of a coexisting pair was 1:1.3. That meant that one was roughly double the volume or weight of the other, because $(1.3)^3 \approx 2$. Hutchinson concluded that this value could tentatively be used "as an indication of the kind of difference necessary to permit two species to co-occur in different niches but at the same level of a food-web" (1959, p. 152). His suggestion was quickly extended to the idea that the sizes of three or more co-occurring species should be regularly spaced along a size axis, with the ratio between consecutively sized species being 1.3 in units of length. Initially, these generalizations were empirical, but soon mathematical models of competition were explored to see if such a "biological constant" could be derived theoretically.

Observations of differences between species that do coexist were complemented by observations that species with very similar morphology fail to coexist. This observation expresses Elton's early idea of "limited membership" in a guild.

Thus, the classic pattern of structure in food-limited guilds consists of regular differences in the morphologies of the species that do coexist, together with the failure to coexist of species having the same morphology.

Now let us consider some examples of these classical patterns and discuss the questions that have been raised about whether such patterns are real.

EXAMPLES OF THE CLASSIC PATTERNS

Diamond (1973, 1975) has studied the birds of the South Pacific for 20 years and has extensively documented their community structure. There are 18 species of fruit pigeons on New Guinea, of which the 12 smaller species are in the genus *Ptilinopus* and the 6 larger in the genus *Ducula*. Diamond reports that all of these species are ecologically similar in being arboreal, living in the crowns rather than in the middle story, being exclusively frugivorous, not taking stones into the gizzard, and hence being restricted to eating soft fruits that can be crushed by the gizzard wall.

Only eight species are present at any particular location in the lowland rain forest. These form a graded size series, with each species weighing approximately 1.5 times as much as the next smaller species. The series starts with a tiny 50-g bird and ends with a huge 800-g form. The size differences indicate resource partitioning because body size influences the choice of the size of the fruits and the locations in which the fruits are collected (see Figure 16–1). Smaller birds take smaller fruits and from thinner perches than do larger birds.

The remaining 10 species are replacements of species that are found in lowland forest but that occur in other habitats; such species are called *habitat vicariants*. Diamond writes that a 76-g species in high-rainfall

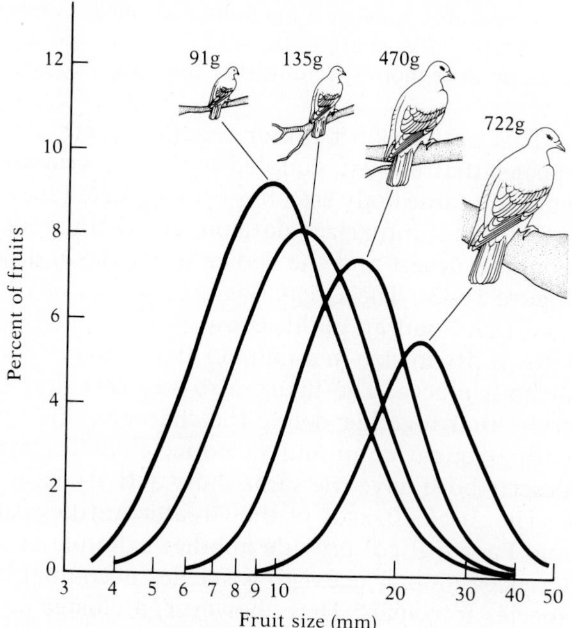

FIGURE 16–1 Percent of fruits of different sizes eaten by fruit pigeon species of different sizes. *(After Diamond 1975.)*

areas is replaced by a different 75-g species in lower-rainfall areas, a 123-g species in forest by a 112-g species in open country, a 592-g species at low elevations by a 613-g species at high elevations, and so forth. Thus, eight size levels are envisioned, corresponding to niches within the fruit pigeon guild. These niches are filled at any location by appropriately sized species.

Diamond suggests that the importance of body size is not restricted to fruit pigeons, but applies to much of the bird fauna of New Guinea. He writes (1973):

> Differences in body size provide the commonest means by which closely related species can take the same type of food in the same place at the same time. . . . Among congeners sorting by size in New Guinea, the ratio between the weights of the larger bird and the smaller bird is on the average 1.90; it is never less than 1.33 and never more than 2.73. Species with similar habits and with a weight ratio less than 1.33 are too similar to coexist locally (that is, to share territories) and must segregate spatially. For instance, the cuckoo-shrikes *Coracina tenuirostris* and *C. papuensis*, segregate by habitat on New Guinea where their average weights are 73 and 74 grams, respectively, but they often occur together in the same tree in New Britain, where their respective weights are 61 g and 101 g. New Guinea has no locally coexisting pairs of species with similar habits and with a weight ratio exceeding 2.73,

presumably because a medium-sized bird . . . can coexist successfully with both the large species and with the small species. Thus one finds a sequence of three or more species rather than just two species of such different sizes.

This pattern of a rather regular separation of body sizes among the species that coexist, coupled with a geographical replacement of species with the same body size, has been reported in other groups as well. Brown (1975) has summarized data on seed-eating rodent communities of the hot Sonoran desert and the cooler Great Basin desert in North America (see Figure 16–2). The Great Basin guild has one less species than the more southern Sonoran guild. Brown also points out that *Peromyscus maniculatus* shifts in size in a manner that leads to regular spacing in the guilds of both places. The figure also suggests that size is coupled with other traits that together define the characteristic types occurring in the seed-eating rodent community. Notice that the largest species from the two deserts both have the same daily activity habits.

The *Anolis* lizards of the Greater Antilles (Cuba, Jamaica, Hispaniola, and Puerto Rico) provide another example of a consistent scheme of differences among coexisting species accompanied by a failure of similar species to coexist. Here, however, a cluster of traits, not only body size, has been used in assaying the similarities and differences. Williams (1983) introduced definitions of *ecomorphs,* a type of lizard with a characteristic size, shape, color, and modal perch position in the vegetation (see Table 16–1). Williams reports that species representing different ecomorphs coexist, whereas species of the same ecomorph do not. In lowland wooded areas and forest in the Greater Antilles, one may find as many as six species of anoles coexisting.

TABLE 16–1 Ecomorphs of *Anolis* lizards in the Greater Antilles

Ecomorph	Size (mm)	Color	Modal perch	Body proportions
Crown giant	>100	Green	High canopy	Massive head, often casqued
Twig dwarf	<50	Gray	Canopy twigs	Long head, short body and legs
Trunk–crown	>70	Green	Canopy, trunk	Long body, short legs
Trunk	<50	Variable	Middle trunk	Short head and body
Trunk–ground	>60	Brown	Lower trunk	Large head, short stocky body, long legs
Grass–bush	<50	Lateral stripe	Grass or bushes	Long head, slender body tail

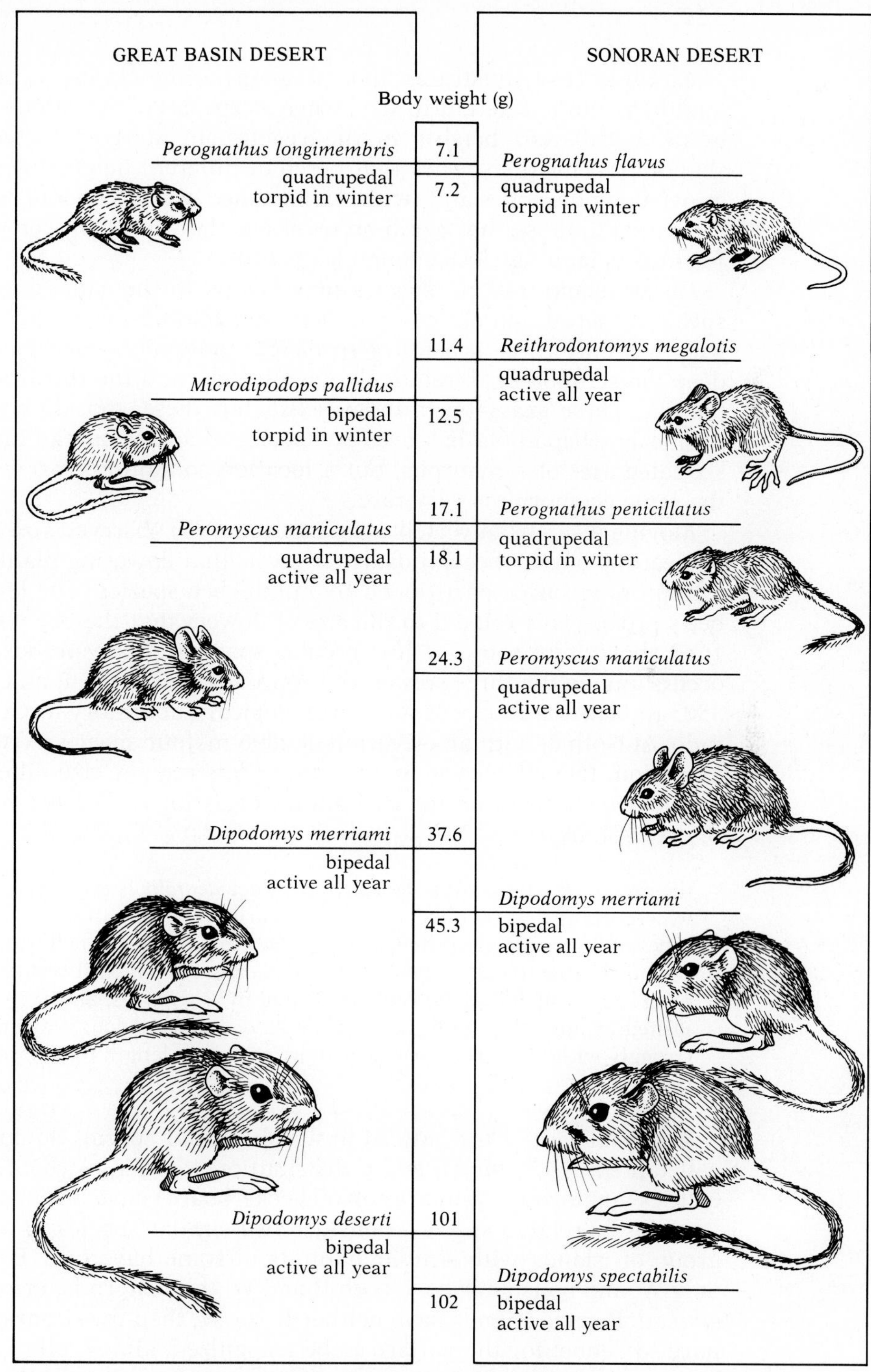

FIGURE 16–2 Schematic representation of niche relations among rodents in the Great Basin Desert *(left)* and the Sonoran Desert *(right)*. *(After Brown 1975.)*

As Table 16–1 illustrates, the species perching closest to one another tend to be quite different in size, whereas species of the same size tend to occur at different heights in the vegetation. Moreover, characteristic shapes exist for the species that perch at different heights: Animals that crawl along branches and twigs, for instance, tend to be long and slender compared to those that perch on trunks so that they can scan the ground for insects landing close enough for capture.

On an island, several species may belong to the same ecotype. Such species generally do not co-occur with one another. For example, Puerto Rico has only three species of trunk–ground anoles—one found only in deep forest, another in relatively open woods, and the third in open arid habitat. Three grass–bush anoles exist, but these, too, do not co-occur. Similarly, Hispaniola has over 36 species of anoles spread among these six categories of ecomorphs, but a location containing two species from the same ecomorph is very rare.

Moving away from vertebrates, Inouye (1977) observed size differences in Nearctic bumblebee guilds. He suggests that flowering plants might be a limiting resource partitioned by bumblebee species. The length of the bee's proboscis is related to the size of flowers that the bee visits. Figure 16–3 shows that more or less regular spacing of bee proboscis lengths occurs among the three species of Virginia Basin, a site at an elevation of 3505 m near the Rocky Mountain Biological Laboratory at Gothic, Colorado. At Gothic, with an elevation of 2896 m, four species exist; with one exception, they also occur with more or less regular size differences. Inouye (1977) writes that the exception, consisting of *Bombus bifarius* and *B. occidentalis,* is

> the exception that proves the rule. . . . *B. occidentalis* is a nectar robber, which means that it collects nectar primarily by biting through the corolla tubes of flowers to gain the nectar. . . . By employing this behavior, *B. occidentalis* is able to collect nectar from flowers from which it would normally be excluded, including flowers not visited by *B. bifarius,* and even flowers not visited by any other bee species. Thus, *B. occidentalis* does not compete strongly with other bee species of short proboscis length for the same floral resources.

A special case of the pattern in which similar species do not coexist is called a *checkerboard pattern,* a distribution pattern in which various localities (analogous to squares on a checkerboard) have one or the other of two closely related species. Consider two similar species, B and W. In a group of islands with similar habitats, if some have only B and others only W, and no islands have both B and W, the pattern is termed a *checkerboard.* If some islands have neither B nor W, then the situation becomes hard to define; for the pattern to be recognized, almost all of the islands have to have either B or W but not both, or no species at all.

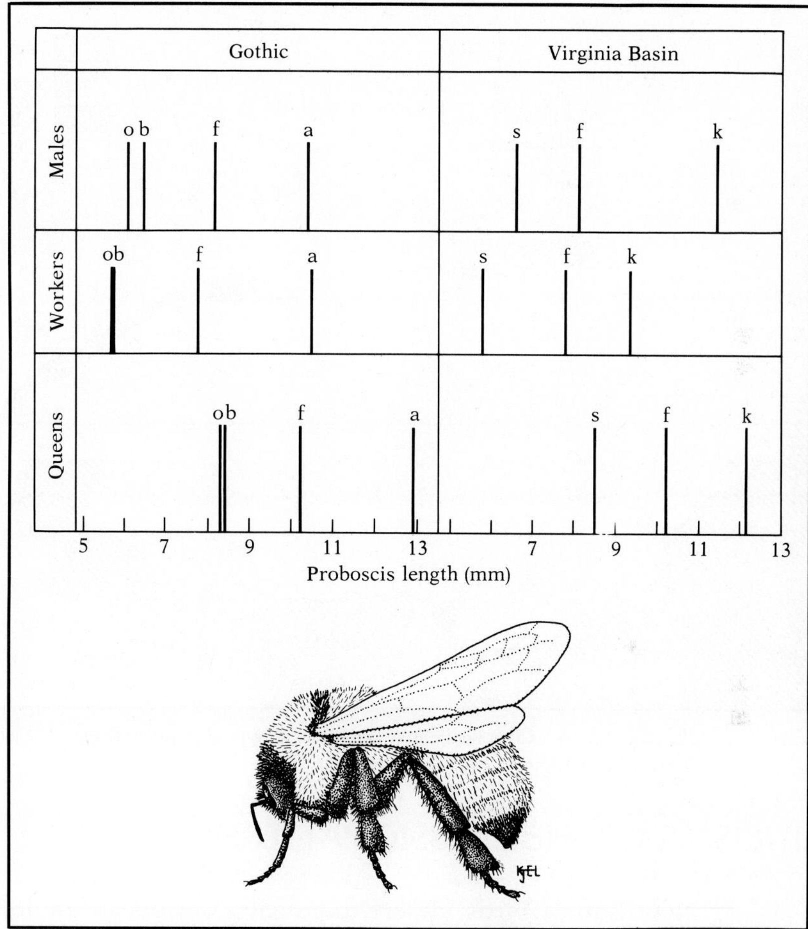

FIGURE 16–3 Schematic representation of niche relations among bumblebees at two sites near the Rocky Mountain Biological Laboratory at Colorado. The species are: o, *Bombus occidentalis;* b, *Bombus bifarius;* f, *Bombus flavifrons;* a, *Bombus appositus;* s, *Bombus sylvicola;* k, *Bombus kirbyellus.* Note the relatively even spacing between the proboscis lengths of the species indicating partitioning of floral resources. Note also how *Bombus occidentalis,* the nectar robber, has converged on *Bombus bifarius,* the exception that proves the rule. *(After Inouye 1977.)*

Diamond (1975) has described checkerboard patterns for several pairs of birds on islands in the neighborhood of New Guinea. Figure 16–4 illustrates a pattern involving two fruit pigeons, *Ptilinopus rivoli* and *P. solomonensis,* which was described as a checkerboard pattern and interpreted as evidence of interspecific competition.

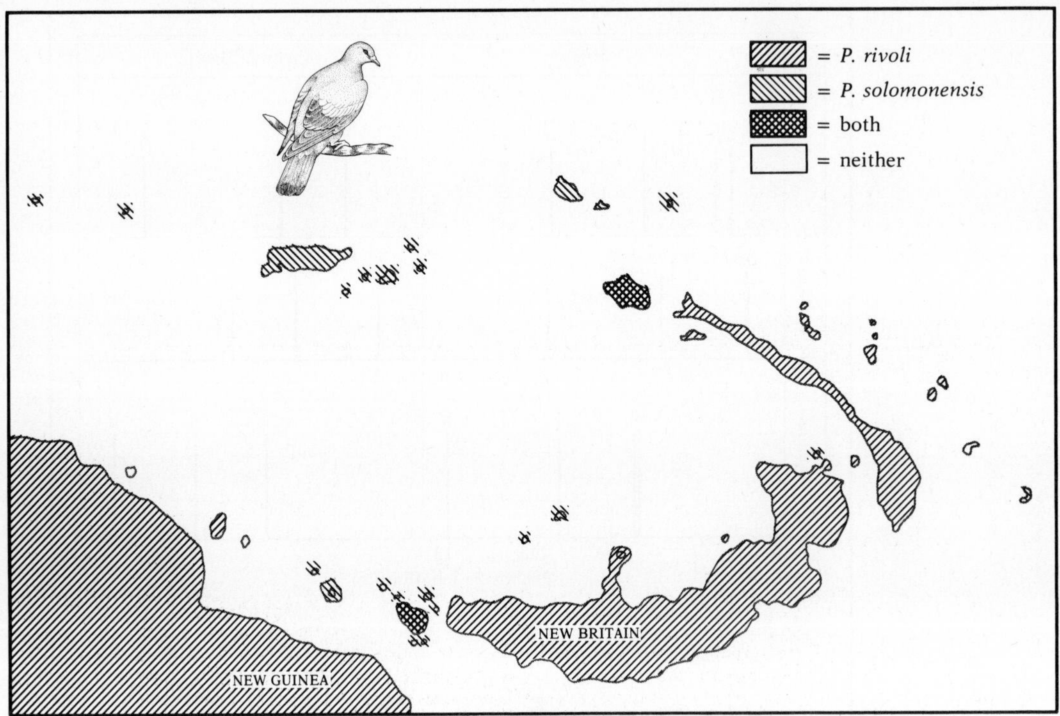

FIGURE 16–4 Checkerboard pattern of two closely related fruit pigeons in the Bismark Islands. *(After Diamond 1975.)*

CRITICISMS OF THE CLASSIC PATTERNS

New Guinea birds, desert mammals, Caribbean anoles, and mountain bumblebees all offer examples of the classic pattern of regular differences among coexisting species coupled with a failure of similar species to coexist. The literature on community ecology includes many other possible examples as well. These observations prompt certain questions: Are the patterns genuine? If so, what can we conclude about biological processes from such patterns? With the first question, we are asking whether the data are complete enough to conclude that the patterns are indeed accurate descriptions of species distributions. With the second question, we are asking whether the patterns, if correct descriptions, are diagnostic of interspecific competition.

Are Classic Patterns Genuine?

Whether all of the examples of the classic patterns proposed in the literature are genuine is not clear. Wiens (1982) has provided important criticisms of the quality of the evidence for the classic patterns, some of which are as follows:

1. The reality of the regular spacing of traits depends on the choice of the trait. The degree of regularity in the spacing between bird species depends on whether body weight, wing length, bill length, bill cross-sectional area, or tarus length is used. Hence, a biological justification should be provided for the choice of the trait. If traits are combined, using multivariate indices, the formula for the index also should be biologically justified.
2. The differences between species are often computed from the average values of the traits among adults, often in the larger sex, for each species. This practice tends to overestimate the differences between species because the juveniles of different species are usually more similar in size than are the adults, and yet much of the population may consist of subadults.
3. Ad hoc assumptions may be made about whether to include "rare" species among those coexisting at a given locality. Again, the criterion used to determine whether two species coexist must have a biological justification. If two species breed at the same locality, then they certainly should be considered to coexist. If one species is represented by an individual that is only migrating through, then they probably should not. However, if the animals are merely trapped, or sighted, then an arbitrary cutoff should be used to discount rare species, and the pattern that is found is sensitive to the choice of the cutoff point.
4. The species differences that are found may not be impressive, but one must ask, what *is* an impressive difference? Several species are usually different in many ways, including size. Does a factor of 1.3 represent a large difference? A factor of 1.1, or 1.01? Moreover, a collection of species will almost inevitably differ in size to some extent simply because they *are* different species. The detection of differences in species size may not be an ecologically important finding; it may be no more than a confirmation of taxonomic distinctness.

The Importance of Patterns in the Distribution of Species

To provide a yardstick against which to measure the importance of patterns in the distribution of species, a group of scientists at Florida State University proposed that the pattern should be compared to patterns that would result in the absence of interspecific competition (Connor and Simberloff 1979, Strong *et al.* 1979, Simberloff and Boecklen 1981). They have called models that predict patterns of species distributions, or body sizes, in the absence of competition *null models*. One type of null model involves data from species lists only; another type involves data on body size.

The main idea of the null models based on presence/absence data is to begin with the actual species lists, to "randomize" the lists, and then to compare the randomized lists with the actual lists. To randomize the species lists, in this context, means to take all the species names from a set of islands, place these names in a container (or *urn*, in the language of statisticians), and then to fabricate a set of island species lists by drawing as many species names from the urn for each island as are actually on that island. In this way, one generates species lists for islands that consist of random rearrangements of the original species lists, but with the same total number of species on each island.

The procedure is then repeated many times in order to generate many rearrangements; the set of all the rearrangements defines a *statistical population* of rearrangements. Finally, the actual species distribution is compared to the statistical population of possible rearrangements. For example, one can determine if the actual species lists contain certain species combinations more often, or lack certain combinations more often, than 95% of the randomly rearranged lists. If so, then the actual species lists are considered significantly different from the statistical population of randomly rearranged species lists. With this approach, Connor and Simberloff (1979) and Strong *et al.* (1979) found that many patterns thought to be present in species lists are not significantly different from random rearrangements of those species lists.

These null models for species distributions have been criticized on three grounds (Grant and Abbott 1980, Diamond and Gilpin 1982, Roughgarden 1983):

1. Testing data against the predictions of null models lacks power. That is, even when competition occurs, the statistical tests lack the ability to discern differences between what happens in the absence of competition and what happens in its presence, unless unreasonably large sample sizes (the species lists from a great many islands) are available.
2. Results of using null models are sensitive to the way the null model is formulated, especially with respect to what is considered to be the *species pool*. Is the species pool the collection of all of the populations that are currently on the islands, or is it the set of species of some nearby continental area that is presumably the source of the populations that colonized the islands? Moreover, are habitat requirements to be taken into account? How should the taxonomic limits of the study be determined—should only cuckoo-doves be considered, for example, or all columbiform birds, or all vertebrates? The choice is critical.
3. What one learns by falsifying a null model is not clear. Perhaps

tests of data against random rearrangements of themselves are irrelevant in determining whether competition is present. One might argue that a more informative alternative model should be constructed by modeling specific biological processes, such as dispersal and demographic stochasticity, while leaving out the other processes, such as interspecific competition, whose importance is being questioned.

In an extension of the null model approach, Simberloff and Boecklin (1981) introduced null models for the distribution of body size to examine whether the regular differences in body size among co-occurring species originally proposed by Hutchinson (1959) are genuine. These null models have in common the assumption that the species in the species pool have body sizes that are uniformly (evenly) distributed between some minimum and maximum. The investigators concluded that many reported instances of regularity among the differences in the body sizes of co-occurring species were not statistically significant. This study is subject to many of the same practical criticisms as those made against null models involving rearrangements of species lists. Is it realistic to assume that, in the absence of competition, body sizes of the species will be evenly distributed between some minimum and maximum?

Schoener (1984) analyzed the regularity in body size differences in the bird-eating hawks, using as a species pool all species of the genus *Accipiter;* as a measure of body size, the wing length was used. Figure 16–5 shows the distribution of wing lengths in the species pool. They are clearly not evenly distributed between a minimum and a maximum; rather, the distribution shows few birds with wing lengths less than 160 mm, most between 160 and 260 mm, and a few more from 260 to over 310 mm. As a result, if pairs were drawn at random from this worldwide species pool, one would find a great many pairs whose members are similar in body size.

Figure 16–5 shows the expected distribution of the ratio of body sizes (larger divided by smaller) for guilds consisting of two species drawn from this pool. Notice that the expected distribution of ratios for randomly drawn pairs is a decreasing curve. The figure also shows the actual distribution of wing-length ratios for the two-species guilds of *Accipiter* in the world. Schoener (1984) reports that "ratios smaller than 1.15 are notably lacking, even though 35% of the possible ratios lie here. This absence suggests a kind of limiting similarity for pairs." Figure 16–5 also shows a curious multimodal distribution. In pairs of *Accipiters,* the ratio of wing lengths is typically about 1.2, 1.5, or 1.8.

Schoener also studied the size ratios between species of adjacent sizes in guilds with more than two species. The results were much the same. He found significantly more species whose adjacent wing-length ratios were greater than expected by assembling the guild at random from all species of *Accipiter.* Schoener also noticed that relatively more guilds with

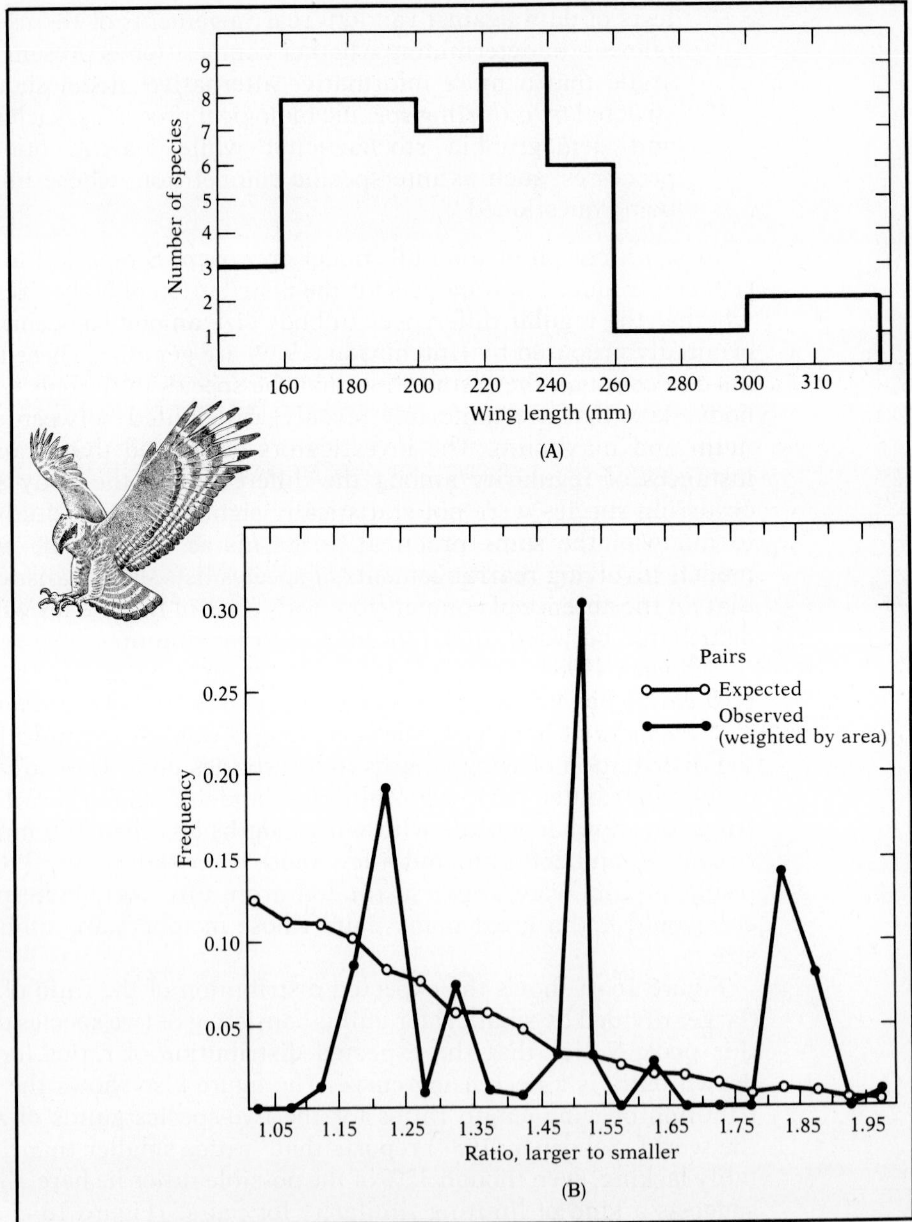

FIGURE 16–5 (A) Distribution of wing lengths in hawks. (B) Distribution of the ratio of larger to smaller wing lengths for pairs of hawk species. *(After Schoener 1984.)*

a few species exist than guilds containing a large number of species, as Table 16–2 illustrates.

Schoener comments that "the small number of realized associations, combined with the decline in the ratio of actual to possible associations

TABLE 16–2 Number of possible and real associations of bird species from the genus *Accipiter*

Number in guild	Number of possible associations	Number of real associations
2	1,081	31
3	8,596	19
4	178,365	9
5	1,533,939	3

Source: After Schoener (1984).

from pairs to quintets, might itself suggest some sort of competitive limitation on the number of species." He also cautions, however, that this hypothesis could not be examined further using the available data.

A statistical examination of the differences in body size among coexisting species was offered by Bowers and Brown (1982) for desert rodents. They tested statistically for independence between body size and co-occurrence in the granivorous rodent guild. They concluded that "species of similar size (body mass ratios <1.5) coexist less frequently in local communities and overlap less in their geographical distributions than expected on the basis of chance, suggesting that coexistence is precluded by interspecific competition."*

The dust is beginning to settle from the recent extensive reexamination of the classical patterns relating body size to species coexistence. Although the matter is still controversial, let us venture two general observations:

- Most ecologists think that the introduction of null models has been constructive. The critiques have led to more rigor and have discouraged a casual and uncritical acceptance of claims about biogeographic patterns.
- Most ecologists believe that the best evidence of competition is direct and that it is based on experiments conducted in the field. Ecologists usually are unwilling to accept patterns relating body size to coexistence—even patterns inconsistent with a null model—as convincing evidence of competition. Although such patterns may suggest competition, few would accept biogeographic patterns as diagnostic of competition.

Thus, let us accept the notion that some guilds occur that exhibit a pattern relating body size and foraging habit to coexistence. Although the details depend on the group being studied, the basic idea is that in *some* taxonomic groups, animals of the same size and habit rarely coexist but animals of sufficiently different size and habit do coexist.

* Still other examinations include those of Case *et al.* (1983) for West Indian birds, Grant and Abbott (1980) for Galapagos finches, Gatz (1979) for fishes, and Diamond and Gilpin (1982) for South Pacific birds.

In the next section, we approach the question of whether such a phenomenon can be caused by competition. We first examine direct evidence that competition is actually present, and then investigate the possible causes of the patterns relating body size and habit to coexistence suggested by models of competition.

DIRECT EVIDENCE FOR COMPETITION

In this section, we describe some case studies of competition to see whether it exists in some of the groups for which size–coexistence patterns have been reported. Let us recall from Chapter 12, in which we introduced the concept of interspecific competition, that experimental evidence for competition now exists in many taxa and ecological situations. Here we are not interested in how widespread competition is so much as whether it has been correctly identified as the cause of size–coexistence patterns. In parallel with the examples of size–coexistence patterns introduced before, we will consider competition in birds, desert mammals, anoles, and bumblebees.

Competition Among Birds

Although direct evidence for competition among birds is slight—perhaps disappointingly so in view of the extent to which bird communities have been used as exemplars of how competition imparts structure to a community—frequent competition may well exist among birds. Perhaps the best-studied group consists of the European tits, which are foliage-gleaning insectivorous birds of the genus *Parus.* They forage almost continuously and are relatively easy to observe. The distribution of these birds and the possible occurrence of a size–coexistence pattern have been examined by Herrera (1981). The evidence for competition between these species has recently been reviewed by Alatalo (1982); also extensive information on the *Parus* of Britain is summarized in Perrins (1979).

Winter is the time of principal resource limitation in these birds. Alatalo (1982) writes: "many studies suggest that foliage-gleaning birds eat only a small part of the available food in summer, . . . whereas in winter they consume a much higher proportion of food resources." Providing extra food during the winter has often increased the size of the breeding populations the following summer. In particular, Jansson *et al.* (1981) showed a doubling of population size for tits with food supplements. Furthermore, they studied individually marked birds and could prove that it was the extra feeding that increased winter survival. Moreover, Askenmo *et al.* (1977) had shown that the presence of the tits causes a great reduction in the number of spiders on spruce branches in the winter, which further supports the premise that winter food is in short supply. In addition, many tit species hoard food during the autumn for use later in the winter, again suggesting that food is limiting in the winter.

The evidence for competition consists principally of changes in the feeding habits and in the abundance of one species when the presumed competitor is removed or otherwise absent. Alerstam *et al.* (1974) compared the foraging of tits on the Swedish mainland and on the island of Gotland. The willow tit (*P. montanus*), crested tit (*P. cristatus*), and coal tit (*P. ater*) occur on the mainland, whereas only the coal tit occurs on Gotland. On the mainland, the coal tit forages on needled twigs of spruce and pine, whereas the willow tit and crested tit forage in the inner part of trees near the tree trunk. However, on Gotland, the coal tit uses the inner parts of trees much more frequently than it does on the mainland. This fact is indirect evidence of competition by the willow and crested tits against the coal tit.

Furthermore, in northern Finland, the coal tit is absent, leaving only the willow and crested tits. Alatalo (1981a) showed that the willow and crested tits use needled twigs more frequently in northern Finland, where the coal tit is absent, than they do in Sweden, where it is present. This observation is indirect evidence of reciprocal competitive interaction. Direct evidence is now also becoming available. Alatalo (1982) reports that during the winter of 1981–1982 most of the willow and crested tits were removed from a site in Sweden, and the remaining coal tits then used the inner areas of the trees more frequently than did the controls. More evidence in the same vein comes from the foraging shifts that occur in mixed-species flocks during the winter. Alatalo (1981b) established that species avoided one another's foraging places when moving in the same flock.

Finally, further evidence of competition is that the coal tit is much more abundant on the island of Gotland than it is on the mainland (Alerstam *et al.* 1974). Also, evidence of an evolutionary effect of the removal of competition is that the foraging shift of the coal tit on Gotland is accompanied by a morphological shift; on Gotland, coal tits are bigger than they are on the mainland, and are closer in size to the crested and willow tits than they are on the mainland.

Thus, a circumstantial case can be offered that interspecific competition is present between the members of a guild of foliage-gleaning birds from northern Europe. Although evidence suggests that food is a limiting resource, the nature and strength of the competition are not yet well characterized. Also, food is almost certainly not the only resource that is limiting. Dhondt and Eyckerman (1980) established that great tits (*P. major*) and blue tits (*P. caeruleus*) compete for roosting sites during winter and that the larger great tit excludes the smaller blue tit from holes. This finding suggests that the competition is not symmetrical.

Many studies illustrate niche shifts and abundance changes in comparisons of island and continental bird faunas (see, e.g., Keast 1968, 1970a). What is unusual in the *Parus* studies is the evidence for resource limitation together with the participation of many workers, which allows corroboration of the data.

Competition Among Desert Mammals

Experimental evidence of competition among desert rodents comes from the work of Price (1978a) with rodents similar to those whose size–coexistence patterns were described earlier. The experiments were carried out in a habitat that is a transition between the lower Sonoran desert and the desert grassland about 50 km south of Tucson, Arizona. The rodents that coexist in the study area are the kangaroo rat, *Dipodomys merriami* (weighing 33 g), and three pocket mice: *Perognathus baileyi* (27 g), *P. penicillatus* (17 g), and *P. amplus* (12 g).

These rodents tend to forage in different microhabitats. In the study, the microhabitats were classified into four categories: (1) "large open space," a patch of bare ground at least 2 m in diameter; (2) "small open space," patches of bare ground between 0.20 and 0.25 m in diameter; (3) "large bush," ground under a shrub that is at least 1 m tall and 2 m in diameter; and (4) "tree," ground under a tree (*Prosopis* or *Cercidium*) that is at least 1.7 m tall. In the summer, the kangaroo rat tends to be found in the large open spaces, *P. amplus* in small open spaces, and both of the other pocket mice under bushes or trees. In the winter, only two of the species are active; the other two are in hibernation.

The first experiment tested whether the range of microhabitats used by these rodents increased if the other species were removed. The experiments were carried out in fenced-in areas to prevent animals from migrating in from surrounding habitats. The range of microhabitats used by each of the three pocket mice increased when the other species were removed relative to their use of microhabitats under natural conditions. The range used by the kangaroo rat also increased, but not enough to be statistically significant.

The second experiment tested whether the microhabitat ranges would contract when other species were added to the enclosures. Kangaroo rats were added in two stages to the enclosures containing pocket mice. The range of microhabitats used by the pocket mice did not contract when only a few kangaroo rats were added, but it did contract when a larger number were added. Conversely, in enclosures that originally had only kangaroo rats, the kangaroo rats also narrowed the range of microhabitats they used when pocket mice were added. All the changes were in the appropriate directions. For example, in the presence of the pocket mice, the kangaroo rats used the open spaces more and were not found on ground under bushes and trees as much as they were in the absence of the pocket mice.

The third experiment involved augmenting the amount of certain habitat types in an unenclosed study area. Bushes were removed to create more of the large open space used by the kangaroo rat. The number of kangaroo rats increased, primarily as a result of immigration from adjacent areas. Also, the numbers of pocket mice decreased. This experiment indicates that the abundance of the animals is influenced by the amount of microhabitat available and suggests that the microhabitat type is, or is correlated with, limiting resources.

The significance of body size in these rodents is a puzzle. An early report by Brown and Lieberman (1973) showed that body size was related to the size of seeds eaten. However, later work has not tended to confirm this finding, and any partitioning of seeds with respect to size is an indirect consequence of microhabitat specializations. What seems to be important is the relation between animal size and the spatial distribution of the seeds that it feeds on. Seeds are more clumped in open spaces than under bushes, and the species differ in their foraging efficiencies for various spatial patterns of seeds (Reichman 1976, Reichman and Oberstein 1977). Moreover, Price (1978b) found different preferences for the spatial pattern of seeds by heteromyid rodents. Thus, it presently appears that patterns of body size and coexistence among desert rodents do not involve resource partitioning of food size in the same way that they do for birds, although microhabitats are partitioned.

A more serious problem in understanding the size structure in the guild of desert seed-eating rodents is that the competitive interaction among the rodents may be weaker than that between rodents and ants. Ant populations also consume seeds and are potential competitors with seed-eating rodents. Studies using enclosures that selectively keep rodents outside of the experimental areas reveal a strong interaction between ants and rodents (Brown *et al.* 1979, Davidson *et al.* 1980).

Competition Among *Anolis* Lizards

Extensive experimental evidence for competition in a size-structured guild comes from studies with *Anolis* lizards on small islands in the Caribbean (Roughgarden *et al.* 1983). The islands that begin near Puerto Rico and extend in an arc toward Central America are much smaller than the Greater Antilles discussed previously, and they support a much smaller fauna. Each of these islands has only one or two species; none has three or more. St. Maarten in particular is unique in having two species whose body sizes are very similar, and this island has been the site of the studies detecting competition.

In *Anolis* lizards, body size is clearly related to prey size. In one of the first studies to show this relation, Schoener and Gorman (1968) found that the two species of anoles on Grenada differ in the size of their prey, with the larger lizard species taking, on the average, larger insect prey than the smaller lizard species. (The average is computed using the distribution of prey volume from various prey-length categories.) Roughgarden (1974a) established further that the size of individuals *within* a species is correlated with the size of their prey.

More recent studies have established, for the anoles of these small islands, that the relation between body size and prey size is not an indirect consequence of foraging in different microhabitats. For example, Adolph and Roughgarden (1983) showed that the anoles of St. Eustatius forage mainly in the same microhabitat. On St. Eustatius, the larger species usually perches higher in the vegetation than does the smaller (typically, the larger is higher about 1 m). This separation in perches presumably reflects the fact that the larger species searches a wider area in order to find large prey, because large prey are rarer than small prey.

St. Maarten has two species whose body size is nearly the same: *Anolis gingivinus* and the slightly smaller *A. pogus*. *A. pogus* is found only in hills in the center of St. Maarten, whereas *A. gingivinus* is found throughout the island, including the central hills, and also on small offshore cays. The abundance of *A. gingivinus* is lower at locations where *A. pogus* is also present than it is at locations where it is the only anole species. Also, *A. gingivinus* perches higher in the vegetation at locations where *A. pogus* is present than it does at locations where it is the only anole. In locations where both species occur, *A. pogus* perches below *A. gingivinus*. These observations suggest that *A. pogus* affects both the abundance and the perch position of *A. gingivinus*.

To see experimentally if competition from *A. gingivinus* affected *A. pogus*, individuals of *A. pogus* were introduced onto a small offshore cay where *A. gingivinus* lives. At some sites on the cay, the resident *A. gingivinus* were removed; at other sites, the residents were left undisturbed. The introduced animals survived about twice as well when the resident animals were removed as when the resident animals were left undisturbed. This experiment shows that the presence of *A. gingivinus* affects the survivorship of introduced individuals of *A. pogus*.

To determine if a reciprocal interaction exists, four experimental sites were prepared consisting of fence-enclosed areas of natural habitat where both species occur, each enclosed area measuring 12 by 12 m. The abundance of anoles in the surrounding habitat was approximately 1 anole per square meter. Sixty individuals of *A. gingivinus* were placed in each enclosure; in addition, 100 individuals of *A. pogus* were added to two of the enclosures. In the enclosures in which *A. pogus* was present, the *A. gingivinus* individuals had less food in their stomachs, grew more slowly, produced fewer eggs, and perched higher in the vegetation than did the individuals in enclosures that did not have *A. pogus* also present. This experiment showed that *A. pogus* has a large effect on *A. gingivinus*.

The experiment directly implicates competition for food as one of the mechanisms of competition because the animals that were observed to have captured less food were the animals that had slower growth and lower egg production. Furthermore, Licht and Gorman (1970) and Stamps (1977) have shown that growth rates of anoles increase when their food supplies are increased.

To examine further whether food is a limiting resource under natural conditions, Pacala and Roughgarden (1984) removed lizards from three fenced enclosures, and three other enclosures were stocked with natural abundances of lizards. The experiment was carried out on the island of St. Eustatius, which is near (20 km from) St. Maarten and has a nearly identical habitat. After 6 months, the enclosures without lizards had two times as many insects on the forest floor and 20 times as many spider webs in the vegetation as the enclosures containing lizards. Also, Toft and Schoener (1983), working on small islands in the Bahamas, discovered a negative statistical correlation between the number of anoles and the number of spiders. Thus, the presence of *Anolis* lizards strongly affects the abundance of arthropods in the forest.

The experiments discussed so far show that strong present-day competition exists on St. Maarten, an island in which the two species have a similar body size and correspondingly high overlap in the sizes of insects that they eat and in the places in the vegetation where they perch. To determine if anoles that differ more in their body sizes also compete, Pacala and Roughgarden (1985) performed similar experiments on St. Eustatius, where the two lizard species differ in body length by a factor of nearly 2. Using the same experimental design as on St. Maarten, they stocked four enclosures with 60 of the larger species of that island (*A. bimaculatus*). In two of these enclosures, they also placed an additional 100 individuals of the smaller species (*A. schwartzi*). Here almost no competition was detectable. The smaller species did not strongly affect the amount of food in the stomachs of individuals of the larger species, or their growth rate, fecundity, or perch position in the habitat. More recently, the reciprocal experiment has also been carried out on St. Eustatius, and the results show that the larger species also has a very small effect on the smaller species there (Rummel and Roughgarden, 1985). These results contrast strongly with the observations of strong competition between the anoles of St. Maarten.

Thus, the studies on *Anolis* from the eastern Caribbean show that strong competition occurs on the only island with two species of similar body size, whereas weak competition occurs on a nearby island with two species of different body size.

Competition Among Bumblebees

Finally, we briefly revisit nectivorous bumblebees, the fourth example of a size-structured guild. Inouye (1978) obtained experimental evidence of competition in a system consisting of the bumblebees (*Bombus appositus* and *B. flavifrons*), using two flowers (*Delphinium barbeyi* and *Aconitum columbianum*) in the Rocky Mountains of Colorado. Normally, each bee species concentrates on a different flower according to the corolla tube length that most closely matches its proboscis length. When each bumblebee was temporarily removed from its preferred flower species, the remaining bee species increased its use of that flower.

The Different Circumstances of Competition

Thus, European insectivorous birds, North American desert seed-eating rodents, eastern Caribbean insectivorous lizards, and North American bumblebees all offer examples of size-structured guilds for which independent experimental evidence of competition now exists. Yet the circumstances of the competition are different among these examples. The kinds of resources that are limiting are different among most of them, and hence the mechanisms of competition also must differ. Also, the experimental studies do not generally indicate if competition is the most important consideration; for example, no research has demonstrated that predation is quantitatively less important than competition for the population dynamics of the species in these examples. Furthermore, in some of these examples, the evidence suggests that competition is continually present or nearly so; it does not represent an extremely rare condition. The studies of *Parus* in Northern Europe and of *Anolis* in the Caribbean have been carried out over several years, in different places, and with many investigators, with generally consistent results. Nonetheless, the studies of Dunham (1980) with desert lizards showing that competition occurs only in exceptional years should be considered an important alternative situation.

COMPETITION THEORY AND SIZE-STRUCTURED GUILDS

The Lotka–Volterra Model Extended to Three Species

As studies to detect the presence of competition experimentally were proceeding, mathematical models for competition were also being analyzed. These models are based on the Lotka–Volterra competition equations introduced in Chapter 12 combined with some additional assumptions that enable the parameters of the Lotka–Volterra equations to be interpreted in terms of resource partitioning.

For two species, recall the Lotka–Volterra equations:

$$\frac{dN_1}{dt} = r_1(K_1 - N_1 - \alpha_{1,2}N_2)\frac{N_1}{K_1}$$

$$\frac{dN_2}{dt} = r_2(K_2 - \alpha_{2,1}N_1 - N_2)\frac{N_2}{K_2}$$

Now for more species, say three, another equation is added, and a term is added to each of the original equations. For example, with three species, dN_1/dt becomes

$$\frac{dN_1}{dt} = r_1(K_1 - N_1 - \alpha_{1,2}N_2 - \alpha_{1,3}N_3)\frac{N_1}{K_1}$$

Notice that the term $\alpha_{1,3}N_3$ has been added. Similarly, the equation for dN_2/dt will have a term added, $\alpha_{2,3}N_3$. The equation for dN_3/dt is cast in the same mold:

$$\frac{dN_3}{dt} = r_3(K_3 - \alpha_{3,1}N_1 - \alpha_{3,2}N_2 - N_3)\frac{N_3}{K_3}$$

In this way, a model for many competing species can be written down, all based on the original Lotka–Volterra equations for two species.

Any large competition model in the Lotka–Volterra mold consists of three groups of coefficients: the r's and K's, and competition coefficients (α's). The α's may be organized as a table called the *α-matrix,* shown in Table 16–3. The α-matrix summarizes the strength of each of the pairwise competitive interactions, with the value of 1 representing the intraspecific competition coefficient.

TABLE 16–3 α-matrix

		Name of species causing competitive effect		
		1	2	3
Name of species receiving competitive effect	1	1	$\alpha_{1,2}$	$\alpha_{1,3}$
	2	$\alpha_{2,1}$	1	$\alpha_{2,3}$
	3	$\alpha_{3,1}$	$\alpha_{3,2}$	1

New Results with Three Species

The first question to ask of a model with three or more competing species is whether the two-species Lotka–Volterra equations are representative of models with more than two species. The answer is no. With two species, let us recall, only three kinds of outcomes are possible: (1) one species always excludes the other, regardless of their initial abundances; (2) either species can exclude the other, provided it is initially the more abundant; (3) both species coexist and achieve steady-state abundances that can be predicted from the α's and K's.

In general, the outcome is a steady state of some sort, with either one or two species present. Also, which outcome results depends only on the size of the α's relative to the ratio of the K's; the r's are unimportant in determining the qualitative outcome with two species, although they do influence the speed with which the outcome will be attained.

With three or more species, the results are different. First, theoretically, species competing according to the three-species Lotka–Volterra equations may oscillate permanently, without ever attaining a steady state. Second, whether the species permanently oscillate or not is determined in part by the size of the r's. The r's not only affect the speed with which an outcome is realized but may also affect the outcome qualitatively.

Figure 16–6 illustrates an example of permanent oscillation among three competing species (May and Leonard 1975). It presents what is called *intransitive competition,* which is the following situation: Species 1

FIGURE 16–6 An example of permanent oscillation among three competing species. *(After May and Leonard 1975.)*

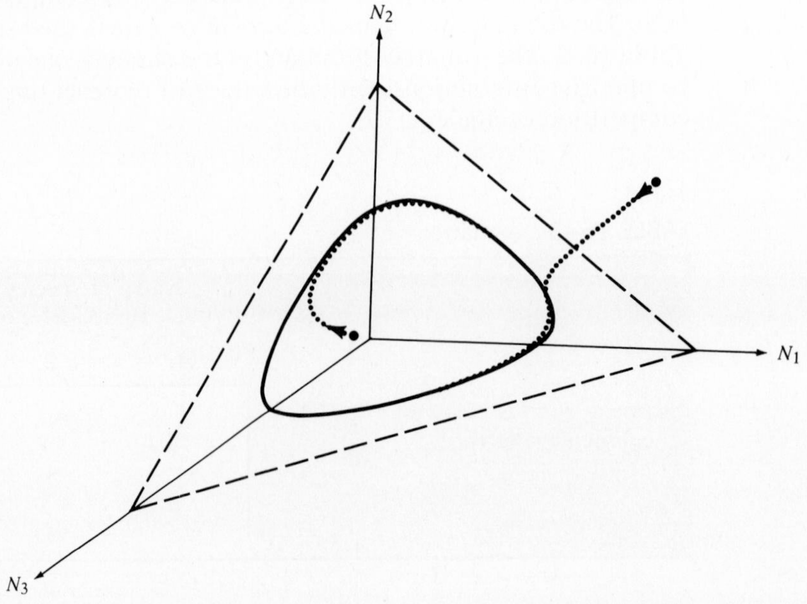

excludes species 2 when both are alone; species 2 excludes species 3 when both are alone; and species 3 excludes species 1 when both are alone. However, when all three are together, a permanent oscillation may result. Whether any examples of these competitively driven oscillations are present in nature is not known. Buss and Jackson (1979) have identified possible examples of intransitive competition among subtidal sessile invertebrates, but the applicability of the three-species Lotka–Volterra competition equations to that system is not yet known. In any case, it is clear theoretically that the presence of competition does not automatically stabilize a system and cause it to attain a steady state. Whether competition leads to a steady state depends on the number of species involved, on how the strength of competition is distributed among all the pairwise interactions, and on the intrinsic rates of increase of the species.

General Two-Species Models

Because the two-species Lotka–Volterra competition equations are not completely representative of the Lotka–Volterra equations with three or more species, we might also ask whether the two-species Lotka–Volterra equations are representative of other possible two-species models of competition. That is, suppose we use equations more complicated than the Lotka–Volterra model to describe the inter- and intraspecific density dependence. Are we likely to get results greatly different from those of the traditional two-species Lotka–Volterra model?

In a short, elegant study of great importance for ecologists, Hirsch and Smale (1974) noted that the two-species Lotka–Volterra model is very robust, provided that it is compared only to other models for *two* competing species. If the right-hand side of the equations for dN_1/dt and dN_2/dt is replaced by an arbitrary continuous function that still represents competition, then oscillations cannot result. The only new possibility that may arise is several simultaneously stable steady states for the system, each attained from a special set of initial abundances. Indeed, Schoener (1976) has introduced some models for competition illustrating this result. However, as the number of species increases beyond two, all bets are off. If many strongly competing species co-exist, one must specify the mathematical form of the density dependence for the competition in detail to obtain a clear-cut prediction of the effect on population dynamics. Also, time lags, if their effect is large enough, theoretically can cause even a two-species competition system to oscillate.

Niche Overlap and Competition

Against this backdrop of diverse results from general competition theory, the application of the Lotka–Volterra equations to competition for food in size-structured guilds appears simple. In size-structured guilds *com-*

petition is assumed to depend on overlap in the use of limiting resources. This restricts the arrangement of competitive strengths among the species and hence reduces the variety of possible outcomes for competition, even with many species. Specifically, the α's are assumed to depend on overlap in resource use as follows: Let $u_i(h)$ be the fraction of the total resource gathered by an individual of species i that comes from resource category h. Similarly, let $u_j(h)$ be the resource use by an individual of species j. Then $\alpha_{i,j}$ is assumed to be proportional to the product of $u_i(h)$ and $u_j(h)$ summed up over all the resource categories:

$$\alpha_{i,j} = c_i[u_i(1)u_j(1) + u_i(2)u_j(2) + \cdots]$$

where c_i is the coefficient of proportionality for species i.

In the literature, various choices for c_i have been discussed. Some (where the c_i's are equal) yield a symmetric α-matrix, others do not. Some yield α values that are always less than or equal to 1; others may produce α values greater than 1.

Overlap is inherently symmetrical. If i overlaps with j, then j automatically overlaps with i. Although reciprocal α's may not be exactly equal, since c_i may differ from c_j, reciprocal α's nonetheless cannot be too different because they still incorporate the same degree of overlap. Here intransitivity and other very asymmetric schemes of competition are ruled out.

The combination of the Lotka–Volterra equations with α's based on overlap, as in the preceding formula, is called *niche theory* (cf. the review in Roughgarden 1979, chap. 24); it may be defined as a body of mathematical competition theory in which niche overlap is the basis of competition. The main mathematical theorem upon which niche theory models rely is that sustained oscillations do not occur; the system always attains a stable steady state.

Limiting Similarity and Realized Similarity

Niche theory competition models have been used to provide hypotheses on how competition can cause the classical size-distribution patterns of size-structured guilds. The observations to be explained, in principle, are the degree of difference between the species that do co-occur and the degree of similarity between species that fail to coexist.

Historically, the first question to be studied was the critical degree of similarity marking the borderline between coexistence and the failure to coexist; this borderline was called the *limiting similarity,* which may be defined as the maximum degree of similarity that two species can have and still coexist. In posing this question, one supposes that if two species are more similar than the limiting similarity, then they cannot coexist and will be found in different places. In contrast, if two species are more different than the limiting similarity, then they may be found together. Moreover, how different are coexisting species? From the answer to the first question, we expect them to be more different than the limiting similarity, but how much so? This second question is concerned with *realized similarity*—that is, the difference that typically is found between competing species that do co-occur.

MacArthur and Levins (1967) were the first to investigate the limiting similarity. They showed that the limiting similarity between two species depends on the ratio of the carrying capacities. Specifically, let species 2 be the species with the lower K. The limiting similarity depends on K_2/K_1. If this ratio is much less than 1, indicating a severe disadvantage to species 2, then species 2 must be quite different from species 1 to coexist with it. However, as K_2/K_1 approaches 1, the limiting similarity approaches 0. Since the ratio of the K's for two species may vary geographically along environmental gradients, this criterion can be used in principle to predict where a species border will occur between two competitors that are adapted to different places along the environmental gradient.

The realized similarity has been studied by Rummel and Roughgarden (1985). According to one hypothesis, the coexisting species represent successful invaders. For example, if, at some place, a guild already contains two different-sized species, then the most likely invader is a species whose size is approximately midway between that of the residents, or much larger or much smaller than either. Because all sizes are not equally likely to invade successfully, the sequential addition of species to a guild can lead to a sequence of sizes. A guild whose sizes represent the result of successive invasions is called an *invasion-structured* guild (Rummel and Roughgarden 1985).

Another hypothesis to explain the realized similarity among members of a size-structured guild involves coevolution leading to resource partitioning. Indeed, this hypothesis is what Lack (1947) proposed for Darwin's finches and what Brown and Wilson (1956) refer to as *character displacement*. In the present context, the hypothesis may be stated as follows: Imagine two species that initially have a similar resource use. Each species may shift its resource use evolutionarily to a different place on the resource axis by, for example, evolving a different body size. As a species shifts away from its competitor, the competition lowers, but it also moves out to resources that yield a lower K. As a result, one might expect each species to shift away from the other up to a point at which the benefits of further shifting to lower competition are matched by the disadvantage of reducing K still further. A guild whose members have sizes determined by coevolution is called a *coevolution-structured* guild (Rummel and Roughgarden 1983).

The kind of character displacement just outlined, in which two initially similar species evolve in opposite directions from each other, may be quite rare. How did two similar species get together in the first place? If the species enter the system one by one, then the most likely additions are species that are already dissimilar from the residents; this is a condition for successful invasion. Only if the habitat is opened up suddenly to simultaneous colonization by several species are we likely to see the initial condition of two similar species in the same guild. Then they may evolve in opposite directions. This appears to have happened with *Hydrobia* mud snails of the Limfjord in Denmark (Fenchel and Christiansen 1977) when the barrier between the ocean and the Limfjord was suddenly destroyed. Perhaps more typically, if one species is present at the center of the resource axis and then another invades, say, to the right of the resident, both will show parallel evolution to the left. The original resident displaces away from the now established invader, and the invader shifts into resources that are now being partially vacated by the original resident and that were inaccessible to it when it first entered the guild.

APPLICATION OF NICHE THEORY TO SIZE-STRUCTURED GUILDS

Freshwater Fish

Werner (1977) used the theory of limiting similarity to account for patterns of coexistence among freshwater fish. The fish communities of small, soft-bottomed glacial lakes in southern Michigan usually contain four to eight species of sunfishes (Centrarchidae) (Keast 1970b, Werner *et al.* 1977). Typically, two of these species are abundant. The bluegill (*Lepomis macrochirus*) generally accounts for about 60% of the fish biomass in a lake and the largemouth bass (*Micropterus salmoides*) for about 20%. The bluegill is a smaller fish than the largemouth bass. The other species are comparatively rare. Specifically, the green sunfish (*Lepomis cyanellus*), which is intermediate in size between the bluegill and the largemouth bass, is a regularly caught but relatively rare component of the fish fauna.

The bluegill's small mouth and compressed body make it efficient at catching small prey. It feeds by sucking in its victims in a current of water, and its form allows quick maneuvering (Alexander 1967). The large mouth of the bass, including its large premaxilla, and the fusiform shape (to minimize drag), allow the bass to catch large prey efficiently, including other fish (Keast and Webb 1966). The green sunfish is intermediate between the bluegill and bass in shape as well as size. As a result, a correlation exists between body size and prey size among these species of fish (Werner 1977).

Werner and Hall (1976, 1977) used a system of identical circular ponds, 29 m in diameter and 1.8 m deep, at the Kellogg Biological Station of Michigan State University to determine that interspecific competition is present among sunfish. Specifically, 900 bluegill fish were placed in one pond, 900 green sunfish in a second, and 900 of still another species, the pumpkinseed sunfish (*Lepomis gibbosus*), in a third. Also, 900 of each species, a total of 2700, were placed in a fourth pond. The species all took larger prey, and had faster growth rates during one season, in the pond in which they were alone than those in which they were present with the congeners.

Werner (1977), using a combination of feeding experiments in the laboratory and data on diets and the distribution of fish lengths in the field, calculated the separation between the niches of the bluegill and the largemouth bass. He predicted, according to the theory of limiting similarity, that a medium-sized species could not coexist in a niche between that of the bluegill and largemouth bass and that a medium-sized species such as the green sunfish must occur in a different habitat from that occupied by the bluegill and the largemouth bass. Surveys using direct observation through scuba diving confirmed that the green sunfish is concentrated in vegetated areas of lakes, whereas the bluegill and largemouth bass occupy the open water (Werner *et al.* 1977).

Anolis Lizards

To explain the biogeographic pattern of body size and species diversity among the anoles of the northeastern Caribbean, Roughgarden *et al.* (1987) hypothesized that invasion and coevolution among eastern Caribbean anoles happen according to the plan outlined in Figure 16–7. This scheme is predicted by coevolutionary niche theory when the competition is asymmetrical in the sense that a large lizard has a greater effect on a smaller lizard than the reverse. According to this hypothesis, an anole species that is a solitary resident of an island evolves a body size that corresponds to the most abundant and usable size class of insects. Most islands in the eastern Caribbean appear to be in this condition. However, if an invader with a larger body size arrives, it can successfully enter the community. The two species then coevolve; both simultaneously become smaller. The invader "overtakes" the original resident and the original resident becomes extinct, leaving the island with one species. This remaining species is temporarily larger than the typical body size, but it eventually converges to the characteristic size of solitary anoles. According to this hypothesis, the typical difference in body size between two co-occurring species in the northeastern Caribbean is not a result of the evolution of character displacement but represents a prerequisite for successful invasion by a second species against an established resident.

By this hypothesis, St. Maarten, the only island with strong present-day competition and with two anoles species of nearly the same size, is derived from the state seen on other islands, with greater differences in

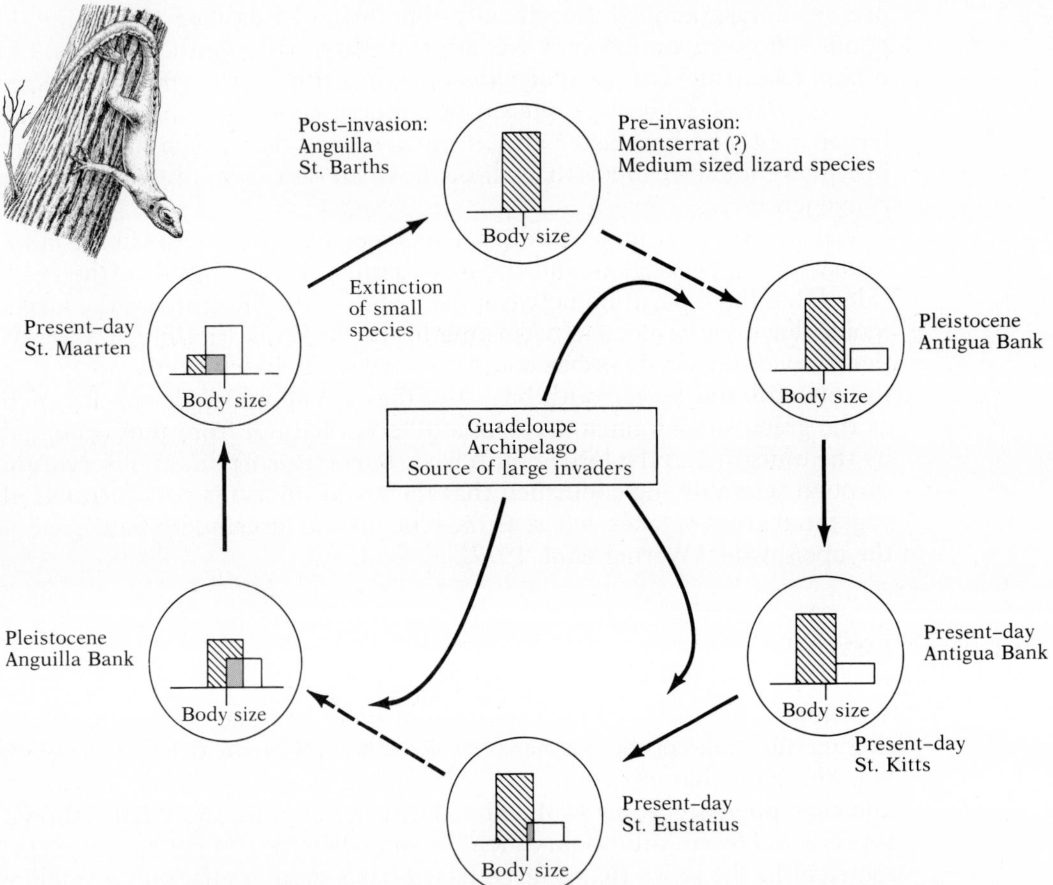

FIGURE 16–7 Taxon cycle among *Anolis* lizards in the northern Lesser Antilles. Different invasions on each of the island banks in the northeastern Caribbean were accomplished by a lizard species larger than the original medium-sized resident. Invader is shown as a clear bar; the original resident as a hatched bar. Within each circle, the horizontal axis is body size; the vertical axis is population size. The larger invader converged on the resident's size, eventually causing the original resident to become extinct. Present-day distributions and fossil evidence support this hypothesis as illustrated. *(After Roughgarden* et al. *1987.)*

body sizes between the co-occurring species. On St. Maarten, the smaller species is hypothesized to be near extinction. This hypothesis for the cyclic plan, illustrated in Figure 16–7, receives support from the ecology of colonization in *Anolis* (Wingate 1965), the fossil record on anoles (Etheridge 1964), from archeological research on land use in the northeastern Caribbean (Rouse and Allaire 1978), and from the fact that a smaller species on Anguilla (which is on the same bank as St. Maarten) has already become extinct since the 1930s (Lazell 1972).

Thus niche theory is beginning to be tested in natural systems where interspecific competition occurs. But what the examples of freshwater fish and eastern Caribbean anoles have in common is that only a few species are involved at any one place and time. The classic patterns involving size-structured guilds that originally motivated the development of niche theory are primarily expressed in complex systems for which not enough is known about whether the species really compete for food and whether the strength of competition varies with the degree of overlap in the way postulated by niche theory models. Further studies are needed to determine whether and how competition has been involved in causing the classic patterns relating size to coexistence in complex food-limited guilds, although important progress has been made.

CHAPTER 17

Substrate-Limited Guilds: Coexistence and Disturbance

Communities on rocky surfaces in the intertidal and shallow subtidal zones are among the most accessible marine communities. Here space, and not food, is generally the limiting resource, and hence the research has tended to focus on mechanisms that either provide or occupy space. The organisms in rocky intertidal and subtidal communities are frequently subject to intense predation, and the composition of these communities is often affected by disturbances from diverse causes, including scraping by ice and damage from logs and other debris that strikes the community.

SYSTEM DEFINITION FOR A HARD-SUBSTRATE SYSTEM

In a community that lives on a substrate covered by water much of the time, immigration and emigration can occur at any location in the system because the dispersing individuals can drop or swim into the system at any point. When birds and insects fly through the air, they typically do so for short periods and at great energetic costs. In contrast, marine and aquatic organisms can remain suspended in the water column almost without energetic costs, and the water column actually becomes a habitat. As a result, the communities on marine substrates are often *open* systems; that is, the recruits into the population are not produced by the adults in the system being studied but are the offspring of adults that live very far away. Moreover, the openness of benthic marine systems is increased by the use of small study sites for ecological research. The small size of sessile invertebrates often leads to the investigation of sites that range from 1 cm^2 to 1 m^2 in size. Also, difficulties with visibility and mobility limit the size of underwater study sites. On a sufficiently large scale, however, marine populations can be considered closed systems, because obviously all offspring in a species are produced by parents in the same species.

THE ROCKY INTERTIDAL ZONE

The Intertidal Zone of Washington State

The community in the high rocky intertidal zone along the coast of Washington State in North America was outlined in a classic study by Dayton (1971). A sketch of a typical rocky intertidal community of the Pacific Coast appears in Figure 17–1. The main features of the community are as follows: Marine plants, including algae and "sea grass" (*Phyllospadix*), are

FIGURE 17–1 Zonation in the rocky intertidal zone on the Pacific coast. *(After Carefoot 1977.)*

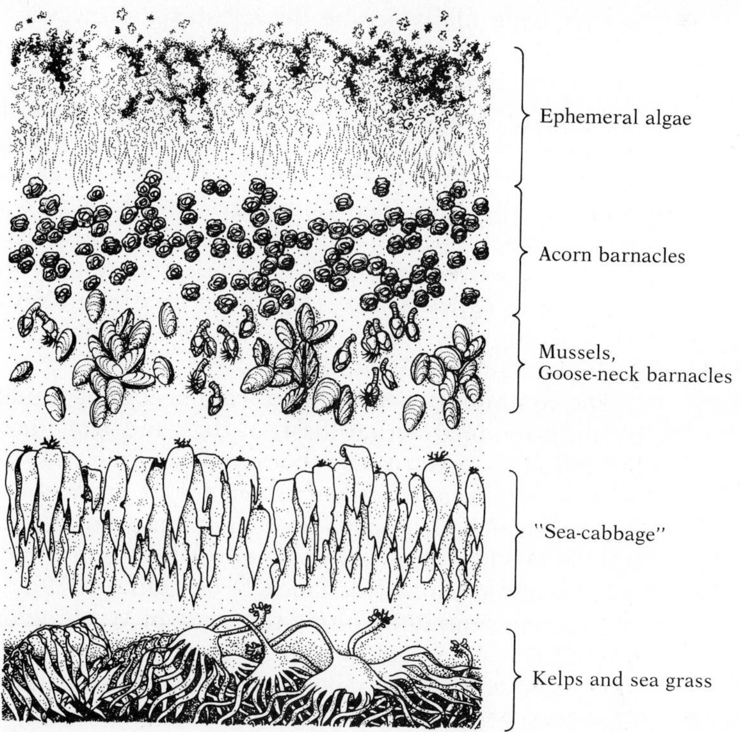

strongly influenced by wave exposure and desiccation. During the winter, many of these plants grow high into the intertidal zone. In the spring, they die back from desiccation. At this time, low tides occur during the day, without coastal fog. On wave-swept coasts, above the algae and sea grass, are beds of mussels. These beds have gaps where logs and other debris have hit the surface (Paine and Levin 1981). Wave action augments the disturbance, ripping out whole clumps of mussels.

Dayton described competition for space by the sessile organisms as a linear hierarchy in which organisms can overgrow, smother, or crush others ranked below them in the hierarchy. The study emphasizes competition for space as involving physical contact between the competing individuals. Dayton reported that barnacles are dominant over algae. Among the barnacles themselves, *Semibalanus cariosus* is dominant over *Balanus glandula,* which in turn is dominant over *Chthamalus dalli.* This dominance order corresponds to the maximum size that the animals can reach; *S. cariosus* reaches about 3 cm in basal diameter, *B. glandula* reaches about 2 cm, and *C. dalli* less than 1 cm. The mussel, *Mytilus californianus,* is at the top of the competitive hierarchy. Its larvae require the prior settlement of certain algae, barnacles, or other mussels, but after successful settlement the mussel can overgrow all other sessile organisms.

Limpets are snails that graze the surface of the rocky intertidal zone. They generally reduce the recruitment of barnacle larvae and algal spores as they continually scrape the substrate. However, the net effect of the limpets depends on another predator, a thaid snail, or whelk. Thaids prey upon the barnacles *B. glandula* and *C. dalli* by drilling through their shells. They prefer *B. glandula,* presumably because of its larger size and more porous shell. If thaids are absent, then limpet grazing leads to a net increase in *C. dalli* and a net reduction in the other barnacles. However, if thaids are present, then limpet grazing leads to a reduction in both of these barnacle species. Adults of both the large barnacle, *S. cariosus,* and the mussel are large enough to avoid predation by thaid snails. This phenomenon is called a *size escape* from predation. Yet another predator exists, the large starfish, *Pisaster ochraceus,* which prefers mussels and can consume adults of *S. cariosus.*

The community is prevented from becoming a monoculture of mussels by the combined action, first, of logs and wave action that lead to the formation of vacant space that can be colonized by algae, barnacles, and other subdominant forms and, second, of predation by the large starfish, *Pisaster,* which also opens up vacant space to colonization. All these features of Dayton's study have generally been confirmed in subsequent studies in the same area (Suchanek 1978, Quinn 1979).

Another community along the coast of Washington State, lower in the intertidal zone, is dominated by algae. Dayton (1975) divided the algae into three categories: canopy species, obligate understory species, and fugitive species. Canopy species grow above the other species and intercept the light. Experiments showed that some of the other algal species undergo a bloom when the canopy species are removed. Obligate understory species die when the canopy species are removed. Fugitive species quickly colonize new space. This study is, in the main, comparable to that of mussel beds in that the canopy species could be classified into a competitive hierarchy at most sites.

Two Principles

The mussel community that Dayton studied has been the source of two principles that apply beyond the immediate context of the intertidal community of the Washington coast: the concept of keystone predator and the concept of intermediate disturbance.

The first concept, that of keystone predator, was proposed by Paine (1966), who demonstrated that if the large starfish *Pisaster* is removed from an area, and if disturbance from waves and logs is not too great, then the area comes to consist of a monoculture of mussels.* The removal of *Pisaster* thus causes the extinction of other species from the system. Paine termed *Pisaster* a *keystone predator.* The key feature of this concept is that the predator prefers the competitively dominant prey species. Other workers since Paine have used the term *keystone species* to refer to any species, not necessarily a predator, whose removal causes the extinction of other species (see, e.g., Gilbert 1980).

According to the second concept, the *intermediate disturbance principle,* the highest number of species are found in a system with an intermediate degree of disturbance. If the combined disturbance rate from predation and wave action is low, then the system approaches a monoculture of the competitive dominant. If the disturbance rate is high, then no opportunity arises for most species to recruit successfully. Therefore, according to this principle, the highest species diversity is found at an intermediate level of disturbance.

A good example of the intermediate disturbance principle comes from the tide pools of New England (Lubchenco 1978), as illustrated in Figure 17–2. The most abundant and important herbivore along the rocky shores of New England is the periwinkle snail, *Littorina littorea.* In tide pools, it prefers the competitive dominants, which are thin, usually green, ephemeral algae. The subdominant is a tough red alga called Irish moss, *Chondrus crispus.* Lubchenco found that when *L. littorea* is absent or rare, an ephemeral green alga, *Enteromorpha,* grows over and dominates other algae, thus reducing diversity. When *L. littorea* is present in intermediate

* *Monoculture* here refers to invertebrates that occupy space within the system. Other species are present, of course, such as those that live in association with mussels.

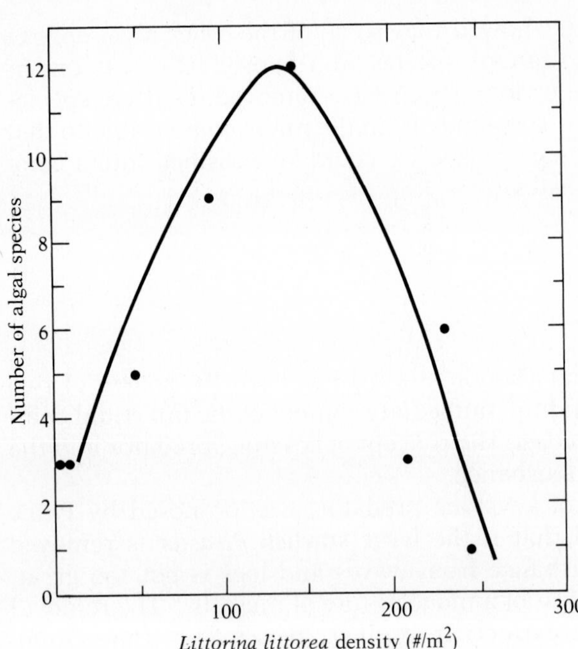

FIGURE 17–2 Number of algal species as a function of the density of snails (*Littorina littorea*)—an illustration of the intermediate disturbance principle. (*After Lubchenco 1978.*)

densities, the abundance of *Enteromorpha* and other ephemeral algae is reduced, local competitive exclusion is prevented, and many algal species are found in the tide pool. At high densities of *L. littorea,* all the species of ephemerals are removed, leaving only the virtually inedible Irish moss, *Chondrus.*

Lubchenco also showed that at other locations, not in tide pools, the periwinkles prefer a competitive subordinate. In these places, algal species diversity decreases monotonically with the number of periwinkles. Paine and Vadas (1969) also found this relationship between species diversity and the degree of disturbance for the sea urchins grazing on algae.

The Rocky Intertidal Zone at Other Places

The rocky intertidal zone on the coast of Washington State is characterized by a competitive hierarchy upon which is superimposed disturbance from predation together with the effects of wave action, logs, and other debris. Is this picture a general picture of the intertidal zone at other places around the world? To some extent, it is. Menge (1976, 1983) has studied the mid-elevation rocky intertidal zone of New England and has found rather similar results at some places there. A sketch of a typical rocky intertidal community of the Atlantic coast appears in Figure 17–3. A competitive hierarchy exists that involves Irish moss (*C. crispus*), barnacles, and mussels. In the absence of predation, mussels dominate. The

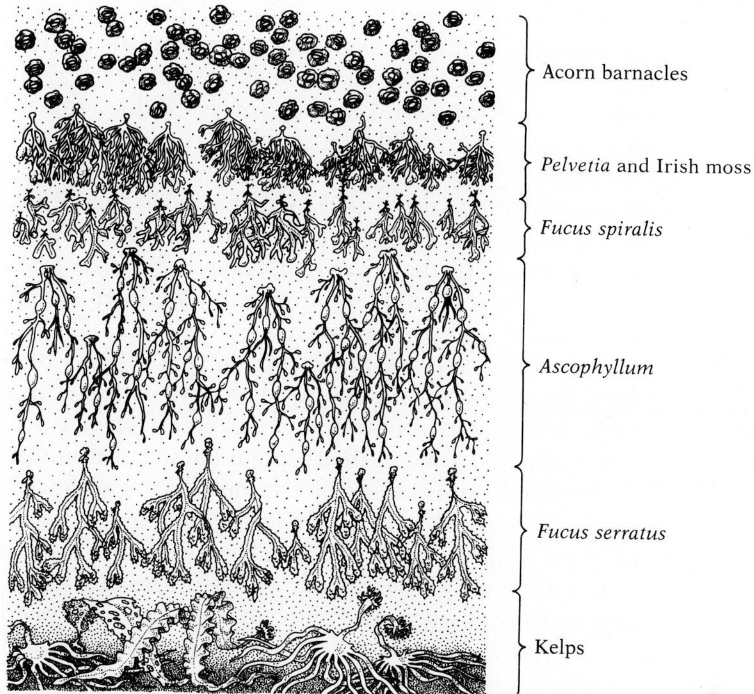

FIGURE 17–3 Zonation in the rocky intertidal zone on the Atlantic coast. *(After Yonge 1949.)*

predators consist of three species of crabs, two species of starfish, and one species of thaid snail. If these predators are present, they consume both mussels and barnacles, allowing the Irish moss to emerge as the major organism occupying space in the mid-elevation to low intertidal zone. This situation is not exactly comparable to that of the Pacific northwest, because the starfish are smaller and tend to be more subtidal than the *Pisaster* of the Pacific, and scraping from ice, not damage from logs, is the common source of abiotic disturbance. Lubchenco and Menge (1978) extended this work to the low intertidal zone in New England, with essentially similar results.

Menge and Lubchenco (1981) have also shown, however, that the rocky intertidal zone along the Pacific coast of Panama is substantially different from that of northern waters. The rocky intertidal zone looks for the most part like bare rock; erect algae and sessile invertebrates are in extremely low abundance, which prevent the development of a distinct pattern of zonation. The algae and sessile invertebrates are concentrated in holes and crevices, whereas in northern temperate communities they also are abundant on the surfaces between crevices. In Panama, the exposed surfaces are bare not because the surface is too hot for organisms to live there, but because of a high predation pressure.

Through experiments, Menge and Lubchenco established that predation from large mobile animals such as grazing fish and crabs remove virtually all the settlers on the surfaces between crevices, while the small predators that are found in the crevices reduce the abundance there, but not to zero. Presumably as a result of the low overall abundance of most invertebrates in the region, few larvae are produced and the settlement rate is low even on areas of rock from which predators are experimentally exluded by cages. Moreover, in Panama the predation pressure is sustained throughout the year. Thus, no escape from predation is possible by attaining a large size or by appearing at times when predators are relatively inactive, as in northern waters. In the Panama system, a major determinant of species diversity along a stretch of the rocky intertidal zone is the configuration of the substrate; in places in which it is complex, it provides various refuges from extremely strong predation pressure.

The intertidal zone of the tropical Pacific, although quite different from the intertidal community of the northern Pacific, is still a community whose composition is strongly influenced by biological interactions, primarily predation. Sousa (1979) has studied an algae-dominated community in a field of boulders that occurs in the low intertidal zone. Here the major form of disturbance that provides vacant space is the overturning of boulders by wave action. Algal populations recolonize cleared surfaces either through vegetative regrowth from surviving individuals or by recruitment from spores. Small boulders turn over more frequently than large boulders. Sousa confirmed the intermediate disturbance principle in this system by finding that medium-sized boulders had a higher species diversity than either small or large boulders. This system differs from the northern Pacific rocky intertidal system in that the mechanism of disturbance is more often abiotic, not the result of predation. Also, the competitive relations cannot be placed in a strict hierarchy, because early-arriving stages have the capability of preempting space from the later stages in the absence of disturbance. This study is also an excellent example of algal succession, the colonization of substrate by plants; we will discuss it again in Part 7 when we introduce the general subject of succession.

Another picture of the community ecology of the intertidal zone has emerged from studies by Underwood *et al.* (1983) on the coast of New South Wales in Australia. The main components of this community include algae, the barnacles *Tesseropora rosea*, a large and a small limpet species, and the predatory thaid *Morula marginalba*. In this system, the alga is not competitively inferior to the barnacle; instead, the alga can preempt the surface and smother established barnacles. The experiments by Underwood *et al.* involve partial as well as total removal of these components. This work is unusual in allowing the possibility of different interactions to be observed, depending on the density of the various components. Indeed, study showed that the larger limpet either increased or decreased the number of barnacles, depending, in part, on the density of the limpets themselves. The presence of many limpets retarded the successful settlement of barnacles; the presence of a few limpets caused a net increase in barnacle survival because the limpets also reduced the hazard from algae through their grazing of algal spores. Furthermore, the number of established barnacles had an effect on the limpets. Limpets migrated away from areas with dense barnacle cover, and if they were confined to such regions they lost weight, suggesting that the presence of adult barnacles could interfere with limpet foraging.

Underwood *et al.* (1983) also reported that the composition of the community is strongly influenced by a great variability in the timing and intensity of larval settlement and also by variability in the number of thaid snails that happen to be present. The settlement levels reported for New South Wales are low even at best; large effects result from fluctuations in what is already a low level of recruitment, a level lower than that which regularly occurs in the intertidal systems of the Pacific northwest, but still higher than that which Menge and Lubchenco (1981) observed along the Pacific coast of Panama.

Roughgarden *et al.* (1984, 1985) offered a model, supported by data on barnacles from the rocky intertidal zone in central California, that seems to reconcile the quite different observations made in the intertidal communities of New South Wales and Washington State. First, they noted that space could be limiting even if the adults were not dense enough to be in physical contact. Space is limiting during the *recruitment* process if the recruitment is only to vacant space. In a given area, the number of recruits continually decreases as the space becomes progressively occupied.

Recruitment, however, is only part of the picture. When vacant space is nearly exhausted, the animals come into close contact with one another. As a result, mortality may increase and growth may decrease. Barnes and Powell (1950) reported that dense clumps (hummocks) of barnacles were more susceptible to removal by wave action than were individuals directly attached to the substrate. Roughgarden *et al.* (1984) reported that the starfish, *Pisaster ochraceus*, preferentially feeds on spots in which the barnacle density is high, rather than on spots in which the barnacles are sparse or young. More generally, as the vacant space drops low enough that animals are physically touching around most of their perimeter, special mechanisms involving crowding, overgrowth, and increased susceptibility to predation and physical disturbance become possible. Thus, space continuously limits recruitment when recruitment is primarily to unoccupied space, and space also limits survival and growth when space is exhausted enough that most individuals are in physical contact with others.

The second point made by Roughgarden *et al.* (1984) is that the average, or overall, level of abundance is set primarily by the settlement rate.

Hence, in locations of low settlement, the densities cannot build up enough for interactions that depend on physical contact to become important. In contrast, high-settlement areas can attain the densities needed to bring contact-mediated interactions into play. Moreover, in the model of Roughgarden *et al.*, the abundance of organisms is directly proportional to the settlement rate, provided settlement is sufficiently low; here settlement is the rate-limiting step. When the settlement rate is high in their model, the abundance becomes independent of the numerical value of the settlement rate and becomes governed by the dynamics of the contact-mediated interactions.

Apparently, the New South Wales system is a low-settlement community. An expected feature of a low-settlement community is that its composition at any time is extremely sensitive to fluctuation in the settlement rate. If, say, an average of 1 larva lands per square centimeter on one rock, whereas 0.5 land per square centimeter on a nearby rock, then after a while the abundance between the rocks will probably differ by a factor of about 2. In a high-settlement situation, 50 larvae may land per square centimeter on one rock and 100 on another. Yet the rocks may be able to support only, say, two or three individuals per square centimeter as adults, and so the composition of the rocks will be primarily determined by postsettlement interactions. It is these postsettlement contact interactions that, in turn, are the basis of a competitive hierarchy, the intermediate disturbance principle, and the keystone predator all of which are observed in the high-settlement intertidal communities of Washington State and New England.

Ultimately, however, some feedback must occur between the abundance and reproductive success of the organisms on the rocks and the concentration of larvae in the adjacent water. Indeed, the low settlement rates Menge and Lubchenco (1981) observed in the intertidal community of the Pacific coast of Panama may be due ultimately to the low abundance of adult invertebrates in that coastal region that result from the high predation there. What could explain the low settlement rate along the coast of New South Wales? Perhaps high predation of larvae exists in the plankton, or regional oceanic currents may generally prevent most larvae from returning to shore. The answer is not known.

THE ROCKY SUBTIDAL ZONE

The marine environment has many types of communities other than those in the rocky intertidal zone, which has been the most studied area primarily because of its accessibility. The community that is physically closest to the intertidal zone is the shallow subtidal zone, and it is receiving increasing attention. Also, the subtidal hard-substrate community is of commercial interest because the organisms from this community are the principal biological agents of marine fouling. *Fouling* refers to the accumulation of organisms on surfaces left in seawater. Fouling organisms settle on surfaces such as the undersides of boats and on wharf pilings. Their presence slows boats by increasing their drag and leads to higher fuel costs. These costs are used to justify the purchase of poisonous antifouling paints for the underside of boats; a huge amount of money is spent annually buying and applying these paints.

Hard-substrate communities in the subtidal zone differ from those in the intertidal zone in three main ways:

1. The competitive relations may not involve a hierarchy.
2. The organisms are frequently colonial.
3. Dispersal often does not involve pelagic larvae and, if so, recruitment is from within local populations.

Sutherland (1974) drew attention to the possibility of nonhierarchical competition in fouling communities with a study of the colonization of artificial ceramic plates suspended horizontally about one-third of a meter below the surface of the water. The organisms that potentially colonize this surface include hydroids, tunicates, bryozoa, sponges, and the organisms that live with them, as shown in Figure 17–4. Sutherland found that

FIGURE 17–4 Subtidal epifaunal community on a settling plate, including bryozoans, tunicates, sponges and anemones. *(S. K. Webster, Monterey Bay Aquarium/BPS.)*

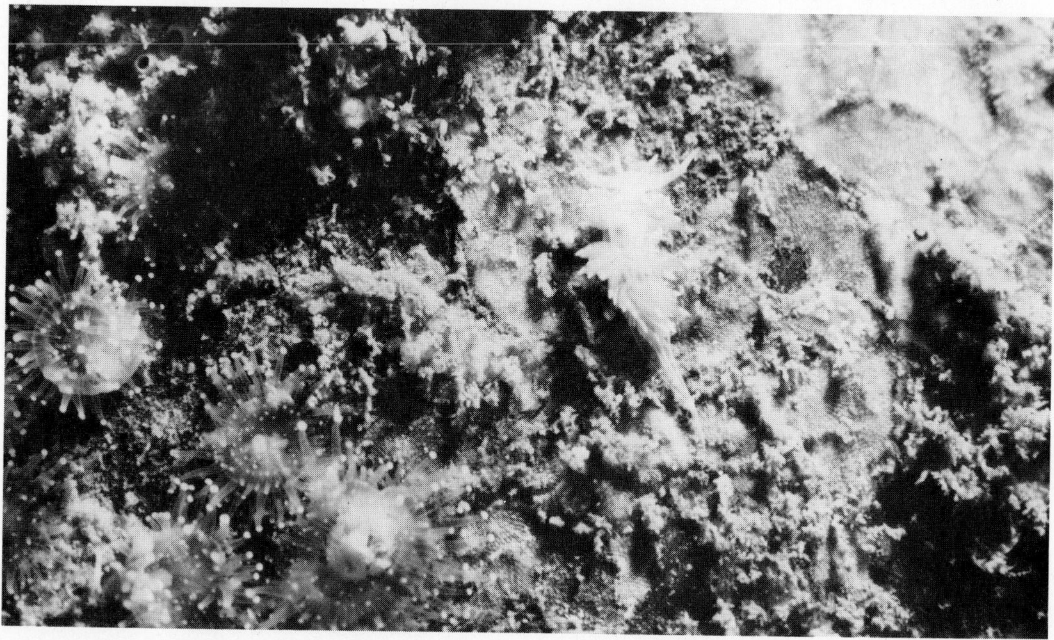

these plates became dominated by different species, depending on the season or year in which they were originally submerged in the water. The species whose larvae were available in sufficient abundance at the time the plate was placed in the water were able to preempt the space. Furthermore, Sutherland observed that the dominant space occupier on a plate could die, presumably from senescence, without any other known cause. This event released vacant space for colonization by the organisms then present in the water. Subsequent studies (Osman 1977) confirmed the highly variable outcome of colonization on artificial substrates and the existence of senescence as a method of producing vacant space.

Further evidence of nonhierarchical competition comes from studies in shaded subtidal coral reef communities. Jackson (1979) and Buss and Jackson (1979) have shown that the competition is actually intransitive under some conditions; that is, A can overgrow B, B can overgrow C, and C can overgrow A. The researchers also found that the competitive relation between two individuals of different species is reversible at times, and not necessarily unidirectional. The angle of encounter, shape, thickness, season, and age can affect which of two organisms overgrows the other. Moreover, Sebens (1982a) has documented that an escape in size from overgrowth competition occurs in the soft coral *Alcyonium*. Only small individuals are susceptible to overgrowth by the ascidian *Aplidium*.

Another feature of subtidal hard-substrate communities is the possibility that competition for food is as important as competition for space. Buss (1979) showed that direct feeding interference occurs between encrusting bryozoa. Sebens (1983), working with the anemones *Anthopleura xanthogrammica* and *A. elegantissima* in the low intertidal zone in Washington State, demonstrated that the size of an individual anemone fluctuated seasonally, increasing during spring and early summer but shrinking or remaining constant during the rest of the year. Moreover, growth was much more rapid in areas in which the abundance of anemones was experimentally reduced relative to control areas, and even in the control areas growth was highest at the sites with the lowest density of anemones. The mechanism controlling growth and shrinkage was identified as intraspecific competition for prey.

The colonization of artificial surfaces such as those used for fouling studies as well as of natural substrate may involve complex behavior by the larvae. Factors influencing the settlement of larvae under laboratory conditions have been reviewed by Crisp (1974, 1976) and Lewis (1978). Gregarious settlement and selectivity for the chemical and physical characters of the substrate have been demonstrated. Even though settlement may be gregarious, spacing behavior has also been observed in the settling larvae of barnacles, although interindividual distances may be reduced in conditions of high crowding. Settlement rates are known to be affected by all of the following for at least some organisms: age and nutritional state of the larvae; color, texture, and rugosity of the substrate; direction and intensity of water movement; color and amount of illumination; and the presence and identity of microflora on the substrate. The significance of these findings for population dynamics under natural conditions is unknown.

Although some organisms of the subtidal zone are short-lived, others may be extremely old. Sebens (1983) estimates that individual *Anthopleura xanthogrammica* anemones persist for "at least several decades." Some individuals of the huge subtidal barnacle *Balanus nubilis* that were marked by the legendary Doc Ricketts of John Steinbeck's *Cannery Row* are known to have lived for over 50 years (G. Haderlie, pers. comm.). These individuals ultimately were killed by boring sponges that weakened the barnacle's test.

The ecological and evolutionary causes of widespread coloniality in subtidal forms are a subject that is beginning to receive attention. Jackson (1977) suggested that colonial animals are better space competitors than are solitary animals for two reasons:

1. Their indeterminate growth through fission allows continuous lateral expansion without requiring intervening steps of mating and recruitment.
2. Large animals are themselves less susceptible to being crowded and overgrown.

He suggests that solitary forms are favored in situations in which predation, disturbance, or competition from plants prevents lateral expansion from monopolizing the surface.

The possible evolutionary and ecological reasons for the long-lived pelagic larval phase of high intertidal organisms such as barnacles and mussels relative to many colonial subtidal forms such as tunicates have received attention. Vance (1973) has focused on the value of feeding during the larval phase. Crisp (1979) has suggested that the long-range dispersal of barnacles is an adaptation to finding vacant space where the habitat is subject to frequent disturbance.

CHAPTER 18

A Diversity of Guilds: Marine, Aquatic, and Terrestrial

CORAL REEF FISH

Coral reefs, which occur in all tropical oceans, teem with life and are among the most complex ecosystems in the world. Some, such as the Great Barrier Reef on the northeast coast of Australia, have a tremendous diversity of fish and invertebrates apparently coexisting. The Caribbean reefs have a lower diversity for most taxa, including fish and stony corals. Reef fish have received particular attention from biologists since scuba apparatus became widely available (see the review in Ehrlich 1975). With scuba equipment, the collection of specimens became possible through spearing, netting, trapping, and use of the respiratory poison rotenone, and the activities of fish can be directly observed.

The community ecology of coral reef fish shares features with size-structured guilds, in which resource partitioning of food is the basis for coexistence, and also with subtidal space-limited invertebrates. Like many invertebrates, coral reef fish generally have eggs and often larvae that disperse for some time as components of the plankton.

Resource Partitioning

The stage for understanding coral reef communities was set by a classic series of studies, not on fish but on a group of predatory snails from the genus *Conus* in the South Pacific. Kohn (1959, 1966, 1967, 1968, 1971) established that about 10 species of *Conus* usually co-occurred and that each species had specific feeding and microhabitat specializations. These findings with *Conus* led to the supposition that resource partitioning was generally the basis for coexistence of coral reef organisms, including fish. Moreover, early descriptions of the fauna of coral reef fish tended to include characteristic "profiles" of the feeding habits and microhabitat use of reef fish based on preliminary information.

Roughgarden (1974b) pointed out, on the basis of Randall's (1967) data on the stomach contents of Caribbean reef fish, that some fish from the Caribbean appeared to show the classical patterns of a size-structured guild. The *Epinephalus* groupers, which are predators, consist of three to four co-occurring species of quite different sizes, and their stomach contents suggest that these species use correspondingly different sizes of prey. Similarly, three species from another genus of groupers, *Serranus,* show partitioning with respect to size. Moreover, in *Epinephalus,* two species of about the same size were found that seemed not to occur together with equal abundance; if one was rare, the other was common, and vice versa. In contrast, Roughgarden (1974b) pointed out that more herbivorous fish, including parrot fish and surgeon fish, lacked evidence of local resource partitioning and geographical replacement. This evidence was preliminary, however, and based on collections made for other purposes.

The Lottery Hypothesis

Sale (1977), in a stimulating and influential paper based mostly on fish from the Great Barrier Reef, suggested that resource partitioning was virtually absent among reef fish and that they are not generally comparable in this regard to *Conus* snails. Sale suggested that fish species diversity could be maintained in the absence of resource partitioning. The idea was basically that space, and not food, is limiting because many species of reef fish, including the damselfish that Sale has primarily studied, are territorial, and many other species, if not territorial, have home ranges. Sale further envisaged that new space opens up on the reef randomly and that fish larvae are constantly available in the plankton to colonize new substrate as it becomes available. Thus, the picture is one of a local system with space-limited recruitment, and Sale analogized its workings roughly to that of a lottery.

It was originally imagined that ample larvae were present in the plankton, so that the total density of adult fish would remain constant while the identity of the species varied at random. Adjacent sites would nonetheless be quite similar because they would be colonized by larvae from the same water mass. Subsequent work has tended to suggest that the settlement is lower than originally imagined, more analogous to that of a low-settlement local system than to that of a high-settlement system (Sale 1980, Williams 1980, Victor 1983). That is, not only the composition but also the density is sensitive to the recruitment rate, but the idea remains essentially the same.

It is difficult to do experiments with reef fish, so little has been known from an experimental standpoint. Recently, however, Thresher (1983) has developed evidence of food limitation and interspecific competition among damselfish and other planktivorous fish. Increased numbers of planktivorous competitors correlate with delayed spawning, smaller clutches, smaller adult size, and possibly also reduced juvenile growth rates for the damselfish, *Acanthochromis polyacanthus*—an unusual reef fish in that it has no pelagic larval stage and the adults care for the young on the reef. This correlational evidence was corroborated by some limited experiments.

It is clear that the lottery hypothesis theoretically can explain *local* coexistence of competitors without resource partitioning. As long as larvae of two or more species are present in the water column and space is available on the reef, then these species will settle there, regardless of how similar they are to each other. The theoretical difficulty arises in accounting for the presence of the larvae to begin with. Chesson and Warner (1981) analyzed a model of random settlement by two species in vacant space and concluded that coexistence on a regional scale was impossible without additional assumptions. They then further supposed that environmental variation was superimposed on the random settlement and found that this additional assumption could theoretically lead to regional coexistence. This assumption leads to a prediction of species diversity increasing with the degree of environmental fluctuation. The prediction runs counter to the species diversity gradient on the Great Barrier Reef, in which more species are found in the tropical north and progressively fewer are found as one proceeds south toward more temperate waters. Abrams (1984) recently also concluded that the process of random settlement in vacant space was insufficient to account for coexistence on a regional scale.

Further Considerations

When one moves to a regional scale, clear patterns emerge of species replacement along transects across various parts of the reef, as Anderson *et al.* (1981) noted and Sale and Williams (1982) confirmed. At least to some extent, the occurrence of species in a local system is a stirring together of species that are based in different areas. If so, local areas of patch reef may have many species as a result of larval dispersal, even though each species does best in a habitat that is special to it. Certainly no single answer is evident that applies to all reef fish, because the degree of resource partitioning related to body size seems to vary with trophic position, and the extent to which food and space are the limiting resources may also vary with trophic level. In addition, the role of predation (whether effectively density independent or not) and other interactions such as mutualism (as in cleaning symbioses) remain to be explored.

MARINE INFAUNAL COMMUNITIES

The majority of the earth is covered by soft sediments, including sand and mud. The deep sea alone—that is, the ocean floor between the continental shelves—covers nearly two-thirds of the planet's surface. Coastal soft sediment communities include sandy beaches, salt marshes, mangroves, and river deltas. Soft sediment communities are the most widespread of all community types.

Yet soft sediment communities are not readily comparable to any others we have considered so far, and new and basic processes are being continually identified in them. Research on soft sediment communities dates to the beginning of community ecology, when the prominent issue was identification and classification of community types. Since then, important experimental studies have been added. Nevertheless, at this time we do not have a good overview of how soft sediment communities work; thus, perhaps the best way to convey an idea of what is happening in these systems is to review some case studies.

Types of Soft Sediment Organisms

The organisms that live in soft sediments may be classified as follows:

- *Infauna* consist of organisms as big as, or bigger than, the grains or particles in the sediment, as shown in Figure 18–1.
- *Microfauna* are the smallest infaunal organisms, ranging from grain size to 0.1 mm.
- *Meiofauna* are infaunal organisms from 0.1 to 0.5 mm.
- *Macrofauna* are infaunal organisms whose size is greater than 0.5 mm.

Most of the organisms can be further classified by their typical feeding habits, including what they eat, where they get it, and whether they are mobile or sedentary. *Suspension feeders* extract particles of food from the water. *Deposit feeders* swallow grains of sediment and assimilate the bacteria and other microorganisms that coat the grains. Members of the infauna also may obtain their food from the surface of the substrate or from within the substrate. The organisms may burrow or may be sedentary. The burrowers include clams, crabs, sea cucumbers, and polychaete

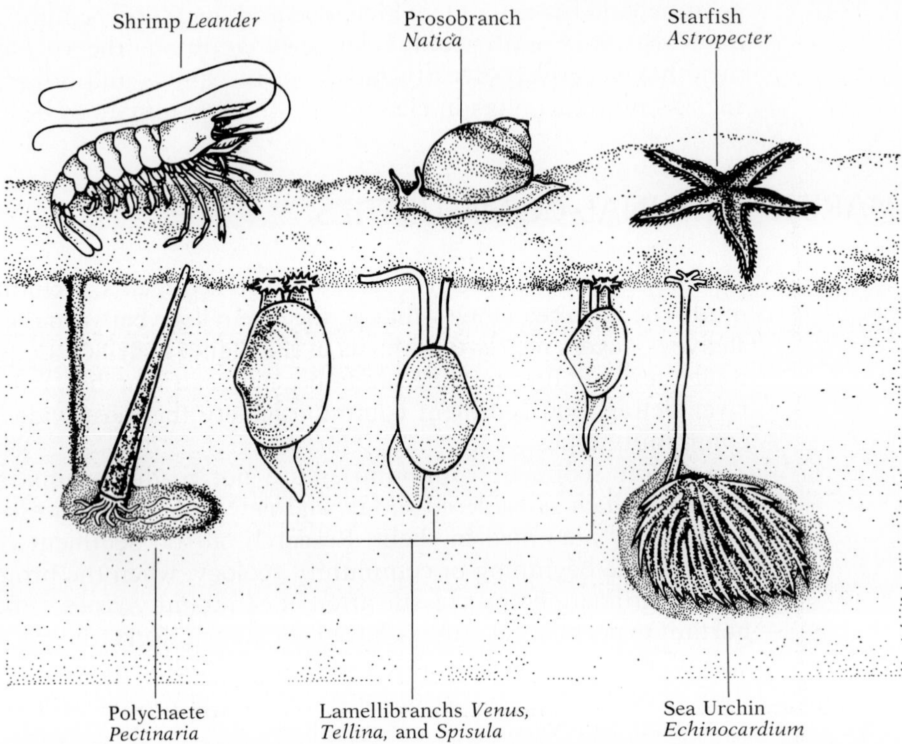

FIGURE 18–1 Sketch of an infaunal community illustrating suspension (clams) and deposit (snail) feeders. *(After Thorson 1957.)*

worms. The sedentary organisms are primarily worms and amphipods that build and live in tubes.

The early investigators of soft sediment communities noticed that discrete packages of species seem to occur together—surprisingly so, in view of the absence of such groupings in the rest of community ecology, as discussed at the beginning of this part.* Woodin (1976) has summarized the community types as (1) collections of burrowing infaunal deposit feeders, (2) collections of suspension feeding clams, and (3) collections of tube-building sedentary forms. Sharp boundaries seem to occur between these three types of communities. Historically, the main question in the community ecology of soft sediment communities has been, why do such discrete communities exist? One factor is the size of the grains in the sediment (Sanders 1958). In Buzzards Bay and Long Island Sound, for example, deposit feeders are more abundant in fine sediments than in coarse sediments.

* See Petersen (1913), Ford (1923), Shelford *et al.* (1935), Thorson (1957), and Sanders (1958).

Another important consideration is that infaunal animals *modify* the infaunal environment itself. Rhoads and Young (1970) experimentally demonstrated that deposit-feeding clams changed the character of the sediment as they processed it, with the result that it became more easily suspended. The consequent increase in turbidity was demonstrated to depress the growth rate of a suspension-feeding clam, *Mercenaria mercenaria.* Thus, infaunal animals can interact by altering the properties of the substrate. In this instance, the interaction is negative and in one direction. This relation would appear to prevent suspension-feeding clams from co-occurring with large numbers of burrowing deposit feeders. However, a reciprocal negative interaction also occurs. Tube-building forms, which are typically suspension feeders, as well as the roots of plants, interfere with the mobility of burrowing deposit feeders (Brenchley 1982). In an influential paper, Woodin (1976) also proposed that the community types are maintained as intact patches by a larva–adult interaction. Specifically, she hypothesized that suspension feeders consume larvae of potential competitors as a result of their filter feeding, thus perpetuating an intact association.

An intact community is also susceptible to disturbance from fish predation, especially from flat fish, and from wading birds. These predators churn up the substrate and allow recolonization. Of course, turbulence from storms and heavy freshwater runoff can also wreak havoc.

Resource Partitioning and Competition

Some resource partitioning also occurs. Peterson (1982) demonstrated that two suspension-feeding clams had strong intraspecific competition but virtually no interspecific competition. These clams occupied less than 10% of the surface area, and space was not considered limiting. They differed greatly in size, but no evidence was found that the species took particles of a different size. Instead, one species seemed to be consuming principally diatoms and other particles from the surface of the mud, whereas the other species was consuming particles from higher levels in the water. Similarly, intraspecific density dependence without much interspecific competition was detected by Wilson (1983) with two species of deposit-feeding polychaete worms. Wilson attributed this occurrence to spacing behavior that leads to little interspecific interaction. The abundance of these worms was on the order of 2 to $10/cm^2$.

Both studies showing low interspecific competition together with high intraspecific competition involve species that normally co-occur. An interesting contrast is provided by two species of mud snails in San Francisco Bay (Race 1982). One species is introduced, the other is native. Race experimentally established that strong competition by the introduced species against the native species caused exclusion of the native species from most of the habitats formerly occupied. The mechanisms of competition primarily involved interference.

A related issue is whether body size in deposit feeders can be used as an indicator of the principal particle size used for feeding. In laboratory studies, Fenchel and Kofoed (1976) found clear evidence of partitioning with respect to the size of diatoms consumed by two snail species of different sizes. As we discussed earlier, these snails in the Limfjord of Denmark are thought to offer an example of the evolution of character displacement.

Facilitation

The effect of tube-building organisms on the recolonization of soft substrate after a disturbance has been studied recently by Gallagher *et al.* (1983). They found that tube-building surface deposit feeders facilitated the recruitment of other taxa. This study is significant because it is the only experimentally documented example of facilitation during succession, as we will discuss in more detail in Part 7.

These studies indicate that extremely high densities of organisms appear in shallow soft bottom communities, that food and not space is often limiting, that resource partitioning can occur, that many interactions are effected through changes in the physical characteristics of the substrate, and that positive as well as negative interactions occur. The recent development of experimental methods for soft substrate communities suggests that many discoveries are near for this important community about which so little is known.

ZOOPLANKTON GUILDS

For our final example of an aquatic community, let us consider the activities in the water column itself. The small organisms inhabiting the water column are called *plankton,* with the algal component being called *phytoplankton* and the animal component the *zooplankton.* The plankton includes organisms whose entire life cycle occurs within the water column, as well as eggs and juvenile phases of organisms that occupy the substrate as adults.

Communities of zooplankton from freshwater lakes, together with freshwater fish, have been the subject of important studies during the last 20 years. Much of the recent work can be traced to a paper by Hutchinson (1951) entitled "Copepodology for the Ornithologist," in which he presented the view that competition plays a major role in limiting the number of zooplankton species that occur in lakes, in a way analogous to the role that competition seems to play in the community ecology of size-structured guilds of birds. In subsequent years, a definite role for competition in freshwater zooplankton was established, although it is by no means the only, or even most important, feature of their community ecology.

Size Patterns and Competition

Hebert (1982) notes that the diversity of species in freshwater lakes in Canada is far less than the number available, in a manner suggesting the concept of limited membership, as discussed earlier in connection with size-structured guilds.

As Table 18–1 shows, Canadian freshwater streams and lakes contain an average of four zooplankton species of cladocerans and calanoids, although within a local region four or five times as many species may be available. Moreover, in Australia and New Zealand, similar-sized species do not coexist, whereas species of different sizes do (Bayly 1964), although this pattern is not found in Ontario lakes. Direct evidence for competition in cladocerans has come from laboratory experiments in which one species is often observed quickly to exclude another (Frank 1957). Good evidence of competition in nature comes from the studies of Hebert (1977) with Australian *Daphnia* species. Egg production in each of two co-occurring species is density dependent both intra- and interspecifically.*

Predation

The next major ingredient of zooplankton communities to receive attention was predation by fish. Brooks and Dodson (1965) showed that freshwater fish prefer larger zooplankton prey and that the death rate of zooplankton in the presence of fish is an increasing function of body size. They showed that the zooplankton community is dominated by large species of zooplankton if fish are absent and by small species if fish are present.

TABLE 18–1 Average number of zooplankton species in Canadian lakes and streams

Region	Locations	Cladocerans[a]	Calanoids[a]
Tundra	135	0.9 (4)	1.9 (5)
Rocky Mountain	340	1.6 (26)	1.1 (21)
Ontario	40	3.7 (13)	2.0 (7)

[a] The number in parentheses is the total number encountered in the survey of the region.

* Further evidence is found in Allan (1973), Jacobs (1977), Lynch (1979), Smith and Cooper (1982), and DeMott and Kerfoot (1983).

Brooks and Dodson interpreted the presence of large zooplankton in the absence of fish as resulting from an inherent competitive superiority of the large zooplankton species over the smaller species. However, subsequent work established that the dominance of large zooplankton species was a result of their lower susceptibility to small invertebrate predators, such as notonectid bugs, and not a result of their competitive superiority (Dodson 1974, Zaret 1980, Grant and Bayly 1981).

Information about species interactions is generally lacking for marine plankton communities, where the main research focus has been on the timing of *blooms* of particular species, on the existence of horizontal and vertical patterns (called *patchiness*), and on the productivity of the plankton community. Blooms are bursts of abundance of species in the phytoplankton. Blooms of phytoplankton often accompany upwelling currents that bring limiting nutrients to the surface. Great population increases in herbivorous zooplankton and fish can follow. The interest in the productivity of plankton communities can be traced to interest in the extent to which the sea can be harvested for food and whether it can be farmed. The subject of productivity is covered further in Part 7.

AMPHIBIAN COMMUNITIES

As we leave aquatic systems and move onto land, we begin with a community that occupies the interface between terrestrial and aquatic systems. Temporary ponds—that is, those that regularly dry out by the end of the summer—typically do not support fish, and the top predators are the larvae of frogs and salamanders. These amphibians usually have an aquatic larval phase characterized by rapid growth but high susceptibility to predators, followed by a terrestrial adult phase when they are less susceptible to predators. Although the adults may live on land, they must return to the ponds to lay eggs. A three-phase life history may even occur, as in the case of the eastern red-spotted newt, *Notophthalmus viridescens*, which returns from its terrestrial phase (when it is called an *eft*) to the pond to reassume an aquatic life. Much is now known about the community dynamics in temporary ponds. Less is known about the dynamics of the terrestrial phase of salamander life histories and about how the community dynamics of temporary ponds is coupled to the population dynamics of the terrestrial phase.

Temporary Ponds

Wilbur (1972) studied temporary ponds in southern Michigan and produced one of the first experimental demonstrations of interspecific competition. The amphibian community consists of four salamanders of the genus *Ambystoma* and the frog *Rana sylvatica*. The larvae of these amphibians are eaten by carnivorous aquatic insects, including notonectids, as well as by larger amphibian larvae themselves. Wilbur showed that both intra- and interspecific density dependence exist among larvae of the *Ambystoma* species under experimental conditions. The assays of density dependence included survivorship, body weight at metamorphosis, and the time to metamorphosis.

More recently, Morin (1983) studied temporary ponds in North Carolina. The technique involves creating artificial temporary ponds that are about 1.5 m in diameter and 0.5 m deep from galvanized steel cattle-watering tanks. The natural ponds in the area are the potential breeding sites for about 25 species of frogs, and up to 17 species may chorus and breed near a single pond during one evening. Little evidence has been found of habitat partitioning among these species. The tadpoles feed by rasping and filtering on suspended and deposited plant material and detritus. The predators on these tadpoles are salamanders, including aquatic adults of the red-spotted newt, *N. viridescens*.

Morin introduced six species of frogs into each of a series of artificial ponds. He varied the number of *Notophthalmus* predators that were introduced in steps of zero, two, four, and eight newts. The results are shown in Figure 18–2. In the absence of salamander predation, the spadefoot toad, *Sciaphiopus holbrooki*, is the most abundant frog to emerge from the ponds, but as the number of newts progressively increases, the most abun-

FIGURE 18–2 Effect of predation intensity on community composition. The number of predators *(Notophthalmus)* is varied in steps of 0, 2, 4, and 8 newts. Species are: S.h., *Scaphiopus holbrooki;* H.c., *Hyla crucifer;* R.s., *Rana spenocephala;* B.t., *Bufo terrestris;* H.ch, *Hyla chrysocelis;* H.g., *Hyla gratiosa. (After Morin 1983.)*

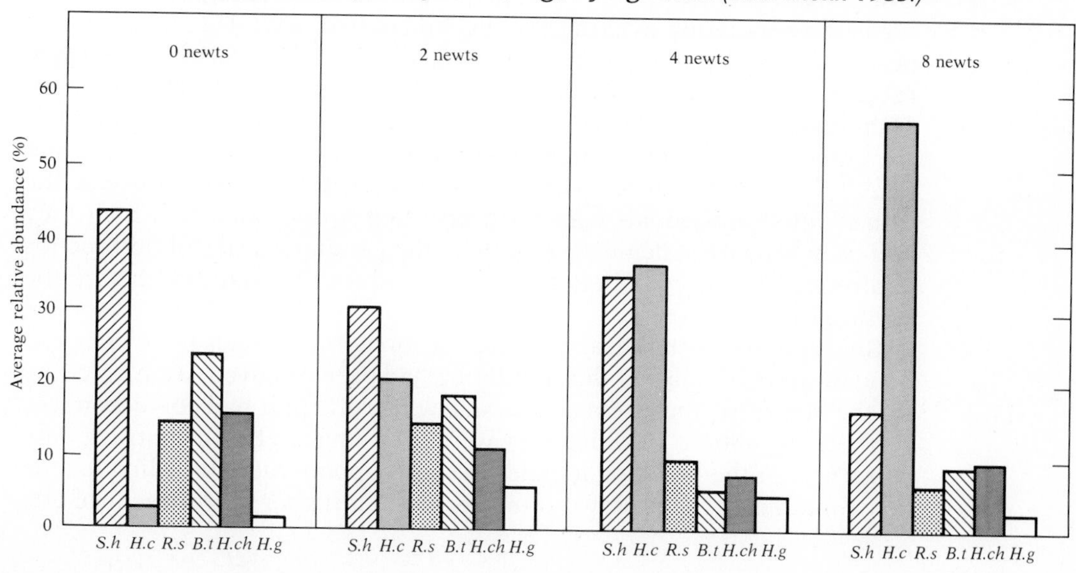

dant frog progressively becomes the treefrog, *Hyla crucifer*. With feeding experiments in the laboratory, Morin also established that the newt selectively preyed on the spadefoot toad (and also the toad *Bufo terrestris*) and indiscriminately on the tadpoles of the other species as they were encountered. Morin concluded that the composition of the pond community resulted from the joint action of predation and competition. Specifically, predation from newts is directly responsible for the decline of the spadefoot toad, but the success of the tree frog, *Hyla crucifer*, is not due to any absence of predation on it relative to that on the other frogs. Instead, it is, for some reason, a superior competitor.

Morin (1983) terms the simultaneous action of predation and competition *predator-mediated competition*. He writes that "density-dependent competition among tadpoles is expected to be less intense where fewer tadpoles escape predation and survive to compete" (p. 131). Predator-mediated competition refers to the joint effects of two or more pairwise interactions—at least one of which is a predator–prey interaction and another a competitive interaction. It does not refer to a higher-order interaction, as might occur, for example, if the competition coefficient between the individuals of different species had been shown to vary as a function of the abundance of a third species. Morin points out that the correlations between growth rates and anuran abundances were very high, between 0.88 and 0.96, so that a simple linear function of competitor densities satisfactorily described most of the variation in mean growth rates among the experiments.

Ponds and the Surrounding Forest

Gill (1978, 1979) took the first steps toward understanding, on a regional scale, the population dynamics of the red-spotted newt. He identified two nearly identical ponds in the Shenandoah Mountains of Virginia, separated by 0.4 km, each pond having populations of over 1000 individuals. Then in 1974 he transferred almost all of the population of one pond to the other, thus reducing the population in one pond to nearly zero and doubling the abundance in the other to over 2000 newts. Most of the transplanted individuals continued breeding in the pond to which they had been brought. The next year, 1975, the population size of both ponds had returned to approximately the same values they had had before the perturbation.

Gill established that nearly 95% of the breeding females in 1975, in the pond from which most of the population had been removed, were breeding for the first time; these females were new recruits that had come from the surrounding habitat in which they were surviving as efts. In contrast, only about 65% of the breeding females in control ponds were breeding for the first time. Thus, the removal of newts led to a higher recruitment of efts that restored the population size to its previous value. Gill's experiments also showed that a majority of the surviving males that were transported in 1974 had returned to their original pond for the breeding season of 1975. This homing behavior is reminiscent of the study by Twitty *et al.* (1964) showing homing in the California red-bellied newt, *Taricha rivularis,* over distances as great as 8 km.

Woodland Salamanders

Salamanders include species that are fully aquatic, partly aquatic and partly terrestrial, and almost fully terrestrial. The plethodontids are lungless salamanders; they breathe through a thin, moist skin. The genus *Plethodon* has many almost completely terrestrial species; all species are called *woodland salamanders.*

Hairston (1980a,b, 1981) has studied the terrestrial salamanders of the Great Smoky Mountains and the Balsam Mountains of North Carolina. In both of these mountain ranges, *Plethodon jordani* occupies a higher altitudinal zone than does *P. glutinosus.* However, in the Great Smoky Mountains, the zone of overlap is very narrow, spanning about 50 m in altitude or less. In the Balsam Mountains, the overlap is extensive, often over 750 m in altitude. Hairston established study sites in the overlap zones on the two mountain ranges for competition experiments. The study plots were not fenced and were approximately circles 24 m in diameter. These plots were censused by counting the number of salamanders seen while walking through the site at night along a predetermined route.

The main experiment, begun in 1974, consisted of removing *P. jordani* at some sites and *P. glutinosus* at others, together with monitoring control sites. By 1976 and continuing through 1978, the number of *P. glutinosus* had increased at all sites at which *P. jordani* was removed, although the effect was less in the Balsam Mountains than in the Smoky Mountains. Hairston noted an indication of a reciprocal interaction in the Smoky Mountains as well, for there removal of *P. glutinosus* led to an increase in *P. jordani* by 1977. In the Balsam Mountains, however, little effect of *P. glutinosus* against *P. jordani* was noted. Hairston concluded that the competition between the two salamander species was less, in both directions, in the Balsam Mountains than in the Smoky Mountains. He also concluded that the lower intensity of competition in the Balsam Mountains permitted the species to overlap more in their altitudinal ranges, whereas the higher intensity of competition in the Smoky Mountains prevented coexistence.

Hairston went on to estimate the competition coefficients and carrying capacities (Table 18–2) of the two *Plethodon* species in both places, in what is one of the first attempts ever to do so with data from experiments with natural populations. The results are roughly consistent with the

TABLE 18–2 Parameters of Lotka–Volterra competition equations for *Plethodon jordani* and *P. glutinosus*

	Great Smoky Mountains	Balsam Mountains
$\alpha_{g,j}$	0.19	0.14
K_g	11.7	11.0
$\alpha_{j,g}$	2.26	0.63
K_j	45.3	47.4
det(α)[a]	0.56	0.91
$\hat{N}_g$[b]	5.2	4.6
$\hat{N}_j$[b]	33.6	44.5

Source: After Hairston (1980a).

[a] The det(α) equals $(1 - \alpha_{g,j} \times \alpha_{j,g})$.

[b] Steady-state abundance predicted by Lotka–Volterra equations based on the α's and K's in the table.

abundance of both control populations during the years 1976–1978. The results should be taken as tentative, however, because the census was done without mark-resighting statistics, and the relation between the counts and the natural population size is unknown. Hairston estimated that only one-quarter to one-tenth of the salamanders were active on even a favorable night, and some migration from the habitat surrounding the study plots also occurred. These considerations introduce an error of unknown magnitude; nevertheless, the study is important as a pioneering attempt to quantify the strength of competition in a situation in which experiments had shown that competition was present.

Hairston (1980b) also transplanted salamanders from the Balsam Mountains to the Great Smoky Mountains and vice versa. These experiments demonstrated that the salamanders of the Great Smoky Mountains, where the competition coefficients are high in both directions, also have a high competitive effect on the salamanders of the Balsam mountains. Conversely, the Balsam Mountain salamanders, with low competition coefficients to one another, also have a low competitive effect on the Smoky Mountain salamanders. Hairston interpreted these findings as showing that evolution had led to an increase in competitive ability in the Great Smoky Mountains and to a decrease in competitive ability in the Balsam Mountains.

Hairston (1981) reported that the experiments showing competition between the two *Plethodon* species failed to show any effect on five other plethodontid salamanders that occurred on the study plots, although in relatively low abundance. The identity of the limiting resource was not clear. Jaeger (1974) indicated that burrowing sites might be limiting, and Hairston suggested that nesting sites might be the resource.

Hairston's studies with *Plethodon* that focus on competition contrast strinkingly with his studies on another plethodontid group, the genus *Desmognathus*. This system is especially puzzling because it initially appeared that these salamanders formed a size-structured guild showing the classic patterns relating size to coexistence. Hairston's (1949) early work with this group was oriented to competition. In the Black Mountains and the adjacent Blue Mountains of North Carolina are six species of *Desmognathus*. The four species that typically coexist in the Black Mountains form a size series. The maximum lengths for males in decreasing order are 70, 63, 46, and 30 mm. The species in this series are *D. quadramaculatus*, *D. monticola*, *D. ochrophaeus*, and *D. wrighti*. The ratios between consecutive sizes are thus about 1.1, 1.3, and 1.5. A pattern of allopatry is also suggested, in which ecological equivalents replace one another along altitudinal gradients.

However, covarying with this sequence in body size is a gradient from aquatic to terrestrial locations. The largest species is found closest to streams, the next largest species is the next closest, the third largest is the third closest, and the smallest is the most terrestrial and found farthest away from streams.

In 1980, Hairston published a hypothesis based on predation, and not competition, as the main consideration for the processes in this system. The key point was established by Organ (1961); a decreasing loss to predation appears along a gradient from an aquatic to a terrestrial habitat. Hairston, following Organ, suggests that the common ancestor of the *Desmognathus* species was a form such as *D. quadramaculatus*, the largest and most aquatic species today. "The differentiation of each species," Hairston writes, "has involved successively greater adaptation to more terrestrial life, smaller size, and decreased length of larval life" (p. 360). If so, the body sizes may be a secondary consequence of the position of a species along the aquatic–terrestrial spectrum. The most aquatic species should be the largest, both because it is a predator itself and because a size escape occurs from predation by other species in the streams.

Clearly, the full picture is far from complete with the *Desmognathus* system. Predation is certainly important, and competition is well documented in other groups of salamanders. Moreover, the species differ in the extent to which the age classes are found in different habitats and using different niches.

Now let us follow the path set by these salamanders—dry our feet and investigate the activities of some completely terrestrial communities.

TERRESTRIAL PLANT GUILDS

Much of community ecology originated with the study of plant communities, especially with their classification and with the question of whether communities are discrete entities. Today plant communities remain the subject of active study, with investigators using new techniques for comparing and contrasting plant and animal communities.

A central question in plant community ecology is, how can we account for the high species diversity of some plant communities? For example, a 0.5-km^2 plot of seasonal moist tropical forest on Barro Colorado Island in central Panama has 380 woody plant species with a stem at least 1 cm diameter at breast height (dbh) and nearly 200 species that achieve diameters of 20 cm dbh or larger (Hubbell and Foster 1983). Explaining how these species coexist poses a problem because the frequent existence of competition among plants for light, water, and nutrients is generally accepted, as was discussed in Chapter 12. Although sites, especially early in succession, may not have enough plants for much competition to occur, the existence of competition in most grasslands and forests is rarely a subject of controversy. However, whether that competition has any role in structuring a plant community is not at all clear.

Resource Partitioning

Resource partitioning is a mechanism that may permit the coexistence of plant species (Harper 1961). For example, Gulmon (1974, 1979) showed that grasses and forbs of an annual grassland in the Mediterranean climate of central California have strategies that lead to success at characteristic times in the seasons and in characteristic places in the habitat. Some species take up nutrients at the beginning of the season, when the weather is cold and the quantity of nutrients in the soil is low. These plants get a head start on the season and so flower early. Other species begin in the spring, when high temperatures together with plentiful rain have promoted the decomposition of the stalks left from the preceding year. This decomposition enriches the soil with nutrients. The species that begin at this time grow quickly and to a larger size than the early-flowering plants.

Grubb (1977) synthesized observations of differing strategies among coexisting plant species in terms of niches with four dimensions. The first dimension is microhabitat—what type of place a plant characteristically occurs in, such as shade or sun, dry or wet soil, and so forth. The second dimension is life form; trees, herbs, shrubs, and vines are separated along this dimension. The third dimension is phenological; this axis separates plants that capture nutrients at different times, as in the grassland example just mentioned. The fourth dimension is the *regeneration niche dimension,* which refers to the type of vacant space, or gap in the canopy left by a tree fall, that a species occupies.

For a community less diverse than a grassland or forest, Cody (1986) has demonstrated that Sonoran and Mojave desert cactus communities have species representing a repeatable array of life forms. His study also showed that on desert islands in the Gulf of California, in which fewer cactus species are found, the species that do occur have a wider variety of forms than species occurring in the more complex communities on the adjacent mainland.

Tilman (1977, 1982) has emphasized that the theoretical possibility of coexistence by niche specialization is high. Models for exploitive competition among plants using different nutrients and light were developed and tested with phytoplankton. These models were extended to perennial grasslands. They show that different plant species can coexist simply by specializing on different ratios of various soil nutrients, water, or light, as well as along the dimensions suggested earlier by Grubb (1977). Thus, it is theoretically possible that coexistence among the species in a diverse plant community is brought about by specialization to the varied circumstances that are present in the area occupied by the community.

Alternative Equilibrium Hypotheses

One alternative to the hypothesis of fine specialization among the plant species in a diverse community is analogous, in population genetics, to maintaining polymorphism by frequency-dependent selection. The idea is that seedlings cannot germinate and grow under the canopy of their own species. Therefore, species can invade the community when rare, but they cannot replace themselves when sufficiently common. Frequency-dependent mechanisms of this type have been termed *compensatory mechanisms* by Connell *et al.* (1984), who found evidence for them in tropical and subtropical forests in Queensland, Australia.

Still another alternative is that a forest is analogous to a section of rocky substrate in the intertidal zone, in which competition is presumed to lead to a dominant tree in the absence of an extrinsic source of mortality that opens vacant space (disturbance). Doyle (1981) has suggested that recurrent hurricanes cause tree falls that maintain species diversity in Puerto Rican rain forest. Doyle's research used the techniques of systems ecology, as discussed at the end of Chapter 23.

A Nonequilibrium Hypothesis

All of the preceding hypotheses are *equilibrium* hypotheses in the sense that the species are viewed as coming to coexist in a steady state. In contrast to these hypotheses, Hubbell (1985) has developed an important *nonequilibrium* hypothesis for tropical forests. The motivation is that more species seem to exist in the forest than there are roles or niches. According to Hubbell's hypothesis, only a few major niches exist for forest trees, and each niche has many species in it. Furthermore, it is hypothesized that the species in the same niche do not coexist; they are equivalent to one another, and their relative abundance changes through time as a random walk. Indeed, the model used by Hubbell is formally equivalent to one of the models that had been proposed for genetic drift in population genetics. The key to this idea is that even if tree species do not coexist in the long run, the time for one of the species to drift to extinction is long. Hubbell calculates that more than a million trees must die and be replaced in a forest that has 2000 trees before a species that was initially present as 200 trees becomes extinct. This time approaches a geological time scale.

The hypothesis of drift in a community's species composition is supported by studies suggesting a competitive equivalence among the plants in a community (Goldberg and Werner 1983). Further support for the hypothesis of community drift comes from a comparison between forests on Barro Colorado and the islands of the Perlas archipelago near Panama. If the forest composition represents a random process involving slow immigration and extinction processes, then Hubbell (1985) predicts that on islands an impoverishment of the tree flora should occur, together with an increase in the apparent dominance of the community by one type of tree, and different species should be dominant on each island. Recent surveys confirm these predictions. Hubbell cautions, however, that equilibrium hypotheses could also be proposed that account for these data, and so the matter is still open.

Plant community dynamics are also influenced by differing dispersal strategies. Rabinowitz *et al.* (1979) and Rabinowitz and Rapp (1981) have discovered that the sparse species within a perennial grassland have greater dispersal abilities than the common species.

Animals can also affect the plant community. To a limited extent, resistance to herbivory, and other traits that affect plant–animal interaction, can be considered as part of the plant's niche, as Tilman (1982) emphasizes. In some circumstances, animals have a large and almost controlling effect on plant community structure. With a sequence of experiments in the plant community of the Sonoran Desert, Inouye *et al.* (1980) showed that ants increase the diversity among annual plants by differentially harvesting seeds of the numerically dominant species, whereas rodents prey selectively on plants that dominate the community in terms of biomass. Furthermore, as we will see, the community ecology of plants may be affected by plant diseases and by pollinators. In particular, the plant–pollinator interaction has been extensively studied in recent years from the standpoint of community ecology.

The Interdependence of Plants and Pollinators

Whenever plants are sympatric, the potential for competing for pollinators exists. If two similar plants flower simultaneously, the rarer one is presumably at a selective disadvantage and tends either to become extinct or to diverge from the commoner species by attracting different pollinators (Levin and Anderson 1970). If the plants are close relatives, they may also show selection against the tendency to hybridize, using different pollinators (Stiles 1975).

At the research site of the Organization of Tropical Studies at Finca La Selva, on the Caribbean coastal plain of Costa Rica, nine species of *Heliconia* (wild plantains, closely related to bananas) are pollinated by nine species of hummingbirds (Stiles 1975). Four of the hummingbirds are "hermit" hummingbirds, with relatively long, down-curved bills. The remainder are short-billed species. The hermits *trapline,* that is, they move over regular routes and primarily visit the five *Heliconia* species with long, down-curved flowers. The four *Heliconia* species with straight or short flowers are most often visited by short-billed hummers, which frequently establish territories around the flowers. A final *Heliconia* species is visited by both kinds of birds.

The partitioning between the territorial and traplining pollinators is based largely on the amount and accessibility of the energy provided in nectar. Large clumps of flowers, copious amounts of nectar, rich nectar, and synchronous flowering all favor the territorial birds, because an adequate resource is available in a defendable space. In contrast, if flower clumps are small, nectar dilute or inaccessible, or flowering asynchronous, traplining seems to be favored. Territoriality among the birds promotes inbreeding in the *Heliconia* plants, because each clump is a single genetic individual; traplining among the birds promotes outbreeding among the plants. Therefore, the strategy for attracting pollinators adopted by a *Heliconia* species may be partly determined by its degree of self-compatibility.

Members of the La Selva *Heliconia* guild are closely related and have various other mechanisms for avoiding hybridization. For example, the hermit-pollinated species, with a single exception, show nonoverlapping peaks of flowering. The two species that flower simultaneously are thought to be genetically isolated by placing their pollen at different places on the birds.

Flowering patterns in the *Heliconia* guild may also be influenced by factors such as competition for dispersal agents (manakins—forest birds of the family Pipridae) and the attacks of diurnal flower-destroying animals that tend to select for early morning flowering.

The idea that plants might compete for insect pollinators was first suggested by Robertson (1895) and continues to attract attention.* Also, plants competing for pollinators may act mutualistically because sequential flowering provides a continuity of resources that maintains nectarivorous species required as pollinators (Heinrich and Raven 1972, Waser and Real 1979).

* See, for example, Mosquin (1971), Reader (1975), Lack (1976), Stiles (1977), and Waser (1978, 1983).

HERBIVORE GUILDS

The best-known and largest of the herbivore guilds consist of small organisms, largely insects, that draw sustenance from leaves. Indeed, they may account for one-fourth or more of all animal species, because herbivorous insects as a whole certainly account for more than half (see, e.g., Gilbert 1979, Erwin 1982, Futuyma 1983, Strong 1983). It is therefore essential that these guilds be understood if ideas about community ecology are to have any generality.

Patterns of Phytophagous Insects

Assemblages of phytophagous insects show some patterns similar to those found in other groups of organisms (Lawton and Strong 1981):

1. Large areas have more species than small ones do; widely distributed plant species are host to more insects than narrowly distributed ones. For instance, the number of pests on tropical crops is strongly correlated with the area under cultivation (Strong 1974, Strong *et al.* 1977), and widespread oak trees in California support more leaf miners than do narrow endemics (Opler 1974).
2. More species are found in complex or persistent habitats than in simple or ephemeral habitats. Large perennial plants are normally attacked by more species than small annuals are; the mowing of grassland greatly depresses populations of leafhoppers and their relatives (Morris 1981). Weeds tend to have fewer enemies than otherwise similar longer-lived species do (Lawton and Schroeder 1978).
3. Niches are narrower in persistent habitats than in ephemeral habitats. Specifically, butterflies that have narrow host ranges tend to feed on "predictable" plants in Scandanavia (Wiklund 1974, 1977, 1982b); those with broad host ranges more frequently eat weedy species.
4. Ecologically significant differences in size occur within and between the species of a guild. For instance, each *instar* (stage between moults) of a growing insect tends to be 1.3 times as large as the previous one, and closely related species occurring together often differ in size (see, e.g., McLure and Price 1975).
5. Communities become saturated with species in ecological time. When a plant is introduced to a new area, the rate at which it seems to acquire new insect herbivores approaches an asymptote in 100–1000 years. For example, no correlation exists between the number of insect species attacking native trees in Britain and the period (4000–13,000 years) during which they have been present (Southwood 1961, Birks 1980).

6. Within a given habitat, species in the same guild often seem to show niche separation. In tropical passionvine (*Passiflora*) butterflies of the genus *Heliconius*, Benson (1978) found resources partitioned along three different niche dimensions—*Passiflora* habitat, *Passiflora* species attacked, and plant part eaten.

Insect ecologists express little disagreement on the reality of these patterns; however, they disagree a great deal on their interpretation.*

Competition Among Insect Herbivores

In some cases, competition among insect herbivores has been clearly demonstrated. Native leafhoppers of the genus *Erythroneura* have been shown to compete at densities that occur naturally in the field (McClure and Price 1975, 1976). The number of male progeny produced per female declined when more than one female of the same species was caged on a single leaf—and quite dramatically when six females were placed on the same leaf (Figure 18–3A). Mixed-species rearings showed similar but somewhat less dramatic declines (Figure 18–3B), indicating that intraspecific competition was more intense than interspecific competition.

Recently, McClure (1983 for summary) has also shown interspecific competition between introduced armored scale insects on eastern hemlocks. The nitrogen and water content of the needles are crucial to the fecundity, development rate, and survival of the scales, and to the density of the scales themselves, so that the phenology of the hemlocks is a major determinant of food quality and quantity. Experiments similar to those done with *Erythroneura* together with geographical evidence indicate that one scale, *Fiorinia externa,* is able competitively to exclude the other, *Tsugaspidiotus tsugae,* from areas in which *Fiorinia* is established. *Fiorinia* colonizes the trees first and reduces the amount of nitrogen available to the later-arriving *Tsugaspidiotus.*

Absence of Competition Among Insect Herbivores

In contrast, careful studies of a guild of 13 stem-boring insects (beetles, flies, and moths) attacking native prairie plants in Illinois showed a great deal of overlap in resource exploitation and little competition (Rathcke 1976). Even when several larvae occupied the same stem, their tunnels rarely met. Overlap in the species attacked was extensive; nine of the insect species showed more than a 70% overlap with at least one other. In addition, little evidence appeared of resource partitioning on the basis of stem diameter or location within the stem. In one case, however, in-

* See, for example, Benson (1978), Lawton and Strong (1981), McClure (1983), Price (1983), Schultz (1983), and Strong *et al.* (1984).

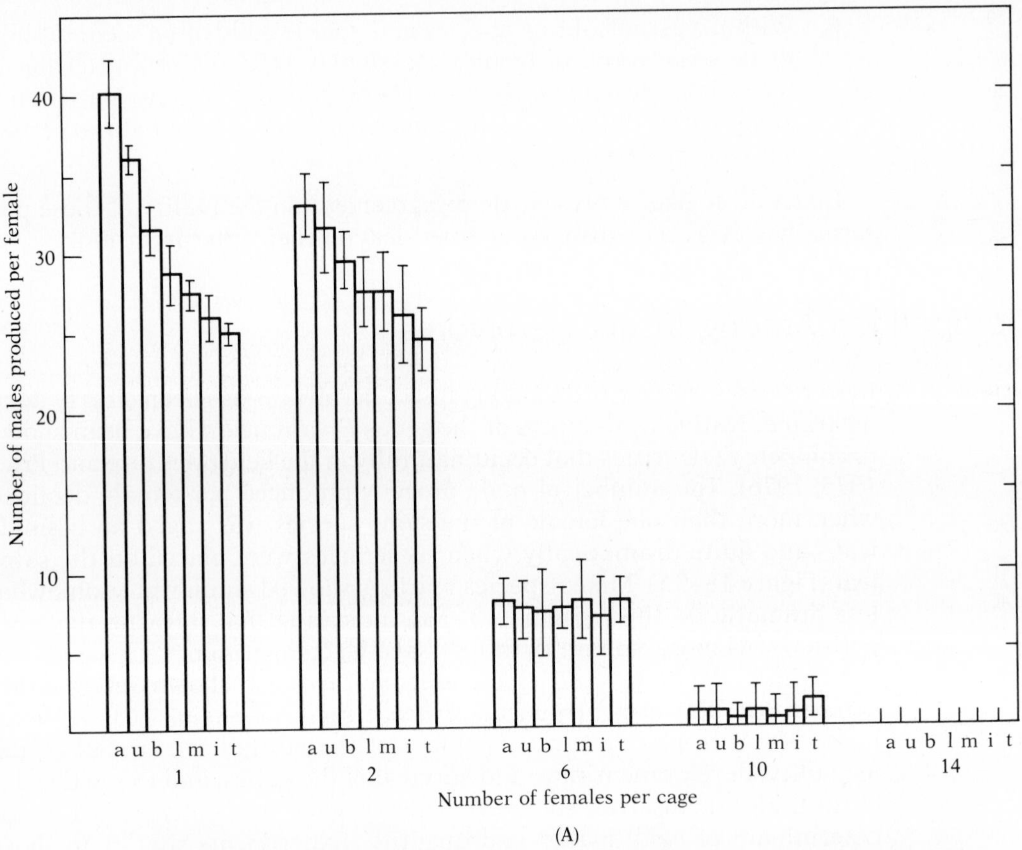

FIGURE 18–3 Leafhopper competition. (A) Mean number of male progeny produced per female in single-species rearings for each of seven *Erythroneura* species *(E. arta, E. usitata, E. bella, E. lawsoni, E. morgani, E. ingrata,* and *E. torella)* from cages containing various numbers of females per leaf. (B) *(Facing page)* Mean number of male progeny produced per female in four mixed-species rearing of pairs of *Erythroneura* species from cages containing various numbers of females per leaf. The unshaded histograms give the means for each species under mixed-species conditions; the shaded histograms give the corresponding results (from A) of single-species rearings for comparison. Error bars show 68 percent confidence interval. *(After McClure and Price 1975.)*

terference competition appeared to be present. Whenever the larva of one species of mordellid beetle met a larva of an *Epiblema* moth (Olethreutidae), the moth was injured and eventually died (see also De Vita 1979). Interspecific aggression with fatal outcomes also occurred among borers in the stems of the intertidal grass *Spartina* in Florida. Here too, however, the competition seemed to be weak and to have little influence on community structure (Stiling and Strong 1983).

Further evidence of a general lack of competition among members of

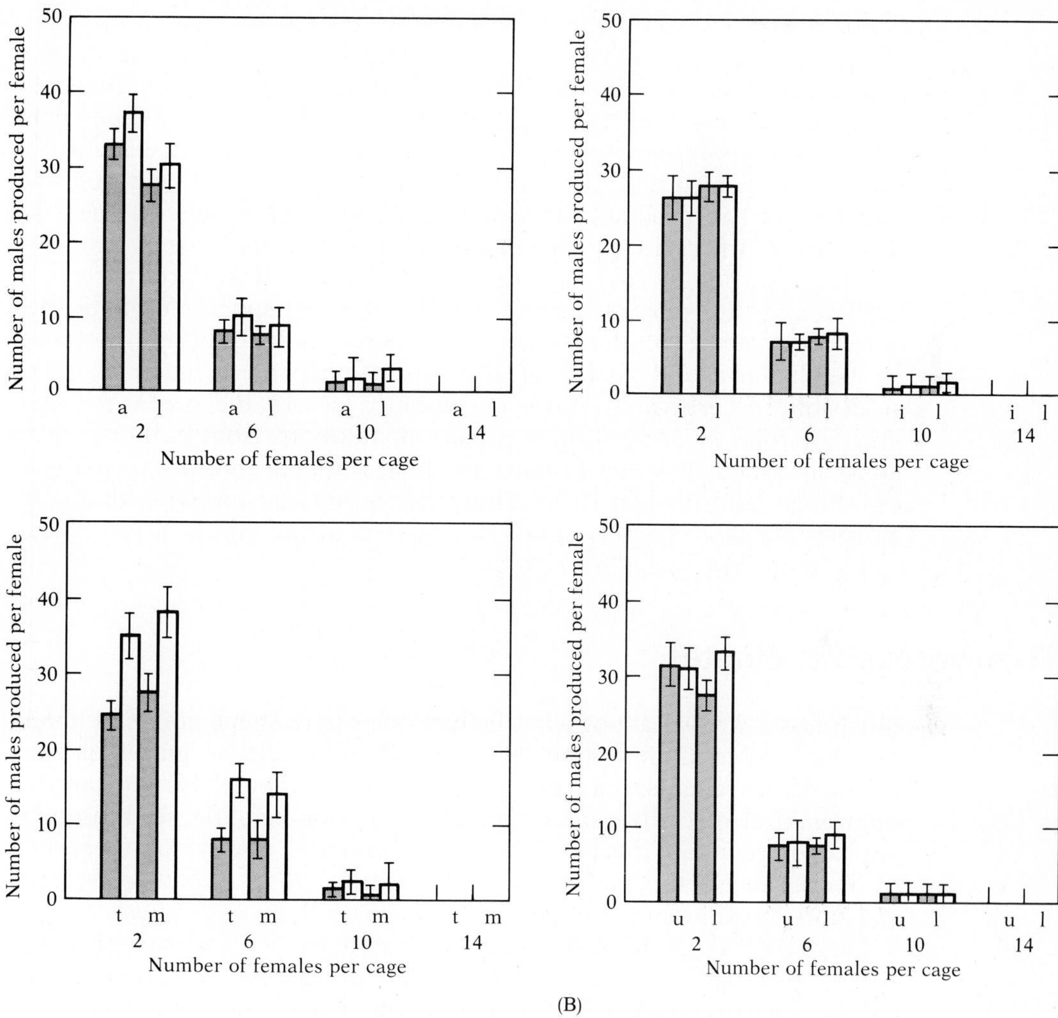

(B)

guilds of insect herbivores is extensive. For instance, studies of the insect communities living in the water-filled floral bracts of two lowland tropical and one mid-altitude tropical species of *Heliconia* (relatives of the banana plant, Seifert and Seifert 1976, 1979a, b) found that most species did not interact significantly with others, and that among those that did, some of the interactions were mutualistic. Only 5 of 40 possible pairwise interactions were competitive, and these only weakly so.

As early as 1889, the great lepidopterist Samuel Hubbard Scudder suggested that the imported cabbage butterfly, *Pieris rapae,* was extirpating the native cabbage butterfly, *P. oleracea,* from New England, and the story of the latter's competitive displacement became widely accepted. However, recently Chew (1981) has shown that the two species partly partitioned food plants and habitats and that the composition and abundance of the crucifer flora were crucial in determining the composition and abundance of the *Pieris* fauna. Chew found little sign that competition was important; changes in land use patterns that altered the crucifer flora were seemingly responsible for the appearance of competitive exclusion. Similarly, Shapiro and Cardé (1970) were unable to find evidence that competition had produced observed partitioning of the environment by three North American satyrine butterflies of the genus *Lethe*.

The suggestion by Hairston *et al.* (1960) that herbivores are rarely food limited and thus must be largely predator limited appears to be headed for a compromise solution.* Food limitation may be important because, for biochemical and phenological reasons, much plant material is not accessible to herbivores. However, climatic factors and predators often play key roles in preventing herbivore populations from building up to levels at which they can exhaust the food resource that does exist (see, e.g., Singer and Ehrlich 1979). Thus, the consensus is growing that competition plays a relatively small role in structuring guilds of insect leaf-eating and stem-boring herbivores.

Herbivorous Vertebrates

Much less is known about what factors operate to shape guilds of noninsect herbivores. Competition operates in size-structured guilds of seed-eating desert rodents that have been studied (c.f. Chapter 16).† Recall that microhabitat specialization, rather than partitioning of the seed resource, appears to structure this herbivore guild. Similarly, five sympatric species of finches in North Carolina do not partition the available seeds, even though larger birds tend to eat larger seeds (Pulliam and Enders 1971).

Large herbivores in the Serengeti ecosystem show distinct partitioning of food resources, and competition has been at least partly responsible, as will be discussed in Part 7. The same is true of many guilds of nectar and pollen eaters. Interference competition has been recorded between members of colonies of stingless bees with rich sources of food, leading to battles that cause the deaths of many individuals (Johnson and Hubbell 1975). North American hummingbirds (nectar feeders) show strong interference competition and partition their flower resources (Armitage 1955, Stiles 1973, Phillips 1975, Carpenter 1978). Bumblebees, in contrast, show mostly exploitation competition but also partition their flower resources (Heinrich 1976, 1979, Inouye 1978, 1980; but see Morse 1977).

Might there be a general reason why competition between insect leaf-

* See, for example, Murdoch (1966), Ehrlich and Birch (1967), and Slobodkin *et al.* (1967).

† See Rosenzweig and Sterner (1970), Brown (1973), Brown and Lieberman (1973), Smigel and Rosenzweig (1974), Price (1978), and Bowers and Brown (1982).

eating herbivores is less frequent than that between either plants or carnivores? One possibility is that the nutritional resources of leaf-eating herbivores are more diverse and on the average require more specialization to use than those of organisms at other trophic levels. Plants, for example, "eat" mostly sunlight and nutrients, the latter extracted from a common pool in the soil. Although carnivores have varied diets, the biochemical characteristics of their foods are quite similar. A lion could benefit from the same diet of flies or mosquitoes that nourishes a lizard or a dragonfly, provided that sufficient quantities could be molded into sizable chunks (and perhaps basted with tasty gravy).

In contrast, the lupine diet of *Glaucopsyche lygdamus* or the milkweeds eaten by monarch butterflies could not support *Pieris rapae* or *Euphydryas editha*. The defensive chemicals of plants present a formidable array of challenges, and relatively few herbivores are able to thrive on a wide variety of plant species.

CHAPTER 19

The Entire Community: From the Parts to the Whole

In this chapter we move, perhaps somewhat ambitiously, from considering pieces of a community to the entire community. Much of what we know in this area is based on the analysis of islands, for islands offer replicated instances of communities that are more or less comparable. Although this chapter is concerned with the entire community, in fact no data actually exist for an entire community. What data do exist pertain only to very big pieces of the community, such as all its plants, all its insects, or all the major compartments through which energy flows; these are pieces nonetheless.

First, let us explore some patterns characteristic of entire communities, or at least big pieces of them. We will then move to the conceptual approaches that have been developed for entire communities. These approaches can be classified into near-equilibrium views, nonequilibrium views, and quasi-equilibrium views, the last being a mixture of the first two.

PATTERNS OF ENTIRE COMMUNITIES

Does any feature of entire communities seem to be general—that is, that occurs in many communities or becomes apparent only when many different communities are compared? The answer is yes. Five basic patterns have been discovered. They are not universal patterns, but they are often present.

Pattern 1: Patterns in the Distribution of Abundance Among Species

The first pattern pertains to how many species are common and how many are rare within a community. Ideally, we could plot an abundance level on the horizontal axis and plot the number of species with that abundance on the vertical axis. For example, if 20 species exist with an abundance between 500 and 1000 individuals, then a point is plotted at, say, 750 on the horizontal axis and 20 on the vertical axis. When the abundance of all the species is taken into account, we obtain a graph of the *distribution* of abundance among the species in the community, as illustrated in Figure 19–1 for British birds (Williams 1964).

Unfortunately, the graph for the distribution of abundance for all the species in a community is rather sensitive to the sampling effort. The data typically come from mass collections, not from mark-recapture or other methods of census focused on each of the populations. For example, a light trap, in which a bulb emitting ultraviolet light attracts many species of moths at night, yields an estimate of the number of animals in each of the moth species. Of course, one must use such data with the hope that the species are more or less equally likely to be caught. However, an even more basic problem concerns the total sample size itself. If the total sample size—that is, the number of individuals of all species combined—is small, then most species will be represented by only one individual, perhaps a few species by a couple of individuals, and maybe only one species by three or more individuals. If that total sample size is quite small, the distribution is a decreasing function; Fisher *et al.* (1943) suggested a formula for this situation.

Preston (1948, 1960, 1962) investigated patterns in the distribution of abundance among species and offered a way of taking account of sample size; more importantly, he produced some generalizations and connections between the distribution of abundance and other community patterns. The main observation is that, as the total sample size increases,

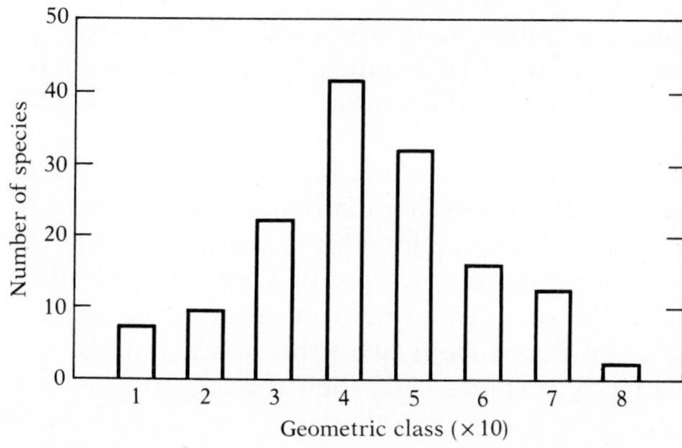

FIGURE 19–1 Distribution of abundance in British birds. Each geometric class represents a factor of 10—e.g., geometric class 1 is for species whose abundance is between 1 and 10, geometric class 2 for species with abundance between 10 and 100, etc. *(After Williams 1964.)*

eventually a point is reached at which the distribution of species abundance appears to be unimodal rather than monotonically decreasing, as before with small total samples. Figure 19–2 illustrates this effect with lepidopteran insects (Williams 1964). The distribution is not symmetrical; the right tail of the distribution is more pronounced than the left. Because of the typical asymmetry, it is conventional to plot the *logarithm* of abundance on the horizontal axis and the number of species with that abundance on the vertical axis. If the base of the logarithm is 2, then the horizontal scale is calibrated in multiples of 2.

Figure 19–2 shows a monotonic decreasing curve for the distribution of abundance among moth species for a total sample size of about 500 individuals. With a sample size of about 3750 individuals, the distribution has a mode. With over 16,000 individuals, the mode has shifted to the right. Indeed, one can imagine a bell-shaped normal distribution that shifts to the right as the total sample is increased. The mode is to the left of the vertical axis when the sample size is small; it becomes "revealed" when the sample size is big enough.

A normal distribution in which the horizontal axis is calibrated in units of the logarithm of a variable is called the *log-normal* distribution. The studies by Preston revealed that many distributions of abundance could be represented as *truncated* log-normal distributions, where the truncation occurs at the vertical axis and depends on the total sample size, as just illustrated.

What is the significance of a log-normal distribution when it is found? To answer this question, we need to digress briefly and consider what is called the *Central Limit Theorem* in statistics and its connection with the usual normal distribution itself. The Central Limit Theorem states that the average of n random variables is approximately normally distributed, even if the distribution of the individual variables is not normal, provided that n is large.

As an example, suppose we use a random number generator to find a number anywhere between 0 and 1 with equal probability. This number is said to be *uniformly distributed* between 0 and 1; its distribution is a flat line between 0 and 1, and so it is not normally distributed between 0 and 1. Next, let us consider a number, Y, that is produced by using the random number generator twice to get two values, X_1, and X_2, which are then combined to get Y according to the formula $Y = (X_1 + X_2)/2$. This new number Y—the average of X_1 and X_2—will itself tend to be more normally distributed than its constituent X's are. Indeed, if on a computer you generate many values for Y by combining many pairs of X's, you will see that Y's distribution looks like a triangle, with its peak at $\frac{1}{2}$. Furthermore, if Y is produced by generating three values of X and then dividing their sum by 3, the distribution of Y would be even more like a normal distribution than when Y is produced from two values of X. Finally, if Y is produced from the average of n values of X, then the distribution of Y

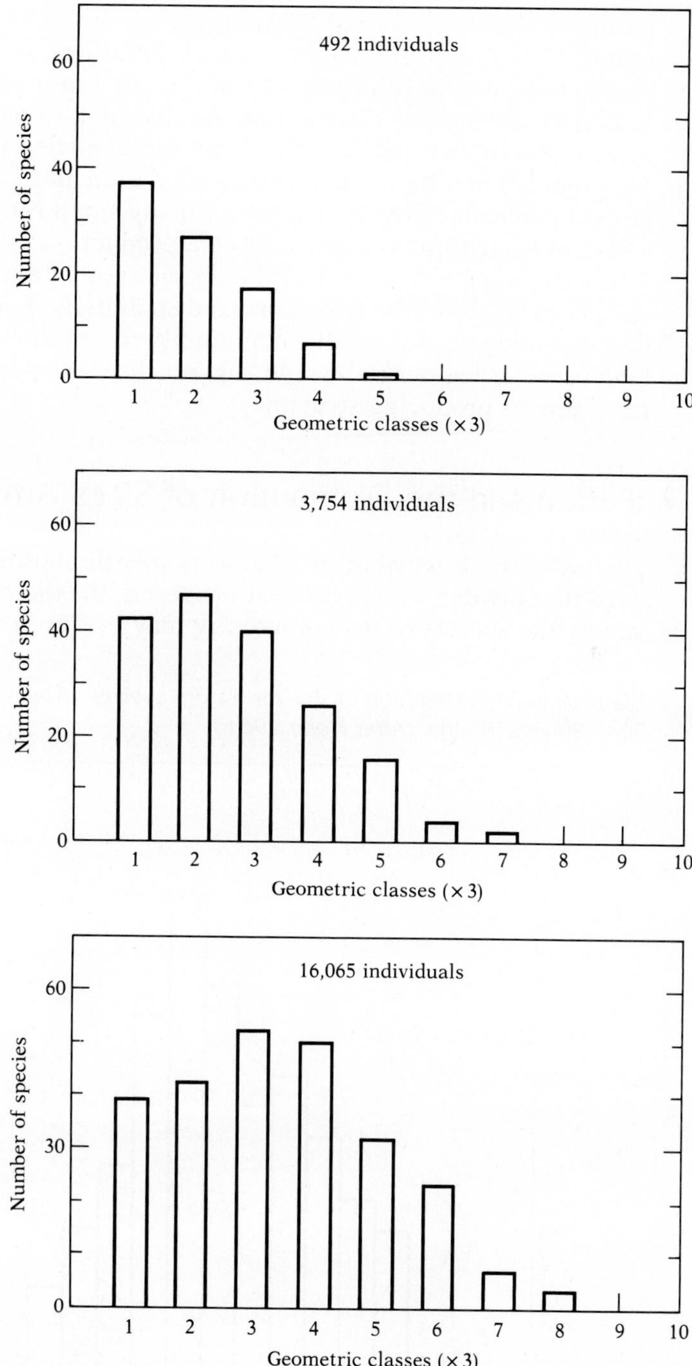

FIGURE 19–2 Distribution of abundance in moths captured by light traps in England. *(After Williams 1964.)*

becomes exactly a normal distribution in the limit as n tends toward infinity. For practical purposes, the distribution may look like a normal distribution if n is 10 or more. The Central Limit Theorem of statistics says that the normal distribution results for a quantity whose value is determined by the sum of many small random effects. In the same spirit, the existence of a log-normal distribution indicates a quantity whose values are determined by the *product* of many small random effects because a sum of logarithms is equivalent to a product.

Let us return to the issue of the distribution of abundance among species. Clearly, observing a log-normal distribution of abundance is not evidence of any structure in the community; if anything, a distribution different from a log-normal would suggest the existence of structure and of mechanisms producing structure.

Pattern 2: Patterns in the Distribution of Sizes Among Species

The log-normal distribution reappears in still another context. In collections that involve a large number of species, the distribution of body sizes among the species of the community may be approximately log-normal.

FIGURE 19–3 Distribution of size for diatom species collected over 155 days from Silver Springs, Florida. *(After Hohn 1961.)*

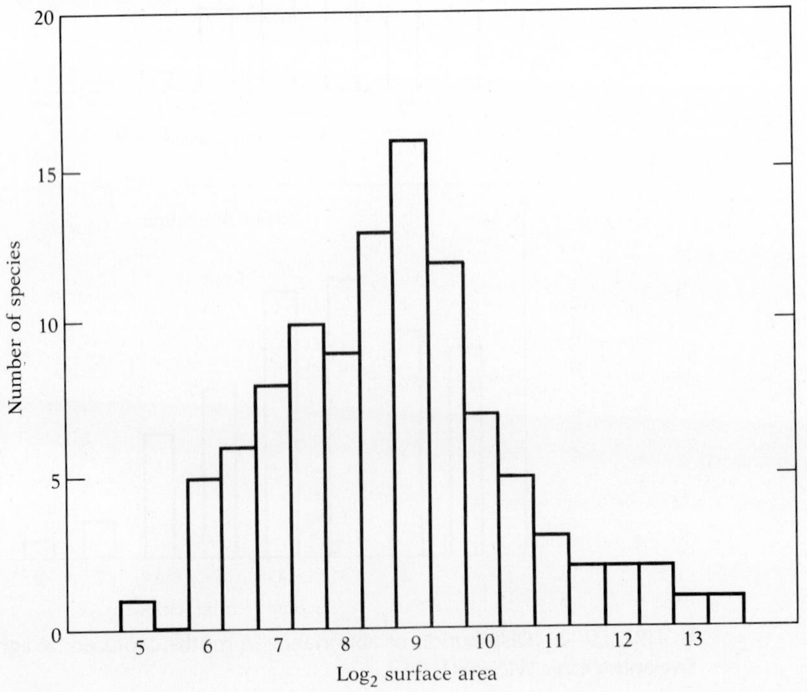

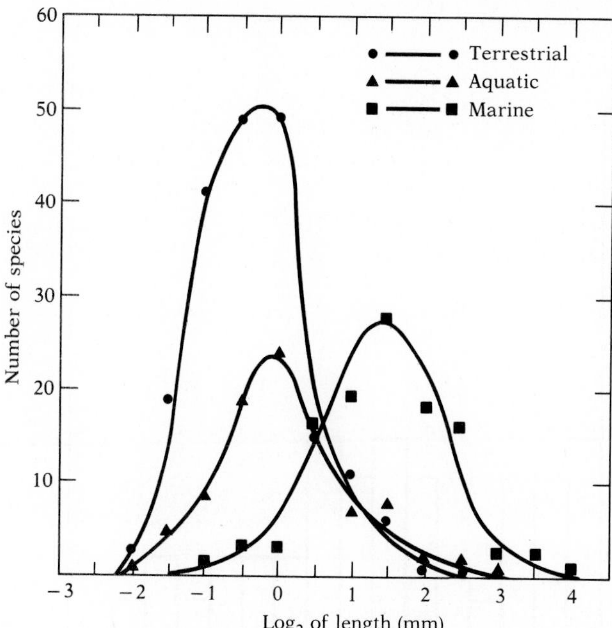

FIGURE 19–4 Distribution of size for nematodes from each of three habitats, according to unpublished data of N. Stanton as summarized by Kirchner (1980).

Figure 19–3 presents an example consisting of the size distribution for diatom species collected from Silver Springs, Florida (Hohn 1961). Figure 19–4 presents the distribution of lengths for nematodes from each of three habitats, according to unpublished data of N. Stanton, as summarized by Kirchner (1980). Also, Schoener and Janzen (1968) documented a log-normal distribution of body sizes for insects.

Kirchner (1980) surveyed the distribution of body sizes among vertebrate species. The distribution for the logarithm of body mass for the mammals of Michigan from the data of Burt (1946) is not log-normal but is skewed to the right (positive skew) (see Figure 19–5). Kirchner found similar results for birds. He concluded that character displacement might lead to a spreading out of body sizes beyond that expected according to a log-normal distribution and that character displacement is apparently more important in vertebrates than in other taxa.

Pattern 3: Relation Between Area and the Number of Species

A third generalization is that the number of species in a site increases with the area of a site. This relation has received attention mostly with respect to the number of species on islands. For example, Darlington (1957) wrote, "division of an area by ten [in going from one island to the next] divides the fauna by two." Since that time, many studies have in-

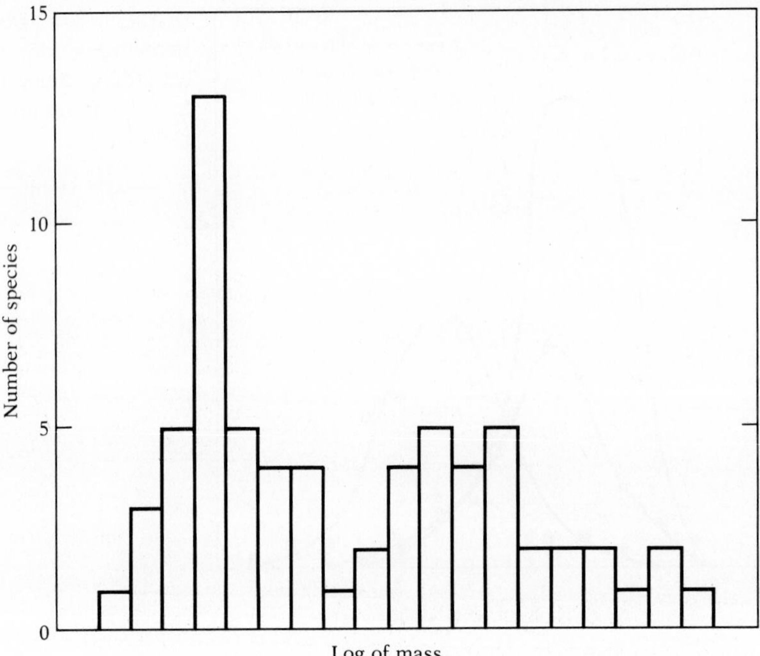

FIGURE 19–5 Distribution of size for the mammals of Michigan. *(After Burt 1946.)*

vestigated the quantitative relationship between island area and the number of species on it. For islands that are all sufficiently close to a source of species, the number of species on the island is approximately proportional to the island's area raised to a value between 0.20 and 0.35. Darlington's rule works out to an exponent of 0.30, which lies in the middle of the range shown by other studies. Figure 19–6 presents a spectacular illustration of the relation between area and species number for islands in the vicinity of New Guinea (Diamond 1973). Here the exponent was 0.22.

Preston (1962) made an important contribution by showing that a typical value of the exponent relating area to species number on an island could be derived by assuming that the relative abundance of species is distributed log-normally and that the absolute abundance of any particular species increases linearly with the island's area. This result, which potentially coupled two of the regularities for whole communities, spawned other investigations of the possible biological basis for the exponent relating species number to island area; this research is synthesized in May (1975a) and Pielou (1975).

The exponent relating species number to area is lower if the area is a nonisolated sample area within an island or continent. The exponent is

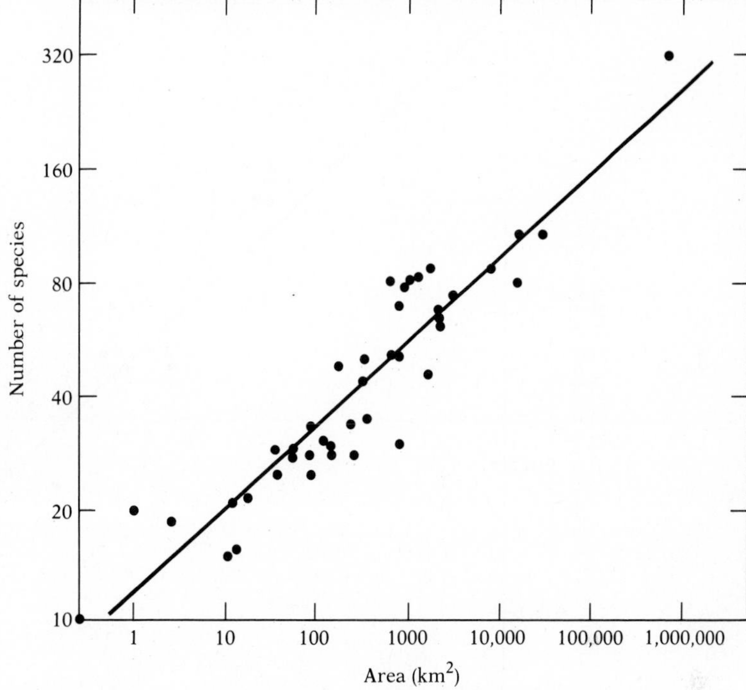

FIGURE 19–6 Relation between area and species number of birds on islands in the vicinity of New Guinea. *(After Diamond 1973.)*

typically between 0.12 and 0.17 in this situation, which MacArthur and Wilson (1967) interpreted as resulting from the presence of rare species in the area that would become extinct if the area were isolated.

Pattern 4: Relation Between the Distance of an Island from Its Source and the Number of Species

Pattern 4 complements the area relation just discussed. Figure 19–7 shows another spectacular example from the islands in the vicinity of New Guinea (Diamond 1973). New Guinea is evidently the source of birds for islands as far away as about 9000 km. The figure shows that the number of species on an island, after taking into account its area, declines as a function of the distance from New Guinea. Diamond also discovered a relation between species number and elevation, such that, on the average, each 1000 m of elevation on an island enriches the avifauna by about 9% of the avifauna at sea level.

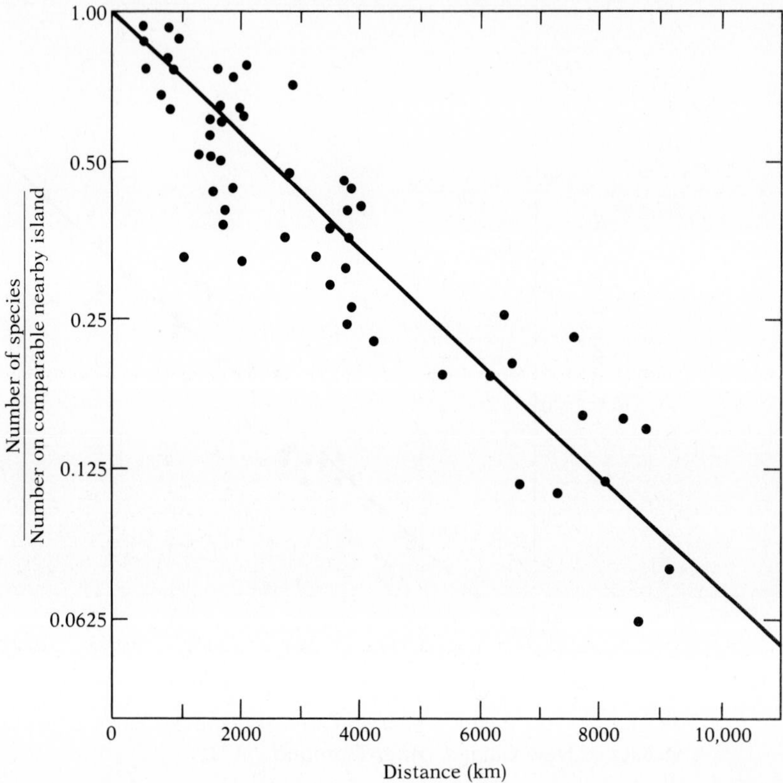

FIGURE 19–7 Relation between distance and species number for birds on islands in the vicinity of New Guinea. A comparable nearby island means an island of the same area that is so close to the source that the distance effect is negligible; the diversity for this nearby island is found using the area relation depicted in Figure 19–6. *(After Diamond 1973.)*

Pattern 5: Relation Between Number of Species and Latitude

The last major pattern affecting whole communities is that the number of species in most groups of organisms increases along a gradient from the temperate zone to the tropics. Simpson (1964) offered an illustration of what has come to be called the *latitudinal gradient in species diversity* with recent mammals of continental North America, as illustrated in Figure 19–8. Similar patterns are found in most taxa (Fischer 1960). One of the few exceptions consists of the brown seaweeds, which have their highest diversity and abundance in temperate waters (Gaines and Lubcheneo 1982).

The latitudinal gradient in species diversity has prompted speculation about its cause (Fischer 1960, Connell and Orias 1964, Pianka 1966). Briefly, the four main conjectures are as follows:

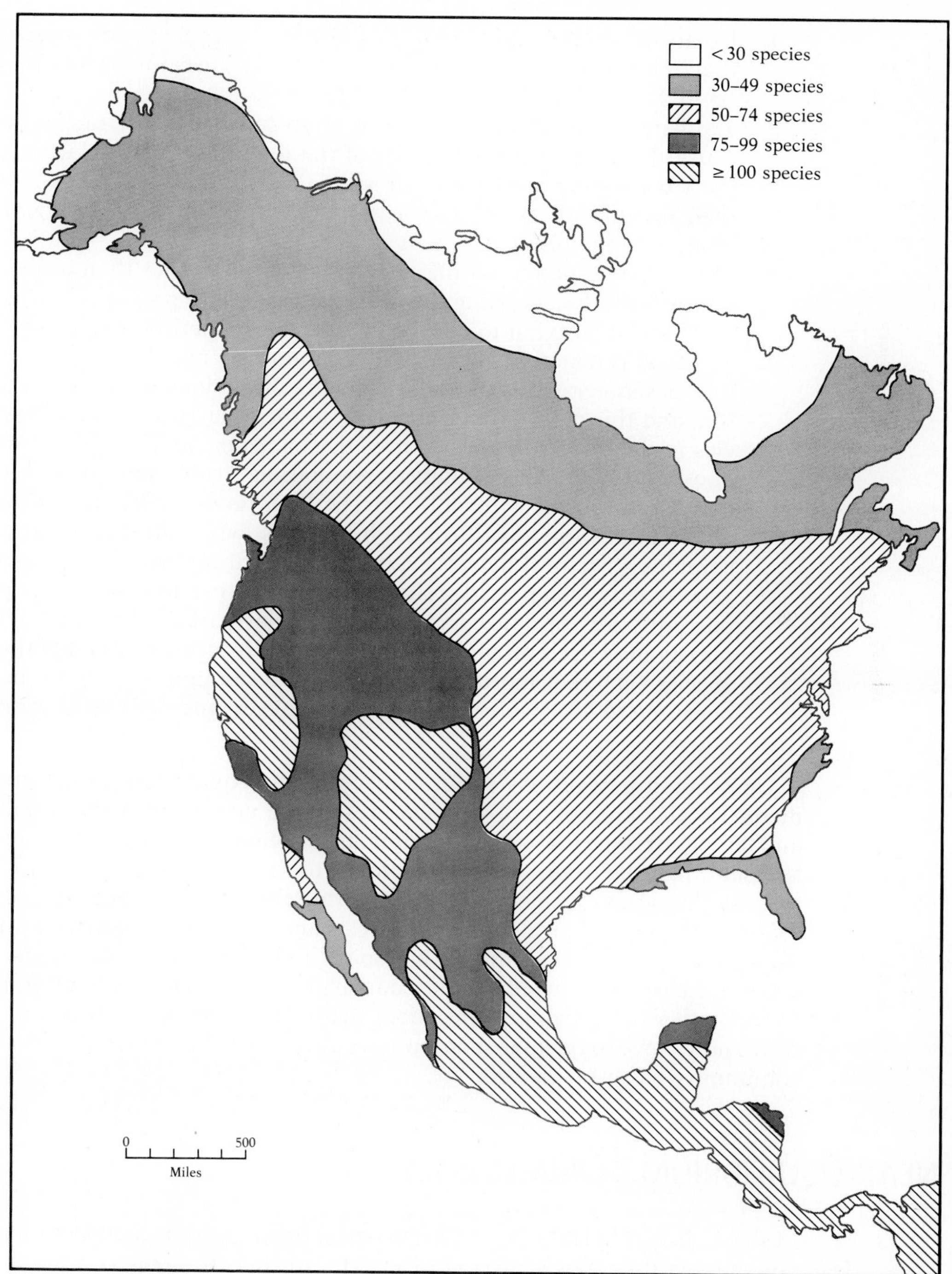

FIGURE 19–8 Latitudinal gradient in species diversity for recent mammals of continental North America. *(After Simpson 1964.)*

1. Because the temperate and arctic latitudes are geologically younger than the tropics, owing to the ice ages, temperate communities may be in a state of accumulating species.
2. Temperate zone species may have, so to speak, broader niches because the fluctuating conditions of the temperate zone relative to the tropics may preclude niche specialization. Fewer species may then be able to "pack" into communities in the temperate zone relative to the tropics.
3. The degree of predation may be higher in the tropics than in temperate latitudes, allowing the coexistence of more prey species that would otherwise fail to coexist because of competitive exclusion.
4. The total productivity of tropical communities may be higher than that of temperate communities because of the longer growing season, and this productivity may lead to the occurrence of more species in the tropics by any of three mechanisms: (a) The higher productivity may allow higher population sizes, thus avoiding extinction (higher K for all). (b) The high productivity of tropical communities may lead to a high biomass, which in turn leads to complex structural heterogeneity in the environment. This heterogeneity may then provide refuges from predation and increased opportunity for specialized niches. (c) Similarly, the structural complexity may itself provide a wider resource spectrum for competing species such that more species may be packed into the community because, so to speak, the "shelf" is longer and will hold more "books."

One should be cautious about conjectures to explain latitudinal gradients because the conjectures often presuppose myths about the tropics and temperate zone that are misleading. Is the temperate zone really more seasonal than the tropics? Many tropical sites have cycles of wet and dry seasons that are as pronounced as temperate zone cycles of hot and cold seasons. The annual productivity of forests and meadows at northern latitudes may easily exceed that of a tropical desert. In short, the communities in a tropical–temperate zone comparison must be matched carefully to each other before conjectures about their differences are well posed.

We now move to the main conceptual approaches with which the entire community has been viewed.

NEAR-EQUILIBRIUM COMMUNITIES

In the near-equilibrium view, an ecological community is a collection of populations that coexist in a steady state. In this view, the community consists of many populations—some competitors with one another, others predators, still others parasites and mutualists, and so forth. The central idea is that community processes consist primarily of all the pairwise interactions among the component populations of the community. Further, these processes are visualized as leading to a steady-state configuration in which the composition of the community and the abundances of the component populations do not change much. What change does occur is hypothesized to be caused by recurring environmental fluctuation. Thus, the community is visualized as a stable entity that is embedded in a fluctuating physical environment. These fluctuations are continually bumping the community, so that its properties are not exactly constant through time, but they are mostly constant.

This view of a community traces to Elton (1927), who emphasized that the number of organisms in each of the populations in a community is continually changing—communities are not static—and that the primary cause of the fluctuations in animal numbers is fluctuation in the environment, as from severe winters and large disturbances such as cyclones. Elton stressed that fluctuations in one population are then transmitted to other populations; as the abundance of prey fluctuates, so does that of their predators. In this view, if the environment should cease fluctuating, the system would settle down to a state of constancy.

In later work, the possibility of internal sources of fluctuation, as from predator–prey limit cycles, was also taken into account. Such cycles would preserve the same species composition through time, even though the abundances of the species would fluctuate. In this view of a community, a key question is how the community responds to the rain of perturbations continually supplied by the environment. Does a little shock cause a big effect, or are environmental shocks absorbed and damped out?

Complexity and Stability

The early thought on how a community responds to fluctuation was that the more complex a community is, the more stable it is. The idea is that a complex community has the possibility of checks and balances, that a perturbation affecting one part of the community will be compensated for by the response in another part. This idea received anecdotal support from the observation that large oscillations, such as the lynx–hare oscillation, occur primarily in the relatively simple animal communities of northern latitudes and that violent insect pest epidemics readily occur in agricultural systems. Little was actually known about oscillations in complex tropical communities, but abundances in tropical communities were presumed to be more constant than those in temperate communities.

Upon reflection, however, we see that the environment itself may fluctuate less at tropical latitudes, and so any lower degree of oscillation there might simply reflect perturbations of lower strength, and not an inherently more stable community. Furthermore, it is now known that substantial oscillations can occur regularly in very complex tropical communities, as documented by outbreaks of the crown-of-thorns starfish (*Acanthaster*) on coral reefs (Glynn 1974, 1976). Nevertheless, the conventional wisdom became that complex communities—that is, those with more species—were more stable than simple ones.

During the last 10 years, this early view has become largely reversed because of mathematical models of communities that originated in studies by May (1973). His studies showed that, in models, a community with many different species is generally *less* stable than a simple one. The reason is easy to understand with a physical analogy. Suppose you are given two wheels from a Tinker Toy set and one stick; obviously, it is easy to connect the two wheels with the stick. Next, suppose you have three wheels and three sticks; again, it is easy to arrange the sticks so that every wheel is connected to every other, the arrangement taking the form of a triangle. Now consider four wheels and six sticks and try to make a structure in which each wheel is connected to all the others. Notice how it becomes increasingly difficult to make the structure; the angles have to be just right or the sticks run into each other, do not insert correctly, and so forth. In short, it becomes extremely difficult to assemble a *completely connected* structure as the number of wheels in it becomes larger.

Moreover, if you do succeed in assembling a complex, completely connected structure, notice that it is rigid. If you push at one end, the push is transmitted across the entire structure. It is not absorbed by the structure, nor is it somehow damped out. Typically, a strongly connected structure communicates perturbations originally delivered at one point to all the other points. In contrast, if you were given many wheels, but had to connect each to only one or two others, you could easily make a spread-out structure. Moreover, a perturbation to one point would have to go a long way before it was felt at the other end. In a loosely connected structure, each point can deal with the perturbations applied specifically to it without also having to contend with perturbations coming to it from many other sources as well. This is the intuition behind the mathematical demonstration that complex communities should not generally fare as well in a fluctuating environment as do simple ones; complex communities are difficult to assemble in the first place and, once assembled, they are generally not more stable to perturbation than simple ones.

The application of this analogy to real communities would suggest that *only* simple communities can persist in highly fluctuating environments. If a complex community were somehow introduced to a highly fluctuating environment, then components of it would become extinct, thereby leaving a simple community. Conversely, in relatively benign environments, the components of a community can accumulate until a rather complex community is attained.

A Test with Food Webs

Food webs are the principal community features that have been analyzed to test whether there is a relation between complexity and stability. A *food web* is a highly stylized description of a community. It pertains only to predator–prey interactions. All the components of the community are scored for the presence or absence of a predator–prey relationship. A table is constructed to indicate the possible pairwise interactions, with the predators ordered along the top row and the prey along the left column. Then a 1 is entered if a predator–prey relation exists and a 0 if it does not. The table of numbers can also be represented as a stick diagram, called a *directed graph,* in which each circle (or rectangle) is a component, and an arrow is drawn from each predator component to each prey component. Figure 19–9 illustrates a food web from a willow forest.

Obviously, a food web barely begins to represent all the interactions in

FIGURE 19–9 Food web from a willow forest. Arrows run from the prey to the predator or from the plant to the herbivore. *(Adapted from Bird, 1930, Cohen 1978.)*

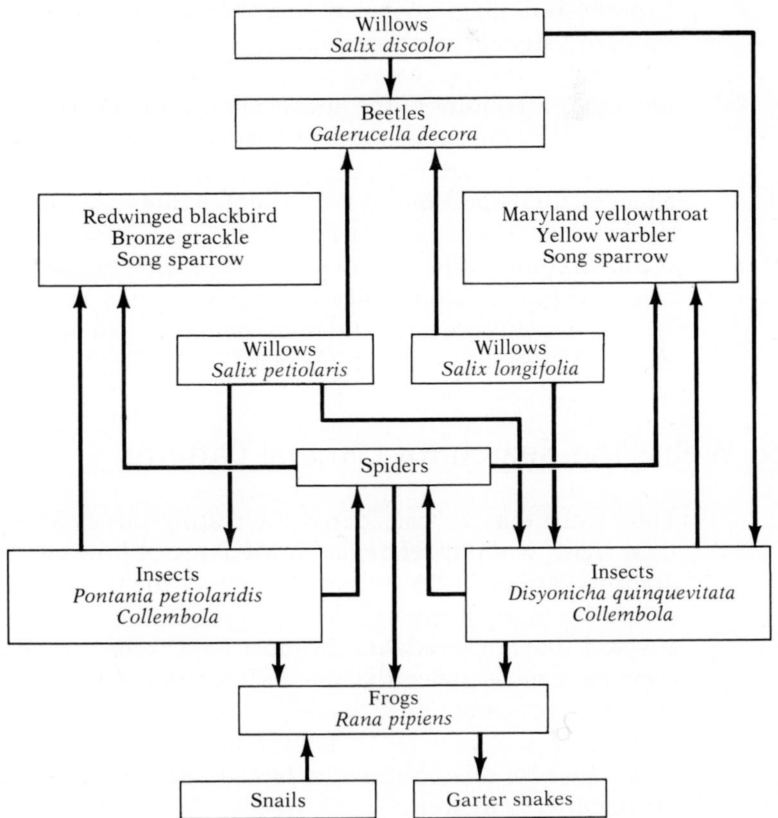

the community, and even the ones it does represent are scored only for presence or absence, and not for strength. Moreover, in practice, no general criteria exist for grouping populations into major components. Some webs have single species as components, and other webs have collections of species as single components; the decision to combine species into a single component is often made on the basis of how much time has been invested in the study of the system, and not on biological grounds. Thus, a food web is an imperfect picture of a community, and its data base is frequently scanty. Nonetheless, one must begin somewhere.

Cohen (1978) and Briand (1983) have collected over 40 published examples of food webs. Briand classified the communities as coming either from greatly fluctuating environments or from rather constant environments. He found that the complexity of webs from the fluctuating environments was much less than that in webs from the constant environments and tentatively interpreted this observation as supporting May's (1983) theoretical prediction that stability decreases with complexity.

The negative relation between diversity and stability discussed earlier pertains specifically to diversity in the species composition of the community. In contrast, McNaughton (1977) demonstrated that the original hypothesis of diversity promoting stability is correct when applied to certain ecosystem properties, not to species composition. As discussed in Part 7, properties of the ecosystem include the total productivity of the plants and other attributes of the flow of energy and nutrients through the entire system. McNaughton showed that perturbations to a community in the Serengeti National Park caused by erratic rainfall produced a smaller effect on total plant biomass in complex communities than on comparable simple ones. The reason for the lower response to the perturbations in the complex communities involved compensation between different plant species; one species would take another's place as conditions changed, preserving a more constant total plant biomass than occurred in communities that lacked such a variety of species.

Food Webs: The Search for General Patterns

Food webs may be considered interesting pieces of information in their own right, without reference to the issue of how complexity is linked to stability. Cohen (1978) initiated much of the recent interest in food webs primarily to see if generalities were present in their architecture. He discovered that the predators in most food webs may be visualized as occupying a niche space that is unidimensional in a special sense. Briand (1983) reports that in 51 webs out of 62, Cohen's observation of unidimensionality is correct. This observation has prompted a search for possible causes of this pattern and has caused investigators to look closer for other patterns as well.

Pimm and Lawton (1983) have summarized the additional patterns that have been detected for food webs as follows:

1. Webs are not too complex. The extent to which species are interconnected in a food web decreases with the number of species in it (Rejmanek and Stary 1979, Yodzis 1980).
2. Food chains are short; typically a food web has three or four trophic levels (Hutchinson 1959).
3. An *omnivore* is defined here as a predator that feeds on more than one trophic level. Omnivores are rare. Scavengers, however, are not rare (Macfadyen 1979, Pimm 1982).
4. A *compartment* is a set of species that interact strongly with each other but not with species of other compartments. Webs are rarely compartmentalized (Pimm and Lawton 1980).

In addition, all food webs satisfy certain obvious requirements—for example, the requirement that every predator have at least one prey. Also, Sugihara (1982) has shown that food webs are "full," in a special sense, and that no "holes" exist in niche space. Discussion about the reality of these patterns, their possible causes, and their biological significance is very active. Nevertheless, these preliminary findings are tantalizing and suggest that major regularities may exist in community organization.

Indirect Effects of Interactions

If a community is a set of interacting populations whose abundances are roughly in equilibrium, then it is possible for pairwise interactions to combine to produce the appearance of interactions between species that are not directly influencing one another. Suppose, for example, that species A interferes with species B, that species B interferes with species C, but that species A does not directly interact with species C at all. Nevertheless, an increase in species A will tend to increase species C because increasing species A will lower the abundance of a species that does negatively affect species C. This case is an instance of the old adage that "he who is the enemy of my enemy is my friend." In terms of community ecology, this type of scheme is an example of what has come to be known as an *indirect effect.*

The concept of indirect effects has become important for three reasons:

1. Indirect effects may be responsible for a spurious indication of competition or other interactions in a perturbation experiment (Holt 1977, Bender *et al.* 1984).
2. Density-dependent natural selection within a species that is a member of a near-equilibrium community may not lead to an increase in the abundance of that species. Indeed, it is theoretically possible

for a species to evolve itself into extinction if it is embedded in a community in which the self-induced indirect effects upon it are strongly negative (Roughgarden 1977).

3. Group selection may act on the indirect effects to produce a closely integrated community (D. S. Wilson 1980). Wilson's examples are based on communities that occur in spatially localized clumps, as in the community that carries out the decomposition of a carcass. It is possible that selection on the level of the entire community can occur according to the picture of group selection that Wilson introduced for single populations, as discussed in Part 2. The intriguing idea emerges that selection at the community level in communities that occur in discrete packages can lead to a functionally integrated set of species. Thus, the early concepts of a community as a *superorganism* might be close to correct for this type of situation.

SPECIES-TURNOVER COMMUNITIES

In 1963, MacArthur and Wilson published a classic paper about the number of species on islands. The paper advanced a view of community structure quite different from the near-equilibrium view just discussed. The authors hypothesized that the number of species on an island remains constant through time as a result of a balance between an immigration rate of new species to the island and the extinction rate for the species already established. According to this picture, the identity of the species continually changes, but the total number of species remains approximately constant. MacArthur and Wilson termed this hypothesis the *equilibrium theory of island biogeography*. The word *equilibrium* here refers to the steady-state value of the *total number* of species on the island; the community is not at, or near, equilibrium with respect to the abundances of the constituent species. This view is essentially a noninteractionist view of a community; each species in the community comes and goes independently of the others, and the community's composition is determined by the rates of arrival and departure.

The Area and Distance Effects

The purpose of the MacArthur–Wilson theory was to explain the relations between the number of species on an island and its area and its distance from the source fauna. Recall the spectacular examples of these relations that have been documented with New Guinea birds by Diamond (1973), which were presented earlier in Figures 19–6 and 19–7. We refer to these relations here as the *area effect* and the *distance effect*.

Here is how the MacArthur–Wilson theory could explain the area effect and distance effect in island biogeography. First, focus on a particular island. Assume that the immigration rate to the island is a decreasing function of the number of species that are already on the island. Indeed, if all the species from the source pool were already on the island, then the immigration rate of new species would be zero by definition. If each species in the source fauna has the same probability of dispersing to the island, then the immigration rate to the island is a linear decreasing function of the number of species already on the island. Second, assume that the extinction rate of species on the island is an increasing function of the number of species on the island. Indeed, if each species has the same probability of becoming extinct, then the rate at which species are becoming extinct is a linear increasing function of the number of species on the island. These curves are drawn in Figure 19–10A. The lines intersect at

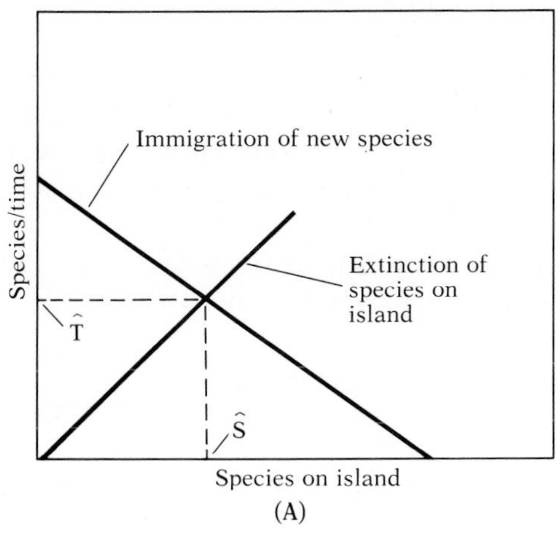

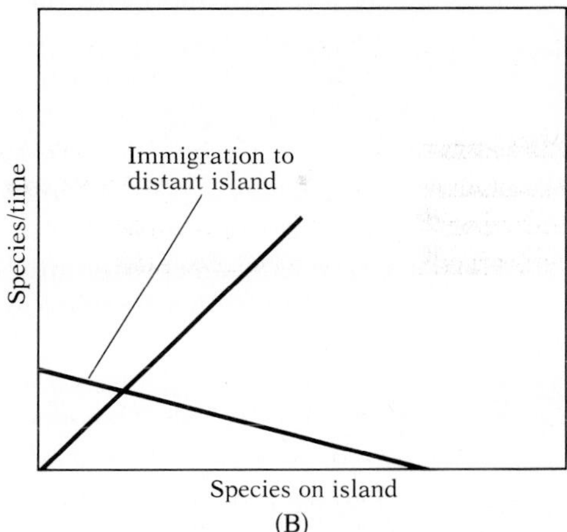

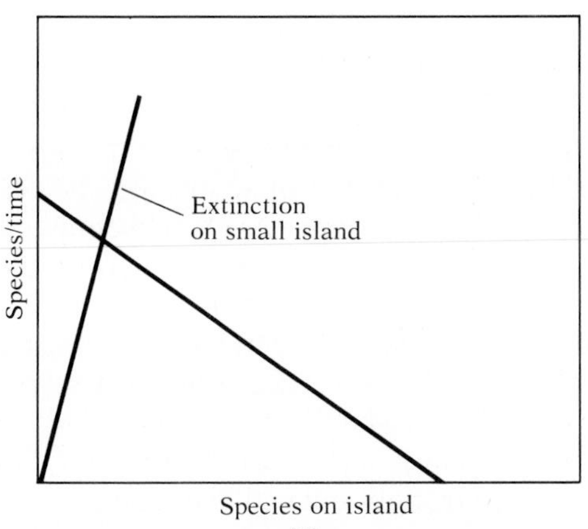

FIGURE 19–10 (A) Hypothetical immigration rate for new species and extinction rate for species already on the island. $\hat{T}$ is the turnover rate at steady state, and $\hat{S}$ is the number of species on the island at steady state. (B) Rates for an island the same size as in (A) but farther from the source. (C) Rates for an island just as far from the source as in (A) but smaller in area.

the point at which the immigration rate equals the extinction rate; the horizontal position of this point is the steady-state species diversity of the island, and the vertical position of this point is called the *turnover rate.* The turnover rate is the value of both the immigration and extinction rates when the steady-state species diversity has been obtained.

The way this model is used to predict the distance effect is to hypothesize that the curve relating immigration to species diversity is lower for distant islands than for islands close to the source fauna, as illustrated in Figure 19–10B. Hence, a distant island attains a lower steady-state species diversity because its input rate is lower. The way to account for the area effect is to hypothesize that the curve relating extinction rate to species diversity is lower for a large island than for a small island, as illustrated in Figure 19–10C. The rationale for this hypothesis is that population sizes should be higher on large islands than on small islands, and so chance extinctions should be less likely on large islands than on small islands. If so, a large island will have higher steady-state species diversity than will a small island because its output rate is lower.

To test this model, one must establish that turnover actually happens on islands. That is, it must be shown that the number of species is remaining approximately constant through time but that the identities of the species are changing. Furthermore, to use the model to explain the area and distance effects, it must also be shown that the immigration curve is lower for distant islands than for islands near the source and that the extinction curve is lower for large islands than for small islands.

Tests of the Turnover Theory

The early tests of the existence of turnover came from experimental systems using microorganisms (Patrick 1967, Cairns *et al.* 1969) and terrestrial arthropods (Simberloff and Wilson 1969, 1970). Simberloff and Wilson used tiny islands formed by mangrove trees in the Florida keys. First, they censused the arthropod fauna on the islands. Then they enclosed each island in a tent, as illustrated in Figure 19–11, and removed all the arthropods with insecticide. Then they observed the recolonization by insects. Figure 19–12 illustrates their data. The number of arthropod species returned to its former level, and the identity of the species clearly changed even after the steady-state species diversity was attained. This study established the reality of turnover in these communities.

The next step was to see what the relation was between island area and distance from the source fauna for the immigration and extinction curves. Gilpin and Diamond (1976) investigated that shape of the immigration and extinction curves for the Solomon Islands and found, first, that the curves were not straight lines but were strongly curved. The rate of extinction on an island as a function of the number of species on the island,

(A)

(B)

FIGURE 19–11 A mangrove island covered by a tent for the experimental application of insecticide. (A) Construction of scaffold around a mangrove island, Florida. (B) Completed tent around the mangrove island. *(Photos by Daniel Simberloff, Florida State University.)*

$E(S)$, varies approximately as $S^{2.5}$. The rate of arrival of new species to the island as a function of the number already on it, $I(S)$, varies approximately as $(1 - S/P)^7$, where P is the number of species in the source fauna. Thus, $E(S)$ rises faster with S than a straight line, and $I(S)$ declines slower with S than a straight line. (These functions are called *convex*.)

Second, Gilpin and Diamond (1976) examined how the $E(S)$ and $I(S)$ curves with these shapes varied among different islands, depending on

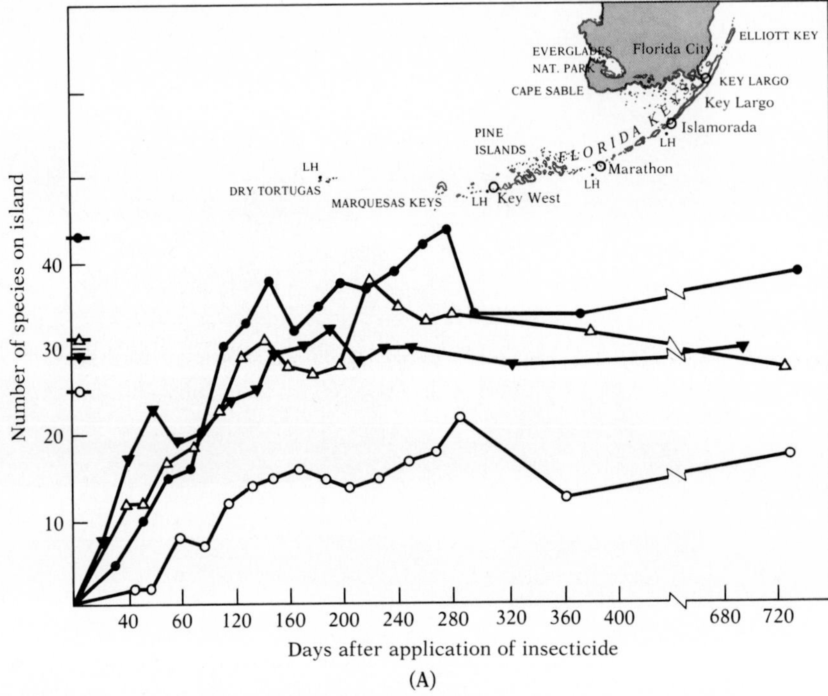

FIGURE 19–12 Recolonization of a mangrove island. (A) Number of species through time on four different tiny islands. Symbols at the vertical axis indicate the abundance on each island before the experiment. (B) (Facing page) Different species come and go during the experiment, demonstrating turnover in the fauna. Each horizontal line shows the history of a particular insect species on an island. *(After Simberloff and Wilson 1970.)*

island area and distance from the source fauna. They found that, for a given S, the $E(S)$ curve varied inversely with island area, A. Also, for a given S, the $I(S)$ curve varied with distance from the source fauna, D, as $\exp(D^{\frac{1}{2}})$. When all this information was combined, the number of bird species on each of the Solomon Islands could be predicted very accurately.

The Significance of Turnover

The finding that $E(S)$ and $I(S)$ are not linear functions of S implies that the species cannot be assumed to have identical and independent rates of arrival and chances of extinction. Two interpretations of the nonlinearity of the extinction and immigration curves are possible.

First, the nonlinearity may indicate that the species differ in their immigration and extinction rates but that turnover still occurs. For example, birds that live primarily on the ground are less likely to disperse than

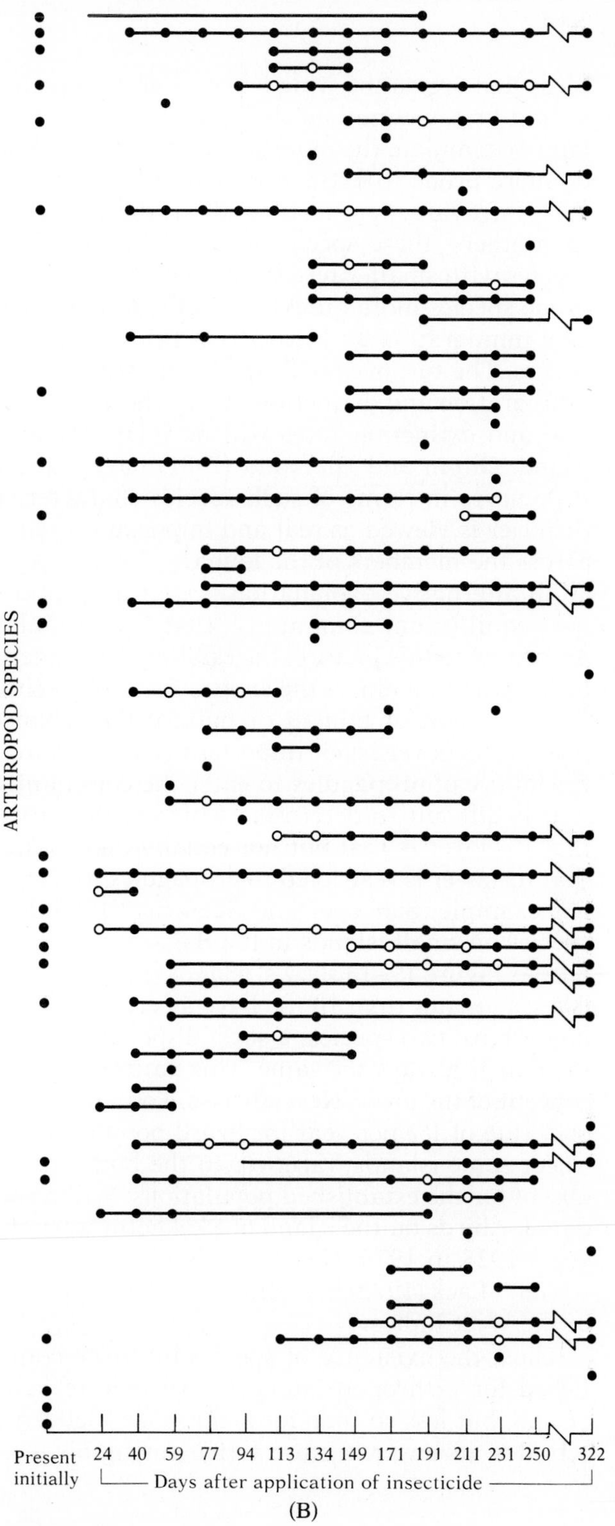
ARTHROPOD SPECIES
Present initially
24 40 59 77 94 113 134 149 171 191 211 231 250 322
Days after application of insecticide
(B)

birds that are on the wing for most of the day. If so, the fauna on an island should first accumulate the species that are prone to dispersal and only later accumulate the more sedentary species. Similarly, some species may be more prone to extinction than others. Some species, such as birds of prey, have large territories and so inherently have lower population sizes than others; these species might be more susceptible to extinction than species with smaller territories. If so, the fauna should lose the extinction-prone species more quickly than the others. The net effect of species-specific immigration and extinction rates is to lead to convex $E(S)$ and $I(S)$ curves. The turnover will involve mostly the species with relatively high immigration and extinction rates, whereas the species with low immigration and extinction rates will be relatively permanent residents on the island. Gilpin and Diamond (1981) have explained the convex $E(S)$ and $I(S)$ curves in terms of such species characteristics. With this approach, turnover is viewed as real and important, though not equally distributed across the members of the fauna.

An alternative explanation is that an island community is actually a near-equilibrium community, that it accumulates fauna until it is full. According to this picture, the established residents are effectively immune from extinction unless the environment itself changes over geological time or as a result of human or natural disturbance of the habitat. In this picture, turnover is not important. What turnover there is represents only the failure of propagules to enter the community after it has filled up.

It is difficult to determine which of these two hypotheses is correct—that turnover is real but not equally distributed over the community or that turnover is restricted to propagules that arrive after the island is full. For example, Schoener and Schoener (1983) have shown that the lizard species on small islands in the Bahamas form what are called *nested subsets.* As Figure 19–13 shows, islands less than 10^4 ft^2 have no lizard species; islands greater than 10^4 ft^2 have a certain species. Islands 6×10^4 ft^2 and larger have two species; a second species has been added to the first, and the pair is always the same. This pattern is difficult to reconcile with the concept of turnover. Nevertheless, Toft and Schoener (1983) report a turnover rate of 1% per year for lizard populations in the Bahamas.

For some islands, turnover in the bird community is almost certainly absent for the established populations. Williamson (1983) presents census data for birds on the island of Skokholm west of Wales in England for the years 1928 to 1979. He shows that what turnover exists is "ecologically trivial." Lack (1976) had offered a similar view for island birds, especially in the West Indies.

Thus, the existence of species-turnover communities has been established for arthropod fauna; for vertebrate fauna, species turnover also occurs, but less so than for arthropods (Schoener 1983b). Among the vertebrates, turnover is more active among birds than among lizards.

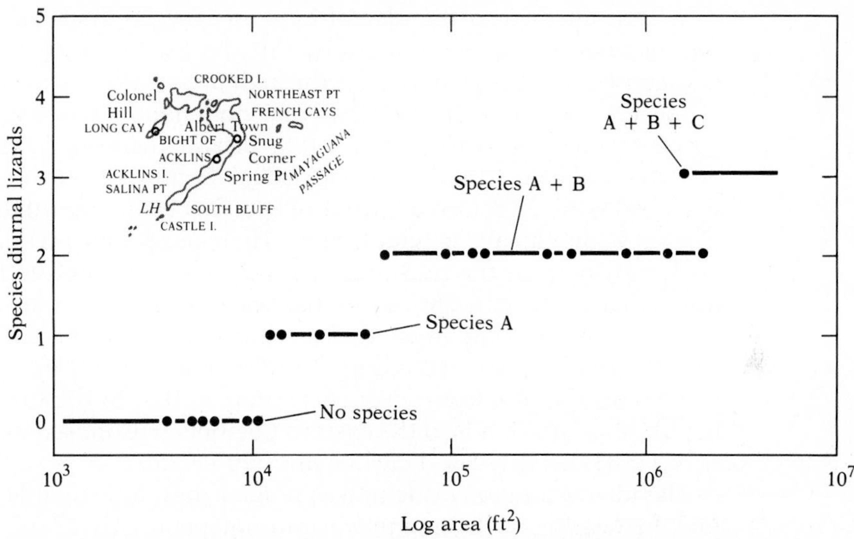

FIGURE 19–13 Relation between island area and number of lizard species on islands in the Bahamas. The species occur as nested subsets. From islands in Lovely Bay on the Crooked-Acklins Bank in the Bahama Islands. *(After Schoener and Schoener 1983.)*

QUASI-EQUILIBRIUM COMMUNITIES: THE TAXON CYCLE

The third view of a community is in effect a mixture of the two views of communities just discussed. Here turnover takes place, but it is slow and is caused by a different mechanism than a balance of immigration and extinction. The idea is that, over evolutionary time, the species in the community are changing, and this coevolution often leads to an extinction. The community can then be reinvaded by a new species, restoring the species diversity to its previous level. Thus, turnover appears in this picture, *but* it is slow and is brought about by the evolutionary change of the species in the community rather than by demographic stochasticity. In this view, at any one time the ecological state of the community is more or less in equilibrium given the current phenotypes of the species in the community. As those phenotypes change, the ecological state changes too; the ecological state dynamically "tracks" the evolutionary state. Thus, in this picture, the ecological state of the community can keep pace with the slow changes that are occurring through evolution. When this evolution gradually results in species phenotypes that are incompatible with coexistence, a species drops out, leaving a community with a lower diversity that is now ripe for invasion by a new species. The cycling of species in this way was termed a *taxon cycle* by Wilson (1961).

Wilson introduced the idea of a taxon cycle to describe the biogeography of ants on Melanesian islands in the South Pacific. He noted that taxonomic lineages seem to pass through phases. At first, a lineage consists of species that occupy primarily a marginal habitat in a source fauna. The species then possess phenotypes that permit their survival in conditions of water and heat stress; they also occupy habitats near the ocean, so their dispersal is likely relative to that of species occupying inland habitats. At this time, the lineage is widespread. Then, as species invade islands, they evolve away from the marginal habitat to a more productive forest habitat. Finally, the lineage comes to consist of species that have relictual distributions, and the lineage as a whole appears to be contracting, even as other lineages are expanding into the now vacated marginal habitat.

The concept of a taxon cycle is similar to that of the turnover of lizard populations on islands in the eastern Caribbean proposed by Roughgarden *et al.* (1983), as discussed earlier in Chapter 16.

The idea of a taxon cycle makes it clear that, in principle, a community may be near or far from equilibrium independently at different levels. A community may be near equilibrium relative to the abundance of its member populations, to the phenotypes in its member populations, and to the ability of other species in the region to invade successfully. These levels correspond, respectively, to local ecology, local evolution, and biogeography.

In Parts 6 and 7, which follow, we move to a still larger scale of investigation—the entire globe. Why are different communities present in various regions of the earth? To answer questions on this scale, we need to think of the entire planet, what determines its climate, how the climate constrains the kind and complexity of a community, and how a community itself can affect the processes that determine the climate in a region.

PART 6

THE DISTRIBUTION OF COMMUNITIES

Communities differ in space and time. Variations in space, as between grasslands and forest or ocean and shore, are obvious. Variations in time, however, have been appreciated only recently; in the mid-nineteenth century, for example, most educated people did not know that the Earth's communities were extremely different 100 million years ago from those of today.

In the three chapters of Part 6, we will discuss the broad geographical and temporal aspects of communities. First, in Chapters 20 and 21, we will consider the distribution of contemporary terrestrial and aquatic communities and the factors responsible for that distribution. Here our knowledge seems reasonably secure, for we can map the occurrence of organisms and study their relationships to various factors in their environments. However, when we turn to the distribution of communities through geological time, as we do in the latter part of Chapter 22, we must rely on inferences from the fossil record and on what geophysicists have discovered about the physical history of the Earth. However, as we shall see, reconstructing the ecology of the past can be a fascinating activity.

CHAPTER 20

Modern Terrestrial Communities: Life on Land

Ecologists divide communities rather arbitrarily into broad structural classes called *biomes*. The basis of this classification is largely the growth form of the dominant plants and the climatic regime to which those plants are adapted (see, e.g., Holdridge 1947, 1967, Lieth 1956, Dansereau 1957). Thus, very warm, moist areas occupied principally by tall evergreen trees are classified as the tropical rain forest biome. Cooler, drier areas dominated by tall deciduous trees form the temperate forest biome. The geographic distribution of major biomes is shown in Figure 20–1A, and the relationship of those biomes to the major climatic variables of temperature and humidity is shown in Figure 20–1B.

CLIMATE

To understand the distribution of biomes, one must know something about the Earth's climatic machinery. That machinery primarily is the blanket of gases, the atmosphere, that surrounds our planet and its oceans. The energy that drives the weather machinery comes from the sun. Differential heating of the atmosphere is responsible for both the short-term changes humans perceive as *weather* and the long-term patterns perceived as *climate*.

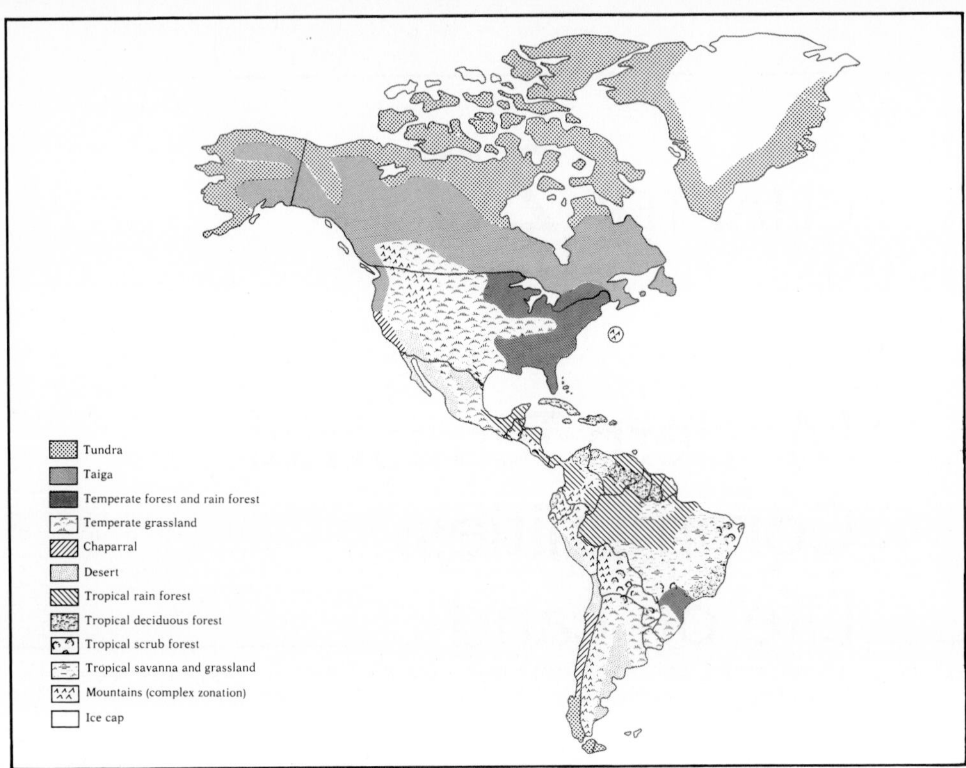
Tundra
Taiga
Temperate forest and rain forest
Temperate grassland
Chaparral
Desert
Tropical rain forest
Tropical deciduous forest
Tropical scrub forest
Tropical savanna and grassland
Mountains (complex zonation)
Ice cap

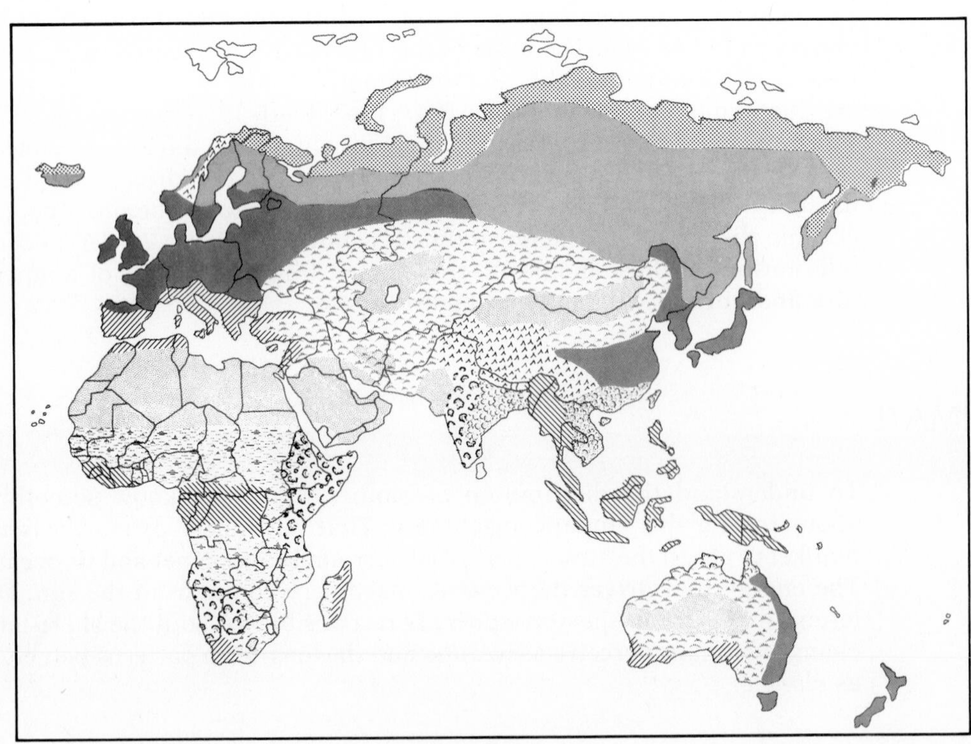

(A)

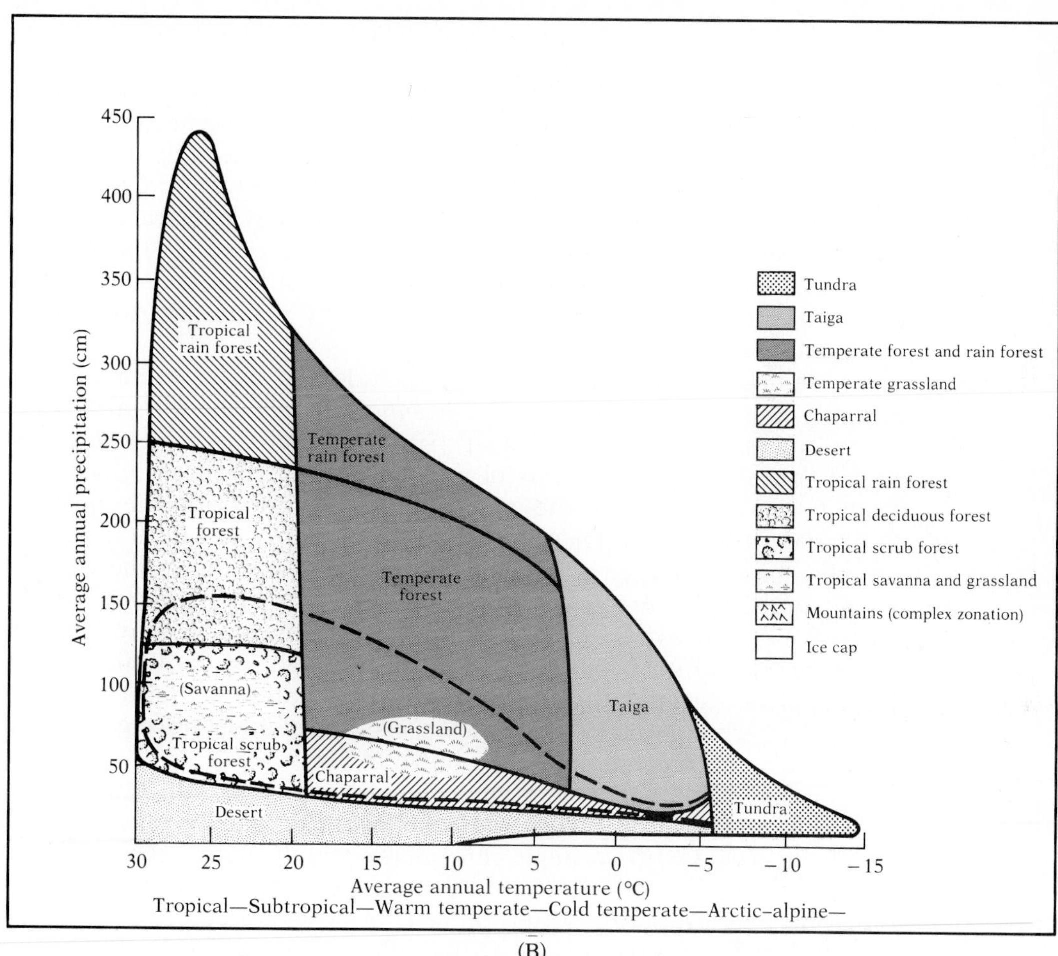

(B)

FIGURE 20–1 (A) *(Opposite)* Distribution of major biomes. (B) The temperature and precipitation conditions that give rise to those biomes. *(After Odum 1971, Whittaker 1975.)*

Climate is not simply a given to which organisms (and thus biomes) react. Earth's biota itself influences the differential heating and thus climatic patterns. Earth's biota affects the composition of the atmosphere, both its component gases and the liquids and solids carried in it. It also affects the extent to which the surface reflects or absorbs solar energy. The technical term for the reflectivity of an object is its *albedo*; thus, the biota influences Earth's albedo.

The interactions of incoming solar energy with the Earth's atmosphere, physical surface, and biota are only partially understood (for more details, see Ehrlich *et al.* 1977a, Schnieder and Londer 1984). Here we present a broad outline and some high points as an introduction.

In a volume of dry air, 78% of the molecules are nitrogen and 21% are oxygen. Climatically, the most important constituents of dry air are the remaining 1%. Carbon dioxide (about 320 ppm), ozone (O_3, about 0.01 ppm), and water content affect the energy balance of the planet, and the concentrations of these key constituents are all influenced by organisms. Carbon dioxide is released into the atmosphere by respiration and decay and removed by photosynthesis. Ozone is produced by physical processes from oxygen, and the atmosphere's oxygen content is entirely the product of eons of photosynthesis. Water vapor is removed from the surface and subsurface and injected into the atmosphere by plants and, to a much smaller degree, by animals.

The climatic importance of carbon dioxide and water vapor may be seen by their roles in the so-called *greenhouse effect,* the formation of a warming "blanket" over the planet's surface. Solar energy reaching the top of the atmosphere is a mix of wavelengths ranging from ultraviolet (0.10–0.40 μm) through the visible (0.40–0.71 μm) to the infrared (0.71–100 μm). The shorter wavelengths of the ultraviolet (0.10–0.30 μm) are absorbed by oxygen and ozone and do not reach the surface. Ultraviolet in the 0.30- to 0.34-μm range is partly absorbed by the ozone, whereas near ultraviolet (0.34–0.40 μm, next to the visible wavelengths) is transmitted almost completely. Most of the light in the visible wavelengths passes through the atmosphere, whereas all of the infrared is absorbed by carbon dioxide, water vapor, ozone and other trace gases. The fate of incoming solar radiation is shown in Figure 20–2.

About half of the solar radiation incident on the top of the atmosphere penetrates to the surface. There the solar radiation may be reflected, and the albedo varies with the angle of incidence and the nature of the surface (water, bare rock, grass, pine needles, and so on). Alternatively, the solar radiation may evaporate water, or melt or sublime ice or snow, or be absorbed and warm the surface. Or the solar radiation may be part of the tiny fraction that is captured by green plants in the process of photosynthesis. All in all, some 69% of the solar energy hitting the planet is absorbed by the Earth–atmosphere system. This absorbed solar energy runs the climatic system, the hydrologic cycle, and the lives of virtually all organisms.

Because the planet is not continually getting hotter, energy must be returning to space as *terrestrial radiation.* It is possible to compute, using the physical laws governing electromagnetic radiation, what the temperature of the planet must be in order to radiate away that 69% of the solar radiation that is absorbed. That temperature is about −18° (roughly 0°F). Why, then, is the average surface temperature about +15°C (59°F)? The answer is that carbon dioxide, water vapor, clouds, and, to a lesser extent, ozone and other trace gases absorb the Earth's infrared radiation. They are themselves warmed and then reradiate the energy—some toward space, but some back toward Earth. Therefore, the atmosphere acts as a

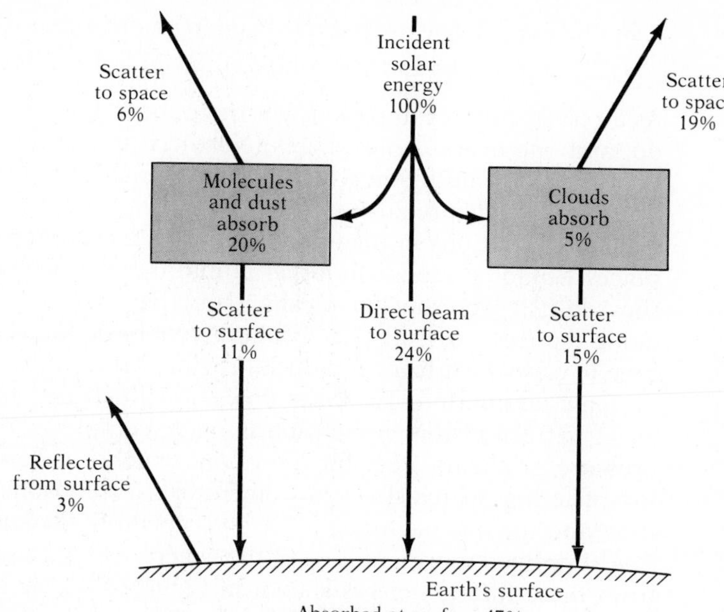

FIGURE 20–2 Fate of incoming solar radiation; percentages are annual global averages. *(After Ehrlich* et al. *1977a.)*

thermal blanket, maintaining the surface at a higher temperature than it would be were there no atmosphere. On the average, the −18° temperature predicted by the calculations is found in the atmosphere at an altitude of about 5 km above the solid surface.

The blanketing effect of atmospheric components is most widely recognized in the commonplace observation that cloudy nights tend to be warmer than clear nights. When the sky is covered with clouds, more of the terrestrial radiation is absorbed and reradiated back toward the surface than when the sky is clear. The term *greenhouse effect* is used for this blanketing phenomenon because glass, like carbon dioxide and water vapor, tends to absorb infrared radiation. However, the term is not entirely appropriate, for most of the heating in a sunny greenhouse results from the inability of wind to remove the heat from inside.

Water vapor plays another significant role in climate. This vapor condenses into clouds, which are critically important in climatic machinery because they account for about two-thirds of Earth's albedo.

Different parts of the planet have different climates, because the incoming solar energy does not heat the Earth–atmosphere system evenly. The amount of solar radiation that strikes the top of the atmosphere varies at different latitudes and at different times of the year. The basic reasons for these differences are as follows:

- The Earth is a sphere.

- The Earth travels around the sun in an orbit that is not quite circular.
- The axis of the Earth's spin is tilted 23.5° from the plane of its orbit.

As a result, equatorial regions get more incoming energy than polar ones do, and one of the hemispheres (Northern or Southern) gets more than the other, depending on which is canted toward the sun at any time, as illustrated in Figure 20–3.

These variations in the amount of solar radiation received at different places would create much larger climatic differences at the surface than they do if it were not for mechanisms that transfer energy horizontally along the surface. In the atmosphere, two basic kinds of movement combine to carry heat from equatorial regions toward the poles: thermal and cyclonic circulation. These air flows are illustrated in Figure 20–4. The exact pattern of this circulation is made complex by the interactions of pressure gradients; coriolis deflection, which is a result of Earth's rotation; friction, which slows movement of the atmosphere close to the surface; and the diverse topography and albedo of the surface.

A simplified representation of the general circulation patterns resulting from these interactions is shown in Figure 20–5. As an example of how these patterns affect the distribution of biomes, note the circulation patterns called *tropical Hadley cells.* They indicate the rising of warm, moist air over the tropics. As the air rises, it cools, and some of its moisture falls as rain. As a result, latent heat of vaporization is released, warming the air again and driving it higher, which leads to further cooling, more rain, reheating, and so on, until most of the moisture has been removed from the air. The result is a very rainy climate in much of the tropics and a high-altitude poleward flow of dry air. As this air moves northward, it radiates its heat into space more rapidly than it acquires it either from solar radiation or from the warmer atmospheric layers below.

At a latitude of about 30° north and south, the now dry, relatively cold air starts to sink into areas of high pressure. In the process, this air is warmed by compression, producing a cascade of warm, dry air about 30° on each side of the equator. The result is a band of deserts at these latitudes; the Sahara of northern Africa, the Kalahari of southern Africa, the Atacama of Chile, and the Sonoran desert of northern Mexico and the southern United States are major examples.

Broad climatic patterns, and thus patterns of biome distribution, can be understood by considering the general circulation pattern, and the ways in which local climates are modified by topographic features and the heat transport activities of the oceans. (The oceans have a large capacity for storing heat, and their currents transfer it horizontally.) For example, areas downwind of mountain ranges receive less precipitation than those on the upwind side. As air flows upslope, it is cooled, and moisture condenses and falls as rain. The air then has less moisture content as it flows over the crest, and a *rain shadow* is created. Sacramento,

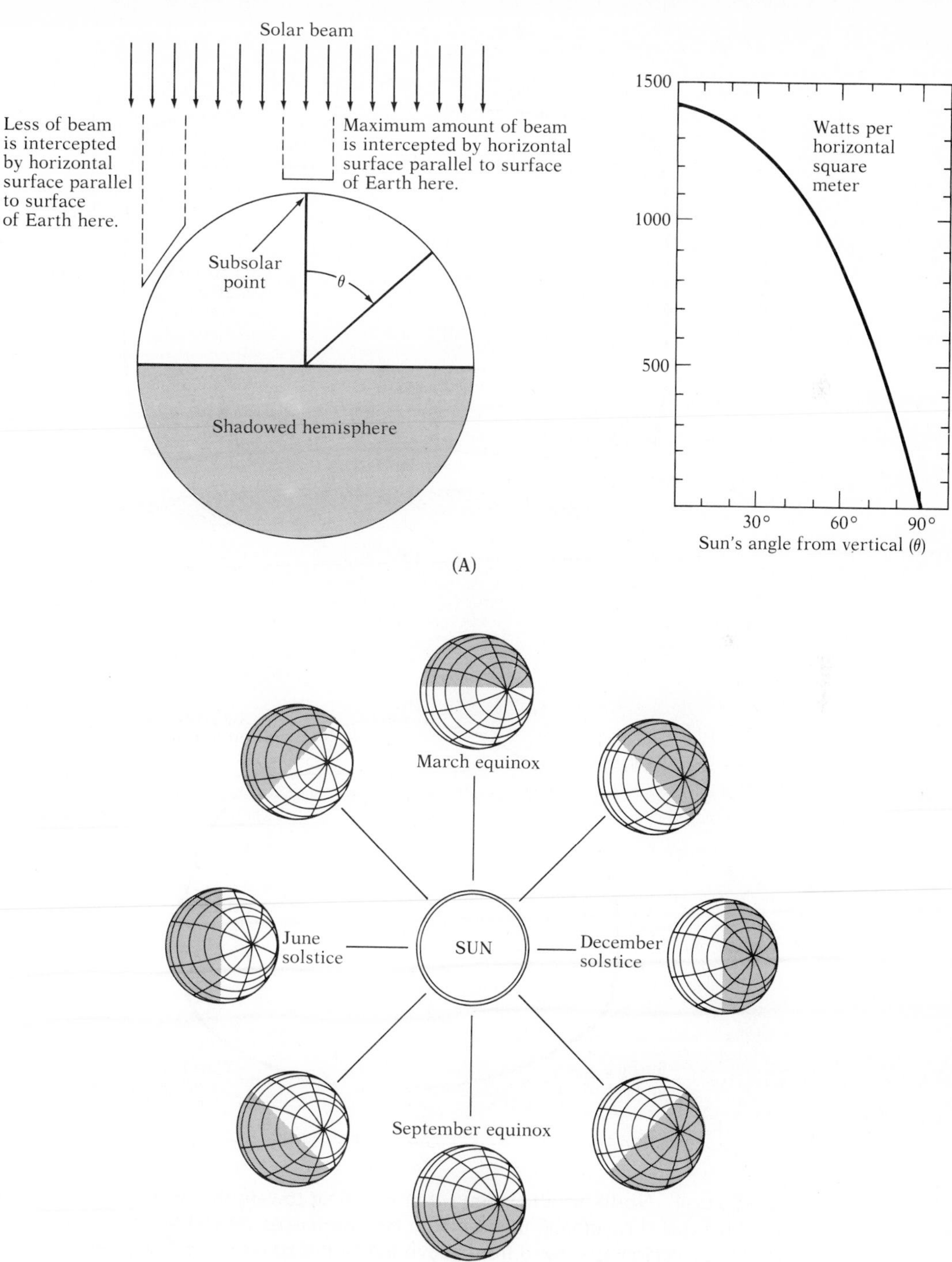

FIGURE 20–3 Differential reception of solar radiation (A) with latitude (angle of sun from vertical); and (B) by seasonal variation. *(After Ehrlich* et al. *1977a.)*

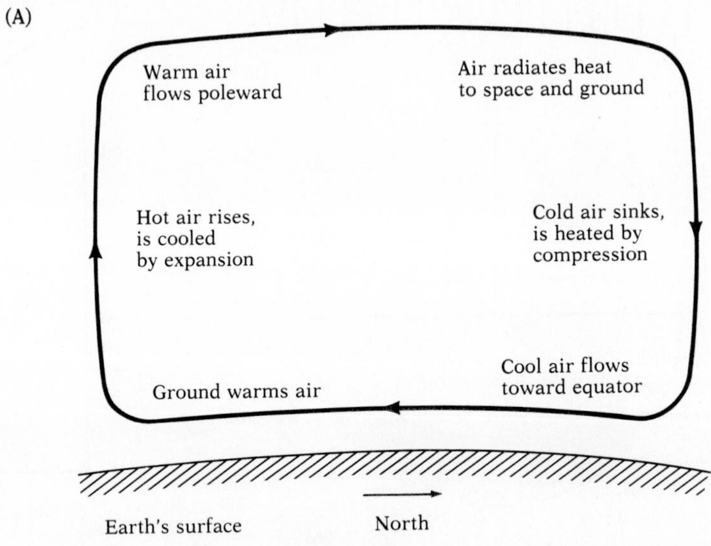

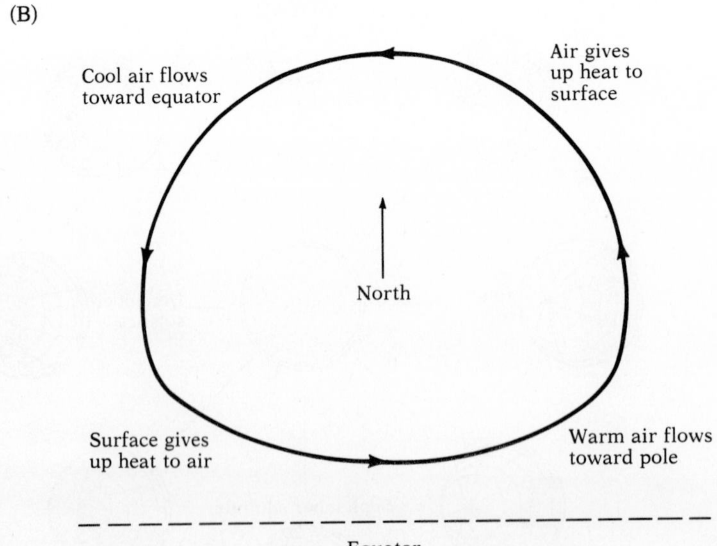

FIGURE 20–4 North–south air flows transport heat toward the poles. (A) Cross section of thermal circulation in the Northern Hemisphere as viewed facing the west. (B) Cyclonic circulation as viewed from above the Earth's surface in the Northern Hemisphere; a low pressure zone is in the center, and circulation is counterclockwise. *(After Ehrlich* et al. *1977a.)*

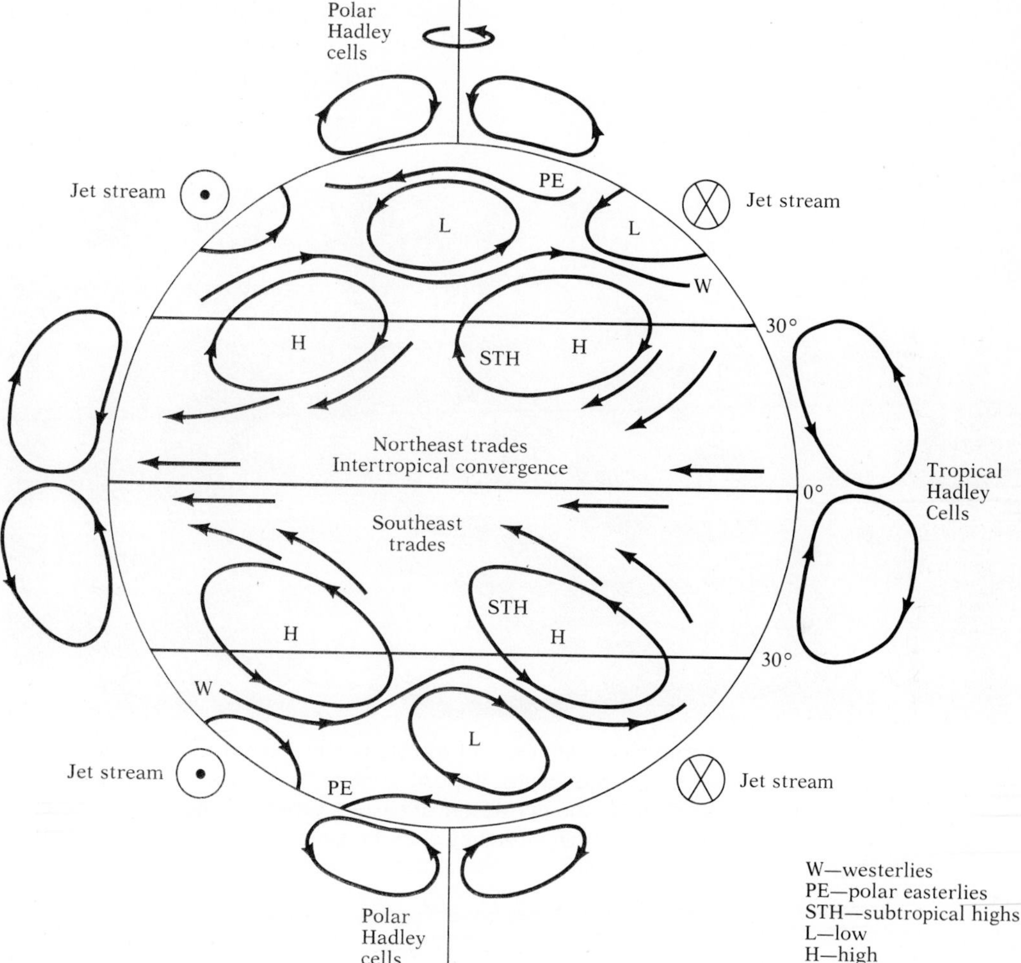

FIGURE 20–5 General air circulation patterns, showing flows near the surface and cross sections of the principal upper troposphere circulations. Circled dot indicates jet stream coming toward viewer; circled X away from viewer. *(After Ehrlich* et al. *1977a.)*

California, on the west (upwind) side of the Sierra Nevada receives about 44 cm of precipitation annually; Reno, Nevada, in the east-side rain shadow, receives less than half of that, about 18 cm (see Figure 20–6).

Referring back to Figure 20–1, note that temperate forests occur much farther north in Europe than they do on the eastern coast of North America (one must remember that England is at the same latitude as Labrador). The reason is that the Gulf Stream, a flow equivalent to some 50 times the flow of all of Earth's rivers, carries heat from the equatorial Atlantic northeastward toward Europe. In contrast, New England is bathed in the cold waters of the Labrador current, which flows southward from between the icy coasts of Greenland and Baffin Island.

FIGURE 20–6 Approaching the crest of the Sierra Nevada from the east in August. Snow can be seen on the mountains, deposited there by moisture-laden air from the Pacific. In the foreground is the dry desert of eastern California, the start of the extensive Sierran rain shadow that stretches over the Great Basin of Nevada and Utah. *(Photo by Paul R. Ehrlich.)*

THE MAJOR BIOMES

The number of major community types (biomes) recognized by ecologists is somewhat arbitrary. Here we will discuss six major terrestrial biomes—ones that each cover substantial portions of Earth's land surface. These biomes are:

- Tropical rain forest
- Temperate forest
- Desert
- Grassland
- Taiga
- Tundra

Tropical Rain Forest

The most species-rich, complex communities are tropical rain forests; a section of one is shown in Figure 20–7. These forests are found in equatorial regions in which the annual rainfull is more than 240 cm, spread relatively evenly throughout the year, and the average annual temperature is more than 17°C, so that neither temperature nor the availability of water is a limiting factor. Tropical rain forests recently covered much of northern South America, Central America, western and central equatorial Africa, the east coast of Madagascar, Southeast Asia, and northwest coastal Australia, as well as various islands in the Indian and Pacific oceans (Richards, 1952). In recent decades, their area has been undergoing rapid reduction (National Research Council, 1980). These changes have been the subject of both great concern and intensified research by biologists (e.g., Leigh *et al.* 1982, Prance 1982, Janzen 1983a, Sutton *et al.* 1983). For example, in Madagascar, only small fragments of once extensive tropical forests persist.

FIGURE 20–7 Interior of tropical rain forest in Corcavado National Park, Costa Rica. Note the relative lack of undergrowth (compare with Figure 20–10). *(Photo by Paul R. Ehrlich.)*

The structural complexity of these forests can be seen in the usual existence of trees of many heights forming several layers (Richards 1952). The tallest trees reach heights of 60 m or so and emerge above the tops of the lower trees, which are intertwined to form a dense *canopy* through which little light penetrates, as depicted in Figure 20–8. The rain forests

FIGURE 20–8 Diagram of the structure of a tropical rain forest . The canopy is essentially closed, greatly reducing the light penetrating to the understory.

FIGURE 20–9 Epiphytes in a Costa Rican rain forest. The two plants with swordlike leaves with the light showing through them are bromeliads—members of the pineapple family, which is rich in epiphytic species. *(Photo by B. J. Miller, Fairfax, Virginia/BPS.)*

are also characterized by a wide variety of epiphytes—plants that grow high on the trees in the sunlit zone and whose roots do not reach the soil, an example of which is shown in Figure 20–9. They also have abundant lianas, woody climbing "vines" whose roots are in the soil but whose leaves are in the canopy.

The undergrowth varies a great deal from place to place but tends to be dense only where a tree fall has created a hole in the canopy or along stream courses—that is, wherever the light-blocking effect of the canopy is removed, as shown in Figure 20–10. Because the trees do not ordinarily branch below the canopy, a person often may walk almost unobstructed through the dim depths of a mature rain forest. In contrast, the forest's edge is usually an impenetrable tangle of shrubs, vines, and young trees.

In many places, the tree diversity—that is, the number of species—is so high that more than 50 species per hectare may be present. They tend to have relatively smooth bark and similar hand-sized oval leaves, often with their midribs prolonged into *drip tips,* as shown in Figure 20–11, which speed the drainage of water and probably help prevent the growth

(A) (B)

FIGURE 20–10 Tropical rain forest in Corcavado National Park, Costa Rica, showing the increased undergrowth in which light penetrates at the edge of a clearing (A) and along a stream (B). *(Photos by Paul R. Ehrlich.)*

of algae and lichens on the leaves. On the other hand, the trees are diverse in flower type and timing, and in pollination and fruit-dispersal mechanisms. Most have shallow roots, and many develop huge buttresses for support, as can be seen in Figure 20–12.

Rain forests tend to resemble each other closely in general botanical features, such as the presence of lianas, the shape of leaves, and the extent of buttressing (Richards, 1973). However, different areas tend to be quite distinct in the details of floristic composition—that is, in the species and genera of plants present and in the relative abundances of genera and families. In these respects, for example, the floras of the tropical rain forests of America and Africa are quite distinct (Smith 1973, Thorne 1973).

Animal life in tropical rain forests is also very diverse, with insects, amphibia, reptiles, and birds especially well represented. For example, about as many butterfly species can be found in a single rain forest locality

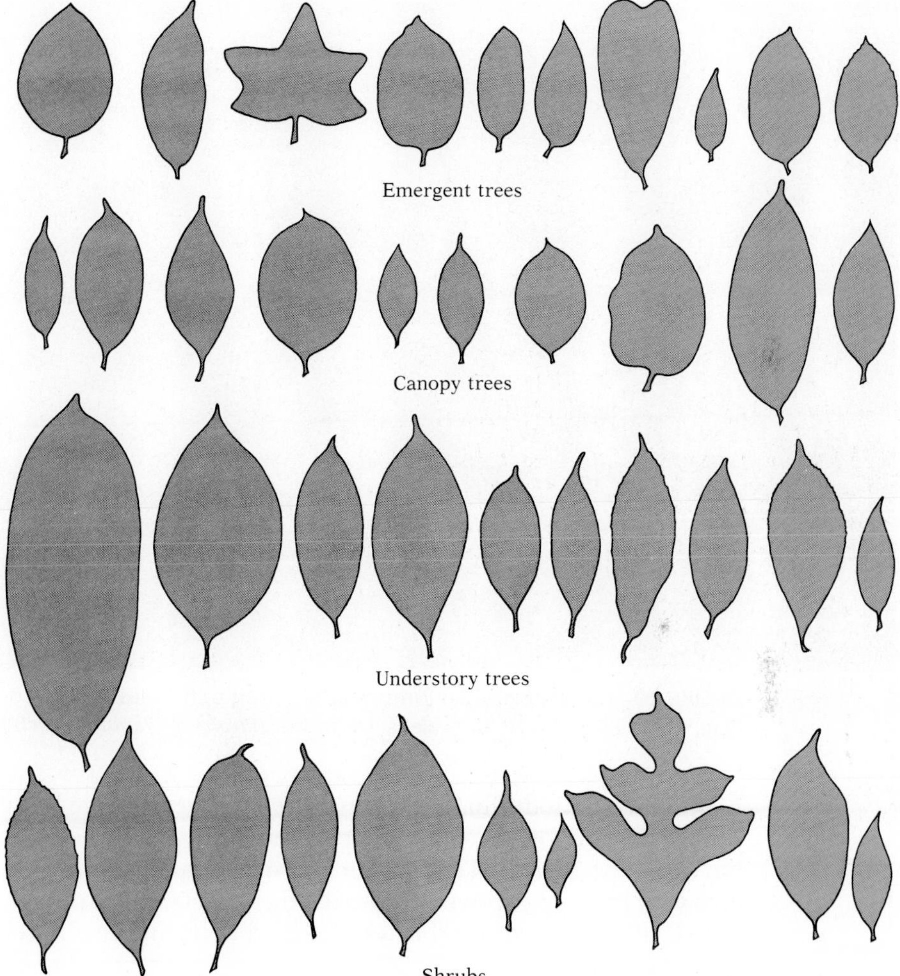

FIGURE 20–11 Leaf outlines of tropical rain forest plants. Note the pointed drip tips on most of them. These features presumably speed the removal of water from the leaves that might otherwise encourage the growth of epiphytes. *(After Richards 1952.)*

as occur in the entire United States—perhaps 500 or 600—including whole groups unknown or poorly represented in the temperate zones. In contrast, large mammals tend to be relatively scarce although monkeys may be important herbivores. Mimicry and protective coloration are pervasive.

Like the floras, the faunas of tropical rain forests in different areas tend to be distinct in details. Thus, although butterflies are abundant (at least in the canopy and forest edge areas), with some exceptions the dominant genera tend to be different in America, Africa, and Asia. For example, the

FIGURE 20–12 Tropical rain forest tree showing buttresses. Corcavado National Park, Costa Rica. Figure for scale is 1.6 m tall. *(Photo by Paul R. Ehrlich.)*

metalmarks (Riodininae) are a major component of the butterfly fauna of American rain forests, perhaps represented by 100 genera; African forests have but two genera. The African and American rain forest ant faunas are also very distinct (Brown 1973). Like the insects, the African and American faunas of terrestrial vertebrates also tend to be highly differentiated (Meggers *et al.* 1973).

In spite of these differences, the faunas of rain forests tend to show much convergent evolution. For example, many different groups of butterflies have evolved iridescent blue upper surfaces combined with cryptically colored under surfaces. A visual predator is confronted with a brilliant, periodically flashing blue surface as the insect flies through the dappled sunlight—a vision that disappears when the butterfly lands and

---

FIGURE 20–13 *(Opposite)* Convergence in mammals of tropical rain forests in South America (*left*) and Africa (*right*): (A) capybara and pigmy hippopotamus; (B) paca and African chevrotain; (C) agouti and royal antelope; (D) brocket deer and yellow-back duiker; (E) giant armadillo and terrestrial pangolin. Note that the members of the pairs are from entirely different groups. For example, the South American mammals in the top three pairs are rodents, whereas the African ones are ungulates. *(After Meggers* et al. *1973.)*

(A)
(B)
(C)
(D)
(E)

(A)

exposes only its underside. In both African and American forests, some lizards have evolved into burrowing creatures, having reduced or lost limbs, elongated bodies, loss of eyesight, and the like, but different lizard groups have evolved in this manner on the two continents (Laurent 1973). Similar trends have occurred throughout rain forest faunas as similar environmental conditions have subjected different animals to many of the same selective pressures; some of these convergences in mammals are shown in Figure 20–13. However, convergences are far from complete. For example, in South America, tree frogs exist that incubate their eggs in pouches, but they do not in Africa. In addition, egg-eating snakes have evolved in the African forests, but not in South America (Laurent 1973).

Lay people see the luxuriant growth of a tropical jungle as implying a rich soil suitable for agriculture, but in fact it usually is not. After forest clearance, the soils generally become thin and nutrient poor. The soils cannot maintain the large reserves of minerals needed for plant growth, such as calcium, phosphorus, and potassium, largely because heavy rainfall and a high rate of water flow through the soil to the water table leach the minerals from the soil. The leaching process leaves behind large residues of insoluble iron and aluminum oxides in the upper soil levels.

(B)

FIGURE 20–14 Destruction of a tropical rain forest about 130 km north of Manaus, Brazil. The forest is first felled and then burned. (A) *(Opposite)* Edge of a clearing, showing intact forest on the left. Light and wind penetrating below the canopy at the edge will change the flora and fauna of the forest for hundreds of meters back from the edge. (B) Center of the clearing after burning. Note the poor, sandy soil and large amounts of unburned material. *(Photos by Paul R. Ehrlich.)*

What nutrients are available in intact tropical rain forest ecosystems are tied up in the lush vegetation itself. Those nutrients that are released by the decay of dropped leaves, fallen trees, and dead animals tend to be taken up by the webwork of shallow roots that laces the nutrient-enriched upper part of the soil. In some intact forests, almost none of them move below the root zone (Stark and Jordan 1978). Because most rain forest trees are evergreen, the process is continuous. A rich layer of humus does not accumulate, as it does in a temperate forest, where, after a general leaf fall, the trees are dormant throughout the winter.

A most interesting feature of some parts of the tropical forest biome is the floodplains that support forests that are periodically inundated. The most extensive such area is 70,000 km^2 of the Amazon basin that is submerged up to depths of 15 m for as many as 8–10 months annually (Goulding 1980). The physiological adaptations of the terrestrial plants for enduring the stress of flooding are not well understood. Some trees have respiratory roots that grow above flood level. At least in the Amazon, these flooded forests play a major role in the fisheries' productivity of the nutrient-poor rivers that drain substantial portions of the forests. The fishes eat fruits and seeds from the forest and in turn may be important in dispersing many of the plant species (Goulding 1980). Thus, apparently coevolution has occurred between fishes and trees.

Tropical rain forests are the great reservoir of species diversity on our planet; as many as one-half of all plants and animals may live in them. Sadly, as Figure 20–14 illustrates, they are also being destroyed more rapidly than any other biome (Ehrlich and Ehrlich 1981, Myers 1983). Indeed, in Central America, the forests are "vaporizing," to borrow a term from tropical ecologist Daniel Janzen. One basic cause is reckless exploitation by commercial interests in rich countries. For example, entire Southeast Asian forests have been converted into wood chips to make cardboard boxes for shipping Japanese stereo sets, and Costa Rican forests have been mowed down to make temporary grazing land to supply cheap beef to fast-food restaurant chains in the United States. Another fundamental cause is runaway human population growth in poor tropical nations, which leads to destruction of forests in order to make farmland and provide firewood. One result of the destruction of Latin American rain forests may already be observable in North America. Various migratory birds that overwinter in South and Central America, such as hooded warblers, Canada warblers, Acadian flycatchers, veeries, and lazuli buntings, are undergoing steady population declines. Most likely this decline is a result of habitat destruction, both in the United States and especially in rain forests at their overwintering grounds. The evidence of an impact on the wintering grounds is not yet conclusive (Wilcove 1985). Nonetheless, because more than one-third of the birds that breed in the United States annually winter down south, the prospect is not cheering.

Temperate Forest

Temperate forest is the forest biome that is familiar to most readers of this book. It is the biome in which Chinese civilization developed (Wang 1961). It also has supported the most recent flowering of Western culture. As defined here, the largest areas occupied by these forests are dominated by broad-leaved deciduous trees, and they are found principally in Western Europe, East Asia, and the eastern United States (Braun 1950). In these regions, temperatures are below freezing each winter, but not usually below −12°C, and annual rainfall is between 75 and 200 cm.

In the eastern United States, this biome is characterized by tall (40–50 m), dense stands of oaks, hickories, maples, and other trees that shed

FIGURE 20–15 Early fall in a temperate deciduous forest, Fairfax County, Virginia. Leaves are just beginning to change color prior to dropping. *(B. J. Miller, Fairfax, Virginia/BPS.)*

their leaves in the fall and regrow them in the spring (see Figure 20–15). Generally, two or three species predominate in any given locality, although several more species may be present and numerous. Unlike the tropical rain forests, temperate forests tend to have a rather rich, shrubby understory, as well as a considerable herbaceous flora. Many of the herbaceous plants take advantage of the abundant sunlight reaching the forest floor in the spring, before the trees leaf out, for growth and reproduction. However, the canopy of a temperate forest is almost never as dense as that of a tropical rain forest, and therefore relatively more light is available to the understory throughout the growing season. Most of the shrubby plants are especially adapted to growth in partial shade.

The epiphytes and lianas that so characterize the rain forest are largely absent from temperate forests, although algae, lichens, and mosses often grow on the tree trunks. The soils also are different from those of rain forests. The annual leaf drop produces soils rich in nutrients, many of

FIGURE 20–16 Southern beech tree, *Nothofagus,* on the shores of Lake Fagnano, Tierra del Fuego, Argentina. The slopes on the shore in the distance are covered with *Nothofagus* forest. *(Photo by Paul R. Ehrlich.)*

which have been extracted from deep in the ground by the penetrating root systems of the trees. In turn, the rich soil supports an extravagant microbiota. In a single gram of forest soil, 650,000 algae have been found; counts of microarthropods, primarily mites and tiny insects, may number in the hundreds of thousands and nematodes (roundworms) in the millions.

In American deciduous forests, characteristic mammals are the white-tailed deer, the black bear, racoons, squirrels, and shrews. Birds such as warblers, woodpeckers, a wide variety of thrushes and flycatchers, wild turkeys, and owls and hawks are common. Snakes such as the copperhead, frogs, and salamanders are well represented, but lizards are not. The variety of understory plants help provide food for the herbivorous vertebrates as well as a great variety of insects. The trees tend to be wind pollinated, but many of their fruits (e.g., acorns, beech nuts) are dispersed by mammals and birds and serve as an important food source for them.

Like the plants, the animals are well adapted to the seasonality of the temperate forest biome. Many of the mammals, such as the bears and the squirrels, are inactive or hibernate during the coldest months. The majority of the birds migrate to warmer climes for the winter. The insects enter diapause, a state of extremely low metabolic activity that occurs at different stages in different species.

When temperate forests are cleared, significant quantities of nutrients may be lost from the soil, as we will see in Part 7. With careful husbandry, however, the richness of soils in this biome can be preserved, as the fertile farms of Bavaria demonstrate.

Not all temperate forests are deciduous. In the southern hemisphere, some of the evergreen *Eucalyptus* forests in Australia and forests of southern beech (*Nothofagus*) in South America (see Figure 20–16), New Zealand, and Australia occur in climatic conditions similar to those of the temperate deciduous forests. More information on temperate forests is available in Reichle (1970).

FIGURE 20–17 An arctic rock desert on Cornwallis Island, Northwest Territories, Canada. Like the fog desert (see Figure 20–20), this rock desert is almost devoid of plants. However, in moist depressions and in areas in which the soil has been fertilized by animal droppings, some sparse vegetation does occur, as can be seen in Figure 20–32(A–B), consisting of photographs taken a few miles from this site. *(Photo by Paul R. Ehrlich.)*

Desert

Deserts are virtually synonymous with lack of water. They are found in areas in which annual rainfall is less than 250 mm. Most deserts are concentrated around latitudes 30° N and 30° S in those regions of descending masses of warm, dry air that we discussed at the beginning of this chapter. However, special conditions have created deserts far from these latitudes, such as in parts of the Canadian arctic, as shown in Figure 20–17.

Besides being dry, most deserts are quite hot in the daytime. Lacking cloud cover, they tend to reradiate their heat rapidly and become quite

FIGURE 20–18 (A) Sonoran semidesert, near Roosevelt Dam, Arizona. Prominent plants are tall, columnar saguaro cactus (*Carnegiea gigantea*), the "fuzzy" cholla cactus (*Opuntia* species), and the spraylike ocotillo (*Fouguieria splendens;* Fouguieriaceae). The saguaro is able to store enough water to fruit even in the driest years; the cholla often reproduces vegetatively as pads washed into stream beds take root. The ocotillo, considered by some a relative of violets, produces a useful wax, is sometimes used in the preparation of native medicines, and is often planted as a fence. (B) *(opposite)* Gila monster (Helodermatidae; beaded lizards), a typical resident of such deserts. It is largely nocturnal, remaining in a burrow during the heat of the day. The tail serves as an organ for storing fat; a plump appendage, as in this individual, indicates a healthy nutritional state. The Gila monster is poisonous, releasing a toxin from grooved fangs, which it chews into prey and enemies. It is not aggressive toward human beings, but its bite can prove fatal. *(Photos by Paul R. Ehrlich.)*

(A)

cold and windy at night. However, temperature regimes show considerable variation. For example, snow is common (but not copious) in the deserts of the northern Great Basin of the United States; it does not occur in the Sahara.

Sometimes the desert biome is subdivided into desert and semidesert areas. In the true deserts, plants cover a tenth or less of the soil surface; in semideserts, they cover between a tenth and about a third. The semidesert landscape is characterized by plants, although they are sparse, as Figure 20–18 makes clear; however, in deserts, plants are so few and small that rock or soil dominates the view, and the most extreme deserts have no vegetation at all (see Figure 20–19).

The need for water conservation dominates the lives of desert organisms. Plants tend to have thick, waterproof outer layers or cuticle, modifications of the breathing pores (stomata), reductions in leaf area, and specialized hairs or outgrowths that reflect light. They often have special water-conserving photosynthetic systems, as we discussed in Part 1. Desert plants are, of course, storage points for water, and they often have evolved impressive arrays of spines for fending off attacks by moisture-

(B)

FIGURE 20–19 Fog desert on the Peruvian coast. (The shoreline is in the distance.) This habitat is devoid of vegetation. *(Photo by Paul R. Ehrlich.)*

seeking animals and, in some cases, for reducing convection and associated water loss. Desert plants also frequently are aromatic, indicating the presence of biochemical defenses against herbivores.

The apparent wide spacing of desert plants is often an illusion. If their roots were visible, they would be seen to occupy all of the subsurface area. Between the plants is a shallow root network that assures that most of the scarce rain water is taken up by the vegetation. In certain soil conditions, desert plants may have extremely long taproots that reach to water supplies deep underground (Figure 20–20).

The seeds of desert plants often contain water-soluble chemicals that inhibit their germination. When these chemicals are leached out by water—that is, after a sufficient amount of rain has fallen—the seeds sprout. Seeds of plants that grow in gravelly arroyos often must be abraded by being washed down the gully before they will germinate. The net result of these mechanisms is that, as Figure 20–21 shows, deserts often break into riotous bloom shortly after heavy rains, as fast-growing annual plants grow, flower, and set seed. In semideserts, after good rains, the temporary annual plant cover can reach over 80%. The pollinators of these plants are usually well synchronized with them, although their timing mechanisms generally are unknown. Indeed, many desert insects are also "annuals," being active only when the annual plants they depend upon are growing.

Annual plants comprise a greater portion of the flora in the desert biome than in any other. When rainfall is plentiful, individual plants may

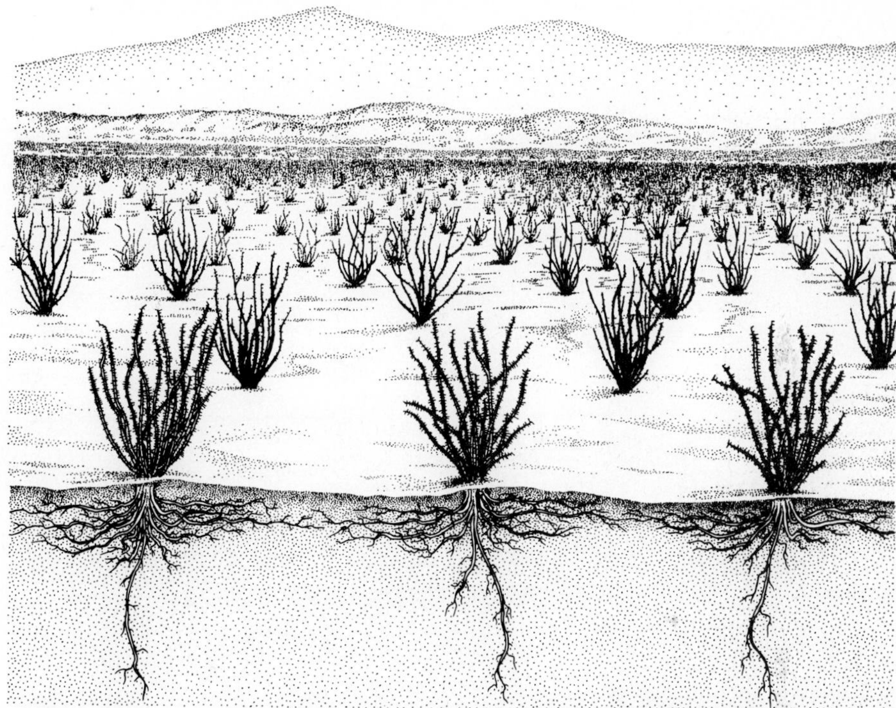

FIGURE 20–20 A stand of desert plants (ocotillo; see Figure 20–19A), showing the characteristic spacing of shoots and network of roots occupying intermediate areas, which allow the plants to maximize their water uptake after rains.

produce numerous flowers; when rainfall is scarce, many individuals will produce only a single bloom. In any case, the flowering season tends to be brief, and many people who travel through deserts have never seen the spectacular floral displays they are capable of producing.

Typical perennial plants of deserts are the succulent and thorny cacti of the Western Hemisphere and succulent thorny members of the milkweed (Asclepiadaceae) and spurge (Euphorbiaceae) families in African deserts, as shown in Figure 20–22. In the cooler semideserts of the northern Great Basin, low shrubs, primarily sagebrush (*Artemisia tridentata*), predominate, as illustrated in Figure 20–23. In the warmer Sonoran desert of southern California, southern Arizona, and northern Mexico, creosote bush (*Larrea divaricata*) replaces the sagebrush. In the hotter North American deserts, other common prominent plants include yuccas and agaves in the lily family. In contrast, the semideserts of South Africa are dominated by ice plants, members of a family (Aizoaceae) that are not prominent in the Western Hemisphere. The Aizoaceae have succulent leaves, whereas the Cactaceae and cactus-like Euphorbiaceae have succulent stems.

FIGURE 20–21 Spring bloom in the Mojave desert of California. The predominant flower is the bright yellow goldfields (*Lasthenia chrysostoma*), an annual composite. *(Photo by John Gerlach/Tom Stack & Associates/BPS.)*

Like the plants, desert animals have also evolved many ways of conserving water (Schmidt-Nielsen 1964, Noy-Meir 1974). One of the most common is for them to be active only in the cool of the night, or at dawn and dusk, and to remain underground in burrows during the heat of midday. The skins and exoskeletons of desert animals tend to be thick barriers against water loss, and their excretory systems are designed to conserve water. The camel, among other adaptations, is able to allow its temperature to rise to 6°C before it must sweat to cool itself (Schmidt-Nielsen 1964). Many desert animals are able to exist without drinking, as they can produce all the water they require metabolically from their food.

Desert soils generally do not contain much organic matter and must be supplied with nitrogen as well as water if they are to be farmed. Much of the uncultivated flatland left today is in this biome, which, in contrast to the tropical rain forest, is *expanding* in response to human activities. The vast Sahara desert was partly the work of humans as well as of climatic changes, the result of millennia of overgrazing, faulty irrigation, and deforestation. Today the Sahara seems to be advancing southward

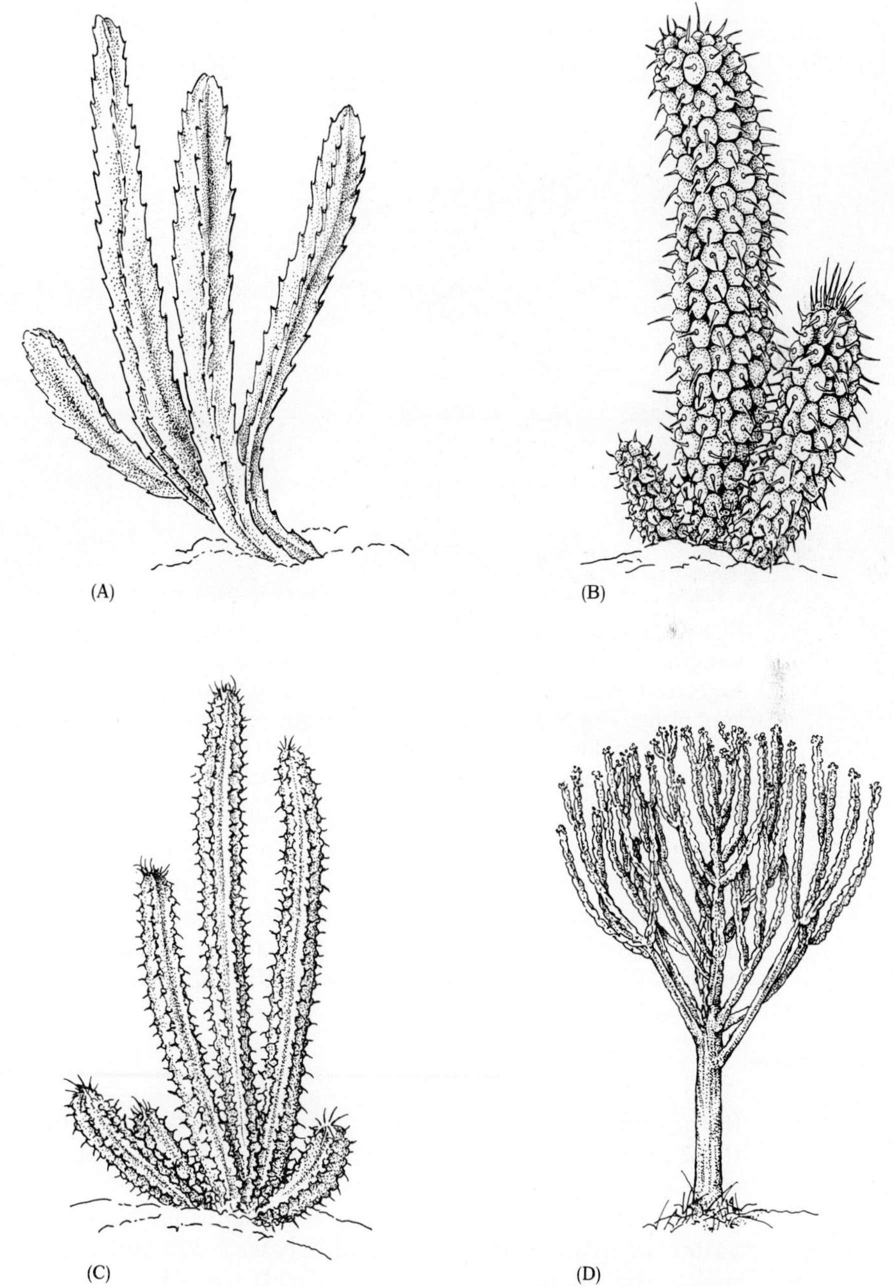

FIGURE 20–22 African desert milkweeds *(Asclepiadaceae)* (A–C) and a spurge *(Euphorbiaceae)* (D). Note the general convergence in form with New World cacti.

FIGURE 20–23 Sagebrush desert in the Great Smokey Valley of Nevada. In the foreground, an area from which cattle have been excluded by a fence, can be seen sagebrush (*Artemisia tridentata*), the dark plants, and a perennial bunch grass, the light plants. The bunch grass has been destroyed by grazing cattle beyond the fence line in the middle distance. Much of the desert area in the western United States has been badly overgrazed. *(Photo by Paul R. Ehrlich.)*

into the drought-plagued sub-Saharan region (the Sahel), its advance abetted by overpopulation of human beings and domestic animals. In the long run, attempts to bring more semidesert areas under cultivation most likely will simply produce more deserts, as irrigation systems fail, soil erosion proceeds, and excessive population growth continues.*

Grassland

In the climatic zone between the conditions favoring temperate forest and those producing desert, *edaphic*, or soil, factors and microclimatic factors, such as those controlled by slope or exposure, determine whether forests grow. Thus, tongues of forest may intrude into areas of desert along river valleys or in other protected, relatively well-watered sites. The forest itself makes the microclimate more moist. If forests are removed from areas otherwise suitable for their growth, a grassland may replace them. In the view of many ecologists, the extensive grasslands of central North Amer-

* For more information on deserts, see Cloudsley-Thompson and Chadwick (1964), Noy-Meir (1973, 1974), and Orians and Solbrig (1977).

FIGURE 20–24 Tall grass prairie. *(David Schimel, Colorado State University.)*

ica, Russia, and parts of Africa are zones between forest and desert in which fire and grazing animals have worked together to prevent the spread of trees (Bragg and Hurlbert 1976, Kucera 1981).

In both North America and Asia, grasslands show differentiation along moisture gradients. In the eastern part of central North America, as far east as Illinois, with about 80 cm annual rainfall, the presettlement vegetation was tall grass prairie, with grasses growing 2 m tall or more, as shown in Figure 20–24. Along the eastern base of the Rocky Mountains, 1300 km to the west, where rainfall is about 40 cm, were the short grass prairies, made up of species that rarely exceed 0.5 m in height, as shown in Figure 20–25. In between, along the gradient, were found the mixed grass prairies. Farther west, in eastern Washington and the Central Valley and coastal valleys of California, were found bunch grass prairies, with distinct clumps of grass between which soil was exposed, as illustrated in Figure 20–26. Today the long grass and mixed grass prairie areas largely are under corn, soybean, and wheat farms. The short grass and bunch grass prairies tend to have been highly modified by heavy grazing. In California, at spots in which the grassland was not farmed, the original prairies have largely been replaced by low, annual, weedy herbs imported from the Mediterranean, a prominent example of the invasion and decimation of a native flora.

FIGURE 20–25 Short grass prairie. *(David Schimel, Colorado State University.)*

FIGURE 20–26 Bunch grass prairie. *(C. W. May/BPS.)*

In Asia, a similar gradient from long grass to short grass occurs, but in this case the trend is from tall grass in the north to short in the south. Other temperate grasslands occur in South Africa (the *veldt*), in Argentina and Uruguay (the *pampas*), and on the fringes of Australia's central desert.

Typical large mammals of the grasslands include the bison (*buffalo*) and pronghorn antelope of North America; wild horses in Eurasia; large kangaroos in Australia; and giraffes, zebras, white rhinoceroses, and a diversity of antelopes in Africa. Other important grassland animals include such predators as coyotes, lions, leopards, cheetahs, and hyenas; a variety of birds, ranging in size from the giant flightless ostriches, rheas, and emus, through vultures and cranes, down to tiny sparrows; rabbits, prairie dogs, aardvarks, and other burrowing mammals; and grasshoppers (or locusts, as they are sometimes known). The plague locusts of biblical fame basically are grassland animals.

Grasslands have higher concentrations of organic matter in their soils than do those of any other biome. A typical grassland soil has about 12 times the humus (dark, partly decomposed organic material) found in a typical forest soil. Grassland soils thus are highly fertile, and it is no coincidence that the area once occupied by temperate grasslands in North America has become the richest agricultural area on Earth. The original soils are quite resistant to erosion by wind or water because of the turf formed by the interlaced roots and creeping underground stems of the grasses. However, when the turf is broken up by plowing, the soils become vulnerable. Careful husbandry is required to prevent gradual loss of soil fertility through erosion. Unfortunately, in most parts of the world, including North America, that husbandry is not being practiced. As a result, humanity's basic source of food is threatened. Agriculturalist Lester Brown (1981) said quite accurately that "Civilization can survive the exhaustion of oil reserves, but not the continuing wholesale loss of topsoil."*

Taiga

North of the temperate zone forests and grasslands and in the mountains of the Northern Hemisphere lies the biome of coniferous forests (Krajina 1969). This biome generally is known by its Russian name, *taiga,* views of which appear in Figure 20–27. Most of the trees have narrow, sclerophyllous (tough, leathery) leaves (*needles*) that are adapted to reduce water loss and persist for 3–5 years. Evergreen trees, especially spruces (*Picea*), firs, (*Abies, Pseudotsuga*), and pines (*Pinus*), are the most prominent trees in this biome, but in local areas, especially disturbed areas, deciduous trees such as aspens (see Figure 20–28), alders, and larches may predominate. Especially along water courses and in boggy ground, willows (*Salix*) often develop dense stands.

* For further information on grasslands, see Costello (1969), McMillan (1959), and French (1979).

(A)

(B)

FIGURE 20–27 Taiga. (A) Along the railroad line to Churchill in northern Manitoba, Canada. Note the continuous forest, largely spruce, on both sides of the track and the tripods supporting telephone lines. The ground is marsh in the summer and frozen in the winter, making it difficult to keep single poles upright. (B) Aerial view of the southern Northwest Territories, Canada, west of the Great Slave Lake. Note the many lakes and large areas of muskeg—open bog with scattered and stunted coni-

(C)

(D)

fers—that are characteristic of large areas of subarctic Canada. (C) Hay River, southern Northwest Territories, Canada, south of the Great Slave Lake. Willows and aspens form a light-colored band along the river bank, where the land is subject to flooding; conifers behind are darker. (D) Plagues of mosquitoes and blackflies are characteristic of taiga in the summer. Note the mosquitoes on the pith helmet of this ecologist on the shore of the Great Slave Lake. *(Photos by Paul R. Ehrlich.)*

FIGURE 20–28 A grove of aspen trees that has grown in an area in which a spruce forest was logged a century before. This taiga location is at an elevation of 3000 m in the Rocky Mountains of Colorado. *(Photo by Paul R. Ehrlich.)*

In general, however, the trees in the taiga are less diverse than they are in temperate forests. They tend to have a conical shape, presumably to reduce branch breakage from heavy loads of snow. Because of the year-round canopy, understory layers tend to be sparse and less diverse than those in other biomes. The gradual decay of the fallen needles in the cold climate produces a characteristic soil that is acidic and much less rich in arthropods and earthworms than are forest or grassland soils.

The taiga is inhabited by bears, lynxes, deer, moose, beavers, squirrels, chipmunks, snowshoe hares, and other mammals, as well as by many birds. However, like the flora, the fauna tends to be depauperate in comparison with those in more temperate biomes. Both the species diversity and the abundance of individuals are reduced. The exception to this rule is the family Mustelidae (weasels, fishers, martins, sables, minks, and wolverines), which is richly represented. The taiga is famous for the cyclic patterns in which the dynamics of hares and lynxes seem to be locked.

Ectothermic vertebrates are even scarcer; snakes are rare and few amphibia exist. As might be expected, insect diversity also is low, but the huge stands of one or a few conifers provide food to support occasional enormous outbreaks of forest pests, such as the spruce budworm, which is the larva of the moth *Choristoneura fumiferana* (Morris 1963). Furthermore, mosquitoes (largely *Aedes*), blackflies (*Simulium*), horseflies (*Tabanus*), and other biting flies are often so abundant as to make life outdoors for human beings a trial during the brief summers.

The generally low diversity of the taiga, the lynx–hare cycles, combined with the occurrence of insect plagues, has been one of the bases of the notion that diversity and stability are related in ecological systems. In the past, the biome has been heavily exploited for furs, but this activity has become less important as fashions in apparel have changed. Now logging is the main form of exploitation, especially for wood pulp. It is often performed with little attempt to protect the soil, threatening the recovery of forests in some areas.

In the Southern Hemisphere, little land area occurs in the latitudes at which one would expect extensive taiga to exist. A similar community

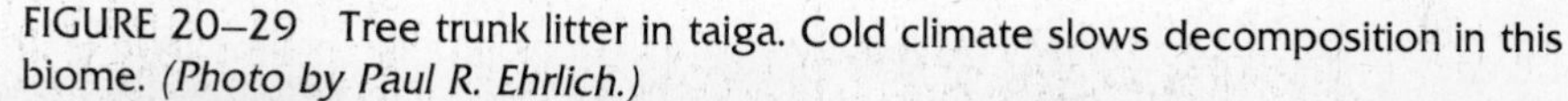

FIGURE 20–29 Tree trunk litter in taiga. Cold climate slows decomposition in this biome. *(Photo by Paul R. Ehrlich.)*

(A)

(B)

FIGURE 20–30 Tundra. (A) Low tundra with willow stands surrounding small lakes on Southampton Island, in northern Hudson Bay, Northwest Territories, Canada. (B) Grassy tundra on the site of an old Eskimo encampment, Duke of York Bay, Southampton Island. The bones are some vertebrae from a white whale. The vegetation is unusually lush here because of the fertilizing effect of the Eskimos' wastes. *(Photos by Paul R. Ehrlich.)*

type, dominated by southern beeches (*Nothofagus*), is found in the cool areas of southern South America and Australia. As in the taiga (see Figure 20–29), these communities have an extensive litter of tree trunks and branches on the forest floor, as the cold climate slows the activities of decomposers.

Tundra

North of the taiga in the Northern Hemisphere are vast, treeless plains also known by a Russian name, *tundra*, examples of which are shown in Figure 20–30. The average annual temperature in the tundra biome is −5°C. In the summer, the ground normally thaws to a depth of only less than 1 m; below that depth, it remains permanently frozen, and this part is called *permafrost*. The permafrost effectively prevents the establishment of trees, even along water courses. However, even without the permafrost the tundra biome would be largely treeless, for the average annual precipitation is usually less than 25 cm. Because of *frost heaving*—the alternate freezing and thawing of the ground—the tundra soil often is patterned into polygons, as shown in Figure 20–31, or hummocks.

Lichens, mosses, grasses, sedges, willows (normally less than 1 m high), and members of the heath family (Ericaceae) are among the dominant plants. In parts of the central arctic archipelago of Canada, however, so

FIGURE 20–31 Soil polygons. *(Photo by Paul R. Ehrlich.)*

little precipitation occurs that a rock desert virtually devoid of vegetation exists.

Tundra mammals include caribou, musk oxen, arctic hares, arctic foxes, wolves, and occasionally grizzly or polar bears. An astonishing diversity of birds, some of them shown in Figure 20–32, migrate to the tundra to breed, feeding on an ephemeral bloom of insects and freshwater invertebrates. Mosquitoes can make human existence nearly as miserable

(A)

(B)

(C)

FIGURE 20–32 Tundra nesting birds. (A) A king eider duck on its nest on the tundra. This species winters off the Atlantic and Pacific coasts of the United States and Canada. (B) A rock ptarmigan on its nest; a chick can be seen next to the mother to her lower left. This species is nonmigratory but changes its feathers to an entirely white set in the winter. Both the duck and the ptarmigan, like most other birds nesting on the treeless tundra, are extremely well camouflaged and therefore difficult to see, even from a few meters away, when on their nests. (C) The same duck as in (A) but seen from about 3 m away. The bird is in the center of the photo. *(Photos by Paul R. Ehrlich.)*

in the tundra as they can in the taiga. Reptiles and amphibians are absent. As in the taiga, animal populations in the tundra may be subject to dramatic oscillations in size. For example, lemmings cycle from scarcity to superabundance and back; in the process, they induce cycles in the animals that feed on them, especially owls and jaegers (predatory gulls).

The tundra has been one of the least exploited biomes; until recently, the taking of furs has been the principal industry. However, that is changing, given mineral exploration and the growing military importance of the north polar region—an unfortunate situation, as the tundra is extremely slow to recover from disruption. Plant growth rates are so slow, in fact, that the tracks of even a single vehicle are likely to persist for decades.*

THE MINOR BIOMES

In addition to the major biomes, some less extensive vegetative formations exist that deserve attention, both because of their intrinsic biological interest and because they emphasize that dividing the terrestrial world into six major biomes is a vast oversimplification. For example, tropical rain forests are difficult to define. In regions in which rainfall is more seasonal, tropical rain forests blend imperceptibly into tropical deciduous forests, in which most of the trees shed their leaves as a water-conservation strategy during the dry season. In some areas, the forests along streams may be rain forests, but back from the streams they may be deciduous.

We will discuss the following minor biomes:

- Tropical savanna
- Tropical scrub and tropical season forest
- Temperate rain forest
- Chaparral

Tropical Savanna

Savannas are tropical grasslands that usually contain scattered, thorny trees, as shown in Figure 20–33. Their large animals are mostly the same as those of adjacent grasslands. Pictures of the animals of the savannas of East Africa will be familiar to viewers of television nature shows: herds of wildebeeste on the Serengeti Plain, a Thompson's gazelle being run down by a cheetah, a black rhino wallowing like a gigantic pig, a pride

* For more information on tundra, see Billings and Mooney (1968), Mani (1968), Bliss *et al.* (1973), Ivens and Barry (1974), and Tieszen (1978).

FIGURE 20–33 African savanna on the western slope of Ngorongoro Crater, Tanzania. Note the giraffes browsing on the widely spaced acacias. *(Photo by Paul R. Ehrlich.)*

of lions resting from the heat of the day under a thorn tree, or a leopard resting in its branches. The concentration and diversity of large mammals in this area are unparalleled and provide us with a hint of what much of our planet must have been like in the Pleistocene, 2 million years ago, before human hunters and climatic change conspired to exterminate most of the large mammals (Martin and Klein 1984).

Savannas generally receive most of their rainfall at one time of the year, which is divided into wet and dry seasons. Birds nest during the wet season, when insects are more common. In the dry season, the larger animals tend to become concentrated in the areas around watering holes. In the savanna of East Africa, rainfall tends to be localized, and animals will travel long distances to graze on the green grasses that result from a rainfall.

A broad band of dry savanna stretches across sub-Saharan Africa, and extensive areas of savanna exist in tropical South America—the *llano* of Brazil and the *campos* of Venezuela. The northern interior of Australia also contains a substantial swathe of savanna.

The debate about the origin of savannas is similar to that about temperate grasslands. Some researchers believe that savannas are local formations maintained by fire, grazing, or special soil conditions (Beard 1953, 1967). For example, many of the South American savannas occur in climates that also support forests. Those in Africa and Australia occur in drier areas than those of the American tropics, and climate probably plays a larger role in maintaining them.

Savannas are presently under assault from overgrazing and, in overpopulated nations such as Kenya, from attempts to farm them. In many cases, the land is too dry for unirrigated agriculture, and yields are disappointing. Some East African savannas, it is hoped, will be preserved as game farms. It has been demonstrated that when herds of native antelopes are mixed, a much higher meat yield can be obtained than by herding cattle. Moreover, unlike cattle, the antelope do not degrade the system by daily trekking to watering holes, because antelopes need less water; indeed, some species do not have to drink at all (Hopcraft 1980).

In moister areas, the trees of savannas become more common and closer together, eventually grading into a different biome—a tropical seasonal woodland.*

Tropical Scrub and Tropical Seasonal Forest

Not all tropical areas dominated by woody plants are rain forests. In the zone between savannas and wet jungles occur a variety of woodlands. The scrub—or, as it is sometimes called, *thorn forest* or *thornwood*—is found in areas in which rainfall is between 50 and 125 cm per year. The scrub is composed of small hardwood trees, very often thorny. Tropical seasonal forests occur in areas in which rainfall is between 125 and 250 cm annually. The seasonal forest, especially with nearly 250 cm of annual rainfall, contains very large trees and is quite dense. Examples are the monsoon forests of Vietnam and India. Indeed the total rainfall in these forests exceed that of some rain forests. As in the savanna, there are distinct wet and dry seasons. Growth of plants occurs predominantly in the wet season, and insects and other animals generally breed then. Teak is an important tree in many tropical seasonal forests, and bamboos are important shrubs.

Temperate Rain Forest

The temperate rain forest biome is found on the northwestern coast of North America, the southeastern coast of Australia, and the southwestern coast of New Zealand in areas that receive 200–380 cm of precipitation a

* For more information on tropical savannas, see Blydenstein (1967), Anderson and Herlocker, (1973), and Webb (1977–1978).

FIGURE 20–34 Olympic rain forest. *(J. N. A. Lott, McMaster University/BPS.)*

year. In America, this biome, an example of which is shown in Figure 20–34, is dominated by coniferous trees that are extremely tall, up to 100 m—primarily redwoods, Douglas fir, and western hemlock. Epiphytic mosses are abundant, and a well-developed understory is present in which both mosses and ferns are prominent. This biome is heavily exploited for timber, the decay-resistant lumber from redwoods being highly valued.

Chaparral

Chaparral is a biome characterized by shrubs or short trees with thick, tough, evergreen leaves, as shown in Figure 20–35. Chaparral is found in maritime climates with winter rainfall—so-called Mediterranean climates—that are cooler than those supporting tropical scrub. In contrast to tropical scrub, chaparral plants are mostly less than 2 m high, and they generally lack spines. The reason for this is unknown, but originally fewer large browsers probably were present in the chaparral, a situation now reversed in many areas because of the importation of goats. However, chaparral shrubs are equipped with prominent chemical defenses, which give them their aromatic character.

Chaparral is found along the southwest coasts of North and South America and Australia, in the Cape Region of Africa, and in the Mediterranean region. The greatest species diversities in chaparral are found in

FIGURE 20–35 Chaparral on a west-facing slope on Stanford University's Jasper Ridge Biological Preserve. This location is a typical habitat for the checkerspot butterfly, *Euphydryas chalcedona,* and its food plant, *Diplacus aurantiacus. (Photo by Paul R. Ehrlich.)*

the Southern Hemisphere, in the rich, low shrubland (*fynbos*) of the Cape province of South Africa and the dry heaths (*kwongan*) of Western Australia. The shrubs in these areas have different evolutionary origins and have been the subject of important studies of biome convergence—the development of similar ecological characteristics by assemblages of organisms from different taxonomic groups.

Chaparral is subject to recurrent burning, which apparently is what permits shrubs to dominate rather than trees. The underground parts of perennial chaparral plants are fire resistant, and the seeds of some annuals must be subjected to fire before they can germinate. People who build their homes in chaparral, as is often done in California, should consider carefully that were it not for periodic fires the chaparral probably would not exist.*

* For more information on chaparral, see Mooney (1971), Thrower and Bradbury (1977), and Cody and Mooney (1978).

VERTICAL ZONATION

Referring back to Figure 20–1, the map of biomes, notice that the mountain ranges are treated separately, with the notation "complex zonation." The reason will become clear if one recalls that climate, especially temperature and precipitation, are the major determinants of what biome is found in an area. Temperatures decline with increasing altitude—about a 6°C drop for every 1000 m in height—and precipitation tends to increase with altitude in desert areas and to be less on the leeward side of mountains (see, e.g., Whittaker and Niering 1965, Terborgh 1971). Therefore, ranges normally cannot be assigned to a biome in their entirety; ranges show altitudinal zones of plant communities, as is apparent in Figure 20–36.

In the Rocky Mountains of Colorado, for example, above about 3500 m, the temperature conditions are similar to those of the tundra biome at sea level north of the Arctic Circle; thus, where soil conditions permit, the mountains are capped by areas of *alpine tundra*. As one proceeds north in the North American mountains, the lower limit of the alpine tundra gets

FIGURE 20–36 Altitudinal zones in the Rocky Mountains of Colorado. The treeline dividing the taiga (middle and foreground) from the alpine tundra is clearly visible on the peak. Note the scattered spruce trees (e.g., left foreground, far right skyline) reinvading the aspen stands (center) that grew after the spruce forest was logged (Figure 20–28). Spruce dominate the forest on the far slope. In the valley to the left, out of sight, is a subalpine meadow community dominated by sagebrush. *(Photo by Paul R. Ehrlich.)*

FIGURE 20–37 Treeline at Churchill, Manitoba, Canada, on Hudson Bay. The treeline here separating the northern tundra from the southern taiga is analogous to the one in Figure 20–36 separating the high-altitude tundra zone from the lower-altitude taiga zone. Note the "one-sided" trees, whose branches grow away from the prevailing cold north winds. *(Photo by Paul R. Ehrlich.)*

progressively lower (as the highest altitude at which trees are found—the *treeline*—also gets lower.) Finally, in northern Canada and Alaska, the tundra is at sea level, and the treeline is a north–south boundary rather than an altitudinal one (see Figure 20–37).

In the northern mountains, tundra occurs as a virtually continous formation. In Colorado, by contrast, not enough area exists above 3500 m for it to be continuous, and the tundra appears as "islands," often as dense archipelagos. However, during recent ice ages, the alpine tundra was directly connected with the lowland tundra, as can be seen by the similarities of their floras and faunas today. Many plants, such as the alpine rose (*Dryas hookeriana*) and the purple saxifrage (*Saxifraga oppositifolia*), occur in both alpine and sea-level tundras. At 3700 m on Cumberland Pass in the Rockies of southwestern Colorado (see Figure 20–38), one can capture the satyrine butterflies *Oeneis polixenes, O. taygete,* and *O. melissa* (the last is shown in Figure 20–39). All of these butterflies are also at home in the arctic tundra surrounding northern Hudson's Bay in Canada.

In contrast, in tropical America and Africa, where the treeline is above 4000 m, the treeless zones of mountains do not show a biotic affinity with the arctic tundra. Instead, the plants and animals are related to those of lower altitudes in the same region. For example, in treeless areas of the high Andes, dark satyrine butterflies appear that are indistinguishable on

FIGURE 20–38 Above the timberline at Cumberland Pass, Colorado, at 3700 m in late June. Note the taiga (dark stands of spruce) below and the abundance of flowers on the tundra. This area is the home of arctic-alpine butterflies, such as those of the genus *Oeneis* (Figure 20–39) that fly in treeless areas high in the Rocky Mountains northward to sea-level tundras on the shores of the Arctic Ocean. *(Photo by Paul R. Ehrlich.)*

the wing from satyrines of the genus *Erebia* of the tundras of Colorado and the Canadian arctic. However, they are members of the genus *Punapedaloides,* which has strictly tropical affinities.

The tundra is representative of the general situation in which vertical ascent of 600 m is roughly equivalent to a trek to the north of about 1000 km. Figure 20–36 shows an altitudinal transect of some 700 m. To find tundra at the 3000-m altitude of the sagebrush of Colorado, one would have to travel almost 2000 km north into the Canadian Rockies.

The complexity of montane zonation in which both altitudinal and moisture gradients occur is well illustrated by a classic study by Whittaker and Niering (1965), the results of which are summarized in Figure 20–40. The various altitudinal zones are displaced upward in the dry end

FIGURE 20–39 The arctic-alpine butterfly, *Oeneis melissa,* photographed at Cumberland Pass (Figure 20–38). Members of the genus *Oeneis* characteristically sun on their sides on rocks, as shown; their color patterns provide fine camouflage against the lichen-covered rocks. (The butterfly is dead center in the photo, with its head to the right. The outer edges of its wings, to the left, can be seen because their fringes are alternating black and white dashes.) *(Photo by Paul R. Ehrlich.)*

of the moisture gradient. Everything else being equal, a biome will occur at a higher altitude where it is dry (xeric) than where it is moist (mesic).

With vertical zonation, as with the horizontal distribution of biomes, the classification is based largely on the structure of plant communities. The distribution of animal communities is often correlated with that of plant communities, but far from perfectly. Apparently sharp discontinuities in plant communities, such as the treeline separating taiga from tundra, do not necessarily also mark boundaries in animal communities. For example, *Erebia* butterflies, though often thought to be characteristic of either tundra or taiga, actually are distributed complexly relative to the edge of the trees, as indicated in Figure 20–41.

Finally, vertical zones differ in one characteristic of the atmosphere that does not change geographically—namely, a decline in atmospheric pressure with altitude. The mass of the atmosphere is concentrated in its lower layers. Some 40% occurs below the altitude of California's Mount

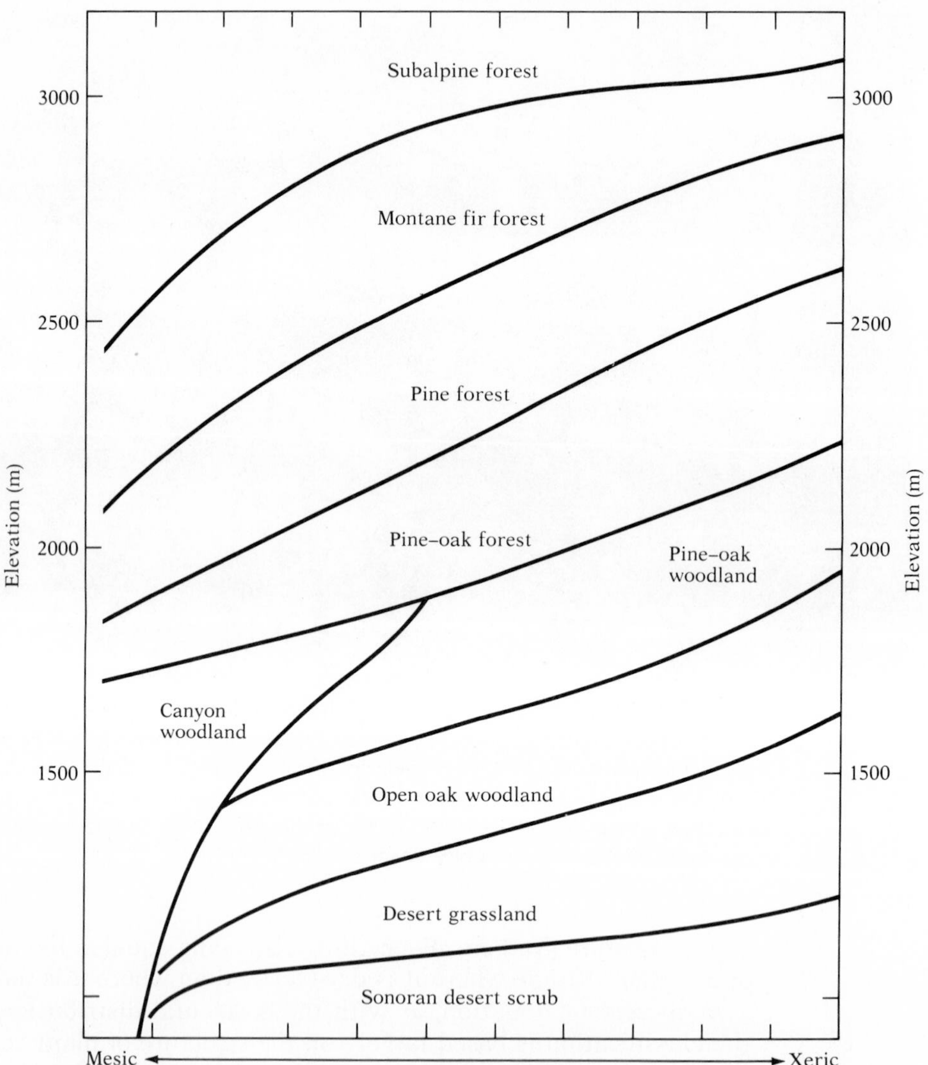

FIGURE 20–40 Effects of altitudinal and moisture gradients on vegetation in the mountains of Arizona. *(After Whittaker and Neiring 1965.)*

Whitney (4400 m), the highest point in the contiguous 48 states. Two-thirds of the atmosphere occurs below the highest bit of land on the planet, at the summit of Mount Everest (8900 m). Organisms that have evolved or developed at high altitudes deal readily with the reduced partial pressure of oxygen, but rapid ascents can have deleterious effects. Human beings born at 3000 m develop large lung capacities and other features that permit them to function at altitudes in ways that genetically identical individuals raised in lowlands cannot.

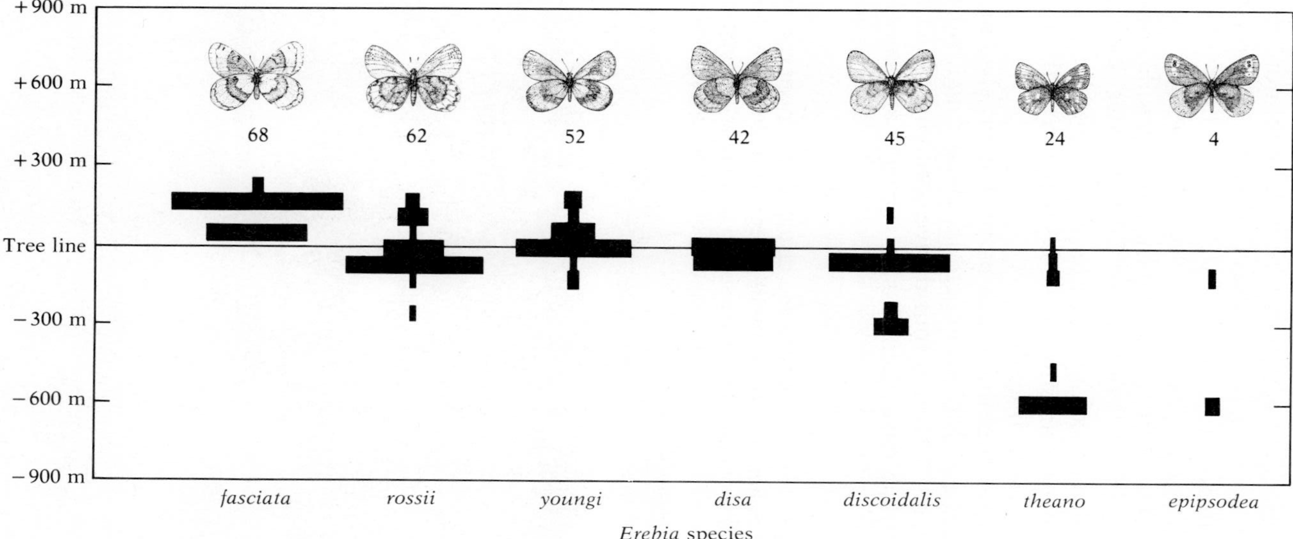

FIGURE 20–41 Distribution of species of *Erebia* butterflies around the tree line of Alaska and the Yukon. Numbers are sample sizes. *(After Ehrlich 1958.)*

SOIL COMMUNITIES

In a sense, the composition of terrestrial communities is determined by the distribution of both climates and soils, as shown in Figure 20–42. However, climates are not just "givens" to which the population in communities must adapt; as we have seen, the communities themselves influence the climate. Even more than the climate, soils are influenced strongly by the communities they support. Like climates, soils are both geographically variable and complicated to understand. Indeed, even to try to define soil as more than simply the layer between the land surface and the solid rock below is difficult.

The soil layer contains much more than crushed or powdered rock; it often includes a rich biota—bacteria, algae, fungi, worms, mites, tiny insects, and so on. Although soils derive nonliving materials from the intact *parent rock* that gave them birth, through a process of fragmentation

FIGURE 20–42 Effect of soils on the distribution of plant communities in the White Mountains of eastern California. In the foreground, the soil is derived from quartzitic sandstone and is occupied by a sagebrush (*Artemisia*) community. In the background is a bristlecone pine (*Pinus longaeva*) woodland occupying soils derived from dolomite. *(Photo courtesy of H. A. Mooney.)*

known as *weathering,* they also derive many inorganic constituents from the living community above and within the soil.

The difficulty in defining soil can be seen from a simple example. A mycorrhizal fungus typically is considered as part of the soil. Is the tree root with which the fungus is complexed also part of the soil? Most people think of the root as part of the tree, but then the soil–plant boundary would be an extraordinarily convoluted surface tracing around the finest root hairs and broken by the fungal mycelia. Consequently, soil scientists operationally consider the soil to begin at the forest floor, and they include the roots as part of it (Jenny 1980).

Soil Chemistry

As one might expect from such a diverse mixture of organic and inorganic substances, the chemistry of soil is not simple. From the point of view of the biological community that it supports, the most important chemical in soil is H_2O. Water is not only essential to the support of the living community in and on the soil, it also acts as a solvent and medium for the chemical reactions in the soil. In turn, soil serves as a crucial reservoir of water, holding it tightly in its pores.

Soils are extremely rich in ions, positively or negatively charged atoms or groups of atoms. Ions in solutions in pure water diffuse readily, but in soils they are often bound to soil particles and are relatively immobile. Plants must obtain the nutrients they require by their uptake of ions in an aqueous solution through root hairs.

Even though soils are rich in ions, an abundance of critical nutrients is not available to plants. The water in the soil tends to have low concentrations of nutrients, and their availability in the water often is a factor limiting plant growth. That availability, in turn, is controlled by factors such as the chemical nature of the parent rock, its degree of weathering—that is, how fine the particles are—the size of pore spaces between the particles, and the amount of humus in the soil. These factors control the size, shape, and chemical composition of the soil particles and the *ion exchange* between the surfaces of the particles and the *soil solution,* the latter defined as water plus what is already dissolved.

Soils are continually evolving. Crystals of rock react with water to produce fine colloidal particles of clay with very different ion exchange capacities than the crystals. These particles, in turn, may interact directly with roots or humus particles or may migrate downward through the soil layers, called *horizons.* Their behavior will be influenced by changes in soil acidity—for instance, owing to changes in humus production or to acid rains. Variation in rainfall can increase or decrease the leaching of nutrients from the upper layers of soil. Formation of an impermeable clay pan by the migration of clay particles can lead to waterlogging, then to changes in the soil's *redox potential*—that is, its potential for oxidation and reduction—and thus to changes in the mobility of important ions, especially iron and manganese.

Classification and Distribution of Soils

Soil classification is a taxonomist's nightmare. Qualities of soils usually show continuous geographic variation; sharp discontinuities between soil types are the exception rather than the rule. Nevertheless, we will find it useful to refer to broad groups of soils that share important attributes. In 1975, the Soil Conservation Service of the United States Department of Agriculture published a new soil taxonomy, which is based on the following mix of characteristics:

- Origin
- Texture
- Chemical properties
- Stratification

Origin. *Origin* refers both to the nature of the parent materials and to whether the soil developed at the source of the parent materials—*residual* soil—or at a distant site—*transported* soil. Two types of soils formed from transported parent materials are often referred to by ecologists. One is *alluvial* soils, in which the parent materials have been deposited by river flow. Alluvial soils are widespread and often very fertile. The second is *loess* (pronounced "less"), which refers to deposits of fertile, unstratified, yellowish soils found in the Mississippi Valley and parts of Europe and Asia. These are believed to have been wind-blown deposits.

Texture. *Texture* refers to the size of individual particles. Particles less than 0.002 mm in diameter are defined as *clay*, between 0.002 and 0.02 mm as *silt*, and between 0.02 and 2 mm as *sand*. *Loam* describes a texture produced by mixed particle sizes. Soil texture classification is illustrated in Figure 20–43. Soil texture greatly influences the water-retaining properties of the soil. Clays, for example, are wetted only slowly but have a large storage capacity, although the stored water may not be readily available to plants. In contrast, sands absorb water quickly, reducing surface runoff relative to clays, but do not retain it well.

Chemical Properties. Many soils are characterized primarily by their chemical characteristics, such as their acidity or an unusual balance of nutrients. For example, *serpentine soils*—so named because of the slippery

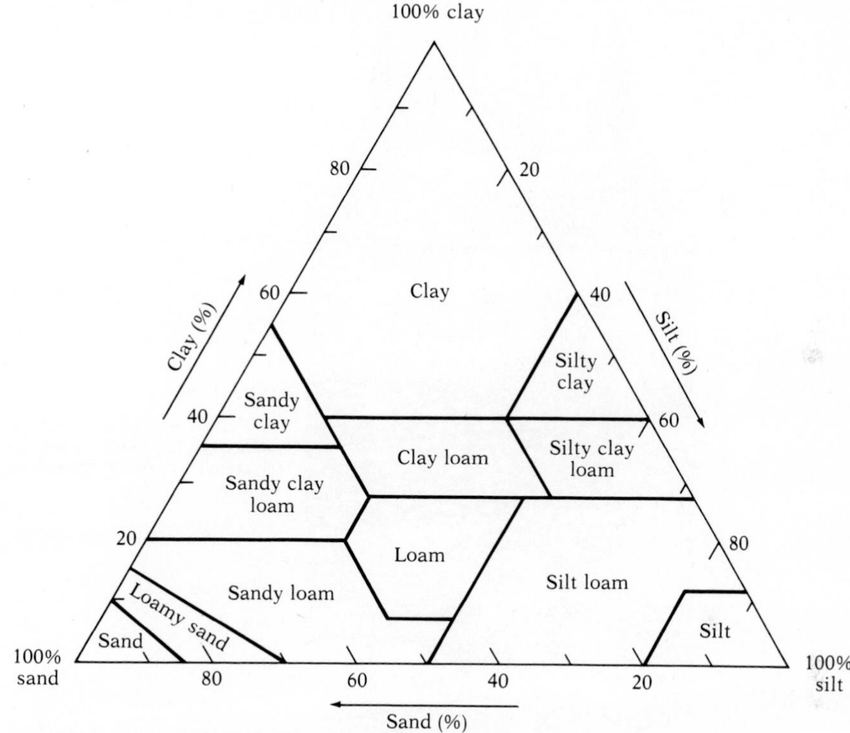

FIGURE 20–43 The classification of soil by texture. *(After Ehrlich* et al. *1977a.)*

character of the parent serpentine rock and the false view that snakes are slippery—tend to have unusually high concentrations of chromium, magnesium, and nickel and relatively low concentrations of calcium, molybdenum, nitrogen, and phosphorus. This combination is lethal to many plants, but a specialized serpentine flora has evolved that consists of plants adapted to the unusual nutrient regime (Proctor and Woodell 1975).

Stratification. Few soils have an undifferentiated vertical profile. (Transported soils such as loess are an exception; they are *azonal.*) Most occur in layers or horizons, as shown in Figure 20–44. Uppermost is the A horizon, in which are found most of the soil organisms and the majority of roots. The water that has passed through this level usually has leached out some of its nutrients. The nutrients tend to collect in the B horizon below it. Because minerals also often migrate up to the B horizon from the parent rock below, the B horizon is sometimes called the *zone of accumulation.* The next horizon, consisting of weathered parent material, is the C horizon. Below that is usually bedrock—the D horizon. Soil scientists often recognize subdivisions within horizons, usually labeled by sub-

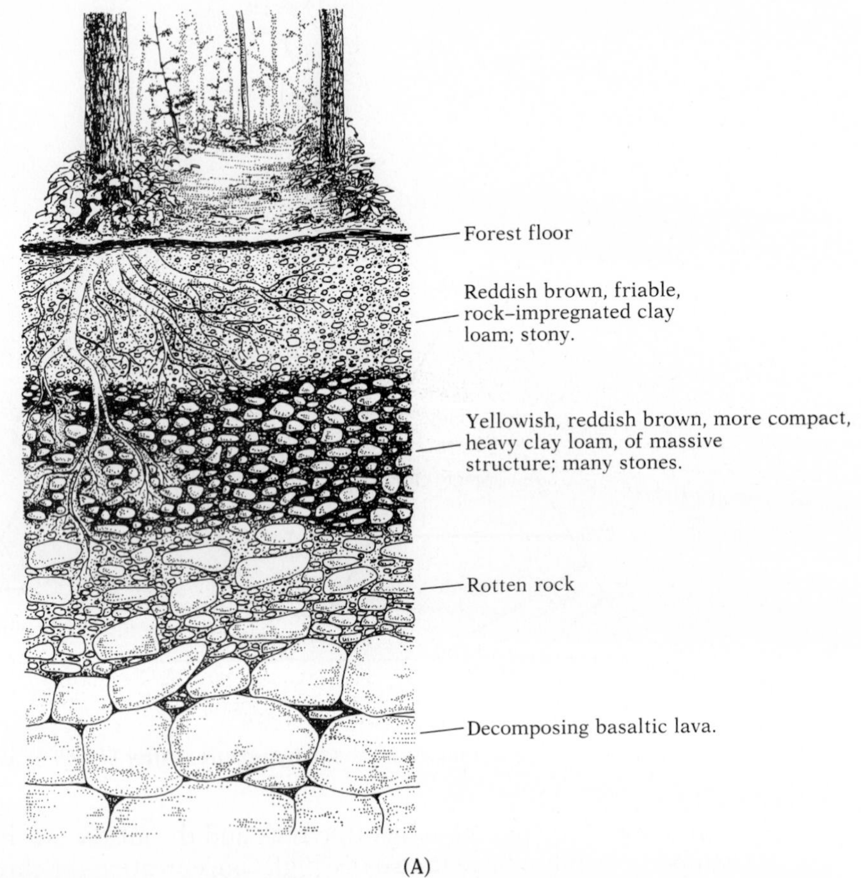

FIGURE 20–44 (A) Soil horizons. (See text for explanation). (B) *(Opposite)* Profile of Alaskan loam, a moderately deep, dark-colored, well-drained soil. *(Photo by Alan E. Aman, USDA—Soil Conservation Service/BPS.)*

(B)

Table 20–1 Major orders of soil recognized by the U.S. Soil Conservation Service

Name	Description
Alfisols	(So named because of their content of aluminum, Al, and iron, Fe.) Quite well developed horizons, with a thin humus layer and bases accumulated in the B horizon. Includes the former gray-brown podzols.
Aridisols	(From the Latin *aridus*, dry.) Water unavailable to plants for long periods; little organic matter; may have clay, gypsum, or carbonate horizons, the carbonate horizon usually within 30 cm of the surface; high base content; reddish to light gray or brown. Includes the former red desert soils and sierozems.
Entisols	(From the Latin *ent-*, existing.) Lack horizons; dominated by mineral materials; young soils. Includes azonal and alluvial soils.
Histosols	(From the Greek *histos*, tissue.) Very high organic matter; brown. Includes half-bog and bog soils, peats, mucks.
Inceptisols	(From the Latin *inceptum*, beginning.) A chemically diverse group containing relatively young soils with little production and movement of clays; horizons not well developed; texture finer than that of loamy sand; humid climates. Includes the former weak podzols, humic gley, and brown forest soils.
Mollisols	(From the Latin *mollis*, soft.) Black or dark brown surface layer of soft or crumbly consistency; often deep; may have a clay or carbonate horizon; rich in bases. Found under semihumid grasslands and forests with well-developed understories. Includes the former chernozems, prairie soils, and rendzinas.
Oxisols	(From the French *oxyde*, oxide.) Old, highly weathered tropical soils; rich in iron oxides and hydeous aluminum silicate; sometimes rich in humus; low in silica. Includes the former laterites and some latosols.
Spodosols	(From the Greek *spodos*, wood ash.) Leaf litter over a humus-rich layer, beneath which is a light (often grayish or whitish) A_2 horizon. Below that is a reddish and black B horizon, the *spodic horizon*, rich in extractable aluminum and iron. Includes well-developed podzols.
Ultisols	(From the Latin *ultimus*, last.) Old soils of humid, usually warm climates; strongly leached, with much clay formed and moved to the B horizon; low base content. Includes the former red and yellow podzols, and some laterites and latisols.
Vertisols	(From the Latin *verto*, to turn.) Clay soils; dark in color; forming characteristic deep, wide cracks on drying. Includes the former grumosol and smolnitza.

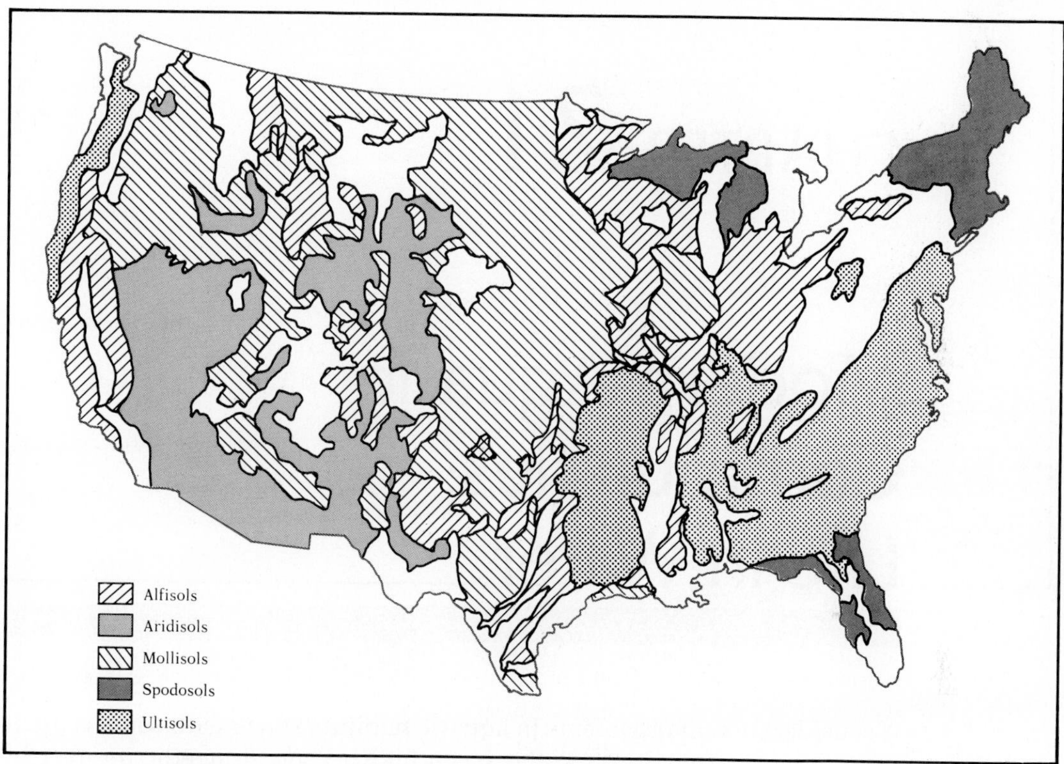

FIGURE 20–45 Map showing the distribution of major soil types in the United States. *(After Jenny 1980.)*

scripts 1, 2, and 3 to indicate top, middle, and bottom. Thus, B_2 would be the middle layer of the B horizon.

The 10 major orders of soil, as recognized by the United States Soil Conservation Service, are shown in Table 20–1. The distribution of major soils is shown in Figure 20–45.

The new terminology shown in Table 20–1 has the advantage of emphasizing the soil as a product rather than the process that formed it. For example, the term *podzol*, frequently found in the older taxonomy, referred to soils formed by the process of *podzolization*. Podzolization, however, simply referred to a movement of aluminum and iron oxides from the A to the B horizon, a process that leads to quite different soils under different conditions. A wide variety of biological communities are found on soils formed by similar processes. In contrast, each of the categories of soil in the new classification tends to support similar communities.

Now we will move on from soils, in which water plays such a key role, to aquatic communities themselves.

CHAPTER 21

Modern Aquatic Communities: Life in Water

A consideration of organisms in aquatic habitats starts with some critical properties of water itself. Water has a high *specific heat*—the number of calories required to raise 1 g of a substance 1°C. Water also has a high *latent heat of fusion* and *latent heat of evaporation*—respectively, the number of calories given up by 1 g of water when it freezes and the number of calories required to evaporate 1 g of water. These properties show that aquatic communities are subject to much less extreme temperature regimes than terrestrial communities in the same geographic area. For example, when ice forms on the surface of a lake, large quantities of heat are given up to the water, retarding further freezing. Conversely, when a hot, dry wind blows over the same lake in the summer, much of the heat it transfers to the water is consumed in evaporating a portion of it rather than in heating the lake as a whole.

Probably little life would exist in fresh water outside of the tropics were it not for another characteristic of water: It reaches its highest density at 4°C—that is, above its freezing point. Without this property, ice would sink and bodies of fresh water outside the tropics would freeze solid.

The amount of salt is the primary basis for classifying aquatic habitats. Fresh water normally has a saltiness or *salinity* of less than 0.5 parts per thousand; sea water has around 35 parts per thousand, at least 70 times as salty. Quite different organisms have adapted to these different salinities, by far the most diverse being found in marine habitats, where life originated.*

* An overview of aquatic systems is given by Barnes and Mann (1980).

SALT WATER

Marine environments are most widespread and most uniform on the planet. Oceans cover about 71% of Earth's surface, but in spite of their vast extent, they tend to be homogenized by currents. For this reason, the vast majority of the oceans contain enough oxygen for heterotrophs and carbon dioxide for autotrophs. However, sunlight and mineral nutrients are unevenly distributed, and variations in the nature of the bottom and shoreline create an environmental complexity to which organisms must adapt (see, e.g., Vermeij 1978).

Sunlight penetrates only far enough to permit photosynthesis in roughly the upper 2% of the ocean's volume, and autotrophs thus are confined to a narrow zone near the surface. Nutrients, especially nitrogen and phosphorus, are limiting in much of the marine environment, and they are found in comparative abundance only in relatively restricted areas in which currents encountering continents create upwellings. In such places, such as along the western coast of South America, *phytoplankton*—free-floating tiny plants—reproduce without nutrient constraints and support a rich fauna. Most of the oceans are resource poor and support so few organisms that they may be thought of as marine deserts.

The oceans are divided into a series of life zones—the most inclusive of which are the *neritic,* near shore, which encompasses the shallow waters over the continental shelves, and the *pelagic,* the open waters beyond the shelves. More detail on the zonation of the oceans is shown in Figure 21–1.

Superimposed on the vertical zonation is geographic variation, based on temperature differences and the configuration of the continental barriers, similar to that which leads to the biome classification of terrestrial areas. Marine biogeographers recognize a variety of *regions* and *provinces* (Briggs 1974, Levinton 1982). Four regions are often recognized in tropical shallow waters: Indo-West Pacific, Eastern Pacific, Western Atlantic, and Eastern Atlantic. These are probably the closest analogues to major biomes found in the oceans.

The Indo-West Pacific region, for example, extends from Africa to east of Tahiti (more than 180° of longitude), is by far the richest in species, and contains groups found nowhere else, such as the giant clams (Tridacnidae) and rabbit fishes (Siganidae). Within that region, a wide variety of provinces—or regions, subregions, or subfaunas—have been recognized on the basis of the different groups of organisms they contain (Briggs 1974).

These provinces can vary dramatically in their faunas. For example, both butterfly fishes (Chaetodonidae), shown in color Plate D, and damselfishes (Pomacentridae) show high diversity near the geographic center of the Indo-West Pacific region at the Australian Great Barrier Reef. This diversity declines as one goes eastward through the region to the Society and Hawaiian Islands (see Table 21–1), the decline being sharper in the pomacentrids (Anderson *et al.* 1981).

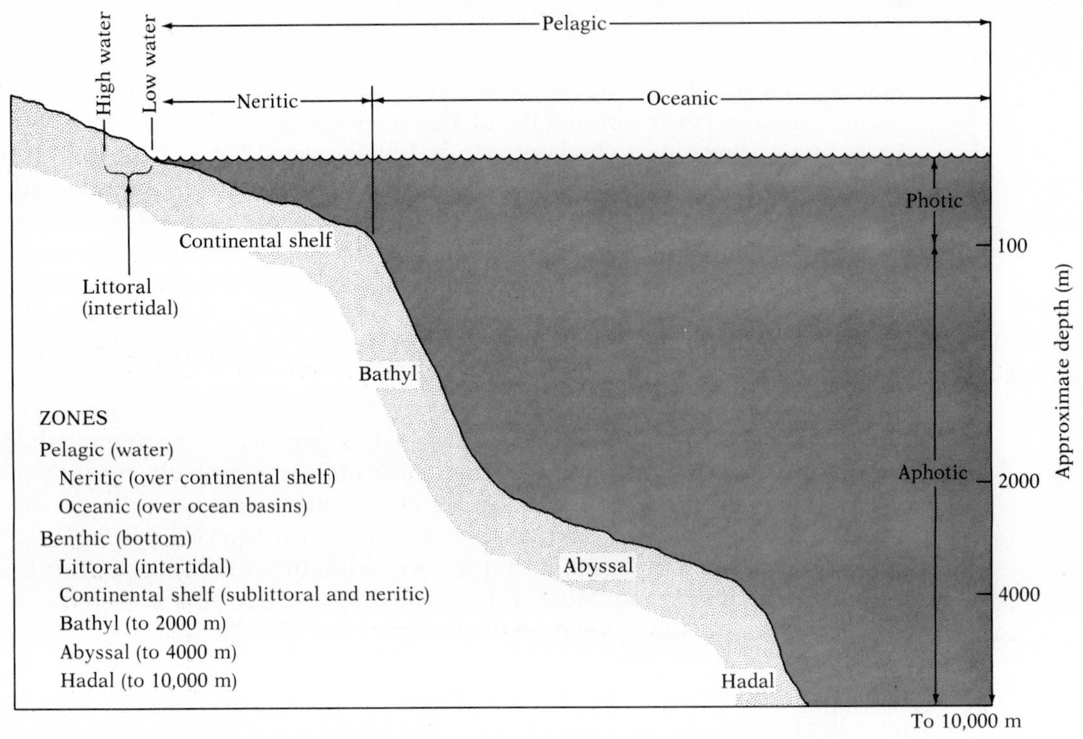

FIGURE 21–1 Cross section of an ocean, showing zonation.

TABLE 21–1 Reef fish diversity on a west–east transect

	Great Barrier Reefs	Fiji	Society Islands	Hawaii
Chaetodontid species	29	23	24	19
Pomacentrid species	91	60	30	15

Source: Anderson *et al.* (1981).

Shallow-Water Habitats

The edge of the sea presents organisms with a challenging array of habitats (Stephenson and Stephenson 1972). There light and nutrients are not normally limiting, but tides and wave action result in gradients of different degrees of immersion and physical disturbance. In the intertidal zone, which is the area between the highest and lowest tides, a complex zonation develops along these gradients, especially on rocky shores (as shown in Fig. 21–2). Below the intertidal zone, a light gradient becomes increas-

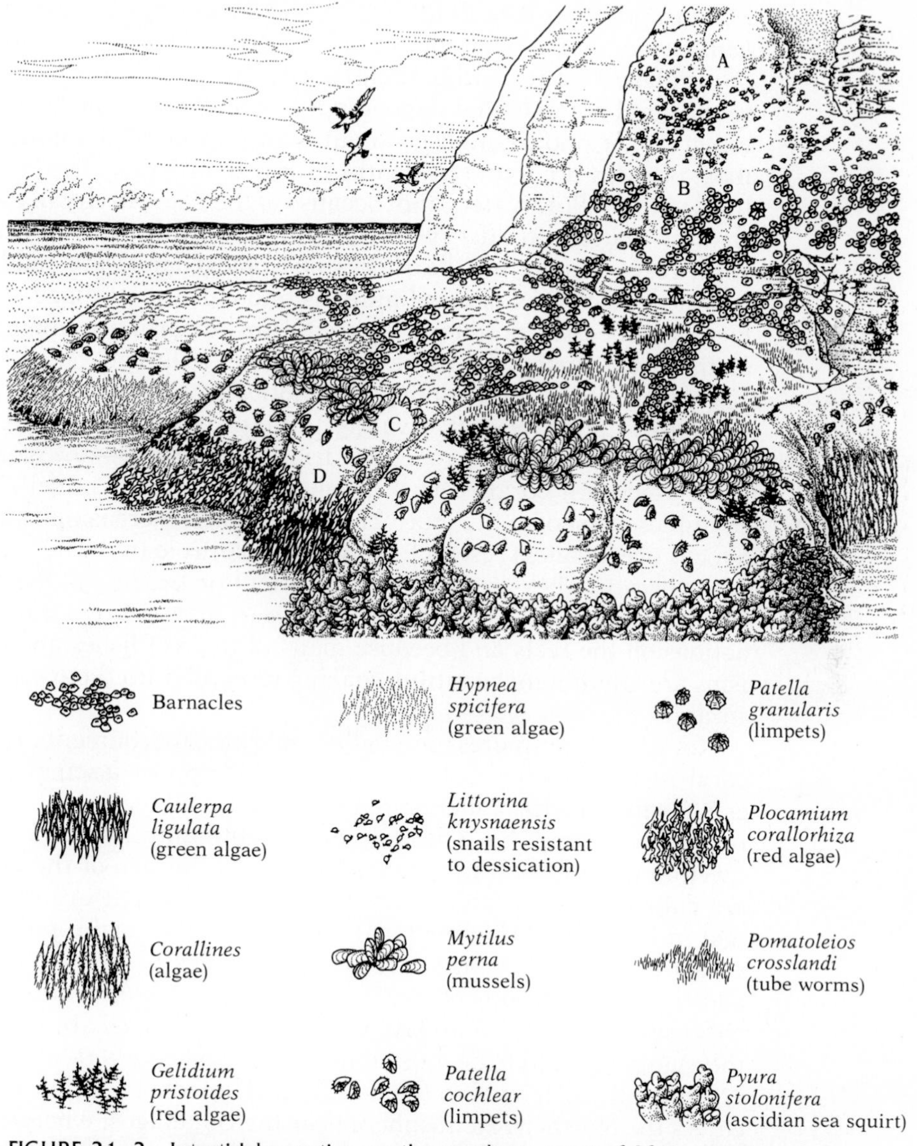

FIGURE 21–2 Intertidal zonation on the southern coast of Africa. (A) The upper edge of the *Littorina* snail zone. (B) The upper edge of the barnacle zone. (C) The upper edge of the zone defined by the limpet *Patella cochlear.* (D) The upper edge of the subtidal zone. The shore is shown at an exceptionally low spring tide on an unusually calm day. The zones are represented in a very diagrammatic fashion, somewhat telescoped; the *Littorina* zone is especially reduced. *(After Stephenson and Stephenson 1972.)*

ingly important. This gradient, for instance, limits the depth at which *hermatypic,* or reef-forming, corals can exist. These animals have symbiotic algae that live within their tissues. The algae transfer photosynthate to the coral animals that depend upon them (Muscatine (1973). At depths too great for photosynthesis, about 90 m, the corals cannot live; indeed, most of them are found at depths of less than 50 m. Coral reefs thus are limited to shallow waters and, because of their sensitivity to cold, to warm shallow waters. In such waters, however, corals thrive. The enormous coral reefs that lie off the shores of tropical lands are made up of the limestone skeletons of coral animals and calcareous algae. They are the largest structures produced by organisms, dwarfing human efforts such as the Great Wall of China or the pyramids of Egypt.

Coral reefs are also among the biotically richest of habitats, with a complexity and diversity rivaling those of tropical rain forests. For example, some 6000–8000 species of fishes, perhaps 30–40% of *all* fishes, are associated with coral reefs (Ehrlich 1975, Lowe-McConnell 1977). Some examples are shown in Figure 21–3. These reefs increasingly are becoming a locus of ecological research because of their role in the nutrient economy of the oceans. They are also drawing attention because of the relative ease with which scuba-equipped scientists can investigate a variety of interactions on the reefs and because many of the reef fishes and other organisms are stunningly beautiful, making research into them especially pleasurable.

Not all of the interest in shallow marine environments is confined to coral reefs, however. Colder waters often support fascinating kelp "forests," which in turn shelter many animals. Virtually all areas of continental shelf ocean bottom maintain a community of *benthic,* or bottom-dwelling, animals; animals that live on the surface of the ocean bottom are called *epifauna,* and those that burrow under it are the *infauna* we met in Chapter 18 (Figure 21–4). Crabs and other arthropods, snails, and a vast diversity of worms are some examples of benthic organisms. In addition, some groups of marine mammals, such as seals, sea lions, and walruses, are confined largely to nearshore areas, as are penguins. All of these animals tend to be most abundant in cooler waters, and the penguins are virtually restricted to the Southern Hemisphere, although one species enters the Northern Hemisphere near the Galapagos, where cool currents travel northward into the equatorial region.

Open Ocean

The producers of the open ocean, the *pelagic zone,* are primarily diverse small phytoplankton, although in some areas floating seaweeds contribute, an example being brown algae such as *Sargassum* of the "Sargasso Sea" near Bermuda. Feeding on the producers are zooplankton, a group representing every major phylum of animals and some minor phyla as

(A)

(B)

FIGURE 21–3 Coral reef fishes. (A) Community in 2–3 m of water at Raiatea in the Society Islands. This community is a typical rich, shallow-water Pacific assemblage, with butterfly fishes, angelfishes, surgeon fishes, and a moorish idol visible. (B) Bluehead wrasse (*Thalassoma bifasciatum*) on a heavily grazed patch of reef in the Grenadines. Caribbean sites like this one have much less diversity than do those of the central and western Pacific and Indian oceans. The small individuals (about 5 cm long) are yellow adults that spawn in groups. Although sexually mature, these individuals continue to grow, and some become "terminal phase" males (8–11 cm) like the large individual shown (which has a blue head). Terminal phase males spawn individually with females. The yellow and bluehead phases of this species were once thought to be different species. *(Photos by Paul R. Ehrlich.)*

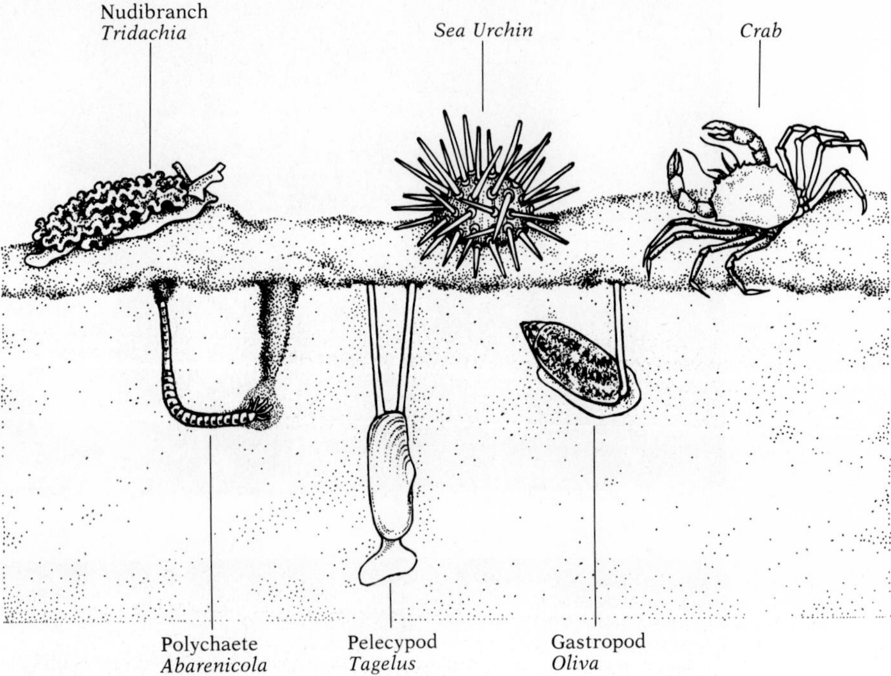

FIGURE 21–4 Examples of benthic organisms—infauna and epifauna.

well. Foraminifera, amoebalike protozoa with calcium carbonate shells, are so abundant that their shells accumulate in sediments called *foraminiferan ooze.* Such oozes from the past are obvious features of geological strata in many areas, and identification of foraminiferan species by the characteristics of their shells is an important tool for recreating the ecology of ancient oceans, with great practical importance in the search for petroleum.

Other zooplankton include radiolarian protozoa, with spectacular spiny silicaceous shells that make them look like miniature "sunbursts," various worms, jellyfishes, shrimplike krill, and numerous other kinds of tiny crustaceans. An important component of open ocean zooplankton is the juvenile forms of many benthic marine animals, ranging from the offspring of sedentary barnacles and oysters to the larvae of reef fishes, the vast majority of which are pelagic.

The *nekton* of the open oceans include those organisms that, unlike the plankton, are strong enough swimmers to overcome currents. They include squid, all manner of fishes, sea turtles, porpoises, and great whales. In addition, sea birds, including albatrosses and petrels, properly are considered denizens of the pelagic zone. Many such birds spend almost their entire lives at sea, feeding on fishes and returning to land only to breed.

Insects, certainly the most successful class of land animals, have barely

(A)

(B)

FIGURE 21–5 Deep-sea fishes. (A) The anglerfish (*Melanocetus johnsoni*), with a large mouth and small eyes, is representative of many deep-sea fishes. The bulb at the end of the stalk projecting from its forehead contains luminous bacteria. Light from the bacteria attracts prey that are then captured by the anglerfish. (B) The anglerfish (*Caulophryne* sp.) ambushes other fish with a quick gulp. The eyes of this fish are vestigial organs; note the rays, however, which are extended and exposed sensory filaments covering the body and which enable the fish to be very sensitive to low-frequency vibrations. *(Photos by B. Robison, University of California, Santa Barbara/BPS.)*

FIGURE 21–6 Black smoker worms. *(Photo by John Edmond, Woods Hole Oceanographic Institution.)*

penetrated the pelagic zone. One group of water striders (hemipterous bugs of the genus *Halobates*) does occur in this zone, as well as in shallower salt water in the tropics and subtropics.

One of the most bizarre faunas on our planet occurs in the eternal darkness of the ocean depths, below the range of light penetration and thus of photosynthesis. These heterotrophs subsist on a steady rain of corpses and feces from the lighted zone above and, of course, on each other. They range from gigantic sharks and a wide variety of grotesque predacious fishes with luminous lures to attract their prey, as shown in Figure 21–5, to sea cucumbers, starfishes, crabs, worms, and other benthic animals.

Strangest of all, entire communities recently have been discovered in the deep ocean around volcanic vents (*black smokers*) in midocean ridges. Near the Galapagos Islands, these vents pour out water that is 20° hotter than the surrounding ocean. Surrounding them live bizarre organisms previously unknown to science (Ballard 1977). The most astonishing, shown in Figure 21–6, is a gutless worm, over 1 m in length, that makes tubes over 3 m long. The worms appear to be nourished by symbiotic chemosynthetic bacteria (Cavanaugh *et al.* 1981) that produce adenosine triphosphate by oxidizing sulfide and reducing CO_2 to organic compounds.

They apparently form the base of the vent community's food chain, making it one of the few that does not use the sun as a basic source of energy. Incredibly, these organisms apparently can grow and reproduce at temperatures well over 200°C. Although this temperature is more than double the boiling point of water, the water remains liquid at these temperatures in the deep seas because of the high pressures. Apparently the proteins, nucleic acids, and other macromolecules of these organisms are also resistant to denaturing at high temperatures as long as they remain under enormous pressure (Baross and Deming 1983).*

FRESH WATER

Like marine communities, freshwater habitats usually have plankton, nekton, and *benthos,* the last a term for bottom-dwelling organisms collectively. An important feature of many freshwater systems is the *periphyton,* communities consisting primarily of tiny algae, bacteria, and protozoa that cling to larger plants, rocks, and the like; sometimes periphyton are called by their German name, *Aufwuchs.*

Freshwater habitats traditionally are divided into standing-water *lentic* habitats, such as swamps, ponds, and lakes, and running-water *lotic* habitats, such as streams and rivers. However, in considering both, it is important to remember their relatively heavy dependence on the *drainage basins* or *catchment areas* that gather in the rain that sustains them.†

Lentic Habitats

The ecology of bodies of standing water is dominated by the consequences of the high *residence time* of the water in them. Water flows in at an inlet, or generally from the surrounding catchment area, and in most cases out at an outlet. However, these flows are usually small in relation to the volume of water in the lake. If a given water molecule remains in the lake a long time, and the lake is not much stirred by the flow, then, in temperate zone lakes during the summer, a characteristic layering or stratification of the water in the lake develops. As water cools, it increases in density until it reaches 4°C. Therefore, as long as no water in the lake is cooler than 4°C, the warmest water will be found at the surface and the temperature will decline with depth. However, the decline is not steady.

* For more information on marine environments, see Steele (1974), Cushing and Walsh (1976), Barnes and Hughes (1982), and Levinton (1982).

† For an overview of freshwater systems, see Bayly and Williams (1973), Hutchinson (1957, 1967, 1975), and Moss (1980).

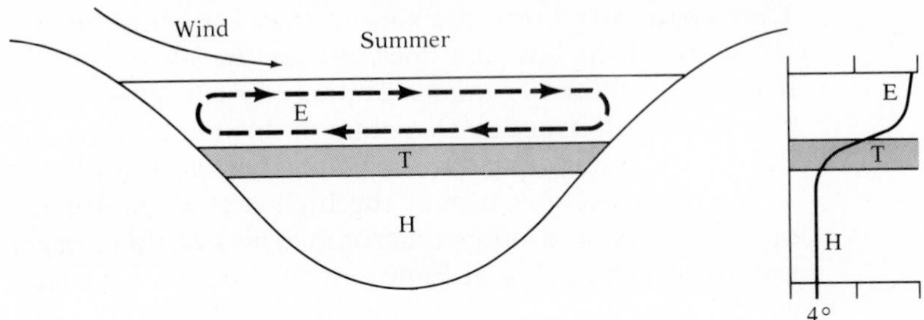

FIGURE 21–7 Cross section of lake stratification and profile of temperature with depth. E = epilimnion; T = thermocline; H = hypolimnion. Temperature scale on the right starts at 0°C and increases to the right. See text for further explanation.

An upper layer called the *epilimnion* is present, consisting of water warmed by the sun and mixed well by the wind. The epilimnion lies over a cool layer too deep to be either warmed or mixed, the *hypolimnion.* As Figure 21–7 indicates, the level at which a rather abrupt transition occurs between the epilimnion and hypolimnion is known as the *thermocline.*

The degree to which light can penetrate the water is important for a reason beyond its warming effect. Light, of course, is required for photosynthesis. If photosynthesis in a community is exactly equal to respiration, that community is said to be at the *compensation point.* In an aquatic community, the depth of light penetration at which the community is able to balance its respiration by photosynthesis, so that the community is at the compensation point, is known as the *compensation level.* Below the compensation level the community becomes dominated by heterotrophs dependent on a rain of processed photosynthate from above.

The compensation level divides an upper autotrophic, or *euphotic,* zone from a lower heterotrophic, or *profundal,* zone. In the summer, the compensation level usually is above the thermocline. Thus, because the green algae that produce oxygen as a by-product of their photosynthesis cannot live below the thermocline, the oxygen content of the waters of the hypolimnion may approach zero. This lack of dissolved oxygen in the hypolimnion is referred to as *summer stagnation.*

The degree of summer stagnation is partly a function of the productivity of the lake. Some lakes have relatively low nutrient contents because of the characteristics of their drainage basins or because of their relative youth—dissolved nutrients have not had a chance to accumulate. Such *oligotrophic,* or poorly fed, lakes tend to have a relatively low density of algae and few rooted plants in their *littoral* zones—that is, the shallows where enough light for photosynthesis can reach the bottom. Therefore, the rain of organic debris into the hypolimnion is relatively sparse, and the population of decomposers to degrade it through oxygen-consuming respiration is relatively small. In oligotrophic lakes, then, relatively little oxygen is consumed in the hypolimnion. Furthermore, many such lakes are sufficiently clear that the compensation level may be *below* the thermocline, and photosynthesis will then occur in the hypolimnion, adding oxygen. As a result, oligotrophic lakes often contain desirable fishes, such as trout, that require both abundant oxygen and cool water.

Most oligotrophic lakes are young in geological terms. As they age, nutrients accumulate, and sediments are deposited that expand the littoral zone. The lakes then support more algae and rooted vegetation, productivity increases, and the lakes may support massive blooms of phytoplankton. The rain of organic matter into the hypolimnion may become heavy, especially as the plankton blooms die out. High levels of respiration in the hypolimnion and higher turbidity of the water, preventing light from penetrating the thermocline, lead to severe oxygen depletion in the depths and the exclusion of fishes such as trout (but not bass, sunfishes, or various "trash" fishes that thrive in warm water and at low oxygen concentrations).

The aged lakes are called *eutrophic,* and the overall process that occurs as they age is called *eutrophication.* Eutrophication can be a natural developmental process in the life of the lake, or it can occur when human intervention rapidly increases nutrient levels, as, for example, when runoff from agricultural areas artificially fertilizes lakes. Such eutrophication often greatly reduces catches of valuable fishes and spoils much of the lake's recreational value. Many other forms of pollution, such as injection of sewage, can also lead to eutrophication.

Because pollution is related to eutrophication, one measure often used as an index to the pollution of lakes and other bodies of water is the *dissolved oxygen concentration.* Another commonly used measure of pollution is the *biological* (or *biochemical*) *oxygen demand* (*BOD*). The BOD is the difference between the production and consumption of oxygen, which may be thought of as the amount of oxygen required for the respiration of the organic materials in the water. The BOD normally is measured in the laboratory as the number of milligrams of O_2 consumed per liter of water in 5 days at 20°C.

As Figure 21–8 indicates, the stratified condition of temperate zone lakes in summer does not persist into the winter. In the fall, the upper layer begins to cool, increase in density, and sink. Fresh oxygen is carried down with it, the thermocline disappears, and the lake becomes thoroughly mixed; the nutrient-rich water of the epilimnion is stirred throughout. At one time, the entire lake is the same temperature; that is, it is *isothermal.*

When winter comes, the surface of the lake freezes, and then a gradient is established from the ice down to the lower part of the lake that contains water at its densest, 4°C. Then in the spring, the ice melts, the water at the surface becomes more dense as it warms to 4°C, and it sinks through the less dense 2–3°C water beneath, producing a *spring overturn* that once again redistributes nutrients and oxygen. This overturn is important to

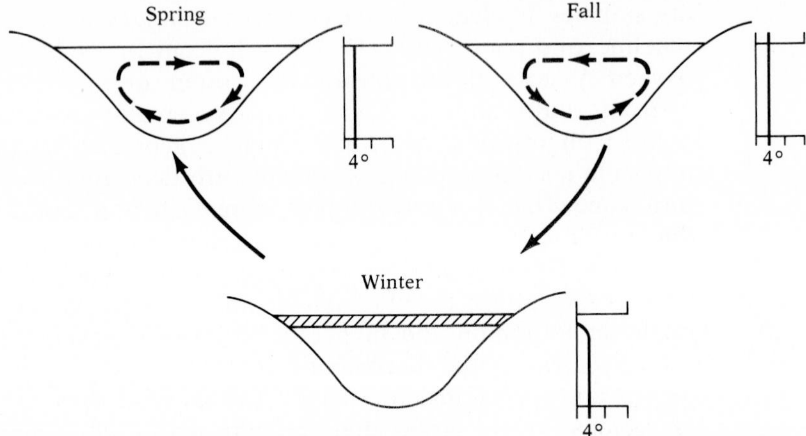

FIGURE 21–8 Annual cycle of a lake. See Figure 21–7 for further explanation.

the continuing productivity of the lake; it often produces a spring plankton bloom as nutrients are brought to the surface at a time of increasing light intensity.

In contrast to temperate lakes, tropical lakes have a rather stable layering created by a weak temperature gradient from the warm surface layer to the only slightly cooler lower layers (and thus slight density differences). Therefore, the ecology of these lakes tends to be quite different from that of temperate lakes. Deep tropical lakes tend to be oxygen poor and unproductive of fishes—as the builders of high dams for tropical development schemes learn to their chagrin.

The community found in any body of fresh water also depends on chemical characteristics of the water other than its oxygen content. For example, as every aquarium enthusiast knows, the pH of the water partly determines which plants and fishes can live in it. Some species, such as neon tetras (*Hyphessobrycon*), come from very acid waters and will breed only at a low pH. Others, such as the livebearing mollies (*Poecilia*), do better in alkaline waters.

The sensitivity of aquatic communities to pH changes and correlated changes in other chemical characteristics has been demonstrated dramatically by the impact of acid rains on lake faunas in the Adirondack Mountains of the northeastern United States. Not only has the water in thousands of lakes been acidified, but the acid rains have mobilized large amounts of aluminum from the soil. The result has been the extermination of the fishes in some 300 lakes and the depression of spotted salamander populations. Similar problems have afflicted lakes subjected to acid rains in Scandanavia (as we will discuss further in Part 7).*

* For an overview of lentic habitats, see Hutchinson (1967).

Lotic Habitats

Lotic habitats differ from lentic habitats primarily in their currents—the low average residence time of a molecule of water—which tend to prevent stratification in either temperature or oxygen concentration. They also differ in their proportionately greater interface with the land—that is, the length of banks in proportion to water volume. Generally, lotic habitats have inadequate production to supply their consumers. Phytoplankton are flushed out by currents, which also, especially in swift waters, discourage the development of rooted plants or algae attached to the bottom. The production deficit is made up by materials swept into the river or stream from the adjacent terrestrial system.

The distinction between lentic and lotic habitats is not as clear as it might first appear. Large lowland rivers may have a quite low flow rate and share many characteristics with lakes, such as major production by phytoplankton. In addition, the shallow waters of wave-scoured shores of lakes tend to resemble fast-running streams in getting more production from algae attached to rocks than from phytoplankton.

Animals that make their homes in streams and rivers have evolved a variety of devices for holding their position against the current. Some, such as trout, are superb swimmers. Various invertebrates burrow into the bottom, and many others have developed holdfasts that anchor them in the stream, as depicted in Figure 21–9.

FIGURE 21–9 Blackfly larva a few mm long, anchored by suckers to a rock on a stream bottom. Note the mouth brushes for filtering food from the flow.

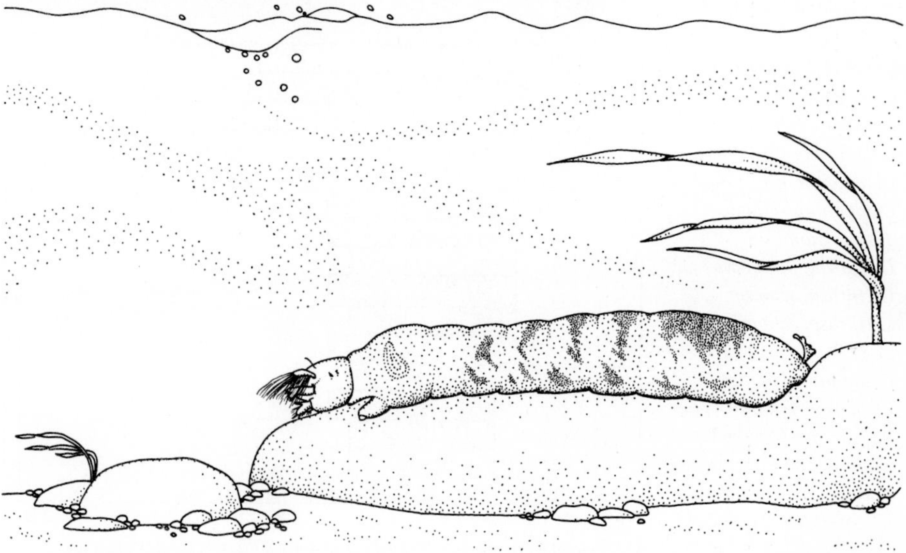

Because the oxygen content of running water normally is high, the animals that inhabit it usually have little tolerance for low-oxygen environments. Therefore, addition of high BOD (oxygen-reducing) pollutants normally has a heavy impact on lotic faunas.*

ESTUARIES

Estuaries are habitats in which fresh and salt waters mingle, as where streams and rivers flow into bays and oceans. Estuaries are characterized by changing salinities as freshwater inputs vary and tides cycle. Organisms that live in them must be especially resistant to osmotic stresses. Because of the general influx of fresh water along oceanic shores, it is not surprising that shallow-water organisms normally show more tolerance to changes in salinity than pelagic organisms (see Figure 21–10). Some animals, such as the brine shrimp *Artemia* and the copepod *Tigriopus*, show enormous tolerance for changes in osmotic pressure.

FIGURE 21–10 Relationship of tolerance to variation in salinity to depth of occurrence of various algae. *(After Biebl 1967.)*

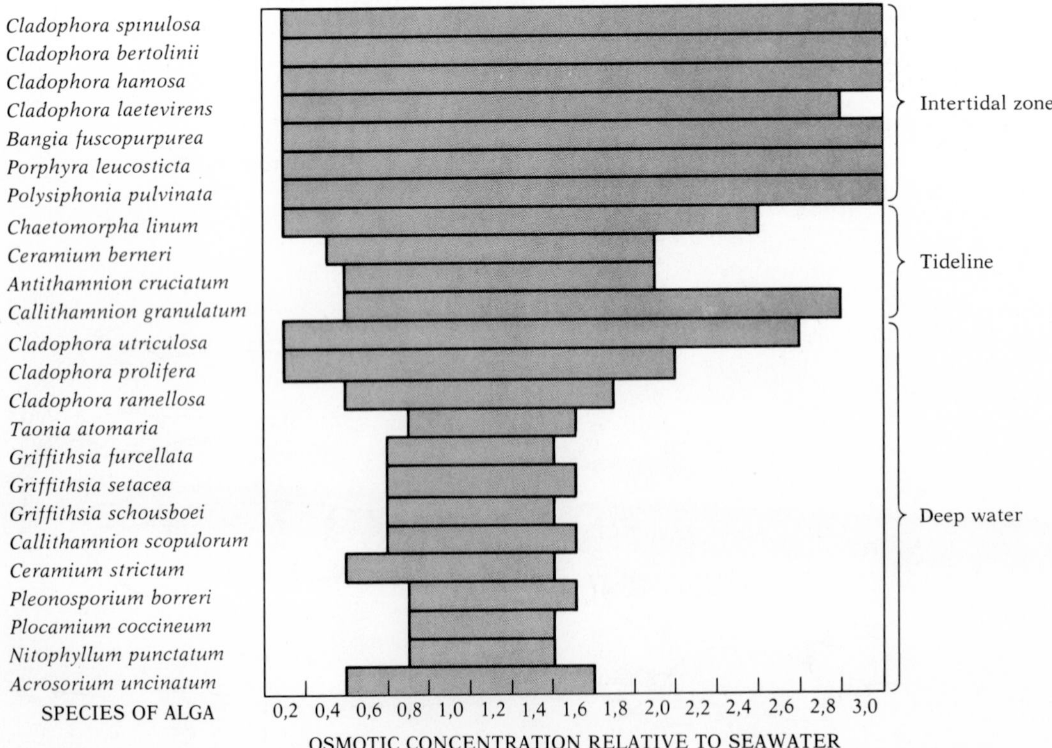

* For an overview of lotic habitats, see Hynes (1970) and Whitton (1975).

Estuaries also tend to have more temperature variation than adjacent oceanic waters. In spite of this and the salinity changes, however, a great many organisms have evolved the capacity to live in them. Apparently the stressful aspects of the estuarine environment are more than compensated for by their function as nutrient traps. The physical circulation patterns tend to retain nutrients that enter estuaries, and the organisms of the benthos tend to retain and recycle nutrients, as well as, in some cases, mobilizing them from deep sediments (Kuenzler 1961, Pomeroy *et al.* 1965, 1969). As a result, productivity is high in estuaries, and they may also play an important role as nurseries for species that are pelagic as adults—possibly including some commercially important fishes and crustaceans.*

* For more information on estuarine systems, see Barnes (1984) and Hedgpeth (1983).

CHAPTER 22

Global Patterns and Communities Through Time: Principles and Reconstructions

In this chapter we continue to enlarge our scale of interest and focus on global patterns and events in geological time.

GLOBAL PATTERNS

Because the physical diversity of our planet has led to a great variety of biological communities, it is appropriate to ask what trends, regularities, or general principles may be perceived in this variety. Is each place unique, or do rules exist that associate one with another?

Productivity

One general characteristic of communities in which trends are easily discernible is their productivity (which is covered in Part 7). For example, although some high-productivity marine communities exist, such as upwelling zones, these communities make up such a small proportion of the oceans that the average marine productivity is less than a quarter of the average terrestrial productivity. Terrestrial communities show a general pattern of high productivity in warmer, moister areas such as tropical forests and swamps. Productivity is lower in cooler, drier communities.

Diversity

The most obvious trend in the diversity of life, a trend that holds for both marine and terrestrial environments, is that species diversity is inversely related to latitude, as we discussed in Chapter 19 and as is illustrated in Figure 22–1. Thus, although about 600 breeding land-bird species are present in Panama, fewer than 50 are present in northern Alaska (MacArthur 1972). Similarly, some 222 ant species are found in São Paulo, Brazil, but only 2 in Tierra del Fuego, 23° farther south. Furthermore, on the order of 35–40% of *all* bony fishes are denizens of tropical coral reefs, and much of the remaining diversity in the group occurs in tropical fresh waters—to the great pleasure of aquarium enthusiasts.

Ecological Equivalents

A casual visitor to an Australian, South African, or Arizonan desert will be impressed by the general similarity of the habitats—dry, with widely spaced, often fleshy-leaved or spiny-leaved plants. However, examination will show that, in spite of their resemblance, the floras and faunas are made up of entirely different species. That resemblance is an important community pattern, testifying to the close similarities of the ecological conditions in these widely separated deserts (just as the convergences shown by the African and South American mammals shown in Figure 20–13 attests to the similarity of rain forest environments on those two continents). The resemblance in the structure of the communities of different organisms in widely separated convergent communities may extend beyond the appearance of the organisms. For example, the relationship between bird species diversity and the diversity in the height of vegetation in Australia and North America is quite similar, as indicated in Figure 22–2.

The degree of convergence of chaparral communities occurring at sites in Mediterranean-type climates in California, Chile, South Africa, and Sardinia has been studied (Cody and Mooney 1978). In the four regions, the dominant plants are evergreen shrubs and trees with sclerophyllous leaves—that is, leaves with high specific weight (weight-to-area ratio)—as shown in Figure 22–3. Furthermore, the relationship between leaf morphology and carbon, water, and nutrient balance is similar in all four sites. However, the taxonomic similarity of the four areas is small. They share no dominant species, and only one genus (*Quercus*—oak) is found in common at two of the four sites. Plant species diversity varies by a factor of four among the sites, with South Africa being floristically richest. Moreover, in spite of the great similarity of the dominant plants in morphology and life-history strategy, the sites show substantial differences in successional relationships and flowering periods, with South Africa again being most different.

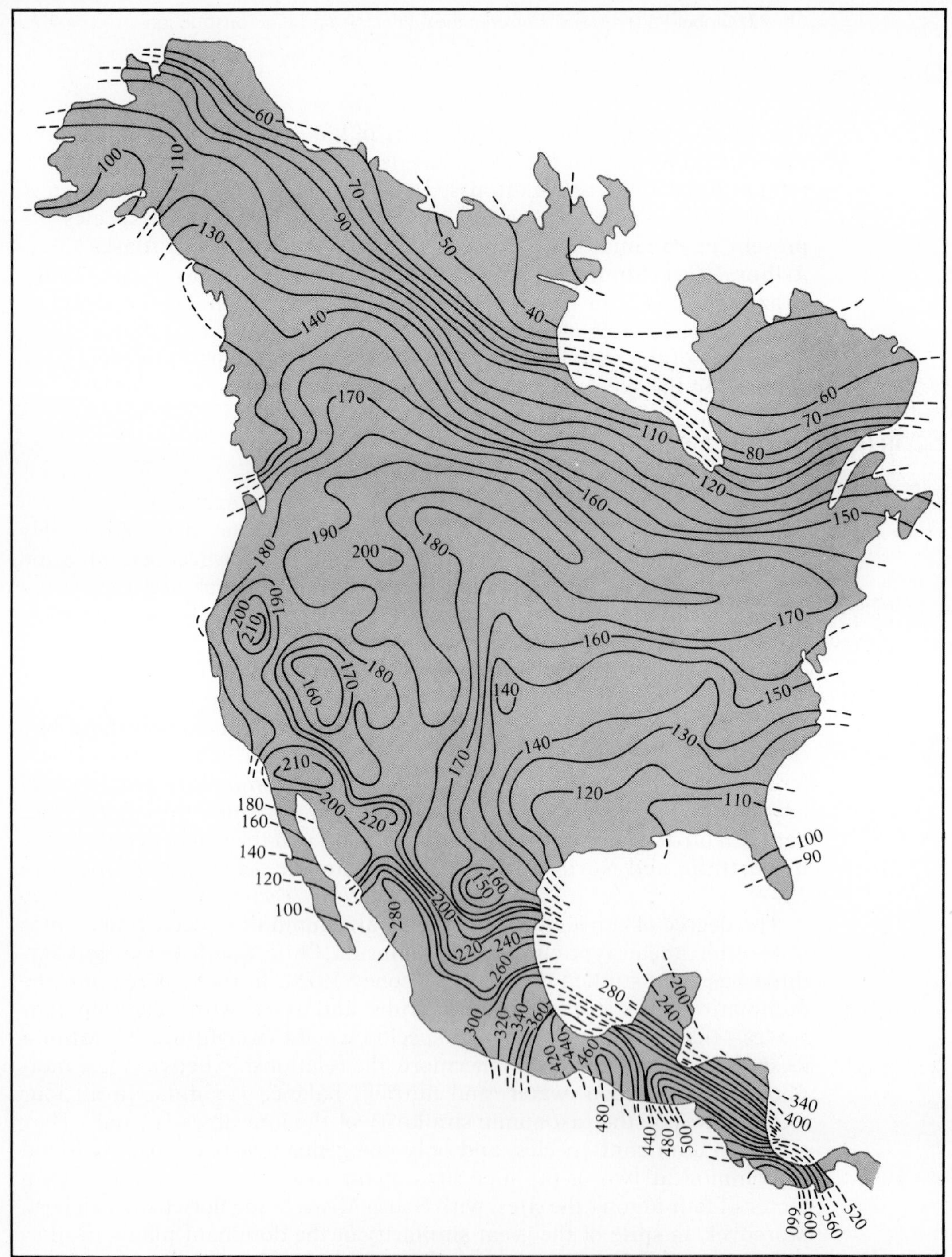

FIGURE 22–1 Species diversity of birds in relation to latitude. Lines connect locations with the same number of species. See also Figure 19–8 for similar data on mammals. *(After Brown and Gibson 1983.)*

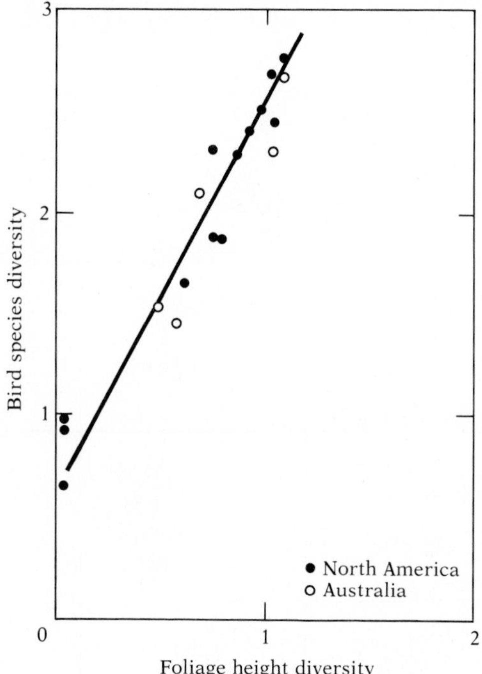

FIGURE 22–2 The relationship of bird species diversity and foliage height diversity in Australia and North America. Scale on axes are in arbitrary units. *(After Recher 1969.)*

These differences, and differences in resources, have led to important divergences in the consumers of the four communities. For example, the *fynbos* of South Africa contains an abundance of *Protea* flowers (Proteaceae). These plants are excellent sources of seeds, nectar, and insects for birds, and they in turn have led to a more divergent bird fauna in the fynbos. Another example is the existence of a well-developed herb layer in the Chilean *matorral,* which accounts for a greater diversity of ground-foraging birds there than at otherwise comparable sites on other continents.

In summary, although substantial structural similarities exist among the four communities investigated, so do substantial differences. As we shall see in the next section, the differences may result partly from the different histories of these four similar areas.

COMMUNITIES THROUGH TIME

Today's organisms are the products of a long, complex evolutionary and ecological history. To understand modern communities fully, one must often look to the past. Sometimes the impact of historical accident is dramatic. It seems clear, for example, that the absence of penguins from

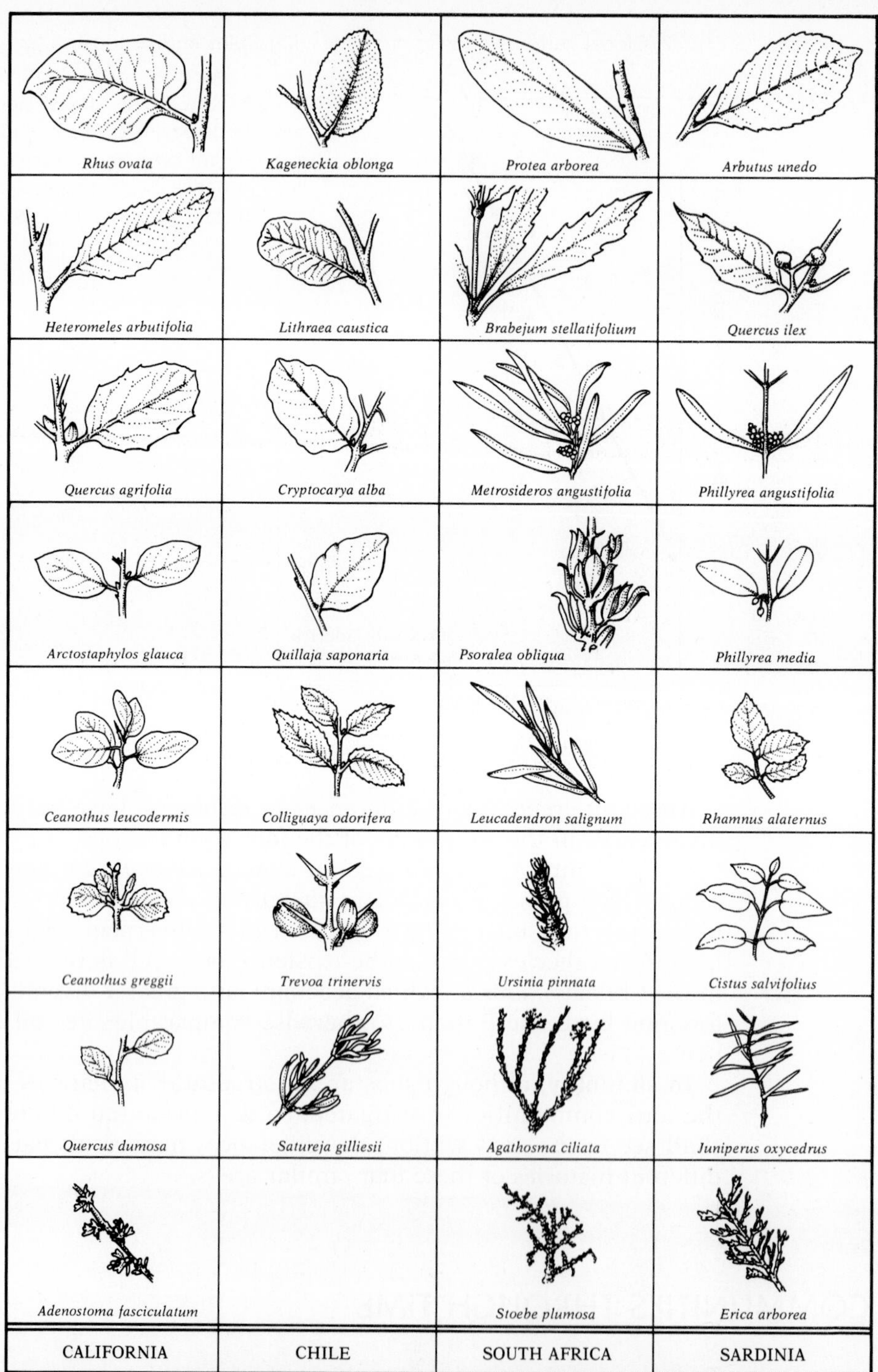

FIGURE 22–3 Leaves of dominant trees and shrubs from four Mediterranean-climate regions. All leaves are one-third of their natural size except for *Protea arborea* and *Brabejum stellatifolium,* which are one-fourth of their natural size. *(After Cody and Mooney 1978.)*

north polar regions and of polar bears from Antarctica results simply from the failure of each to reach the other polar region—not from an inability to adapt to the conditions at the other pole. In this section, then, we will take a brief look at historical ecology—both because of its intrinsic interest and because of the light it can throw on the present.

THE NEW GEOLOGY

Since the first maps of the globe were drawn, people undoubtedly have noted the resemblance between the S-shaped curve of the eastern edge of the Western Hemisphere continents and the S-shaped curve of the western edge of the Eastern Hemisphere continents. Benjamin Franklin even speculated that beneath an outer shell Earth could be a dense fluid and that the shell would be "capable of being broken and disordered by the violent movements of fluids on which it rested" (quoted in Sullivan 1974, p. 75).

In the early 1900s, German meteorologist Alfred Wegener proposed that the continents moved around, because they were made of lighter rocks than were the ocean basins. He developed a complete scheme of *continental drift,* postulating the existence of an ancient land mass that broke up to form the present continents. However, physicists calculated that no conceivable force could cause the granitic continents to plow like ships through the basaltic ocean basins. Thus, until the 1960s, Wegener's ideas of continental drift languished, and geologists believed that the continents were where they had always been, that the Earth's crust had solidified just once, and that movements since had been largely vertical "breathing" rather than horizontal drifting.

Then, in the 1960s, evidence began to accumulate to the contrary. The past orientation of the Earth's magnetic field, preserved in old rocks when they first solidified, was found to indicate that the north magnetic pole was far from where it should have been. Either the pole had migrated more than was thought possible, or the continents had moved relative to it.

Next, a ridge almost 50,000 miles long was discovered wandering through the ocean basins. The section in the Atlantic Ocean had an S shape paralleling that of the coastlines, as can be seen in Figure 22–4. It was then hypothesized that the ridges were the sites of upwellings of molten rock, or *magma,* that were continually producing new ocean bed (Dietz 1961, Hess 1962). An elegant proof of the correctness of this view was provided by studies of the patterns of magnetism (see Figure 22–5) in the ocean bed on either side of the ridge (Vine and Matthews 1963). Other evidence quickly accumulated that supported the idea that ocean floor is being continuously created.

If the ocean floors were steadily growing, this meant either that Earth as a whole was growing or that ocean beds somewhere were being de-

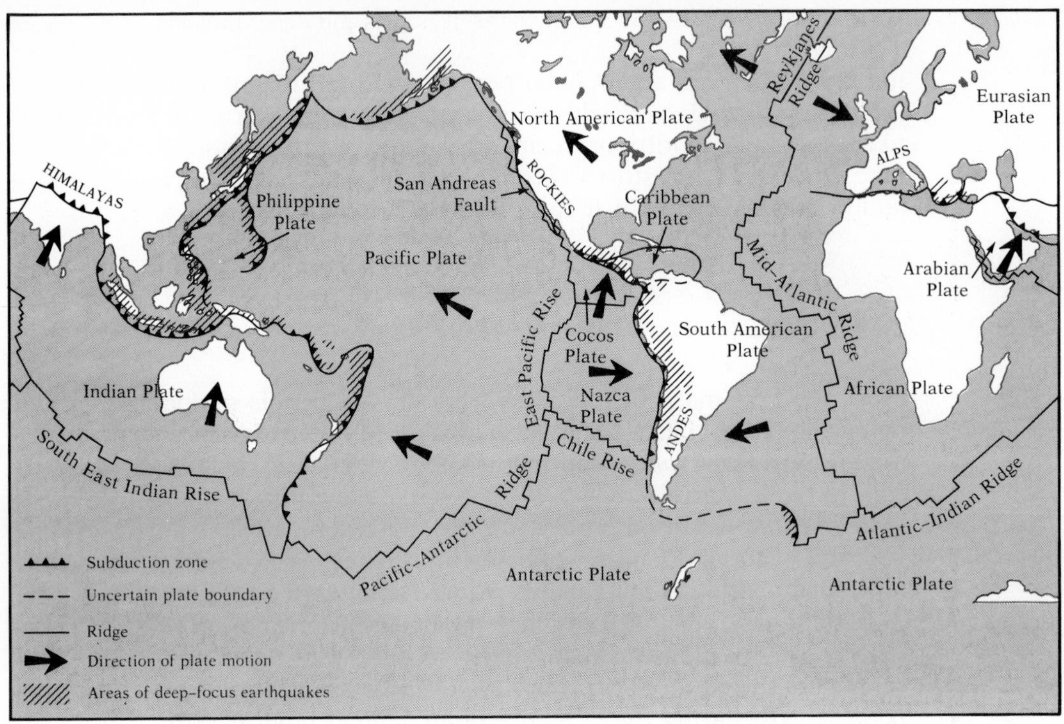

FIGURE 22–4 Midocean ridges and tectonic plates. See text for explanation. *(After Uyeda 1978.)*

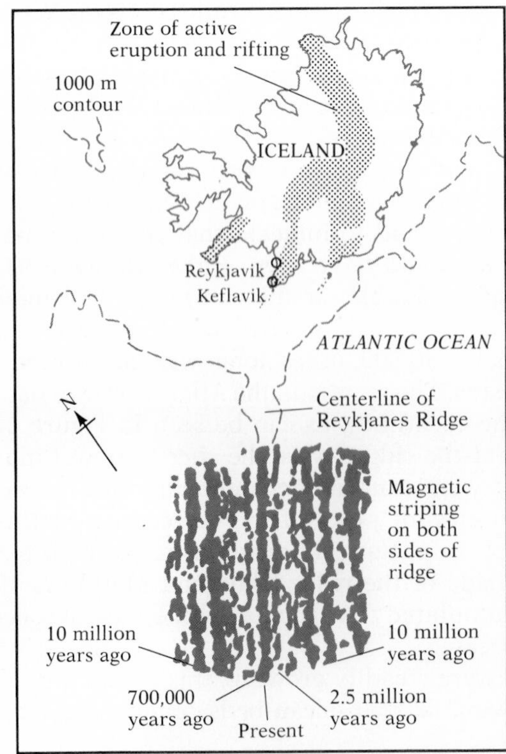

FIGURE 22–5 Magnetic "striping" along the mid-Atlantic ridge south of Iceland. The striping records periodic reversals of Earth's magnetic field. The orientation of the field is preserved in the magma as it solidifies; the symmetrical pattern confirms that new sea bed is formed by magma pushing out both directions from the ridge. Dark striping represents sea floor with "normal" (present day) polarity; light striping reversed polarity. When Earth's polarity reverses again, material solidifying along the ridge will be striped "white," dividing the present dark stripe. *(After Sullivan 1974.)*

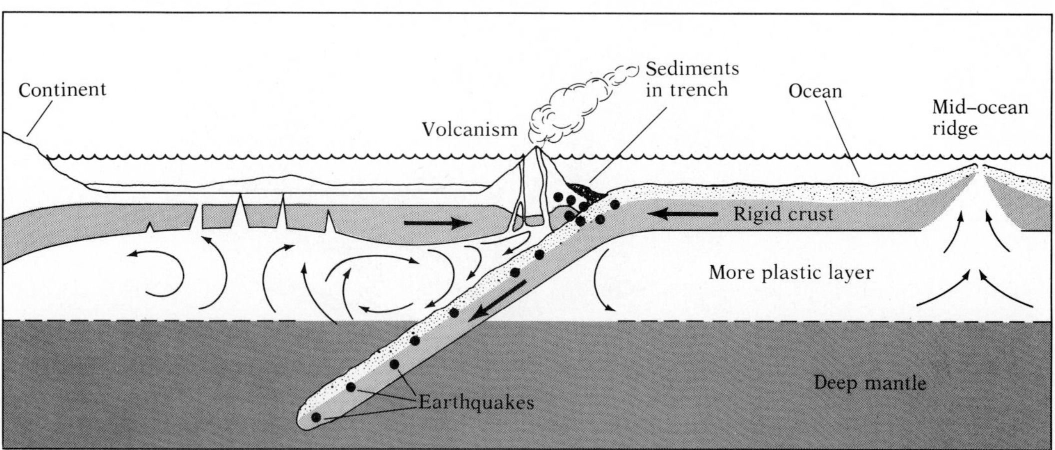

FIGURE 22–6 Subduction. The rigid crust of one tectonic plate is overridden by the edge of another, and a trench forms where they meet. Arrows in the more plastic layer show possible patterns of convective flow. *(After Toksöz 1975.)*

stroyed. The location of the destruction was provided by seismic studies in which geophysicists were able to plot closely the distributions of earthquakes. They showed, for example, a zone of quakes angling down from the Pacific shore under the Andes. These quakes were interpreted as being the grumblings of Earth as it digested ocean floor beneath the west coast of South America.

Based on these investigtions, and on other seismic studies showing Earth to contain several plastic layers under a crust 60 to 100 mi thick, a new picture of an ever-changing surface emerged. Earth's crust was fractured into a series of plates, proportionately thinner than an eggshell, moving about on the hot, more fluid layers below, as represented in Figure 22–6. At certain places, such as the midocean ridges, new crust was being formed from molten material coming up from below. At others, such as the deep ocean trenches, one plate was being overridden by another. The edge of the lower plate was being shoved back progressively into the hot layers—being *subducted*—and being melted down and returned to magma (see Figure 22–6).

Wegener had been largely right. The continents do move, but not *through* the crust—rather, *along with it.* The pace is stately, only an inch or so a year. Some 200 million years ago, all the continents were jammed together into a single land mass, now called Pangaea, the name originally proposed by Wegener. Since then, they have drifted to their present positions, as shown in the sequence of illustrations in Figure 22–7. In 50 million more years, they will have drifted into the configuration shown in Figure 22–7F.

Continental drift, of course, has had dramatic effects on Earth's climatic machinery. It is thought that 500 million years ago, Pangaea was

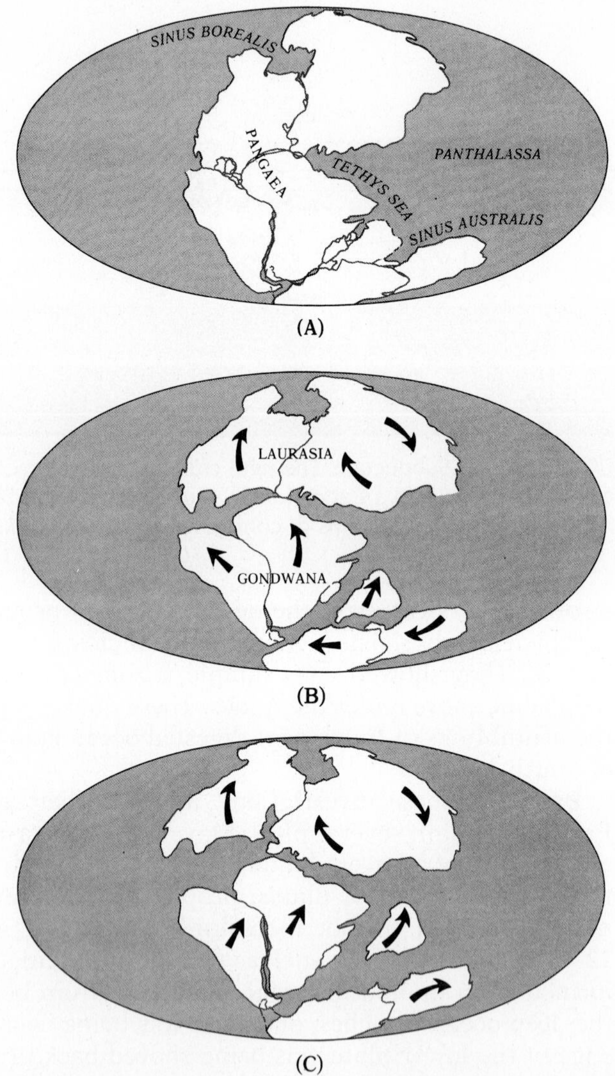

FIGURE 22–7 Continental drift. Arrows show general directions of movements; the tectonic plates that bear the continents do the actual drifting. For simplicity, the plate boundaries, which are largely under the oceans, are not shown here. (A) The supercontinent of Pangaea, 200 million years ago. (B) Pangaea breaking up into a northern supercontinent, Laurasia, and a southern supercontinent, Gondwana, 100 million

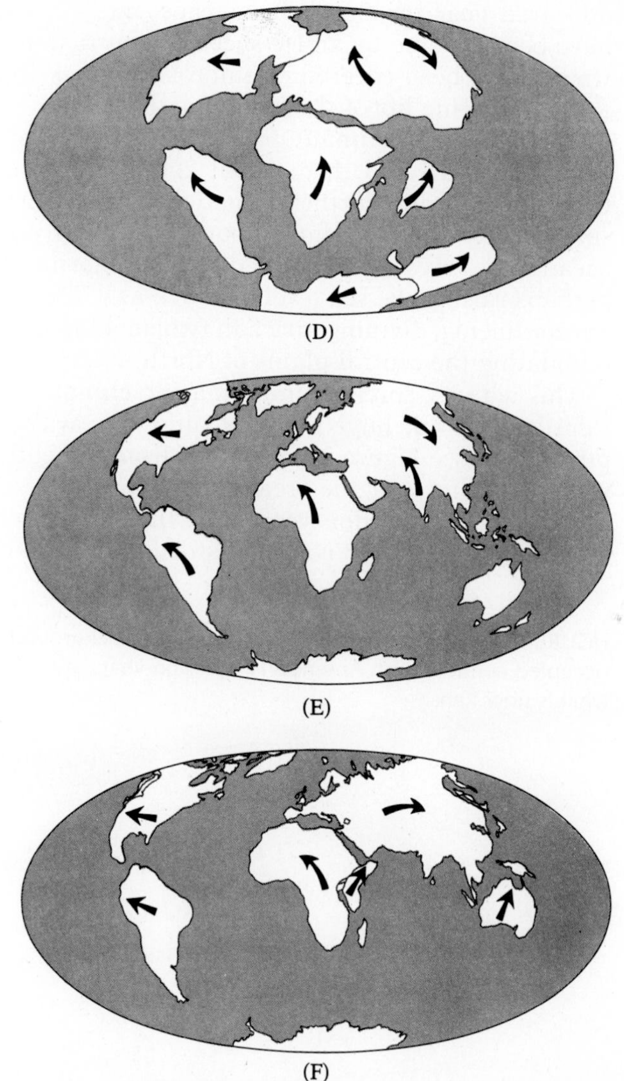

years ago. (C) The breakup continues 135 million years ago. (D) As the dinosaurs disappear, the South Atlantic opens up, 65 million years ago. (E) The Earth today. (F) The future: 50 million years from now, the Atlantic widens, the Pacific shrinks, and the west coast of the United States today's city of Los Angeles has drifted to the north of San Francisco. *(After Uyeda 1978.)*

appeared near the South Pole. Thus, most of Earth's land mass would have been covered by an ice sheet. In contrast, the Northern Hemisphere would have been covered by unbroken warm oceans—habitats in which the rich marine life of the Paleozoic thrived.

The terrestrial climate gradually ameliorated as Pangaea inched its way northward and then, about 200 million years ago in the Triassic, began to break up. Further climatic change occurred during a widespread surge of sea bed spreading about 110 million years ago. The broad mid-ocean ridges created and then displaced enormous quantities of seawater, perhaps raising the sea level as much as 475 m. Shallow seas penetrated the continents, dividing Africa in two, making Europe an archipelago, and inundating the central plains of North America.

This activity spread benign marine climates over much of the globe. Reptiles thrived; huge herds of dinosaurs walked the Earth, and giant pterosaurs (see Figure 22–8) soared from the cliffs bordering the shallow seas stretching over the area that is now Texas, Oklahoma, and Kansas. Then, about 85 million years ago, the spreading subsided. The shallow seas receded, and the continents drifted farther toward the North Pole.

FIGURE 22–8 Pterosaurs soaring from cliffs (background) over the shallow sea that occupied central North America 100 million years ago; the cliffs were located in what is now Kansas.

These events, combined with the creation of new mountain ranges by the surge—the Himalayas, Alps, Andes, and Rockies—chilled Earth as a whole and increased the temperature differential between the equator and poles and between the seasons.

These climatic changes caused the disappearance of many groups of organisms and a general decline of the reptiles. Whether they were responsible for the ultimate extinction of the dinosaurs (as we discuss later in this chapter) remains controversial, but they probably helped them on their way. Also, it is significant that the two main groups to expand after the demise of the dinosaurs, birds and mammals, are both endotherms. Whether changes in the sun's output or in the composition of the atmosphere have added to the climatic variation also is not known.

We have just painted with a broad brush what is now sometimes called the "new view of the Earth." It is a view of great significance to ecologists and evolutionists because it pictures not a static planet but one continuously swept by great environmental change. Thus, both the dynamics and genetics of populations have changed in response not just to the kinds of short-term changes experienced in historic times but to enormously greater changes in geologic time.

BROAD HISTORICAL PATTERNS

The outstanding events in the history of life, as reconstructed from the fossil record, are displayed in Table 22–1. Note, for example, that *all* ecology for the first 4 billion years of Earth's history was aquatic ecology. The

TABLE 22–1 The geological time scale

Era	Period	Epoch	Millions of years from start to present	Major events
	Quaternary	Recent	0.01	Repeated glaciations of the northern hemisphere;
		Pleistocene	2	extinctions of many large mammals; evolution of modern people
Cenozoic	Tertiary	Pliocene	12	Proliferation of mammals,
		Miocene	25	birds, snakes, and lizards;
		Oligocene	36	insects flourish and diversify
		Eocene	58	
		Paleocene	65	

TABLE 22–1 The geological time scale *(continued)*

Era	Period	Epoch	Millions of years from start to present	Major events
	Cretaceous		144	Angiosperms become dominant over gymnosperms; mass extinctions at end
Mesozoic	Jurassic		213	Dinosaurs abundant; appearance of birds, mammals, and angiosperms; gymnosperms dominant
	Triassic		248	First dinosaurs; gymnosperms rise to dominance
	Permian		286	Adaptive radiation of reptiles; extensive glaciations; extinction of many marine groups, including trilobites; amphibians; early plants decline
	Carboniferous		360	Great coal forests of club mosses, horestails, and ferns; age of amphibians; first reptiles; early radiation of insects
Paleozoic	Devonian		408	Age of fishes; first amphibians and seed plants
	Silurian		438	First plants and arthropods invade land
	Ordovician		505	First vertebrates; sponges and reef-building cnidarians abundant
	Cambrian		590	Appearance of most invertebrate phyla and plants
			1500	Origin of eukaryotes
Precambrian			2500	Origin of modern photosynthetic bacteria
			4000?	Evolution of anaerobic bacteria
Origin of Earth			4600	

first plants did not invade the land until the Silurian, some 425 million years ago. Note also that not until 35 million years ago or so, in the Tertiary, did communities began to take on their modern form. At that time, flowering plants, the most prominent feature of the majority of today's communities, were developing their modern diversity, mammals had diversified to replace the dinosaurs, and our own ancestors had reached the status of ape.

Considerable controversy exists over the pace at which the ecology of the past was transformed. Recall from Chapter 6 that the gradualist school of thought holds that evolution has occurred more or less continuously, although at varying rates. The evolution of higher taxa—genera, families, orders, classes, and so on—is viewed as simply the summation of millions of years of microevolution. Recently, however, a group of paleontologists has challenged this view. They claim that the gaps in the fossil record are too frequent to be explained simply as the result of accidents of sampling—that is, failure of fossils to be formed or found. Their interpretation of that record is that it is primarily one of the sudden appearance of species or higher categories, followed by a period of stasis in which little or no evolution occurs, followed in turn by a rapid disappearance. This theory of punctuated equilibria thus has a quite different emphasis from that of phyletic gradualism.

However, much of the paleontological evidence does not fit the punctuated equilibrium model (see, e.g., Gingerich 1977). Neither does the fossil record of our own species. Furthermore, considerable evidence indicates that speciation is going on continuously at the present time. Group after group of living organisms shows populations in all stages of the speciation process—from ones that have small statistical genetic differences to ones so distinct that they are verging on becoming full species. Still, these facts also may be consistent with the punctuationist position. The present could be an episode of unusually rapid speciation. Perhaps more important, species now being formed may have low survivorship, especially as they deviate from ancestral forms. The pruning action of *species selection*—that is, the differential survival of species—eventually might produce a macroevolutionary picture of relative stasis (Stanley 1975).

All biologists are agreed that the fossil record presents prominent examples of relatively sudden, perhaps catastrophic environmental changes leading to mass extinctions (Elliot 1986). For example, one, the biggest of all, occurred at the end of the Permian, when over 95% of marine species disappeared (Newell 1967, Lewin 1983a). Another occurred at the end of the Cretaceous, as we will see when we discuss the ecology of dinosaurs. The degree to which the pace of evolution between these spectacular events is stately is likely to remain controversial (Gould 1977). The answer has implications both for the rates of change of environments and for the coupling of those rates to changes in species and communities in the past and in the future.

THE ECOLOGY OF DINOSAURS

In recent years, a dramatic and controversial reappraisal of the paleo-ecology of dinosaurs has taken place. Since before Darwin's time, dinosaurs had been viewed as being gigantic versions of lizards; indeed, the word *dinosaur* means "terrible lizard." Gigantic herbivorous dinosaurs of the infraorder Sauropoda (sauropods), such as *Apatosaurus* (also known as *Brontosaurus*—see Figure 22–9) and *Diplodocus*, were thought to have lived mostly under water to support their 30- to 50-ton bulks. When they did move onto the land, they were pictured as lumbering slowly and stupidly over the landscape, dragging their 10-m tails behind them.

The monstrous carnivores that fed on the sauropods were equally presumed to be sluggish creatures. Some paleontologists deemed *Tyrannosaurus* (see Figure 22–10) and *Allosaurus* as capable of little more than lying in wait for their robotlike prey and then leaping on them with a short burst of activity, much as a sunning lizard lunges quickly at a passing insect.

This view of the dinosaurs, which can be traced back to the great French and British anatomists of the early nineteenth century, Baron Georges

FIGURE 22–9 *Apatosaurus*, one of the sauropods, a group of dinosaurs that included the largest animals ever to walk on land. The sauropods thrived 65–150 million years ago. *Apatosaurus* is better known as *Brontosaurus*.

FIGURE 22–10 *Tyrannosaurus rex*, an impressive carnivore that presumably preyed upon sauropods (Figure 22–9). Its remains have been found in sediments 70–75 million years old in Mongolia and North America. This animal measured some 13 m from the nose to the tip of the tail; its jaws, armed with 15 cm saw-edged teeth, were more than 1 m in length.

Cuvier and Richard Owen, is based on the notion that dinosaurs were simply enlarged versions of modern reptiles. Reptiles have metabolisms very different from those of birds and mammals, and moreover, they are not very intelligent when chilled. The dinosaur-as-giant-lizard view proved compelling and enduring—so much so that some paleontologists claimed that the sauropods could not have stood erect but must have had their bodies suspended between bent and splayed-out legs, like modern lizards and crocodilians. This belief was manifestly preposterous, for the anatomy of the legs showed clearly that they had the same design as an elephant's legs, straight and pillarlike for supporting great weight. Besides, the rib cage of a *Brontosaurus* was sufficiently deep that, if its legs really were articulated like those of a crocodile, the gigantic creature would have required a groove in the ground in order to move. Still, that serious scientists would propose such a scheme in the face of all of the anatomical evidence shows how deeply rooted the view was.

In retrospect (but only in retrospect) it is difficult to understand why the dinosaur-as-giant-lizard notion was so entrenched. The sauropods were upright and quadripedal, and their predators were upright and bipedal. Clearly, unlike contemporary crocodiles, they did not spend most of their time resting on their bellies. Therefore, if they did not, they may have required a more efficient metabolic system to permit the intense bursts of energy and rapid recovery from oxygen deficits that are characteristic of large, fast, highly maneuverable mammals today (Bakker 1972, 1975). Perhaps dinosaurs and birds, which share key anatomical features with them, should even be joined in a single taxonomic class; birds, in essence, should be viewed as surviving dinosaurs (Bakker and Galton 1974).

The gigantic bulk of many dinosaurs precludes the possibility that they, like small lizards, were warmed to operating temperature by the sun each morning. Their surface areas are too small relative to their volume to permit basking to raise their body temperatures even a small amount in less than a few hours. (The exact times are a matter of some dispute and would vary with size, shape, color, routing of the vascular system, and so on.) Furthermore, experiments with alligators, as well as theoretical calculations, indicate that in order to warm its viscera by basking in the sun, a large dinosaur would have to be subjected to a skin-to-guts temperature gradient that would cook the tissues near the surface (Colbert *et al.* 1946, 1947). Equally, as may be recalled from Part 1, should an alligator overheat, it would have great difficulty cooling rapidly.

For these and other reasons, dinosaurs are now viewed by some paleontologists as having regulated their body temperature much as modern mammals do (reviewed by Ostrom 1980). Support for this hypothesis has been claimed on the basis of attempts to reconstruct food chains that included dinosaurs. Because modern reptiles do not use nearly as much energy per unit of time as mammals of the same size, reptilian predators can be much more abundant relative to their prey than equivalent mammalian predators. However, in the late Cretaceous, for example, remains of the herbivorous duck-billed hadrosaurs and rhinolike ceratopsians vastly outnumbered the fossils of the carnivorous dinosaurs that presumably fed on them (Bakker 1972, 1980). The predator/prey biomass ratio appears more appropriate for the lions and antelopes of the African *veldt* than for a lizard population and its insect prey (as we will see in Part 7).

Of course, serious problems arise when one attempts to treat the fossils associated in a geological stratum as a random sample of a community. This difficulty has led other scientists to conclude that the food-chain evidence does not support the "warm-blooded dinosaurs" hypothesis (Beland and Russell 1980). For example, herbivorous dinosaurs may have spent more of their time in habitats in which they were more likely to be fossilized than the carnivores that preyed upon them. It is also possible that the largest herbivorous dinosaurs, like the elephants that eventually evolved into a similar niche, were relatively free of predation.

The view that dinosaurs were, in fact, mostly ectotherms is still held by many paleontologists (e.g., Beland and Russell 1979, Benton 1979, Regal and Gans 1980). They argue, for instance, that large size alone could produce high and relatively constant body temperatures in ectotherms living in an equitable climate. Moreover, they hypothesize, large dinosaurs could have remained permanently in such climates by seasonal migrations, whereas small ones hibernated or aestivated. Furthermore, many paleontologists claim that mammal-like temperature regulation would have been too "costly" a trait, requiring more food than could be provided by the arid Triassic environments in which dinosaurs thrived (Benton 1979). Indeed, some believe that ectothermy would be a *preferred* trait in very large animals in tropical climates (Regal and Gans 1980). Even the mammals show variation in their degree of temperature regulation (Eisenberg 1981).

The dispute over the metabolic patterns of dinosaurs has led to several other interesting avenues of speculation. Rather than being sluggish, stupid creatures, as had been hypothesized, dinosaurian predators—especially the smaller ones—may well have been swift, well-coordinated, relatively clever hunters. Recent studies of the brains of smaller predatory dinosaurs indicate a capability for rapid motor coordination and social behavior more like that of an ostrich than of a reptile. Such organisms could have placed considerable predation pressure on the small, nocturnal Mesozoic mammals. Indeed, some dinosaurs may well have made their living as nocturnal or crepuscular hunters, snatching up our tiny ancestors and, for more than 130 million years, successfully barring the mammals from most of the niches mammals now occupy.

Perhaps the greatest ecological puzzle posed by the dinosaurs is what caused the extinction of such an incredibly successful group of organisms. They had been gradually declining in the late Mesozoic; at the end of the Cretaceous, the survivors disappeared from the fossil record in a geological twinkling of an eye. Many other groups, such as the ammonites (mollusks with spiral, nautiluslike shells) and various marine phytoplankton (Smit and Hertogen 1980), bowed out with them.

Few, if any, features of the geological record have been the subject of more speculation and controversy. Some of the speculation has been nonsensical, such as the notion that dinosaurs as a group became "senile" and died of old age or that mammals wiped them out by eating their eggs. Other speculation relates to the change in floras at the end of the Mesozoic. It has even been suggested that the decline of ferns changed the diets of herbivorous dinosaurs and that they died out of constipation. More reasonably, the extinctions have been blamed on the failure of the herbivores to coevolve rapidly enough with the flowering plants as the plants developed a vast array of antiherbivore chemicals.

None of the plant-related theories, however, explain why other groups, such as some marine phytoplankton, disappeared simultaneously with the dinosaurs, why some groups of predacious dinosaurs did not persist by preying on mammals, or why at least some dinosaurs did not manage to keep up in the coevolutionary race with their food plants. Furthermore, inasmuch as flowering plants appeared about 120 million years ago, why did their chemical warfare take 50 million years to conquer the dinosaurs and then do the deed with catastrophic suddenness?

More tenable explanations of the sharp faunal and floral punctuation at the end of the Mesozoic involve planetwide climatic change. In our discussion of continental drift, we mentioned a possible cooling trend developing at that time. Such a cooling might have had deleterious consequences for large, warm-blooded animals that, in an equitable climate, had not evolved insulating fur or feathers, just as today's elephants, rhinos, and hippos are essentially hairless. A cooling trend might also have posed difficulties for ectotherms; recall the scarcity of amphibia and reptiles in the taiga and their absence from the tundra.

Various suggestions also have been made regarding an extremely sudden event causing the extinctions, such as a radiation "storm" created by a supernova in our portion of the galaxy or a catastrophe caused by collision of the Earth with an asteroid or a giant fragment of a comet's tail (Hsü 1980, Kerr 1980, Clube and Napier 1982a,b). A collision might have killed many organisms outright with blast and heat, then lofted huge amounts of dust into the atmosphere, which in turn might have eliminated most photosynthesis for months. This event would have left as survivors primarily animals in the arctic and temperate zones that had been in diapause or hibernation in the winter hemisphere.

The collision hypothesis is supported by some unusual thin geological strata, 65 million years old, containing high concentrations of iridium. They have been found in the Apennine mountains, near Gubbio, Italy, and elsewhere (Alvarez *et al.* 1980, Ganapathy 1980). Normally, iridium is found in only trace amounts in the Earth's crust. That metal has a strong affinity for iron, and most terrestrial iridium presumably sank into the core of our planet when heavy and light elements sorted out billions of years ago. However, iridium is relatively abundant in meteorites. The thin, iridium-rich deposits form a boundary between layers below containing marine organisms typical of the Cretaceous and layers above containing Paleocene fossils.

The matter has been heavily debated.* For example, recent discoveries in New Mexico of strata of terrestrial origin with an iridium-rich layer have been interpreted as showing that land organisms did not suffer catastrophic extinction at precisely the same time as marine organisms. The view that a sharp discontinuity occurred at the Cretaceous–Tertiary boundary and that it was caused by an asteroid impact continues to re-

* Papers dealing with this controversy include Mclean (1980, Smit (1980), Hickey (1981), Archibald and Clemens (1982), Hsü *et al.* (1982), O'Keefe and Ahrens (1982), Officer and Drake (1983), Alvarez *et al.* (1984a,b), Berggren and Van Colvering (1984), Hallam (1984), Russell (1984), Smit and van der Kaars (1984).

ceive support. Some of the most impressive recent evidence is that quartz grains from the iridium-rich boundary layer show features of *shock metamorphism*—changes in their crystal structure such as those found in grains from the sites of meteor impacts (Bohor *et al.* 1984). The logical conclusion is that the quartz grains were part of the fallout from an asteroid impact. However, whether that impact actually caused the extinctions remains uncertain (Officer and Drake 1985).

Some evidence exists that catastrophic meteor impacts have occurred quite regularly in the history of life.* Some researchers believe that they happen, on the average, about every 26 million years. Some believe that these events can be explained by the passage of an as yet undiscovered companion star of Earth (already named Nemesis) through the Oort comet cloud. The passage of the star, or periodic oscillations of the solar system about the plane of the galaxy, perturbs the orbits of the comets and leads to a comet shower, which in turn leads to large-body impacts with Earth.

Like the endothermy controversy, the debate over the extinctions at the end of the Cretaceous highlights both the great fascination and the difficulties of reconstructing the ecology of the past. Scientists have long been hard pressed to determine what had wiped out one of the largest and most successful group of organisms—a group well represented in the fossil record. Similarly, a debate has long raged over whether or not human hunters played a major role in the extermination of many species of large mammals during the Pleistocene (see, e.g., Martin and Wright 1967, Lewin 1983c, Martin and Klein 1984). If such major events are difficult to reconstruct, however, it seems likely to be some time before more subtle aspects of Earth's ecological past can be illuminated.

GEOGRAPHIC OPPORTUNITY

Biogeography is the branch of ecology that attempts to explain the large-scale geographical distribution of organisms. Many aspects of biogeography already have been discussed, from differential vagility and island biogeographic theory to the climatic machinery that is critical in determining where different organisms can survive. It is clear that an additional major factor determining what organism lives in a given place is simply what organism has had an opportunity to arrive there. We know, for instance, from the successful transplantation of organisms as diverse as English sparrows and *Opuntia* cactus between continents that many can flourish in habitats that they had previously not reached. Similarly,

* Key references on the periodic meteor shower hypothesis are Raup and Sepkoski (1982, 1984), Sepkoski (1982), Quinn (1983), Raup *et al.* (1983), Davis *et al.* (1984, 1985), Kerr (1984), Rampino and Stothers (1984a,b), Whitmire and Jackson (1984), Clube and Napier (1984, 1985).

the penguins apparently evolved in the Southern Hemisphere, for all known fossil penguins—some dating back to the late Eocene 45 million years ago—have been found within the range of living species (Simpson 1976). Through time they have not crossed the gigantic barrier of the tropical seas. Presumably, penguins could make a living in arctic regions if they successfully reached them. Other diving birds do or did flourish in the Arctic; indeed, the term *penguin* was first applied to a Northern Hemisphere flightless marine bird, the now extinct great auk, a creature remarkably like medium-sized penguins (Halliday 1978). However, the critical experiment of attempting to transplant penguins to the Arctic has never been properly performed, although a small release of king penguins in Norway resulted in survival for several years (Long 1981).

Accidental and purposeful transfers of organisms to places outside of their original range often have had unfortunate side effects. Goats, for example, have been moved to many parts of the planet far from their native haunts near the Mediterranean, where they thrive in a great variety of communities. However, the communities into which they are introduced are often all but destroyed by the activities of these voracious herbivores. For example, the island of Saint Helena in the south Atlantic was once covered by heavy forests. Following timbering, goats that had been introduced ate the seedlings, prevented regrowth of the trees, and converted the island into a rocky wasteland. A large literature has been compiled on the history of introductions (see, e.g., Elton 1958). A recent fine addition is a comprehensive book by Long (1981) on introduced birds.

Sometimes safe biogeographic experiments are possible, although on a less spectacular scale. The rare checkerspot butterfly *Euphydryas gillettii* occurs in the Rocky Mountains north of the Wyoming Basin, a low area splitting the range in southern Wyoming and northwestern Colorado. However, apparently a suitable habitat, containing both larval food plants and adult nectar sources, exists in central Colorado. In 1977, eggs and larvae were moved from Granite Creek, Wyoming, to Gothic, Colorado; the locations appear in Figure 22–11. A population was established that in 1986, after nine generations, was still persisting (Holdren and Ehrlich 1981 and unpublished). Clearly, the butterfly's inability to cross the barrier, not the suitability of the habitat, was the reason for the southern limit of the range of *E. gillettii.*

Most biogeographic questions are not amenable to such experimental tests. Biogeographers largely rely on analysis of modern patterns of geographic distribution, combined with hypotheses about evolutionary relationships, in order to recreate the distributional history of Earth's biota. This activity can be traced back to Darwin and especially to Alfred Russell Wallace, cofounder of evolutionary theory. Wallace's classic works (1869, 1876, 1880) made him the father of zoogeography. Among other contributions, he defined the major biogeographic regions still recognized today:

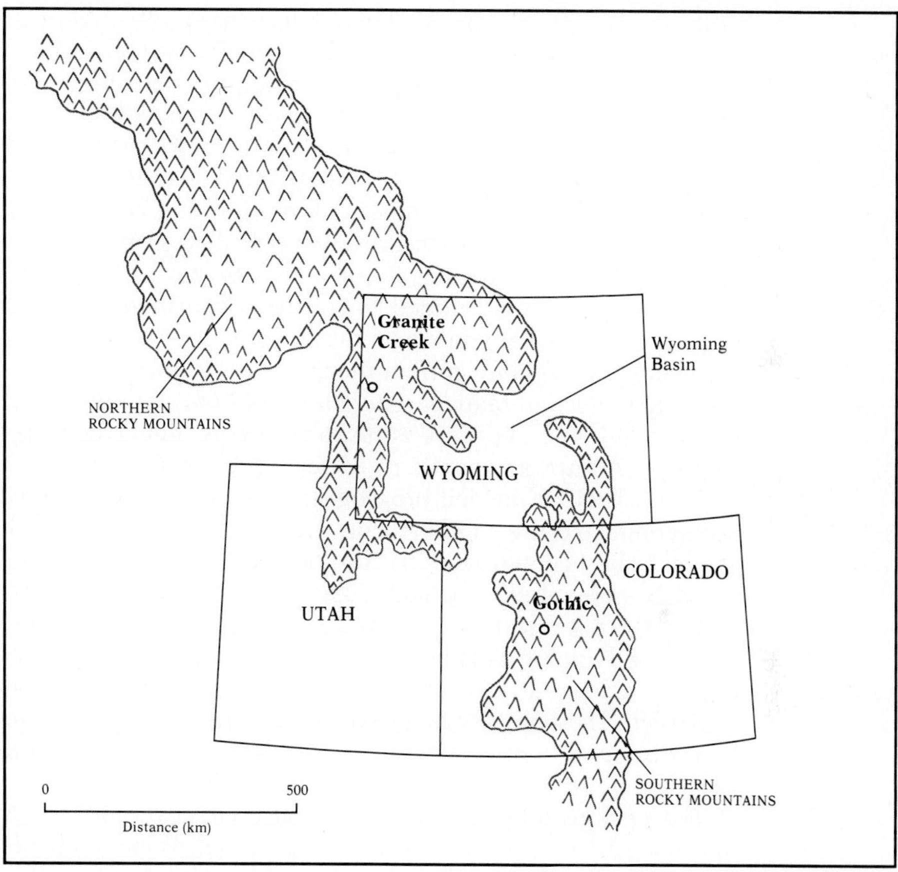

FIGURE 22–11 Locations of the donor population and recipient site for an *E. gillettii* transplant experiment. Granite Creek: doner; Gothic: recipient. *(After Holdren and Ehrlich 1981.)*

- Neartic—North America north of Mexico
- Neotropical—Central and South America
- Palearctic—Europe, North Africa, and northern Asia
- Oriental—Southeast Asia
- Australian—Australia, New Guinea, the Celebes, and some other islands of the southwestern Pacific

Indeed, the boundary between Wallace's Oriental and Australian regions, which lies between the Celebes and Lombock on the east and Borneo and Bali on the west, is still known as *Wallace's Line* (Darlington 1957, Brown and Gibson 1983). The line marks an area of great faunal change. For example, few Australian marsupials penetrate west of it, and few Oriental mammals are found east of it. Significant changes also appear in the bird, reptile, amphibian, and freshwater fish faunas in this area (Darlington, 1957). These differences can undoubtedly be traced to the arrival of the Australian tectonic plate at the continental shelf of southeast Asia only about 15 million years ago (Brown and Gibson 1983).

Plate tectonics also explains the southern distribution of a variety of organisms thought to be closely related to one another, a distribution that long presented biogeographers with a dilemma. How did southern beech trees (*Nothofagus*) come to grow in such distant places as New Zealand and Chile? Why are ratites, flightless birds that are often quite large, and their fossil remains confined to southern continents—ostriches in Africa and southwest Asia, emus and cassowaries in Australia and New Guinea, kiwis in New Zealand, extinct moas in New Zealand, and extinct elephant birds in Africa and Madagascar? How does one explain fossils of the mammallike Triassic reptile *Lystrosaurus* being unearthed in such diverse places as Antarctica, southern Africa, and India?

Such distributions led biogeographers to hypothesize *land bridges* that once connected the southern continents and have since sunk beneath the oceans. They encouraged ornithologists to assume that the ratites were actually a polyphyletic group—that is, an assemblage of species that independently became flightless and were actually more closely related to groups of flying birds than to each other. Once continental drift was shown to have occurred, however, these hypotheses were demonstrated to be incorrect. These southern distributions are explained by ordinary overland dispersal of ancestral forms within Pangaea prior to its breakup in the Triassic (Colbert 1971, Cracraft 1973, Pielou 1979).

The new geology has had a dramatic effect on biogeographic thought and has fueled the development of a school of *vicariance biogeographers* (Croizat *et al.* 1974, Nelson and Platnick 1981, Nelson and Rosen 1981). This school focuses on the development of barriers—caused by plate tectonics, changes in sea level, and the like—within already established broad ranges of organisms as virtually the sole explanation for present discontinuities in related groups of organisms. As previously indicated, such *secondary discontinuities* clearly do explain many biogeographic patterns. However, many others, such as the establishment of the unique fauna of the Galapagos islands, clearly are equally the result of *primary discontinuities*—the establishment of populations on both sides of barriers by successful dispersal across them. In fact, many dispersal events have been observed; the repopulation of Krakatoa after all of its biota was exterminated by a volcanic explosion is just one example. Today's distributions are the result of both the fragmentation of previously continuous ranges *and* dispersal across established barriers (Mayr 1982a,b).

The problems of sorting out these processes and of explaining phenomena such as centers of origin—that is, geographic centers of species richness within a genus or genus richness within a family—are challenges for a revitalized field of historical biogeography (Vermeij 1978, Pielou 1979,

Stott 1981, Endler 1982, Brown and Gibson 1983). So are predicting the results of human interventions that could give organisms access to new habitats. Substantial debate has already taken place about one possible such intervention—the creation of a sea-level canal across Central America that would permit the distinct marine biotas on Atlantic and Pacific sides of the isthmus to invade each other (Briggs 1968, 1974, Rubinoff 1968, Topp 1969, Vermeij 1974). The issue is complex, and results for one taxonomic group are likely to be different from those for another (Vermeij 1978).

COMMUNITIES OF THE FUTURE

Biogeography deals with the past distribution of life. Can anything be said about its future? One thing is clear: *Unless current trends are altered,* Earth's biota will be much less diverse than it is now. By 2100, more than half of the plants and animals that exist today almost certainly will be extinct (National Research Council 1980). The consequences of this enormous change for humanity may be catastrophic (Raven 1976, Myers 1979, Ehrlich and Ehrlich 1981, Lewin 1983c, Simberloff 1986). This loss of diversity will not affect all communities equally; the tropics, with their disproportionate richness, will suffer correspondingly significant losses.

Increasingly, *Homo sapiens* will have to manage other organisms to preserve what diversity is left. Wilderness will decline, to be "replaced" by parks, zoos, and botanical gardens. Whether ecologists will have learned enough by then about the properties of natural communities to allow remnant communities to be successfully managed in the long term remains to be seen.

COMMUNITIES OF THE FUTURE

PART 7

ECOSYSTEMS

The most comprehensive type of system studied by ecologists is an ecosystem—all of the organisms in an area together with the physical environment with which they interact. For example, a lake ecosystem includes the plants, animals, and microorganisms of the lake, the water of the lake, and the rocks and sediments of the lake bottom.

The concept of an ecosystem is essential for two reasons. First, many kinds of organisms modify the physical environment and affect each other through that modification. Second, global geochemical cycles, which are essential parts of the machinery that supports life on Earth, are strongly influenced by the rates of flow of nutrients and energy through local ecosystems. Indeed, those cycles are part of the largest ecosystem, the biosphere—the "thin film" of life that occurs near the surface of Earth and the physical-chemical environment in which it is embedded.

In this part, Chapter 23 will first describe the functions, structure, and control of ecosystems. Chapter 24 will describe biogeochemical cycles. Chapter 25 covers a natural ecosystem and, finally, Chapter 26 discusses the concept of ecosystem services, which relate ecosystem ecology to the maintenance of human society.

CHAPTER 23

Ecosystem Functions and Structure: Energy Flows and Successional Changes

An *ecosystem,* as mentioned, consists of all of the organisms in an area and the physical environment with which they interact. An ecosystem is described by the quantity of material and energy that is contained in each of its major compartments together with the rates of flow of energy and materials among those compartments. The flow of energy is unidirectional, whereas the flow of materials may be cyclic.

The laws of thermodynamics govern ecosystems, just as they govern the physical universe. According to the first law of thermodynamics, energy is conserved. Energy is not created or destroyed, only changed in form. If energy seems to disappear at one place, it must turn up in some form somewhere else. According to the second law of thermodynamics, every spontaneous process that involves the transformation of energy from one form to another proceeds in a certain direction, as in an exothermic chemical reaction or the movement of material down a concentration gradient.

At the end of the reaction, less usable energy is present than was there at the start; some of the energy, though not destroyed, has been degraded to a less useful form. In particular, the products of a spontaneous reaction do not have as much useful energy as the original reactants, for if they did, charcoal could turn back spontaneously into wood and oxygen. Similarly, materials do not move spontaneously up a concentration gradient; otherwise, a drop of ink once dispersed in a glass of water might spontaneously reconcentrate. The ecological importance of these laws is that energy transformations in an ecosystem are unidirectional and that along a chain of successive energy transformations less usable energy is available at each step.

ENERGY TRANSFER IN A SIMPLE FOOD CHAIN

In one of the pioneering studies of energy flow in an ecosystem, Golley (1960) examined a vegetation→mouse→weasel→food chain in an abandoned farm field (*old field*) ecosystem in Michigan; the results are shown in Figure 23–1. Of 94.2×10^8 kcal of solar energy falling on each hectare of the study plot, only about half of it (47.1×10^8 kcal) lies in the portion of the spectrum that is usable by plants in photosynthesis. Of that amount, as can be seen from the figure, only a little over 1% is actually used in photosynthesis.

The plants themselves use some 15% of the photosynthate to drive their own life processes; thus, in this system, almost 1% of the solar energy ends up as energy stored in the chemical bonds of the plants. Of that amount stored in the plants, about 30% is available to mice. Of this available amount, mice consume only 1.5%, and of that, most is used in respiration or unused. Only 2.0% is production—that is, energy incorporated into the mouse tissue. Thus, only about 0.03% (0.015×0.02) of the chemical energy theoretically available to the mice ends up incorporated into mouse tissue.

At the third stage of the food chain, Golley found that the weasels consumed more than the mice produced. The difference was presumably made up by immigrating mice and other prey species. Again, at this trophic level only about 2% of consumption becomes production. A tiny proportion of the solar energy incorporated into the plants, roughly 0.00026%, ends up in the chemical bonds of weasel bodies. The rest either leaves this food chain—for example, plant material is not consumed by mice but by insects, or mice are consumed by hawks or by scavengers—or is degraded into heat energy in the process of respiration.

Because the total amount of energy available to the entire chain is that originally fixed by plants, the amount available to other trophic levels declines as energy moves up the food chain. More energy is available to herbivores, for example, than to carnivores that feed on them.

ENERGETICS OF ENTIRE ECOSYSTEMS

Energy flows can be traced in entire ecosystems in a manner similar to that in the old field food chain. The total energy fixed by all of the plants in the ecosystem is called the *gross primary production (GPP)*. The plants themselves use part of this energy to maintain their life process. The re-

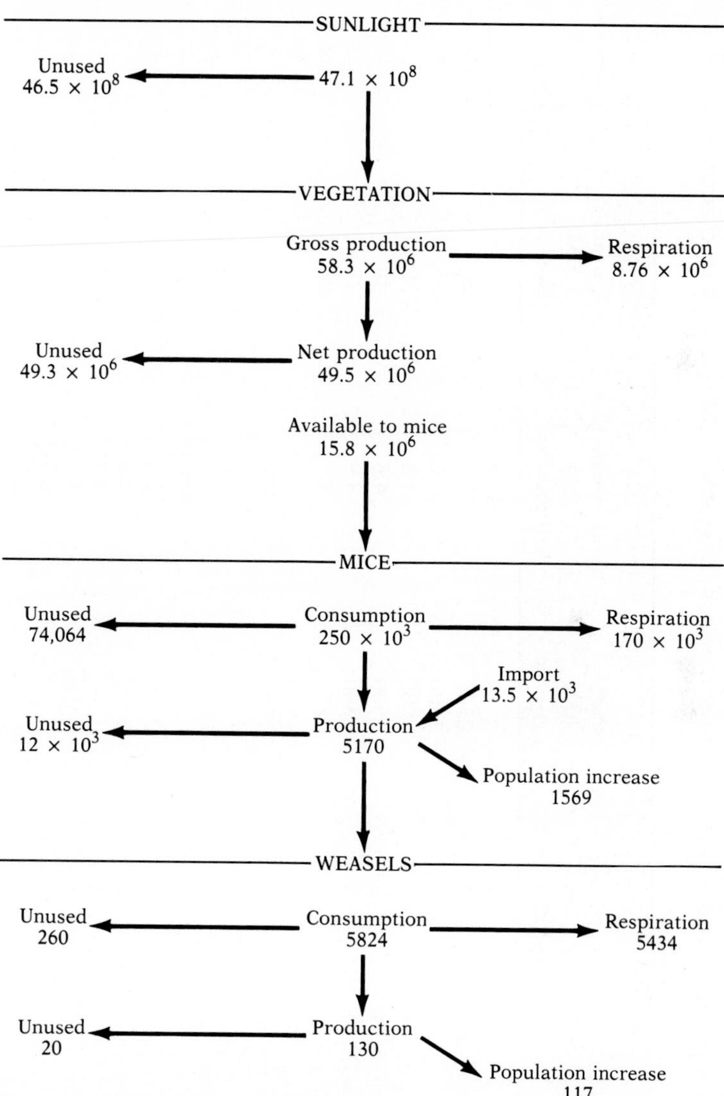

FIGURE 23–1 Energy flow (kcal/hectare/year) through the food chain on 1 hectare of an old field in Michigan, May 1956 to May 1957. *(After Golley 1960.)*

mainder, the *net primary production (NPP)*, is the energy that is available to the heterotrophs in the system. NPP has been calculated for the largest ecosystem, the biosphere. The *biosphere*, as mentioned, is the thin layer of life that occurs near the surface of the Earth and the physical-chemical environment in which it is embedded. Global NPP is estimated to be about 11.2×10^{17} kcal/year (Ajtay *et al.* 1979, DeVooys 1979). This figure is not based on direct measurements of energy, but on estimates, so it may be in error by 50% or more.

TABLE 23–1 Surface areas, net primary productivity, and plant biomass (phytomass) of terrestrial and aquatic ecosystems. DM = dry matter

Ecosystem type	Surface area (million km^2)	NPP DM g/m^2 yr	Total production DM (billion tons)	Living phytomass DM (kg/m^2)	Total living phytomass DM (billion tons)
1. Forests	31.3		48.68		950.5
Tropical humid	10	2300	23	42	420
Tropical seasonal	4.5	1600	7.2	25	112.5
Mangrove	0.3	1000	0.3	30	9
Temperate evergreen/coniferous	3	1500	4.5	30	90
Temperate deciduous/mixed	3	1300	3.9	28	84
Boreal coniferous (closed)	6.5	850	5.53	25	162.5
Boreal coniferous (open)	2.5	650	1.63	17	42.5
Forest plantations	1.5	1750	2.62	20	30
2. Temperate woodlands (various)	2	1500	3	18	36
3. Chaparral, maquis, brushland	2.5	800	2	7	17.5
4. Savanna	22.5		39.35		145.7
Low tree/shrub savanna	6	2100	12.6	7.5	45
Grass dominated savanna	6	2300	13.8	2.2	13.2
Dry savanna thorn forest	3.5	1300	4.55	15	52.5
Dry thorny shrubs	7	1200	8.4	5	35
5. Temperated grassland	12.5		9.75		20.25
Temperated moist grassland	5	1200	6	2.1	10.5
Temperated dry grassland	7.5	500	3.75	1.3	9.75
6. Tundra arctic/alpine	9.5		2.12		13.05
Polar desert	1.5	25	0.04	0.15	0.23
High arctic/alpine	3.6	150	0.54	0.75	2.7
Low arctic/alpine	4.4	350	1.54	2.3	10.12

7. Desert and semidesert scrub	21		3		16.5
Scrub dominated	9	200	1.8	1.1	9.9
Irreversible degraded	12	100	1.2	0.55	6.6
8. Extreme deserts	9		0.13		0.78
Sandy hot and dry	8	10	0.08	0.06	0.48
Sandy cold and dry	1	50	0.05	0.3	0.3
9. Perpetual ice	15.5	0	0	0	0
10. Swamps and marshes	2		7.25		26.25
Temperate	0.5	2500	1.25	7.5	3.75
Tropical	1.5	4000	6	15	22.5
11. Bogs, unexploited peatlands	1.5	1000	1.5	5	7.5
12. Cultivated land	16		15.05		6.64
Temperate annuals	6	1200	7.2	0.1[a]	0.6
Temperate perennials	0.5	1500	0.75	5	2.5
Tropical annuals	9	700	6.3	0.06[a]	0.54
Tropical perennials	0.5	1600	0.8	6	3
13. Human area	2[b]	500	0.4	4	3.2
Terrestrial subtotal	147.3	897[a]	132.2	8.44[a]	1243.9
14. Lakes and streams	2	400	0.8	0.02	0.04
15. Marine[c]	361	254	91.6	0.001	3.9
Aquatic subtotal	363	255[a]	92.4	0.001[a]	3.94
GRAND TOTAL	510.3	440	224.6	2.46	1247.8

Sources: After Ajtay *et al.* (1979) and De Vooys (1979).

[a] Average values.

[b] Of which only 40% (or 0.8×10^{12} m^2) productive.

[c] Studies currently in progress are demonstrating that marine productivity has been underestimated. See Brewer *et al.* 1986 and Martin *et al.* 1986.

The global averages conceal enormous local differences in primary production, as shown in Table 23–1, as well as in the *standing crop* of plants; standing crop is defined as the weight of plants present at any given time. All of the consumers in each ecosystem depend on the NPP for their supply of energy. (For simplicity we are ignoring material that is either imported into a system or exported from it.) However, ecosystems differ in the *efficiency* with which consumers at different trophic levels are able to acquire energy from the level below them and to incorporate that energy into *biomass,* the weight of living organisms. Many kinds of ecological efficiencies have been defined and given diverse names (Kozlovsky 1968). Some of the most important are illustrated in Figure 23–2.

Lindeman's efficiency, also called *ecological efficiency,* is the ratio of energy acquired by one trophic level (N + 1) to that acquired by the previous (N). It is named after ecologist Raymond L. Lindeman in recognition of his classic work (1942) on ecological energetics. *Assimilation efficiency* is the fraction of light striking plants that is actually used in photosynthesis, or it is the fraction of energy consumed—taken in or *ingested*—by animals that is actually absorbed from the food. *Growth efficiency* pertains to the splitting of the assimilated energy into the portion degraded through the metabolic activities of organisms or lost through excretion and the portion incorporated into new tissues. *Exploitation efficiency* pertains to the fraction of those tissues that are consumed by the

FIGURE 23–2 Details of energy pathways in a link between consumer trophic levels (herbivore → predator or predator → predator) in a food chain, illustrating how efficiencies are calculated. *(After Kozlovsky 1968.)*

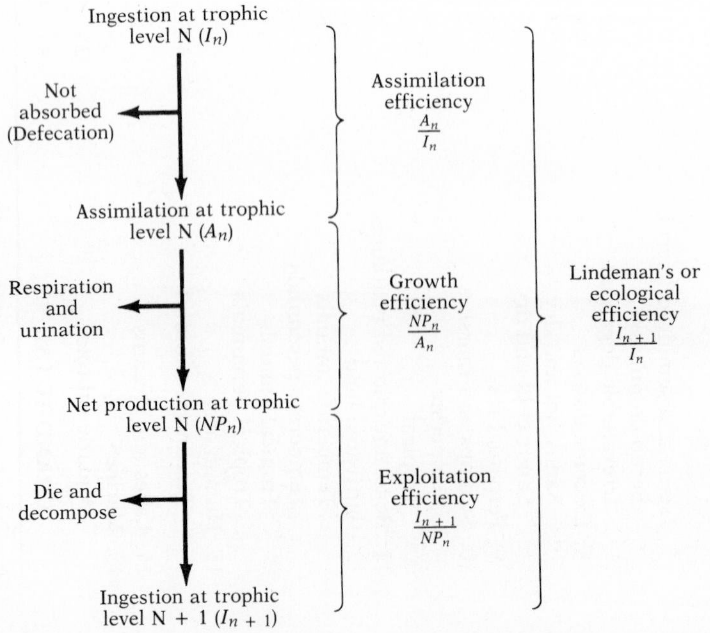

next trophic level. Lindeman's efficiency is the *product* of assimilation efficiency, growth efficiency, and exploitation efficiency.

At the producer level, as we saw in the vegetation to mouse to weasel food chain, assimilation efficiency is extremely low. Assimilation efficiencies in animals vary with diets. For example, a millipede feeding on decaying wood may have an assimilation efficiency of 15% (O'Neill 1968). An elephant munching on vegetation may have about 30% (Petrides and Swank 1966). A seed-eating kangaroo rat may have about 80% (Chew and Chew 1970). The weasel in our food chain was about as efficient as one can get—96%. In general, carnivores are more efficient than herbivores, which is not surprising, considering that so much relatively indigestible material such as cellulose occurs in plant tissues.

Growth efficiency at the producer level can be quite high, in contrast to assimilation efficiency. Plants do not spend a lot of energy in locomotion and thus are able to incorporate a high percentage of their intake into their tissues. Efficiencies above 60–70% are common, especially outside of the tropics. An efficiency of 30%, for example, has been estimated for a tropical rain forest (Golley and Misra 1972).

Generally, ectotherms have higher growth efficiencies than endotherms do because they do not divert energy toward maintaining their body temperatures. Insects tend to use about 33% of their assimilated energy in tissue building and reproduction, birds and mammals less than 1–5%. Growth efficiencies tend to be higher in young than in old animals (Phillipson 1966). For example, growth efficiency in chickens drops sharply at the age of 3 or 4 months. The poultry industry maximizes its meat production by slaughtering broilers at that point.

Lindeman's efficiencies are often said to be about 10% at levels above the producer level. This 10% rule of thumb is convenient for thinking about many ecological problems. We can picture each successive trophic level as having roughly one-tenth of the energy of the preceding level available to it. On this basis and with the same primary production, about 10 times more people who eat corn could be supported than people who eat corn-fed beef.

This progressive reduction of available energy as it passes along food chains, slavishly following the dictates of the second law of thermodynamics, results in a pyramidal structure of ecosystems. Figure 23–3 shows an energy-flow pyramid for the aquatic ecosystem at Silver Springs, Florida (H. T. Odum 1957). The pyramidal structure may also appear when

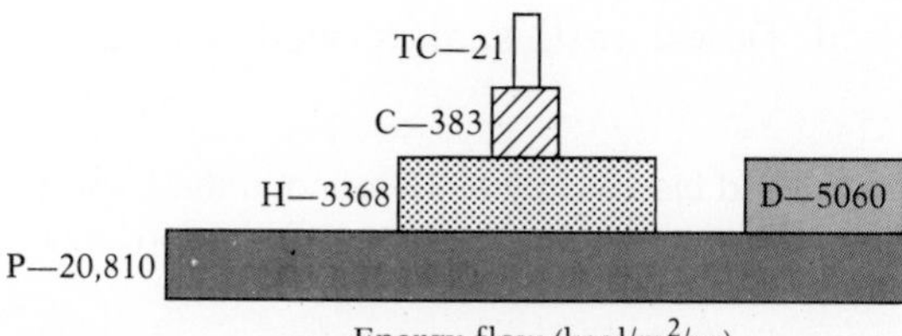

FIGURE 23–3 Energy-flow pyramid for Silver Springs, Florida. P = producers; H = herbivores; C = carnivores; TC = top carnivores; D = decomposers. *(Data from H. T. Odum 1957.)*

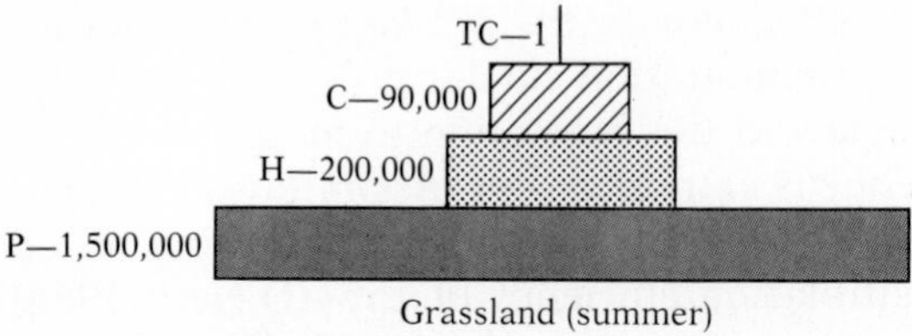

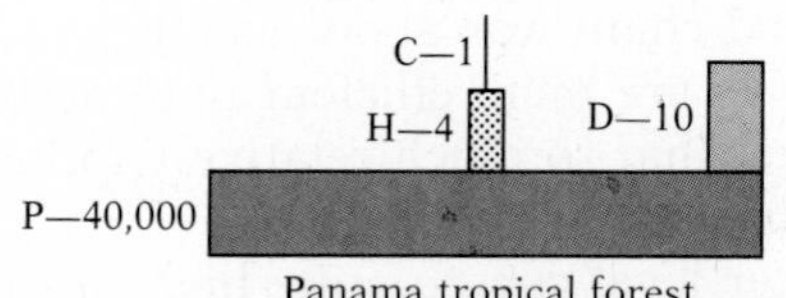

FIGURE 23–4 Pyramid of numbers for a grassland and biomass for a tropical forest. Numbers represent individuals per 1000 m^2 and grams of dry weight per square meter, respectively. See Figure 23–3 for an explanation of abbreviations and numbers. *(After E. Odum 1971.)*

the number of individuals or biomass at each trophic level is graphed (see Figure 23–4), although pyramids may also be inverted (see Figure 23–5).

The reason a pyramid may be inverted is obvious if the pyramid represents *numbers* of organisms. A few producer individuals can support a great many consumers, if the producers are trees and the consumers are bugs. But how can more biomass of consumers exist than of producers? An inverted pyramid of *biomass* occurs when the *turnover* of the producers is much more rapid than that of the consumers. Generally, small organisms have high metabolic rates; that is, they use energy faster per unit weight than larger ones, as we discussed in Part 1. Thus, the biomass that a given flow of energy supports tends to be inversely related to the size of the organisms. As an extreme example, consider a population of plankton with a generation time of 1 week supporting a population of blue whales with a generation time of 20 years. At any given moment, the total weight of whales may be considerably higher than that of plankton, just as you might always weigh more than the food in your refrigerator, provided that you restock it once a week.

Metabolic rates can strongly influence terrestrial biomass pyramids as well. For example, a much higher biomass of reptiles than of mammals can be supported by a given food supply. Mammals use a substantial portion of the calories they ingest in maintaining their body temperatures; reptiles utilize solar energy directly for this purpose. One consequence is the high population densities achieved by reptilian herbivores such as the Galapagos marine iguana and by reptilian carnivores in deserts. The iguana maintains large populations, as shown in Figure 23–6, on relatively sparse, low-calorie seaweed. Desert rattlesnakes need only catch

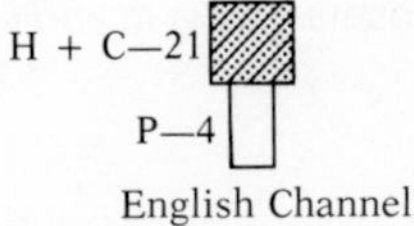

FIGURE 23–5 Inverted biomass pyramid found in the English Channel. H + C = herbivores plus carnivores; P = producers. *(After E. Odum 1971.)*

(A)

(B)

FIGURE 23–6 Marine iguanas, Galapagos Islands. (A) Close-up of individuals. (B) Aggregation. The iguanas are excellent swimmers and feed underwater on seaweed. *(Photos by Paul R. Ehrlich.)*

and eat a few rodents annually. The large monitor lizard known as the *Komodo dragon* eats the equivalent of about a pig a month—its own weight every 2 months—whereas a cheetah consumes something like four times its own weight in the same period (Kruuk and Turner 1967).

SUCCESSION

The study of how ecosystems are organized can be introduced with the topic of *succession,* which was originally defined as the sequential change in plant communities after disturbance. Succession is a topic that straddles community ecology, where it has historical roots, and ecosystem ecology, where it now resides more naturally as a result of recent interest in the fate of nutrient stocks and nutrient flows during succession. Succession also has been studied in animal communities (see, e.g., Shelford 1913, Southwood *et al.* 1979, Day and Osman 1981). However, most of our emphasis will be on plant communities and their physical substrates.

PRIMARY AND SECONDARY SUCCESSION: THE CLASSIC VIEW

In tropical rain forests, human beings have long practiced what is variously called *shifting, slash-and-burn, swidden,* or *milpa* agriculture. As still practiced in some areas, a farmer cuts a small clearing in the forest during the driest season of the year. After the felled trees have dried out, they are burned, and desired crops are planted in the nutrient-rich ashes. Pests are suppressed by the hot fire. The next season, the crop residues are burned, but the fire is less hot, the resultant ash less nutritious, and pest control less successful. Most important, soil nutrients are "mined" as they are taken up by the crops, which are then removed from the clearing.

After a season or two, the clearing must be abandoned, and the farmer moves and starts the process again at another site. At the original site, various grasses and other herbaceous plants invade, followed by shrubs and saplings. As the saplings mature into trees, the dense, leafy canopy characteristic of tropical moist forests is reestablished (Nye and Greenland 1960, Wrigley 1982). Many of the earlier occupants of the clearing die because they are unable to get enough light for photosynthesis; that is, they are "shaded out." Over time, ecological succession has returned the original clearing to tropical forest.

Two kinds of succession have long been recognized: primary and secondary. *Primary succession* is the development of a biological community where one did not previously exist. In terrestrial ecosystems, primary succession ordinarily begins with the colonization by *pioneer* plants of bare rock created by cooling volcanic magma or exposed by a landslide or retreating glacier, or of a newly formed sand dune or another inorganic substrate. Succession, then, proceeds with the establishment of a series of plant communities, each contributing to soil formation and replacing the other, each step paving the way for the next, until a relatively stable community is established that is characteristic of the climatic regime of the area and the parent rock from which the soil was formed. That relatively stable community is called a *climax*. This so-called classic view of primary succession, as we will see, has been the center of considerable dispute.

Secondary succession occurs on substrate that has previously supported a community, as when the tropical forest climax is destroyed by a milpa farmer. Secondary succession is usually thought of as a shorter process than primary succession because the laborious process of soil building does not have to occur.

Three crucial aspects of the classic view are the following:

- A specific sequence of plant communities occurs on a given site.
- Each community prepares the site for the subsequent invaders.
- The end of the sequence is a stable climax community.

EXAMPLES OF SUCCESSION

Botanist W. S. Cooper (1923) described primary succession on moraines—the debris deposited by glaciers—at Glacier Bay, Alaska (see Figure 23–7). There glaciers have retreated as much as 100 km in the past 200 years or

FIGURE 23–7 Retreating glacier at Glacier Bay. *(Photo by B. F. Molnia, Terraphotographics/BPS.)*

so, exposing large areas of bare moraines to colonization. Through the use of tree rings and historical records, the ages of a series of experimental plots were determined. By comparing plots of various ages, Cooper pieced together the sequence of events on a typical plot.

The first pioneers to colonize the bare moraines are primarily mosses and a perennial herb, *Epilobium latifolium,* both in a thin layer of soil resting on the surface of polished rock. The next invaders are a horsetail (a reedy plant) in finer, moister soils and a mat-forming, nitrogen-fixing herb, *Dryas,* in coarser, drier soils. The *Dryas* are soon joined by prostrate willows that have thick, woody stems growing along the ground and are anchored to the surface by adventitious roots. Adventitious structures on plants are those growing from unusual places—in this case, roots growing along the length of the prostrate stems. The *Dryas*–willow mat, which is shown in Figure 23–8, traps humus and begins to build the nitrogen-rich soil on which later successional plants grow (Crocker and Major 1955).

The prostrate willows are followed by erect willows, and these trees are joined by alders that form thickets. By now the soil pH that was originally about 8.0 has declined to less than 5.0. Various herbs and shrubs, such as strawberries, bearberries, yarrows, and arnicas, form an understory in the thickets. Gradually Sitka spruce, some individuals of which were present in the pioneer stage, begin to grow above (overtop) the willows and alders. The alders die as they are gradually shaded out. A solid stand of spruces, with a rich flora of mosses and a few hemlocks in the understory, as shown in Figure 23–9, develops after about 170 years. This stand is the most mature forest seen at Glacier Bay. Given enough

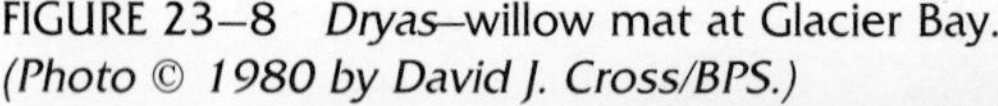

FIGURE 23–8 *Dryas*–willow mat at Glacier Bay. *(Photo © 1980 by David J. Cross/BPS.)*

FIGURE 23–9 Spruce forest at Glacier Bay.
(Photo by B. F. Molnia, Terraphotographics/BPS.)

time, Cooper hypothesized, the hemlocks would increase until the spruce-hemlock climax characteristic of southeastern Alaska developed.

Eventually, however, on level and gently sloping land, the forests themselves disappear. First, for reasons as yet unknown, new kinds of mosses, water-absorbent *sphagnum* mosses, appear on the forest floor. As the sphagnum layer grows, the forest floor becomes increasingly waterlogged, and roots cannot get the oxygen they require. The trees begin to die from the top down (Lawrence 1958). As the trees die, succession occurs in the sphagnum mosses, as species that can grow in the shade of the forest are replaced by those adapted to life in the open. Gradually the forest is replaced by boggy land, some 50% of which is eventually occupied by ponds and in which accumulated sphagnum mosses start to form peat—land often referred to as *muskeg*. Whether the muskeg is a climax that will persist until major geologic changes occur is not known.

Change is more rapid in the secondary succession that follows the burning of chaparral vegetation in southern California than in the primary succession following glacial retreat (Hanes 1971). Most of the plant species are dormant during the dry season. After a fire, the species that will once again become dominant in the climax, especially chamise (*Adenostoma fasciculatum*) and scrub oak (*Quercus dumosa*), sprout from burned stumps. Their seeds, in contrast, remain dormant until the rains come in the fall and then germinate. The combined result is an abundance of new vegetative growth and seedlings of the previous climax dominants, along with species such as deerweed (*Lotus scoparius*), which were present only as seeds prior to the fire. The latter come from abundant seeds that survive in the soil for long periods and germinate only after a fire.

Succession in such chaparral communities consists primarily of a sorting out of species that were present right after the fire. The chamise and scrub oak in low-altitude coastal chaparral are largely shaded out in the first decade after burning. Within 30 to 50 years, depending on the site, a climax of mature shrubs is established. If another fire does not occur and a stand ages beyond that period, mature shrubs start to die and sage (*Salvia mellifera*), a short-lived smaller shrub, fills the gaps.

Beyond the age of 60 years the stands of chaparral tend to become "senile." Species diversity declines, little annual growth takes place, and a leafy lichen grows on the shrubs, rocks, and leaf litter. It has been suggested that toxic substances originating in the chamise interfere with normal decomposition and suppress the germination of grass seedlings. Thus, fire not only may remove the decadent chaparral but also may destroy the toxic substances and permit a vigorous stand to regenerate.

Succession is usually studied as a sequence of communities of higher plants. However, in marine environments, it may be viewed in terms of changes in the communities of sessile animals. Successional sequences in patch reefs formed around single colonies of the scleractinian coral *Montastrea annularis* have been studied by Smith and Tyler (1975). As the colony grows from a small, rounded head to about 1 m in diameter, the edge begins to grow beyond the base, and an overhanging lip is formed. The undersurface erodes, forming numerous chambers, other sessile organisms invade dead areas, and what was a single *Montastrea* colony becomes a patch reef. As growth continues to several meters in diameter, the weight of the growing overhang, combined with wave action, eventually leads to its partial or complete collapse. After the collapse, colonies of other corals, originally attached to the *Montastrea* base, spread over the limestone platform, that is, the remains of the *Montastrea* dome, forming an even more complex patch reef. The hypothetical stages of development are shown in Figure 23–10. The sequence may take 500–1000 years or more.

During this sequence, the diversity of reef fish species generally tends to increase as the original colony grows and as more refuges and diverse microhabitats are provided (see Figure 23–11). Partial collapse leads to an increase in diversity by adding to the heterogeneity of the reef, but full collapse reduces the amount of cave shelter from predators that is available and thus leads to a small decline in diversity. That decline is accompanied by a shift from species that hover in caves during the day (e.g.,

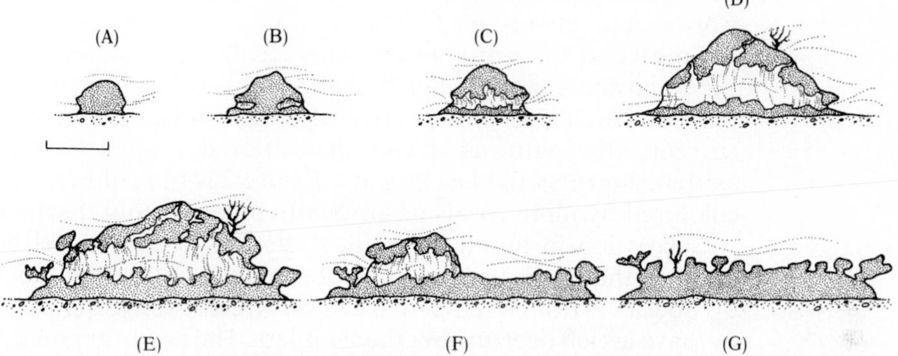

FIGURE 23–10 Proposed successional sequence as *Montastrea* coral dome begins to build (A–B), becomes more complex (C–E), and collapses (F–G). Scale line is one meter. *(After Smith and Tyler 1975.)*

FIGURE 23–11 Development of fish communities associated with reefs thought to be at successional stages similar to those shown in Figure 23–10. Stages of succession go from a simple single *Montastrea* colony (1) through a complex dome reef with a large overhang and caves (3) to a collapsed complex dome reef (6). Volume is the volume of a cylinder that would enclose the reef; diversity is measured by the number of species. *(After Smith and Tyler 1975.)*

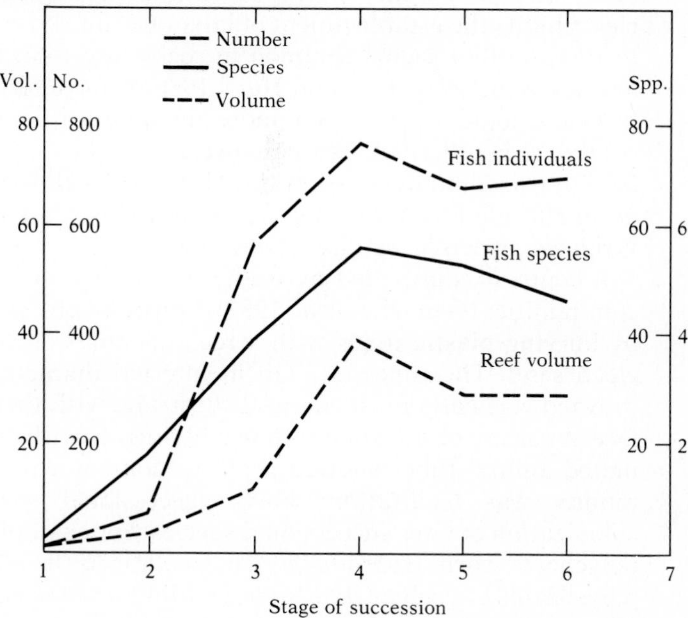

cardinalfishes) to those that live in such places as sponges or abandoned worm tubes (gobies, blennies).

Smith and Tyler did not say whether they considered the final stage of the sequence a climax, but it seems unlikely that the final stage is permanent. Patch reefs almost certainly come and go as storms and changing currents alter patterns of sand deposition and other factors. A reef such as that shown as the last stage in Figure 23–10 could either grow as it is colonized by more corals or gradually break up and disappear.

Succession in another marine system, a boulder field dominated by algae in the low intertidal zone of southern California, has been studied by Sousa (1979). Here open space on which succession occurs is created by wave action overturning the boulders. The newly exposed bare surfaces are recolonized first by algae, either through vegetative regrowth of surviving individuals or by recruitment of spores. The first pioneer is sea lettuce, a green alga (*Ulva*). It is followed by a sequence of red algae, culminating in populations of the red alga *Gigartina canaliculata*.

Sousa showed, however, that algal populations earlier in the sequence are not replaced by later forms unless the former are removed or damaged by the grazing crab, *Pachygrapsus crassipes*. Not only do the earlier successional species not prepare the way for the later, the earlier ones—as long as they are undisturbed—prevent the later ones from occupying space. Here the possibility of succession is traced to the organisms at the different stages being differentially susceptible to disturbance and predation. Fragile forms, such as the sea lettuce, are easily damaged or devoured; eventually the tougher red algae dominate. The presence of pioneer species inhibits the establishment of later ones until the former are damaged, a situation that seems common in rocky intertidal habitats (Sutherland and Karlson 1977, Dean and Hurd 1980, Paine and Levin 1981).

Small boulders turn over more frequently than large boulders. As a result, the boulder field is a patchwork of many stages in algal succession. Medium-sized boulders show a higher species diversity than either large or small boulders, and thus the system illustrates the intermediate disturbance principle, as discussed in Chapter 17.

A contrast is provided by studies of succession in soft-bottom benthic communities (Gallagher *et al.* 1983). Empty 10-cm^2 patches were produced by burying plastic tubes with screen openings on their sides filled with clean sand. The tubes (3.7 cm in internal diameter, 15 cm long) were inserted vertically in the natural substrate, with their lips below the surface. A variety of transplantation experiments with animals and with simulated animal tubes showed that the dominant interaction in the community was facilitation: Early successional species facilitated the colonization of later successional species. For example, the presence of the polychaete worm *Hobsonia florida* encouraged the recruitment of another polychaete, oligochaetes, a tube-building crustacean, and two bivalves. The simulated tubes facilitated the establishment of the crustacean and oligochaete worms.

SUCCESSION IN SPACE AND TIME

An entire successional sequence, from bare substrate to climax, is called a *sere*. The sequences at Glacier Bay, from moraine to spruce-hemlock forest or to muskeg, are seres. Perhaps the most famous of all seres were those studied by early ecologists H. C. Cowles (1899) and V. E. Shelford (1913) and later by J. S. Olson (1958) in the sand dunes along the shore of Lake Michigan in Indiana. The seres stretched from newly formed dunes along the lake edge that were stabilized by the pioneer grass *Ammophila breviligulata* to a forest inland. The development of these seres is shown diagrammatically in Figure 23–12. Note the complex patterns that depend primarily on the initial conditions.

All of these seres are actually *sequences in space* that are interpreted as *sequences in time*. The seres are considered to be representative of se-

FIGURE 23–12 Lake Michigan dune succession (simplified). Note the varying sequences that depend heavily on the initial conditions. Undercover types: c = chokecherry/poison ivy; p = "prairie"; b = blueberry-huckleberry; h = other herbs that thrive in places with moderate moisture. *(After Olson 1958.)*

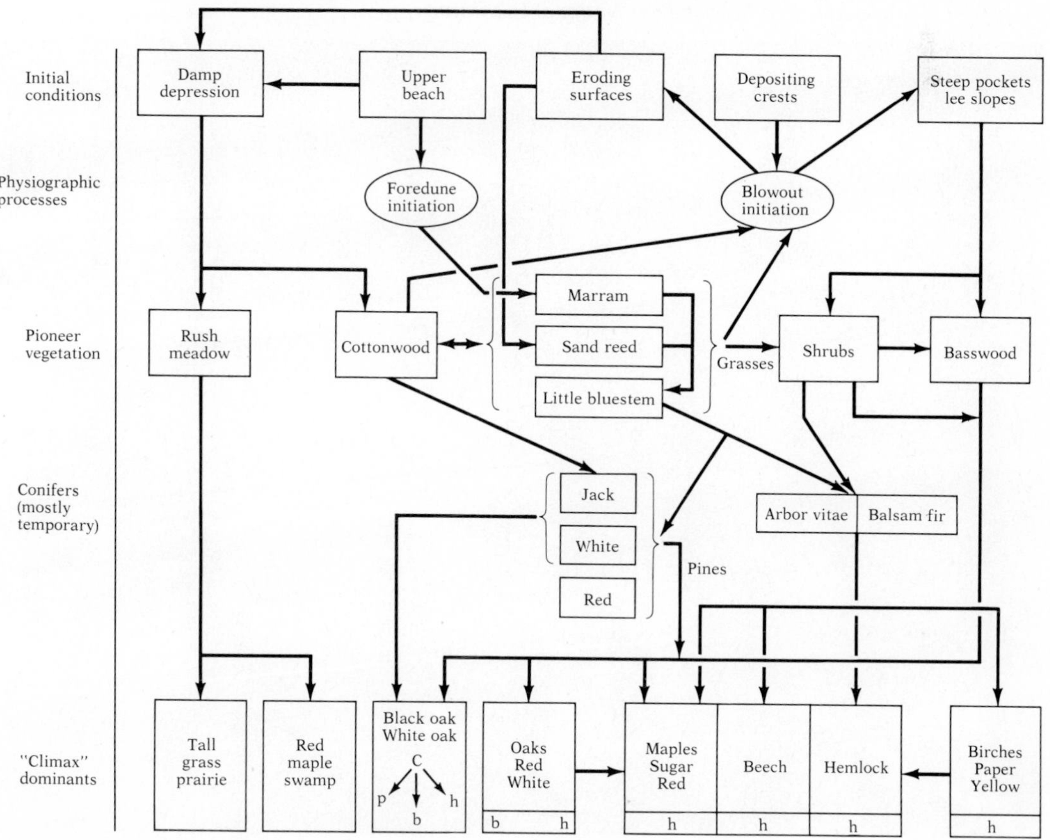

quences in time because they stretch from conditions that are stressful—for example, without soil or with a harsh microclimate—to those that are less so. Because of the slow pace of most successional changes, ecologists usually must infer the properties of seres in time from those observed in seres in space. This fact means that differences from place to place in a spatial sequence must be assumed to be minor, so that seres can represent changes in the *same place* through time.

Woodwell (1970) examined the patterns imposed on terrestrial ecosystems by various kinds of stresses, using an artificial stress gradient. The system was an oak-pine forest at Brookhaven National Laboratory in New York, in which the most abundant trees were white, scarlet, black oak (*Quercus alba, Q. coccinea,* and *Q. velutina,* respectively) and pitch pine (*Pinus rigida*). For 7 years the forest was exposed to chronic irradiation from 9500 curies of radioactive cesium, ^{137}Cs, the effects of which are shown in Figure 23–13. During that time, five different zones of modified vegetation were established, proceeding outward from the source (Woodwell 1967). These five zones were as follows:

FIGURE 23–13 Effects of irradiation on a forest ecosystem. *(Photo courtesy G. M. Woodwell at Brookhaven National Laboratory.)*

1. A devastated zone adjacent to the source, in which no higher plants survived. That zone was exposed to more than 200 roentgens per day (R/day). (This is about half of the dose that would be required to kill about half of the human beings exposed to it for a single day.) Certain lichens and mosses survived in this zone, some close enough to the source to receive more than 1000 R/day.
2. A zone receiving more than 150 R/day, in which the sedge *Carex pennsylvanica* ultimately formed a continuous cover.
3. A shrub zone receiving more than 40 R/day, in which several species survived, including the shrub oak (*Quercus ilicifolia*) near its outer margin.
4. A zone receiving more than 16 R/day, in which the pitch pine (*Pinus rigida*) was exterminated but the oaks persisted.
5. A zone receiving less than 2 R/day, in which no disturbance was evident, although the rate of growth in the height of both oaks and pines was measurably reduced at exposures as small as 1 R/day.

The pattern caused by radiation stress is similar to that caused by other stresses. The plants found in the more stressful radiation conditions are the pioneer types that are also the first invaders of highly disturbed sites. For example, Buell and Cantlon (1953) found that oak-pine forests in New Jersey reacted to the stress of frequent winter burning of the understory by a loss of taller shrubs first and by a substantial increase in *Carex pennsylvanica,* the plant that formed the "sedge zone" at Brookhaven.

Woodwell hypothesized that one factor leading to a higher stress sensitivity of larger plants is their lower ratio of photosynthetic structures to supportive structures. Indeed, it is possible that the size of trees is limited by the decline of this ratio with growth (Horn 1971). Stresses also apparently operate to reduce the ratio, because photosynthetic structures (leaves, leaf buds) on larger plants are much more likely to be damaged by wind drying, ice, herbivore attack, or radiation than bark or supporting structures that only respire. However, additional work on this problem is needed, for one supporting structure, the heartwood, supports but does not respire (Sprugel 1984).

Successional changes on the same plots were followed over long periods of time under relatively controlled conditions in a grassland ecosystem (Silvertown 1980). At Rothamsted in England, the Park Grass Experiment, started in 1856, subjected a series of plots to various fertilizer regimes. Sections of each plot were sampled at irregular intervals, and the absolute and percentage compositions of each species in the cropped biomass were determined. The data showed that the plots reached equilibria with respect to the main floral components—grasses, legumes, and miscellaneous other species. However, these equilibria of floral components concealed changes in the relative abundance of species within each component.

Different fertilizer regimes produced different equilibria, but important regularities were observed. For example, the more biomass a plot generated, the lower the species diversity that occurred on that plot. This result conforms to the observations in nature by Green (1972) that successional changes in grasslands tend to increase productivity and decrease diversity and that the most productive grasslands show the smallest array of species (McNaughton 1968). The result also supports the idea stated by Grime (1973) that, up to a point, various kinds of stress, such as lack of nutrients, tend to increase diversity. The stress limits the vigor of the good competitors, encouraging other species, especially those adapted to the particular stress imposed, as represented in Figure 23–14. Of course, too much stress will again decrease diversity because fewer and fewer species can survive the conditions. One might call this an *intermediate stress hypothesis*.

It is sometimes possible to determine in detail what the natural successional sequence has been. Henry and Swan (1974) reconstructed the history of woody plants from before 1665 to 1967 on a small (20 × 20 m) plot in a virgin forest in southwestern New Hampshire. They identified all the living stems ("trunks"), dead unburied stems and stem fragments, and woody remnants buried in the forest floor. Wood and charcoal could be identified by species microscopically. Size, age, and growth rates could be determined from the growth rings in living and relatively intact dead stems.

Henry and Swan discovered that three forests of quite different composition had occupied the site. Prior to about 1665, the trees were largely coniferous, and the plot was dominated by white pine (*Pinus strobus*). That forest was destroyed by a fire. Between the fire and 1897, a stand developed that was dominated by hemlock (*Tsuga canadensis*) and beech (*Fagus grandifolia*) but also contained white pine and white birch (*Betula papyrifera*). The hemlock was represented by 2 out of 16 identified trees that existed prior to 1665; the beech and birch were not represented in the earlier stand. Between 1897 and 1938, the hemlock-beech stand was vir-

FIGURE 23–14 Impact of environmental stress on species diversity. C = species with high competitive ability; R = species highly resistant to prevailing stresses; O = other species *(After Grime 1973.)*

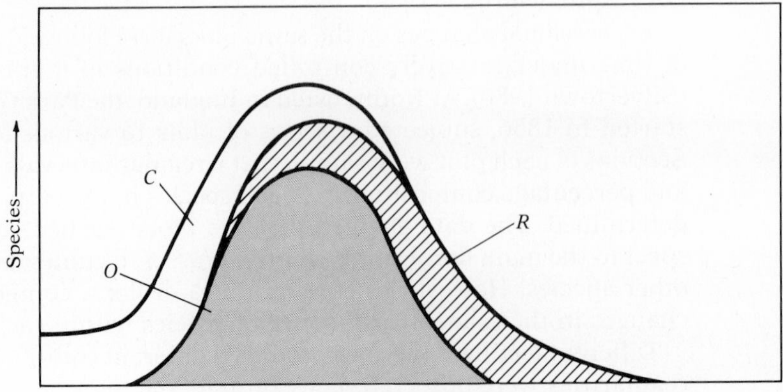

tually destroyed by four windstorms that culminated in the extremely destructive hurricane of 1938. Since that time, the forest that has regenerated on the plot again has hemlock as the most common tree, but the second most common in terms of total number of individuals is black birch (*Betula lenta*), a tree not previously represented. Another newcomer, red maple (*Acer rubrum*), was almost as common as black birch. Beech is in fourth place, and white birch is represented by just two small trees. Two other species not previously found in the history of the plot appeared: sugar maple (*Acer saccharum*) and striped maple (*Acer pensylvanicum*). Using the same reconstruction technique on a forest in Massachusetts, Oliver and Stephens (1977) also found varying vegetational compositions developing in response to periodic disturbances.

In the tropics, the course of succession in a cleared forest plot is conjectured often to be influenced by the mycorrhizal fungus content of the soil, which in turn depends on the previous floral composition. Different stable communities may develop, depending on the fungal biota of the soil (Janos 1980). In addition, the species composition of the mature trees that provides the relatively stable community at the end of a successional sequence in a rain forest gap depends both on the original size of the gap and, more interestingly, on the frequency distribution of natural gaps generated by tree falls in the particular forest. Gaps are a spatial resource for tree species. For example, about 75% of the canopy tree species in a rain forest in Costa Rica are estimated to be dependent on gaps to reach reproductive size (Hartshorn 1978), and two-thirds of the important canopy species on the Solomon Island of Kolombangara also are dependent on gaps (Whitmore 1974).

According to Denslow (1980), tree-fall gaps are a resource that the forest trees appear to partition; large gap, small gap, and understory guilds exist with different life-cycle strategies. The frequency distribution of gaps in a forest is a function of topography, local climatic conditions, edaphic factors, and the tree species composition. All of these factors are thought to play a critical role in succession. Although the statistical distribution of gap sizes has not yet been documented for any tropical forest, Denslow argues that the distributions are important in producing, for example, a roughly twofold difference in turnover times between forest stands in Costa Rica and Malaysia (Poore 1968, Hartshorn 1978). The key factor in succession in tropical forests appears to be the differences in the life cycles of the species involved in forest regeneration (Gómez-Pompa and Vázquez-Yanes 1974, 1981).

CLASSIC AND INDIVIDUALISTIC VIEWS OF SUCCESSION

Early workers on succession believed that they had found an important regularity among the complexities of nature. Indeed, Clements thought of a vegetational community as a superorganism. He wrote: "each climax formation is able to reproduce itself, repeating with essential fidelity the stages of its development. The life-history of a formation is a complex but definite process, comparable in its chief features with the life-history of an individual plant" (Clements 1916, p. 3).

No modern ecologist would take such an extreme view of succession, but two schools of thought still exist—the classic, tracing back to Clements, and the individualistic, tracing back to Henry Gleason, whose view was that communities in essence were simply collections of individuals that shared common physiological tolerances, as discussed in Part 5.

The influence of the Clementsian school can be seen today in the views of E. P. Odum, one of the leading ecosystem ecologists of our times. In a much-cited paper, "The Strategy of Ecosystem Development" (1969), Odum presented a model of succession, a process that "has many parallels in the developmental biology of organisms, and also in the development of human society." He defined succession as having three basic characteristics:

1. It is orderly, reasonably directional, and predictable.
2. It is controlled by the biological community, even though the basic constraints upon it are set by the physical environment.
3. It culminates in a community in which the "maximum biomass . . . and symbiotic function between organisms are maintained per unit of energy flow."

Odum concluded that the "strategy" of ecosystems is to develop in such a way as to be maximally protected against disturbance by fluctuations in the physical environment. He stated that this activity is also an evolutionary goal of the biosphere as a whole, a viewpoint that has been expanded into the Gaia hypothesis, to be discussed in Chapter 24. His model of the trends to be expected during succession is reproduced as Table 23–2, and his figure illustrating parallel energetic trends during succession in a forest ecosystem and a laboratory culture of microorganisms is shown in Figure 23–15.

The Odum model amounts to a series of predictions about changes in ecosystems through time: Biomass will increase, photosynthesis (in nonpolluted ecosystems) will at first exceed respiration, and then their ratio will converge on 1, species diversity will increase, food chains will intertwine more complexly into food webs, *r*-selected organisms will be replaced by *K*-selected organisms, and nutrient conservation and system resistance to perturbation will be increased. In addition, others have suggested that succession leads to an increase in primary production (Margalef 1968), an increase in soil depth and differentiation (Whittaker 1970), and a progressive inhibition of nitrification (Rice and Pancholy 1972).

How well do data from actual ecosystems fit these predictions? Modern members of the individualistic school think they do not fit well at all. For example, Drury and Nisbet (1973) examined the literature for evidence of

TABLE 23–2 A model of ecological succession: trends to be expected in the development of ecosystems

Ecosystem attributes	Developmental stages	Mature stages
Community energetics		
Gross production/community respiration (P/R ratio)	Greater or less than 1	Approaches 1
Gross production/standing crop biomass (P/B ratio)	High	Low
Biomass supported/unit energy flow (B/E ratio)	Low	High
Net community production (yield)	High	Low
Food chains	Linear, predominantly grazing	Weblike, predominantly detritus
Community structure		
Total organic matter	Small	Large
Inorganic nutrients	Extrabiotic	Intrabiotic
Species diversity	Low	High
Stratification and spatial heterogeneity	Poorly organized	Well-organized
Life history		
Niche specialization	Broad	Narrow
Size of organism	Small	Large
Life cycles	Short, simple	Long, complex
Nutrient cycling		
Mineral cycles	Open	Closed
Nutrient exchange rate, between organisms and environment	Rapid	Slow
Role of detritus in nutrient regeneration	Unimportant	Important
Selection pressure		
Growth form	For rapid growth (r-selection)	For feedback control (K-selection)
Production	Quantity	Quality
Overall homeostasis		
Nutrient conservation	Poor	Good
Stability (resistance to external perturbations)	Poor	Good

Source: After Odum (1969).

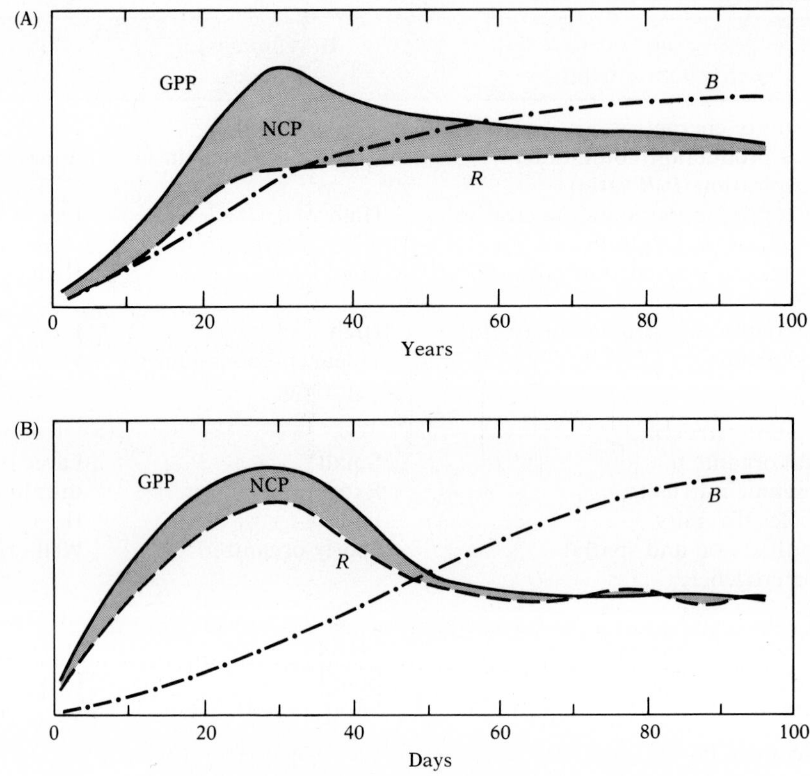

FIGURE 23–15 Parallel energetic trends in succession (A) in a forest and (B) in a laboratory aquatic microcosm. GPP = gross primary production; NCP = net community production (shade—the difference between GPP and *R*); R = total community respiration; *B* = total biomass. *(After Odum 1971.)*

the trends predicted by Odum. They quoted, for instance, Odum's own (1960) classic study of old field succession, claiming that it did not show the regular changes in energetics illustrated in Figure 23–15, but rather "a steady state in productivity . . . even though species composition was changing from year to year." They cited examples in which soil development is not progressive and in which climaxes developed on "immature soils."

Two fundamental tenets of the classical view of succession are that sequences of distinct plant associations occur and that each association facilitates the establishment of the next. Drury and Nisbet concluded that neither tenet was well supported by evidence from actual systems. They gave examples providing partial support to what Egler (1952), an early critic of classical succession, called the *initial floristic composition* hypothesis. Under that hypothesis, the propagules (seeds, living roots) of the entire successional sequence are present at the start, and succession merely represents the unfolding of that initial flora, as appears to be true of the secondary succession of chaparral we discussed earlier. Such unfolding of the initial composition also seems to occur in the classic Hubbard Brook study of succession following the clear-cutting of a northern hardwood forest, where "although the relative importance of individual species will change as their life strategies unfold and as shifting stages of dominance occur in the ecosystem (Harper and Ogden 1970), the great bulk of the species that will participate in the changes that follow clear-cutting are present, in active or dormant form, in the intact forest" (Bormann and Likens 1979).

In the Hubbard Brook forest, a major portion of the new growth following clear-cutting is from advance regeneration—that is, growth of established seedlings and roots sprouting from stumps and from the germination of buried seeds. The buried-seed strategy is especially important because in many northern hardwood forests millions of seeds are buried in the soil of each hectare, some of which may remain viable for a century or more, such as the seeds of the pin cherry (Marks 1974).

These situations cast doubt on both tenets of the classical view of succession, for one distinct association cannot be said to give way to another if the other is already there, and the necessity of site preparation for the climax species is dubious if individuals of the climax are present in the pioneer community to begin with. Indeed, early successional species may *inhibit* the development of later species. Olson (1958) showed, for example, that once blueberries and black oaks became established on a site in the Lake Michigan sand dunes, they tended to reproduce themselves rather than give way to the beech-maple climax. Apparently, these species tie up nutrients and produce an acid leaf litter that discourages the beeches and maples. Other examples are cited by Drury and Nisbet: Removal of pioneer trees in Michigan speeded the appearance of later successional species; early successional plants in old fields may modify their environments chemically so as to inhibit later species; shrubs discourage the development of mycorrhizal fungi whose presence is required by climax trees.

However, some cases have been found in which plants apparently make an area less inhabitable for their own species and, in this sense, prepare the way for continued succession. After investigating old fields in central Oklahoma and southeastern Kansas, Booth (1941) described a four-stage successional sequence. The first was a weedy stage that lasted for 2 or 3 years; sunflowers (*Helianthus annuus*), other composites, and crabgrass (*Digitaria sanguinalis*) were prominent. This stage was followed for about a decade by a stage involving annual grasses, dominated by *Aristida oligantha*. In the third and fourth stages, perennial grasses dominated. Laboratory experiments indicate that the sunflowers, crabgrass, and other species of the weedy stage produce toxins that tend to suppress their own seedlings and other early weeds (Wilson and Rice 1968, Parenti and Rice 1969).

It would be a mistake, of course, to conclude that the weedy species are in some sense "sacrificing" themselves so that succession can proceed. The behavior of each species is, presumably, the result of selection that maximizes the number of offspring left by the average individual. As Drury and Nisbet point out, the self-inhibition must be a side effect of the production of toxins evolved to inhibit competitors. In plants such as crabgrass with widely dispersed seeds, the loss from inhibited seedlings near the parent plant may be more than compensated for by the additional seed production permitted by freedom from competition.

In a synthesis of the theory of succession, Connell and Slatyer (1977) outline three alternative models of succession in plants and sessile marine animals, as depicted in Figure 23–16. They are the facilitation, tolerance, and inhibition models.

The *facilitation model,* which is closely related to the classical model, assumes that early successional species prepare the site for later species—that each stage facilitates the establishment of the next.

The *tolerance model* depends on different resource-exploitation strategies of species to produce sequences of site occupancy. The first species to appear have a life history featuring rapid dispersal of propagules and rapid growth; they either migrate to the site or are already there, contained in a seed bank. These species neither help nor hinder other species that disperse and grow more slowly. The slowly growing and dispersing species are more tolerant of shortages of resources such as light and nutrients and can mature in the presence of the early successional species that they replace.

The *inhibition model* is the antithesis of the first. It holds that each species tends to inhibit the invasion of others. Those that invade or sprout first occupy the territory; later species get a chance only when the early successional individuals die or are damaged.

How well do these models fit the available data? As you can see, the facilitation model seems to apply best to examples of primary succession, such as the soil development in the *Dryas*–willow mat at Glacier Bay and the stabilization of sand dunes by *Ammophila* that permit later successional species to become established. It also seems to describe soft-bottom benthic succession (Gallagher *et al.* 1983). The facilitation model also applies to *heterotrophic successions* in corpses and logs, in which decomposers at one stage of decay often produce the conditions required by their successors. For example, certain insects must bore through the bark of logs before others can attack their inner wood.

Experiments in one terrestrial system support the facilitation model. Seeds and seedlings of the saguaro cactus survive only if they are shaded (Niering *et al.* 1963, Steenbergh and Lowe 1969, Turner *et al.* 1969). The shade in which they do survive is usually provided by "nurse plants." The nurse plants not only provide an appropriate microclimate, as described by Turner *et al.* (1966), but also present a physical (and perhaps psycho-

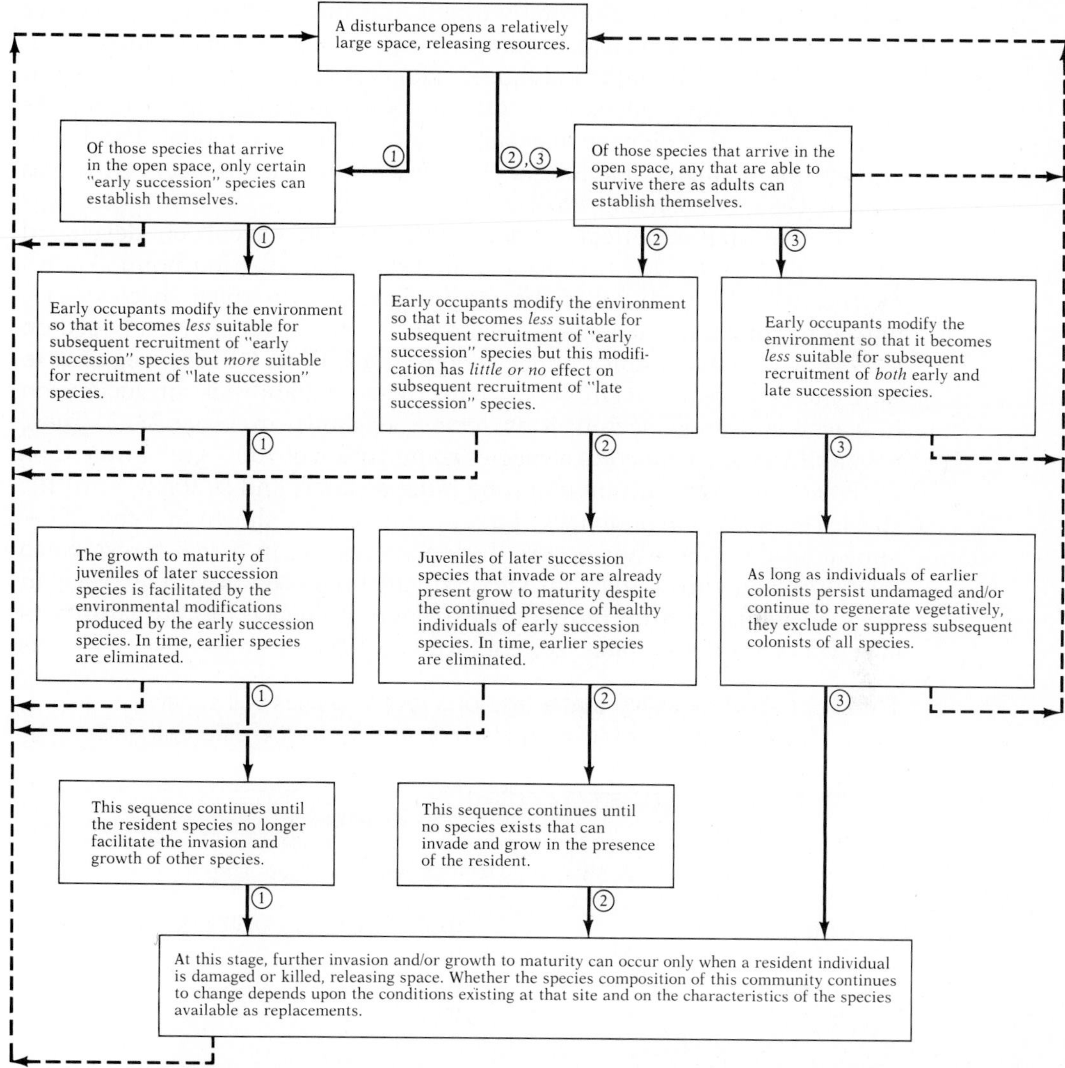

FIGURE 23–16 Alternative models (1, 2, and 3) of succession. Broken lines indicate interruptions in the process. *(After Connell and Slatyer 1977.)*

logical) barrier to grazing rodents that cause heavy seedling mortality. Except for the benthic infauna, pioneer marine species only rarely help make sites more hospitable for those later in the successional sequence (Connell 1972). Indeed, Sutherland (1974) found that in hard-substrate subtidal communities, additional individuals could colonize only after the death and disintegration of early occupants.

Both the tolerance and inhibition models predict that late successional species can become established without any site preparation by pioneer species, as is often true of secondary terrestrial succession. Abundant evidence exists for the inhibition model: Pioneers can prevent the establishment of later ones, reduce their growth rates, and lower the probability of their survival. Evidence also supports the tolerance model: Later species are more efficient exploiters of resources than the pioneers (Hils and Vankat 1982, Parrish and Bazzaz 1982).

Recently, mathematical models of succession have been developed (Botkin and Miller 1974, Horn 1975b, 1981). The starting point is with measures of the probability that an individual of a given species is replaced by another of the same species or of some other species. Based on the occurrence of saplings under canopy trees, Horn (1975b) was able to estimate a 50-year *transition matrix* of these probabilities for succession in a woods near Princeton, New Jersey, as shown in Table 23–3. These probabilities are applied to a species composition of 100% grey birch, then to the species composition one time interval later, and so forth, until the species composition no longer changes. The results, shown in Table 23–4, appear to describe what occurs during succession in a real forest. Such modeling can throw light on successional processes in species such as trees, in which mature individuals strongly influence the probability of survival of immature ones (see e.g., Runkle 1981).

TABLE 23–3 Probability that a tree of a given species will survive or be replaced by one of its own kind (diagonal) or of a different kind (off-diagonal)

	50 years hence			
Now	**Gray Birch**	**Blackgum**	**Red Maple**	**Beech**
Gray Birch	0.5	0.36	0.50	0.09
Blackgum	0.1	0.57	0.25	0.17
Red Maple	0	0.14	0.55	0.31
Beech	0	0.01	0.03	0.96

Source: Horn (1976).

TABLE 23–4 Predicted composition of a succession

Age of forest (years)	**0**	**50**	**100**	**150**	**200 ...**	**∞**	**Very old forest**
Gray Birch	100	5	1	0	0	0	0
Blackgum	0	36	29	23	18	5	3
Red Maple	0	50	39	30	24	9	4
Beech	0	9	31	47	58	86	93

Source: Horn (1976).

CLIMAX ECOSYSTEMS

Pollen studies show that dramatic changes occurred in forest and prairie ecosystems over thousands of years as continental glaciers waxed and waned; these changes are represented in Figure 23–17. If the climate in an area is altered, one would expect its biological community, and thus its ecosystem, to be replaced by another. However, close examination shows that all climax ecosystems are in a perpetual state of alteration (see e.g., Davis 1981). The time scales of significant change vary with factors such as the generation time of the dominant organisms and the frequency of disturbance.

Some of the changes appear to be more or less cyclic, with periodic natural or human-induced changes playing a major role in determining ecosystem structure and function. Periodic toppling of trees by hurricanes plays such a part in forests in New England. Fires do the same in northern forests (see, e.g., Loucks 1970, Heinselman 1973, Wright 1974, Tande 1979). Fires are also cyclic in chaparral and other Mediterranean-type communities (reviewed by Mooney *et al.* 1981c). Cyclic changes on a scale of decades have been recorded in the Western English Channel ecosystem (Southward 1980). Finally, long-term change is superimposed upon all other scales of change because of the slowly changing geology and climate of the planet. We see that a climax can be no more than a period of short-term constancy in an ecosystem, because of the gradual changes of plate tectonics.

SYSTEMS ECOLOGY

Mathematical models for the dynamics of an ecosystem often are based on simulation methods, using a digital computer. This approach has been called *systems ecology* because it is focused on the understanding of large, complex systems. Systems ecology has been pioneered with models about the dynamics of forests. These models shed light on how succession occurs, how nutrients are exchanged among the components of a forest ecosystem, and how the forest responds to perturbations such as the spread of tree diseases (e.g., chestnut blight), forest fires, and various techniques of logging (Botkin *et al.* 1972, Shugart and West 1977, Shugart *et al.* 1981, Shugart 1984). These models aid in understanding both what the processes in a forest ecosystem are and how to manage a forest as a resource.

Figure 23–18 presents a flow diagram for a computer program called FORET that simulates the dynamics of a stand of deciduous forest in the southern Appalachian mountains; it includes 33 tree species. This model is similar to others that have been developed since then that apply, for example, to *Eucalyptus*-dominated forests of the Australian Alps (Noble *et*

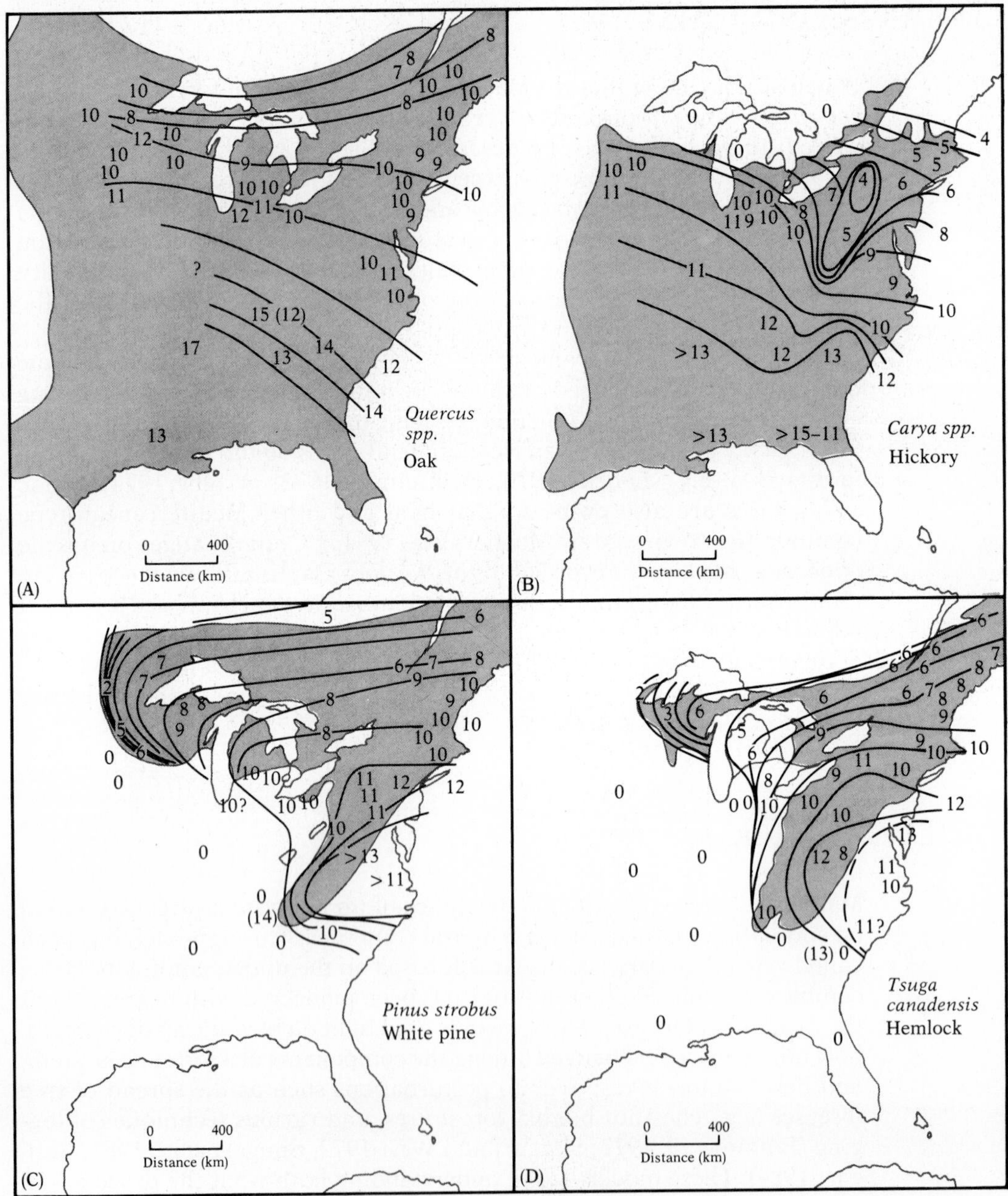

FIGURE 23–17 Migration of climax communities as indicated by various tree species. Numbers indicate date in thousands of years before present based on pollen records. (*After Davis 1981.*)

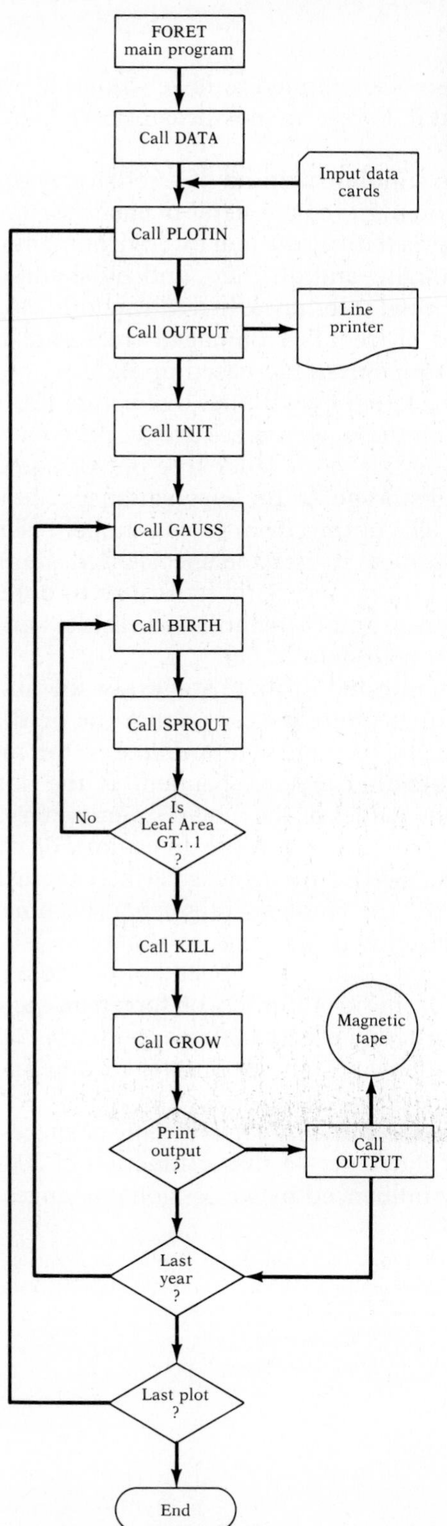

FIGURE 23–18 Flow diagram of the FORET model. Each rectangle represents a computer subroutine, and each diamond-shaped area represents a decision point in the program. *(After Shugart and West 1977.)*

al. 1980). Models have also been applied to lower montane rain forests in the Luquillo Experimental Forest in northeastern Puerto Rico (Doyle 1981).

All these forest models simulate a stand of trees in a specified area by computing the annual growth in the diameter of each tree in the site. The growth of each tree is assumed to be affected by environmental conditions, such as temperature, drought, and altitude, and by shading from other trees. In the model, individual trees are killed at random according to the life table for each species of tree. Fire-damaged trees and trees that are suppressed are assumed to incur an increased probability of mortality as a result. Trees may also be killed by wildfire, by hurricanes, and by being hit by other trees falling on them. New trees are recruited at random into the vacant area left by the fall of an older tree or, for some species, by vegetative sprouting, seed storage, or the survival of special reproductive organs, as in *Eucalyptus.* The output from such a model resembles a field data sheet from a sample plot; it lists the species and diameter of each tree. Typically, the model is run for many iterations to obtain the mean of the overall forest response, and the simulation results can be analyzed in the same way as actual field data.

A systems model is often tested in three stages. In the first stage, *verification,* the model's parameters are chosen so that the predictions of the model fit certain data. In the second stage, *validation,* the model is compared with additional data that are independent of the data originally used when determining the values of the model's parameters. In the third stage, the model is *applied* to basic questions in ecology or resource management. For example, the FORET model was verified against data on the composition of forests after the chestnut blight had become endemic. It was validated by comparison with the composition of forests prior to the spread of the chestnut blight. It has since been applied to a prediction of the effects of chronic atmospheric pollutants on forests in the southeastern United States, the reconstruction of a 16,000-year pollen chronology under changing climatic conditions, and the response of certain forest birds following the harvesting of selected trees.*

Now that we have introduced the important topic of succession, we can pass on to the cycling of nutrients—a major function of ecosystems and one that itself is strongly influenced by successional changes.

* The techniques of systems ecology have now been applied to a variety of ecosystems. Further important reading includes Holling (1973), Patten *et al.* (1975), and Watt (1975).

CHAPTER 24

Biogeochemical Cycles: Connecting the Physical and the Biotic

Earth is an open system with respect to energy; it is a closed system with respect to materials. The amounts of the elements do not change with time; with trivial exceptions, such as the iridium in meteorites, nothing but energy enters or leaves the planet.

However, the elements of Earth are not static. They are continually changing position, phase, and combinations with one another. The cycling of elements makes nutrients continuously available to organisms and thus makes life possible. This cycling also prevents the accumulation of elements in forms, quantities, and locations that are deleterious to organisms. Nutrients flow along paths that bind together the living and physical portions of the biosphere; these paths are called *biogeochemical cycles*.

MECHANISMS OF NUTRIENT CYCLING

The major *pools* of nutrients are the atmosphere, the oceans, soils, and Earth's crust. The major pathways of rapid nutrient transport among these pools are in movements of water and air. Nutrients are carried by rivers, streams, groundwater, rain, glaciers, moving organisms, ocean currents, and winds. Nutrients enter the atmosphere directly as gases, such as CO_2 released by animals, or as tiny particles, or they can be carried

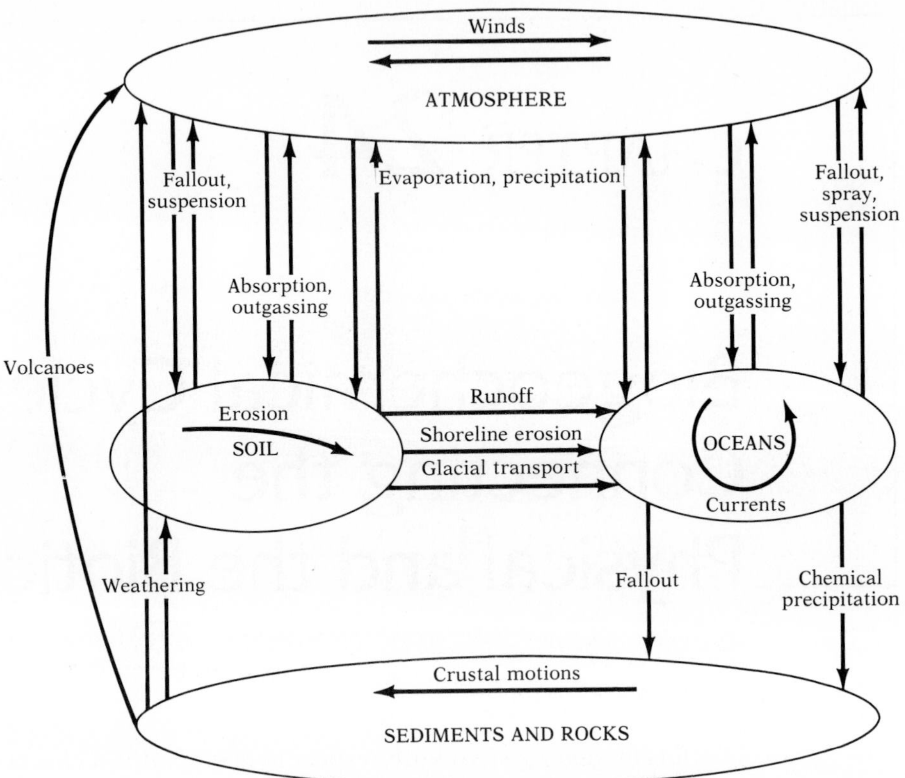

FIGURE 24–1 Transport among compartments in the physical world. *(After Ehrlich et al. 1977b.)*

along by evaporating water. Nutrients can leave the atmosphere as gases, such as CO_2 taken up by plants, or as solid "fallout," or can be carried out by rain or snow. Geological movements provide much slower transport of nutrients. Because the transport mechanisms are so similar, and so often dependent on the movement of water, the cycles of various nutrients tend to be linked with one another. The pathways among pools are shown schematically in Figure 24–1; the hydrologic (water) cycle is diagrammed in Figure 24–2.

The magnitude of some global materials flows are shown in Table 24–1. Note that the oceans lose much more water by evaporation than they receive in flows off the land. The reason, of course, is that over 90% of the water evaporated from the oceans returns to them directly as rain or snow.

When the size of a pool is constant, one can calculate how long an average atom or molecule of nutrient spends in the pool. The result is known as the *residence time*. The residence time is simply the size of the pool divided by the sum of the outflows. For example, about 4.8×10^{21} g

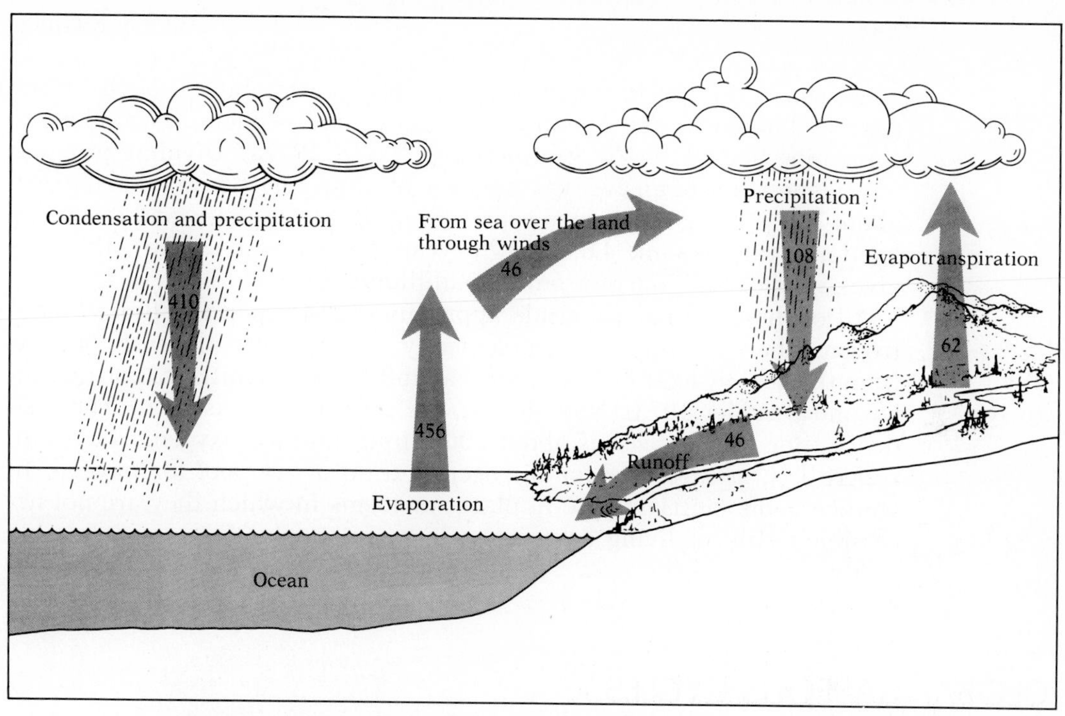

FIGURE 24–2 The hydrologic cycle. Flows are in units of 1000 km^3/yr. *(After Ehrlich et al. 1977b.)*

TABLE 24–1 Some global materials flows[a]

Pathway	Magnitude of flow (billion MT/yr[b])
To oceans	
River discharge (H_2O)	32,000
Groundwater flow (H_2O)	4,000
River-borne suspended solids (today)	18
River-borne suspended solids (prehuman)	5(?)
River-borne dissolved solids	4
Ice-sheet transport of solids	2
Shoreline erosion	0.3
To and from atmosphere	
Evaporation (H_2O)	526,000
CO_2 exchange in photosynthesis	220
CO_2 from burning fossil fuels	16
Dust and smoke from land	1(?)
Volcanic gases and debris	0.1

Source: After Ehrlich *et al.* (1977b).

[a] Uncertainty in most estimates is roughly ±20 percent.
[b] MT = metric tons (10^6 g).

of oxygen is present in the atmosphere. Flows into and out of the atmosphere—biological and chemical oxidations and reductions—amount to something like 2.4×10^{17} g/year (Ehrlich *et al.* 1977a). Dividing provides a residence time of an average molecule of some 20,000 years. In contrast, about 7.6×10^{21} g of nitrogen is present in the atmosphere, and it is being fixed at a rate of some 4.0×10^{14} g/year (Ehrlich *et al.* 1977a). This rate gives a residence time of around 20 million years.

All of these figures are crude approximations, especially those for nitrogen, where the pool is not in a steady state because presently outflow from the pool by fixation is greater than inflow from denitrification (Strahler and Strahler 1973). Nonetheless, the critical point is clear. The residence time for nitrogen is about 6000 times that for oxygen. The atmosphere is a *storage pool* for nitrogen and an *active pool* for oxygen. In storage pools, nutrients are in places or forms in which they are not accessible easily to living organisms; in active pools, they are accessible easily.

GLOBAL GASEOUS CYCLES

Both oxygen and nitrogen participate in *gaseous cycles,* as characteristic of elements that form gaseous compounds at temperatures normally encountered in the biosphere. These elements therefore can diffuse into the atmosphere and move with its circulation. All of the major building blocks of living organisms—carbon, hydrogen, oxygen, and nitrogen—move in gaseous cycles. The cycles of carbon, hydrogen, and oxygen especially are tightly linked, and we will discuss them together.

Carbon, Hydrogen, and Oxygen

Net primary production adds O_2 to the atmosphere and subtracts CO_2 and H_2O. Why, then, does not the oxygen content of the atmosphere continually increase and its store of carbon dioxide and water continually decrease? The reason is that plant material is not accumulating continuously but is being decomposed into CO_2 and H_2O by the respiration of autotrophs, respiration that withdraws O_2 from the atmosphere about as rapidly as photosynthesis adds it.

How, then, did gaseous oxygen accumulate in the atmosphere in the first place, inasmuch as the evidence indicates that it was all produced by photosynthesis in the last 2 billion years (Cloud and Gibor 1970)? The answer lies in the coupling of the O_2 cycle with the carbon cycle, as represented in Figure 24–3. The flows of carbon follow pathways similar to those taken by energy in ecological systems. Carbon enters a food chain through the autotrophs and then moves up the chain, with some returning

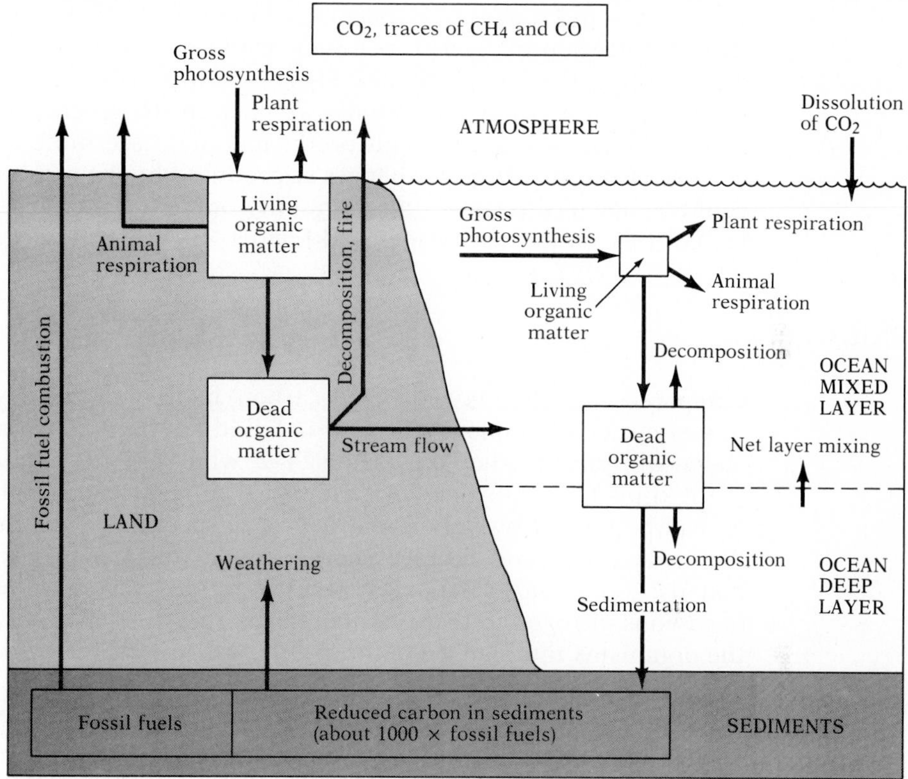

FIGURE 24–3 The carbon cycle. *(After Ehrlich* et al. *1977b.)*

to the atmospheric pool at every step as both autotrophs and heterotrophs produce some CO_2 from respiration. Over time, however, some of the carbon has left the cycle and escaped oxidation. For every atom of carbon removed from the carbon cycle, the two atoms of oxygen that were released when that carbon atom was originally fixed—connected to a solid or nonvolatile form—remained in the atmosphere. Carbon that has not been oxidized to carbon dioxide is the stuff from which fossil fuel deposits are made. These concentrations of carbon in Earth's crust originated as plant materials buried in swamps and the sediments of shallow seas from which both anaerobic decomposers and oxygen were absent. However, the vast majority of unoxidized carbon is not in fossil fuels but in sediments in which the carbon concentration is less than one-tenth of 1%.

Because of carbon storage in the crust, the atmospheric CO_2 pool is some three orders of magnitude smaller than that of O_2. As carbon and O_2 are tightly linked, the residence time of CO_2 is correspondingly shorter—less than 10 years—and any withdrawal of carbon from the unoxidized pool has a much greater effect on the atmospheric concentration of CO_2 than of O_2. A rate of burning of fossil fuels that would lower the O_2 concentration by 0.001% per year would increase the CO_2 concentration by 0.7%. Burning of fossil fuels is now increasing the concentration of CO_2 which will have a dramatic impact on the greenhouse effect (described in Chapter 20) without producing any discernible change in the O_2 concentration. Climatic changes as produced by the resultant warming could be one of this most important environmental problems around the turn of the century (Schneider and Londer, 1984; Keeling *et al.*, 1984).

Nitrogen

Compared with the movements of carbon, hydrogen, and oxygen, the nitrogen cycle is complex. Nitrogen is found in the cycle in a number of chemical forms, as shown in Table 24–2, which are not available equally to all kinds of organisms. For example, the gigantic atmospheric pool of N_2 can be tapped by only a relatively few species of bacteria and algae. Plants and animals require more complex nitrogen-containing compounds: ammonia (NH_3) and especially nitrates (NO_3^-). The conversion of gaseous nitrogen to these compounds is known as *nitrogen fixation,* and the organisms that can accomplish this task are called *nitrogen fixers,* as discussed in Part 1. Nitrogen fixers are the link between the main accessible pool of nitrogen and all other organisms. Thus, they are responsible for continuation of life as we know it.

The best-known nitrogen fixers are bacteria of the genus *Rhizobium,* which live as symbionts in nodules on the roots of legumes. They obtain

TABLE 24–2 Chemical forms of nitrogen

Formula	Name	Oxidation number[a]	Comments
NH_3	Ammonia	−3	Major nutrient form
NH_4^+	Ammonium ion	−3	From NH_3 dissolved in water
NH_2^+	Amino group	−1	Constituent of protein
N_2	Nitrogen gas	0	Bulk of atmosphere
N_2O	Nitrous oxide	+1	Laughing gas, controls natural ozone cycle
NO	Nitric oxide	+2	Combustion product
NO_2^-	Nitrite ion	+3	Link in N cycle
NO_2	Nitrogen dioxide	+4	From NO oxidized in atmosphere
NO_3^-	Nitrate ion	+5	Principal nutrient form

Source: After Ehrlich *et al.* (1977b).
[a] Negative oxidation numbers denote more reduced forms, and positive numbers, more oxidized forms.

TABLE 24–3 Some chemistry of nitrogen in organisms

Step	Reaction	Energy	Organism
Fixation	$2N_2 + 6H_2O \longrightarrow 4NH_3 + 3O_2$	In	*Rhizobium, Azotobacter, Gloeocapsa, Plectonema*
Amino acid synthesis (ammonification is the reverse)	$2NH_3 + 2H_2O + 4CO_2 \longrightarrow 2CH_2NH_2COOH + 3O_2$	In	Many, bacteria and others
Nitrification	$2NH_4^+ + 3O_2 \longrightarrow 2NO_2^- + 4H^+ + 2H_2O$	Out	*Nitrosomonas, Nitrosospira, Nitrosolobus*
Nitrification	$2NO_2^- + O_2 \longrightarrow 2NO_3^-$	Out	*Nitrobacter*
Denitrification	$4NO_3^- + 2H_2O \longrightarrow 2N_2 + 5O_2 + 4OH^-$	Out	*Pseudomonas*
Denitrification	$5S + 6KNO_3 + 2CaCO_3 \longrightarrow 3K_2SO_4 + 2CO_2 + 3N_2$	Out	*Thiobacillus denitrificans*
Denitrification	$C_6H_{12}O_6 + 6NO_3^- \longrightarrow 6CO_2 + 3H_2O + 6OH^- + 3N_2O$	Out	Many

Sources: After Ehrlich *et al.*, 1977b.

energy from the plant and use the energy, as well as part of the plant's metabolic machinery, to carry out the nitrogen-fixing reaction:

$$2N_2 + 6H_2O \longrightarrow 4NH_3 + 3O_2$$

In turn, the legume is able to use some of the ammonia to synthesize amino acids, such as glycine:

$$2NH_3 + 2H_2O + 4CO_2 \longrightarrow 2CH_2NH_2COOH + 3O_2$$

The nitrogen-fixing activities of *Rhizobium* have made legumes a key tool for farmers to use in restoring the fertility of nitrogen-depleted soils—that is, as "green manures."

Rhizobium is the most common nitrogen-fixing organism, and various *Rhizobium* species each associate with a single species of legume. The details of how the symbionts recognize each other are now under intensive investigation. Another important symbiotic nitrogen-fixing bacterium is *Anabaena*, a cyanobacterium or blue-green alga that lives in the leaves of the floating water fern *Azolla* and that plays a major role in fertilizing rice

FIGURE 24–4 The nitrogen cycle. *(After Ehrlich* et al. *1977b.)*

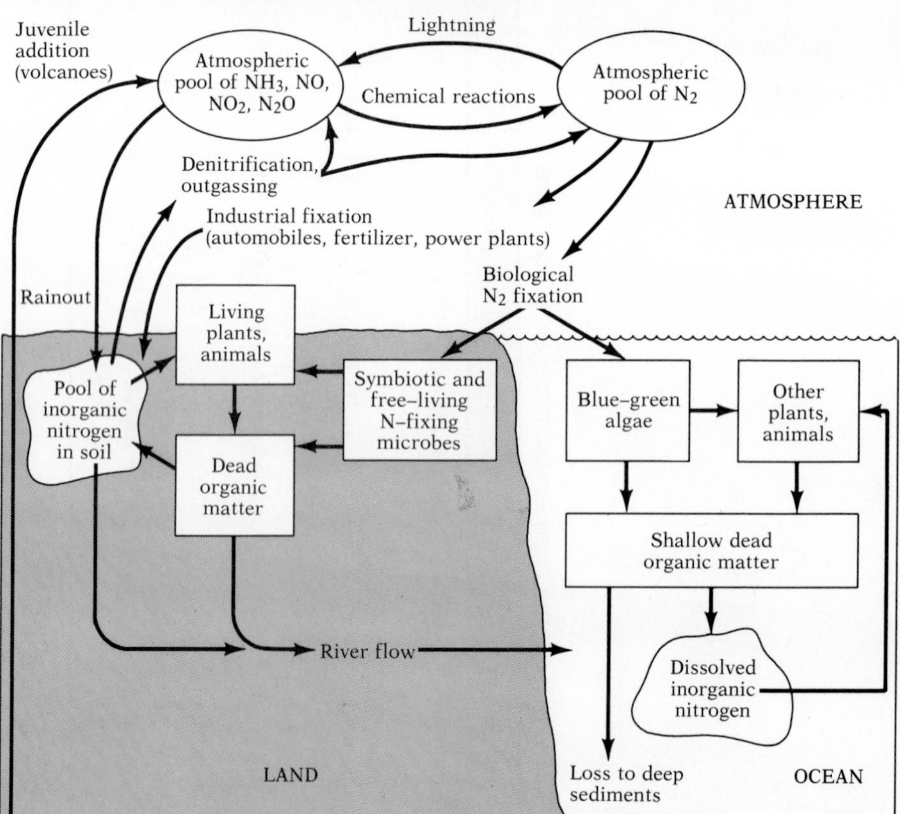

paddies. Such symbiotic relationships with nitrogen fixers are scattered throughout the plant kingdom. Nitrogen fixation is also accomplished by free-living bacteria, such as the aerobic *Azotobacter* and the anaerobic *Clostridium,* both of which are common decomposers in soils.

Amino acids are synthesized not only from ammonia but also from nitrate. Nitrate is produced by oxidation of ammonia in two processes known collectively as *nitrification.* As Table 24–3 indicates, nitrification yields energy to the several kinds of soil bacteria—*Nitrosospira, Nitrosomonas, Nitrosolobus,* and *Nitrobacter*—that carry it out (Belser and Schmidt 1978, Belser 1979). The nitrogen incorporated into living organisms returns to the soil when organisms die, in animal wastes (which are rich in urea), or in fallen leaves. Decomposers break down the proteins of dead organisms. The nitrogen-containing residues and urea are converted to ammonia by certain soil bacteria and fungi in an oxidation process that yields them energy—the reverse of the amino acid synthesis process previously shown.

FIGURE 24–5 The sulfur cycle. *(After Ehrlich* et al. *1977b.)*

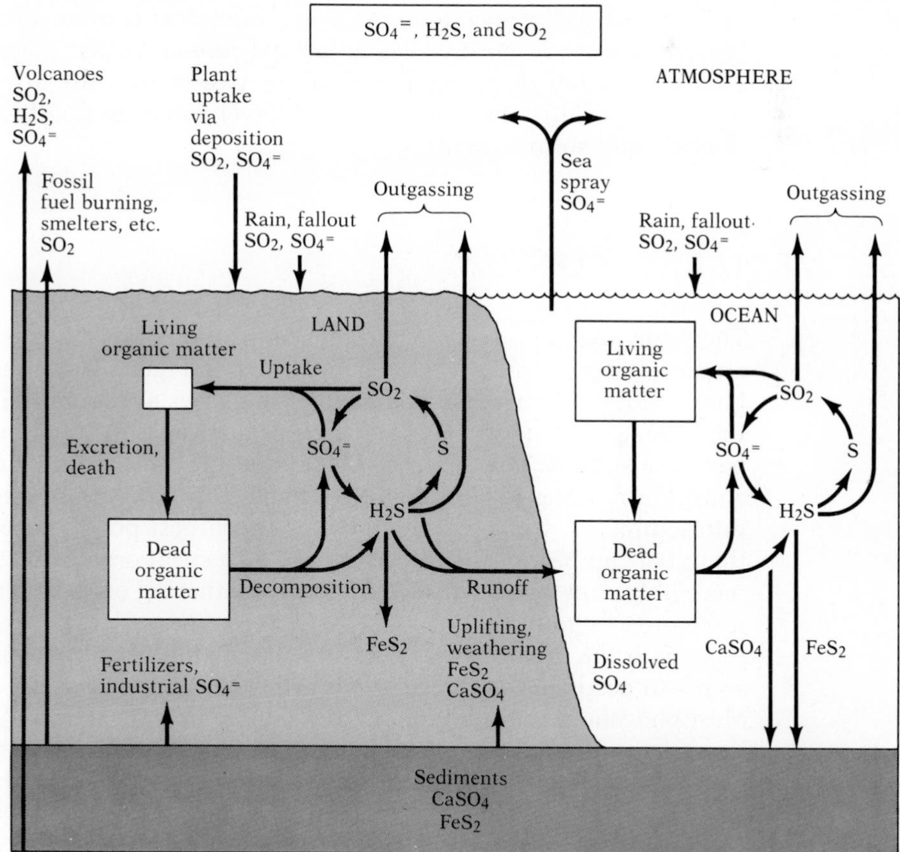

If the story ended here, no nitrogen cycle (see Figure 24–4) would exist, and NO_3 would build in the soil until it reached toxic levels. To close the loop, *denitrification* must be able to occur—that is, some way for volatile forms of nitrogen (N_2, N_2O) to be formed and released back to the atmospheric pool. Many microorganisms do just that, as may be seen in Figure 24–4. Nonetheless, as is evident from the heavy lines in the figure, most of the action in the nitrogen cycle involves internal loops rather than exchange with the atmospheric pool.

The nitrogen cycle has some pathways that do not involve fixation or denitrification by organisms (see Figure 24–4). Most interesting, perhaps, is *industrial fixation*. In the *Haber process*, nitrogen and hydrogen gases are combined with a catalyst at high temperature and pressure to yield ammonia, which is then used as a base for the manufacture of fertilizers. The Haber process now accounts for a substantial portion of all nitrogen fixation.

Sulfur, which is critical to the structure of proteins, has a gaseous cycle, as Figure 24–5 indicates, that is similar in many respects to that of nitrogen, except for the absence of a gigantic atmospheric pool; however, this cycle is less well understood (Missouri Botanical Garden 1975). One important feature of both cycles is labeled *rainout* in the diagrams of the cycles. Injection of nitrogen and sulfur oxides into the atmosphere by industrial societies has increased the travel along that pathway by rains of nitric and sulfuric acids.

SEDIMENTARY CYCLES

The pathways followed by phosphorus in the ecosphere are a good example of a sedimentary cycle, as illustrated in Figure 24–6. Because phosphorus does not form compounds that are gases at normal temperatures, the closing of the cycle proceeds at a geological rather than a biological pace. Phosphorus flowing from the land to the sea cannot be returned by a quick atmospheric path. Instead, almost all phosphorus is incorporated into sediments, which make up by far the largest pool. Phosphorus from that pool can become available only to terrestrial organisms through the incredibly slow processes of uplift and weathering or through human intervention by mining.

As with the nitrogen cycle, most of the action in the phosphorus cycle seems to occur in the internal loops (those not involving the sediments). Most phosphorus salts are not highly soluble in water, so that much more is carried into the sea in particles eroded from the land than in solution. The only situations in which rapid return from sea to land occurs are when seabirds eat fish and deposit their guano on land and when marine organisms are harvested.

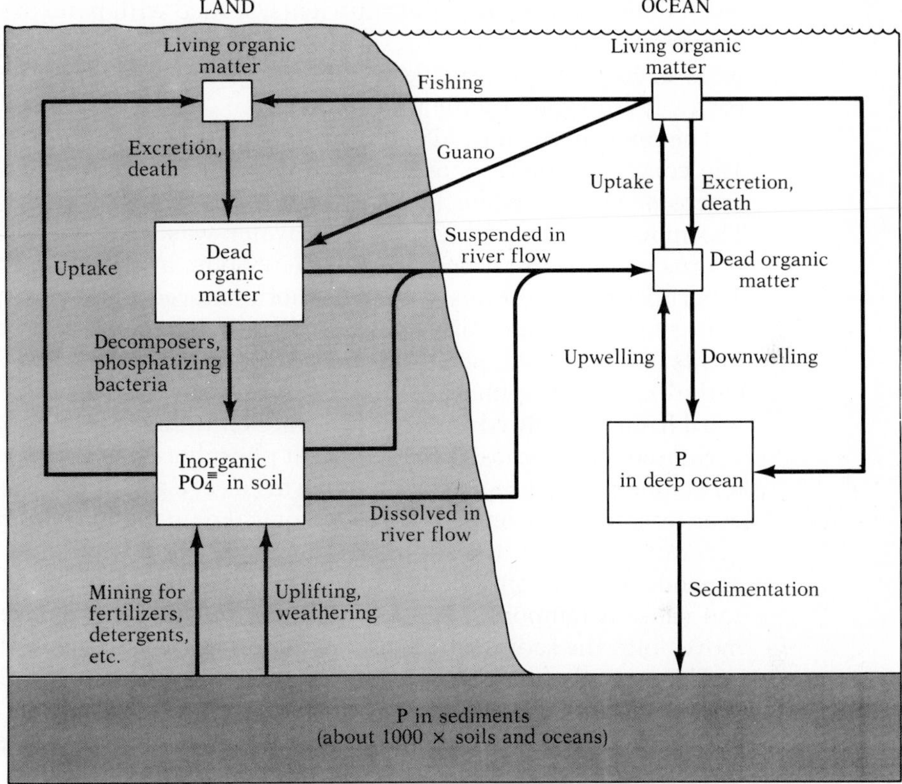

FIGURE 24–6 The phosphorus cycle. *(After Ehrlich et al. 1977b.)*

The internal loops of the phosphorus cycle in aquatic systems have been subject to intensive studies using a *tracer,* a radioactive isotope of phosphorus, ^{32}P, that behaves chemically just like nonradioactive phosphorus. Pioneering tracer studies of the phosphorus cycle in lakes by Coffin and his colleagues (1949), Hutchinson and Bowen (1947, 1950), and Rigler (1956) and in marine systems by Pomeroy (1960) revolutionized the view of the movements of that element. Prior to the 1940s, ecologists focused on gradual seasonal changes in the availability of phosphorus. In spring, plankton used the nutrient and bloomed, exhausting the supply and causing nutrient-poor water in the summer. Then decay during the winter slowly reestablished the supply. The tracer studies, in contrast, showed that phosphorus in aquatic ecosystems could have turnover times of days, hours, or even minutes.

The cycling of phosphorus was examined in detailed experiments at the Hanford Laboratories in Richland, Washington, in which ^{32}P was introduced into 200-L aquarium microcosms (Whittaker 1961). A *microcosm* is an artificial system set up to resemble a larger natural system. The microcosms had been established with either deionized water plus nutrients or Columbia River water plus nutrients, seeded with pond organisms, and allowed to develop for 6 weeks. After the tracer was introduced, samples were taken of water, plankton, and filamentous algae to determine the pathways followed by the phosphorus.

The phosphorus moved rapidly out of the water and into the plankton. The concentration of phosphorus in the plankton of low-phosphorus microcosms became as much as 2 million times as high as in the water itself. The movement of ^{32}P back into the water was equally rapid. The phosphorus completely turned over in a few hours at most, depending on the experimental conditions. The phosphorus moved more slowly into algae, animals, mud, and sediments. Some turnover took place between sediments and mud and water, but no equilibrium was reached in the course of the experiments. Instead, a steady movement was observed of ^{32}P downward from the lighted water and surfaces into the depths of the substrate.

Subsequent studies have found that phosphorus behaves in natural systems much as it does in the microcosms (see, e.g., Rigler 1964). The turnover time tends to depend on the biological demand for phosphorus, being generally higher in the summer than in the winter. In lakes, the biota is dependent on a continuous input of phosphorus. A pulse of phosphorus can cause a temporary plankton bloom, but the phosphorus inexorably moves into the sediment, where it is no longer available to the plankton.

The mining of phosphate rock (see Figure 24–6) constitutes a substantial human intervention in the natural phosphate cycle. That intervention is required because of human interference in the internal loops of the cycle on farms and pastures. Crops and animals are removed from the system, and residues and sewage are not returned to it. This activity amounts to "mining" the soil of its nitrogen, phosphorus, and other nutrients and results in the need for fertilizer.

NUTRIENTS IN LOCAL COMMUNITIES

In describing nutrient cycles so broadly, we have concealed the basis on which the global picture has been assembled, the magnitude of variation in biogeochemical cycles from ecosystem to ecosystem, and the variation in successional stages of the same ecosystem. The kind of study that provides the basis for understanding overall cycling of nutrients in the biosphere is typified by classic investigations of six contiguous, small (12- to 43-ha) watershed ecosystems in the Hubbard Brook Valley of New Hampshire (Likens *et al.* 1977). The system was chosen because it was underlain by a watertight metamorphic bedrock, so that little chance existed of water loss by deep seepage. Each watershed had nutrient inputs from the movement of animals, from wind, rain, and snow, and from nutrient outputs from wind and stream flow. However, most of the nutrient input and output could be determined by monitoring the nutrient content of precipitation and streams.

Precipitation collectors measured rain and snowfall throughout the experimental area. Stream-measuring stations, using small dams built on bedrock, measured the outflow from each of the watersheds. These stations provided data on the volume of inflows and outflows. Rain, snow, and stream-water samples were taken weekly for chemical analysis. Combining all the data led to the precipitation input, the stream-water output budgets shown in Table 24–4.

Considerable annual variation in the flows results largely from annual changes in the amount of precipitation. Precipitation inputs are important nutrient sources for the system. For example, sulfur and nitrogen oxides in rain and snow provide a substantial portion of the uptake of sulfur and nitrogen by the green plants. Overall the data tended to show a net loss of calcium, magnesium, potassium, sodium, sulfate, aluminum, and dissolved silica. A net gain occurred in ammonium, hydrogen ion, nitrate, chloride, and phosphate.

The losses of dissolved inorganic substances, amounting to some 74.7 kg/ha/yr, are accounted for by weathering of rocks within the system, an unmeasured "input" to the active pools of the substances lost. In addition, the accumulation of nitrogen indicated by the figures for ammonium and nitrate are insufficient to account for the increase of nitrogen recorded in living and dead biomass in the system. The calculations of Likens and his colleagues suggest that about 14 kg/ha/yr may be entering the system in gaseous form or as dry deposition of aerosols.

VARIATION IN ENERGY–NUTRIENT FLUXES AMONG ECOSYSTEMS

Many factors influence patterns of nutrient circulation in terrestrial systems, including temperature, precipitation patterns, dominant vegetation, fire regime, and underlying base rock. Data from some forest systems and trees are shown in Table 24–5. Note that even within temperate deciduous forests, differences can be striking (Duvigneaud and Denaeyer-de Smet 1970). For example, a German beech forest growing on diorite takes up, adds to the biomass, and returns to the soil about twice the quantity of nutrients annually of another German beech forest growing on granite (Klausing 1956).

In general, tundra, taiga, arid, and semiarid ecosystems have low productivity; subtropical and tropical forests, bogs, and floodplains are one to two orders of magnitude more productive (refer back to Table 23–1). The biomass per unit area also varies greatly from ecosystem to ecosystem. Both productivity and biomass affect the position of pools of nu-

TABLE 24–4 Annual budgets of precipitation inputs and stream-water outputs of dissolved substances for undisturbed watersheds of the Hubbard Brook Experimental Forest (alternate years only shown)

Substance (kg/ha)	1963–1964	1965–1966	1967–1968	1969–1970	1971–1972	1973–1974	Total 1963–1974 (kg/ha)	Annual mean (kg/ha)
Calcium								
Input	3.0	2.7	2.8	2.3	1.2	2.0	23.8	2.2
Output	12.8	11.5	14.2	16.7	12.4	21.7	151.2	13.7
Net	−9.8	−8.8	−11.4	−14.4	−11.2	−19.7	−127.4	−11.5
Magnesium								
Input	0.7	0.7	0.7	0.5	0.4	0.4	6.3	0.6
Output	2.5	2.9	3.7	3.5	2.8	4.6	34.6	3.1
Net	−1.8	−2.2	−3.0	−3.0	−2.4	−4.2	−28.3	−2.5
Potassium								
Input	2.5	0.6	0.7	0.8	0.3	0.8	9.8	0.9
Output	1.8	1.4	2.2	2.1	1.6	2.8	20.7	1.9
Net	0.7	−0.8	−1.5	−1.3	−1.3	−2.0	−10.9	−1.0
Sodium								
Input	1.0	2.0	1.7	1.9	1.5	1.4	17.4	1.6
Output	5.9	6.9	9.1	8.0	5.8	9.8	79.5	7.2
Net	−4.9	−4.9	−7.4	−6.1	−4.3	−8.4	−62.1	−5.6
Aluminum								
Input	[a]	[a]	[a]	[a]	[a]	[a]	[a]	[a]
Output	1.6[b]	1.7	2.1	2.2	1.7[b]	3.2[b]	21.9	2.0
Net	−1.6	−1.7	−2.1	−2.2	−1.7	−3.2	−21.9	−2.0
Ammonium								
Input	2.6[b]	2.6	3.2	2.7	2.8	3.7	31.6	2.9
Output	0.27[b]	0.92	0.24	0.51	0.05	0.42	3.7	0.34
Net	2.3	1.7	3.0	2.2	2.8	3.3	27.9	2.6

Hydrogen								
Input	0.85[b]	0.85	0.96	0.93	0.97	1.14	10.62	0.96
Output	0.08[b]	0.05	0.06	0.09	0.13	0.20	1.13	0.10
Net	0.77	0.80	0.90	0.84	0.84	0.94	9.49	0.86
Sulfate								
Input	33.7[b]	41.6	46.7	29.3	33.0	52.8	418.3	38.0
Output	42.7[b]	47.8	58.5	48.1	46.8	84.7	580.3	52.8
Net	−9.0	−6.2	−11.8	−18.8	−13.8	−31.9	−162.0	−14.8
Nitrate								
Input	12.8[c]	17.4	22.3	14.9	21.4	30.9	209.5	19.0
Output	6.7[c]	6.5	12.7	29.6	18.7	34.8	177.5	16.1
Net	6.1	10.9	9.6	−14.7	2.7	−3.9	32.0	2.9
Chloride								
Input	5.5[b]	2.6	5.0	5.3	5.5	4.4	67.9	6.2
Output	3.7[b]	4.3	5.3	4.2	3.6	5.9	50.4	4.6
Net	1.8	−1.7	−0.3	1.1	1.9	−1.5	17.5	1.6
Phosphate								
Input	0.09[b]	0.10[b]	0.11[b]	0.10[b]	0.10[b]	0.13	1.16	0.11
Output	0.02[b]	0.02[b]	0.02[b]	0.02[b]	0.02[b]	0.029	0.225	0.020
Net	0.07	0.08	0.09	0.08	0.08	0.10	0.94	0.09
Bicarbonate								
Input	[a]	[a]	[a]	[a]	[a]	[a]	[a]	[a]
Output	6.2[b]	6.2	9.6	6.0	6.6[b]	12.5[b]	84.2	7.7
Net	−6.2	−6.2	−9.6	−6.0	−6.6	−12.5	−84.2	−7.7
Dissolved silica (SiO_2)								
Input	[a]	[a]	[a]	[a]	[a]	[a]	[a]	[a]
Output	30.5[b]	36.1	42.1	41.4	30.6	56.7	414.3	37.7
Net	−30.5	−36.1	−42.1	−41.4	−30.6	−56.7	−414.3	−37.7

Source: After Likens *et al.* (1977).

[a] Not measured, but trace quantities.

[b] Calculated value, based on weighted average concentration during years when chemical measurements were made and on amount of precipitation or streamflow during the specific year.

[c] Calculated from weighted average concentration for 1964–1966 times precipitation or streamflow for 1963–1964.

TABLE 24–5 Annual uptake, retention, and release of nutrients by the trees of woodland ecosystems (kg/ha)

		Chemical Elements (kg/ha)				
		K	Ca	Mg	P	N
Spruce on podsol soil U.S.S.R.	Uptake	18	56	9	3.2	61
	Retained	6	8	2	0.6	9
	Released by litter fall	12	48	7	2.6	52
Spruce on rich soil U.S.S.R.	Uptake	20	52	7	2.7	62
	Retained	3	6	1	0.3	8
	Released by litter fall	17	46	6	2.4	54
Birch U.S.S.R.	Uptake	30	107	29	11	111
	Retained	17	53	10	6	45
	Released by litter fall	13	54	19	5	66
Mixed oakwood U.S.S.R.	Uptake	85	102	16	7	92
	Retained	23	16	3	4	33
	Released by litter fall	62	86	13	3	59
Mixed oakwood Belgium	Uptake	69	201	19	6.9	92
	Retained	16	74	6	2.2	30
	Released by litter fall	53	127	13	4.7	62
Beech forest Germany on diorite	Uptake	8.5	72.4	9.8	9.4	—
	Retained	4.5	49	5.3	6.1	—
	Released by litter fall	4.0	23.4	4.5	3.3	—
Beech forest Germany on granite	Uptake	4.7	24.5	4.6	5.6	—
	Retained	2.5	23.8	2.5	2.8	—
	Released by litter fall	2.2	10.7	2.1	2.8	—
Nothofagus truncata New Zealand	Uptake	34	84	12	3.3	10
	Retained	4	10	1	0.7	3
	Released by litter fall	30	74	11	2.6	37

Sources: After Ovington (1962) and Duvioneaud and Denaeyer-DeSmet (1970).

trients and their movement along pathways. Superficially similar terrestrial systems can be quite different. In relatively infertile forests, including many tropical rain forests, many of the nutrients are stored in the biomass; in fertile ones, the nutrients tend to be concentrated in the soil.

There are fundamental differences between terrestrial and aquatic ecosystems, many of which are traced directly to the buoyancy of aquatic organisms and to the absorption of light in water. Aquatic organisms generally spend much less energy than terrestrial ones do holding themselves up. They usually do not have to make a continuous effort to remain upright or to devote resources to rigid structures (Parsons 1976). In addition, aquatic animals use an order of magnitude less energy to move than do terrestrial animals (Gold 1973). Thus, terrestrial ecosystems are characterized by an enormously greater standing crop of strong structural materials—primarily wood, bones, and insect cuticle—than is found in most aquatic systems. In aquatic systems, where the majority of producers are short-lived phytoplankton, turnover times are generally much more rapid, and inverted biomass pyramids are common. Much is still to be learned about flows of energy and nutrients in aquatic systems. For example, Pomeroy (1979) extended the work of Steele (1974; see also Landry 1976) in modeling energy flux in marine systems. His simulations of a continental shelf community, as represented in Figure 24–7, cast strong doubt on the assumption that a 10% efficiency could be assumed in marine food webs and in the levels of primary productivity that had been generally imputed to marine systems (again, refer back to Table 23–1). Pomeroy concluded that either "conventional assumptions about efficiency must be low by a factor of 2 to 3, or measurements of photosynthesis must be low by a factor of 5 to 10."

The accuracy of measurements of photosynthesis also has been open to question (Peterson 1980). Thus, the standard view of the primary productivity of the seas, including the idea that much of the open ocean is a "biological desert," may be erroneous. Moreover, the discrepancy may be caused by underestimates of *both* efficiency and productivity. The results

FIGURE 24–7 Results of a run of a simulation model of a continental shelf community. Letters indicate compartments in the model; numbers directly under the compartments are the stocks (in kcal/m^2). Energy flows are indicated by arrows, and their magnitudes are given in the numbers next to the arrows (in kcal/m^2/yr). In this run, grazing zooplankton dominate the energy flow. Pomeroy could not design a model with 10 percent transfer efficiency that would carry enough energy to the top consumers to support them. *(After Pomeroy 1979.)*

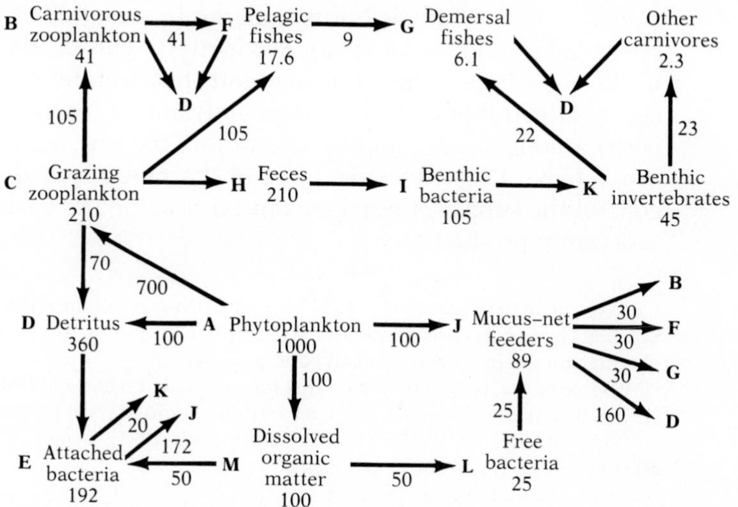

of further investigations of this problem in energetics may have important implications for understanding nutrient cycling in the ocean. They also, of course, could influence strategies for harvesting edible marine organisms in an increasingly food-short world.

ENERGY–NUTRIENT FLUXES AND SUCCESSION

Patterns of productivity and nutrient cycling also change with succession. Generally, in primary succession, nutrients are thought to flow rapidly through the system in the early stages and then begin to accumulate as they are incorporated into biomass and detritus (Vitousek and Reiners 1975). If net ecosystem production* declines during the approach to a relatively permanent state, the rate of output of nutrients converges with the rate of input (Coats *et al.* 1976, Vitousek 1977). In the secondary succession, the period immediately after disturbance can be a time at which the output of nutrients exceeds the input.†

The details of the processes—such as the weathering of soils and rocks, erosion, nitrogen fixation, net production, and decomposition that interact to create these trends—have become the subject of an increasingly rich literature (see, e.g., Turner 1975, Gorham *et al.* 1979, Vitousek 1983). Let us consider some examples. Rapid net production by fast-growing plants has been shown to preserve pools of nutrients that might otherwise be lost to streams following clear-cutting of New England and southern Appalachian forests (Marks and Bormann 1972, Boring *et al.* 1981). The state of the soil nitrogen pool in relation to the requirements of the vegetation is a critical factor in understanding the influence of harvesting forests (Swank and Waide 1980). In primary succession, both the amount and the chemical form of phosphorus undergo irreversible changes as soil develops (Walker and Syers 1976). Productivity can decline late in succession if phosphorus becomes limiting, and this decline in productivity in turn can result in higher loss of other nutrients. Clearly, reciprocal reactions occur among the developing vegetation, the soil, and nutrient fluxes in an ecosystem.‡ These investigations of the biotic and abiotic processes that control the supply of nutrient capital provide the basis for understanding ecosystem productivity.

* Net primary productivity (NPP) is gross primary productivity (GPP) minus plant respiration. Net ecosystem productivity (NEP) is GPP minus total ecosystem respiration, including storage in detritus (mostly) and animals.

† Some of the key references on this topic include Likens and Bormann (1974), Woodwell (1974), Swank and Douglas (1977), and Vitousek and Matson (1985).

‡ See Vitousek and White (1981), Walker *et al.* (1981), Swank *et al.* (1984), Vitousek and Matson (1984), and Wood *et al.* (1984).

THE GAIA HYPOTHESIS

The average temperature of the planet has remained at between 10 and 20°C throughout the last 3.5 billion years, even though the heat output of the sun has increased over that period and the chemical composition of the atmosphere has changed and thus so has the greenhouse effect. Similarly, the salinity of the sea has remained quite constant for the last hundreds of millions or even billions of years, after building to its present level in perhaps 80 million years of the washing of salts from the land into the sea (Lovelock 1979). Some scientists, such as Margulis and Lovelock (1974) and Lovelock and Epton (1975), have hypothesized that a giant system, made up of all organisms, atmosphere, seas, and the land surface, controls the temperature and chemical composition of the biosphere in ways that result in the Earth's continuing to remain hospitable to life. They point out that the temperature of the Earth's surface has not dropped to −50°C since life first appeared. They also note that the oceans did not turn to heavy brine until multicellular organisms evolved. On the basis of physical processes alone, both changes might well have happened. These scientists see in this gigantic system properties characteristic of an organism, and the system has been named *Gaia,* after the ancient Greek Earth goddess.

EMERGENT PROPERTIES OF ECOSYSTEMS

The Gaia hypothesis represents one extreme view of Earth's ecosystems, the ultimate development of the early Clementsian superorganism concept of the biological community. Although most ecologists today are doubtful about the Gaia hypothesis and about the utility of the superorganism analogy, many believe that ecosystems have *emergent properties,* properties unique to the ecosystem level of biological organization that are wholly unpredictable from observation of the components of that unit (Salt 1979). Under this definition, the species diversity of a community is not an emergent but a *collective* property, being the summation of characteristics of components of the community. If one knows all of the species present, one can predict this property of the community. On the other hand, the propensity of ecosystems to retain nutrients after fires and of early seral stages to be efficient at nutrient retention seem to be emergent properties (Vitousek and Reiners 1975, Woodmansee 1978, Woodmansee and Wallach 1981). Knowledge of the individual species probably would not have permitted predictions of resistance of the system to nutrient leakage.

CHAPTER 25

A Natural Ecosystem: The Serengeti

Let us now examine some of the features of a single large ecosystem. We have chosen a system that has been relatively thoroughly studied and may already be familiar to you from nature films—the Serengeti–Mara ecosystem of northern Tanzania and southern Kenya, East Africa (Sinclair and Norton-Griffiths 1979).

THE PHYSICAL SETTING

The Serengeti–Mara ecosystem (see the map in Figure 25–1) is more or less self-contained, a plateau of some 25,000 km^2 bounded on the east by hills, on the south and southwest by rocky woodlands and cultivated areas, on the west by Lake Victoria, and on the northwest and north by cultivation and escarpment. The plateau lies just south of the equator, and it reaches almost 1800 m in the east, sloping downward to about 1200 m at Lake Victoria. Low hills covered with sparse *Acacia* woodland occur in the western portion of the system, with extensive plains between them and the bordering volcanic mountains (which include the famous Ngorongoro Crater) in the east. However, this ecosystem is best defined not physically but as the area influenced by large migratory herds of one species of antelope, the wildebeeste or white-bearded gnu (see Figure

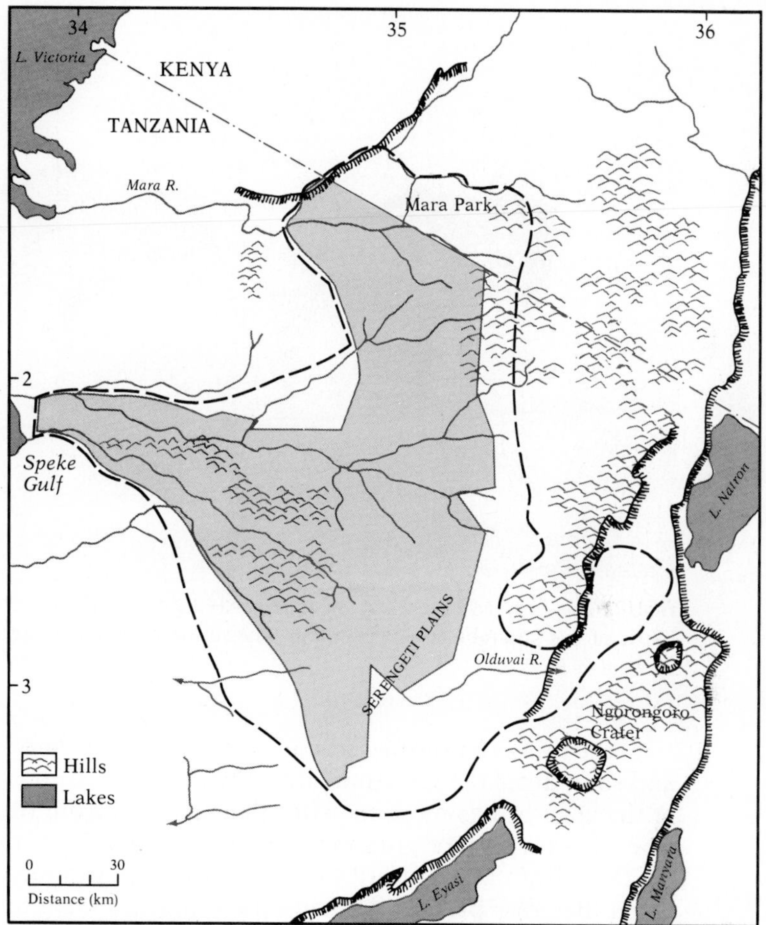

FIGURE 25–1 Map of the Serengeti–Mara ecosystem. The shaded area is the boundary of the Serengeti National Park; the dashed line is the area used by wildebeeste. *(After Sinclair, 1979a.)*

25–2). This area is characterized by the presence of the last great assemblage of ungulate (hoofed mammal) species anywhere on this planet.

As with other ecosystems, the broad constraints within which the Serengeti–Mara system operates are set by its physical environment (Sinclair 1979b). In its equatorial setting, the major factor limiting plant growth is moisture, which is provided by an uneven and seasonal rainfall. The rain, in turn, is controlled by movements of the Intertropical Convergence Zone, which travels back and forth across the equator, lagging behind the sun's movements by some 6 weeks. When the zone moves south, it brings relatively dry winds from the northeast and, starting around November, a small amount of rain.

FIGURE 25–2 Wildebeeste, or white-bearded gnu, on the Serengeti plain. Vast herds of this strange-looking antelope shape the entire Serengeti ecosystem. *(Photo by Paul R. Ehrlich.)*

These rains end the dry season; sometimes they last only until January, and sometimes they continue into March. Then the northward movement of the zone brings more moisture-laden winds from the southeast. Water from these winds, originating over the Indian Ocean, then provides heavier rains from March until May. July through October is the dry season, when the grass produced during the wet season can no longer grow. The eastern hills produce a rain-shadow effect, so that the wet-season rainfall increases from a mean of some 500 mm in the southeast to 800 mm in the northwest. Dry season rainfall shows a similar geographic pattern, increasing from 100 to 300 mm.

On top of this annual and spatial variation in rainfall is superimposed longer-term temporal variation in the physical environment—that of climate change. For at least a decade before the 1971 dry season, average dry season rainfall in the northern woodlands of the Serengeti was some 100 mm less than in 1971–1976. Spatial variation on a smaller scale is provided by the differential water-holding capacity of soils. Along ridgetops, the soils can store relatively little water, and these soils tend to have nutrients leached out of them. The ridgetop soils are linked into a *soil catena,* a connected series of soils that leads down the slopes to rich alluvial soils in the valleys, to which nutrients flow.

The term *catena* is also used to describe the broad soil trends in the Serengeti. The plains were formed by airborne deposits of volcanic materials from the eastern range, which includes Ngorongoro. The fallout was coarsest in the plains near the base of the volcanoes and has formed leached porous soils there. Farther west in the plains, and in the northwest, to which the finer materials were blown, the smaller particles form a more closely packed soil that holds water better. Thus, an animal can "move down the catena," either by traveling from a ridgetop to a valley or by migrating north and west—in either case, moving toward soils that hold more water.

THE MIGRATIONS

More than a million wildebeeste migrate around the Serengeti in order to find the highest-quality forage. During the dry season, they tend to be concentrated in thorn woodland in the relatively moist northwestern areas that, in the wet season, will be infested with mosquitoes and tsetse flies. At the beginning of the wet season, the wildebeeste migrate from poor pasture toward thunderstorms, which they can detect at a distance of 25 km by sound and at 100 km by perceiving the anvil-topped storm clouds, in order to take advantage of the grass that will sprout after the rains (Baker 1978). At times, the wildebeeste move as much as 80 km if food and water are widely separated, as they cannot go more than 5 days without drinking. The animals move eastward into the plains; then from January to May they move in a generally clockwise gyre in the central and eastern plains of the Serengeti. At the end of the wet season, as the grass on the plains dries up, the wildebeeste return to the moist northwest.

In these movements, the wildebeeste are generally accompanied by zebras and Thomson's gazelles. However, the three species do not follow precisely the same routes and timetable; they tend to partition the resources available to them. During the wet season, all species are concentrated on the upper catena of the plains, feasting on high-protein ephemeral grasses. When the wet season is drawing to a close, the zebras are the first to move off down the catena. As illustrated in Figure 25–3, they preferentially eat the protein-poor stems of the drying grasses (Bell 1970, 1971).

Next come the wildebeeste, which feed more on leaves and the sheathes at the base of the leaves—plant parts containing more protein than the stems and made more accessible by their removal by the zebras. Then the Thomson's gazelles take their turn at occupying the now-short grassy turf, as illustrated in Figure 25–4. They feed heavily on protein-rich forbs (herbacious dicotyledons), made accessible by the previous grazers, in addition to grass sheathes and leaves (Jarman and Sinclair 1979, Maddock 1979). The Thomson's gazelles, with their slender muzzles, are able to feed much more selectively than either the zebras or the wildebeeste. They

FIGURE 25–3 Zebras grazing on dry grass. *(Photo by Paul R. Ehrlich.)*

FIGURE 25–4 Thomson's gazelle grazing in short grass. Note the slender muzzle. *(Photo by Paul R. Ehrlich.)*

generally avoid areas of long grass, perhaps because their small size (about 67 cm at the shoulder, as opposed to 130 cm for wildebeeste and zebra) prevents them from easily spotting predators there.

The sequence of utilization raises an interesting question regarding the physiological ecology of ungulates. How can the zebra, a nonruminant, exist on a diet lower in protein than that of the wildebeeste and gazelle, both members of the ruminant cattle family?

All large animals probably benefit digestively from microbial fermentation in their digestive tracts. However, mutualistic digestive relationships with bacterial and protozoan symbionts reach an especially high degree of development in herbivores. Only with the aid of microorganisms can the insoluble cellulose and related compounds in plants be broken down to obtain energy and to make accessible the protein within the plant cells. The ruminants' digestive tracts have an enlarged section, the rumen, in which ground-up plant materials and symbionts are mixed together and held for up to several days.

Ruminants get their name from their habit of "chewing their cuds," returning food from the rumen to the mouth for further mastication. This chewing increases the surface area of food exposed to enzymes in the rumen, and the saliva helps neutralize acids produced by the digestive process there. The plant cells are thoroughly crushed and their contents made available for absorption in the intestine—an efficient, if slow, system.

The nonruminant zebra, however, can extract enough protein and food energy to survive from cellulose-rich, low-protein forage that could not support the ruminant wildebeeste. Microbial breakdown of cellulose in horses (the zebra is basically a horse) takes place in the intestine *after* the extraction of protein in the simple stomach. However, in the intestine, digestion is roughly twice as fast as in the rumen of a cow, and food passes through horses at roughly twice the speed at which it moves through cattle. Thus, a zebra is able to process the low-protein grass stems at a rate sufficiently high to permit it to maintain itself, even though it extracts less protein per kilogram of food eaten (Bell 1970, 1971). The zebra's system is so much faster that it more than compensates for its lower efficiency, and it can thrive on a diet that would starve a wildebeeste.

NONMIGRANT UNGULATES

Nonmigrant ungulates also tend to partition their environment and resources. For example, two antelopes, topi and impala, and the African buffalo have broadly overlapping ranges. However, the topi, shown in Figure 25–5, dominate in the short-grass open plains; the buffalo, shown

FIGURE 25–5 Topi in the Serengeti. The dark legs are characteristic of this large antelope, which is primarily a grazer. *(Photo by Paul R. Ehrlich.)*

in Figure 25–6, prevail in long-grass areas; and the impala, shown in Figure 25–7, predominate in *Acacia* woodlands. At times, all three species graze together, the buffalo mowing the grass relatively unselectively, the topi eating leaves from grass of medium height, and the impala picking out green leaves from bushes and short grasses and eating seeds and fruits.

The migratory and resident herbivores in the Serengeti also interact. For example, as they pass, the zebras and wildebeeste mow the long grass.

FIGURE 25–6 African buffalo. Note their broad muzzles. *(Photo by Paul R. Ehrlich.)*

FIGURE 25–7 Male impala in *Acacia* woodland. *(Photo by Paul R. Ehrlich.)*

This mowing probably benefits the impalas, which normally avoid long grass, but the migrants also probably reduce the total grass available to the impala, causing them to start eating the limited supplies of dicot leaves, seeds, seedpods, and fruits sooner than otherwise would be necessary.

HERBIVORE IMPACT ON PLANTS

The impact of the herbivore community on the Serengeti grassland has been the subject of intensive study (Sinclair 1975, McNaughton 1976, 1979a,b, 1983). On the average, the ungulates require about 20% of the yearly production of the some 6 tons/ha of plants in the tall grasslands of the Serengeti. However, averages can be misleading (remember the statistician who drowned in a lake that averaged only 2 ft deep). During the wet season, food is superabundant; in April, some 10 times the normal requirement is produced, as Table 25–1 makes clear. However, in July through September, production ranges from none to slightly over half of the normal requirement (which, in these figures, is adjusted for the movement of migrants). Thus, at some times of year, food is in quite short

TABLE 25–1 Mean monthly food availability and requirement in the tall grassland (kg/ha) with the approximate range in values in parentheses. Available food is monthly production plus food remaining from previous month.

Month	Monthly production	Food requirement				Food available	Food remaining
		Invertebrates	Ungulates	Small mammals	Total		
Nov.	555	34.90	111.33	3.33	149.56	589.8	440.3
Dec.	781	56.50	29.22	8.87	94.59	1277.8	1183.3
Jan.	668	56.50	45.04	8.87	110.41	1907.8	1797.3
Feb.	633	56.50 (23–140)	48.07 (32–80)	8.87	113.44 (64–229)	2486.8	2373.4
Mar.	857	56.50	48.07	8.87	113.44	3286.9	3173.4
Apr.	1154	56.50	48.07	4.95	109.52	4384.0	4274.4
May	706	56.50	43.95	4.95	105.40	5036.9	4931.5
Jun.	340	56.50	129.50	4.95	190.95	5328.0	5137.6
Jul.	0	10.00	124.60	4.95	139.55	0	0
Aug.	37.5	5.90 (2.3–14.8)	164.60 (110–248)	3.33	173.83 (116–266)	38.4 (0–80)	0
Sep.	92	4.60 (1.8–11.5)	164.60 (110–248)	3.33	172.53 (115–263)	92.9 (0–149)	0
Oct.	154	4.60 (1.8–11.5)	164.60 (110–248)	3.33	172.53 (115–263)	154.9 (30–216)	0
Total	5978	455.60	1121.65	68.60	1645.25		

Source: After Sinclair (1975).

supply—in July, only growth from the previous wet season is available—and the pressure on both animals and plants is severe.

Studies using exclosures consisting of fenced areas from which large grazers are excluded have shown that grazing increased the diversity of grasses while reducing the average height of their foliage. Grazing also causes individual plants to reallocate resources from roots to shoots and increases above-ground productivity. However, the true total impact of grazing on primary productivity remains a controversial matter.

Herbivores usually change the competitive relationships among various plants. For example, grasses are generally more grazing resistant than dicotyledonous plants. In the dicots, the primary meristem (the permanently embryonic tissue responsible for the elongation of roots and shoots) is at the ends of the shoots, where it is vulnerable to removal by grazers. In contrast, grasses have meristems just above each node of the stem, called *intercalary meristems,* where they are much less likely to be eaten. These relatively soft areas are protected and supported by the sheathes at the base of the leaves. The blade of the leaf also has a meristematic area at its base; thus, unlike dicot leaves, it can continue to grow if the distal part of the leaf is eaten. Furthermore, branching, or tillering, in plants occurs primarily at ground level, and the resulting rosette is often extended and reproduced by rhizomes (stems growing horizontally underground) or by stolons (stems growing along the ground surface).

If a grass plant is smashed flat, differential growth at the intercalary meristem occurs and once again aims the shoot upward. This vertical regrowth, in combination with the ability of chewed leaves to continue to elongate and the basal tillering with the spread by horizontal stems, makes grasses extremely resistant to trampling and feeding by ungulates. The presence of these animals in large numbers favors the continued dominance of grasses and the perpetuation of grasslands.

Plant populations outside of permanent exclosures on the Serengeti proved to be different genetically from those inside, being dwarfed and more prostrate in growth form as a result of the selection pressures applied by the grazers. Indeed, McNaughton (1979a) believes the Serengeti system is one of the few remaining great reservoirs of grazing-resistant plant genotypes, a "genetic library" of potentially incalculable value to humanity in a world in which overgrazing is one of the most critical environmental problems.

THE PREDATORS

Just as the large herbivores on the Serengeti tend to partition their resources, so do the predators that prey on those herbivores (Bertram 1979). Lions, which weigh 100–200 kg, feed primarily on zebras and wildebeeste

FIGURE 25–8 Courting lions in the Serengeti. *(Photo by Paul R. Ehrlich.)*

when they are within the territory of the pride. Lions (shown in Figure 25–8) are found in both plains and woodlands. Their prey among the nonmigratory ungulates includes giraffes and buffalo, which no other predator can kill, as well as resident antelope and warthogs when the migrants are not around. Hunting is performed usually by a group of lionesses hunting mostly at night by stalking their prey, then making a short sprint (Schaller 1972).

The other lionlike cat, the leopard, which weighs 35–60 kg and is shown in Figure 25–9, is confined to the woodlands and generally takes smaller prey than the lion does. Leopards' choice of prey overlaps with that of the lions at the upper end of the size range—for example, Thomson's gazelles, topi, and occasional zebras—but leopards feed extensively on small antelope such as Kirk's dik-dik (38 cm at the shoulder), small carnivores, hares, and birds. Leopards are solitary, nocturnal stalk-and-sprint predators.

Cheetahs (see Figure 25–10) are as large and heavy as leopards but much more slender in build than the other big felines. They hunt in the daytime, primarily on the plains, taking small antelopes and hares, which they stalk and then run down at speeds of up to 95 km/hr (60 mph). The classic long-distance sprint predator, a cheetah may chase its prey for as much as 350 m, as opposed to a maximum of 50 m for a lion (McLaughlin 1970, Schaller 1972).

Hyenas, such as that shown in Figure 25–11, hunt the plains at night and in the early hours of the morning, in groups of 1–3 for wildebeeste and other antelopes and in groups of 4–20 for the larger zebra (Kruuk

FIGURE 25–9 Leopard resting in the Serengeti. Note how spots help to camouflage it in dappled sunlight. *(Photo by Paul R. Ehrlich.)*

FIGURE 25–10 A cheetah. Notice the slender build of this long-distance sprint predator. *(Photo by Paul R. Ehrlich.)*

FIGURE 25–11 Hyena resting in the Serengeti. *(Photo by Paul R. Ehrlich.)*

1972). As pursuit predators, they may harry their prey for up to 3 km. These fierce predators are involved in one of the more interesting coevolutionary interactions in East Africa. Hyenas appear to serve as models in a mimicry complex with the smaller hyaenid known as the *aardwolf,* the subspecies of which each closely resemble sympatric forms of hyena (Gingerich 1975). Leopards prey regularly on jackals, which are similar in size to the aardwolf. However, the aardwolf, by erecting its mane, can assume the appearance of the larger and more dangerous hyena, thus presumably giving even a leopard second thoughts.

The other Serengeti pursuit predator, wild dogs, shown in Figure 25–12, take an array of prey similar to those of the hyenas (Estes and Goddard 1967, Schaller 1972, Malcolm and Van Lawick 1975). However, wild dogs hunt mostly by day, the entire pack of 2–19 individuals participating in chases over 2 km long. Both of the pursuit predators specialize more in the old, young, sick, and wounded in their prey populations than do the sprint predators, even though the sprint predators probably kill animals less fit than the average.

The available prey seem to be partitioned even more finely than indicated here (Bertram 1979). For example, hyenas feed heavily on young wildebeeste calves on the plains, away from the territories of most lion prides. By the time the wildebeeste move down the catena to the woodlands, the calves are not as easy to catch and the lions feed mostly on adults.

The extent of partitioning implies that relatively little present-day exploitation competition is occurring among these predators. However, in-

(A)

(B)

FIGURE 25–12 Wild dogs. (A) Adults hunting in the Serengeti. (B) Adults and pups at their den. *(Photos by Paul R. Ehrlich.)*

terference competition is another matter. Only the cheetah does not add to its diet by scavenging, where interspecific conflicts often occur. Hyenas obtain about a third of their food by scavenging, lions obtain 10–15%, leopards 5–10%, and wild dogs 3%. Only the lions do not experience significant interference from other predators. Cheetahs lose 10–12% of their prey to hyenas and occasionally to lions, wild dogs lose about half to hyenas, and hyenas and leopards are thought to lose 5% or more to lions.

THE DECOMPOSERS

The final major element in the Serengeti system consists of the decomposers. Many of these decomposers are similar to those of other ecosystems—flies whose maggots help break down cadavers, beetles that feed on hides or dung, bacteria and fungi that do much of the fine work of reducing organic compounds in dead plants and animals, and in feces, to inorganic compounds. However, the Serengeti ecosystem is unique today in having a great variety of large scavenger-decomposers—including lions, leopards, hyenas, and wild dogs, all of which play this ecological role at least part of the time. Indeed, of some 40 million kg worth of ungulates that die annually, these animals—along with the cheetahs, which eat only their own kills—devour some 14 million kg (Houston 1979). Interestingly, the three species of jackals in the Serengeti (Figure 25–13) prey mostly on insects and small mammals. Carrion makes up only some 3% of their diets.

The remaining 26 million kg of dead animals contain about 14 million kg of soft tissues. Of these, an estimated 12 million kg are eaten by the most abundant large decomposers in the system—seven species of vul-

FIGURE 25–13 Black-backed jackal in the Serengeti. *(Photo by Paul R. Ehrlich.)*

FIGURE 25–14 Vultures in the Serengeti. These scavengers are major decomposers in the Serengeti ecosystem. *(Photo by Paul R. Ehrlich.)*

tures, some of which are shown in Figure 25–14, and the marabou stork, shown in Figure 25–15. The remainder is consumed by insects, other invertebrates, and bacteria. The carrion-feeding birds, like the herbivores and predators, tend to have only partly overlapping use of resources and habitats.

The large white-backed and Ruppell's vultures have long, sharp bills and barbed tongues adapted for cutting and gripping the soft muscles and viscera on which they feed. White-backs are primarily lowland birds; Ruppell's prefer the hills. Large vultures also include the less common white-headed and lappet-faced (Nubian) vultures. They have deep, strong, hooked bills well adapted for dealing with tough meat, skin, and tendons. Both nest in isolated trees in the savanna. The last large vulture in the Serengeti, the bearded vulture (lammergeyer), is rare and found only near the eastern mountains. Its strong beak is adapted for tough tearing and its tongue for removing the marrow from long bones, which it shatters by dropping from a height onto rocks.

Egyptian and hooded vultures weigh only one-third to one-half as much as the larger vultures and have long, slender, comparatively weak bills, which they use to glean scraps of carrion, carnivore dung, and to catch a variety of small animals. The Egyptian vulture is famous for having

FIGURE 25–15 Marabou storks, important scavenger-decomposers in the Serengeti. The smaller birds in the foreground are the sacred ibis. *(Photo by Paul R. Ehrlich.)*

learned to crack ostrich eggs by throwing rocks at them (van Lawick-Goodall and van Lawick 1966). The Egyptian vulture is largely restricted to the eastern plains in the Serengeti; the hooded vulture is common in plains, savanna, and woodlands. The marabou stork in the Serengeti is almost exclusively a scavenger. Its deep, broad, lengthy bill cannot be used to tear meat from carcasses; it must snatch what it can away from its fellow scavengers.

Thus, we can see the major outlines of the Serengeti–Mara system. It is a temporally and geographically varying, semiarid physical setting, with energy and nutrients entering grasses, herbs, bushes, and trees. Both pass through an array of herbivores that includes an unusual abundance and variety of ungulates and through a carnivore trophic level that has an equally unusual number of big predators. In their turn, the decomposers—including many large ones—make their living on what remains of the available energy and recycle the nutrients.

UNGULATE SOCIAL SYSTEMS

Of course, this major outline omits important detail about how the components of the system function. We will look briefly at just one example: how the combined selective pressures of getting food and avoiding predators has helped to shape the social systems of the ungulates as it has those in other areas (Geist 1974).

The Serengeti antelopes range in size from the 4-kg dik-dik, shown in Figure 25–16, to the 700-kg eland, shown in Figure 25–17. Dik-diks live in pairs with their young on permanent territories in woodland with thick cover. They use olfactory signaling, because visual or auditory signaling would reveal their positions to predators. The territories, from which all other dik-diks are excluded, presumably are large enough to contain sufficient food items, even during the leanest time of year. The tiny dik-diks have higher metabolic requirements per unit of body weight than do larger species. As a result, they must feed quite selectively on tender young leaves, buds, fruits, and seeds from a variety of species. Both the need to search out special food items and their small size presumably select against herding and favor territoriality in dik-diks.

In contrast, larger antelopes, such as impala, are not as picky about their food; their greater food requirements, and mouths too large to make fine discriminations among foods, mandate a lower quality of intake. Impala cannot remain hidden; they must move about, seeking the best graze and browse; herds of up to 100 or more shift their habitat preference

FIGURE 25–16 Kirk's dik-dik. This antelope is the size of a jackrabbit. *(Photo by T. W. Ransom/BPS.)*

FIGURE 25–17 The cowlike eland is the largest of the Serengeti antelopes. It does not migrate, although herds gradually change location as they graze. *(Photo by T. W. Ransom/BPS.)*

seasonally. Roughly a third of the males hold territories wherever the females are found at a given time; the remainder of the males form a bachelor herd. A male marks his territory by his presence (visual signal), roaring (auditory), and urinating, defecating, and rubbing secretions from forehead glands on plants (all chemical signals) (Jarman and Jarman 1979). Females and young wander over larger areas than the individual male territories. If an attack occurs, the entire impala herd reacts as a coordinated unit, and members of the herd also benefit from group alertness and communication about predators.

Other antelopes, such as waterbuck (Figure 25–18), kongoni (Figure 25–19), sedentary populations of wildebeeste, Grant's gazelles (Figure 25–20), and Thomson's gazelles, have social systems with many of the same features. An especially notable characteristic is the holding by males of spatially defined territories within the region occupied by the population at a given time; these territories are abandoned when the herd moves on. Migratory wildebeeste no longer have spatially defined territories. Instead, they set up tiny, temporary, sometimes mobile territories wherever the females are, defending them during the short rutting period. This pattern allows the males to maintain exclusive sexual access to a group of females in spite of their having to travel constantly in search of food.

FIGURE 25–18 Waterbuck, a large Serengeti antelope that does not migrate but remains near standing water. This animal feeds mostly on grasses, especially those not preferred by other grazers, but will also browse on trees and shrubs. This adult male will hold a territory within the area occupied by the herd at any given time. *(Photo by Paul R. Ehrlich.)*

FIGURE 25–19 Kongoni or Coke's hartebeeste. Note the fusion of the horns at the base into a "pedicel," which is characteristic of this nonmigratory antelope, which feeds mostly on grasses. *(Photo by Paul R. Ehrlich.)*

FIGURE 25–20 Grant's gazelle. This larger, nonmigratory relative of the Thomson's gazelle is distinguished from the latter by the way the white on the rump extends above the tail. Like the Thomson's gazelle, in nondrought conditions it gets all of its water from the food it eats. *(Photo by Paul R. Ehrlich.)*

In buffalo, and perhaps in eland (the largest of the antelopes), there is a different system. Both form large herds that move a great deal. Rather than defending territories, the males simply exercise individual dominance to keep subordinate males away from a female detected to be in estrus. Like the wildebeeste, both the buffalo and the eland males solve the problem of monopolizing access to females simply by staying with them. Neither of these species has a short rutting period, so attempting a brief defense of a group of females would be to no avail; rivals must be perpetually excluded.

Buffalo and eland herds cooperate in the defense of their young (Kruuk 1972, Sinclair 1974). Buffalo calves appear to be less able to fend for themselves than wildebeeste calves, and Jarman and Jarman (1979) suggest that attempts by males to cut out groups of females from the herd would expose the calves to predation. The formation of cohesive groups by buffalo and other large species is facilitated by their ability to tolerate lower-quality food, because of their smaller metabolic requirements per unit body weight. They can mow down abundant, evenly dispersed, relatively protein-poor food, such as a lawn of grass. They do not have to disperse to find patches of high-quality food.

PERTURBATIONS OF THE SYSTEM

Finally, what can be said of the stability of the Serengeti–Mara ecosystem? If the system is disturbed, does it return to its previous state? Two large-scale perturbations of the system have been observed. The first was a great epidemic of rinderpest, a virus disease of ruminants native to the steppes of Asia. The disease may have been introduced with cattle brought by the British from Russia in 1884 during their incompetent and unsuccessful attempt to relieve the beseiged General "Chinese" Gordon at Khartoum.

Whatever its origin, the disease first decimated the cattle of the Masai and other herding tribes. The loss of cattle led to horrible famines between the 1890s and 1920, with at least two-thirds of the Masai dying. By 1890, the disease had moved to native ruminants, and the buffalo, wildebeeste, and giraffes were disappearing. One result was that predators starved, and lions switched to eating people. Thus, man-eating lions of Tsavo became famous in 1898; in 1920, one lion in Kenya is reported to have devoured 84 people (Sinclair 1979b).

The appearance of the man-eaters led to the abandoning of land by cultivators. This situation, combined with the disappearance of the pastoralists and the native grazers and browsers, led to a recapturing of plains areas by woodland, as secondary succession resumed in areas in which it once had been arrested by agriculture and grazing. By 1910 or so, the wild ruminants were becoming resistant to the disease, and these ruminants fed tsetse flies, which expanded their range into the new brushy areas. The sleeping sickness that the tsetse carried reduced the human populations further.

Gradually, in the 1930s, brush-control programs began to suppress the tsetse fly populations, and vaccination began to reduce the impact of rinderpest on cattle populations. Slowly the rinderpest began to fade from the native ruminants, first producing mortality only in yearlings that had not yet acquired immunity and finally disappearing in the early 1960s. The results were dramatic. The survival of wildebeeste yearlings increased from 25% to 50%, and their population shot up from a quarter million individuals in 1961 to half a million in 1967. Buffalo went from 30,000 to 50,000 in the same period. Interestingly, the nonruminant zebras, immune to rinderpest, showed no such response.

The second perturbation was the increase in dry season precipitation in 1971–1976, as previously mentioned. This precipitation increased the productivity of the grasslands at the time when it is usually at a low point. That productivity led to a further increase in wildebeeste and buffalo populations, the wildebeeste reaching about 1.3 million by 1977. For reasons that are not entirely clear, no such increase occurred in other grazers, such as the zebras. Some evidence appeared that interspecific competition from the wildebeeste slowed the increase of the buffalo, but no sign appeared of a reverse effect (Sinclair 1979b). Competition may exist between wildebeeste and zebras, with a possible small decline in zebras, but census data are not adequate to document it. The capacity of zebras to survive on low-quality forage may have reduced the potential impact of the wildebeeste eruption on them. Sinclair believes that interspecific competition, acting periodically, could be the reason for much of the ungulate habitat partitioning, especially that of wildebeeste and buffalo, but that it may also be somewhat fortuitous or caused by factors unrelated to competition.

The increase in wildebeeste may have had a positive effect on the population size of Grant's gazelles by altering the competitive relationships among the plants. Herbs are favored over grasses by the intense grazing pressure on grasses, and the gazelles eat the herbs preferentially. This change, in turn, may have been responsible for an increase in the population of cheetahs, which feed heavily on the gazelles.

The increase in wildebeeste also might have been expected to benefit their major predators, the lions and hyenas; however, that was not the case. The postrinderpest population explosion of wildebeeste was not accompaned by similar eruptions in numbers in the lions and hyenas. Apparently the reason is that their territorial populations are controlled by the food supply in the scarce period, when the migratory wildebeeste are absent. The predators did respond to the rainfall-induced increases in resident ungulates such as the topi, kongoni, and warthog (see Figure 25–21). In the early 1970s, lions almost doubled in numbers, and the hyena

FIGURE 25–21 Warthog, a nonmigratory herbivore of the Serengeti that prefers habitats with watering and wallowing sites. Warthogs feed primarily on short grass, but in times of shortage they will use their tusks to dig for roots and tubers. *(Photo by Paul R. Ehrlich.)*

population increased by almost 50% (Sinclair 1979a). These increases had a negative impact on the population size of wild dogs, both from increased interference competition and from predation on their young by hyenas.

Other elements of the system changed as well. The increased rainfall reduced the frequency of dry season fires; that in turn helped the survival of small *Acacia* trees, leading to increased numbers of giraffe. Giraffes tend to keep small trees from escaping into the mature class, as shown in Figure 25–22. The mature trees are killed by elephants, which push them over, resulting in a complex, dynamic, fire–giraffe–elephant–tree system (see, e.g., Pellew 1983). This system may cycle with a long periodicity because of the generation times of the organisms involved.

The major linkages in the complex Serengeti ecosystem are shown in Figure 25–23. The illustration shows how physical factors—spatial and temporal heterogeneity—can mediate the degree to which competitive and predator–prey interactions play significant roles in the functioning of the system. Successional change is a constant feature of African savannas, but it is multidirectional (Walker 1981). The paleontological record indicates that the major features of the Serengeti ecosystem—including the

FIGURE 25–22 Giraffe browsing on a small *Acacia* tree. Almost all the giraffes' fodder comes from trees and bushes. *(Photo by Paul R. Ehrlich.)*

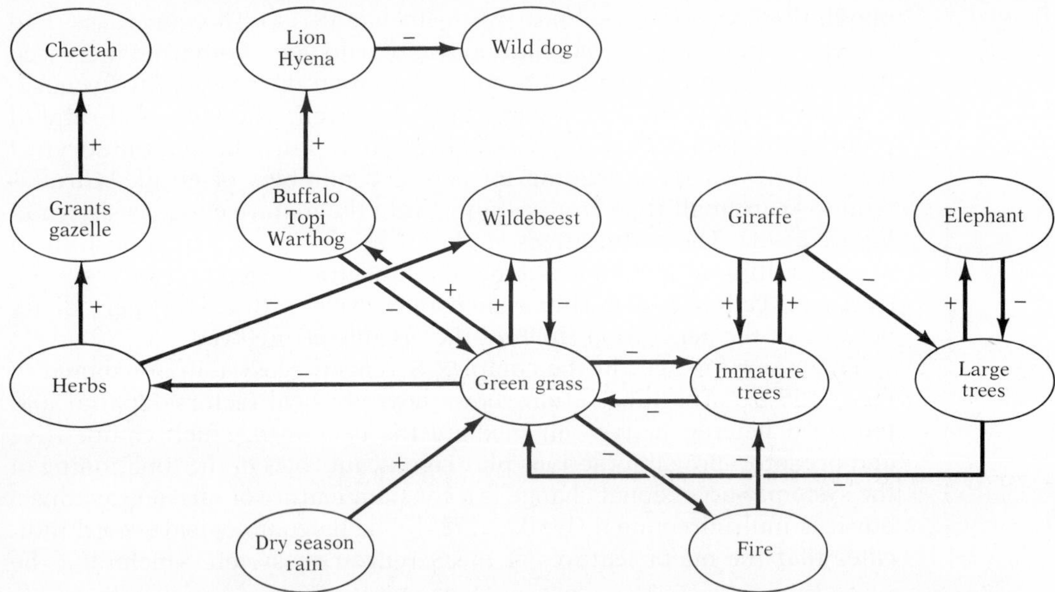

FIGURE 25–23 Major linkages in the Serengeti ecosystem. Minus signs indicate that higher values in the preceding oval result in lower values in the succeeding oval. Plus signs result in higher values in the succeeding oval. *(After Sinclair 1979a.)*

semiarid climate—have been relatively stable for at least a million years and possibly for much longer. Nonetheless, if a change in climate brought an increase in rainfall to the point where the major migrants became sedentary, the entire structure of the system would change dramatically.

CHAPTER 26

Ecosystem Services: Human Dependence on Natural Systems

Although most people do not realize it, human society receives an array of free services from natural ecosystems. These include amelioration of the weather, maintenance of the quality of the atmosphere, supplying of fresh water through control of the hydrologic cycle, generation and maintenance of soils, disposal of wastes, supplying of food from the sea and inland waters, recycling of the nutrients essential to agriculture, control of the vast majority of crop pests and organisms that could carry disease to *Homo sapiens,* and the maintenance of a vast genetic "library" from which humanity has already withdrawn, in the form of foods, medicines, and industrial products, the very basis of civilization (Ehrlich and Ehrlich 1981).

Let us consider what changes in these services might mean for human beings and for the Earth itself.

LOSS OF ECOSYSTEM SERVICES

Perturbations both natural and *anthropogenic*—human caused—can result in an alteration or even a loss of ecosystem services. One example of an anthropogenic perturbation is the increasing injection of oxides of sulfur and nitrogen into the atmosphere from power plant and factory smokestacks and from automobile exhausts. These compounds undergo chemical reactions that convert them into sulfuric and nitric acids. Because of the presence of these powerful acids in the atmosphere, rains over much of the Northern Hemisphere are 10 to 1000 or more times as acidic as rain from unpolluted skies (Likens and Bormann 1974).

Acid Deposition

Until recently, the record for acidity was held by a rainfall at Pitlochry, Scotland, which had a pH of 2.4, as acidic as vinegar (Sage 1980). However, in 1981, precipitation samples were gathered from the Tibetan Plateau in Qinghai Province, China. One, from the city of Machin, had a pH of 2.25, roughly that of lemon juice (Harte 1983). Machin, with a population of 20,000 to 40,000, is heated by coal stoves, and coal fires also are used for cooking and boiling water. (All drinking water in China must be boiled; no potable water is available.)

The impacts of acid rains on ecosystem services are not fully understood, but what is known is ominous. For example, in the Adirondack Mountains of New York, not only have lake waters been acidified, but the sulfuric acid and, to a lesser extent, the nitric acid have been reacting with the soil to release large amounts of aluminum, which is washed into the lakes. As a result, one ecosystem service, provision of lake fishes, has been terminated. All fish populations in 300 Adirondack lakes are extinct, and one of the most valuable species, brook trout, may have been wiped out over the entire area.

The Adirondack situation is not unique. Canadian scientists have identified 48,000 lakes that, if current trends continue, within two decades will not be able to support life. In Nova Scotia, acid rains have already made a third of the spawning rivers uninhabitable for Atlantic salmon, combining with the dams, pesticides and other pollution, overfishing, and poaching to reduce greatly the populations of this commercially valuable species.

Evaluating the overall impact of acid deposition on ecosystems especially terrestrial systems, is difficult, in part because a great deal depends on the buffering capacity of soils. Enough is known about the effects of lowered pH on biological systems to draw some preliminary conclusions. An important one is that the nitrogen cycle, which provides an absolutely essential ecosystem service, is susceptible to disruption by acid rains. Both nitrogen fixation and nitrification may be inhibited at lowered pH (Babich *et al.* 1980). In general, enzymes are designed to work at a specific pH, and they tend to lose their activity when the pH changes. At the cellular level, lowered pH level causes proteins to denature. At the ecosystem level, acid rains tend to mobilize cations, including toxic metals.

These impacts still are observed largely in aquatic systems or inferred from knowledge accumulated in the laboratory. However, a body of information is growing on the impacts of acid rains on terrestrial systems in nature. For example, in the Green Mountains of Vermont, in the period 1965–1980, acid rains apparently were responsible for a decline in the coverage of mosses, organisms that play a part in two key ecosystem services—water metering and erosion control (Vogelmann 1982). In addition, although the issue is disputed hotly, acid rains may be implicated in a decline of red spruce in the same area during the same period (Siccama *et al.* 1982).

There is growing evidence that forests in Europe are in grave peril from ozone and SO_3, dry deposition of SO_3, and acid rains. The West German government estimates that some 1500 ha of forest were killed by acid deposition in Bavaria between 1977 and 1982 and another 80,000 ha have been seriously damaged (Pearce 1982). German soil biochemist Bernhard Ulrich stated at a 1982 conference on acid rains in Stockholm that damage to soils was far reaching and possibly irreversible. The buffering capacity of many soils is becoming exhausted, he claimed, and acid waters are leaching nutrients such as magnesium and calcium from the soil. Worse yet, toxic aluminum, normally bound harmlessly to organic matter in soils, is mobilized by sulfate from acid rains. As the amount of aluminum approaches that of calcium, toxicity becomes a problem; when the ratio of calcium to aluminum approaches 1:7, all tree growth stops. At the forest research station at Solling, this ratio fell from 1:1 to 1:5 in 1982. Trees in 1982 were dying from the top down, but the ultimate demise of the plant is initiated when the aluminum starts to inhibit cell division in its roots. Deterioration of the roots means loss of control of water and nutrient uptake and an opening of an invasion route for pathogens. Ulrich claims that the aluminum also damages soil bacteria, leading to the soil system's generation of acid on its own, a lethal positive-feedback mechanism. He contends that 2 million ha of German forest are close to the threshold of root damage and that forest damage will greatly accelerate in central Europe in the next decade. Whether these mechanisms actually are responsible is highly controversial. What is not in dispute is that forests are dying, and virtually all scientists believe that atmospheric pollution is the overall cause.

Forest Clearing and Nutrient Export

Less dramatic assaults than acid rains can interfere with the services provided by forest ecosystems. The classic series of studies of succession in the Hubbard Brook system by Bormann and Likens (1979) provide insight into the loss of ecosystem services that can accompany clear-cutting of a northern hardwood forest. In 1965, one Hubbard Brook watershed was clear-cut, but the felled trees were left in place, minimizing soil disturbance. Further, regrowth was suppressed for 3 years by herbicides in order to permit nutrient losses from deforestation to be measured without the interfering factor of nutrient uptake by new growth. One result was a weakening of the flood-control service of the system. Lack of transpiration increased the water content of the soil, and peak flow rates increased after storms. The nutrient-recycling and erosion-control services also suffered; the export of both nutrients and soil particles in stream flow increased. In this system, a relatively short period, less than a decade, appears to be required before the ecosystem can restore those services. Control over nutrient export seems to be reestablished more rapidly than that over water flow. Other services, such as filtering of pollutants from the air and production of timber, of course, will be restored much more slowly.

What if the forest is not allowed to regenerate? For example, what if livestock grazing, gathering of firewood, extinction of key species, or climatic change caused by widespread deforestation prevent the system from reestablishing itself? What if perturbation of ecosystems reduces or eliminates other services?

These are not hypothetical questions. Massive assaults on ecosystems by humanity produce well-known syndromes that are accompanied by severe degradation of their public-service functions. By far the best documented is desertification (United Nations 1977). Deserts are expanding almost everywhere, and even some areas that were once tropical rain forest have been desertified. The syndrome affects most ecosystem services, most obviously climate amelioration, supplying of fresh water, generation and maintenance of soil, and maintenance of the genetic library.

Desertification

In Africa, cattle play a major role in causing desertification, as is evident in Figure 26–1. Cattle are viewed as a source of wealth by many African peoples, such as the Masai. From an ecological point of view, however, they are a source of poverty in hot, semiarid climates. In order to drink, cattle, goats, and sheep must walk daily to a water supply. This activity consumes a good deal of energy and slows the rate at which they gain weight. It also results in their trampling valuable grasses and compacting the soil surface along the paths to the water. In contrast, wild herbivores have much less need to drink. Some, such as eland, oryx, and Grant's gazelle, probably do not have to drink at all, obtaining all the water they require from the vegetation they eat. Others, such as impala, do drink some water, but they still require much less than cattle do. Most native

FIGURE 26–1 Cattle and desertification in the Sahel. *(Agency for International Development/BPS.)*

African herbivores conserve water much more efficiently in digestion; for example, nearly all of the moisture is extracted from the intestinal contents of gazelles, and dry feces are released.

In contrast, cow pats are produced moist; thus, they rapidly lose ammonia, and hence the vital nutrient nitrogen, to the atmosphere. They then rapidly dry in the sun and heat up, which kills the bacteria and fungi that might speed their decomposition. The flat, dried cow pat also kills the grass beneath it. In contrast, the dry fecal pellets of antelopes are roughly spherical. They fall between the grass blades, do not heat up, and retain their nitrogen, as can be seen in Figure 26–2. Rather than tending to create a "fecal pavement," as cattle droppings do, they break down readily and return nutrients to the soil.

Moreover, like most grazers, cattle have quite specific food preferences; they graze some grass species heavily and others not at all. In cattle-raising areas, the species composition of the forage changes, with the grass species not eaten by cattle becoming increasingly common. However, native herbivores partition the available plant resources, their diets to varying degrees complementing each other. Thus, not only are the water-conserving native herbivores better adapted to the semiarid habitats of the African savannas, they also do not degrade it physically or chemically.

(A) (B) (C)

FIGURE 26–2 Differences between cattle and antelope feces. (A) dry cowpad; (B) bare soil under cowpad; (C) antelope feces that drop between grass blades. See text for details. *(Photos by Paul R. Ehrlich.)*

Cattle, on the other hand, parading back and forth to watering holes and producing their destructive droppings, have been a major engine of desertification on the continent.

Fighting Desertification: Game Ranching

All of these differences led wildlife biologist David Hopcraft to conclude that the soundest way to exploit many African grasslands—and maintain their ecosystem services—was not to graze cattle on them but to organize ranches to raise and harvest native herbivores (Hopcraft 1970). The idea is not unique.* However, the experiment that Hopcraft and his wife are running is unusual; since 1978, they have been transforming ecological theory into practice on their 20,000-acre ranch on the Athi Kapiti plains near Nairobi.

* See, for example, Dasmann and Mossman (1961) and Field (1979). See also Coe *et al.* (1976) and Lewis (1977) for related issues.

The Hopcraft ranch is stocked with a variety of grazers and browsers, including antelopes, zebra, giraffe, and ostrich. Cattle are being phased out and one day may be replaced by a closely related bovine, the African buffalo. For the present, however, the cattle serve as a valuable "control" for comparing costs and meat yields on the same land with those of the native animals. A principal tool of the Hopcrafts' operation is a perimeter fence 30 mi long especially designed not to injure animals that run into it. A great deal of research is being carried out at the ranch. The dynamics of the various populations are being monitored and the food preferences of the different animals are recorded, and veterinarians at the ranch are studying the parasites of the animals harvested.

So far, the results of the experiment are exceeding early hopes and expectations. The herds of native herbivores have been steadily increasing as well as yielding meat. Simultaneously, the condition of the range has been *improving*—and the combined biomass of cattle and native herbivores has increased by some 35% in the past few years. Harvesting is efficient and more humane than in an abattoir. One night each week, men in Land Rovers spotlight surplus male animals and kill them instantly with a high-velocity bullet to the brain. Each carcass is rapidly processed under the scrutiny of a government inspector in a modern facility (see Figure 26–3).

The Hopcrafts are experimenting with various ways of marketing the meat and seem to have established the basic economic feasibility of African game ranching. The costs are substantially lower than those of raising cattle in that region. Much less water has to be supplied to the animals, so that less capital must be sunk into bores, dams, piping, and so on. Moreover, unlike cattle, game animals tend to be resistant to local parasites and diseases and do not need to be dipped or inoculated. Herding and corraling are unnecessary; the native herbivores handle their own predator protection. Indeed, the Hopcrafts' herds have been expanding in spite of almost no predator control. Lions, cheetahs, hyenas, and jackals all inhabit the ranch, and they get a small share of the game.

The exact economic advantage of the game ranch over a cattle ranch has not yet been determined. One reason is that the eventual carrying capacity of the range and the sustainable yield are not yet known. However, the biomass of native herbivores that can live on the ranch without degrading it seems to be considerably higher than that of cattle. The percentage annual increase in the stock also seems to be appreciably higher, running well over 35% per year in the last few years for the fast-maturing game animals. Data gathered so far suggest that the yield of lean meat from an operation such as the game ranch will be at least twice the poundage per acre that is taken from the best cattle ranch in the region.

In addition to rapid maturation rates in the game species, a major reason for this higher productivity is that a mix of herbivores, partitioning their resources, can use much more of the vegetation than cattle can. For

FIGURE 26–3 Meat-processing facility on the Hopcraft game ranch. The large carcasses in the foreground are wildebeeste, in the center distance is a kongoni, and in the right foreground is a Thomsen's gazelle. *(Photo by Paul R. Ehrlich.)*

example, giraffes gain a great deal of their sustenance from the tops of *Acacia* trees, a food source out of reach of the other large herbivores and totally unexploited by cattle. The ultimate yield from the ranch will depend, in part, on the mix of animals finally established. For example, if the emphasis is put on harvesting small animals such as Thomson's gazelles rather than large ones such as eland, production may be lower. The reason is that the smaller animals have higher metabolic rates and thus require more forage per pound of meat produced. Calculations are further complicated because the optimal mix of herbivores will itself depend on their exact food preferences and on the precise mix of edible plants available.

An additional advantage of the game ranch is the potential for selling the hides. Game hides should have a much higher value than cowhides, but at the moment the sale of game hides is prohibited by the Kenyan government—quite properly, inasmuch as most game hides are obtained by poaching. Possibly a licensing system could be devised to allow game ranchers to market their hides, as mink ranchers do in the United States. Combining the higher meat yields with the lower costs of production and adding in the possible profits from hides, one can foresee a bright economic future for game ranching. Of course, some problems are still to be overcome. The main ones involve breaking traditions—traditions of what meat is good to eat, traditions among scientists in animal husbandry for whom the idea of game ranching is too novel, and traditions among African pastoral peoples to whom cattle are the main symbol of wealth.

In Africa, however, the benefits of breaking with these traditions would be enormous. In the face of extremely rapid human population growth, game animals are fast disappearing from the continent; even national parks are under intense pressure from expanding agriculture and poaching. Hunger already is widespread on the continent, which has been stricken by disastrous droughts and famines in the last few years. Food supplies per person in most African countries south of the Sahara desert have declined by well over 10% since 1970 as population growth has outstripped the gains in food production. A major contributing factor to this continent-wide tragedy has been desertification caused in significant part by overgrazing of semiarid lands by traditional domestic animals. Clearly, game ranching could help both to preserve Africa's unique large animals and to contribute substantially to its food supply. It illustrates one of many ways that a knowledge of basic ecosystem ecology can greatly benefit humanity.

Other Assaults

Related to desertification, and possibly an even more serious threat to ecosystem services, is soil erosion. Recent estimates indicate that unless current rates of erosion are slowed, a monotonic global decline will occur in agricultural productivity (Brown 1981). Equally ominous is the trend toward *toxification* of the entire planet, a trend in which the increase in acid precipitation is just one factor (see, e.g., Woodwell 1983). Finally, we have the *decay of biological diversity,* the accelerating trend of extinction of populations and species (Council on Environmental Quality 1980, National Research Council 1980, Ehrlich and Ehrlich 1981). One species of Earth's millions of kinds of animals, *Homo sapiens,* is now using nearly 40% of the potential terrestrial NPP (Vitousek *et al.* 1986), and that species is "planning" to double its population size in a half century or so. This bodes ill both for other organisms and for humanity. The loss of diversity will lead to the progressive failure of the genetic library service and threatens the other services because living organisms play vital roles in the delivery of all of them. All of these trends are lowering the long-term carrying capacity of Earth for *H. sapiens.*

SUBSTITUTION AND ECOSYSTEM SERVICES

One way that humanity can seek to avoid this lowering of the carrying capacity is to find ways to substitute for the lost services. Deserts can be irrigated. Dams can be built to control floods. Exotic ground covers can be planted to slow erosion. Timber or firewood can be harvested from replanted forests. Insect pests can be suppressed with synthetic organic pesticides. Honeybees can be introduced to take over pollination services once supplied by native insects.

Generally, however, the replacements are less satisfactory than the original service. Consider, for example, F. H. Bormann's description of the problems of abiotic substitution for the services lost through deforestation:

> We must find replacements for wood products, build erosion control works, enlarge reservoirs, upgrade air pollution control technology, install flood control works, improve water purification plants, increase air conditioning, and provide new recreational facilities. These substitutes represent an enormous tax burden, a drain on the world's supply of natural resources, and increased stress on the natural system that remains. Clearly, the diminution of solar-powered natural systems and the expansion of fossil-powered human systems are currently locked in a positive feedback cycle. Increased consumption of fossil energy means increased stress on natural systems, which in turn means still more consumption of fossil energy to replace lost natural functions if the quality of life is to be maintained [Bormann 1976, p. 759].

Similar statements could be made about other abiotic substitutions, such as pesticides for natural pest control, synthetic fertilizers for natural soil maintenance, and chlorination for naturally purified water. All of these substitutions are less desirable than the original service because they require a continual energy subsidy that in turn has deleterious impacts on remaining natural systems. Moreover, as evidenced by phenomena such as insecticide resistance, eutrophication from fertilizer runoff, and bad taste and a possible cancer risk from chlorinated water, the substitutes are less desirable on other grounds as well.

Biotic substitutions—replacing extinct populations with exotic ones of the same or other species—also are usually unsatisfactory (Ehrlich and Mooney 1983). Populations often cannot be satisfactorily transplanted. Either the transplants die out, rapidly or gradually, or they persist but lead to degradation of the ecosystem. The failure of large copper butterflies of Dutch stock to become established in England is an example of a failed transplant (Duffy 1977). The reduced energy flow and the export of soil nutrients in exotic Monterey pine plantations in Australia is an example of ecosystem degradation (Springett 1976, Feller 1978).

Nevertheless, both biotic and abiotic substitutions will be increasingly necessary in the future. Unfortunately, satisfactory substitutes are unlikely to be found at anything like the rate at which ecosystems are being degraded. With advancing knowledge of ecosystem ecology, that situation may brighten somewhat, and ecosystems synthesized by humanity may substitute more successfully for natural systems (Wilson and Willis 1975). It would seem desirable, therefore, to preserve natural ecosystems and thus the populations and species that function within them. The ways of accomplishing this preservation are the focus of a new ecological discipline, *conservation biology* (Soule and Wilcox 1980). It is a discipline the roots of which go back to the mid-twentieth century and earlier, to men such as George Perkins Marsh, who in 1874 published a classic account of human impact on the biosphere. The roots go back also to that hero of modern conservationists, Aldo Leopold, who put the extinction crisis in a nutshell (1953) when he wrote: "If the biota, in the course of aeons, has built something we like but do not understand, then who but a fool would discard seemingly useless parts? To keep every cog and wheel is the first precaution of intelligent tinkering."

EPILOGUE

Where to Now?

Considering how complex the relationships of organisms are to their physical environments and to one another, the increase in human understanding of ecological processes is gratifying. We have begun to gain more knowledge about the basics of such diverse phenomena as the capture of energy from the sun by plants, the control of changes in the size of populations, the evolution of social behavior, the role of competitive and predatory interactions and disturbance in shaping communities, and the control of planetary cycles of nutrients by ecosystems. The past few decades have been a period of great excitement and progress in ecology. What does the future hold?

Since the mid-1960s, ecology has rapidly become one of the sciences most in the public eye. As the population-resource-environment crisis has unfolded, ecological, or "environmental," questions increasingly have become the center of political debates. What should the population policies of the United States be—or those of Russia, Mexico, China, or India? What is the maximum yield from fisheries? How will the extinction of the natural relatives of crop plants affect agricultural productivity? To what degree should the use of synthetic organic pesticides or fertilizers be restricted? Is the use of the poison 1080 likely to reduce coyote populations, and what will be its side effects? Are the possible negative impacts of acid rains on ecosystem functioning severe enough to justify expensive controls on the release of sulfur and nitrogen oxides? What will be the ecological consequences of depletion of the ozone layer, or of the deforestation of the Amazon basin, or of an all-out nuclear war?

Such questions will become even more central to the political process in the next few decades as humanity attempts to accommodate a growing human population and supply each individual with an increasing level of material affluence. The pressure on ecologists to supply answers will increase apace, and this pressure will probably increase more rapidly than will scientific knowledge of the issues involved. Therefore, one likely demand on our discipline in the future will be an escalating need to make judgments in the face of uncertainty and to communicate the nature of those judgments to decision makers and to the general public.

In many cases, ecologists will be forced to give opinions based on first principles and inference rather than on experimentation. For example, the crucial question of how the destruction of tropical rain forests would change global climatic patterns ideally might be settled by running experiments, which, in a sense, are now well under way. However, from what we already understand about the basic mechanisms that drive Earth's climate system, we can say that destruction of those forests probably will negatively affect humanity.

Still, a maximum effort should now be made to increase our ecological knowledge in order to reduce the uncertainty in making policy recommendations. Perhaps equally important, increasing ecological knowledge would increase humanity's understanding of vital aspects of the world we live in. Knowledge that helps us to survive is necessary, but *Homo sapiens* also highly values knowledge for its own sake.

In what areas should future efforts be concentrated? In actuality, the direction the science takes results from the collective decisions of all ecologists, and the direction changes more or less continuously. However, at the moment, several trends and needs seem clear. Solving current problems is unnecessarily difficult because of a lack of *baseline data*. Estimating the impacts of acid rains is hindered because the length of time they have been falling in various areas is not known. "Normal" dynamic patterns in populations cannot be described because few populations have systematically been censused for many generations. Successional changes have been followed for only relatively short periods of time, which is why some ecologists have been relocating sites of old photographs in order to document vegetational changes, as demonstrated by the photographs shown in Figure E–1. In order to minimize such problems in the future, an increasing effort, recently funded by a special program of the National Science Foundation, has gone into establishing sites and programs for long-term ecological research (LTER). It is critical that this program be extended to marine and tropical sites.

Related to the need for LTER programs is that for biological survey programs. To an astonishing degree, ecologists do not know what organ-

(A)

(B)

FIGURE E–1 West slope of the Stansbury Mountains, south of the Great Salt Lake, Utah. (A) 1901 photo by G. K. Gilbert. (B) 1980 photo by Garry F. Rogers. Succession over this 79-year period has restored a juniper woodland that, according to the numerous stumps in the older photograph, may have been destroyed by fire. Note that the new woodland is denser. *(Photos courtesy of Garry F. Rogers.)*

isms occur where today, so that changes in distributions and abundances often are difficult to document. This need is especially severe in the fast-disappearing tropical rain forests, where the majority of organisms have not even been named and described (National Research Council 1980).

Even in the rich United States, which seemingly would have been explored thoroughly, biological inventories are sadly lacking. For example, no national flora* has been produced, a work that would give a complete description of the plant species of the nation and their distribution. An attempt to get funds to start one failed in the 1970s. The passage of the Endangered Species Act by the United States Congress in 1973 underlined the paucity of information on many groups. For example, some species thought to be endangered have, when investigated, turned out to be widespread and common. Other populations and species have disappeared simply because no one knew they were endangered in the first place. The interest in endangered organisms has stimulated some work on the distribution of the American biota, especially state inventories promoted by a private group, the Nature Conservancy, but much more effort is needed.

Clearly, then, one immediate need is to stimulate more concentration on the critical task of documenting the state of populations, communities, and ecosystems and recording changes in a representative sample over the long term. In short, ecologists must have a more accurate idea of the phenomena they are trying to explain. Trying to develop national policies for environmental management without knowing the overall state of the environment, and whether it is changing, is as foolish as trying to determine national policy about bank interest rates without knowing the size of the money supply and whether it has been increasing or decreasing.

Beyond this basic need, we see certain directions in which the explanatory process should move. At present, ecologists are somewhat divided between those interested in mathematical theory and those oriented primarily toward field or laboratory observation and experimentation. As we hope this book demonstrates, each approach enriches the other. Doing careful, insightful field work is no more or less a rigorous and intellectually challenging activity than developing an analytical or computer model of an ecological system or process. As we have seen, the models can be enormously helpful in thinking about and guiding the investigation of complex natural situations. However, they would be sterile without the knowledge of nature that gave birth to them and the possibility of testing them in natural situations. Field, laboratory, and theoretical ecology are natural allies, and in the future we hope and expect that they will be even more closely integrated with one another.

Similarly, we foresee a trend toward better integration between the different levels of ecological organization. For example, elucidation of

* *Flora* is a term used both for an entire plant community and for a book (or books) describing such a community. Here we are using it in the latter sense.

some nutrient cycling problems will likely involve knowledge of the dynamics of populations of microorganisms. The answers to many questions about the dynamics of populations appear to require insights into the physiological ecology of the organisms involved and of others with which they interact.

Overall we expect that ecological research will increasingly involve a *multivalent approach* in which teams of people with different backgrounds, interested in different levels of organization, will collaborate. However, such team research will supplement, not supplant, the single investigator who, armed with field glasses, population cage, pencil, or computer, so often has had the insights that have led to great advances in ecology.

All in all, progress in ecology will not accelerate significantly unless society invests more in the subject—and that investment must first of all be monetary, for in today's world human talent tends to flow toward financial resources. More graduate students must be trained, more jobs must be created, and more money must be made available to provide ecologists with the opportunity and tools to do the monitoring, inventorying, and research that is needed. In a quite real sense, *H. sapiens* is betting its future on being able to deal rationally with its environment. However, it is investing only a few tens of millions of dollars per year in basic research to learn how to do this. In contrast, our species spends on weapons roughly $1.4 billion *per day*. It is not a priority system to give one great confidence in human judgment.

Yet all is not dark. Even with limited resources, ecology has come a long way, and we do not see signs of a slackening of the pace. In fact, we find it enormously exciting to be involved in a discipline that is both highly challenging and extremely important to our society. We hope that in this book we have been able to share some of that excitement with you.

GLOSSARY

Abiotic Physical characteristic of the place in which an organism lives, such as temperature of the air or water.

Acacia Trees of the genus *Acacia* (Leguminosae).

Acclimate To change phenotypically in a new environment. An individual acclimates; adaptation occurs over many generations. *See* Adaptation.

Adaptation Evolutionary change that makes an organism function better in a given environment. An adaptation is a structure, function, or behavior of an organism that aids its survival or reproduction.

Additive Genetic Variance The degree of phenotypic variation in the population caused by genetic variation, provided that the heterozygote has a phenotype intermediate between those of the two homozygotes.

Adenosine Triphosphate *See* ATP.

African Buffalo *Syncerus cafer,* large wild relatives of domestic cattle, also known as *Cape buffalo.*

Aggressive Mimicry Situation in which a predator mimics a harmless model in order not to alert potential prey.

Albedo The proportion of light reflected by an object, as in the extent the Earth's surface reflects solar energy.

Alga A photosynthesizing plant that normally lacks multicellular sex organs; that is, egg and sperm are not surrounded by layers of sterile nonreproductive cells. Many algae are single-celled phytoplankton.

Alkaloids Bitter, nitrogen-containing compounds that tend to be poisonous to animals and are important defensive chemicals in many plants. They include caffeine, nicotine, morphine, and strychnine.

Allee Effect Positive density dependence, usually found only when the population size becomes quite low.

Allele One of two or more alternate forms of a gene. Alleles occur at the same locus (physical position) of a chromosome.

Allopatric Having nonoverlapping distributions; polar bears and lions are allopatric. *See also* Sympatric.

Allopatric Speciation The evolutionary divergence of geographically isolated populations sufficient to prevent cross breeding between them.

Alluvial Soils Soils that have been deposited by running water.

Altricial Birds Birds whose young are unable to move about and feed when they hatch. *Compare* Precocial birds.

Altruism Performance by one individual, at its expense, of actions that benefit another individual.

Alula Structure in birds, consisting of a few feathers on the "thumb," which forms a leading-edge slot that helps to maintain smooth air flow and allows flight at slow speeds.

American Jacana *Jacana spinosa,* also known as the northern jacana. Jacanas are pantropical, moorhenlike birds that have long toes and are able to walk on floating lily leaves. Several species of jacanas are polyandrous, a rare condition in birds that makes jacanas of special interest in behavioral ecology.

Amino Acid Small organic molecules containing an $-NH_2$ group and a $-COOH$ group; some 20 different kinds serve as "building blocks" of protein molecules. In the process of making a protein, amino acids hook together chemically, producing a string of amino acids.

Amino Acid Sequence The linear arrangement of amino acids in a protein.

Ammonite An extinct relative of the octopus with a flat, spiral shell.

Androgen A sex hormone responsible for the development of male secondary sexual characteristics such as body hair and a deep voice.

Androgenital Syndrome Production of too much androgen by a female fetus.

Angelfish A crescent-shaped cichlid fish (freshwater) of the genus *Pterophyllum*, either *P. scalare* or *P. eimekei*, which are commonly kept in home aquariums. Also, poster-colored saltwater fishes of the family Chaetodontidae, subfamily Pomacanthinae.

Angle of Attack Angle at which birds intercept air flow or relative wind with their wings in order to generate lift.

Anoles Members of the lizard genus *Anolis*.

Anolis Genus of about 300 species primarily in tropical Central and South America and in the Caribbean. In the West Indies, they are the principal ground-feeding insectivorous vertebrates.

Anopheles The genus of mosquitoes that carries malaria.

Anthropocentric Focused on human values; considering humanity to be the standard or frame of reference.

Anthropogenic Caused by humans.

Apatosaurus A giant herbivorous dinosaur, more commonly known as *Brontosaurus*.

Aposematic Being conspicuous and serving to warn—for example, when toxic animals advertise their poisonousness by aposematic black and red coloration.

Artificial Selection Selection in which an experimenter or plant or animal breeder creates a differential reproduction of genotypes by choosing the parents of each generation.

Asexual Reproduction Any reproduction not involving the fusion of gametes, such as dividing in two or budding.

Aspect Ratio Ratio of length to width; birds with long, slender, wings, such as albatrosses, have a high aspect ratio.

Assimilation Efficiency The fraction of light striking plants that actually is used in photosynthesis or the fraction of energy consumed (e.g., ingested) by animals that actually is absorbed from the food.

Association Analysis Investigation of the extent to which species co-occur.

ATP Adenosine triphosphate, a phosphorus-containing compound that is the "energy currency" of all organisms.

Australopithicines Upright, small-brained relatives of modern human beings of the genus *Australopithecus* that lived roughly 1–4 million years ago. Some australopithicines were undoubtedly ancestors of *Homo sapiens*.

Automimicry Phenomenon by which a set of different individuals of the same species mimic each other.

Autotroph An organism that derives its energy directly from the physical environment, such as sunlight or inorganic chemical reactions. *See* Heterotroph.

Batesian Mimicry Mimicry of a dangerous or distasteful organism by a harmless or tasty one.

Bearded Vulture *Gypaetus barbatus*, also known as the *lammergeyer*.

Benthic Living on the bottom of a body of water.

Biogeochemical Cycles Cycling or flow of nutrients along paths that bind together the living and physical portions of the biosphere.

Biogeography The study of present and past distributions of organisms and their causes.

Biological Amplification The tendency for chemicals, especially toxic ones, to become more concentrated in organisms than they are in the environment.

Biological Clock An internal mechanism that permits an organism to tell the time without external cues.

Biological Community All of the organisms that live in a given area; sometimes a more restricted group is designated, such as the "bird community"—all of the birds found in a given area.

Biological Control The use of predators (or pathogens) to control a pest.

Biological Oxygen Demand (BOD) A measure of the difference between the production and consumption of oxygen, a common measure of pollution. The BOD in a laboratory is normally the number of milligrams of oxygen consumed per liter of water in 5 days at 20°C.

Biomass The weight of living organisms in an area.

Biome A major ecological community type, such as grassland or desert.

Biometrical An approach predicting evolutionary change based on parent–offspring correlation as a description of inheritance.

Biosphere The thin layer of the life that occurs near the surface of the Earth and the physical-chemical environment in which it is embedded.

Biota All the living creatures, plants, animals, and microorganisms in an area.

Birth Rate The number of individuals born in a given period of time divided by the number in the population at the midpoint of the period.

Biston betularia The peppered moth, famous as the subject of studies of natural selection in the field.

Black Death *See* Bubonic plague.

Black Smokers Hot water springs in deep ocean trenches that support small ecosystems based on bacteria that derive energy from inorganic chemical reactions.

Blennies A family of small, slender marine fishes that include cleaner mimics and dash-and-slash predators.

Bloom The sudden explosive growth of a plankton population.

Blue-Green Algae Photosynthesizing bacteria of the group known as *cyanobacteria*.

Bombardier Beetles Certain bettles of the genus *Brachinus* (Carabidae) that defend themselves with a chemical spray from the tip of the abdomen.

Bower Birds Perching birds from Australia and New Guinea (Ptilonorhynchidae); the males of some species build elaborate, often decorated structures for courtship displays.

Brontosaurus *See Apatosaurus.*

Brood Parasites Animals that lay their eggs in the nests of other species, which then serve as foster parents and raise the young. Cuckoos are the classic example.

Bubonic Plague An extremely serious disease caused by the bacterium *Yersina pestis*. It is carried from rats to human beings by fleas, and in epidemics is transmitted directly from person to person.

Bulk Flow The movement of liquid or gas caused by pressure differences.

Bushmaster *Lachesis mutus*, the largest poisonous snake in the Western Hemisphere. The bushmaster is a pit viper (Crotalinae), a member of the same group that includes the rattlesnake.

C_3 Pathway The capture of carbon dioxide and its incorporation into a sugar, in which the product is a molecule with three carbon atoms.

C_4 Pathway The capture of carbon dioxide and its incorporation into a sugar, in which the product is a molecule with four carbon atoms.

Cabbage Butterfly A white butterfly of the genus *Pieris* that feeds on plants in the cabbage family, Cruciferae (Brassicaceae).

Cactoblastis cactorum A small moth introduced into Australia that successfully controlled the plague of *Opuntia* cactus there.

CAM Pathway The capture of carbon dioxide and its incorporation into a sugar, in which CO_2 is taken up during the day and fixed using the C_3 pathway at night. CAM stands for crassulacean acid metabolism, from a family of succulent plants, Crassulaceae.

Canalization Production of the same phenotype by several different genotypes.

Canopy The overlapping, intertwined crowns (tops) of forest trees. In a tropical rain forest, the canopy is normally so dense as to make the interior of the forest quite dark, and an entire distinct biota occupies the canopy.

Capybara A dog-sized South American rodent, *Hydrochaeris hydrochaeris*.

Carbon Fixation The capture of carbon dioxide and its incorporation into a sugar.

Carnivore A flesh eater; any animal that attacks and eats other living animals (scavengers are usually not included).

Carrying Capacity The maximum size of a population that can be supported by the resources of the area it occupies.

Cassowary A large, flightless bird of Australia and New Guinea in the genus *Casuarius*, related to the emu and the ostrich.

Caste A structurally and usually functionally distinct group of individuals found in a social insect colony such as workers, soldiers, or reproductives.

Catena A connected series of soil types.

Cenozoic Era The major division of geological time stretching from the end of the Mesozoic, some 65 million years ago, to the present.

Ceratopsians Large herbivorous dinosaurs with prominent horns; the dinosaur equivalent of rhinos.

Chaparral A vegetation type characteristic of Mediterranean climates that is dominated by shrubs with thick, small leaves. Chaparral is maintained by recurrent burning; in the absence of fires, it is replaced by other vegetation types.

Character Displacement The phenomenon whereby two species are more different from each other where they are sympatric than where they are allopatric.

Checkerboard Pattern A distribution pattern in which various localities (analogous to squares on a checkerboard) have one or the other of two closely related species.

Checkerspot Butterflies Members of the family Nymphalidae, subfamily Nymphalinae, tribe Melitaeini. Small to medium-sized butterflies, usually with a black, white, and orange checkered pattern on the wings. In this book, members of the genus *Euphydryas*.

Chemical Defenses Toxic or repellent chemicals used by an organism to discourage predators.

Chemical Potential Energy required to convert water from some state—for example, as present in soil, in air, or in a leaf—to a reference state, such as pure liquid water under 1 atm and at 20°C.

Chlorinated Hydrocarbons Synthetic organic poisons containing chlorine that are often used as pesticides. They tend to be persistent and to move around in the environment; they often accumulate in fatty tissues of organisms that feed high on food chains. DDT is the best-known example; polychlorinated biphenyls (PCBs) are also chlorinated hydrocarbons.

Chloroquine A synthetic antimalarial drug.

Cichlid Fishes Freshwater fishes of the family Cichlidae, which includes angelfishes, mouthbreeders, and Jack Dempseys. A large number of species of cichlids live in each of the great lakes of East Africa, occupying a great variety of niches.

Cleaner Wrasses Indo-Pacific fishes of the genus *Labroides* of the family Labridae, which survive by eating parasites and dead tissues of other fishes that permit themselves to be "cleaned."

Climate Space The set of environmental conditions in which an animal can maintain a steady-state temperature that is between its lethal limits. *See* Steady-state temperature.

Climax Community The most stable community on a given soil under a given set of climatic conditions; the end of a successional sequence. *See* Succession.

Cline A gradual change in the characteristics of a geographically contiguous set of populations; a gradient in a given character.

Clone A group of individuals produced asexually from a single ancestor.

Coarse-Grained Environment An environment in which resources occur in patches sufficiently large with respect to an organism's activities that it can select among them. *Compare* Fine-grained environment.

Codynamic Describes population interactions which involve reciprocal influences on population size. *See* Coevolutionary.

Coevolution The reciprocal evolution of two or more interacting populations. The effect of a predator on the evolution of a prey species, and vice versa, is an example.

Coevolutionary Describes population interaction that involves two species affecting each other's evolution. *See* Codynamic.

Coevolution-Structured Guild A guild whose members have body sizes determined by coevolution.

Cohort A group of individuals of the same age.

Cold Hardening Preparing physiologically to undergo the stress of winter.

Commensalism Relationship in which one species benefits and the other apparently suffers no harm.

Communal Courtship Display A display of male animals in the same place, competing for females.

Communication An action by an animal that creates a response in another, thought to be, on the average, to the advantage of the communicator.

Community *See* Biological community.

Community Diversity Usually the number of different species in a community.

Community Stability The speed with which the community, if disturbed, returns to its original state.

Compensation Level In an aquatic community, the depth of light penetration at which the plants are able to balance their respiration by photosynthesis.

Compensation Point *See* Light compensation point.

Competition The use of a limited resource by two or more different individuals, or the interference by one individual with the resource use of others.

Competitive Exclusion A situation in which one species eliminates another as a result of competition.

Conductivity Constant of proportionality of the pathway along which water is flowing, relating the speed of flow to the pressure difference.

Conservation Biology The discipline concerned with the scientific aspects of the conservation of biological resources, especially its diversity of species.

Conspecifics Members of the same species.

Consumer Any organism that lives by feeding on other organisms, dead or alive. The term includes all animals—herbivores, carnivores, and decomposers; parasitic and decomposer plants; and most microorganisms. Those left are producers. *Compare* Producer.

Continential Drift The continuous movement of the continents and oceans as they are carried along on the tectonic plates.

Control Group In an experiment, the group that is left alone to see what would happen without intervention (e.g., the flies that are treated just like those exposed to DDT in a selection experiment, except that they are not exposed to DDT).

Convective Heat Transfer Coefficient The constant of proportionality relating the speed of heat flow to the difference between body temperature and air temperature.

Convergent Evolution Development of a similar form or function in unrelated organisms living in the same environment, such as the shapes of whales and fishes.

Convict Tang A common Pacific Ocean surgeonfish, *Acanthuris triostegus* (Acanthuridae).

Cooperation Performance of actions by one individual, but without cost to itself, that benefit another individual.

Crepuscular Active at dawn or dusk.

Cretaceous Period The last period of the Mesozoic Era, stretching from roughly 135 million years ago to 65 million years ago. The Cretaceous saw the origin of the flowering plants and the demise of the dinosaurs.

Cretaceous–Tertiary Boundary The boundary between the last period of the Mesozoic and the first period of the Cenozoic, some 65 million years ago. The boundary is marked by the extinction of many groups of organisms, including the dinosaurs, thought by some to have been caused by the collision of an asteroid with Earth.

Crocodilians Crocodiles, alligators, gavials, and their extinct relatives.

Cuckoo Dove Fruit-eating pigeon that lives in

the middle story of shaded forests from India to Melanesia.

Cyanobacteria *See* Blue-green algae.

Cyclops A tiny freshwater crustacean.

Dark Reaction The second part of photosynthesis; adenosine triphosphate (ATP) and nicotinamide adenosine dinucleotide phosphate (NADPH) are used to capture carbon dioxide and incorporate it into sugar. *See* Light reaction.

Death Rate The number of individuals dying in a given period of time divided by the total in the population at the midpoint of the period.

Deciduous Forest A forest in which the dominant trees shed their leaves before an unfavorable season (normally in the fall in the temperate zone and before a dry season in the tropics).

Decomposer An organism that survives by feeding on the carrion or wastes of other organisms. Decomposers play a crucial role in nutrient cycles by breaking down complex organic molecules into simpler constituents.

Decomposition The breaking down of complex organic molecules into simpler constituents, often to the elements themselves.

Defaunate To remove all animals from.

Demographic Stochasticity Fluctuation in population size caused by random variation in offspring number that inevitably occurs even when the environment is practically uniform.

Demographic Unit A population sufficiently isolated from others that its dynamics are independent of migration.

Demography Study of the dynamics and age structure of populations.

Denitrification The conversion of nitrate and nitrite to elemental nitrogen by microorganisms.

Density Dependent Influenced by the size of the population, usually through resource depletion or increased social interaction. *Compare* Density independent.

Density Independent Not influenced by the size of the population, usually caused by weather-related changes. Compare Density dependent.

Deoxyribonuceic Acid *See* DNA.

Deposit Feeders Organisms that swallow grains of sediment and assimilate the bacteria and other microorganisms that coat the grains.

Desertification The creation of desert as a result of human activities.

Diapause A resting state of an insect in which development is suspended and little energy is used. Insects often enter diapause prior to a period of unfavorable environmental conditions.

Didinium A barrel-shaped predatory protozoon.

Differential Reproduction Reproduction of some individuals more frequently than others.

Diffuse Coevolution Coevolution in which many organisms are interacting so that pairwise coevolution is difficult to detect.

Dik-Dik A jack-rabbit sized African antelope (*Madoqua*).

Dioecious Plants Plants with male and female flowers on different plants, in which case the entire plant may be called male or female. *Compare* Monoecious plants.

Diplacus A plant of the snapdragon family that serves as the larval food for some checkerspot butterflies.

Diplodocus A large, *Apatosaurus*-like dinosaur.

Diploid Having two sets of chromosomes in undifferentiated cells.

Directional Selection Selection leading to an increase or decrease in the average value of a trait.

Dispersal Movement away from the place of birth or origin.

Dispersion The spatial pattern of individuals in a population.

Dissolved Oxygen Concentration Measure used as an index of the pollution of lakes because it is a measure of eutrophication. *See* Eutrophication.

Distribution The geographic area occupied by a taxonomic group—for example, a species or genus.

DNA Deoxyribonucleic acid, the chemical that carries the genetic code in the sequence of the four kinds of subunits (nucleotides) of which it is made; the genetic material.

Dominance Hierarchy Linear arrangement of individuals, each either dominant or submissive to others, determined by the results of aggressive interactions.

Drag Force resulting from the components of pressure and friction.

Drosophila The genus of fruit flies (sometimes called *vinegar flies*) that are used extensively in genetic research.

Dung Flies Flies that survive as larvae living in decomposing feces, especially *Scatophaga stercoraria* (Scatophagidae).

Dutchman's Pipe Vine in the genus *Aristolochia*, family Aristolochiaceae.

Ecological Efficiency *See* Lindeman's efficiency.

Ecology The science that deals with the relationship between organisms and their physical and biological environments.

Ecomorph An animal with a characteristic size, shape, color, and modal location in the habitat.

Ecosystem The biological community in an area and the physical environment with which it interacts.

Ecosystem Services Services provided to humanity by natural ecosystems such as amelioration of weather, control of water flows, maintenance of soils, disposal of wastes, and recycling of nutrients.

Ectotherm *See* Heliothermic thermoregulator.

Egyptian Vulture *Neophron percnopterus*, also known as *Pharaoh's chicken*.

Eighty-Nine Butterfly Various species of the genera *Callicore* and *Diaethria* (Nymphalidae) that have a pattern on the underside of the hind wings that resembles the number 89.

Eland *Tragelaphus oryx*, a cowlike large antelope of southern Africa.

Electrophoresis A procedure for separating different proteins on the basis of their electric charge.

Emu Australian flightless bird (*Dromiceius novae-hollandiae*) related to, and almost as large as, the ostrich.

Endemic Native to a particular region.

Endothermic Thermoregulator An animal, usually bird or mammal, that uses its metabolism as a primary source of body heat and uses physiological mechanisms to hold its body temperature nearly constant.

Environment All of the elements in an organism's surroundings that can influence its behavior or survival.

Environmental Variance Phenotypic variance among individuals with the same genotype.

Enzyme A protein that serves as a biological catalyst, speeding the rate of a chemical reaction.

Eocene Period The second period of the Cenozoic era, about 45 million years ago.

Epifauna Animals that live on the surface of a soft sediment.

Epilimnion Upper layer in water, which is warmed by the sun and mixed by the wind. *See* Hypolimnion, Thermocline.

Escherichia coli A bacterium isolated from human feces that has been a key tool for geneticists working on the biochemical mechanisms of heredity.

Estrus The stage of the sexual cycle of a female mammal when it is both receptive to intercourse and able to conceive.

Ethology Study of animal behavior in natural situations.

Euphydryas Genus of checkerspot butterflies, restricted to North America and Eurasia, that has been the subject of intensive ecological investigations.

Eutrophic Rich in dissolved nutrients. *Compare* Oligotrophic.

Eutrophication Nutrient enrichment of a lake or stream, often by runoff of fertilizer or addition of sewage.

Evolution The process of change and diversification that led over billions of generations from one or a few simple protoorganisms to the rich array of life forms that have populated Earth.

Evolutionarily Stable Strategy A strategy that, when common, cannot be displaced by another other strategy that is rare.

Evolutionary Biology The branch of biology that focuses on evolutionary processes and evolutionary history.

Exothermic Process A chemical reaction or change of phase that releases energy to the environment.

Exploitation Competition Competition in which two or more organisms consume the same limited resource.

Exploitation Efficiency The fraction of tissues at one trophic level consumed by organisms at the next trophic level.

Exponential Growth Population size conforming to an exponential function of time; the population increase in a period is a fixed percentage of the size of the population at the beginning of the period.

Family (Taxonomic) A taxonomic category between genus and order. Dogs and their relatives (wolves, jackals, foxes, etc.) are in the family Canidae; cats and their relatives (lynxes, lions, tigers, cheetahs, etc.) are in the family Felidae. The human family is Hominidae; the rose family is Rosaceae. Animal family names always end in *-idae*; plant family names almost always end in *-aceae*.

Fauna The animals of an area.

Faunal Collapse A dramatic loss of animal diversity, often resulting from the isolation of an area or the extinction of a keystone species.

Fecundity The rate of production of offspring by a female.

Field Capacity Capacity of soil to hold water; any water beyond this maximum leaves the soil as runoff.

Fine-Grained Environment An environment in which resources occur in patches so small relative to the activities of an organism that it cannot selectively forage in the patches. *Compare* Coarse-grained environment.

Fish Lice Crustacean parasites of fishes.

Fitness Net reproductive output of an individual; the number of offspring it places in the next generation.

Fixation Conversion of an element or compound from an unavailable to an available form.

Fixed Said of a gene pool, in which all individuals have the same allele at a particular locus.

Flora The plants of an area.

Flour Beetles Small beetles of the genus *Tribolium*.

Flycatchers A variety of birds that perches on limbs and flies out to catch insects in midair or perch.

Foliage Height Diversity A measure of the vertical structure of the vegetation in an area.

Food Chain A feeding sequence such as grass → zebra → lion; used to describe the flow of energy and materials in an ecosystem.

Food-Limited Guild Guild of species that compete with one another for food.

Food Web Intertwined food chains.

Foraging Strategies Ways in which animals choose their diets and allocate their time when seeking, catching, and eating food.

Forb An herbaceous plant other than a grass.

Founder Effect Said of colonization events and experiments in which the outcome is determined by the genetic attributes of the colonists.

Frequency-Dependent Selection Differential reproduction of individuals in which the fitness of a type of individual varies with its frequency in the population.

Functional Response The rate at which an individual predator consumes prey as a function of the number of prey available to it; the reaction of an individual to the availability of prey. *See* Numerical response.

Fundamental Theorem of Natural Selection The average fitness of the individuals in a population increases each generation with a speed that is proportional to the variance of fitness among the individuals in the population.

Fungi A group of organisms distinct from both plants and animals, but traditionally considered with plants (and studied by botanists). Fungi are largely decomposers; mushrooms and molds are among the most familiar forms. Some are parasitic; a few, such as the fungi that cause athlete's foot and valley fever (a serious lung infection), attack human beings. Others, such as *Penicillium*, the source of penicillin, have been a great boon to humans.

Game Ranching Herding native game animals instead of the standard domestic animals, especially cattle.

Gamete A haploid reproductive cell that fuses with another to produce a diploid gamete; a sperm or an egg.

Gause's Principle The proposition that two species cannot coexist if their niches (occupations) are the same. Replaced with the concept of "limiting similarity."

Gel A gelatinlike material, used in electrophoresis as a medium for protein migration.

Gene The basic unit of Mendelian heredity.

Gene Frequencies In genetics, the proportions of the different alleles in the gene pool; the frequency of A_1 is the fraction of the gene pool at the A locus that is type A_1.

Gene Pool The collection of all the genes from all the individuals in the population.

Gene Splicing An array of techniques by which genes from one organism can be inserted into another organism.

Genet A genetic individual; all the tissue that grows from a single fertilized egg.

Genetical Theory of Evolution An approach to predicting evolutionary change based on specified genetic mechanisms for inheritance of a trait.

Genetic Code The arrangement of the four DNA subunits (a codon) that designate which of 20 amino acids is to be added to the end of a growing protein molecule. Combinations of three of the four (triplets) code for the amino acids and for other instructions such as "stop making the protein now."

Genetic Drift Random changes in gene frequency caused by using a finite "sample" of the gene pool of each generation. Drift is the process by which the gene pool's composition fluctuates from one generation to another solely because the alleles drawn for a particular generation have slightly different frequencies than they did in the preceding generation, even without selection.

Genetic Engineering Altering the genetic makeup of organisms by using selection or gene-splicing techniques.

Genetic Variability Genetic differences among individuals in a population.

Genetics The branch of biology that deals with heredity.

Genome All the genes in an organism.

Genotype An individual's genetic endowment, as contrasted with its phenotype. Often restricted to the portion of the genotype affecting a single characteristic. *Compare* Phenotype.

Genotype Frequencies In genetics, the proportions of the different genotypes in the population; the frequency of A_1A_1 is the fraction of individuals with the genotype A_1A_1.

Genus A set of one or more species; the main taxonomic category between species and family. The checkerspot species *Euphydryas editha* belongs to the genus *Euphydryas*; humans belong to the genus *Homo* (along with some of their fossil ancestors such as Peking, Java, and Heidelberg men, all *Homo erectus*). The generic name forms the first half of the full binomial name for a species. The plural of genus is *genera*.

Geographic Variation The almost ubiquitous phenomenon of organisms of the same species being genetically different in different parts of the geographic range.

Geometric Average The *n*th root of a product of *n* numbers.

Geometric Growth Exponential growth with a discrete time interval.

Giant Armadillo *Priodontes maximus*. Found throughout most of South America east of the Andes, this armadillo reaches almost 5 ft in length, including the tail.

Glucose A sugar involved in energy metabolism.

Glucose-6-Phosphate Glucose with a phosphate group. Formation of this molecule is the first step in glycolysis, the metabolic breakdown of glucose to release energy.

Glycerol An alcohol that serves as antifreeze in some animals.

Goby A fish of the family Gobiidae. Gobies often sit on the bottom of a body of water and often have fins on their underside modified into a sucking disk that helps hold them in position.

Gondwanaland The giant southern continent that resulted when Pangaea split into two fragments some 200 million years ago. The northern fragment was Laurasia.

Gooseneck Barnacle A group of somewhat flattened barnacles that usually have a shell made up of five limy plates. In spite of their superficial resemblance to clams, barnacles are marine crustaceans related to stony barnacles, acorn barnacles, and more distantly to crabs and lobsters.

Gradual View View of evolution as continuing gradual process of change.

Grant's Gazelle *Gazella granti.* A graceful small African antelope with lyre-shaped horns. It is slightly larger than the similar Thomson's gazelle, with which it often occurs; however, it is easily distinguished because its white rump markings extend above the base of the tail, whereas those of the Thomson's gazelle do not.

Great Auk *Alca impennis*, a penguinlike diving bird of the northern oceans that became extinct in the 1840s.

Greenhouse Effect Warming of Earth caused by water vapor, carbon dioxide, and some other substances in the atmosphere that are largely transparent to incoming solar radiation but absorb the outgoing long wavelength infrared that is radiated by Earth. Without the greenhouse effect, Earth's surface would, on the average, be about 30°C colder.

Gross Primary Production The rate at which energy is bound (and organic matter created) per unit area by the photosynthesis of green plants.

Group Selection Differential reproduction of populations, as contrasted with individual selection, which is the differential reproduction of individuals within populations.

Grouper A large predacious marine fish of the family Serranidae.

Growth Efficiency The portion of the energy assimilated by an organism that is incorporated into new tissues.

Grunt A marine fish of the family Pomadasyidae. Resting schools of these fishes, which hunt invertebrates in sandy areas at night, are a prominent feature of coral reefs in the Caribbean.

Guild A group of species that makes a living in the same way, such as fruit-eating birds or ambush predators.

Haber Process Industrial process in which nitrogen and hydrogen gases are combined with a catalyst at high temperature and pressure to yield ammonia, which is then used as a base for the manufacture of fertilizers.

Habitat An organism's home, the place where it lives, such as in deep woods, running streams, coral reefs, or the human bloodstream (for certain life stages of certain species of malaria parasite).

Habitat Vicariants Species that are replacements of a form found in one habitat but that occur in other habitats; they occupy the same niche and do not coexist in the same habitat.

Haplodiploidy The existence of haploid males and diploid females in the same species; the sex-determination system in bees, wasps, and ants, in which males develop from unfertilized eggs (and thus remain haploid) and females from fertilized eggs (which are diploid).

Haploid Having only one set of chromosomes.

Hardy–Weinberg Law The proposition that the genotype ratios produced from random mating remain unchanged through time; only true in the absence of natural selection, genetic drift, and mutation.

Hardy–Weinberg Ratios The genotypic ratios produced by random mating or random union of gametes; at a locus with two alleles, the ratios are $p^2{:}2pq{:}q^2$, where p is the frequency of one allele and q is the frequency of the other.

Harem A group of females controlled by one male.

Heliconia A South American genus of bananalike herbaceous plants (Strelitziaceae).

Heliconius The genus of longwing butterflies (Nymphalidae).

Heliothermic Thermoregulator An animal whose body temperature is roughly equal to that of its surroundings and that controls its body temperature during the day by changing its position throughout the day so as to adjust the amount of solar radiation it receives.

Herbivore An organism that eats plants.

Heritability The slope of the regression line between the phenotype of an offspring and the average phenotype of its two parents.

Hermaphrodite An organism having both male and female reproductive organs.

Hermatypic Reef-forming; said of corals.

Heterosis The condition in which the heterozygote has a higher fitness than either homozygote.

Heterotroph A plant, animal, or microorganism that cannot extract energy from inorganic reactions. *See* Autotroph.

Heterozygote A diploid individual having two different alleles at a locus, one derived from the father and one from the mother.

Heterozygous Adjective of heterozygote; describes an individual with different alleles at the same locus on each of its chromosomes.

Hexokinase The enzyme responsible for adding a phosphate group to glucose to form glucose-6-phosphate in glycolysis. *See* Glucose-6-phosphate.

Hibernate To pass the winter in torpor, that is, with a greatly lowered body temperature.

Home Range Area through which mobile animals regularly move in the course of their normal activities.

Homo erectus The immediate ancestor of *Homo sapiens*; Java and Peking people.

Homo habilis Link between australopithicines and *Homo erectus*.

Homozygous Describes an individual with the same alleles at a particular locus on both of its matching chromosomes.

Honeycreepers A small family (Drepaniidae) of Hawaiian birds that provides a classic example of adaptive radiation. A single ancestral species that reached the Hawaiian Islands evolved into a series of descendant species showing a diversity of bill shapes and sizes.

Honeyeaters A large family (Meliphagidae) consisting of some 170 species of birds of the southwest Pacific and Australia. They have rather long bills and brush-tipped tongues, and feed on nectar (hence their name), insects, and fruit.

Horizons Soil layers.

Host An organism that can be infected with a specific parasite.

Humus A dark, decaying organic substance found in soils.

Hypolimnion Cool layer in water that is too deep to be warmed by the sun or mixed by the wind. *See* Epilimnion, Thermocline.

Hypothesis A proposed explanation for a phenomenon.

Ichthyology The study of fishes.

Ichthyosaurs Porpoiselike relatives of dinosaurs.

Impala *Aepyceros melampus*. One of the most elegant of African antelopes. The males have graceful, lyre-shaped horns; the females are hornless.

Inbreeding Mating system in which adults mate with relatives more often than is expected by chance.

Inclusive Fitness The number of copies of an individual's genes that are placed in the next generation's gene pool as a result of its reproductive activities combined with the reproductive activities of its relatives.

Industrial Melanism Evolution of dark coloration in moths in heavily industrialized areas.

Inertial Thermoregulators Animals such as large reptiles that develop body temperatures above the air temperature because their low surface-to-volume ratio limits cooling.

Infauna Animals that live in sediments on the ocean bottom.

Inflorescence A cluster of flowers with a definite arrangement; inflorescences of some plant families, such as the Compositae, look like individual flowers to the uninitiated.

Infrared Radiation Heat radiation of too long a wavelength to be visible to the human eye.

Insectivore An organism that eats insects.

Insularization Division into islands or island-like units.

Integrated Pest Management The use of a variety of ecologically sound techniques to keep pest populations below the size at which they cause serious economic damage.

Interference Competition Competition in which one species prevents the other from having access to a limiting resource.

Intermediate Disturbance Principle The proposition that the highest species diversity in a guild of space-limited competitors occurs at an intermediate degree of disturbance.

Intersexual Selection The situation in which members of one sex create a reproductive differential in the other by preferring some individuals as mates. *Compare* Intrasexual selection.

Interspecific Competition Competition between individuals of different species.

Interspecific Territoriality Defense of a territory against individuals of other species.

Intrasexual Selection The situation in which some members of one sex have more offspring than others because they compete with one another for opportunities to mate with the other sex. *Compare* Intersexual selection.

Intraspecific Competition Competition between individuals of the same species.

Intrinsic Rate of Increase The difference between the density-independent components of the birth and death rates.

Invasion-Structured Guild A guild of species whose body sizes represent the result of successive invasions.

Invertebrates Animals without backbones—for example, insects, worms, and mollusks.

Ionizing Radiation Radiation capable of causing mutations.

Island Biogeography The study of the distributions of organisms on islands.

Isolation Absence of migration.

Isotope One of two or more varieties of the same element with virtually identical chemical properties and different atomic weights.

Jawless Fishes Hagfishes and lampreys, fishes with sucking mouths and no bones (class Agnatha).

Keystone Mutualist A species upon which other species depend for their persistence at a site.

Keystone Predator A predator whose removal leads to reduced species diversity among the prey.

Kilocalorie Roughly the amount of heat energy required to raise the temperature of 1 kg (2.2 lb) of water 1°C (1.8°F).

Kin Selection Selection favoring an altruistic trait because the survival and reproduction of close relatives of the altruist are sufficiently enhanced.

Kirk's Dik-Dik *Madoqua kirki* of east and southwestern Africa. *See* Dik-dik.

Kiwi *Apteryx australia*. A virtually wingless nocturnal bird from New Zealand. Less than 2 ft in length, the kiwi is a distant relative of the ostrich and other Southern Hemisphere flightless birds. The kiwi nearly became extinct after Europeans settled New Zealand, but now it is totally protected and apparently thriving.

Kongoni *Alcelaphus buselaphus*. An African antelope with short horns that appear to be fused at the base. Also called *Coke's hartebeeste*.

Kudzu Vine *Pueraria thunbergiana*, a leguminaceous vine from China and Japan that is widely used for control of erosion and that has become a pest in the southeastern United States.

Lammergeyer *See* Bearded vulture.

Lappet-Faced Vulture *Aegypius tracheliotus*. Also called *Nubian vulture*.

Larva An early, free-living developmental stage of an animal that does not resemble the adult. A caterpillar is the larva of a butterfly or moth, a maggot the larva of a fly, a tadpole the larva of a frog or toad.

Latent Heat of Evaporation Number of calories required to evaporate 1 g of water.

Latent Heat of Fusion Number of calories given up by 1 g of water when it freezes.

Laurasia The northern fragment of Pangaea; the southern fragment was Gondwanaland.

Laws of Thermodynamics Laws that describe energy. The first says that energy is conserved. The second law says that as a result of all spontaneous processes, the availability of energy decreases.

Leach Dissolve out by water percolating through the soil.

Leaf Cutter Ants Members of the ant tribe Attini that eat fungi grown on a mulch of cut leaves in special chambers of their nests called *fungus gardens*. Parades of workers of these ants, each carrying a cut piece of leaf like an umbrella, are a common sight in the tropics of the Western Hemisphere, where they are major herbivores in tropical forests.

Lek An area used for communal courtship displays by males attempting to attract and mate with females.

Lentic habitats Standing-water habitats, such as ponds and lakes. *See* Lotic habitat.

Lepidoptera The insect order containing butterflies and moths.

Light Compensation Point The light level at which photosynthesis just balances respiration.

Light Reaction The first part of photosynthesis; adenosine triphosphate (ATP) stores energy in phosphate bonds, and nicotinamide adenosine dinucleotide phosphate (NADPH) stores energy as the ability to reduce other compounds. See Dark reaction.

Limit Cycle A stable, sustained oscillation in the population sizes of species, as in models of predator and prey.

Limiting Resource A resource that is in short supply compared with the demand for it.

Limiting Similarity The maximum degree of similarity that two species can have and still coexist.

Lincoln Index Formula used to estimate the population size from mark-recapture data.

Lindeman's Efficiency The ratio of energy acquired by one trophic level to that acquired by the next; also called *ecological efficiency*. Lindeman's efficiency is the product of the assimilation efficiency, the growth efficiency, and the exploitation efficiency. *See* Assimilation efficiency, Exploitation efficiency, Growth efficiency.

Littoral Zone Near-shore zone of sea or lake, where enough light for photosynthesis can reach the bottom.

Lizard Fish A predaceous marine fish of the family Synodidae. Lizard fishes have large heads and well-armed jaws, and sit on the bottom (or largely buried in sand) waiting to dash out at their prey.

Logistic Equation The equation for population dynamics in which the birth and death rates depend linearly on population size. Because the population growth rate declines as the carrying capacity is approached, a population growth curve shaped like an S stretched left to right.

Log-Normal Distribution Normal distribution in which the horizontal axis is calibrated in units of the logarithm of a variable.

Lotic Habitat Running-water habitat, such as a stream or river. *See* Lentic habitat.

Lystrosaurus A fossil, swamp-dwelling reptile whose remains have been found in Antarctica, India, and southern Africa.

Macroevolution Evolution above the population level, involving speciation and the divergence and history of major groups.

Maggot The larva of a fly.

Magma Molten rock from the interior of the Earth.

Malathion An organophosphate insecticide with relatively low mammalian toxicity.

Marabou Stork *Leptoptilos crumeniferus*, a carrion-eating African stork.

Mark-Release-Recapture A technique for estimating population size and determining movement patterns by finding the proportions and locations of previously marked and released animals. See also Lincoln index.

Mating System The pattern of matings within a population, including how mates find one another, how many mates individuals have, and how long mates remain associated.

Medfly Mediterranean fruit fly, *Ceratitis capitata*, an economic pest of the family Trypetidae whose larvae feed on a vast variety of cultivated and wild fruits.

Mediterranean Climate A maritime climate with winter rain and summer drought.

Mediterranean Fruit Fly *See* Medfly.

Meiosis Two successive divisions of cell nuclei, resulting in the diploid chromosome number being reduced to the haploid number, and in the course of which recombination occurs.

Melanic Very dark colored or black.

Mendelian Genetics The transmission of traits according to the rules first discovered by the Austrian monk Gregor Mendel.

Mesic An environment with ample rainfall and a well-drained soil.

Mesozoic Era The geological era 225–65 million years ago; the age of the dinosaurs.

Microclimate The climate in the immediate vicinity of an organism or in a specific habitat.

Microcosm A tiny ecosystem in a container explored as an analogy to large ecosystems.

Microevolution Evolution within populations, involving primarily changes in the frequency of different alleles.

Microhabitat The part of the more general habitat frequented by an organism. One habitat in which California thrashers are found is chaparral; their microhabitat is close to the ground in or near bushes.

Midparent The average of the two parents.

Migration The movement of individuals; it may change the frequency of genes in a population.

Milpa Agriculture Slash and burn farming.

Mimic An organism whose appearance resembles that of another kind of organism.

Mimicry Phenomenon in which one organism evolves a resemblance to another.

Mirid Bugs Insects of the order Hemiptera, family Miridae. Largely herbivores with sucking mouthparts, although some prey on other insects.

Mitosis Division of a cell nucleus in which the chromosomes divide and each daughter nucleus ends up with exactly the same chromosome complement as the original nucleus.

Moa One of a variety of species of wingless birds from New Zealand, related to ostriches, kiwis, emus, and cassowaries. The last moas were killed off many centuries ago by the Polynesians when they arrived in New Zealand.

Model In mimetic associations, an organism that is mimicked by another kind of organism.

Molecular Biology The branch of biology that attempts to understand the chemical mechanisms of life.

Molly A small fish of the genus *Poecilia (Mollienesia)* whose eggs hatch internally so that the young are born alive. They are popular with aquarists and thrive in alkaline water.

Momentum In physics, the mass times the velocity of an object.

Monoecious Plants Plants in which flowers of both sexes—male flowers that produce pollen and female flowers that bear ovules—are found on the same plant. *Compare* Dioecious plants.

Monogamy A mating system in which one male and one female mate with one another exclusively.

Mortality Death rate.

Mouthbrooder A fish in which the female carries the eggs and young in her mouth to protect them.

Müllerian Mimicry The mutual resemblance of distasteful organisms that presumably makes it easier for unintelligent predators to learn what patterns to avoid. In Müllerian mimicry complexes, all members are both models and mimics.

Mutant An individual carrying a mutation.

Mutation An allele that has changed into another, ordinarily recognized when the common allele in the population changes to a less common one. Also used for major changes in the structure of chromosomes.

Mutualism A situation in which an association between two or more species is beneficial to all involved.

Mycorrhizae Fungi that form a mutualistic relationship with the roots of plants, aiding them in the uptake of nutrients in return for energy.

Nagana A disease of ungulates in Africa, related to human sleeping sickness, to which domestic cattle are susceptible. The protozoan agent, like that of the human disease, is a trypanosome.

Naked Mole-Rats *Heterocephalus glaber* (Bathyergidae); also known as *sand puppies*.

These burrowing, almost hairless rodents live in colonies organized much like those of social insects; they occur in northeast Africa.

Natality Birthrate.

Natural Selection The differential reproduction of genotypes, where the differential is large enough to exceed the effect of genetic drift.

Nectivore An organism that feeds on nectar.

Nekton Organisms of the open oceans that are strong enough swimmers to overcome currents, such as squid, sea turtles, and great whales.

Neon Tetra *Hyphessobrycon innesi* (Characidae). A small, slender, brilliantly colored fish that lives in acid fresh waters in South America; it is popular with aquarists.

Neritic Zone Life zone in the ocean near shore, encompassing the shallow waters over the continental shelves. *See* Pelagic zone.

Net Primary Production Gross primary production minus that respired by the photosynthesizing plants themselves. *Compare* Gross primary production.

Neutralists Geneticists who believe that most naturally occurring variation in enzymes does not affect their functioning enough for selection to differentiate among them. *Compare* Selectionists.

Neutrality Controversy The argument between selectionists and neutralists over the significance of the large amount of genetically based variation in enzymes that occurs in natural populations. *See* Neutralists, Selectionists.

Niche The way in which an organism obtains its resources; the "occupation" of an organism as contrasted with its "address" or habitat.

Niche Theory A body of mathematical competition theory in which niche overlap is the basis of competition.

Nitrification Production of nitrites and nitrates as microorganisms break down organic compounds that contain nitrogen.

Nonregulators Animals that exert no physiological or behavioral control over their body temperature, such as animals that burrow in soil and very small invertebrates.

Nonrenewable Resource A resource that is used up over time.

Nothofagus A southern beech tree (Fagaceae) with a fragmented distribution in the southern Andes, New Guinea, and Australia.

Novel Mutation A change to an allele that is not already in the population.

Nubian Vulture *See* Lappet-faced vulture.

Nucleotides The building blocks of deoxyribonucleic acid (DNA).

Null Hypothesis The hypothesis that the effect one is looking for has not occurred.

Numerical Response The rate at which predators are born as a function of the number of prey available; the reaction of the predator population to the availability of prey. *See* Functional response.

Nutrient A substance necessary for the normal growth, development, and reproduction of an organism. Often used in the more restricted sense of inorganic nutrients taken up by plants from air or water.

Nutrient Cycles The mostly closed pathways followed by nutrients as they move through ecosystems.

Oligotrophic Deficient in plant nutrients. *Compare* Eutrophic.

Omnivores Animals that eat both plant and animal materials.

Operational Sex Ratio Ratio of sexually ready males to fertilizable females.

Opuntia Prickly pear cactus, which became a plague when introduced into Australia.

Order (Taxonomic) A major category above the family and below the phylum. Butterflies and moths belong to the order Lepidoptera, wolves and lions to the order Carnivora.

Ordination Investigation of the distribution of species along environmental gradients.

Organic Diversity The variety of plants, animals, and microorganisms that inhabit Earth. Often confined to the diversity of species, although diversity within species is also important ecologically and evolutionarily.

Origin In soils, refers to nature of the parent materials and to whether the soil is developed at the source of the parent materials or at a distant site.

Orthocarpus A secondary larval food plant for some populations of the Bay checkerspot butterfly; Owl's clover, in the family Scrophulariaceae.

Oryx *Oryx gazella*, a medium-sized African antelope with long, slender, rather straight horns.

Outbreeding Mating system in which adults mate with relatives less often than expected by chance. *Compare* Inbreeding.

Owl's Clover *See Orthocarpus.*

Paca *Agouti paca*; a large rodent, more than 2 ft long, of the rain forests of tropical America.

Pairwise Coevolution Coevolution in which interaction between organisms is sufficiently close that each species is the primary actor in the evolution of the other.

Paleontology The study of fossil organisms.

Paleozoic Era The geological era roughly 620–230 million years ago, during which the main groups of organisms differentiated.

Pangaea The giant single continent that existed prior to some 200 million years ago, at which time it gradually broke up and plate tectonics began to generate the configuration of continents that exist today.

Pangolin A superficially anteaterlike mammal of the genus Manis found in Asia and Africa. Its body is covered with overlapping scales, making it resemble walking pine cones. It feeds mainly on termites and ants.

Paradise Fish The most common cleaner wrasse, *Labroides dimidiatus*.

Paramecium A genus of protozoa (single-celled organisms) common in pond water, sometimes called *slipper animalcules* because of their rounded, elongated shape and the presence of a deep groove leading to the "mouth."

Parasite A predator that feeds on its prey, usually called a *host*, without killing it immediately (if ever). Parasites are usually much smaller than their prey and may live inside them, on their surface, or (in the case of brood parasites) in the same nest with the host's young.

Parasitoid An insect predator that develops inside its host, eventually and inevitably killing it.

Parasol Ants Leaf-cutter ants.

Parental Investments The resources individuals devote to offspring.

Pascal A unit of pressure; 10^6 Pascals or 1 Mega Pascal (1 MPa) equal 9.87 atm.

Passion Fruit Vine A vine of the genus *Passiflora* (Passifloraceae); larval food plant of longwing butterflies.

Pathogenic Organism An organism that causes a disease.

Pecking Order *See* Dominance hierarchy.

Pelagic Zone Life zone of the ocean consisting of the open waters beyond the continental shelves. *See* Neritic zone.

Pentatomid Bug An insect in the family Pentatomidae of the order Hemiptera.

Peppered Moth *Biston betularia* in the family Lymantridae.

Perennial An organism whose life cycle is longer than 1 year.

Perfect Flower One that is hermaphroditic. See Hermaphrodite.

Phenotype The appearance and behavior of an organism; results from the expression of a genotype in a given environment. *Compare* Genotype.

Pheromone Chemical messages used for communication among individuals.

Phosphorus An element essential to life because of its role in energy metabolism.

Photosynthesis The process by which green plants take up carbon dioxide and convert it to carbohydrates.

Photosynthetic Action Spectrum Graph of how fast 1 cm^2 of leaf photosynthesizes as a function of the color (wavelength) of light shining on it.

Phyletic Evolution Change within an evolutionary line. *Compare* Speciation.

Phylogenetic Tree A treelike diagram showing the time of splitting of various evolutionary lines and sometimes expressing their degree of divergence; a pictorial representation of a group's phylogeny.

Phylogeny The evolutionary history of a group.

Physiological Ecology The subdiscipline of ecology concerned with the dynamic relationship of individuals to their physical environments and resources.

Phytophthora Infestans Potato blight; the fungus responsible for the Irish potato famine in the nineteenth century.

Phytoplankton Plankton that photosynthesize.

Pit Viper A poisonous snake of the Western Hemisphere in the subfamily Crotalinae of the family Viperidae. It includes the rattlesnake, cottonmouth, bushmaster, and fer-de-lance. The pits are sensory organs on the face that detect the infrared (heat) radiation of the prey.

Plankton Mostly tiny organisms that are so small as to be carried with the currents.

Plantago Erecta An annual, native plantain of the west coast of the United States that serves as the primary larval food plant of some populations of the Bay checkerspot butterfly.

Plantain A plant of the family Plantaginaceae. Some of these plants are widespread weeds. Also, a wild banana.

Plasmodium The genus of protozoa that cause malaria. Four species infect human beings, all causing somewhat different symptoms: *P. malariae, P. falciparum, P. vivax,* and *P. ovale.*

Plate Tectonics The movement of crustal plates of Earth, carrying with them continents that thus drift. *See* Continental drift.

Polyandry A mating system in which one female mates with more than one male.

Polygyny A mating system in which one male mates with more than one female.

Polymorphism The presence in a population of two or more distinct forms of a gene.

Population A collection of organisms of the same species.

Population Biology The subdiscipline of biology that deals with populations of organisms.

Population Dynamics The study of constancy and change in population size.

Poster Colors Bright colors of many reef fishes, once thought to be territorial warnings.

Precocial Birds Birds whose young are able to move about and feed upon hatching. *See* Altricial birds.

Predator An organism, usually an animal, that obtains most of its food by eating other animals. The term is sometimes extended to consider herbivores as predators on plants.

Prey What a predator eats.

Prickly Pear Cactus *See Opuntia.*

Pride A family group of lions, defined by the continued presence of certain females, males being more or less temporary adjuncts controlling access to the females until ousted by other males.

Primary Consumer An herbivore.

Primary Succession Succession on a newly exposed site, such as land exposed by a retreating glacier or an emerging volcanic island.

Producer A green plant (almost always) or other organism that can bind energy from inorganic sources into the chemical bonds of organic compounds. *Compare* Consumer.

Productivity Roughly the dry weight of plant material produced in an area per unit time.

Pterosaurs Gliding reptiles that were contemporaneous with dinosaurs.

Punctuated Equilibrium The view that in phylogenies there have been intense bursts of speciation punctuated by long periods with little significant evolution.

Pupa The resting stage of an insect in which the structure of a larva is broken down and an adult assembled. Found in insects with complete metamorphosis, primarily the orders containing beetles; butterflies and moths; bees, wasps and ants; and flies.

Pyrethrum A plant chemical defense that is

also used by humans as an insecticide. It is extracted from daisylike flowers and is relatively nontoxic to warm-blooded animals.

Queen In Hymenoptera and termites, the individual that lays eggs; contrasted with other individuals that are sterile and that provision the nest and carry out other tasks that assist the queen.

Quinine An alkaloid extracted from cinchona bark; it is an effective treatment for malaria.

Radioisotope A radioactive isotope.

Rain Forest Tropical woodland with an annual rainfall of at least 100 in. and marked by lofty broad-leaved evergreen trees forming a continuous canopy. *See* Temperate rain forest.

Ramet An ecological unit; the entity noticed in the field as an individual—for example, a bamboo shoot from an underground runner.

Random Mating Mating system whereby diploid adults mate at random independently of their genotype at specific loci.

Random Union of Gametes Mating system whereby gametes unite at random independently of their genotype at specific loci.

Recombination Mixing of genes from paternal and maternal chromosomes during meiosis.

Recurrent Mutation A mutation from an allele to another allele that is already in the gene pool.

Relative Fitness Fitness coefficient scaled so that the largest equals 1.

Relative Wind Airflow; like airplanes, birds generate lift by having their wings intercept the airflow at an angle, known as the *angle of attack*. *See* Angle of attack.

Renewable Resource A resource that is replenished. *Compare* Nonrenewable resource.

Residual Soil Soil developed at the source of its parent materials. *See* Transported soil.

Resource Any substance, object, or position consumed or occupied by an organism during its growth, maintenance, or reproduction.

Resource Partitioning The use of different, or mostly different, types of limiting resources by two or more species of organisms.

Resting Schools Schools of fishes that rest on coral reefs during the day and feed elsewhere at night.

Reynolds Number An index to describe the relative amount of the pressure and friction components in drag, used for objects in any fluid, including water and air.

Rinderpest A serious viral disease of ruminants that decimated herds of cattle and native ruminants in Africa.

Root Nodules Tiny bulbous structures on the roots of legumes that harbor nitrogen-fixing bacteria.

Rule of Constant Yield Observation that the yield from a plot is independent of the sowing densities; applies to a range of intermediate sowing densities.

Ruminant An ungulate with a complex three- or four-chambered stomach with a large first chamber called the rumen. In the rumen, cellulose is broken down with the aid of mutualistic microorganisms. Ruminants repeatedly regurgitate and rechew the contents of the rumen to speed the digestive process—a behavior known as "chewing the cud." Ruminants include cattle, antelopes, sheep, and deer.

Rutting Period Estrus period. *See* Estrus.

Sabre-Toothed Blennies A group of marine fishes of the family Blenniidae that includes cleaner mimics and slash-and-dash predators.

Savanna A tropical grassland containing widely spaced trees.

Scatophaga A dung fly of the family Scatophagidae.

Schistosomiasis A serious human disease caused by parasitic trematode worms that live inside the blood vessels. The disease is contracted by contact with water containing the immature forms of the worms, which have aquatic snails as intermediate hosts. The African variety of schistosomiasis is called *bilharzia*.

Search Image The image presumed to be formed in the minds of predators, in response to which they tend to concentrate on a common type of prey. People searching for a cryptic animal often quickly find additional individuals after the first is spotted, presumably because they form search images also.

Secondary Succession Succession following the disturbance of a preexisting successional stage or climax.

Selectionists Geneticists who believe that most naturally occurring variation in enzymes causes changes that affect the functioning of those enzymes sufficiently to be "detectable" by selection. *Compare* Neutralists.

Selfing Producing zygotes by self-fertilization.

Sere An entire successional sequence, from bare substrate to climax.

Serpentine Soil A soil with an unusual mineral composition characteristic of sites occupied by Bay checkerspot butterfly populations. It is a refuge for native plants, apparently because the European weeds that have covered most of California seem to have difficulty invading it.

Sex Ratio Proportion of males to females. *See* Operational sex ratio.

Sexual Reproduction Reproduction that involves the fusion of gametes.

Sib Selection A pattern of artificial selection in which the breeding stock for each generation consists of the brothers and sisters of the individuals that showed the character to be selected for.

Sickle-Cell Anemia A fatal disease caused by distortion of the red blood cells. It occurs in individuals, usually of African origin, who are homozygous for a gene that changes one amino acid residue in the protein hemoglobin. Heterozygotes, however, are resistant to a dangerous kind of malaria found in Africa.

Silurian Period A period of the mid-Paleozoic era, about 430–395 million years ago. During the Silurian, vascular plants and modern algae and fungi first appeared in the fossil record.

Sitka Spruce *Picea sitchensis*, the largest of the spruces, at home in the Pacific Northwest, where the summers are cool and damp.

Size Escape Attaining a size large enough so that predators cannot be effective.

Sleeping Sickness A serious human disease found in Africa and caused by protozoa called *trypanosomes* that are transmitted by tsetse flies. It is related to nagana. *See* Nagana.

Slipper Limpet A marine mollusk of the genus *Crepidula*, which sometimes is a serious pest in oyster beds because it competes with the oysters for sitting space.

Social Behavior Behavior involving unrelated or distantly related individuals of the same population; it excludes interactions between close relatives such as parents and offspring or siblings.

Social Insects Wasps, bees, ants, and termites, which often live in complex societies, with individuals performing specialized tasks, such as defending the nest or caring for the young.

Sociobiology The evolutionary study of social behavior, including human behavior.

Solar Tracking Trait of certain plants in which plants orient their leaves to face the sun throughout the entire day.

Southern Beech Various species of the genus *Nothofagus*.

Speciation The splitting process of evolution that creates new kinds of organisms. *Compare* Phyletic evolution.

Species A distinct kind of organism. When populations of two kinds occur together without interbreeding, they are considered different species. When the populations do not occur together, the judgment of whether they belong to different species or are just geographic varieties of the same species can be arbitrary.

Species Diversity The number of species in an area.

Species Selection The differential diversification of species; some species become extinct or do not continue to diversify, whereas others continue to split into new species.

Specific Heat Number of calories required to raise 1 g of a substance 1°C.

Sperm Precedence Event in which sperm from later matings tend to displace those from earlier matings.

Stabilizing Selection Selection in which individuals near the average for a trait reproduce more than those far from the average.

Stable Age Distribution The age distribution ultimately attained by a population in a constant environment.

Stasis A condition in which species remain essentially unchanged for stretches of geological time—often millions of years.

Steady State A condition where an input rate balances an output rate; also called an *equilibrium* or *fixed point*.

Steady-State Temperature A constant temperature that results when the input of heat in an animal balances its output rate of heat.

Stomate A tiny opening in the underside and sometimes the top of leaves and stems through which gases enter and leave a plant.

Strain The stretching of a material, such as a tendon; the change in length caused by the force.

Stress The force on an object, tending to produce deformation, as in the force on a tendon.

Subduction Condition in which one tectonic plate plunges under another and is melted down into magma again.

Succession A regular progression of communities replacing each other on a site until a relatively permanent climax community is established.

Survivorship Curve A curve describing the probability of living to age x or more.

Suspension Feeders Organisms that extract particles of food from the water above the substrate.

Symbiosis A physically close relationship between individuals of two different species; includes parasitism (in which one is helped and the other harmed), commensalism (in which one is unaffected and the other aided), and mutualism (in which both are helped).

Sympatric Having overlapping distributions; occurring together in the same area.

Sympatric Speciation Speciation without geographic isolation.

Syngamy The fusion of sex cells.

Systems Ecology A focus on understanding large and complex systems, such as the dynamics of forests, with mathematical models based on simulation models, using a digital computer.

Taiga Biome of coniferous forests in the mountains of the Northern Hemisphere.

Taxon (*plural:* Taxa) A taxonomic group that is distinct enough to be assigned a name and placed in an official taxonomic category.

Taxon Cycle A postulated series of changes in the phyletic evolution of colonizing species on isolated islands.

Taxonomy The science of the classification of organisms.

Temperate Rain Forest Woodland of temperate but usually rather mild climatic areas with heavy rainfall, usually including numerous kinds of trees and distinguished from tropical rain forest by the presence of a dominant tree species. *See* Rain forest.

Tensile Strength Force required to break material, such as a bone or tendon.

Territorial Signals The ways in which territory holders announce their ownership.

Territory A defended portion of an organism's home range.

Tertiary Period The first and longest period of the Cenozoic, starting some 65 million years ago and ending 2 million years ago at the start of the Quaternary Period (Pleistocene epoch).

Tetraploid Having four sets of chromosomes in undifferentiated cells.

Thermocline The level in the water at which a rather abrupt transition occurs between the epilimnion and the hypolimnion, marked by a change in temperature.

Thomson's Gazelle *Gazella thomsoni.* A small, abundant East African antelope, similar to Grant's gazelle. *See* Grant's gazelle.

3/2 Thinning Rule The observation that the average weight of a plant varies inversely with the sowing density raised to the 3/2 power.

Topi *Damaliscus lunatus*; a medium-sized antelope with dark, often black, upper legs, which is widely distributed in sub-Saharan Africa. Also known as *sassaby, hirola,* and *Hunter's antelope*.

Transpiration The loss of water vapor from plants, primarily through the small pores (stomates) in the leaves through which carbon dioxide enters.

Transported Soil Soil developed at a distant site from its parent materials. *See* Residual soil.

Trapline To move along a regular route while harvesting scattered resources.

Tribolium The genus of flour beetles (family Tenebrionidae) that have been used extensively in laboratory studies of competition.

Trophic Level Feeding level—for example, producer, herbivore, and first-level carnivore.

True Bugs Members of the insect order Hemiptera.

Trypanosome Parasitic protozoans of the genus *Trypanosoma*. At one stage of their life cycle they are slender, more or less pointed at both ends, and on one end bear a long slender process known as a *flagellum*, with which they pull themselves along.

Tsetse Fly A fly of the genus *Glossina* (Muscidae). Endemic to Africa, tsetse flies are ferocious, tough bloodsuckers that transmit human sleeping sickness and nagana, both caused by trypanosomes.

Tundra Treeless plains north of the taiga in the Northern Hemisphere.

Turgor Pressure Pressure of the cell against its cell wall as the result of water moving by osmosis into the cell from outside; this pressure gives nonwoody parts of a plant most of their rigidity.

Twin-Berry *Lonicera involucrata* (Caprifoliaceae), the bushy relative of the honeysuckle that serves as larval food plant for Gillette's checkerspot.

Tyrannosaurus The largest and most impressive predatory dinosaur. This creature, which stood 6 m high and weighed about 8 tons, lived some 70 million years ago.

Understory The undergrowth in a forest.

Ungulate A hooved animal. Ungulates include horses, cattle, elephants, rhinos, and pigs.

Vicariance Biogeography Biogeography focused on the development of barriers primarily by plate tectonic processes.

Viscosity The resistance of a liquid to flowing over itself.

Vital Rates Birth and death rates.

Volterra Principle The proposition that if some factor causes equal rises in death rates in both predators and prey in a predator–prey system, the predator population size will drop disproportionately.

Water Chevrotain *Hyemoschus aquaticus* (Tragulidae). Rabbit-sized relatives of deer found in the rain forests of central Africa.

Water Potential The chemical potential of water.

Waterbuck *Kobus ellipsiprymnus*. A heavy bodied, medium-sized antelope widespread in sub-Saharan Africa. Only males are horned; the rump is characteristically whitish or white-ringed.

Whiptail Lizard A New World lizard of the genus Cnemidophorus (Teiidae), characterized by a slender, whiplike tail.

White-Backed Vulture *Gyps africanus*. Also known as the *griffon vulture*.

White-Eyes Old World birds of the genus *Zosterops* (Zosteropidae), characterized by a ring of white surrounding the eye.

White-Headed Vulture *Aegyius occipitalis*.

Wild Type The commonest phenotype, genotype, or allele in a population.

Xeric An environment in which production by green plants is limited by lack of water.

Zebu Cattle An Asiatic breed of cattle with a humped back and large dewlap. It is quite resistant to heat and insect attack.

Zygote The diploid cell resulting from the fusion of two haploid gametes; the fertilized egg.

BIBLIOGRAPHY

Abele, L. G., and S. Gilchrist, 1977. Homosexual rape and sexual selection in acanthocephalan worms. Science 197:81–83.

Abrams, P. 1984. Recruitment, lotteries, and coexistence in coral reef fish. Am. Nat. 123:44–55.

Adolph S., and J. Roughgarden. 1983. Foraging by passerine birds and *Anolis* lizards on St. Eustatius (Neth. Antilles), with implications for interclass competition. Oecologia 56:313–317.

Aker, C. L., and D. Udovic. 1981. Oviposition and pollination behavior of the yucca moth, *Tegeticula maculata* (Lepidoptera: Prodoxidae), and its relation to the reproductive biology of *Yucca whipplei* (Agavaceae). Oecologia 49:96–101.

Alatalo, R. V. 1981a. Habitat selection of forest birds in the seasonal environment of Finland. Annales Zoologici Fennici 18:103–114.

———. 1981b. Interspecific competition in tits *Parus* spp. and the goldcrest *Regulus regulus:* foraging shifts in multispecific flocks. Oikos 37:335–344.

———. 1982. Evidence for interspecific competition among European tits *Parus* spp.: a review. Annales Zoologici Fennici 19:309–317.

Alcock, J. 1979. Multiple mating in *Calopteryx maculata* (Odonata: Calopterygidae) and the advantage of non-contact guarding by males. J. Nat. Hist. 13:439–446.

———. 1982. Postcopulatory mate guarding by males of the damselfly *Hetaerina vulnerata* Selys (Odonata: Calopterygidae). Anim. Behav. 30:99–107.

Alerstram, T., S. Nilsson, and S. Ulfstrand. 1974. Niche organization during winter in woodland birds in southern Sweden and the island of Gotland. Oikos 25:321–330.

Alevizon, W. S. 1976. Mixed schooling and its possible significance in a tropical western Atlantic parrotfish and surgeonfish. Copeia 4:796–798.

Alexander, R. 1967. Functional design in fishes. Hutchinson, London.

Alexander, R. D. 1961. Aggressiveness, territoriality and sexual behavior in field crickets (Orthoptera: Gryllidae). Behaviour 17:131–217.

———. 1974. The evolution of social behavior. Annu. Rev. Ecol. Syst. 5:325–383.

Alexander, R. D., and P. W. Sherman. 1977. Local mate competition and parental investment in social insects. Science 196:494–500.

Alexander, R. M. 1968. Animal mechanics, University of Washington Press, Seattle.

Allan, J. 1973. Competition and the relative abundance of two cladocerans. Ecology 54:484–498.

Allee, W. C., A. E. Emerson, O. Park, T. Park, and K. P. Schmidt. 1949. Principles of Animal Ecology. W. B. Saunders & Co., Philadelphia.

Allen, G. R. 1975. The anemonefishes: their classification and biology. 2d ed. T. F. H. Publications, Neptune City, N.J.

Allison, A. C. 1982. Coevolution between hosts and infectious disease agents and its effects on virulence. Pages 245–267 *in* R. M. Anderson and R. M. May, eds. Population biology of infectious diseases. Springer-Verlag, New York.

Alvarez, L. W., W. Alvarez, F. Asaro, and H. V. Michel. 1980. Extraterrestrial cause for the Cretaceous-Tertiary extinction. Science 208:1095–1108.

Alvarez, W., L. W. Alvarez, F. Asaro and H. V. Michel. 1984a. The end of the Cretaceous: sharp boundary or gradual transition? Science 223:1183–1186.

Alvarez, W., E. G. Kauffman, F. Surlyk, L. W. Alvarez, F. Asaro, and V. Michel. 1984b. Impact theory of mass extinctions and the invertebrate fossil record. Science 223:1135–1141.

Anderson, D. J. 1967. Studies on structure in plant communities. IV. Cyclical succession in *Dryas* communities from North-West Iceland. J. Ecol. 55:629–635.

Anderson, G., A. Ehrlich, P. Ehrlich, J. Roughgarden, B. Russell, and F. Talbot. 1981. The community structure of coral reef fishes. Am. Nat. 117:476–495.

Anderson, G. D., and D. J. Herlocker. 1973. Soil factors affecting the distribution of the vegetation types and their utilization by wild animals in the Ngorongoro Crater, Tanzania. J. Ecol. 61:627–653.

Anderson, M. 1982. Female choice selects for extreme tail length in a widowbird. Nature 299:818–820.

Anderson, R. M., and R. M. May. 1982. Coevolution of hosts and parasites. Parasitology 85:411–426.

Anderson, W. 1971. Genetic equilibrium and population growth under density-dependent selection. Am. Nat. 105:489–498.

Andersson, M., and C. G. Wiklund. 1978. Clumping versus spacing out: experiments on nest predation in fieldfares *(Turdus pilaris)*. Anim. Behav. 26:1207–1212.

Andrewartha, H. G., and L. C. Birch. 1954. The distribution and abundance of animals. University of Chicago Press, Chicago.

Aneshansley, D., T. Eisner, J. M. Widom, and B. Widom. 1969. Biochemistry at 100°C: the explosive discharge of bombardier beetles *(Brachinus)*. Science 165:61–63.

Archibald, J. D., and W. A. Clemens. 1982. Late Cretaceous extinctions. Am. Scientist 70:377–385.

Archley, W. R., and D. Woodruff. 1981. Evolution and speciation. Cambridge University Press, Cambridge.

Armitage, K. B. 1955. Territorial behavior in fall migrant Rufous hummingbirds. Condor 57:239–240.

———. 1962. Social behaviour of a colony of the yellow-bellied marmot *Marmota flaviventris*. Anim. Behav. 10:319–331.

———. 1975. Social behavior and population dynamics of marmots. Oikos 26:341–354.

———. 1981. Sociality as a life history tactic of ground squirrels. Oecologia (Berl.) 48:36–49.

———. 1982a. Marmots and coyotes: behavior of prey and predator. J. Mammal. 63(3):503–505.

———. 1982b. Social dynamics of juvenile marmots: role of kinship and individual variability. Behav. Ecol. Sociobiol. 11:33–36.

———. 1984. Recruitment in yellow-bellied marmot populations: kinship, philopatry, and individual variability. Pages 377–403 *in* J. O. Murie and G. R. Michener, eds. The biology of ground-dwelling squirrels: annual cycles, behavioral ecology, and sociality. University of Nebraska Press, Lincoln.

Armitage, K. B., and D. W. Johns. 1982. Kinship, reproductive strategies and social dynamics of yellow-bellied marmots. Behav. Ecol. Sociobiol. 11:55–63.

Askenmo, C., A. von Brömssen, A. Ekman, and C. Jansson. 1977. Impact of some wintering birds on spider abundance in spruce. Oikos 28:90–94.

Askew, R. R. 1971. Parasitic insects. American Elsevier, New York.

Atchley, W. R., and D. Woodruff, eds. Evolution and Speciation: Essays in Honor of M. J. D. White. Cambridge Univ. Press, New York, 1981.

Atjay, G. L., P. Ketner, and P. Duvigneaud. 1979. Terrestrial primary production and phytomass. Pages 129–181 *in* B. Bolin, E. T. Degens, S. Kempe, and P. Ketner, eds. The carbon cycle. Wiley, New York.

Ayala, F. 1965. Evolution of fitness in experimental populations of *Drosophila serrata*. Science 150:903–905.

———. 1968. Evolution of fitness. II. Correlated effects of natural selection on the productivity and size of experimental populations of *Drosophila serrata*. Evolution 22:55–65.

———. 1969. Evolution of fitness. V. Rate of evolution of irradiated populations of *Drosophila*. Proc. Natl. Acad. Sci. USA 63:790–793.

Babich, H., D. L. Davis, and G. Stotzky. 1980. Acid precipitation. Environment 22:6–41.

Bailey, N. T. J. 1964. The elements of stochastic processes with applications to the natural sciences. Wiley, New York.

Baker, H. G., and P. D. Hurd. 1968. Intrafloral ecology. Annu. Rev. Entomol. 13:385–415.

Baker, M. C. 1975. Song dialects and genetic differences in white-crowned sparrows *(Zonotrichia leucophrys)*. Evolution 29:116–244.

———. 1982a. Genetic population structure and vocal dialects in *Zonotrichia* (Emberizidae). Pages 209–235 *in* D. E. Kroodsma and E. H. Miller, eds. Acoustic communication in birds. Vol. 2. Academic Press, New York.

———. 1982b. Vocal dialect recognition and population genetic consequences. Am. Nat. 22:561–569.

———. 1983a. Genetic population structure and vocal dialects in *Zonotrichia* (Emberizidae). Pages 209–235 *in* D. E. Kroodsma and E. H. Miller, eds. Acoustic communication in birds. Vol. 2. Academic Press, New York.

———. 1983b. The behavioral response of female Nuttall's white-crowned sparrows to male song of natal and alien dialects. Behav. Ecol. Sociobiol. 12:309–315.

Baker, M. C., L. R. Mewaldt, and R. M. Stewart.

1981. Demography of the white-crowned sparrow *(Zonotrichia leucophrys nuttallii).* Evolution 62:636–644.

Baker, M. C., D. B. Thompson, G. L. Sherman, M. A. Cunningham, and D. F. Tomback. 1982a. Allozyme frequencies in a linear series of song dialect populations. Evolution 36:1020–1029.

Baker, M. C., G. L. Sherman, T. C. Theimer, and D. C. Bradley. 1982b. Population biology of white-crowned sparrows: residence time and local movements of juveniles. Behav. Ecol. Sociobiol. 11:133–137.

Baker, M. C., M. A. Cunningham, and A. D. Thompson, Jr. 1984. Cultural and genetic differentiation of two subspecies of white-crowned sparrow. Condor 86:359–367.

Baker, R. R. 1978. The evolutionary ecology of animal migration. Holmes and Meier, New York.

Bakker, R. T. 1972. Anatomical and ecological evidence of endothermy in dinosaurs. Nature 238:81–85.

———. 1975. Dinosaur renaissance. Sci. Am. April:58–78.

———. 1980. Dinosaur heresy—dinosaur renaissance. Pages 351–462 *in* R. D. K. Thomas and E. Olson, eds. A cold look at the warm-blooded dinosaurs. Westview Press, Boulder, Colo.

———. 1983. The deer flees, the wolf pursues: incongruencies in predator–prey coevolution. Pages 350–382 *in* D. J. Futuyma and M. Slatkin, eds. Coevolution. Sinauer, Sunderland, Mass.

Bakker, R. T., and P. M. Galton. 1974. Dinosaur monophyly and a new class of vertebrates. Nature 248:168–172.

Balfour, D. 1983. Infanticide in the Columbian ground squirrel, *Spermophilus columbianus.* Anim. Behav. 31:949–950.

Ballard, R. D. 1977. Notes on a major oceanographic find. Oceanus 20:35–44.

Baptista, L. F. 1975. Song dialects and demes in sedentary populations of the white-crowned sparrow *(Zonotrichia leucophrys nuttalli).* Univ. Calif. Publ. Zool. 105:1–52.

Barash, D. P. 1973. The social biology of the Olympic marmot. Anim. Behav. Monogr. 6(3):171–245.

———. 1974. The evolution of marmot societies: a general theory. Science 185:415–420.

Barclay-Estrup, P., and C. H. Gimingham. 1969. The description and interpretation of cyclical processes in a heath community. J. Ecol. 57:737–758.

Barducci, T. B. 1972. Ecological consequences of pesticides used for the control of cotton insects in Cañete Valley, Peru. *In* M. T. Farvar and J. P. Milton, eds. The careless technology: ecology and international development. Natural History Press, Garden City, N.Y.

Barlow, G. W. 1974. Extraspecific imposition of social grouping among surgeonfishes (Pisces: Acanthuridar). J. Zool. (Lond.) 174:333–340.

Barnes, H., and H. Powell. 1950. The development, general morphology and subsequent elimination of barnacle populations of *Balanus crenatus* and *B. balanoides* after a heavy initial settlement. J. Anim. Ecol. 19:175–179.

Barnes, R. S. K. 1984. Estuarine biology. 2d ed. Edward Arnold, London.

Barnes, R. S. K., and R. N. Hughes. 1982. An introduction to marine ecology. Blackwell, Oxford.

Barnes, R. S. K., and K. H. Mann. 1980. Fundamentals of aquatic ecosystems. Blackwell, Oxford.

Baross, J. A., and J. W. Deming. 1983. Growth of "black smoker" bacteria at temperatures of at least 250°C. Nature 303:423–426.

Barrett, G. W., and R. Rosenberg, eds. 1981. Environmental monographs and symposia. Wiley-Interscience, New York.

Barrett, J. A. 1983. Plant-fungus symbioses. Pages 137–160 *in* D. J. Futuyma and M. Slatkin, eds. Coevolution. Sinauer, Sunderland, Mass.

Bateman, A. J. 1948. Intra-sexual selection in *Drosophila.* Heredity 2:349–368.

Bateson, P., ed. 1983. Mate choice. Cambridge University Press, Cambridge.

Bawa, K. S. 1980. Evolution of dioecy in flowering plants. Annu. Rev. Ecol. Syst. 11:15–39.

———. 1982. Seed dispersal and evolution of dioecism in flowering plants—a response to Herrera. Evolution 36:1322–1325.

Bayly, I. A. E., and W. D. Williams. 1973. Inland waters and their ecology. Longman, London.

Bayly, J. 1964. A revision of the Australasian species of the freshwater genus *Boeckella* and *Hemoboeckella* (Copepoda: Calanoida). Aust. J. Marine and Freshwater Res. 15:180–238.

Bazzaz, F. A. 1979. The physiological ecology of plant succession. Annu. Rev. Ecol. Syst. 10:351–371.

———. 1983. Characteristics of populations in relation to disturbance in natural and man-modified ecosystems. Pages 259–275 *in* H. A. Mooney and M. Godron, eds. Disturbance and ecosystems: components of response. Springer-Verlag, Berlin.

Bazzaz, F. A., and S. T. A. Pickett. 1980. Physiological ecology of tropical succession: a comparative review. Annu. Rev. Ecol. Syst. 11:287–310.

Beard, J. S. 1945. The progress of plant succession on the soufriere of St. Vincent. J. Ecol. 33:1–9.

———. 1953. The savanna vegetation of northern tropical America. Ecol. Monogr. 23:149–215.

———. 1967. Some vegetation types in tropical Australia in relation to those of Africa and America. J. Ecol. 55:271–290.

Beattie, A. J. 1983. The nest chemistry of two seed-dispersing ants species. Oecologia 56:99–103.

———. 1985. The evolutionary ecology of ant–plant mutualisms. Cambridge University Press, Cambridge.

Beattie, A. J., and D. C. Culver. 1982. Inhumation: how ants and other invertebrates help seeds. Nature 297:627.

Beattie, A. J., C. L. Turnbull, R. B. Knox, and T. Hough. n. d. The vulnerability of pollen and fungal spores to ant secretions: some evolutionary implications. (In preparation.)

Beattie, A. J., C. Turnbull, R. B. Knox and E. G. Williams. 1984. Ant inhibition of pollen function: a possible reason why ant pollination is rare. Am. J. Bot. 71:421–426.

Béland, P., and D. A. Russell. 1980. Dinosaur metabolism and predator/prey ratios in the fossil record. Pages 85–102 *in* R. D. K. Thomas and E. Olson, eds. A cold look at the warm-blooded dinosaurs. Westview Press, Boulder, Colo.

Bell, G. 1982. The masterpiece of nature: the evolution and genetics of sexuality. University of California Press, Los Angeles.

Bell, R. H. V. 1970. The use of the herb layer by grazing ungulates in the Serengeti. Pages 111–124 *in* A. Watson, ed. Animal populations in relation to their food sources. Blackwell, Oxford.

———. 1971. A grazing ecosystem in the Serengeti. Sci. Am. 225:86–94.

Belser, L. W. 1979. Population ecology of nitrifying bacteria. Annu. Rev. Microbiol. 33:309–333.

Belser, L. W., and E. L. Schmidt. 1978. Diversity in the ammonia-oxidizing nitrifier population of a soil. Appl. Envir. Microbiol. 34:403–410.

Bender, E. A., T. J. Case, and M. E. Gilpin. 1984. Perturbation experiments in community ecology: theory and practice. Ecology 65:1–13.

Benson, W. W. 1978. Resource partitioning in passion vine butterflies. Evolution 32:493–518.

Benton, M. J. 1979. Ectothermy and the success of dinosaurs. Evolution. 33:983–997.

Berenbaum, M. R. 1980. Adaptive significance of midgut pH in larval lepidoptera. Am. Nat. 115:138–146.

———. 1981. Patterns of furanocoumarin production and insect herbivory in a population of wild parsnip *(Pastinaca sativa L.)*. Oecologia 49:236–244.

Berenbaum, M. R., and P. Feeny. 1981. Toxicity of angular furanocoumarins to swallowtail butterflies: escalation in a coevolutionary arms race? Science 212:927–929.

Berger, J. 1978. Group size, foraging and antipredator ploys: an analysis of bighorn sheep decisions. Behav. Ecol. Sociobiol. 4:91–99.

Berggren, W. A., and J. A. Van Couvering, eds. 1984. Catastrophes and earth history. Princeton University Press, Princeton, N.J.

Bernabo, J. C. 1981. Quantitative estimates of temperature changes over the last 2700 years in Michigan based on pollen data. Quat. Res. 15:143–159.

Bernabo, J. C., and T. Webb III. 1977. Changing patterns in the Holocene pollen record of northeastern North America: a mapped summary. Quat. Res. 8:64–96.

Bernays, E. A. 1978. Tannins: an alternative viewpoint. Ent. Exp. Appl. 24:44–53.

Bertram, B. C. R. 1975. Social factors influencing reproduction in wild lions. J. Zool. 177:462–482.

———. 1976. Kin selection in lions and in evolution. Pages 281–301 *in* P. Bateson and R. Hinds, eds. Growing points in ethology. Cambridge University Press, Cambridge.

Bertram, B. C. R. 1979. Serengeti predators and their social systems. Pages 221–248 *in* A. R. E. Sinclair and M. Norton-Griffiths, eds. Serengeti: dynamics of an ecosystem. University of Chicago Press, Chicago.

Biebl, R. W. 1967. Protoplasmatische Okologie. Naturw. Rund. 20:248–252.

Bierzychudek, P. 1982. The demography of Jack-in-the-pulpit, a forest perennial that changes sex. Ecology 52:335–351.

Billings, W. D., and H. A. Mooney. 1968. The ecology of arctic and alpine plants. Biol. Rev. 43:481–529.

Birch, L. C., and P. R. Ehrlich. 1967. Evolutionary history and population biology. Nature 214:349–352.

Bird, R. D. 1930. Biotic communities of the aspen parkland of central Canada. Ecology 11:356–442.

Birks, H. J. B. 1980. British trees and insects: a test of the time hypothesis over the last 13,000 years. Am. Nat. 115:600–605.

Bligh, J., J. L. Cloudsley-Thompson, and A. G. MacDonald, eds. 1976. Environmental physiology of animals. Wiley, New York.

Bliss, L. C., G. M. Courtin, D. L. Pattie, R. R. Riewe, D. W. A. Whitfield, and P. Widden. 1973. Arctic tundra ecosystems. Annu. Rev. Ecol. Syst. 4:359–399.

Blydenstein, J. 1967. Tropical savanna vegetation of the llanos of Colombia. Ecology 48:1–15.

Boake, C. R. B., and R. R. Capranica. 1982. Aggressive signal in "courtship" chirps of a gregarious cricket. Science 218:580–582.

Boerner, R. E. J. 1982. Fire and nutrient cycling in temperate ecosystems. BioSci. 32:187–192.

Bohart, G. E. 1955. Gradual nest supersedure within the genus *Osmia*. Proc. Entomol. Soc. Wash. 57:203–204.

Bohor, B. F., E. E. Foord, P. J. Modreski, and D. M. Triplehorn. 1984. Mineralogic evidence for an impact event at the Cretaceous–Tertiary boundary. Science 224:867–869.

Bolin, B., P. J. Crutzen, P. M. Vitousek, R. G. Woodmansee, E. D. Goldberg, and R. B. Cook. 1983. Interactions in biogeochemical cycles. Pages 1–39 *in* B. Bolin and R. B. Cook, eds. The major biogeochemical cycles and their interactions. Wiley, New York.

Boltzmann, L. 1905. Populäre Schriften. Leipzig.

Booth, W. E. 1941. Revegetation of abandoned fields in Kansas and Oklahoma. Am. J. Bot. 28:415–422.

Boring, L. R., C. D. Monk, and W. T. Swank. 1981. Early regeneration of a clear-cut southern Appalachian forest. Ecology 62:1244–1253.

Bormann, F. H. 1976. An inseparable linkage: conservation of natural ecosystems and the conservation of fossil energy. BioSci. 26:754–760.

Bormann, F. H., and G. E. Likens. 1977. The fresh air-clean water exchange. Nat. Hist. Nov:63–71.

———. 1979. Catastrophic disturbance and the steady state in northern hardwood forests. Am. Sci. 67:660–669.

Bormann, F. H., G. E. Likens, D. W. Fisher, and R. S. Pierce. 1968. Nutrient loss accelerated by clear-cutting of a forest ecosystem. Science 159: 882–884.

Botkin, D. B. 1980. A grandfather clock down the staircase: stability and disturbance in natural ecosystems. Pages 1–9 *in* R. H. Waring, ed. Forests: fresh perspectives from ecosystem analysis. Proc. 40th Ann. Bio. Coll. Oregon State University Press, Corvallis.

Botkin, D. B., J. F. Janak, and J. R. Wallis. 1972. Some ecological consequences of a computer model of forest growth. J. Ecol. 60:849–872.

Botkin, D. B., and R. S. Miller. 1974. Complex ecosystems: models and predictions. Am. Sci. 62:448–453.

Bowers, M. A., and J. H. Brown. 1982. Body size and coexistence in desert rodents: chance or community structure? Ecology 63:391–400.

Bowers, M. D., I. L. Brown, and D. Wheye. 1985. Bird predation as a selective agent in a butterfly population. Evolution 39:93–103.

Bowman, R. I. 1961. Morphological differentiation and adaptation in the Galapagos finches. Univ. Calif. Publ. Zool. 58:1–302.

Boyce, M. 1984. Restitution of *r*- and *K*-selection as a model of density-dependent natural selection. Annu. Rev. Ecol. Syst. 15:427–447.

Boyce, M. S. 1979. Seasonality and patterns of natural selection for life histories. Am. Nat. 114:569–583.

Boyce, M. S., and D. J. Daley. 1980. Population tracking of fluctuating environments and natural selection for life histories. Am. Nat. 115:480–491.

Bragg, T. B., and L. C. Hurlbert. 1976. Woody plant invasion of unburned Kansas bluestem prairie. J. Range Mgmt. 29:19–29.

Branch, G. M. 1975a. Intraspecific competition in *Patella cochlear* Born. J. Anim. Ecol. 44:263–281.

———. 1975b. Mechanisms reducing intraspecific competition in *Patella* spp.: migration, differentiation and territorial behaviour. J. Anim. Ecol. 44:575–585.

Braun, E. L. 1950. Deciduous forests of eastern North America. Blakiston, Philadelphia.

Bray, J. R. 1961. Measurement of leaf utilization as an index of minimum level of primary consumption. Oikos 12:70–74.

Bray, J. R., and J. T. Curtis. 1957. An ordination of the upland forest communities of southern Wisconsin. Ecol. Monogr. 27:325–349.

Breedlove, D. E., and P. R. Ehrlich, 1968. Plant–herbivore coevolution: lycaenids and lupines. Science 162:671–672.

———. 1972. Coevolution: patterns of legume predation by a lycaenid butterfly. Oecologia 10:98–104.

Brenchley, G. 1982. Mechanisms of spatial competition in marine soft-bottom communities. J. Exp. Mar. Biol. Ecol. 60:17–33.

Brewer, P. G., K. W. Bruland, R. W. Eppley, and J. J. McCarthy. The global ocean flux study (GOFS): Status of the U.S. GOFS program. EOS 67:827–832.

Briand, F. 1983. Environmental control of food web structure. Ecology 64:253–263.

Briggs, J. C. 1968. Panama sea-level canal. Science 162:511–513.

———. 1974. Marine zoogeography. McGraw-Hill, New York.

Brockmann, H. J. 1984. The evolution of social behavior in insects. Pages 340–361 *in* J. R. Krebs and N. B. Davies, eds. Behavioral ecology: an evolutionary approach. 2d ed. Sinauer, Sunderland, Mass.

Brooks, J., and S. Dodson. 1965. Predation, body size and composition of plankton. Science 150:28–35.

Brooks, R. J., and J. B. Falls. 1975. Individual recognition by song in white-throated sparrows. I. Discrimination of songs of neighbors and strangers. Can. J. Zool. 53:879–914.

Brooks, R. R., R. D. Reeves, Y. Xing-Hua, D. E. Ryan, J. Holzbecher, J. D. Collen, V. E. Neall, and J. Lee. 1984. Elemental anomalies at the Cretaceous-Tertiary boundary, Woodside Creek, New Zealand. 1984. Science 226:539–542.

Brower, J. VZ. 1958a. Experimental studies of mimicry in some North American butterflies. l. *Danaus plexippus* and *Limenitis archippus archippus*. Evolution 12:32–47.

———. 1958b. Experimental studies of mimicry in some North American butterflies. 2. *Battus philenor* and *Papilio troilus, P. polyxenes* and *P. glaucus*. Evolution 12:123–136.

———. 1958c. Experimental studies of mimicry in some North American butterflies. 3. *Danaus gilippus berenice* and *Limenitis archippus floridensis*. Evolution 12:273–285.

———. VZ. 1960. Experimental studies of mimicry. IV. The reactions of starlings to different proportions of models and mimics. Am. Nat. 94:271–282.

Brower, L. P. 1970. Plant poisons in a terrestrial food chain and implications for mimicry theory. Pages 69–82 *in* K. L. Chambers, ed. Biochemical coevolution. Proc. 29th Ann. Biol. Coll. Oregon State University Press, Corvallis.

Brower, L. P., and J. VZ. Brower. 1962. The relative abundance of model and mimic butterflies in natural populations of the *Battus philenor* mimicry complex. Ecology 43:154–158.

Brower, L. P., F. H. Pough, and H. R. Meck. 1970. Theoretical investigations of automimicry. I. Single trial learning. Proc. Natl. Acad. Sci. USA 66:1059–1066.

Brower, L. P., W. N. Ryerson, L. L. Coppinger, and S. C. Glazier. 1968. Ecological chemistry and the palatability spectrum. Science 161:1349–1351.

Brown, I. L., and P. R. Ehrlich. 1980. The population biology of the butterfly, *Euphydryas chalcedona*. The structure of the Jasper Ridge colony. Oecologia (Berl.) 47:239–251.

Brown, J. 1975. Geographical ecology of desert rodents. Pages 315–341 *in* M. Cody and J. Diamond, eds. Ecology and evolution of communities. Harvard University Press, Cambridge, Mass.

Brown, J., Jr., D. Davidson, and O. Reichman. 1979. An experimental study of competition between seed-eating desert rodents and ants. Am. Zool. 19:1129–1143.

Brown, J. H., and A. C. Gibson. 1983. Biogeography. Mosby, St. Louis, Mo.

Brown, J. H., and A. Kodric-Brown. 1979. Convergence, competition and mimicry in a temperate community of hummingbird pollinated flowers. Ecology 60:1022–1035.

———. 1981. Reply to Williamson and Black's comment. Ecology 62:497–498.

Brown, J. H., and G. A. Lieberman. 1973. Resource utilization and coexistence of seed-eating desert rodents in sand dune habitats. Ecology 54:788–797.

Brown, J. L. 1974. Alternate routes to sociality in jays—with a theory for the evolution of altruism and communal breeding. Amer. Zool. 14:63–80.

Brown, J. L., and G. H. Orians. 1970. Spacing patterns in mobile animals. Annu. Rev. Ecol. Syst. 1:239–262.

Brown, L. 1981. Building a sustainable society. Norton, New York.

Brown, L. R., ed. 1984. State of the world, 1984. Norton, New York.

———. 1985. State of the world, 1985. Norton, New York.

Brown, R. H. S. 1961. Flight. Pages 289–305 *in* A. J. Marshall, ed. Biology and comparative physiology of birds. Vol. 2. Academic Press, New York.

Brown, W. L., Jr. 1973. A comparison of the Hylean and Congo-West African rain forest ant faunas. Pages 160–185 *in* B. J. Meggers, E. S. Ayensu, and W. D. Duckworth, eds. Tropical forest ecosystems in Africa and South America: a comparative review. Smithsonian Institution Press, Washington, D.C.

Brown, W. L., Jr., and E. Wilson. 1956. Character displacement. Syst. Zool. 7:49–64.

Buchmann, S. L., and J. P. Hurley. 1978. A biophysical model for buzz pollination in angiosperms. J. Theor. Biol. 72:639–657.

Buechner, H. K., and R. Schloeth. 1965. Ceremonial mating behavior in Uganda kob (*Adenota kob thomasi* Neumann). Z. Tierpsychol. 22:209–225.

Buell, M. F., and J. E. Cantlon. 1953. Effects of prescribed burning on ground cover in the New Jersey pine region. Ecology 34:520–528.

Bull, J. J. 1983. Evolution of sex determining mechanisms. Benjamin/Cummings, Menlo Park, Calif.

Burnet, M., and D. O. White. 1972. Natural history of infectious disease. 4th ed. Cambridge University Press, Cambridge.

Burt, W. 1946. The mammals of Michigan. University of Michigan Press, Ann Arbor.

Burton, R., and M. Feldman. 1983. Physiological effects of an allozyme polymorphism: Glutamate-pyruvate transaminase and response to hyperosmotic stress in the copepod, *Tigriopus californicus*. Biochem. Genetics 21:239–251.

Bush, G. L. 1969. Sympatric host race formation and speciation in fridivorous flies of the genus *Rhagoleris* (Diptera, Tephritidae). Evolution 23:237–251.

———. 1975. Modes of animal speciation. Ann. Rev. Ecol. Syst. 6:339–364.

Bush, G. L., S. M. Case, A. C. Wilson, and J. L. Patton. 1977. Rapid speciation and chromosomal evolution in mammals. Proc. Natl. Acad. Sci. USA 74:3942–3946.

Buss, L. 1979. Bryozoan overgrowth interactions—the interdependence of competition for space and food. Nature 281:475–477.

Buss, L., and J. Jackson. 1979. Competitive networks: nontransitive competitive relationships in cryptic coral reef environments. Am. Nat. 113:223–234.

Buzzati-Traverso, A. 1955. Evolutionary changes in components of fitness and other polygenic traits in *Drosophila melanogaster* populations. Heredity 9:153–186.

Bygott, J., B. Bertram, and J. Hanby. 1979. Male lions in large coalitions gain reproductive advantages. Nature 282:839–841.

Cade, T. J. 1982. The falcons of the world. Cornell University Press, Ithaca, N.Y.

Cain, A. J., and P. M. Sheppard. 1950. Selection in the polymorphic land snail, *Capaen nomoralis*. Heredity 4:275–294.

Cairns, J., Jr., M. L. Dahlberg, K. L. Dickson, N. Smith, and W. Waller. 1969. The relationship of freshwater protozoan communities to the MacArthur–Wilson equilibrium model. Am. Nat. 103:439–454.

Calvin, M., and J. A. Bassham. 1962. The photosynthetic carbon compounds. Benjamin, New York.

Camin, J. H., and P. R. Ehrlich. 1958. Natural selection in water snakes (*Natrix sipedon* L.) on islands in Lake Erie. Evolution 12:504–511.

Campanella, P. J. 1975. The evolution of mating systems in temperate zone dragonflies (Odonata: Anisoptera) II. *Libellula luctuosa* (Burmeister). Behaviour 54:278–310.

Campanella, P. J., and L. L. Wolf. 1974. Temporal leks as a mating system in a temperate zone dragonfly (Odonata: Anisoptera) I: *Plathemis lydia* (Drury). Behaviour 51:48–87.

Campbell, G. S. 1977. An introduction to environmental biophysics. Springer-Verlag, New York.

Caraco, T. 1982. A remark on Markov models of behavioural sequences. Anim. Behav. 30:1255–1256.

Caraco, T., and L. L. Wolf. 1975. Ecological determinants of group sizes of foraging lions. Am. Nat. 109(967):343–352.

Carefoot, T. 1977. Pacific seashores. University of Washington Press, Seattle.

Carey, F. G. 1982. A brain heater in the swordfish. Science 216:1327–1329.

Carey, M., and V. Nolan, Jr. 1975. Polygyny in indigo buntings: a hypothesis tested. Science 190:1296–1297.

Carpenter, F. L. 1978. A spectrum of nectar-eating communities. Am. Zool. 18:809–819.

Carpenter, F. L., and R. E. MacMillen. 1976. Threshold model of feeding territoriality and test with a Hawaiian honeycreeper. Science 194:639–642.

Case, T. J., J. Faaborg, and R. Sidell. 1983. The role of body size in the assembly of West Indian bird communities. Evolution 37:1062–1074.

Caughly, G. 1966. Mortality patterns in mammals. Ecology 47:906–918.

———. 1977. Analysis of vertebrate populations. Wiley, London.

Cavanaugh, C. M., S. L. Gardiner, M. L. Jones, H. W. Jannasch, and W. B. Waterbury. 1981. Procaryotic cells in the hydrothermal vent tubeworm *Riftia pachyptila* Jones: possible chemoautotrophic symbionts. Science 213:340–342.

Cavener, D., and M. Clegg. 1981. Evidence for biochemical and physiological differences between enzyme genotypes in *Drosophila melanogaster*. Proc. Natl. Acad. Sci. USA 78:4444.

Chadab, R., and C. W. Rettenmeyer. 1975. Mass recruitment by army ants. Science 188:1124–1125.

Chapman, H. H. 1945. The effect of overhead shade on the survival of loblolly pine seedlings. Ecology 26(3):274–282.

Charlesworth, B. 1977. Population genetics, demography, and the sex ratio. Pages 345–363 *in* F. Christianson and T. Fenchel, eds. Measuring selection in natural populations. Springer-Verlag, Berlin.

———. 1980. Evolution in age-structured populations. Cambridge University Press, Cambridge.

———. R. Lande, and M. Slatkin. 1982. A neo-Darwinian commentary on macroevolution. Evolution 36:474–498.

Charnov, E. L. 1976a. Optimal foraging: attack strategy of a mantid. Am. Nat. 110:141–151.

———. 1976b. Optimal foraging: the marginal value theorem. Theor. Pop. Biol. 9:129–136.

———. 1978. Sex-ratio selection in eusocial hymenoptera. Am. Nat. 112(984):317–326.

———. 1982. The theory of sex allocation. Princeton University Press, Princeton, N.J.

Charnov, E. L., and J. Bull. 1977. When is sex environmentally determined? Nature 266:828–830.

Charnov, E. L., and J. R. Krebs. 1975. The evolution of alarm calls: altruism or manipulation? Am. Nat. 109:107–112.

Charnov, E. L., J. M. Smith, and J. J. Bull. 1976. Why be an hermaphrodite? Nature 263:125–126.

Chesson, P., and R. Warner. 1981. Environmental variability promotes coexistence in lottery competitive systems. Am. Nat. 117:923–943.

Chew, F. S. 1981. Coexistence and local extinction in two pierid butterflies. Am. Nat. 118:655–672.

Chew, R. M., and A. E. Chew. 1970. Energy relationships of the mammals of a desert shrub (*Larrea tridentata*) community. Ecol. Monogr. 40:1–21.

Christian, K. A., and C. R. Tracy. 1981. The effect of thermal environment on the ability of hatchling Galapagos land iguanas to avoid predation during dispersal. Oecologia 49:218–223.

Clarke, B. 1969. The evidence for apostatic selection. Heredity 24:347–352.

Clausen, J., D. Keck, and W. Hiesey. 1941. Regional differentiation in plant species. Am. Nat. 75:231–250.

Clements, F. E. 1916. Nature and structure of the climax. J. Ecol. 24:252–284.

Cloud, P., and A. Gibor. 1970. The oxygen cycle. Sci. Am. Sept.:110–123.

Cloudsley-Thompson, J. L., and M. J. Chadwick. 1964. Life in deserts. Foulis, London.

Clube, S. V. M., and W. M. Napier. 1982a. Spiral arms, comets and terrestrial catastrophisms. Q. J. R. Astron. Soc. 23:45–66.

———. 1982b. Close encounters with a million comets. New Scientist July 15:148–151.

———. 1984. Terrestrial catastrophism—nemesis or galaxy? Nature 311:635–636.

———. 1985. Terrestrial catastrophism—nemesis or galaxy? Nature 313:503.

Clutton-Brock, T. H., and P. H. Harvey. 1977. Primate ecology and social organization. J. Zool. (Lond.) 183:1–39.

———. 1978. Mammals, resources and reproductive strategies. Nature 273:191–195.

Clutton-Brock, T. H., F. E. Guinness, and S. D. Albon. 1982. Red deer, behavior and ecology of two sexes. University of Chicago Press, Chicago.

Coats, R. N., R. L. Leonard, and C. R. Goldman. 1976. Nitrogen uptake and release in a forested watershed: Lake Tahoe Basin, California. Ecology 57:995–1004.

Cody, M. L. 1971a. Ecological aspects of reproduction. Avian Biol. 1:461–512.

———. 1971b. Finch flocks in the Mohave desert. Theor. Popul. Biol. 2:142–148.

———. 1974. Competition and the structure of bird communities. Monographs in population biology. Princeton University Press, Princeton, N.J.

———. 1979. Habitat selection and interspecific territoriality among the sylviid warblers of England and Sweden. Ecol. Monogr. 48:351–396.

———. 1986. Structural niches in plant communities. Pages 381–405 *in* J. Diamond and T. Case, eds. Community ecology. Harper & Row, New York.

Cody, M. L., and H. A. Mooney. 1978. Convergence versus nonconvergence in Mediterranean-climate ecosystems. Annu. Rev. Ecol. Syst. 9:265–321.

Coe, M. J., D. H. Cumming, and J. Phillipson. 1976. Biomass and production of large African Herbivores in relation to rainfall and primary production. Oecologia (Berl.) 22:341–354.

Coffin, C. C., F. R. Hayes, L. H. Jodrey, and S. G. Whiteway. 1949. Exchange of materials in a lake as studied by the addition of radioactive phosphorus. Can. J. Res. D, 27:207–222.

Cohen, D. 1966. Optimizing reproduction in a randomly varying environment. J. Theor. Biol. 12:119–129.

Cohen, J. 1977. Ergodicity of age structure in populations with Markovian vital rates. II. General states. Adv. Appl. Probab. 9:18–37.

———. 1978. Food webs and niche space. Princeton University Press, Princeton, N.J.

Colbert, E. H. 1971. Tetrapods and continents. Q. Rev. Biol. 46:250–269.

Colbert, E. H., R. B. Cowles, and C. M. Bogert. 1946. Bearing of temperature tolerances in the alligator on habits, evolution, and extinction of the dinosaurs. Bull. Amer. Mus. Nat. Hist. 86:365–366.

———. 1947. Rates of temperature increase in the dinosaurs. Copeia 1947(2):141–142.

Colwell, R. 1973. Competition and coexistences in a simple tropical community. Am. Nat. 107:737–760.

———. 1979. The geographical ecology of hummingbird flower mites in relation to their host plants and carriers. Pages 461–468 *in* J. G. Rodriguez, ed. Recent advances in acarology. Vol. II. Academic Press, New York.

———. 1981. Group selection is implicated in the evolution of female-biased sex ratios. Nature 290:401–403.

———. 1982. Female-biased sex ratios. Nature 298:495–496.

Connell, J. H. 1972. Community interactions on marine rocky intertidal shores. Annu. Rev. Ecol. Syst. 3:169–172.

———. 1980. Diversity and the coevolution of competitors, or the ghost of competition past. Oikos 35:131–138.

———. 1983. On the prevalence and relative importance of interspecific competition: evidence from field experiments. Am. Nat. 122:661–696.

Connell, J. H., and E. Orias. 1964. The ecological regulation of species diversity. Am. Nat. 98:399–414.

Connell, J. H., and R. O. Slatyer. 1977. Mechanisms of succession in natural communities and their role in community stability and organization. Am. Nat. 111:1119–1144.

Connell, J. H., J. G. Tracey, and L. J. Webb. 1984. Compensatory recruitment, growth, and mortality as factors maintaining rain forest tree diversity. Ecol. Monogr. 54:141–164.

Conner, W. E., T. Eisner, R. K. Vander Meer, A. Guerrero, D. Ghiringelli, and J. Meinwald. 1980. Sex attractant of an arctiid moth (*Utetheisa ornatrix*): a pulsed chemical signal. Behav. Ecol. Sociobiol. 7:55–63.

Connor, E., and E. D. McCoy. 1979. The statistics and biology of the species-area relationship. Am. Nat. 113:791–833.

Connor, E., and D. Simberloff. 1979. The assembly of species communities: chance or competition? Ecology 60:1132–1140.

Conrads, K., and W. Conrads. 1971. Regional dialekte des Ortlans (*Emberiza hortulana*) in Deutschland. Vogelwelt 92:81–100.

Cook, E. 1971. The flow of energy in an industrial society. Sci. Am. 224:135–144.

Cook, R. B. 1983. The impact of acid deposition on the cycles of C, N, P, and S. Pages 345–364 *in* B. Bolin and R. B. Cook, eds. The major biogeochemical cycles and their interactions. Wiley, New York.

Cooke, G. D. 1967. The pattern of autotrophic succession in laboratory microcosms. BioSci. 17:717–721.

Cooke, W. B. 1979. The ecology of fungi. CRC Press, Boca Raton, Fla.

Cooper, W. S. 1923. The recent ecological history of Glacier Bay, Alaska: II. The present vegetation cycle. Ecology 4:223–246.

Costello, D. F. 1969. The prairie world. Crowell, New York.

Cott, H. B. 1940. Adaptive coloration in animals. Methuen, London.

Council on Environmental Quality. 1980. Global 2000: entering the 21st century. U. S. Government Printing Office, Washington, D.C.

Cowles, H. C. 1899. The ecological relations of the vegetation of the sand dunes of Lake Michigan. Bot. Gaz. 27:95–117; 167–202; 281–308; 361–391.

Cowling, E. B., and R. A. Linthurst. 1981. The acid precipitation phenomenon and its ecological consequences. BioSci. 31:649–654.

Cox, D. R., and H. D. Miller. 1965. The theory of stochastic processes. Wiley, New York.

Cox, P. A. 1981. Niche partitioning between sexes of dioecious plants. Am. Nat. 117:295–307.

———. 1982. Vertebrate pollination and the maintenance of dioecism in Freycinetia. Am. Nat. 120(1):65–80.

Cracraft, J. 1973. Continental drift, paleoclimatology and the evolution and biogeography of birds. J. Zool. (Lond.) 169:455–545.

Crespigny, C. C. de. 1869. Notes on the friendship existing between the malacopterygian fish *Premnas bimaculeatus* and the *Actinia crassicornis*. Proc. Zool. Soc. Lond. 1869:248–249.

Crisp, D. 1974. Factors influencing the settlement of marine invertebrate larvae. Pages 177–265 *in* P. T. Grant and A. M. Mackie, eds. Chemoreception in marine organisms. Academic Press, New York.

———. 1976. Settlement responses in marine organisms. Pages 83–124 *in* R. C. Newell, ed. Adaptations to environment: essays on the physiology of marine organisms. Butterworths, London.

———. 1979. Dispersal and re-aggregation in sessile marine invertebrates, particularly barnacles. Pages 319–327 *in* G. Larwood and B. R. Rosen, eds. Biology and systematics of colonial organisms. Academic Press, New York.

Crocker, R. L., and J. Major. 1955. Soil development in relation to vegetation and surface age at Glacier Bay, Alaska. J. Ecol. 42:427–448.

Croizat, L., G. J. Nelson, and D. E. Rosen. 1974. Centers of origin and related concepts. Syst. Zool. 23:265–287.

Crook, J. H. 1964. The evolution of social organization and visual communication in weaver birds (Ploceinae). Behavior. Suppl. 10:1–178.

———. 1965. The adaptive significance of avian social organizations. Symp. Zool. Soc. Lond. 14:181–218.

———. 1970. Social organization and the environment: aspects of contemporary social ethology. Anim. Behav. 18:197–209.

Crow, J., and M. Kimura 1965. Evolution in sexual and asexual populations. Am. Nat. 99:439–450.

———. 1970. An introduction to population genetics theory. Harper & Row, New York.

Cullen, E. 1957. Adaptations in the kittiwake to cliff-nesting. Ibis 99(2):275–302.

Culver, D. C., and A. J. Beattie. 1980. The fate of *Viola* seeds dispersed by ants. Am. J. Bot. 67:710–714.

Cummins, K. W. 1974. Structure and function of stream ecosystems. BioSci. 24:631–641.

Cunningham, M. A., and M. C. Baker. 1983. Vocal learning in white-crowned sparrows: sensitive phase and song dialects. Behav. Ecol. Sociobiol. 13:259–269.

Cushing, D. H., and J. J. Walsh., eds. 1976. The ecology of the seas. Saunders, Philadelphia.

Dansereau, P. 1957. Biogeography—an ecological perspective, Ronald Press, New York.

Dantzler, W. H. 1982. Renal adaptations of desert vertebrates, BioSci. 32:108–113.

Darling, F. Fraser. 1937. A herd of red deer. Oxford University Press, London.

Darlington, P. J. 1957. Zoogeography: The geographical distribution of animals. Wiley, New York.

Darwin, C. 1859. On the origin of species. John Murray, London.

Dasmann, R. F., and A. S. Mossman. 1961. Commercial use of game animals on a Rhodesian ranch. Wild Life September/December:7–14.

Davenport, D., and K. S. Norris, 1958. Observations on the symbiosis of the sea anemone *Stoichaetes* and the pomacentrid fish *Amphiprion percula*. Biol. Bull. 115:397–410.

Davidson, D., J. Brown, Jr., and R. Inouye. 1980. Competition and the structure of granivore communities. BioSci. 30:233–238.

Davies, N. B. 1978. Ecological questions about territorial behaviour. Pages 317–350 *in* J. R. Krebs and N. B. Davies, eds. Behavioral ecology: an evolutionary approach. Blackwell, Oxford.

Davies, N. B., and T. R. Halliday. 1979. Competitive mate searching in male common toads, *Bufo bufo*. Anim. Behav. 27:1253–1267.

Davies, M., P. Hut, and R. A. Muller. 1984. Extinction of species by periodic comet showers. Nature 308:715–717.

———. 1985. Terrestrial catastrophism: nemesis or galaxy? Nature 313:503.

Davis, M. B. 1981. Quaternary history and the stability of forest communities. Pages 132–153 *in* D. C. West, H. H. Shugart and D. B. Botkin, eds. Forest succession: concepts and application. Springer-Verlag, New York.

Dawkins, R., and T. R. Carlisle. 1976. Parental investment, mate desertion and a fallacy. Nature 262:131–133.

Dawkins, R., and J. R. Krebs. 1978. Animal signals: information or manipulation? Pages 292–309 *in* J. R. Krebs and N. B. Davies, eds. Behavioral ecology. Blackwell, Oxford.

———. 1979. Arms races between and within species. Proc. R. Soc. Lond. 205:489–511.

Dawson, W. R., and G. A. Bartholomew. 1958. Metabolic and cardiac responses to temperature in the lizard, *Dipsosaurus dorsalis*. Physiol. Zool. 31:100–111.

Day, R. W., and R. W. Osman. 1981. Predation by *Patiria miniata* (Asteroidea) on bryozoans: prey diversity may depend on the mechanism of succession. Oecologia 51:300–309.

Day, T., P. Hillier, and B. Clarke. 1974. Properties of genetically polymorphic isozymes of alcohol dehydrogenase in *Drosophila melanogaster*. Biochem. Genetics 11:141–153.

Dayton, P. 1971. Competition, disturbance and community organization: the provision and subsequent utilization of space in a rocky intertidal community. Ecol. Monogr. 41:351–389.

———. 1975. Experimental evaluation of ecological dominance in a rocky intertidal algal community. Ecol. Monogr. 45:137–159.

Dean, T. A., and L. E. Hurd. 1980. Development in an estuarine fouling community: the influence of early colonists on later arrivals. Oecologia (Berl.) 46:295–301

DeAngelis, D. L. 1980. Energy flow, nutrient cycling and ecosystem resilience. Ecology 61:764–771.

DeBach, P. 1974. Biological control by natural enemies. Cambridge University Press, London.

Deevey, E. S., Jr. 1947. Life tables for natural populations of animals. Q. Rev. Biol. 22:283–314.

Delcourt, H. R., D. C. West, and P. A. Delcourt. 1981. Forests of the southeastern United States: quantitative maps for above ground woody biomass, carbon and dominance of major tree taxa. Ecology 62:879–887.

Delyamure, S. L. 1955. The helminthofauna of marine animals in the light of their ecology and phylogeny. Izd. Akad. Nauk SSSR, Moscow. (Trans. TT67-51202, U.S. Dept. of Commerce, Springfield, Va.)

DeMott, W., and C. Kerfoot. 1983. Competition among cladocerans: nature of the interaction between *Bosmina* and *Daphnia*. Ecol. Monogr.

den Boer, P. J. 1981. On the survival of populations in a heterogeneous and variable environment. Oecologia (Berl.) 50:39–53.

Denno, R. F., and M. S. McClure. 1983. Variability: a key to understanding plant-herbivore interactions. Pages 1–12 *in* R. F. Denno and M. S. McClure, eds. Variable plants and herbivores in natural and managed systems. Academic Press, New York.

Denny, M. W., T. L. Daniel, and M. A. R. Koehl. 1985.

Mechanical limits to size in wave-swept organisms. Ecol. Monogr. 55:69–102.

Denslow, J. S. 1980. Gap partitioning among tropical rainforest trees. Biotropica 12 (suppl.):47–55.

Desowitz, R. S. 1981. New Guinea tapeworms and Jewish grandmothers: tales of parasites and people. Norton, New York.

De Vita, J. 1979. Niche separation and the broken-stick model. Am. Nat. 114:171–178.

De Vooys, C. G. N. 1979. Primary production in aquatic systems. Pages 259–292 *in* B. Bolin, E. T. Degens, S. Kempe, and P. Ketner, eds. The carbon cycle. Wiley, New York.

De Vos, A., R. H. Manville, and G. van Gelder. 1956. Introduced mammals and their influence on native biota. Zoologica 41:163–194.

Dhondt, A., and R. Eyckerman. 1980. Competition between the great tit and the blue tit outside the breeding season in field experiments. Ecology 61:1291–1298.

Diamond, J. 1973. Distributional ecology of New Guinea birds. Science 179:759–769.

———. 1975. Assembly of species communities. Pages 342–444 *in* M. Cody and J. Diamond, eds. Ecology and evolution of communities. Harvard University Press, Cambridge, Mass.

———. 1981. Birds of paradise and the theory of sexual selection. Nature 293:257–258.

———. 1982a. Evolution of bowerbirds' bowers: animal origins of the aesthetic sense. Nature 297:99–102.

———. 1982b. Mimicry of friarbirds by orioles. Auk 99:187–196.

———. 1982c. Rediscovery of the yellow-fronted gardener bowerbird. Science 216:431–434.

Diamond, J., and M. Gilpin. 1982. Examination of the "null" model of Connor and Simberloff for species co-occurences on islands. Oecologia 52:64–74.

Didden-Zopfy, B., and P. S. Nobel. 1982. High temperature tolerance and heat acclimation of *Opuntia biglovii.* Oecologia (Berl.) 52:176–180.

Dietz, R. S. 1961. Continent and ocean basin evolution by spreading of the sea floor. Nature 190:854–857.

Dietz, R. S., and J. C. Holden. 1970. The breakup of pangaea. Sci. Am. October:30–41.

Dingle, H., and J. P. Hegmann, eds. 1982. Evolution and genetics of life histories. Springer-Verlag, New York.

Dirzo, R. 1984. Herbivory: a phytocentric overview. Pages 141–165 *in* R. Dirzo and J. Sarukhan, eds. Perspectives on plant population ecology. Sinauer, Sunderland, Mass.

Dixey, F. A. 1897. Mimetic attraction. Trans. Ent. Soc. London 1897:317–332.

———. 1909. On Müllerian mimicry and diaposematism. Trans. Ent. Soc. London 1909:559–583.

Dixon, A. F. G. 1975. Effect of population density and food quality on autumnal reproductive activity in the sycamore aphid, *Drepanosiphum platanoides* (Schr.) J. Anim. Ecol. 44:297–303.

Dobzhansky, T. 1954. Evolution as a creative process. Caryologia, suppl. 435–449.

Dodson, C. H. 1975. Coevolution of orchids and bees. Pages 91–99 *in* L. E. Gilbert and P. H. Raven, eds. Coevolution of animals and plants. University of Texas Press, Austin.

Dodson, S. 1974. Zooplankton competition and predation: an experimental test of the size–efficiency hypothesis. Ecology 55:605–613.

Doherty, P. J. 1983. Tropical territorial damselfishes: is density limited by aggression or recruitment? Ecology 64:176–190.

Dolinger, P., P. R. Ehrlich, W. L. Fitch, and D. E. Breedlove. 1973. Alkaloid and predation patterns in Colorado lupine populations. Oecologia (Berl.) 13:191–204.

Downhower, J. F., and K. B. Armitage. 1971. The yellow-bellied marmot and the evolution of polygamy. Am. Nat. 105:355–370.

———. 1981. Dispersal of yearling yellow-bellied marmots *(Marmota flaviventris).* Anim. Behav. 29:1064–1069.

Doyle, T. W. 1981. The role of disturbance in the gap dynamics of a montane rain forest: an application of a tropical forest succession model. Pages 56–73 *in* D. C. West, H. H. Shugart, and D. B. Botkin, eds. Forest succession. Springer-Verlag, New York.

Drury, W. H., and I. C. T. Nisbet. 1973. Succession. J. Arnold Arbor. 54:331–368.

Duffy, E. 1977. The reestablishment of the large copper butterfly, *Lycaena dispar batava* Obth., on Woodwalton Fen Nature Reserve, Cambridgeshire, England, 1969–73. Biol. Cons. 12:143–157.

Duncan, T. 1973. Forty years of succession on an abandoned building site at the University of Michigan Biological Station. Mich. Bot. 12:167–177.

Dunford, C. 1977. Kin selection for ground squirrel alarm calls. Am. Nat. 11:782–785.

Dunham, A. 1978. Food availability as a proximate factor influencing individual growth rates in the iguanid lizard *Sceloporus merriami.* Ecology 59:770–778.

———. 1980. An experimental study of interspecific competition between the iguanid lizards *Scelo-*

porus merriami and *Urosaurus ornatus*. Ecol. Monogr. 50:309–330.

Duvigneaud, P., and S. Denaeyer-de Smet, 1970. Biological cycling of minerals in temperate deciduous forests. Pages 199–225 *in* D. E. Reichle, ed. Analysis of temperate forest ecosystems. Springer-Verlag, New York.

Edson, K. M., S. B. Vinson, D. B. Stoltz, and M. D. Summers. 1981. Virus in a parasitoid wasp: suppression of the cellular immune response in the parasitoid's host. Science 211:582–583.

Egler, F. E. 1952. Vegetation science concepts I. Initial floristic composition, a factor in old-field vegetation development. Vegetation 4:412–417.

Ehleringer, J., and O. Bjorkman. 1977. Quantum yields for CO_2 uptake in C_3 and C_4 plants: dependence on temperature, carbon dioxide and oxygen concentration. Plant Physiol. 56:86–90.

Ehleringer, J., and I. Forseth. 1980. Solar tracking by plants. Science 210:1094–1098.

Ehrlich, P. R. 1958. Problems of arctic-alpine insect distribution as illustrated by the butterfly genus *Erebia* (Satyridae). Proc. 10th Intl. Cong. Entomol. 1:683–686.

———. 1961a. Intrinsie barriers to dispersal in the checkerspot butterfly *Euphydryas editha*. Science 134:108–109.

———. 1961b. Has the biological species concept outlived its usefulness? Syst. Zool. 10:167–176.

———. 1965. The population biology of the butterfly, *Euphydryas editha*. II. The structure of the Jasper Ridge Colony. Evolution 19:327–336.

———. 1970. Coevolution and the biology of communities. Pages 1–11 *in* K. L. Chambers, ed. Proc. 29th Annu. Biol. Colloq. 1968. Oregon State University Press, Corvallis.

———. 1975. The population biology of coral reef fishes. Annu. Rev. Ecol. Syst. 6:213–247.

———. 1984. The structure and dynamics of butterfly populations. In the biology of butterflies. Symp. Brit. Ent. Soc. 11:25–40.

Ehrlich, P. R., and L. C. Birch. 1967. The "balance of nature" and "population control." Am. Nat. 101:97–107.

Ehrlich, P. R., and S. E. Davidson. 1960. Techniques for capture-recapture studies of Lepidoptera populations. J. Lepidop. Soc. 14:227–229.

Ehrlich, P. R., and A. H. Ehrlich. 1961. How to know the butterflies. Brown, Dubuque, Iowa.

———. 1973. Coevolution: heterotypic schooling in Caribbean reef fishes. Am. Nat. 107:157–160.

———. 1981. Extinction: The causes and consequences of the disappearance of species. Random House, New York.

———. 1982. Social behavior of butterfly and surgeonfishes on coral reefs: some mirror experiments. Oecologia (Berl.) 54:138–140.

Ehrlich, P. R., A. H. Ehrlich, and J. P. Holdren. 1977b. Ecoscience: population, resources, environment. Freeman, San Francisco.

Ehrlich, P. R., and L. E. Gilbert. 1973. The population structure and dynamics of a tropical butterfly, *Heliconius ethilla*. Biotropica 5(2):69–82.

Ehrlich, P. R., and R. W. Holm. 1962. Patterns and populations. Science 137:652–657.

Ehrlich, P. R., R. W. Holm, and I. L. Brown. 1976. Biology and society. McGraw-Hill, New York.

Ehrlich, P. R., R. W. Holm, and D. R. Parnell. 1974. The process of evolution. McGraw-Hill, New York.

Ehrlich, P. R., A. E. Launer, and D. D. Murphy. 1984. Can sex ratio be defined or determined? The case of a population of checkerspot butterflies. Am. Nat. 124(4):527–539.

Ehrlich, P. R., and H. A. Mooney. 1983. Extinction, substitution and ecosystem services. BioSci. 33:248–254.

Ehrlich, P. R., and D. D. Murphy. 1981. The population biology of checkerspot butterflies (*Euphydryas*). Biologisches Zentralblatt 100:613–629.

Ehrlich, P. R., D. D. Murphy, and A. E. Launer. 1983. The role of adult feeding in egg production and population dynamics of the checkerspot butterfly, *Euphydryas editha*. Oecologia 56:257–263.

Ehrlich, P. R., D. D. Murphy, M. C. Singer, C. B. Sherwood, R. R. White, and I. L. Brown. 1980. Extinction, reduction, stability and increase: the responses of checkerspot butterfly (*Euphydryas*) populations to the California drought. Oecologia (Berl.) 46:101–15.

Ehrlich, P. R., and P. H. Raven. 1964. Butterflies and plants: a study in coevolution. Evolution 8:586–608.

———. 1969. Differentiation of populations. Science 65:1228–1232.

Ehrlich, P. R., F. H. Talbot, B. C. Russell, and G. R. V. Anderson. 1977a. The behavior of chaetodontid fishes with special reference to Lorenz's "poster coloration" hypothesis. J. Zool. (Lond.). 183:213–228.

Ehrlich, P. R., and D. Wheye. 1984. Some observations on spatial distribution in a montane population of *Euphydryas editha*. J. Res. Lep. 23:143–152.

———. 1986. "Nonadaptive" hill topping behavior

in male checkerspot butterflies (*Euphydryas editha*). Am. Nat. 127:477–483.

Ehrlich, P. R., and R. R. White. 1980. Colorado checkerspot butterflies: isolation, neutrality and the biospecies. Am. Nat. 115:328–341.

Ehrlich, P. R., R. R. White, M. C. Singer, S. W. McKechnie, and L. E. Gilbert. 1975. Checkerspot butterflies: a historical perspective. Science 188:221–228.

Ehrman, L. 1983. Endosymbioses. Pages 128–136 *in* D. J. Futuyma and M. Slatkin, eds. Coevolution. Sinauer, Sunderland, Mass.

Eisenberg, J. F. 1981. The mammalian radiations: an analysis of trends in evolution, adaptation and behavior. University of Chicago Press, Chicago.

Eisenberg, J. F., N. A. Muckenhirn, and R. Rudran. 1972. The relation between ecology and social structure in primates. Science 176:863–874.

Eisner, T. 1958. The protective role of the spray mechanism of the bombardier beetle, *Brachynus ballistarius* Lec. J. Insect Physiol. 2:215–220.

———. 1960. Defense mechanisms of arthropods. II. The chemical and mechanical weapons of an earwig. Psyche 67:62–70.

———. 1970. Chemical defense against predation in arthropods. Pages 157–217 *in* E. Sondheimer and J. B. Simeone, eds. Chemical ecology. Academic Press, New York.

———. 1981. Leaf folding in a sensitive plant: a defensive thorn-exposure mechanism? Proc. Natl. Acad. Sci. USA 78(1):402–404.

Eisner, T., and D. J. Aneshansley. 1982. Spray aiming in bombardier beetles: jet deflection by the coanda effect. Science 215:83–85.

Eisner, T., D. Hill, M. Goetz, S. Jain, D. Alsop, S. Camazine, and J. Meinwald. 1981. Anti-feedant action of Z-dihydromatricarin acid from soldier beetles (*Chauliognathus* spp.). J. Chem. Ecol. 7:1149–1158.

Eisner, T., and J. Meinwald. 1966. Defensive secretions of arthropods. Science 153:1341–1350.

Elbadry, E. A., and M. S. F. Tawfik. 1966. Life cycle of the mite *Adactylidium* sp. (Acarina: Pyemotidae), a predator of thrips eggs in the United Arab Republic. Annu. Ent. Soc. Am. 59(3):458–461.

Eldredge, N., and S. J. Gould. 1972. Punctuated equilibria: an alternative to phyletic gradualism. Pages 82–115 *in* T. J. M. Schopf, ed. Models in paleobiology. Freeman, Cooper, San Francisco.

Elliott, D. K., ed. 1986. Dynamics of extinction. Wiley, New York.

Ellison, L., and W. R. Houston. 1958. Production of herbaceous vegetation in openings and under canopies of Western aspen. Ecology 39(2):337–345.

Elton, C. 1927. Animal ecology. Reprinted 1966 by Science Paperbacks and Methuen, London.

———. 1933. The ecology of animals. Reprinted 1966 by Science Paperbacks, London.

———. 1958. The ecology of invasions by animals and plants. Methuen, London.

Elton, C., and M. Nicholson. 1942. The ten-year cycle in numbers of the lynx in Canada. J. Anim. Ecol. 11:215–244.

Emlen, S. T. 1976. Altruism in mountain bluebirds? Science 191:808–809.

———. 1978. The evolution of cooperative breeding in birds. Pages 245–281 *in* J. R. Krebs and N. B. Davies, eds. Behavioral Ecology: An Evolutionary Approach. Blackwell Scientific Publications, Oxford.

———. 1981. Altruism, kinship, and reciprocity in the whitefronted bee-eater. Pages 245–281 *in* R. Alexander and D. Tinkle, eds. Natural selection and social behavior: recent research and new theory. Chiron, New York.

———. 1984. Cooperative breeding in birds and mammals. Pages 305–339 *in* J. R. Krebs and N. B. Davies, eds. Behavioral ecology: an evolutionary approach. 2d ed. Blackwell, Oxford.

Emlen, S. T., and L. W. Oring. 1977. Ecology, sexual selection and the evolution of mating systems. Science 197:215–223.

Emlen, S. T., and S. Vehrencamp. 1983. Cooperative breeding strategies among birds. Pages 93–120 *in* A. Brush and G. A. Clark, Jr., eds. Perspectives in ornithology. Cambridge University Press, Cambridge.

Emmel, T. C. 1975. Butterflies. Knopf, New York.

Endler, J. 1977. Geographic variation, speciation and clines. Princeton University Press, Princeton, N.J.

Endler, J. A. 1982. Alternative hypotheses in biogeography: introduction and synopsis of the symposium. Amer. Zool. 22:349–354.

Erickson, E., and S. L. Buchmann. 1983. Electrostatics and pollination. Pages 173–184 *in* C. E. Jones and R. J. Little, eds. Handbook of experimental pollination biology. Van Nostrand Reinhold, New York.

Erwin, T. L. 1982. Tropical forests: their richness in Coleoptera and other arthropod species. Coleopt. Bull. 36:74–75.

Eschel, I. 1972. On the neighbor effect and the evolution of altruistic traits. Theor. Popul. Biol. 3(3):258–277.

Estes, R. D., and J. Goddard. 1967. Prey selection and hunting behavior of the African wild dog. J. Wildl. Mgmt. 31:52–70.

Etheridge, R. 1964. Late Pleistocene lizards from Barbuda, British West Indies. Bull. Florida State Mus. 9:43–75.

Ewel, J., ed. 1980. Tropical succession. Biotropica (suppl.) 12:1–95.

Ewel, J., C. Berish, B. Brown, N. Price, and J. Raich. 1981. Slash and burn impacts on a Costa Rican wet forest site. Ecology 62:816–829.

Fager, E. W., and J. A. McGowan. 1963. Zooplankton species groups in the North Pacific. Science 140:453–460.

Fagerstrom, T. 1982. Maternal investment, female rivalry and a fallacy. Oikos 39(1):116–118.

Falconer, D. S. 1981. Introduction to quantitative genetics, 2nd ed. Longman's, New York.

Farner, D. S. 1955. Bird banding in the study of population dynamics. Pages 397–449 *in* A. Wolfson, ed. Recent studies in avian biology. University of Illinois Press, Urbana.

Feder, H. M. 1966. Cleaning symbiosis in the marine environment. Pages 327–380 *in* S. M. Henry, ed. Symbiosis. Vol. 1. Academic Press, New York.

Feeny, P. P. 1975. Biochemical coevolution between plants and their insect herbivores. Pages 1–19 *in* L. E. Gilbert and P. H. Raven, eds. Coevolution of animals and plants. University of Texas Press, Austin.

———. 1976. Plant apparency and chemical defense. Pages 1–40 *in* J. M. Wallace and R. L. Mansell, eds. Biochemical interaction between plants and insects. Plenum, New York.

Feinsinger, P. 1976. Organization of a tropical guild of nectarivorous birds. Ecol. Monogr. 46:257–291.

———. 1983. Coevolution and pollination. Pages 282–310 *in* D. J. Futuyma and M. Slatkin, eds. Coevolution. Sinauer, Sumderland, Mass.

Feldman, M. W., F. B. Christiansen, and L. D. Brooks. 1980. Evolution of recombination in a constant environment. Proc. Natl. Acad. Sci. USA 77(8): 4838–4841.

Feller, M. C. 1978. Nutrient movement in soils beneath eucalypt and exotic conifer forests in southern central Victoria. Aust. J. Ecol. 3:357–372.

Felsenstein, J. 1974. The evolutionary advantage of recombination. Genetics 78:737–756.

Fenchel, T., and F. Christiansen. 1977. Selection and interspecific competition. Pages 477–498 *in* F. Christiansen and T. Fenchel, eds. Measuring selection in natural populations. Vol. 19. Lecture notes in biomathematics. Springer-Verlag, New York.

Fenchel, T., and L. Kofoed. 1976. Evidence for exploitative interspecific competition in mud snails (Hydrobiidae). Oikos 27:367–376.

Fenner, F., and F. Ratcliffe. 1965. Myxomatosis. Cambridge University Press, Cambridge.

Fenton, M. B., and J. H. Fullard. 1981. Moth hearing and the feeding strategies of bats. Am. Sci. 69:266–275.

Ferguson, G. W., C. H. Bohlen, and H. P. Woolley. 1980. *Sceloporus undulatus:* comparative life history and regulation of a Kansas population. Ecology 61:313–322.

Field, C. 1983. Allocating leaf nitrogen for the maximization of carbon gain: leaf age as a control on the allocation program. Oecologia 56:341–347.

Field, C. and H. A. Mooney. 1986. The photosynthesis-nitrogen relationship in wild plants. Pages 25–55 *in* P. J. Givinish, On the economy of plant form and function. Cambridge University Press, Cambridge.

Field, C. R. 1974. Scientific utilization of wildlife for meat in East Africa: a review. J. Sth. Afr. Wildl. Mgmt. Ass. 4:177–183.

Fischer, A. 1960. Latitudinal variations in organic diversity. Evolution 14:64–81.

Fisher, J., and R. Hinde. 1949. The opening of milk bottles by birds. Brit. Birds 42:347–358.

Fisher, R. A. 1927. On some objections to mimicry theory: statistical and genetic. Trans. Ent. Soc. London 1927:269–278.

———. 1930. The genetical theory of natural selection. Clarendon, Oxford.

———. 1958. The genetical theory of natural selection. 2nd ed. Dover, New York.

Fisher, R., A. Corbet, and C. Williams. 1943. The relation between the number of species and the number of individuals in a random sample of an animal population. J. Anim. Ecol. 12:42–58.

Fitch, H. S. 1947. Predation by owls in the Sierran foothills of California. Condor 49:137–151.

Floody, O. R., and A. P. Arnold. 1975. Uganda kob (*Adenota kob thomasi*): territoriality and the spatial distributions of sexual and agonistic behaviors at a territorial ground. Z. Tierpsychol. 37:192–212.

Florence, R. G. 1965. Decline of old-growth redwood forests in relation to some soil microbiological processes. Ecology 46:52–64.

Ford, E. 1923. Animal communities of the level seabottom in the waters adjacent to Plymouth. J. Mar. Biol. Ass. U.K. 13:164–222.

Ford, E. B. 1945. Butterflies. Collins, London.

Fossey, D. 1983. Gorillas in the mist. Houghton Mifflin, Boston.

Fox, L. R. 1981. Defense and dynamics in plant–herbivore systems. Amer. Zool. 21:853–864.

Fox, L. R., and B. J. MacCauley. 1977. Insect grazing on *Eucalyptus* in response to variation in leaf tannins and nitrogen. Oecologia 29:145–162.

Fox, S. F. 1978. Natural selection on behavorial phenotypes of the lizard, *Uta stansburiana*. Ecology 59:834–847.

Frank, P. 1957. Coactions in laboratory populations of two species of *Daphnia*. Ecology 38:510–519.

———. 1965. The biodemography of an intertidal snail population. Ecology 46:831–844.

Frankel, E. 1977. Previous *Acanthaster* aggregations in the Great Barrier Reef. Proc. 3rd Int. Coral Reef Symp. May, pp. 201–208.

Freeman, D. C., L. G. Klikoff, and K. T. Harper. 1976. Differential resource utilization by the sexes of dioecious plants. Science 193:597–599.

French, N., ed. 1979. Perspectives in grassland ecology: results and implications of the US/IBP grasslands biome study. Ecol. Stud. 32. Springer-Verlag, New York.

Fricke, H., and S. Fricke. 1977. Monogamy and sex change by aggressive dominance in coral reef fish. Nature 266:830–832.

Frisch, K. von. 1950. Bees, their vision, chemical senses, and language. Cornell Univ. Press, Ithaca.

———. 1967. The dance language and orientation of bees. Harvard Univ. Press, Cambridge.

Fry, C. H. 1972. The social organisation of bee-eaters (Meropidae) and co-operative breeding in hot-climate birds. Ibis 114:1–114.

———. 1977. The evolutionary significance of cooperative breeding in birds. Pages 127–136 *in* B. Stonehouse and C. M. Perrins, eds. Evolutionary ecology. Macmillan, London.

Fullard, J. H. 1977. Phenology of sound-producing arctiid moths and the activity of insectivorous bats. Nature 267:42–43.

Fullard, J. H., and R. M. R. Barclay. 1980. Audition in spring species of arctiid moths as a possible response to differential levels of insectivorous bat predation. Can. J. Zool. 58:1745–1750.

Fullard, J. H., M. B. Fenton, and J. A. Simmons. 1979. Jamming bat echolocation: the clicks of arctiid moths. Can. J. Zool. 57:647–649.

Futuyma, D. J. 1976. Food plant specialization and environmental predictability in Lepidoptera. Am. Nat. 110:285–292.

———. 1979. Evolutionary biology. Sinauer, Sunderland, Mass.

———. 1983. Evolutionary interactions among herbivorous insects and plants. Pages 207–231 *in* D. J. Futuyma and M. Slatkin, eds. Coevolution. Sinauer, Sunderland, Mass.

Futuyma, D. J., and F. Gould. 1979. Association of plants and insects in a deciduous forest. Ecol. Monogr. 49:33–50.

Gaines, S. D. and J. Lubchenco. 1982. A unified approach to marine plant-herbivore interactions. II. Biogeography. Ann. Rev. Ecol. Syst. 13:111–138.

Gallagher, E. D., P. A. Jumars, and D. D. Trueblood. 1983. Facilitation of soft-bottom benthic succession by tube builders. Ecology 64:1200–1216.

Ganapathy, R. 1980. A major meteorite impact on the Earth 65 million years ago: evidence from the Cretaceous–Tertiary boundary clay. Science 209:921–923.

Garrity, S. D., and S. C. Levings. 1981. A predator–prey interaction between two physically and biologically constrained tropical rocky shore gastropods: direct, indirect, and community effects. Ecol. Monogr. 51:267–286.

Gaston, A. J. 1978. The evolution of group territorial behavior and cooperative breeding. Am. Nat. 112:1091–1100.

Gates, D. 1965. Energy, plants and ecology. Ecology 46:1–14.

Gatz, A., Jr. 1979. Community organization in fishes as indicated by morphological features. Ecology 60:711–718.

Gause, G. F. 1934. The struggle for existence. Williams & Wilkins, Baltimore.

Geist, V. 1974. On the relationship of social evolution and ecology in ungulates. Am. Zool. 14:205–220.

Ghilarov, M. S. 1967. Abundance, biomass and vertical distribution of soil animals in different zones. Pages 611–629 *in* K. Petrusewicz, ed. Secondary productivity of terrestrial ecosystems. Warszawa, Krakow.

Gilbert, L. E. 1971. Butterfly–plant coevolution: has *Passiflora adenopoda* won the selectional race with heliconiine butterflies? Science 172:585–586.

———. 1972. Pollen feeding and reproductive biology of *Heliconius* butterflies. Proc. Natl. Acad. Sci. USA 69:1403–1407.

———. 1975. Ecological consequences of a coevolved mutualism between butterflies and plants. Pages 210–240 *in* L. E. Gilbert and P. H. Raven, eds. Coevolution of animals and plants. University of Texas Press, Austin.

———. 1979. Development of theory in the analysis of insect-plant interactions. Pages 117–154 *in* D. Horn, R. Mitchell, and G. Stairs, eds. Analysis of ecological systems. Ohio State University Press, Columbus.

———. 1980. Food web organization and the conservation of neotropical diversity. Pages 11–33 *in* M. E. Soulé and B. A. Wilcox, eds. Conservation biology. Sinauer, Sunderland, Mass.

———. 1982. The coevolution of a butterfly and a vine. Sci. Am. August:110–121.

———. 1983. Coevolution and mimicry. Pages 263–281 *in* D. J. Futuyma and M. Slatkin, eds. Coevolution. Sinauer, Sunderland, Mass.

Gill, D. 1978. The metapopulation ecology of the red-spotted newts, *Notophthalmus viridescens* (Rafinesque). Ecol. Monogr. 48:145–166.

———. 1979. Density dependence and homing behavior in adult red-spotted newts *Notophthalmus viridescens* (Rafinesque). Ecology 60:800–813.

Gill, F. B., and L. L. Wolf. 1975. Economics of feeding territoriality in the golden-winged sunbird. Ecology 56:333–345.

Gilpin, M., and J. Diamond. 1976. Calculation of immigration and extinction curves from the species–area–distance relation. Proc. Natl. Acad. Sci. USA 73:4130–4134.

———. 1981. Immigration and extinction probabilities for individual species: relation to incidence functions and species colonization curves. Proc. Natl. Acad. Sci. USA 78:392–396.

Gingerich, P. D. 1975. Is the aardwolf a mimic of the hyaena? Nature 253: 191–192.

———. 1977. Patterns of evolution in the mammalian fossil record. Pages 469–500 *in* A. Hallam, ed. Patterns of evolution as illustrated by the fossil record. Elsevier, Amsterdam.

Gittleman, J. L., and P. H. Harvey. 1982. Carnivore home-range size, metabolic needs and ecology. Behav. Ecol. Sociobiol. 10:57–63.

Givnish, T. J. 1980. Ecological constraints on the evolution of breeding systems in seed plants: dioecy and dispersal in gymnosperms. Evolution 34:959–972.

Glasser, J. W. 1979. The role of predation in shaping and maintaining the structure of communities. Am. Nat. 113:631–641.

Gleason, H. A. 1926. The individualistic concept of the plant association. Bull. Torrey Bot. Club 53:1–20.

———. 1953. Dr. H. A. Gleason, distinguished ecologist. Bull. Ecol. Soc. Amer. 34:40–42.

Glutz von Blotzheim, V., K. Bauer, and E. Bezzel. 1971. Handbuch der vogel mitteleuropas. Vol. 4. Akad. Verlag, Frankfurt.

Glynn, P. 1974. The impact of *Acanthaster* in corals and coral reefs in the eastern Pacific. Envir. Conser. 1:295–304.

———. 1976. Some physical and biological determinants of coral community structure in the eastern Pacific. Monogr. 46:431–456.

Goel N. S., and N. Richter-Dyn. 1974. Stochastic models in biology. Academic Press, New York.

Gold, A. 1973. Energy expenditure in animal locomotion. Science 181:275–276.

Goldberg, D. E., and P. A. Werner. 1983. Equivalence of competitors in plant communities: a null hypothesis and a field experimental approach. Am. J. Bot. 70:1098–1104.

Golley, F. B. 1960. Energy dynamics of a food chain of an old-field community. Ecol. Monogr. 30:187–206.

Golley, F. B., and J. B. Gentry. 1965. A comparison of variety and standing crop of vegetation on a one-year and a twelve-year abandoned field. Oikos 15:185–199.

Golley, F. B., and R. Misra. 1972. Organic production in tropical ecosystems. BioSci. 22:735–736.

Gómez-Pompa, A., and C. Vásquez-Yanes. 1974. Studies on the secondary succession of tropical lowlands: the life cycle of secondary species. Proc. 1st Int. Cong. Ecol., The Hague, pp. 336–342.

———. 1981. Successional studies of a rain forest in Mexico. Pages 246–266 *in* D. C. West, H. H. Shugart, and D. B. Botkin, eds. Forest succession: concepts and application. Springer-Verlag, New York.

Gómez-Pompa, A., C. Vázquez-Yanes, and S. Guevara. 1972. The tropical rain forest: a nonrenewable resource. Science 177:762–765.

Gorham, E., P. M. Vitousek, and W. A. Reiners. 1979. The regulation of chemical budgets over the course of terrestrial ecosystem succession. Ann. Rev. Ecol. Syst. 10:53–84.

Gosz, J. R., R. T. Holmes, G. E. Likens, and F. H. Bormann. 1978. The flow of energy in a forest ecosystem. Sci. Am. March:93–102.

Gould, F. 1983. Genetics of plant-herbivore systems: interactions between applied and basic study. Pages 599–653 *in* R. F. Denno and M. S. McClure, eds. Variable plants and herbivores in natural and managed systems. Academic Press, New York.

Gould, S. J. 1977. Eternal metaphors of paleontology. Pages 1–26 *in* A. Hallam, ed. Patterns of evolution as illustrated by the fossil record. Elsevier, Amsterdam.

———. 1980. Is a new and general theory of evolution emerging? Paleobiol. 6:119–130.

Goulding, M. 1980. The fishes and the forest: explorations in Amazonian natural history. University of California Press, Berkeley.

Grant, J., and J. Bayly. Predator induction of crests in morphs of the *Daphnia carinata* King complex. Limnol. Oceanogr. 26:201–218.

Grant, P., and I. Abbott. 1980. Interspecific competition, island biogeography and null hypotheses. Evolution 34:332–341.

Grant, P. R., 1983. The role of interspecific competition in the adaptive radiation of Darwin's finches. Pages 187–199 *in* R. I. Bowman, M. Berson, and A. E. Leviton, eds. Patterns of evolution in Galapagos organisms. Spec. Pub. 1, American Association for the Advancement of Science, Pacific Division. Washington, D.C.

Grant, V. 1982. Punctuated equilibria: a critique. Biol. Zentralbl. 101:175–184.

Green, B. H. 1972. The relevance of seral eutrophication and plant competition to the management of successional communities. Biol. Cons. 4:378–384.

Green, R. H. 1968. The estimation of density dependence. Ecology 49:555–556.

Greenberg, L. 1979. Genetic component of bee odor in kin recognition. Science 206:1095–1097.

Greene, H. W., and R. W. McDiarmid. 1981. Coral snake mimicry: does it occur? Science 213:1207–1212.

Greenwald, O. E. 1974. Thermal dependence of striking and prey capture by gopher snakes. Copeia 1974:141–148.

Greenwood, P. H. 1974. The cichlid fishes of Lake Victoria, East Africa: the biology and evolution of a species flock. Bull. Brit. Mus. Nat. Hist. (Zool.) 6(suppl.):1–134.

Gregg, K. B. 1975. The effect of light intensity on sex expression in species of *Cycnoches* and *Catasetum* (Orchidaceae). Selbyana 1:101–113.

Griffin, D. R., F. A. Webster, and C. R. Michael. 1960. The echolocation of flying insects by bats. Anim. Behav. 8:141–154.

Grime, J. P. 1973. Competitive exclusion in herbaceous vegetation. Nature 242:344–347.

Grubb, P. 1977. The maintenance of species richness in plant communities: the importance of the regeneration niche. Biol. Rev. 52:107–145.

Gulmon, S. L. 1974. The relationship between species diversity and fluctuating climate in California annula grassland. Ph.D. diss. Stanford University, Stanford, Calif.

———. 1979. Competition and coexistence: three annual grass species. Am. Mid. Naturalist 101:403–416.

Gurtin, M., and R. MacCamy. 1974. Nonlinear age-dependent population dynamics. Arch. Rat'l. Mech. Anal. 54:281–300.

Gwynne, D. T. 1982. Mate selection by female katydid (Orthoptera: Tettigoniidae, *Conocephalus nigropleurum*). Anim. Behav. 30:734–738.

Hadley, N. F. 1972. Desert species and adaptation. Am. Sci. 60:338–346.

Hairston, N. G. 1949. The local distribution and ecology of the plethodontid salamanders of the southern Appalachians. Ecol. Monogr. 19:47–73.

———. 1980a. The experimental test of an analysis of field distributions: competition in terrestrial salamanders. Ecology 61:817–826.

———. 1980b. Evolution under interspecific competition: field experiments on terrestrial salamanders. Evolution 34:409–420.

———. 1980c. Species packing in the salamander genus *Desmognathus:* what are the interspecific interactions involved? Am. Nat. 115:354–366.

———. 1981. An experimental test of a guild: salamander competition. Ecology 62:65–72.

Hairston, N. G., F. E. Smith, and L. B. Slobodkin. 1960. Community structure, population control and competition. Am. Nat. 94:421–425.

Haldane, J. B. S., and S. D. Jayakar. 1963. Polymorphism due to selection of varying direction. J. Genetics 58:237–242.

Hallam, A. 1984. The causes of mass extinction. Nature 308:686–687.

Halliday, T. 1978. Vanishing birds: their natural history and conservation. Holt, Rinehart and Winston, New York.

Hamilton, W. D. 1964. The genetical evolution of social behavior. J. Theor. Biol. 7:1–52.

———. 1966. The moulding of senescence by natural selection. J. Theor. Biol. 12:12–45.

———. 1967. Extraordinary sex ratios. Science 156:477–488.

———. 1971. Geometry for the selfish herd. J. Theor. Biol. 31:295–311.

———. 1980. Sex versus non-sex versus parasite. Oikos 35:282–290.

Hamilton, W. D., and M. Zuk. 1982. Heritable true fitness and bright birds: a role for parasites? Science 218:384–387.

Hamilton, W. J., III, R. E. Buskirk, and W. H. Buskirk. 1976. Defense of space and resources by chacma (*Papio ursinus*) baboon troops in an African desert and swamp. Ecology 57:1264–1272.

Hanes, T. L. 1971. Succession after fire in the chaparral of Southern California. Ecol. Monogr. 41:27–52.

Hansell, M. H. 1982. Colony membership in the wasp, *Parischnogaster striatula* (Stenogastrinae). Anim. Behav. 30:1258–1259.

Haper, J. L. 1961. The evolution and ecology of closely related species living in the same area. Evolution 15:209–227.

Harley, J. L. 1939. The early growth of beech seedlings under natural and experimental conditions. J. Ecol. 27:384–400.

Harper, J. 1977. Population biology of plants. Academic Press, New York.

Harper, J., and J. Ogden. 1970. The reproductive strategy of higher plants: I. The concept of strategy with specific reference to *Senecio vulgaris*. L. J. Ecol. 58:681–689.

Harper, J., and J. White. 1974. The demography of plants. Annu. Rev. Ecol. Syst. 5:419–463.

Harris, H. 1966. Enzyme polymorphism in man. Proc. R. Soc. Lond. B. 164:298–310.

Harte, J. 1982. Modeling lake-water mineralization processes. J. Theor. Biol. 99:553–569.

———. 1983. An investigation of acid precipitation in Qinghai Province, China. Atmos. Envir. 17:403–408.

Hartl, D. L. 1980. A primer of population genetics. Sinauer, Sunderland, Mass.

Hartshorn, G. S. 1978. Tree falls and tropical forest dynamics. Pages 617–628 *in* P. B. Tomlinson and M. H. Zimmerman, eds. Tropical trees as living systems. Cambridge University Press, Cambridge.

———. 1980. Neotropical forest dynamics. Biotropica 12 (suppl.):23–30.

Hassell, M. P. 1975. Density-dependence in single-species populations. J. Anim. Ecol. 44:283–295.

———. 1978. The dynamics of arthropod predator–prey systems. Princeton University Press, Princeton, N.J.

———. 1985. Insect natural enemies as regulating factors. J. Anim. Ecol. 54:323–334.

Hatch, M. D., and C. R. Slack. 1970. Photosynthetic CO_2—fixation pathways. Annu. Rev. Plant Physiol. 21:141–162.

Haxo, F. T., and L. R. Blinks. 1950. Photosynthetic action spectra of marine algae. J. Gen. Physiol. 33:389–435.

Hazlett, B. A. 1982. Chemical induction of visual orientation in the hermit crab *Clibanarius vittatus*. Anim. Behav. 30:1259–1260.

Heatwole, H., and R. Levins. 1972. Trophic structure stability and faunal change during recolonization. Ecology 53:531–534.

Hebert, P. 1977. Niche overlap among species in the *Daphnia carinata* complex. J. Anim. Ecol. 46:399–409.

———. 1982. Competition in zooplankton communities. Annales Zoologici Fennici 19:349–356.

Herbert, P. D. N., P. S. Ward, and R. Harmsen. 1974. Diffuse competition in *Lepidoptera*. Nature 252: 389–391.

Heckel, D., and J. Roughgarden. 1979. A technique for estimating the size of lizard populations. Ecology 60:966–975.

Hedgpeth, J. W. 1983. Brackish waters, estuaries and lagoons. Pages 739–757 *in* O. Kinne, ed. Marine ecology: a comprehensive integrated treatise on life in oceans and coastal waters. Vol. V, Pt. 2. Wiley, New York.

Hedrick, P. W. 1983. Genetics of populations. Van Nostrand Reinhold, New York.

Heinrich, B. 1976. Resource partitioning among some eusocial insects: bumblebees. Ecology 57: 874–889.

———. 1979. Resource heterogeneity and patterns of movement in foraging bumblebees. Oecologia 40:235–246.

———. 1983. Do bumblebees forage optimally and does it matter? Amer. Zool. 23:273–281.

Heinrich, B., and P. H. Raven. 1972. Energetics and pollination ecology. Science 176:597–602.

Heinrich, B., and S. L. Collins. 1983. Caterpillar leaf damage and the game of hide-and-seek with birds. Ecology 64(3):592–602.

Heinselman, M. L. 1973. Fire in the virgin forests of the Boundary Waters Canoe Area, Minnesota. Quat. Res. 3:329–382.

———. 1981. Fire and succession in the conifer forests of northern North America. Pages 374–405 *in* D. C. West, H. H. Shugart, and D. B. Botkin, eds. Forest succession: concepts and application. Springer-Verlag, New York.

Heisler, I. L. 1981. Offspring quality and the polygyny threshold: a new model for the 'sexy son' hypothesis. Am. Nat. 117:316–328.

Hendrickson, J. A., Jr., and P. R. Ehrlich. 1971. An expanded concept of "species diversity." Notulae Naturae October 12:439:1–6.

Henry, J. D., and J. M. A. Swan. 1974. Reconstructing forest history from live and dead plant material—an approach to the study of forest succession in southwest New Hampshire. Ecology 55:772–783.

Herrera, C. M. 1974. Trophic diversity of the barn

owl *Tyto alba* in continental western Europe. Ornis. Scand. 5:181–191.
———. 1981. Combination rules among western European *Parus* species. Ornis Scand. 12:140–147.
———. 1982. Breeding systems and dispersal-related maternal reproductive effort of southern Spanish bird-dispersed plants. Evolution 36:1299–1314.
Hess, H. H. 1962. History of ocean basins. Pages 599–620 *in* A. E. J. Engel, H. L. James, and B. F. Leonard, eds. Petrological studies: a volume in honor of A. F. Buddington. Geological Society of America, Boulder, Colo.
Hickey, L. J. 1981. Land plant evidence compatible with gradual, not catastrophic, change at the end of the Cretaceous. Nature 292:529–531.
Hill, R. W. 1976. Comparative physiology of animals: an environmental approach. Harper & Row, New York.
Hils, M. W., and J. L. Vankat. 1982. Species removal from a first-year old-field plant community. Ecology 63:705–711.
Hinde, R. A., and J. Fisher. 1951. Further observations on the opening of milk bottles by birds. British Birds 44:393–396.
Hirsch, M., and S. Smale. 1974. Differential equations, dynamical systems and linear algebra. Academic Press, New York.
Hobson, E. S. 1969. Comments on certain recent generalizations regarding cleaning symbiosis in fishes. Pac. Sci. 23:35–39.
Holdren, C. E., and P. R. Ehrlich. 1981. Long range dispersal in checkerspot butterflies: Transplant experiments with *E. gillettii*. Oecologia 50:125–129.
———. 1982. Ecological determinants of food plant choice in the checkerspot butterfly *Euphydryas editha* in Colorado. Oecologia 52:417–423.
Holdridge, L. R. 1947. Determination of world plant formations from simple climatic data. Science 105:367–368.
———. 1967. Life zone ecology. Tropical Science Center, San Jose, Costa Rica.
Holldobler, B., and C. Michener. 1980. Mechanisms of identification and discrimination in social hymenoptera. Pages 35–58 *in* H. Markl, ed. Evolution of social behavior: hypotheses and empirical tests. Verlag Chemie, Deerfield Beach, Fla.
Holling, C. S. 1965. The functional response of predators to prey density and its role in mimicry and population regulation. Mem. Ent. Soc. Canad. 45:1–60.
———. 1973. Resilience and stability of ecological systems. Annu. Rev. Ecol. Syst. 4:1–23.
Holm, C. H. 1973. Breeding sex ratios, territoriality and reproductive success in the red-winged blackbird *(Agelaius phoeniceus)*. Ecology 54(2): 356–365.
Holmes, J. C. 1983. Evolutionary relationships between parasitic helminths and their hosts. Pages 161–185 *in* D. J. Futuyma and M. Slatkin, eds. Coevolution. Sinauer, Sunderland, Mass.
Holmes, R. T. 1970. Differences in population density, territoriality and food supply of dunlin on arctic and subarctic tundra. Pages 303–319 *in* A. Watson, ed. Animal populations in relation to their food resources. Blackwell, Oxford.
Holn, M. 1961. The relationship between species diversity and population density in diatom populations from Silver Springs, Florida. Trans. Amer. Microsc. Soc. 80:140–165.
Holt, R. 1977. Predation, apparent competition and the structure of prey communities. Theor. Popul. Biol. 12:197–229.
Hopcraft, D. 1970. East Africa: the advantages of farming game. Span 13:11–14.
———. 1980. Nature's technology: the natural land-use system of wildlife ranching. Vital Speeches 46:465–469.
Horn, H. S. 1968. The adaptive significance of colonial nesting in the Brewer's blackbird (Euphagus cyanocephalus). Ecology 49:682–694.
———. 1976. The adaptive geometry of trees. Princeton University Press, Princeton, N.J. pp. 187–204.
———. 1975a. Forest succession. Sci. Am. 232:90–98.
———. 1975b. Markovian properties of forest succession. Pages 196–211 *in* M. L. Cody and J. M. Diamond, eds. Ecology and evolution of communities. Harvard University Press, Cambridge, Mass.
———. 1976. Succession. Pages 253–271 *in* R. M. May, ed. Theoretical ecology: principles and applications. Sinauer, Sunderland, Mass.
———. 1981. Some causes of variety in patterns of secondary succession. Pages 24–35 *in* D. C. West, H. H. Shugart, and D. B. Botkin, eds. Forest succession: concepts and application. Springer-Verlag, New York.
Horvitz, C., and D. W. Schemske. 1984. Effects of ants and an ant-tended herbivore on seed production of a neotropical herb. Ecology 65:1369–1378.
Houston, A., and J. McNamara. 1982. A sequential approach to risk-taking. Anim. Behav. 30:1260–1261.
Houston, D. C. 1979. The adaptions of scavengers. Pages 263–286 *in* A. R. E. Sinclair and M. Norton-

Griffiths, eds. Serengeti: dynamics of an ecosystem. University of Chicago Press, Chicago.

Howard, D. J., and R. G. Harrison. 1983. Habitat segregation in ground crickets: the role of interspecific competition and habitat selection. Ecology 65:69–76.

Howe, H. F. 1976. Egg size hatching asynchrony, sex and brood reduction in the common grackle. Ecology 57:1195–1207.

———. 1977a. Bird activity and seed dispersal of a tropical wet forest tree. Ecology 58(3):539–550.

———. 1977b. Sex ratio adjustment in the common grackle. Science 198:744–746.

———. 1979a. Evolutionary aspects of parental care in the common grackle, *Quiscalus quiscula* L. Evolution 33(1):41–51.

———. 1979b. Fear and frugivory. Am. Nat. 114:925–931.

———. 1982. Fruit production and animal activity at two tropical trees. Pages 189–199 *in* E. Leigh, Jr., A. S. Rand, and D. Windsor, eds. The ecology of a tropical forest: seasonal rhythms and long-term changes. Smithsonian Institution Press, Washington, D.C.

Howe, H. F., and G. F. Estabrook. 1977. On intraspecific competition for avian dispersers in tropical trees. Am. Nat. 111(981):817–832.

Howe, H. F., and J. Smallwood. 1982. Ecology of seed dispersal. Annu. Rev. Ecol. Syst. 13:201–228.

Howe, H. F., and G. A. Vande Kerckhove. 1979. Fecundity and seed dispersal of a tropical tree. Ecology 60(1):180–189.

Hrdy, S. B. 1979. Infanticide among animals: a review, classification and examination of the implications for reproductive strategies of females. Ethol. Sociobiol. 1:13–40.

Hsü, K. J. 1980. Terrestrial catastrophe caused by cometary impact at the end of Cretaceous. Nature 285:201–203.

Hsü, K. J. (and 19 coauthors). 1982. Mass mortality and its environmental and evolutionary consequences. Science 216:249–256.

Hubbell, S. P., and R. B. Foster. 1983. Diversity of canopy trees in a neotropical forest and implications for conservation. Pages 25–41 *in* S. Sutton, T. C. Whitmore, and A. Chadwick, eds. Tropical rain forest: ecology and management. Blackwell, Oxford.

———. 1986. Biology, chance, and history and the structure of tropical rain forest tree communities. Pages 314–330 *in* J. Diamond and T. Case, eds. Community ecology. Harper & Row, New York.

Huey, R. B. 1983. Natural variation in body temperature and physiological performance in a lizard *(Anolis cristatellus)*. Pages 484–490 *in* A. Rhodin and K. Miyata, eds. Advances in herpetology and evolutionary biology. Museum of Comparative Zoology, Cambridge, Mass.

Hughes, A. L. 1982. Confidence of paternity and wife-sharing in polygynous and polyandrous systems. Ethol. Sociobiol. 3:125–129.

Hunter, G. W., W. W. Frye, and J. C. Swartzwelder. 1966. A manual of tropical medicine. Saunders, Philadelphia.

Hutchinson, G. E. 1951. Copepodology for the ornithologist. Ecology 32:571–577.

———. 1957. A treatise on limnology. I. Geography, physics and chemistry. Wiley, New York.

———. 1959. Homage to Santa Rosalia, or why are there so many kinds of animals? Am. Nat. 93:145–159.

———. 1967. A treatise on limnology. II. Introduction to lake biology and the limnoplankton. Wiley, New York.

———. 1975. A treatise on limnology. III. Limnological botany. Wiley, New York.

Hutchinson, G. E., and V. T. Bowen. 1947. A direct demonstration of the phosphorus cycle in a small lake. Proc. Natl. Acad. Sci. USA 33:148–153.

———. 1950. Limnological studies in Connecticut. IX. A quantitative; radio-chemical study of the phosphorus cycle in Linsley Pond. Ecology 31:194–203.

Huxley, J. S. 1914. The courtship-habits of the great crested grebe *(Podiceps cristatus)*; with an addition to the theory of sexual selection. Proc. Zool. Soc. Lond. 35:491–562.

Hynes, H. B. N. 1970. The ecology of running waters. Liverpool University Press, Liverpool.

Imboden, D. M. 1974. Phosphorus model of lake eutrophication. Limnol. Oceanogr. 19:297–304.

Inouye, D. W. 1977. Species structure of bumblebee communities in North America and Europe. Pages 35–40 *in* W. Mattson, ed. The role of arthropods in forest ecosystems. Springer-Verlag, New York.

———. 1978. Resource partitioning in bumblebees: experimental studies in foraging behavior. Ecology 59:672–678.

———. 1980. The effect of proboscis and corolla tube lengths on patterns and rates of flower visitation by bumblebees. Oecologia 45:192–201.

Inouye, D. W., and O. R. Taylor, Jr. 1979. A temperate region plant–ant seed predator system: consequences of extrafloral nectar secretion by *Helianthella quinquenervis*. Ecology 60:1–7.

———. 1980. Variation in generation time in *Frasera speciosa* (Gentianaceae), a long-lived perennial monocarp. Oecologia 47:171–174.

Inouye, R. S., G. S. Byers, and J. H. Brown. 1980. Effects of predation and competition on survivorship, fecundity, and community structure of desert annuals. Ecology 61:1344–1351.

Irvine, D. n. d. Rain forest adaptations: patch management through succession. ms. 9pp.

Itô, Y. 1980. Comparative ecology. Cambridge University Press, London.

Ivens, J. D., and R. G. Barry. 1974. Arctic and alpine environments. Methuen, London.

Ivlev, V. S. 1961. Experimental ecology of the feeding of fishes. (Trans. from Russian.) Yale University Press, New Haven, Conn.

Iwasa, Y. 1981. Role of sex ratio in the evolution of eusociality in haplodiploid social insects. J. Theor. Biol. 93:125–142.

Jackman, R., S. Nowicki, D. J. Aneshansley, and T. Eisner. 1983. Predatory capture of toads by fly larvae. Science 222:515–516.

Jackson, J. 1977. Competition on marine hard-substrata: the adaptive significance of solitary and colonial strategies. Am. Nat. 111:743–767.

———. 1979. Overgrowth competition between encrusting cheilostome ectoprocts in a Jamaican cryptic reef environment. J. Anim. Ecol. 48:805–823.

———. 1981. Interspecific competition and species distributions: the ghosts of theories and data past. Amer. Zool. 21:889–901.

Jacobs, J. 1977. Coexistence of similar zooplankton species by differential adaptation to reproduction and escape in an environment with fluctuating food and enemy densities. II. Field data analysis of *Daphnia*. Oecologia (Berl.) 30:303–329.

Jaeger, R. 1974. Interference or exploitation? A second look at competition between salamanders. J. Herpetol. 8:191–194.

Jain, S., and A. Bradshaw. 1966. Evolutionary divergence among adjacent plant populations. Heredity 21:407–441.

Jaksic, F. 1981. Abuse and misuse of the term "guild" in ecological studies. Oikos 37:397–400.

James, F. C. 1971. Ordinations of habitat relationships among breeding birds. Wilson Bull. 83: 215–236.

Janos, D. P. 1980. Mycorrhizae influence tropical succession. Biotropica 12 (suppl.):56–64.

Jansson, C., J. Ekman, and A. von Brömssen. 1981. Winter mortality and food supply in tits, *Parus* spp. Oikos 37:313–322.

Janzen, D. H. 1966. Coevolution of mutualism between ants and acacias in Central America. Evolution 20:249–275.

———. 1969. Seed-eaters versus seed size, number, toxicity and dispersal. Evolution 23:1–27.

———. 1971. Seed predation by animals. Annu. Rev. Ecol. Syst. 2:465–492.

———. 1979. How to be a fig. Annu. Rev. Ecol. Syst. 10:13–51.

———. 1980. When is it coevolution? Evolution 34:611–612.

———. 1983a. Dispersal of seeds by vertebrate guts. Pages 232–262 *in* D. J. Futuyma and M. Slatkin, eds. Coevolution. Sinauer, Sunderland, Mass.

———, ed. 1983b. Costa Rican natural history. University of Chicago Press, Chicago.

Jarman, P. J. 1974. The social organisation of antelope in relation to their ecology. Behaviour 48:216–266.

Jarman, P. J., and M. V. Jarman. 1979. The dynamics of ungulate social organization. Pages 185–220 *in* A. R. E. Sinclair and M. Norton-Griffiths, eds. Serengeti: dynamics of an ecosystem. University of Chicago Press, Chicago.

Jarman, P. J., and A. R. E. Sinclair. 1979. Feeding strategy and the pattern of resource-partitioning in ungulates. Pages 130–163 *in* A. R. E. Sinclair and M. Norton-Griffiths, eds. Serengeti: dynamics of an ecosystem. University of Chicago Press, Chicago.

Jarvis, J. V. M. 1981. Eusociality in a mammal: cooperative breeding in naked mole-rat colonies. Science 212:241–250.

Jarvis, J. V. M., and J. B. Sale. 1971. Burrowing and burrow patterns of East African mole rats *Tachyoryctes, Heliophobius*, and *Heterocephalus*. J. Zool. 163:451–479.

Jenni, D. A. 1974. Evolution of polyandry in birds. Amer. Zool. 14:129–144.

———, and G. Collier. 1972. Polyandry in the American Jacana (*Jacana spinosa*). Auk 89:743–769.

Jenny, H. 1980. The soil resource: origin and behavior. Springer-Verlag, New York.

Johns, D. W., and K. B. Armitage. 1979. Behavioral ecology of alpine yellow-bellied marmots. Behav. Ecol. Sociobiol. 5:133–157.

Johnson, D. W., G. S. Henderson, D. D. Huff, S. E. Lindberg, D. D. Richter, D. S. Shriner, D. E. Todd, and J. Turner. 1982. Cycling of organic and inorganic sulphur in a chestnut oak forest. Oecologia (Berl.) 54:141–148.

Johnson, E. A. 1981. Vegetation organization and dynamics of lichen woodland communities in the Northwest Territories, Canada. Ecology 62:200–215.

Johnson, L. K., and S. P. Hubbell. 1975. Contrasting foraging strategies and coexistence of two bee species on a single resource. Ecology 56:1398–1406.

Johnson, N. D., C. C. Chu, P. R. Ehrlich, and H. A.

Mooney. 1984. The seasonal dynamics of leaf resin, nitrogen and herbivore damage in *Eriodictyon californicum* and their parallels in *Diplacus aurantiacus*. Oecologia (Berl.) 61:398–402.

Johnson, P. L., and W. T. Swank. 1973. Studies of cation budgets in the Southern Appalachians on four experimental watersheds with contrasting vegetation. Ecology 54:70–80.

Jones, C. D., and R. J. Little, eds. 1983. Handbook of experimental pollination biology. Van Nostrand Reinhold, New York.

Jordan, C. F., and G. Escalante. 1980. Root productivity in an Amazonian rain forest. Ecology 61:14–18.

Kadlec, J. A., and W. H. Drury. 1968. Structure of the New England herring gull population. Ecology 49:644–676.

Kareiva, P. 1983. Influence of vegetation texture on herbivore populations: resource concentration and herbivore movement. Pages 259–289 *in* R. F. Denno and M. S. McClure, eds. Variable plants and herbivores in natural and managed systems. Academic Press, New York.

Karlin, S. 1966. A first course in stochastic processes. Academic Press, New York.

———. 1973. Sex and infinity: a mathematical analysis of the advantages and disadvantages of recombination. Pages 155–194 *in* M. Bartlett and R. Hiorns, eds. The mathematical theory of the dynamics of biological populations. Academic Press, New York.

———. 1976. Population subdivision and selection-migration interaction. Pages 617–657 *in* S. Karlin and E. Nevo, eds. Population genetics and ecology. Academic Press, New York.

———. 1979. Models for multifactorial inheritance. I. Multivariate formulations and basic convergence results. Theor. Popul. Biol. 15:308–355.

Karlin, S., and E. Nevo, eds. 1976. Population genetics and ecology. Academic Press, New York.

Kaufman, D. W. 1973. Was oddity conspicuous in prey selection experiments? Nature 244:111.

Kawai, M. 1965. Newly-acquired pre-cultural behavior of the natural troop of Japanese monkeys on Koshima Island. Primates 6:1–30.

Kawamura, S. 1963. The process of sub-culture propagation among Japanese macaques. Pages 82–90 *in* C. H. Southwick, ed. Primate social behavior: an enduring problem. Van Nostrand, Princeton, N.J. [Originally published in Japanese in J. Primatology. 1959. 2(1):43–60.]

Keast, A. 1968. Competitive interactions and the evolution of ecological niches as illustrated by the Australian honeyeater genus *Melithreptus* (Meliphagidae). Evolution 22:762–784.

———. 1970a. Adaptive evolution and shifts in niche occupation in island birds. Biotropica 2:61–75.

———. 1970b. Food specialization and bioenergetic interrelations in the fish faunas of some small Ontario waterways. Pages 377–411 *in* J. Steele, ed. Marine food chains. Oliver and Boyd, Edinburgh.

Keast, A., and D. Webb. 1966. Mouth and body form relative to feeding ecology in the fish fauna of a small lake, Lake Opinicon, Ontario. J. Fisheries Board Canad. 23:1845–1874.

Keeling, C. D., A. F. Carter, and W. G. Mook. 1984. Seasonal, latitudinal, and secular variations in the abundance and isotope ratios of atmospheric CO_2. J. Geophys. Res. 89(D3):4615–4628.

Keith, L. B. 1963. Wildlife's ten-year cycle. University of Wisconsin Press, Madison.

Kelly, C. D., and T. F. Hourigan. 1983. The function of conspicuous coloration in chaetodontid fishes: a new hypothesis. Anim. Behav. 31:615–617.

Kendall, D. G. 1948. A form of wave propagation associated with the equations of heat conduction. Proc. Cambridge Phil. Soc. 44:591–594.

Kennedy, M. K., and R. W. Merritt. 1980. Horse and buggy island. Nat. Hist. May:34–40.

Kerr, R. A. 1980. Asteroid theory of extinctions strengthened. Science 10:514–517.

———. 1984. Periodic impacts and extinctions reported. Science 223:1277–1279.

Kettlewell, H. B. D. 1958a. Industrial melanism in the Lepidoptera and its contribution to our knowledge of evolution. Proc. 10th Int. Cong. Entomol. 2:831–841.

———. 1958b. A survey of the frequencies of *Biston betularia* L. (Lep.) and its melanic forms in Britain. Heredity 12:51–72.

———. 1973. The evolution of melanism. Clarendon, Oxford.

Kettlewell, H., and R. Berry. 1961. The study of a cline. *Amathes glareosa* Esp. and its melanic f. *edda* Stand. (Lep.) in Shetland. Heredity 24:1–14.

Keyfitz, N., and W. Flieger. 1971. Population: facts and methods of demography. Freeman, San Francisco.

Kiddle, M. 1961. Men of yesterday: a social history of the western district of Victoria. Melbourne University Press, Melbourne.

King, A. W., and S. L. Pimm. 1983. Complexity, diversity, and stability: a reconciliation of theoretical and empirical results. Am. Nat. 122:229–239.

King, C. E., and P. S. Dawson. 1971. Population biology and the *Tribolium* model. Evol. Biol. 5:133–227.

King, D. 1981. Tree dimensions: maximizing the rate of height growth in dense stands. Oecologia (Berl.) 51:351–356.

King, D., and J. Roughgarden. 1982. Graded allocation between vegetative and reproductive growth for annual plants in growing seasons of random length. Theor. Popul. Biol. 22:1–16.

———. 1983. Energy allocation patterns of the California grassland annuals, *Plantago erecta* and *Clarkia rubicunda*. Ecology 64:16–24.

King, J. R. 1972. Variation in the song of the rufous-collared sparrow, *Zonotrichia capensis*, in northwestern Argentina. Z. Tierpsychol. 30:344–373.

Kirchner, T. 1980. Community structure in relation to body size of species. Ph.D. diss. Colorado State University, Fort Collins, Col.

Kitchell, J. F., R. V. O'Neill, D. Webb, G. W. Gallepp, S. M. Bartell, J. F. Koonce, and B. S. Ausmus. 1979. Consumer regulation of nutrient cycling. BioSci. 29:28–34.

Klausing, O. 1956. Untersuchungen über den Mineralumsatz in Buchenwäldern auf Granit und Diorit. Forstwiss. Cbl. 75:18–32.

Kleiman, D. G., and J. F. Eisenberg. 1973. Comparisons of canid and felid social systems from an evolutionary perspective. Anim. Behav. 21:637–659.

Klopfer, P. H. 1982. Mating types and human sexuality. BioSci. 32(10):803–806.

Kluyver, H. N., and L. Tinbergen 1953. Territory and the regulation of density in titmice. Arch Neerland Zool. 10:265–289.

Koehl, M. A. R. 1984. How do benthic organisms withstand moving water? Amer. Zool. 24:57–70.

Koehl, M. A. R., and J. R. Strickler. 1981. Copepod feeding currents: food capture at low Reynolds number. Limnol. Oceanogr. 26:1062–1073.

Koehn, R. 1978. Physiology and biochemistry of enzyme variation: the interface of ecology and population genetics. Pages 51–72 *in* P. F. Brussard, ed. Ecological genetics: the interface. Springer-Verlag, New York.

Koenig, W. D. 1981. Reproductive success, group size and the evolution of cooperative breeding in the acorn woodpecker. Am. Nat. 117:421–443.

———. 1984. Coalitions of male lions: making the best of a bad job? Nature 293(5831):413.

Koenig, W. D., and F. A. Pitelka. 1981. Ecological factors and kin selection in the evolution of cooperative breeding in birds. Pages 261–280 *in* R. D. Alexander and D. W. Tinkle, eds. Natural selection and social behavior: recent research and new theory. Chiron, New York.

Kohn, A. 1959. The ecology of *Conus* in Hawaii. Ecol. Monogr. 28:47–90.

———. 1966. Food specialization in *Conus* in Hawaii and California. Ecology 47:1041–1043.

———. 1967. Environmental complexity and species diversity in the gastropod genus *Conus* on Indo–West Pacific reef platforms. Am. Nat. 101:251–259.

———. 1968. Microhabitats, abundance and food of *Conus* on atoll reefs in the Maldive and Chagos Islands. Ecology 49:1046–1061.

———. 1971. Diversity, utilization of resources and adaptive radiation in shallow-water marine invertebrates of tropical oceanic islands. Limnol. Oceanogr. 16:332–348.

Koivisto, I. 1965. Behavior of the black grouse, *Lyrurus tetrix* (L.), during the spring display. Finnish Game Res. 26:1–60.

Konishi, M., and F. Nottebohm. 1969. Experimental studies in the ontogeny of avian vocalizations. Pages 29–48 *in* R. A. Hinde, ed. Bird vocalizations: their relation to current problems in biology and psychology. Cambridge University Press, Cambridge.

Kozlovsky, D. G. 1968. A critical evaluation of the trophic level concept. I. Ecological efficiencies. Ecology 49:49–60.

Krajina, V. J. 1969. Ecology of forest trees in British Columbia. Ecol. Western North America 2:1–147.

Krebs, J. R. 1971. Territory and breeding density in the granttite, *parus major* L. Ecology 52:2–22.

———. 1974. Colonial nesting and social feeding as strategies for exploiting food resources in the great blue heron (*Arden herodias*). Behavior 51:99–134.

———. 1977. The significance of song repertoires: the Beau Geste hypothesis. Anim. Behav. 25:475–478.

Krebs, J. R., and N. B. Davies, eds. 1978. Behavioral ecology: an evolutionary approach. Sinauer, Sunderland, Mass.

Krebs, J. R., J. T. Erichsen, M. I. Webber, and E. L. Charnov. 1977. Optimal prey selection in the great tit (*Parus major*). Anim. Behav. 25:30–38.

Krebs, J. R., J. C. Ryan, and E. L. Charnov. 1974. Hunting by expectation or optimal foraging? a study of patch use by chickadees. Anim. Behav. 22:953–964.

Kreitman, M. 1983. Nucleotide polymorphism at the alcohol dehydrogenase locus of *Drosophila melanogaster*. Nature 304:412–417.

Krischik, V. A., and R. F. Denno. 1983. Individual, population and geographic patterns in plant defense. Pages 463–512 *in* R. F. Denno and M. S. McClure, eds. Variable plants and herbivores in natural and managed systems. Academic Press, New York.

Kroodsma, D. E. 1974. Song learning, dialects and dispersal in Bewick's wren. Z. Tierpsychol. 35: 352–380.

Kruijt, J. P., and J. A. Hogan. 1967. Social behaviour

on the lek in black grouse, *Lyrurus tetrix tetrix* (L.). Ardea 55:203–240.

Kruuk, H. 1972. The spotted hyena. University of Chicago Press. Chicago.

Kruuk, H., and M. Turner. 1967. Comparative notes on predation by lion, leopard, cheetah, and wild dog in the Serengeti area, East Africa. Mammalia 31:1–27.

Kucera, C. L. 1981. Grasslands and fire. Pages 9–111 *in* H. A. Mooney, T. M. Bonnicksen, N. L. Christensen, J. E. Lotan, and W. A. Reiners, eds. Fire regimes and ecosystem properties. USDA Forest Service General Technical Report WO-26.

Kuenzler, E. J. 1961. Phosphorus budget of a mussel population. Limnol. Oceanogr. 6:400–415.

Kullenberg, B. 1961. Studies in *Ophrys* pollination. Zool. Bidr. Upps. 34:1–340.

Lack, A. 1976. Competition for pollinators and evolution in *Centaurea*. New Phytol. 77:787–792.

Lack, D. 1947. Darwin's finches. Cambridge University Press, Cambridge.

———. 1954. The natural regulation of animal numbers. Oxford University Press, Oxford.

———. 1966. Population studies of birds. Clarendon, Oxford University Press, London.

———. 1968. Ecological adaptations for breeding in birds. Methuen, London.

———. 1976. Island biology illustrated by the land birds of Jamaica. University of California Press, Berkeley.

Lamprecht, J. 1978. The relationship between food competition and foraging group size in some larger carnivores. Z. Tierpsychol. 46:337–343.

Landry, M. R. 1976. The structure of marine ecosystems: an alternative. Mar. Biol. 35:1–7.

Lange, O. L. 1959. Untersuchungen uber Wurmehaushalt und Hitzeresistenz mauretanischer Wusten und Savannenpflanzen. Flora (Jena) 147: 595–651.

Larcher, W. 1975. Physiological plant ecology. M. A. Biederman-Thorson, trans. Springer-Verlag, New York.

Laurent, R. F. 1973. A parallel survey of equatorial amphibians and reptiles in Africa and South America. Pages 259–266 *in* B. J. Meggers, J. E. S. Ayensu, and W. D. Duckworth, eds. Tropical forest ecosystems in Africa and South America: a comparative review. Smithsonian Institution Press, Washington, D.C.

Law, R., A. D. Bradshaw, and P. D. Putwain 1977. Life-history variation in *Poa annua*. Evolution 31:233–246.

Lawrence, D. B. 1958. Glaciers and vegetation in southeastern Alaska. Am. Sci. June: 89–124.

Lawton, J. H., and S. McNeill. 1979. Between the devil and the deep blue sea: on the problem of being a herbivore. Pages 223–244 *in* R. M. Anderson, B. D. Turner, and L. R. Taylor, eds. Population dynamics. Blackwell, Oxford.

Lawton, J. H., and D. Schroder. 1978. Some observations on the structure of phytophagous insect communities: the implications for biological control. Pages 57–63 *in* Proc. 4th Int. Symp. Biol. Control Weeds. Institute for Food and Agriculture Science, University of Florida, Gainesville.

Lawton, J. H., and D. R. Strong, Jr. 1981. Community patterns and competition in folivorous insects. Am. Nat. 118:317–338.

Lazell, J. 1972. The *Anolis* (Sauria, Iguanidae) of the Lesser Antilles. Bull. Mus. Comp. Zool. 143:1–115.

Le Boeuf, B. J. 1974. Male-male competition and reproductive success in elephant seals. Amer. Zool. 14:163–176.

Leigh, E. G., Jr., A. S. Rand, and D. M. Windsor, eds. 1982. The ecology of a tropical forest: season rhythms and long-term changes. Smithsonian Institution Press, Washington, D.C.

Lenski, R. E., and P. Service. 1982. The statistical analysis of population growth rates calculated from schedules of survivorship and fecundity. Ecology 63:655–662.

Leopold, A. 1953. Round River. Oxford Univ. Press, Oxford.

Lerner, I. M. 1958. The genetic basis of selection. Wiley, New York.

Lerner, I. M., and F. K. Ho. 1961. Genotype and competitive ability of *Tribolium* species. Am. Nat. 95:329–343.

Leslie, P. H. 1945. On the use of matrices in certain population mathematics. Biometrika 33:183–212.

———. 1948. Some further notes on the use of matrices in population mathematics. Biometrika 35:213–245.

Lessells, C. M. 1985. Parasitoid foraging: should parasitism be density dependent? J. Anim. Ecol. 54:27–41.

Levene, H. 1953. Genetic equilibrium when more than one ecological niche is available. Am. Nat. 87:311–313.

Levin, D. A., and W. W. Anderson. 1970. Competition for pollinators between simultaneously flowering species. Am. Nat. 104:345–354.

Levinton, J. S. 1982. Marine ecology. Prentice-Hall, Englewood Cliffs, N. J.

Lewin, R. 1983a. No dinosaurs this time. Science 221:1168–1169.

———. 1983b. Santa Rosalia was a goat. Science 221:636–639.

———. 1983c. What killed the giant mammals? Science 221:1168–1169.

Lewis, C. 1978. A review of substratum selection in free-living and symbiotic cirripeds. Pages 207–218 *in* F. Chia and M. Rice, eds. Settlement and metamorphosis of marine invertebrate larvae. Elsevier, New York.

Lewis, J. G. 1977. Game domestication for animal production in Kenya: activity patterns of eland, oryx, buffalo and zebu cattle. J. Agric. Sci., Camb. 89:551–563.

Lewis, W. M., Jr. 1974. Primary production in the plankton community of a tropical lake. Ecol. Monogr. 44:377–409.

Lewontin, R. C. 1974. The genetic basis of evolutionary change. Columbia University Press, New York.

Lewontin, R. C., and J. L. Hubby. 1966. A molecular approach to the study of genic heterozygosity in natural populations of *Drosophila pseudo-obscura.* Genetics 54:595–609.

Licht, P., and G. Gorman. 1970. Reproductive and fat cycles in Caribbean *Anolis* lizards. Univ. Calif. Publ. Zool. 95:1–52.

Lieth, H. 1956. Ein Beitrag zur Frage der Korrelation zwischen mittleren Klimawerten und Vegetationsformationen. Berl. Deutsch. Bot. Gesell. 69:169–176.

———, ed. 1978. Patterns of primary production in the biosphere. Dowden, Hutchinson and Ross, Stroudsburg, Pa.

Ligon, J. D., and S. H. Ligon. 1982. The cooperative breeding behavior of the green woodhoopoe. Sci. Am. July:26–134.

Likens, G. E., and F. H. Bormann. 1974. Acid rain: a serious regional environmental problem. Science 184:1176–1179.

Likens, G. E., F. H. Bormann, R. S. Pierce, J. S. Eaton, and N. M. Johnson. 1977. Biogeochemistry of a forestall ecosystem. Springer-Verlag, New York.

Likens, G. E., F. H. Bormann, R. S. Pierce, and W. A. Reiners. 1978. Recovery of a deforested ecosystem. Science 199:492–495.

Lincoln, D. E. 1980. Leaf resin flavonoids of *Diplacus aurantiacus.* Biochem. Syst. Ecol. 8:347–400.

Lincoln, D. E., T. S. Newton, P. R. Ehrlich, and K. S. Williams. 1982. Coevolution of the checkerspot butterfly *Euphydryas chalcedona* and its larval food plant *Diplacus aurantiacus:* larval response to protein and leaf resin. Oecologia (Berl.) 52:216–223.

Lindman, R. L. 1942. The trophic-dynamic aspect of ecology. Ecology 23:399–417.

Little, R. J. 1983. A review of floral food deception mimicries with comments on floral mutualism. Pages 294–309 *in* C. E. Jones and R. J. Little, eds. Handbook of experimental pollination biology. Van Nostrand Reinhold, New York.

Livingston, R. J., ed. 1979. Ecological processes in coastal and marine systems. Plenum, New York.

Lloyd, D. G. 1980. Sexual strategies in plants. III. A quantitative method for describing the gender of plants. N. Z. J. Bot. 18:103–108.

Lloyd, J. E. 1966. Studies on the flash communication system in *Photinus* fireflies. Miscellaneous Publications. Museum of Zoology. University of Michigan Press, Ann Arbor.

———. 1975. Aggressive mimicry in photuris fireflies: signal repertoires by femme fatales. Science 187:452–453.

Lloyd, M. 1968. Self-regulation of adult number by cannibalism in two laboratory strains of flour beetles *(Tribolium castaneum).* Ecology 49:245–259.

Loiselle, P. V., and G. W. Barlow, 1978. Do fishes lek like birds? Pages 31–75 *in* E. S. Reese and F. J. Lighter, eds. Contrasts in behavior: adaptations in the aquatic and terrestrial environments. Wiley, New York.

Long, J. L. 1981. Introduced birds of the world. Universe, New York.

Lorenz, K. 1962. The function of colour in coral reef fishes. Proc. R. Inst. Gt. Br. 39:282–296.

Lotka, A. J. 1925. Elements of physical biology. Reprinted 1956 by Dover, New York.

———. 1932. The growth of mixed populations: two species competing for a common food supply. J. Wash. Acad. Sci. 22:461–469.

———. 1945. Population analysis as a chapter in the mathematical theory of evolution. Pages 355–384 *in* W. E. Le Gros Clark and P. B. Medawar, eds. Essays on growth and form, presented to D'Arcy Wentworth Thompson, Oxford University Press, Oxford.

Loucks, O. L. 1962. Ordinating forest communities by means of environmental scalars and phytosociological indices. Ecol. Monogr. 32:137–166.

———. 1970. Evolution of diversity, efficiency and community stability. Amer. Zool. 10:17–25.

Louda, S. M. 1982a. Distribution ecology: variation in plant recruitment over a gradient in relation to insect seed predation. Ecol. Monogr. 52(1): 25–41.

———. 1982b. Inflorescence spiders: a cost/benefit analysis for the host plant, *Haplopappus venetus* Blake (Asteraceae). Oecologia 55:185–191.

Lovelock, J., and S. Epton. 1975. The quest for Gaia. New Scientist Feb. 6:304–306.

Lovelock, J. E. 1979. Gaia: a new look at life on Earth. Oxford University Press, Oxford.

———. 1974. Atmospheric homeostasis by and for the biosphere: the Gaia hypothesis. Tellus 26:2–10.

Low, R. M. 1971. Interspecific territoriality in a pomacentid reef fish, *Pomacentrus flavicauda* Whitley. Ecology 52:648–654.

Lowe, V. P. W. 1969. Population dynamics of the red deer (*Cervus elaphus* L.) on Rhum. J. Anim. Ecol. 38:425–457.

Lowe-McConnell, R. H. 1977. Ecology of fishes in tropical waters. Edward Arnold, London.

Lubchenco, J. 1978. Plant species diversity in a marine intertidal community: importance of herbivore food preference and algal competitive abilities. Am. Nat. 112:23–39.

Lubchenco, J., and B. Menge. 1978. Community development and persistence in a low rocky intertidal zone. Ecol. Monogr. 59:67–94.

Luckinbill, L. S., R. Arking, M. Clare, W. Cirocco, and S. Buck. 1985. Selection for delayed senescence in *Drosophila melanogaster.* Evolution 38:996–1003.

Lumpkin, S., K. Kessel, P. G. Zenone, and C. J. Erickson. 1982. Proximity between the sexes in ring doves: social bonds or surveillance? Anim. Behav. 30:506–513.

Lutz, H. J. 1945. Vegetation on a trenched plot twenty-one years after establishment. Ecology 26(2):200–202.

Lynch, M. 1979. Predation, competition and zooplankton community structure: an experimental study. Limnol. Oceanogr. 24:253–272.

MacArthur, R. H. 1962. Some generalized theorems of natural selection. Proc. Natl. Acad. Sci. USA 231:123–138.

———. 1972. Geographical ecology: Patterns in the distribution of species. Harper & Row, New York.

MacArthur, R. H., and J. H. Connell. 1967. The biology of populations. John Wiley & Sons, New York.

MacArthur, R. H., and R. Levins. 1967. The limiting similarity, convergence and divergence of coexisting species. Am. Nat. 101:377–385.

MacArthur, R. H., and E. O. Wilson. 1963. An equilibrium theory of insular zoogeography. Evolution 17:373–387.

———. 1967. The theory of island biogeography. Princeton University Press, Princeton, N.J.

Macfadyen, A. 1979. The role of the fauna in decomposition processes in grassland. Sci. Proc. R. Dublin. Soc. Sers. A 6:197–206.

MacLulich, D. A. 1937. Fluctuations in the numbers of the varying hare *(Lepus americanus)*. Univ. Toronto Studies; Biol. Ser., no. 43.

MacMahon, J., D. Schimpf, D. Anderson, K. Smith, and R. Bayn. 1981. An organism-centered approach to some community and ecosystem concepts. J. Theor. Biol. 88:287–307.

MacMahon, J. A. 1980. Ecosystems over time: succession and other types of change. Pages 27–57 *in* R. H. Waring, ed. Forests: fresh perspectives from ecosystem analysis. Proc. 40th Annu. Bio. Coll. Oregon State University Press, Corvallis.

MacNally, R. C. 1979. Social organisation and interspecific interactions in two sympatric species of *Ranidella* (Anura). Oecologia 42:293–306.

Maddock, L. 1979. The "migration" and grazing succession. Pages 104–129 *in* A. R. E. Sinclair and M. Norton-Griffiths, eds. Serengeti: dynamics of an ecosystem. University of Chicago Press, Chicago.

Maddox, J. 1984. Extinctions by catastrophe? Nature 308:685.

Malcolm, J. R., and H. van Lawick. 1975. Notes on wild dogs hunting zebras. Mammalia 39:231–240.

———, and K. Marten 1982. Natural selection and the communal rearing of pups in African wild dogs *(Lycaon pictus)*. Behav. Ecol. Sociobiol. 10:1–13.

Malloch, D. W., K. A. Pirozynski, and P. H. Raven. 1980. Ecological and evolutionary significance of mycorrhizal symbioses in vascular plants (a review). Proc. natl. Acad. Sci. USA 77(4):2113–2118.

Mani, M. S. 1968. Ecology and biogeography of high altitude insects. Ser. Ent. 4. Junk, The Hague.

Manley, G. V., J. W. Butcher, and J. E. Cantlon. 1975. Relationship of insects to distribution and abundance of *Melampyrum lineare* (Serophulariaceae). Pedobiologia 15:385–404.

Maranto, G., and S. Brownlee. 1984. Why sex? Discover February:24–28.

Margalef, R. 1968. Perspectives in ecological theory. University of Chicago Press, Chicago.

Margulis, L., and J. E. Lovelock. 1974. Biological modulation of the Earth's atmosphere. Icarus 21:471–478.

Mariscal, R. N. 1972. Behavior of symbiotic fishes and sea anemones. Pages 2327–2360 *in* H. E. Winns and B. L. Olla, eds. Behavior of marine animals. Vol. 2. Plenum, New York.

Marks, P. L. 1974. The role of pin cherry (*Prunus pensylvanica* L.) in the maintenance of stability in northern hardwood ecosystems. Ecol. Monogr. 44:73–88.

Marks, P. L., and F. H. Bormann. 1972. Revegetation following forest cutting: mechanisms for return for steady-state nutrient cycling. Science 176:914–915.

Marler, P. R. 1952. Variation in the song of the chaffinch. Ibis 94:458–472.

———. 1955. Characteristics of some animal calls. Nature (Lond.) 176:6–8.

———. 1968. Visual systems. Pages 103–126 *in* T. A. Sebeck, ed. Animal communications. University of Indiana Press, Bloomington.

———. 1970. A comparative approach to vocal learning: song development in white-crowned sparrows. J. Comp. Physiol. Psychol. 71 (suppl.): 1–25.

———. 1976. On animal aggression: the roles of strangeness and familiarity. Amer. Psych. March:239–246.

Marler, P. R., and P. Mundinger. 1971. Vocal learning in birds. Pages 389–450 *in* H. Moltz, ed. The ontogeny of vertebrate behavior. Academic Press, New York.

Marler, P. R., and M. Tamura. 1962. Song "dialects" in three populations of white-crowned sparrows. Condor 61:368–377.

Marsh, D., J. S. Kennedy, and A. R. Ludlow. 1978. An analysis of anemotactic zigzagging flight in male moths stimulated by pheromone. Physiol. Entomol. 3:226–240.

Marshall, G. A. K. 1908. On diaposematism, with reference to some limitations of the Müllerian hypothesis of mimicry. Trans. Ent. Soc. London 1908:93–142.

Martin, J. H., G. A. Knauer, D. M. Karl and W. W. Broenkow, 1986. VERTEX: Carbon cycling in the northeast Pacific. Deep Sea Res. In press.

Martin, L. D. 1983. The origin of birds and avian flight. Pages 105–129 *in* R. F. Johnston, ed. Current ornithology. Plenum, New York.

Martin, P. S., and R. G. Klein, eds. 1984. Quaternary extinctions: a prehistoric revolution. University of Arizona Press, Tucson.

Martin, P. S., and H. E. Wright, Jr. 1967. Pleistocene extinctions: the search for a cause. Yale University Press, New Haven, Conn.

Mason, L. G., P. R. Ehrlich, and T. C. Emmel, 1968. The population biology of the butterfly, *Euphydryas editha.* VI. Phenetics of the Jasper Ridge Colony, 1965–1966. Evolution 22:46–54.

Mattson, W. J., and N. D. Addy. 1975. Phytophagous insects as regulators of forest primary production. Science 190:515–522.

Mautz, W. J. 1980. Factors influencing evaporative water loss in lizards. Comp. Biochem. Physiol. 67:429–437.

May, R. M. 1973. Stability and complexity in model ecosystems. Princeton University Press, Princeton, N.J.

———. 1974. Historic first with community ecology. Nature 251:376–377.

———. 1975a. Patterns of species abundance and diversity. Pages 81–120 *in* M. Cody and J. Diamond, eds. Ecology and evolution of communities. Harvard University Press, Cambridge, Mass.

———. 1975b. Successional patterns and indices of diversity. Nature 258:285–286.

———. 1982. The impact, transmission and evolution of infectious diseases. Nature 297:539.

———. 1983. Parasitic infections as regulators of animal populations. Am. Sci. 71(1):36–45.

May, R. M., and R. M. Anderson. 1983. Parasite-host coevolution. Pages 186–206 *in* D. J. Futuyma and M. Slatkin, eds. Coevolution. Sinauer, Sunderland, Mass.

May, R. M., and W. Leonard. 1975. Nonlinear aspects of competition between three species. SIAM J. Appl. Math. 29:243–253.

May, R. M., and R. H. MacArthur. 1972. Niche overlap as a function of environmental variability. Proc. Natl. Acad. Sci. USA 69:1109–1113.

Maynard Smith, J. 1968. Evolution in sexual and asexual populations. Am. Nat. 102:469–473.

———. 1978. The Evolution of Sex. Cambridge Univ. Press, Cambridge.

———. 1982. Evolution and the theory of games. Cambridge University Press, Cambridge.

Maynard Smith, J., and G. R. Price. 1973. The logic of animal conflict. Nature 248:15–18.

Mayr, E. 1963. Animal species and evolution. Harvard University Press, Cambridge, Mass.

———. 1969. The biological meaning of species. Biol. J. Linn. Soc. 1:311–320.

———. 1982a. Vicariance biogeography (review). Auk 99:618–620.

———. 1982b. Systematics and biogeography: cladistics and vicariance (review). Auk 99:621–622.

McClure, M. S. 1983. Competition between herbivores and increased resource heterogeneity. Pages 125–153 *in* R. F. Denno and M. S. McClure, eds. Variable plants and herbivores in natural and managed systems. Academic Press, New York.

McClure, M. S., and P. W. Price. 1975. Competition among sympatric *Erythroneura* leafhoppers (Homoptera: Cicadellidae) on American sycamore. Ecology 56:1388–1397.

———. 1976. Ecotope characteristics of coexisting *Erythroneura* leafhoppers, Homoptera: Cicadellidae, on sycamore. Ecology 57:928–940.

McFarland, W. N., F. H. Pough, T. J. Cade, and J. B.

Heiser. 1979. Vertebrate life. Macmillan, New York.

McIntosh, R. P. 1980. The background and some current problems of theoretical ecology. Synthesis 43:195–255.

McKaye, K. R. 1977. Defense of a predator's young by a herbivorous fish: an unusual strategy. Am. Nat. 111(978):301–315.

———. 1979. Defense of a predator's young revisited. Am. Nat. 113:595–601.

———. 1981a. Field observation on death feigning: a unique hunting behavior by the predatory cichlid, *Haplochromis livingstoni* of Lake Malawi. Envir. Biol. Fish. 6(3/4):361–365.

———. 1981b. Natural selection and the evolution of interspecific brood care in fishes. Pages 173–183 *in* R. Alexander and J. Tinkle, eds. Natural selection of social behavior. Chiron, New York.

———. 1983. Ecology and breeding behavior of a cichlid fish, *Cyrtocara eucinostomus,* in a large lek in Lake Malawi, Africa. Envir. Biol. Fish. 8:81–96.

McKaye, K. R., and T. Kocher. 1983. Head ramming behaviour by three paedophagous cichlids in Lake Malawi, Africa. Anim. Behav. 31:206–210.

McKaye, K. R., and N. M. McKaye. 1977. Communal care and kidnapping of young by parental cichlids. Evolution 31:674–681.

McKaye, K. R., and M. K. Oliver. 1980. Geometry of a selfish school: defense of cichlid young by bagrid catfish in Lake Malawi, Africa. Anim. Behav. 28:4.

McKechnie, S. W., P. R. Ehrlich, and R. R. White. 1975. Population genetics of *Euphydryas* butterflies. I. Genetic variation and the neutrality hypothesis. Genetics 81:571–594.

McKey, D. 1975. The ecology of coevolved seed dispersal systems. Pages 159–191 *in* L. E. Gilbert and P. H. Raven, eds. Coevolution of animals and plants. University of Texas Press, Austin.

McLaughlin, R. 1970. Aspects of the biology of cheetahs *Acinonyx jubatus* (Schreber) in Nairobi National Park. M.Sc. thesis. University of Nairobi, Nairobi.

McLean, D. M. 1980. Terminal Cretaceous catastrophe. Nature 287:760.

McLean, I. G. 1983. Paternal behaviour and killing of young in Arctic ground squirrels. Anim. Behav. 31:32–44.

McMillan, C. 1959. The role of ecotypic variation in the distribution of the central grasslands of North America. Ecol. Monogr. 29:285–308.

McNaughton, S. J. 1968. Structure and function in California grasslands. Ecology 49:962–972.

———. 1976. Serengeti migratory wildebeest: facilitation of energy flow by grazing. Science 191:92–94.

———. 1977. Diversity and stability of ecological communities: a comment on the role of empiricism in ecology. Am. Nat. 111:515–525.

———. 1979a. Grazing as an optimization process: grass–ungulate relationships in the Serengeti. Am. Nat. 113:691–703.

———. 1979b. Grassland-herbivore dynamics. Pages 46–81 *in* A. R. E. Sinclair and M. Norton-Griffiths, eds. Serengeti: dynamics of an ecosystem. University of Chicago Press, Chicago.

———. 1983. Serengeti grassland ecology: the role of composite environmental factors and contingency in community organization. Ecol. Monogr. 53:291–320.

Mech, L. D. 1966. The wolves of Isle Royale. Fauna of the National Parks of the U.S. Fauna sers. 7. U.S. Government Printing Office, Washington, D.C.

Medawar, P. B. 1946. Old age and natural death. Mod. Quart. 1:30–56.

Meentemeyer, V., E. O. Box, and R. Thompson. 1982. World patterns and amounts of terrestrial plant litter production. BioSci. 32:125–128.

Meggers, B. J., E. S. Ayensu, and W. D. Duckworth, eds. 1973. Tropical forest ecosystems in Africa and South America: a comparative review. Smithsonian Institution Press, Washington, D.C.

Melampy, M. N., and H. F. Howe. 1977. Sex ratio in the tropical tree *Triplaris americana* (Polygonaceae). Evolution 31:867–872.

Menge, B. A. 1972. Foraging strategy of a starfish in relation to actual prey availability and environmental predictability. Ecol. Monogr. 42:25–50.

———. 1976. Organization of the New England rocky intertidal community: role of predation, competition and environmental heterogeneity. Ecol. Monogr. 46:355–393.

———. 1983. Components of predation intensity in the low zone of the New England rocky intertidal region. Oecologia (Berl.) 58:141–155.

Menge, B. A., and J. Lubchenco. 1981. Community organization in temperate and tropical rocky intertidal habitats: prey refuges in relation to consumer pressure gradients. Ecol. Monogr. 51:429–450.

Mertz, D. 1975. Senescent decline in flour beetles selected for early adult fitness. Physiol. Zool. 48:1–23.

Mertz, D. B., D. A. Cawthon, and T. Park. 1976. An experimental analysis of indeterminacy in *Tribolium.* Proc. Natl. Acad. Sci. USA 73:1368–1372.

Michelsen, A. 1979. Insect ears as mechanical systems. Am. Sci. 67:696–706.

Michener, C. D. 1962. An interesting method of pollen collecting by bees from flowers with tubular anthers. Rev. Biol. Trop. 10:167–175.

———. 1964. Reproductive efficiency in relation to colony size in hymenopterous societies. Insects Sociaux 11:317–342.

———. 1974. The social behavior of the bees. Belknap Press, Cambridge, Mass.

———. 1982. Early stages in insect social evolution: individual and family odor differences and their functions. Bull. Entomol. Soc. Amer. March:7–11.

Michener, G. R. 1982. Infanticide in ground squirrels. Anim. Behav. 30:936–938.

Michod, R. E. 1980. Evolution of interactions in family-structured populations: mixed mating models. Genetics 96:275–296.

———. 1982. The theory of kin selection. Annu. Rev. Ecol. Syst. 13:23–55.

Michod, R. E., and W. W. Anderson. 1980. On calculating demographic parameters from age frequency data. Ecology 61:265–269.

Miller, R. S. 1967. Pattern and process in competition. Adv. Ecol. Res. 4:1–74.

Milstead, W. W., A. S. Rand, and M. M. Stewart. 1974. Polymorphism in cricket frogs: an hypothesis. Evolution 28:489–491.

Missouri Botanical Garden. 1975. Sulfur in the environment. Missouri Botanical Garden, St. Louis.

Mitchell, R. 1983. Effects of host-plant variability on the fitness of sedentary herbivorous insects. Pages 343–370 *in* R. F. Denno and M. S. McClure, eds. Variable plants and herbivores in natural and managed systems. Academic Press, New York.

Mitter, C., and D. J. Futuyma. 1983. An evolutionary-genetic view of host-plant utilization by insects. Pages 427–459 *in* R. F. Denno and M. S. McClure, eds. Variable plants and herbivores in natural and managed systems. Academic Press, New York.

Moehlman, P. D. 1979. Jackal helpers and pup survival. Nature 277:382–383.

———. 1981. Why do jackals help their parents? Nature 289:824–825.

———. 1983. Socioecology of silverbacked and golden jackals (*Canis mesomelas* and *Canis aureus*). Pages 423–453 *in* J. F. Eisenberg and D. B. Kleiman, eds. Recent advances in the study of mammalian behavior. Spec. Pub. No. 7, American Society of Mammalogists.

Mohr, H. 1960. Zum Erkennen von Raubvogeln, insbesondere von Sperber und Baumfalk, durch Kleinvogeln. Z. Tierpsychol. 17(6):686–699.

Montgomerie, R. D. 1981. Why do jackals help their parents? Nature 289:824.

Mooney, H. A., ed. 1977. Convergent evolution in Chile and California: Mediterranean climate ecosystems. Dowden, Hutchinson and Ross, Stroudsburg, Pa.

———. 1981. Primary production in Mediterranean-climate regions. Pages 249–255 *in* F. di Castri, D. W. Goodall, and R. L. Specht, eds. Mediterranean-type shrublands. Elsevier, Amsterdam.

Mooney, H. A., T. M. Bonnicksen, N. L. Christensen, J. E. Lotan, and W. A. Reiners. 1981a. Fire regimes and ecosystem properties. U.S. Forest Service General Technical Report WO-26.

Mooney, H. A., and J. R. Ehleringer. 1978. The carbon gain benefits of solar tracking in a desert annual. Plant Cell Envir. 1:307–311.

Mooney, H. A., P. R. Ehrlich, D. E. Lincoln, and K. S. Williams. 1980a. Environmental controls on the seasonality of a drought deciduous shrub, *Diplacus aurantiacus,* and its predator, the checkerspot butterfly, *Euphydryas chalcedona.* Oecologia (Berl.) 45:143–146.

Mooney, H. A., C. Field, S. L. Gulmon, and F. A. Bazzaz. 1981b. Photosynthetic capacity in relation to leaf position in desert versus old-field annuals. Oecologia (Berl.) 50:109–112.

Mooney, H. A., and M. Godron, eds. 1983. Disturbance and ecosystems: components of response. Springer-Verlag, Berlin.

Mooney, H. A., and S. L. Gulmon. 1983. Determinants of plant productivity—natural versus man-modified communities. Pages 146–158 *in* H. A. Mooney and M. Godron, eds. Disturbance and ecosystems: components of response. Springer-Verlag, Berlin.

Mooney, H. A., S. L. Gulmon, P. W. Rundel, and J. Ehleringer. 1980b. Further observations on the water relations of *Prosopis tamarugo.* Oecologia (Berl.) 44:177–180.

Mooney, H. A., K. S. Williams, D. E. Lincoln, and P. R. Ehrlich. 1981c. Temporal and spatial variability in interaction between the checkerspot butterfly *E. chalcedona* and its principal food source, the California shrub, *Diplacus aurantiacus.* Oecologia 50:195–198.

Moore, P. D. 1980. The slow process of succession. Nature 288:436–437.

Morin, P. 1983. Predation, competition and the composition of larval anuran guilds. Ecol. Monogr. 53:119–138.

Morris, M. G. 1981. Responses of grassland invertebrates to management by cutting. III. Adverse effects on *Auchenorrhyncha.* J. Appl. Ecol. 18:107–123.

Morris, R. F., ed. 1963. The dynamics of epidemic

spruce budworm populations. Mem. Ent. Soc. Canad. 31:1–332.

Morse, D. H. 1977. Resource partitioning in bumblebees: the role of behavioral factors. Science 197:678–680.

Morton, E. S. 1973. On the evolutionary advantages and disadvantages of fruit-eating in tropical birds. Am. Nat. 107:8–22.

Mosquin, T. 1971. Competition for pollinators as a stimulus for the evolution of flowering time. Oikos 22:398–402.

Moss, B. 1980. Ecology of fresh waters. Wiley, New York.

Mueller, H. C. 1968. Prey selection: oddity or conspicuousness? Nature 217:92.

Mueller, L., and F. Ayala. 1981. Trade-off between *r*-selection and *k*-selection in *Drosophila* populations. Proc. Natl. Acad. Sci. USA 78:1303–1305.

Mueller, L. D., B. A. Wilcox, P. R. Ehrlich, D. G. Heckel, and D. D. Murphy, 1985. A direct assessment of the role of genetic drift in determining allele frequency variation in populations of *Euphydryas editha*. Genetics. 110:495–511.

Muller H. 1932. Some genetic aspects of sex. Am. Nat. 66:118–138.

Murdoch, W. W. 1966. Community structure, population control and competition—a critique. Am. Nat. 100:219–226.

Murie, A. 1944. The wolves of Mount McKinley. Fauna Nat. Parks. U. S. Fauna Ser. 5

Murphy, G. 1968. Pattern in life history phenomena and the environment. Am. Nat. 102:52–64.

Muscatine, L. 1973. Nutrition of corals. Pages 77–115 *in* O. A. Jones and R. Endean, eds. The biology and geology of coral reefs. Vol 2. Academic Press, New York.

Muth, F. A. 1980. Physiological ecology of desert iguana *(Dipsosaurus dorsalis)* eggs: temperature and water relations. Ecology 61:1335–1343.

Myers, J. P., P. G. Conners, and F. A. Pitelka. 1981. Optimal territory size and the sanderling: compromises in a variable environment. Pages 135–158 *in* A. C. Kamil and T. D. Sargent, eds. Foraging behavior: ecological, ethological and psychological approaches. Garland STPM Press, New York.

Myers, K. 1971. The rabbit in Australia. Pages 478–506 *in* P. J. den Boer and G. Gr. Gradwell, eds. Dynamics of numbers in populations. Proc. Adv. Study Inst. Dynamics Numbers Popul. (Oosterbeek. 1970). Center for Agric. Pub. and Doc. (Pudoc), Wageningen, The Netherlands.

Myers, N. 1979. The sinking ark. Pergamon, New York.

———. 1983. The primary source: tropic forests and our future. Norton, New York.

Nadkarni, N. M. 1981. Canopy roots: convergent evolution in rainforest nutrient cycles. Science 214:1023–1024.

Nanney, D. L. 1982. Genes and phenes in *Tetrahymena*. BioSci. 32(10):783–788.

National Research Council. 1980. Research priorities in tropical biology. National Academy of Sciences, Washington, D.C.

Nedrow, W. W. 1937. Studies on the ecology of roots. Ecology 18(1):27–52.

Nei, M., P. A. Fuerst, and R. Chakraborty. 1976. Testing the neutral mutation hypothesis by distribution of single loci heterozygosity. Nature 262:491–493.

Nelson, G. J., and N. Platnick. 1981. Systematics and biogeography: cladistics and vicariance. Columbia University Press, New York.

Nelson, G. J., and D. E. Rosen, eds. 1981. Vicariance biogeography. Columbia University Press, New York.

Nelson, K., and D. Hedgecock. 1980. Enzyme polymorphism and adaptive strategy in the decapod crustacea. Am. Nat. 116:238–280.

Nevo, E. 1978. Genetic variation in natural populations: patterns and theory. Theor. Popul. Biol. 13:121.

Nevo, E., A. Beiles, and R. Ben-Shlomo. 1984. The evolutionary significance of genetic diversity: ecological, demographic and life history correlates. Pages 13–213 *in* G. S. Mani, ed. Evolutionary dynamics of genetic diversity. Lecture notes in biomathematics 53. Springer-Verlag, New York.

Newell, N. D. 1967. Revolutions in the history of life. Geol. Soc. Amer. Spec. Paper 89:63–91.

Newell, S. J., and E. J. Tramer. 1978. Reproductive strategies in herbaceous plant communities during succession. Ecology 59:228–234.

Newton, I. 1985. Lifetime reproductive output of female sparrowhawks. J. Anim. Ecol. 54:241–253.

Newton, I., and M. Marquiss. 1979. Sex ratio among nestlings of the European sparrowhawk. Am. Nat. 113:309–315.

Nicholson, A. J. 1958. The self-adjustment of populations to change. Cold Spring Harbor Symp. Quant. Biol. 22:153–173.

Niering, W. A., R. H. Whittaker, and C. H. Lowe. 1963. The saguaro: a population in relation to environment. Science 142:15–23.

Nobel, G. K. 1939. The role of dominance in the social life of birds. Auk 56:263–273.

Nobel, P. S. 1983. Biophysical plant physiology and ecology. Freeman, San Francisco.

Noble, I. R., H. H. Shugart, and J. S. Schauer. 1980. A description of BRIND, a computer model of succession and fire response of the high-altitude *Eucalyptus* forests of the Brindabella Range, Australian Capital Territory. ORNL/TM-7041. Oak Ridge National Laboratory, Oak Ridge, Tenn.

Noble, I. R., and R. O. Slatyer. n.d. The effect of disturbance on plant succession. ms. 15pp.

Norris, K. S. 1967. Color adaptation in desert reptiles and its thermal relationships. Pages 162–229 *in* W. W. Milstead, ed. Symposium on lizard ecology, a symposium. University of Missouri Press, Columbia.

Norton, H. 1928. Natural selection and mendelian variation. Proc. Lond. Math. Soc. 28:1–45.

Nottebohm, F. 1969. The song of the Chingolo *(Z. capensis)* in Argentina: description and evaluation of a system of dialects. Condor 71:299–315.

Noy-Meir, I. 1973. Desert ecosystems: environment and producers. Annu. Rev. Ecol. Syst. 4:25–51.

———. 1974. Desert ecosystems: higher trophic levels. Annu. Rev. Ecol. Syst. 5:195–214.

Nye, P. H., and D. J. Greenland. 1960. The soil under shifting cultivation. Comm. Bur. Soils Tech. Comm. No. 51.

O'Neill, R. 1976. Paradigms of ecosystem analysis. Pages 16–19 *in* S. A. Levin, ed. Ecological theory and ecosystem models. Institute of Ecology, Indianapolis, Ind.

Odendaal, F., Y. Iwasa, and P. Ehrlich. 1985. Duration of female availability and its effect on butterfly mating systems. Am. Nat. 125:673–678.

Odum, E. G. 1971. Fundamentals of ecology. 3d ed. Saunders, Philadelphia.

Odum, E. P. 1960. Organic production and turnover in old field succession. Ecology 41:34–49.

———. 1969. The strategy of ecosystem development. Science 164:262–270.

———. 1977. The emergence of ecology as a new integrative discipline. Science 195:1289–1293.

Odum, H. T. 1957. Trophic structure and productivity of Silver Springs, Florida. Ecol. Monogr. 27:55–112.

Officer, C. B., and C. L. Drake. 1983. The Cretaceous–Tertiary transition. Science 219:1383–1390.

———. 1985. Terminal Cretaceous environmental events. Science 227:1161–1167.

Ogden, J. C., and P. R. Ehrlich. 1977. The behavior of heterotypic resting schools of juvenile grunts (Pomadasyidae). Mar. Biol. 273–280.

O'Keefe, J. D., and T. J. Ahrens. 1982. Impact mechanics of the Cretaceous-Tertiary extinction bolide. Nature 298:123–127.

Oliver, C. D., and E. P. Stephens. 1977. Reconstruction of a mixed-species forest in central New England. Ecology 58:562–572.

Olson, J. S. 1958. Rates of succession and soil changes on Southern Lake Michigan sand dunes. Bot. Gaz. 119:125–170.

O'Neill, R. V. 1968. Population energetics of the millipede, *Narceus americanus* (Beauvois). Ecology 49:803–809.

O'Neill, R. V., and D. L. DeAngelis. 1981. Comparative productivity and biomass relations of forest ecosystems. Pages 411–449 *in* D. E. Reichle, ed. Dynamic properties of forest ecosystems. Cambridge University Press, Cambridge.

Opler, P. A. 1974. Oaks as evolutionary islands for leaf-mining insects. Am. Sci. 62:67–73.

Organ, J. 1961. Studies of the local distribution, life history and population dynamics of the salamander genus *Desmognathus* in Virginia. Ecol. Monogr. 31:189–220.

Orians, G. 1966. Food of nestling yellow-head blackbirds, Caribou Parklands, British Columbia. Condor 68:321–337.

Orians, G., and M. Willson. 1964. Interspecific territories in birds. Ecology 45:736–745.

Orians, G. H. 1969. On the evolution of mating systems in birds and mammals. Am. Nat. 103(934):589–603.

Orians, G. H., and O. T. Solbrig, eds. 1977. Convergent evolution in warm deserts. Dowden, Hutchinson and Ross, Stroudsburg, Pa.

Osman, R. 1977. The establishment and development of a marine epifaunal community. Ecol. Monogr. 47:37–63.

Ostrom, J. H. 1974. Archeopteryx and the origin of flight. Q. Rev. Biol. 49:27–47.

———. 1980. The evidence for endothermy in dinosaurs. Pages 15–54 *in* R. D. K. Thomas and E. Olson, eds. A cold look at the warm-blooded dinosaurs. Westview Press, Boulder, Colo.

Otis, D. L., K. P. Burnham, G. C. White, and D. R. Anderson. 1978. Statistical inference from capture data on closed animal populations. Wildl. Monogr. 62:1–135.

Otte, D. 1975. Plant preference and plant succession. Oecologia 18:129–144.

Ovington, J. D. 1962. Quantitative ecology and the woodland ecosystem concept. Advan. Ecol. Res. 1:103–192.

———. 1965. Organic production, turnover and mineral cycling in woodlands. Biol. Rev. 40:295–336.

Owen, D. F. 1980. Camouflage and mimicry. Oxford University Press, Oxford.

Owen-Smith, R. N. 1971. Territoriality in the white

rhinoceros *(Ceratotherium simum)* Burchell. Nature 231:294–296.

———. 1975. The social ethology of the white rhinoceros *Ceratotherium simum* (Burchell 1817). Z. Tierpsychol. 38:337–384.

Pacala, S., and J. Roughgarden. 1982. Resource partitioning and interspecific competition in two two-species insular *Anolis* lizard communities. Science 217:444–446.

———. 1984. Control of arthropod abundance by *Anolis* lizards on St. Eustatius (Neth. Antilles). Oecologia 64:160–162.

———. 1985. Population experiments with the *Anolis* lizards of St. Maarten and St. Eustatius. Ecology 66:128–141.

Packard, G. C., C. R. Tracy, and J. J. Roth. 1977. The physiological ecology of reptilian eggs and embryos and the evolution of viviparity within the class Reptilia. Biol. Rev. 52:71–105.

Packer, C. 1977. Reciprocal altruism in *Papio anubis*. Nature 265:441–443.

Packer, C., and A. D. Pusey. 1982. Cooperation and competition within coalitions of male lions: kin selection or game theory? Nature 296:740–742.

Page, G., and D. F. Whitacre. 1975. Raptor predation on wintering shorebirds. Condor 77:73–83.

Paine, R. 1966. Food web complexity and species diversity. Am. Nat. 100:65–75.

Paine, R., and S. Levin. 1981. Intertidal landscapes: disturbance and the dynamics of pattern. Ecol. Monogr. 51:145–178.

Paine, R., and R. Vadas. 1969. The effects of grazing by sea urchins, *Strongylocentrotus* spp., on benthic algal populations. Limnol. Oceanogr. 14:710–719.

Palmblad, I. G. 1968. Competition in experimental populations of weeds with emphasis on the regulation of population size. Ecology 49:26–34.

Parenti, R. L., and E. L. Rice. 1969. Inhibitional effects of *Digitaria sanguinalis* and possible role in old-field succession. B. Torrey Bot. Club. 96:70–78.

Park, T. 1954. Experimental studies of interspecific competition. II. Temperature, humidity and competition in two species of *Tribolium*. Physiol. Zool. 27:177–238.

———. 1962. Beetles, competition and populations. Science 138:1369–1375.

Parker, G. A. 1970a. The reproductive behaviour and the nature of sexual selection in *Scatophaga stercoraria* L. (Diptera: Scatophagidae). VII. The origin and evolution of the passive phase. Evolution 24:774–788.

———. 1970b. Sperm competition and its evolutionary consequences in the insects. Biol. Rev. 45:525–567.

———. 1971. The reproductive behaviour and the nature of sexual selection in *Scatophaga stercoraria* L. (Diptera: Scatophagidae). VI. The adaptive significance of emigration from the dropping during the phase of genetical contact. J. Anim. Ecol. 40:215–233.

———. 1974. The reproductive behaviour and the nature of sexual selection in *Scatophaga stercoraria* L. (Diptera: Scatophagidae). IX. Spatial distribution of fertilization rates and evolution of male search strategy within the reproductive area. Evolution 28:93–108.

Parker, M. A., and R. B. Root. 1981. Insect herbivores limit habitat distribution of a native composite, *Machaeranthera canescens*. Ecology 62:1390–1392.

Parrish, J. A. D., and F. A. Bazzaz. 1982. Responses of plants from three successional communities to a nutrient gradient. J. Ecol. 70:233–248.

Parsons, T. R. 1976. The structure of life in the sea. Pages 81–97 *in* D. H. Cushing and J. J. Walsh, eds. The ecology of the seas. Blackwell, Oxford.

Patrick, R. 1967. The effect of invasion rate, species pool and size of area on the structure of a diatom community. Proc. Natl. Acad. Sci. USA 58:1335–1342.

Patten, B. C., D. A. Egloff, and T. H. Richardson. 1975. Total ecosystem model for a cove in Lake Texoma. Pages 205–421 *in* B. C. Patten, ed. Systems analysis and simulation in ecology. Vol. 3. Academic Press, New York.

Payne, I. 1980. Trees and disease. New Scientist. 3 January:12–14.

Payne, R. B. 1981. Population structure and social behavior: models for testing the ecological significance of song dialects in birds. Pages 108–120 *in* R. D. Alexander and D. W. Tinkle, eds. Natural selection and social behavior. Chiron, New York.

Pearce, F. 1982. The menace of acid rain. New Scientist 95:419–424.

Pearson, O. P. 1964. Carnivore-mouse predation: an example of its intensity and bioenergetics. J. Mammal. 45:177–188.

Peckarsky, B. L. 1980. Predator–prey interactions between stoneflies and mayflies: behavioral observations. Ecology 61(4):932–943.

———. 1982. Aquatic insect predator–prey relations. BioSci. 32:261–266.

———. 1984. Predator–prey relations among aquatic insects. Pages 196–254 *in* V. H. Resh and D. M.

Rosenberg, eds. The ecology of aquatic insects. Praeger, New York.

Pellew, R. A. P. 1983. The impacts of elephant, giraffe and fire upon the *Acacia tortilis* woodlands of the Serengeti. Afr. J. Ecol. 21:41–74.

Pennycuick, C. J. 1975. Mechanisms of flight. Pages 1–75 *in* D. S. Farner and J. R. King, eds. Avian biology, Vol. 5. Academic Press, New York.

Perrill, S. A., H. C. Gerhardt, and R. Daniel. 1978. Sexual parasitism in the green tree frog *(Hyla cinerea)*. Science 200:1179–1180.

Perrin, W. F., and J. E. Powers. 1980. Role of a nematode in natural mortality of spotted dolphins. J. Wildl. Manag. 44:960–963.

Perrins, C. M. 1968. The purpose of high-intensity alarm calls in small passerines. Ibis 110:200–201.

———. 1979. British tits. Collins, London.

Petersen, C. 1913. Valuation of the sea II. The animal communities of the sea bottom and their importance for marine zoogeography. Report Danish Biol. Stn. 21:1–44.

Peterson, B. J. 1980. Aquatic primary productivity and the ^{14}C-CO_2 method: a history of the productivity problem. Annu. Rev. Ecol. Syst. 11:359–385.

Peterson, C. 1982. The importance of predation and intra- and interspecific competition in the population biology of two infaunal suspension-feeding bivalves, *Protothaca staminea* and *Chione undatella*. Ecol. Monogr. 52:437–475.

Petrides, G. A., and W. G. Swank. 1966. Estimating the productivity and energy relations of an African elephant population. Pages 831–842 *in* Proc. 9th Int. Grassland Cong., Sao Paulo.

Petrinovich, L., T. Patterson, and L. T. Baptista. 1981. Song dialects as barriers to dispersal: a reevaluation. Evolution 35:180–188.

Pfeifer, S. 1982. Disappearance and dispersal of *Spermophilus elegans* juveniles in relation to behavior. Behav. Ecol. Sociobiol. 10:237–243.

Phillips, A. R. 1975. The migrations of Allen's and other hummingbirds. Condor 77:196–205.

Phillips, J. C. 1928. Wild birds introduced or transplanted in North America. Tech. Bull. U.S. Dept. Agric. 61:1–63.

Phillipson, J. 1966. Ecological energetics. Arnold, London.

Pianka, E. 1966. Latitudinal gradients in species diversity: a review of concepts. Am. Nat. 100:33–46.

———. 1970. On *r*- and *K*-selection. Am. Nat. 104:592–597.

Pickett, S. T. A., and F. A. Bazzaz. 1978. Organization of an assemblage of early successional species on a soil moisture gradient. Ecology 59:1248–1255.

Pielou, E. C. 1974. Population and community ecology: principles and methods. Gordon & Breach, New York.

———. 1975. Ecological diversity. Wiley, New York.

———. 1979. Biogeography. Wiley, New York.

Pierce, N. E., and S. Easteal. 1986. The selective advantage of attendant ants for larvae of a lycaenid butterfly, *Glaucopsyche lygdamus*. J. Anim. Ecol. 55:451–462.

Pierce, N. E., and P. S. Mead. 1981. Parasitoids as selective agents in the symbiosis between lycaenid butterfly larvae and ants. Science 211: 1185–1187.

Pierce, N. E., and W. R. Young. 1986. Lycaenid butterflies and ants: two species stable equilibria in mutualistic, commensal, and parasitic interactions. Am. Nat. In press.

Pimm, S. L. 1980. Food web design and the effect of species deletion. Oikos 35:139–149.

———. 1982. Food webs. Chapman and Hall, London.

———. 1984. The complexity and stability of ecosystems. Nature 307:321–326.

Pimm, S., and J. Lawton. 1980. Are food webs compartmented? J. Anim Ecol. 49:879–898.

———. 1983. The causes of food web structure: energy flow and natural history. Pages 45–49 *in* D. L. DeAngelis, W. M. Post, and G. Sugihara, eds. Current trends in food web theory. Oak Ridge National Laboratory, Oak Ridge, Tenn.

Pitelka, F. A., R. T. Holmes, and S. F. MacLean, Jr. 1974. Ecology and evolution of social organization in arctic sandpipers. Amer. Zool. 14:185–204.

Place, A., and D. Powers, 1979. Genetic variation and relative catalytic efficiencies: lactate dehydrogenase B allozymes of *Fundulus heteroclitus*. Proc. Natl. Acad. Sci. USA 76:2354–2358.

Pleasants, J. M. 1980. Competition for bumblebee pollinators in rocky mountain plant communities. Ecology 61:1446–1459.

Pleszczynska, W. K. 1978. Microgeographic prediction of polygyny in the lark bunting. Science 201:935–936.

Pollak, E. I., and T. Thompson. 1982. Multiple matings and sexual dichromatism in the dwarf Gourami, *Colisa lalia*. Anim. Behav. 30:1257–1258.

Pomeroy, L. R. 1960. Residence time of dissolved phosphate in natural waters. Science 131:1731–1732.

———. 1979. Secondary production mechanisms of continental shelf communities. Pages 163–186 *in*

R. J. Livingston, ed. Ecological processes in coastal and marine systems. Plenum, New York.

Pomeroy, L. R., R. E. Johannes, E. P. Odum, and B. Roffman. 1969. The phosphorus and zinc cycles and productivity of a salt marsh. Pages 412–419 *in* D. J. Nelson and F. C. Evans, eds. Proceedings of the Second Symposium on Radioecology. Clearinghouse for Federal and Scientific Technical Information, Springfield, Va.

Pomeroy, L. R., E. E. Smith, and C. M. Grant. 1965. The exchange of phosphate between estuarine water and sediment. Limnol. Oceangr. 10:167–172.

Pomeroy, L. R., and R. G. Wiegert, eds. 1981. The ecology of a salt marsh. Springer-Verlag, New York.

Poore, M. E. D. 1968. Studies in Malaysian rainforest. I. The forest on triassic sediments in Jengka Forest Reserve. J. Ecol. 56:143–196.

Porter, W. P., and D. Gates. 1969. Thermodynamic equilibria of animals with the environment. Ecol. Monogr. 39:227–244.

Porter, W. P., J. W. Mitchell, W. A. Beckman, and C. B. DeWitt. 1973. Behavioral implications of mechanistic ecology, thermal and behavioral modeling of desert ectotherms and their microclimate. Oecologia (Berl.) 13:1–54.

Post, W. M., W. R. Emanuel, P. J. Zinke, and A. G. Stangenberger. 1982. Soil carbon pools and world life zones. Nature 298:156–159.

Pough, F. H. 1976. Acid precipitation and embryonic mortality of spotted salamanders, *Ambystoma maculatum.* Science 192:68–70.

Powell, J. A., and R. A. Mackie. 1966. Biological interrelationships of moths and *Yucca whipplei.* Univ. Calif. Publ. Entomol. 42:1–46.

Prance, G. T., ed. 1982. Biological diversification in the tropics. Columbia University Press, New York.

Preston, F. 1948. The commonness and rarity of species. Ecology 29:254–283.

———. 1960. Time and space and the variation of species. Ecology 41:611–627.

———. 1962. The canonical distribution of commonness and rarity. Ecology 43:185–215, 410–432.

Price, M. V. 1978a. The role of microhabitat in structuring desert rodent communities. Ecology 59:910–921.

———. 1978b. Seed dispersion preferences of coexisting desert rodent species. J. Mammal. 56:731–751.

Price, P. W. 1977. General concepts on the evolutionary biology of parasite. Evolution 31:405–420.

———. 1980. Evolutionary biology of parasites. Princeton University Press, Princeton, N.J.

———. 1983. Hypotheses on organization and evolution in herbivorous insect communities. Pages 559–596 *in* R. F. Denno and M. S. McClure, eds. Variable plants and herbivores in natural and managed systems. Academic Press, New York.

Price, P. W., C. E. Bouton, P. Gross, B. A. McPherson, J. N. Thompson, and A. E. Weiss. 1980. Interactions among three trophic levels: influence of plants on interactions between insect herbivores and natural enemies. Annu. Rev. Ecol. Syst. 11:41–65.

Proctor, J., and S. Proctor. 1978. Color in plants and flowers. Everest, New York.

Proctor, J., and S. R. J. Woodell. 1975. The ecology of serpentine soils. Adv. Ecol. Res. 9:255–366.

Proctor, M., and P. Yeo. 1972. The pollination of flowers. Taplinger, New York.

Prosser, C. L. 1973. Comparative animal physiology. Vol. 1. Environmental physiology. 3d ed. Saunders, Philadelphia.

Pulliam, H. R. 1974. On the theory of optimal diets. Am. Nat. 108:59–74.

———. 1983. Ecological community theory and the coexistence of sparrows. Ecology 64:45–52.

Pulliam, H. R., and F. Enders. 1971. The feeding ecology of five sympatric finch species. Ecology 52:557–566.

Pusey, A., and C. Packer. 1983. Once and future kings. Nat. Hist. August:55–63.

Pyke, G. H. 1978. Optimal foraging in bumblebees and coevolution with their plants. Oecologia 36:281–292.

———. 1984. Optimal foraging theory: A critical review. Annu. Rev. Ecol. Syst. 15:523–575.

Pyke, G. H., H. R. Pulliam, and E. L. Charnov. 1977. Optimal foraging: A selective review of theory and tests. Q. Rev. Biol. 52:137–154.

Quinn, J. 1979. Disturbance, predation and diversity in the rocky intertidal zone. Ph.D. diss. University of Washington, Seattle.

Quinn, J. F. 1983. Mass extinctions in the fossil record. Science 219:1239–1240.

Rabinowitz, D., B. K. Bassett, and G. E. Renfro. 1979. Abundance and neighborhood structure for sparse and common grasses in a Missouri prairie. Am. J. Bot. 66:867–869.

Rabinowitz, D., and J. K. Rapp. 1981. Dispersal abilities of seven sparse and common grasses from a Missouri prairie. Am. J. Bot. 68:616–624.

Rabinowitz, D., J. K. Rapp, V. L. Sork, B. J. Rathcke, G. A. Reese, and J. C. Weaver. 1981. Phenological properties of wind- and insect-pollinated prairie plants. Ecology 62:49–56.

Race, M. S. 1982. Competitive displacement and predation between introduced and native mud snails. Oecologia 54:337–347.

Ralls, K. 1971. Mammalian scent marking. Science 171:443–449.

———. 1977. Sexual dimorphism in mammals: avian models and unanswered questions. Am. Nat. 111(981):917–938.

Rampino, M. R., and R. B. Stothers. 1984a. Terrestrial mass extinctions, cometary impacts and the sun's motion perpendicular to the galactic plane. Nature 308:709–712.

———. 1984b. Geologic rhythms and cometary impacts. Science 226:1427–1431.

Rand, A. S. 1967. Predator–prey interactions and the evolution of aspect diversity. Atlas do Simpósio sobre a Biotica Amazonica 5:73–83.

Randall, J. 1967. Food habits of reef fishes of the West Indies. Stud. Trop. Oceanogr. 5:665–847.

Randall, J., and G. Helfman. 1972. *Diproctacanthus xanthurus,* a cleaner wrasse from the Palau Islands, with notes on other cleaning fishes. Trop. Fish Hobbyist 20:87–95.

Raschke, K. 1956. Uber die Physikalischen Beziehungen zwischen Wamm evbergangzahl Strahlung saustausch, Temperatur und Transpiration eines Blatres. Planta 48:200–239.

Ratcliffe, D. 1979. The end of the large blue butterfly. New Scientist November 8:

Rathcke, B. J. 1976. Competition and coexistence within a guild of herbivorous insects. Ecology 57:76–87.

Raup, D. M., and J. J. Sepkoski, Jr. 1982. Mass extinctions in the marine fossil record. Science 215:1503.

———. 1984. Periodicity of extinctions in the geologic past. Proc. Natl. Acad. Sci. USA 81:801–805.

Raup, D. M., J. J. Sepkoski, Jr., and S. M. Stigler. 1983. Mass extinctions in the fossil record. Science 219:1240–1241.

Rausher, M. D. 1978. Search image for leaf shape in a butterfly. Science 200:1071–1073.

———. 1979. Larval habitat suitability and oviposition preference in three related butterflies. Ecology 60:503–511.

———. 1980. Host abundance, juvenile survival and oviposition preference in *Battus philenor.* Evolution 34:342–355.

———. 1982. Population differentiation in *Euphydryas editha* butterflies: larval adaptation to different hosts. Evolution 36:581–590.

———. 1983a. Alteration of oviposition behavior by *Battus philenor* butterflies in response to variation in host-plant density. Ecology 64:1028–1034.

———. 1983b. Ecology of host-selection behavior in phytophagous insects. Pages 223–257 *in* R. F. Denno and M. S. McClure, eds. Variable plants and herbivores in natural and managed systems. Academic Press, New York.

Rausher, M. D., and P. Feeny. 1980. Herbivory, plant density and plant reproductive success: the effect of *Battus philenor* on *Aristolochia reticulats.* Ecology 61:905–917.

Rausher, M.D., D. A. Mackay, and M. C. Singer. 1981. Pre- and post-alighting host discrimination by *Euphydryas editha* butterflies: the behavioral mechanisms causing clumped distributions of egg clusters. Anim. Behav. 29:1220–1228.

Rausher, M. D., and D. R. Papaj. 1983a. Demographic consequences of discrimination among conspecific host plants by *Battus philenor* butterflies. Ecology 64:1402–1410.

———. 1983b. Host plant selection by *Battus philenor* butterflies: evidence for individual differences in foraging behaviour. Anim. Behav. 31: 341–347.

Raven, P. H. 1976. Ethics and attitudes. Pages 155–179 *in* J. B. Simmons, R. I. Beyer, P. E. Brandham, G. L. Luchs, and V. H. T. Parry, eds. Conservation of threatened plants. Plenum, New York.

Reader, P. M., and T. R. E. Southwood. 1981. The relationship between palatability to invertebrates and the successional status of a plant. Oecologia 51:271–275.

Reader, R. J. 1975. Competitive relationships of some bog ericads for major insect pollinators. Canad. J. Bot. 53:1300–1305.

Real, L. 1981. Uncertainty and pollinator-plant interactions: The foraging behavior of bees and wasps on artificial flowers. Ecology 62:20–26.

———. 1983. Pollination biology. Academic Press, Orlando, Fla.

Real, L., J. Ott, and E. Silverfine. 1982. On the tradeoff between the mean and the variance in foraging: Effect of spatial distribution and color preference. Ecology 63:1617–1623.

Recher, H. F. 1969. Bird species diversity and habitat diversity in Australia and North America. Am. Nat. 103:75–80.

Regal, P. J., and C. Gans. 1980. The revolution in thermal physiology: implications for dinosaurs.

Pages 167–188 *in* R. D. K. Thomas and E. Olson, eds. A cold look at the warm-blooded dinosaurs. Westview Press, Boulder, Colo.

Reichle, D. E., ed. 1970. Analysis of temperate forest ecosystems. Springer-Verlag, New York.

———. ed. 1981. Dynamic properties of forest ecosystems. Cambridge University Press, Cambridge.

Reichman, O. 1976. Effects of rodents on germination of desert annuals. United States/International Biological Program. Desert Biome Res. Memo. 76-20. Utah State University, Logan.

Reichman, O., and D. Oberstein. 1977. Selection of seed distribution types of *Dipodomys merriami* and *Perognathus amplus*. Ecology 58:636–643.

Reiners, W. A., and N. M. Reiners. 1970. Energy and nutrient dynamics of forest floors in three Minnesota forests. J. Ecol. 58:497–519.

Reiss, M. 1982. Males bigger, females biggest. New Scientist October 28:226–229.

Rejmanek, M., and P. Stary. 1979. Connectance in real biotic communities and critical values of stability in model ecosystems. Nature 280:311–313.

Rhoades, D. F., and R. G. Cates. 1976. A general theory of plant antiherbivore chemistry. Pages 168–213 *in* J. W. Wallace and R. L. Mansell, eds. Biochemical interaction between plants and insects. Plenum, New York.

Rhoads, D., and D. Young, 1970. The influence of deposit-feeding organisms on sediment stability and community trophic structure. J. Mar. Res. 28:150–178.

Rice, E. L., and S. K. Pancholy. 1972. Inhibition of nitrification by climax ecosystems. Am. J. Bot. 59:1033–1040.

Rice, J., R. D. Ohmart, and B. W. Anderson. 1983. Habitat selection attributes of an avian community: a discriminant analysis investigation. Ecol. Monogr. 53:263–290.

Richards, P. W. 1952. The tropical rain forest. Cambridge University Press, London.

———. 1973. Africa, the "odd man out." Pages 21–26 *in* B. J. Meggers, J. E. S. Ayensu, and W. D. Duckworth, eds. Tropical forest ecosystems in Africa and South America: a comparative review. Smithsonian Institution Press, Washington, D.C.

Richardson, J. L. 1980. The organismic community: resilience of an embattled ecological concept. BioSci. 30:465–472.

Ricklefs, R. E. 1975. The evolution of cooperative breeding in birds. Ibis 117:531–534.

Ricklefs, R. E., and K. O'Rourke. 1975. Aspect diversity in moths: a temperate–tropical comparison. Evolution 29:313–324.

Rigler, F. H. 1956. A tracer study of the phosphorus cycle in lake water. Ecology 37:550–556.

———. 1964. The phosphorus fractions and the turnover time of inorganic phosphorus in different types of lakes. Limnol. Oceanog. 9:511–518.

Roach, B., T. Eisner, and J. Meinwald. 1980. Defensive substances of opilionids. J. Chem. Ecol. 6:511–516.

Robertson, C., 1895. The philosophy of flower seasons and the phaenological relations of the entomophilous flora and the anthophilous insect fauna. Am. Nat. 29:97–117.

Robertson, D. R. 1972. Social control of sex reversal in a coral-reef fish. Science 177:1007–1009.

Robertson, D. R., H. P. A. Sweatman, E. A. Fletcher, and M. G. Cleland. 1976. Schooling as a mechanism for circumventing the territoriality of competitors. Ecology 57:1208–1220.

Robertson, G. P., and P. M. Vitousek, 1981. Nitrification potentials in primary and secondary succession. Ecology 62:376–386.

Rockwood, L. L. 1973. The effect of defoliation on seed production of six Costa Rican tree species. Ecology 54:1363–1369.

Rodin, L. E., and N. I. Bazilevich. 1967. Production and mineral cycling in terrestrial vegetation. Oliver and Boyd, Edinburgh.

Roeder, K. D. 1962. The behavior of free-flying moths in the presence of artificial ultrasonic pulses. Anim. Behav. 10:300–304.

———. 1964. Aspects of noctuid tympanic nerve response having significance in the avoidance of bats. J. Ins. Physiol. 10:529–546.

Roeder, K. D., and A. E. Treat. 1960. The acoustic detection of bats by moths. Proc. 11th Int. Cong. Entomol. 3:7–11.

———. 1961. The detection and evasion of bats by moths. Am. Sci. 49:135–148.

Rogers, G. F. 1982. Then and now: a photographic history of vegetation change in the Central Great Basin Desert. University of Utah Press, Salt Lake City.

Rollins, R. 1963. The evolution and systematics of *Leavenworthia* (Cruciferae). Contrib. Gray Herbarium, Harvard Univ. 192:3–198.

Romans, R. C., ed. 1981. Vegetation maps for eastern North America: 40,000 years B.P. to the present. Pages 123–165 *in* Geobotany. II. Plenum, New York.

Root, R. B. 1967. The niche exploitation pattern of the blue-gray gnatcatcher. Ecol. Monogr. 37:317–350.

———. Organization of a plant–arthropod association in simple and diverse habitats: the fauna of collards (*Brassica oleracea*). Ecol. Monogr. 43:95–124.

Rose, M. R. 1979. Quantitative genetics of adult female life-history in *Drosophila melanogaster*. Ph.D. diss. University of Sussex.

———. 1984. Laboratory evolution of postponed senescence in *Drosophila melanogaster*. Evolution 38:1004–1010.

Rose, M. R., and B. Charlesworth. 1980. A test of evolutionary theories of senescence. Nature 287:141–142.

Rosenzweig, M. L. 1971. Paradox of enrichment: destabilization of exploitation ecosystems in ecological time. Science 171:385–387.

Rosenzweig, M. L., and P. W. Sterner. 1970. Population ecology of desert rodent communities: body size and seed-husking as bases for heteromyid coexistence. Ecology 51:217–224.

Rothstein, S. I. 1971. Observation and experiment in the analysis of interactions between brood parasites and their hosts. Am. Nat. 105(941):71–74.

———. 1975. Evolutionary rates and host defenses against avian brood parasitism. Am. Nat. 109(966): 161–176.

Roubik, D. W. 1982. Obligate necrophagy in a social bee. Science 217:1059–1060.

Roughgarden, J. 1971. Density-dependent natural selection. Ecology 52:453–468.

———. 1972. Evolution of niche width. Am. Nat. 106:683–718.

———. 1974a. Niche width: Biogeographic patterns among *Anolis* lizard populations. Am. Nat. 108:429–442.

———. 1974b. Species packing and the competition function with illustrations from coral reef fish. Theor. Popul. Biol. 5:163–186.

———. 1975. Evolution of marine symbiosis—a simple cost–benefit model. Ecology 56:1201–1208.

———. 1976. Resource partitioning among competing species—a coevolutionary approach. Theor. Popul. Biol. 9:388–424.

———. 1977. Coevolution in ecological systems: results from "loop analysis" for purely density-dependent coevolution. Pages 499–518 *in* F. Christiansen and T. Fenchel, eds. Lecture notes in mathematics. Vol. 19. Springer-Verlag, New York.

———. 1979. The theory of population genetics and evolutionary ecology: an introduction. Macmillan, New York. (Reprinted 1987)

———. 1983. Competition and theory in community ecology. Am. Nat. 122:583–601.

Roughgarden, J., S. Gaines, and Y. Iwasa. 1984. Dynamics and evolution of marine populations with pelagic larval dispersal. Pages 111–128 *in* R. M. May, ed. Exploitation of marine communities. Dahlem Konferenzem. Springer-Verlag, Berlin.

Roughgarden, J., S. Gaines, and S. Pacala. 1987. Supply side ecology: the role of physical transport processes. *In* P. Giller and J. Gee, eds. Organization of communities: past and present. Blackwell Scientific Publications. London.

Roughgarden, J., D. Heckel, and E. Fuentes. 1983. Coevolutionary theory and the biogeography and community structure of *Anolis*. Pages 371–410 *in* R. Huey, E. Pianka, and T. Schoener, eds. Lizard ecology: studies on a model organism. Harvard University Press, Cambridge, Mass.

Roughgarden, J., Y. Iwasa, and C. Baxter. 1985. Demographic theory for an open marine population with space-limited recruitment. Ecology 66:54–67.

Roughgarden, J., S. Pacala, and J. Rummel. 1984. Strong present-day competition between the *Anolis* lizard populations of St. Maarten (Neth. Antilles). Pages 203–220 *in* B. Shorrocks, ed. Evolutionary ecology. Blackwell, England.

Rouse, I., and L. Allaire. 1978. Caribbean. Pages 431–481 *in* R. Taylor and C. Meighan, eds. Chronologies in New World archeology. Academic Press, New York.

Rowe, J. S. 1961. The level-of-integration concept and ecology. Ecology 42:420–427.

Rubinoff, I. 1968. Central American sea-level canal: possible biological effects. Science 161:857–861.

Rummel, J., and J. Roughgarden. 1983. Some differences between invasion-structured and coevolution-structured competitive communities: a preliminary theoretical analysis. Oikos 41:477–486.

———. 1985. The effects of reduced perch-height separation on the competition between two Lesser-Antillean *Anolis* lizards. Ecology 66:430–444.

Runkle, J. R. 1981. Gap regeneration in some old-growth forests of the eastern United States. Ecology 62:1041–1051.

Russell, D. A. 1984. The gradual decline of the dinosaurs—fact or fallacy? Nature 307:360–361.

Ryan, D. F., and F. H. Bormann. 1982. Nutrient resorption in northern hardwood forests. BioSci. 32:29–32.

Ryther, J. H. 1969. Photosynthesis and fish production in the sea. Science 166:72–76.

Ryther, J. H., and W. M. Dunstan. 1971. Nitrogen, phosphorus and eutrophication in the coastal marine environment. Science 171:1008–1013.

Sage, B. 1980. Acid drops from fossil fuels. New Scientist 6 March: 743–745.

Sale, P. 1977. Maintenance of high diversity in coral reef fish communities. Am. Nat. 111:337–359.

———. 1978. Reef fishes and other vertebrates: a comparison of social structures. Pages 313–346 *in* E. S. Reese and F. J. Lighter, eds. Contrasts in behavior: adaptations in the aquatic and terrestrial environments. Wiley, New York.

———. 1980. The ecology of fishes on coral reefs. Annu. Rev. Oceanogr. Mar. Biol. 18:367–421.

Sale, P., and D. Williams. 1982. Community structure of coral reef fishes: are the patterns more than those expected by chance? Am. Nat. 120:121–127.

Salisbury, E. J. 1929. The biological equipment of species in relation to competition. J. Ecol. 17(2):197–222.

Salisbury, F. B., and C. W. Ross. 1978. Plant physiology. 2d ed. Wadsworth, Belmont, Calif.

Salt, G. 1979. A comment on the use of the term Emergent Properties. Am. Nat. 113:145–148.

———. 1983. Roles: their limits and responsibilities in ecological and evolutionary research. Am. Nat. 122:697–706.

Sanders, H. 1958. Benthic studies in Buzzards Bay I. Animal–sediment relationships. Limnol. Oceanogr. 3:245–258.

Sandler, L., Y. Hiraizumi, and I. Sandler. 1959. Meiotic drive in natural populations of *Drosophila melanogaster*. I. The cytogenetic basis of segregation distortion. Genetics 44:233–250.

Sargent, T. D. 1978. On the maintenance of stability in hindwing diversity among moths of the genus *Catocala* (Lepidoptera: Noctuidae). Evolution 32:424–434.

Sarukhan, J. 1974. Studies on plant demography: *Ranunculus repens* L., *R. bulbosus* L. and *R. acris* L. II. Reproductive strategies and seed population dynamics. J. Ecol. 62:675–716.

———. 1978. Studies on the demography of tropical trees. Pages 163–184 *in* P. B. Tomlinson and M. H. Zimmermann, eds. Tropical trees as living systems. Cambridge University Press, Cambridge.

Schall, J. J., and E. R. Pianka. 1980. Evolution of escape behavior diversity. Am. Nat. 115:551–556.

Schaller, G. B. 1972. The Serengeti lion. University of Chicago Press, Chicago.

Schemske, D. W., and N. Brokaw. 1981. Treefalls and the distribution of understory birds in a tropical forest. Ecology 62:938–945.

Schjelderup-Ebbe, T. 1922. Beiträge zur Sozialpsychologie des Haushuhns. Zeit. f. Psychologie 88:225–252.

Schlesinger, W. H., and D. S. Gill. 1980. Biomass, production and changes in the availability of light, water and nutrients during the development of pure stands of the chaparral shrub, *Ceanothus megacarpus* after fire. Ecology 61:781–789.

Schluter, D., T. D. Price, and P. R. Grant. 1985. Ecological character displacement in Darwin's finches. Science 227:1056–1059.

Schmidt-Nielsen, K. 1964. Desert animals: physiological problems of heat and water. Clarendon, Oxford.

Schnieder, S. H., and R. Londer. 1984. The coevolution of climate and life. Sierra Club, San Francisco.

Schoener, T. 1968. Sizes of feeding territories among birds. Ecology 49:123–141.

———. 1971. Theory of feeding strategies. Annu. Rev. Ecol. Syst. 2:369–404.

———. 1973. Population growth regulated by intraspecific competition for energy or time: some simple representations. Theor. Popul. Biol. 4:56–84.

———. 1974. Competition and the form of habitat shift. Theor. Popul. Biol. 6:265–307.

———. 1976. Alternatives to Lotka-Volterra competition: models of intermediate complexity. Theor. Popul. Biol. 10:309–333.

———. 1982. The controversy over interspecific competition. Am. Sci. 70:586–595.

———. 1983a. Field experiments on interspecific competition. Am. Nat. 122:240–285.

———. 1983b. Rate of species turnover decreases from lower to higher organisms: a review of the data. Oikos 41:372–377.

———. 1984. Size differences among sympatric, bird-eating hawks: a worldwide survey. Pages 254–281 in D. R. Strong, D. Simberloff, L. G. Abele, and A. B. Thistle, eds. Ecological communities: conceptual issues and the evidence. Princeton University Press, Princeton, N.J.

Schoener, T., and G. Gorman. 1968. Some niche differences in three Lesser-Antillean lizards of the genus *Anolis*. Ecology 49:819–830.

Schoener, T., and D. Janzen. 1968. Notes on environmental determinants of tropical versus temperate insect size patterns. Am. Nat. 102:207–224.

Schoener, T., and A. Schoener. 1983. Distribution of vertebrates on some very small islands II. Patterns in species number. J. Anim. Ecol. 52:237–262.

Schultz, J. C. 1983. Habitat selection and foraging tactics of caterpillars in heterogeneous trees. Pages 61–90 *in* R. F. Denno and M. S. McClure, eds. Variable plants and herbivores in natural and managed systems. Academic Press, New York.

Schulze, E. D., and D. J. Bloom. 1984. Relationship between mineral nitrogen influx and transpiration in radish and tomato. Plant Physiol. 76:827–828.

Scott, J. R., C. R. Tracy, and D. Pettus. 1982. A biophysical analysis of daily and seasonal utilization of climate space by a montane snake. Ecology 63:482–493.

Scriber, J. M. 1983. Evolution of feeding specialization, physiological efficiency and host races in selected Papilionidae and Saturniidae. Pages 373–412 *in* R. F. Denno and M. S. McClure, eds. Variable plants and herbivores in natural and managed systems. Academic Press, New York.

Searcy, W. A. 1983. Response to multiple song types in male song sparrows and field sparrows. Anim. Behav. 31:948–949.

Sebens, K. 1982a. Competition for space: growth rate, reproductive output and escape in size. Am. Nat. 120:189–197.

———. 1982b. The limits to indeterminate growth: an optimal size model applied to passive suspension feeders. Ecology 63:209–222.

———. 1983. Population dynamics and habitat suitability of the intertidal sea anemonies *Anthopleura elegantissima* and *A. xanthogrammica.* Ecol. Monogr. 53:405–433.

Seifert, R. P., and F. H. Seifert. 1976. A community matrix analysis of *Heliconia* insect communities. Am. Nat. 110:461–483.

———. 1979a. A *Heliconia* insect community in a Venezuelan cloud forest. Ecology 60:462–467.

———. 1979b. Utilization of *Heliconia* (Musaceae) by the beetle *Xenarescus monocerus* (Oliver) (Chrysomelidae:Hispinae) in a Venezuelan forest. Biotropica 11:51–59.

Sepkoski, J. J., Jr. 1982. Mass extinctions in the Phanerozoic oceans: a review. Geologic Society of America, Spec. Paper 190:283–289.

Seyfarth, R. M., and D. L. Cheney. 1984. Grooming, alliances and reciprocal altruism in vervet monkeys. Nature 308:541–543.

Shapiro, A. M. 1974. Partitioning of resources among lupine-feeding Lepidoptera. Am. Mid. Naturalist 91:243–248.

Shapiro, A. M., and R. T. Cardé. 1970. Habitat selection and competition among sibling species of satyrid butterflies. Evolution 24:48–54.

Shelford, V. E. 1911. Ecological succession: pond fishes. Biol. Bull. 21:127–151.

———. 1913. Animal communities in temperate America. University of Chicago Press, Chicago.

Shelford, V., A. Weese, L. Rice, D. Rasmussen, and A. MacLean. 1935. Some marine biotic communities of the Pacific coast of North America. Part I. General survey of the communities. Ecol. Monogr. 5:249–332.

Sherman, K. J. 1983. The adaptive significance of postcopulatory mate guarding in a dragonfly, *Pachydiplax longipennis.* Anim. Behav. 31:1107–1115.

Sherman, P. W. 1977. Nepotism and the evolution of alarm calls. Science 197:1246–1253.

———. 1981. Reproductive competition and infanticide in Belding's ground squirrels and other animals. Pages 311–331 *in* R. Alexander and D. Tinkle, eds. Natural selection and social behavior: recent research and new theory. Chiron, New York.

———. 1982. Infanticide in ground squirrels. Anim. Behav. 30:938–939.

Shorrocks, B., ed. 1984. Evolutionary ecology. Blackwell, Oxford.

Shugart, H. H. 1984. A theory of forest dynamics. Springer-Verlag, New York.

Shugart, H. H., and D. C. West. 1977. Development of an Appalachian deciduous forest succession model and its application to the assessment of the impact of the chestnut blight. J. Environ. Mgmt. 5:161–179.

Shugart, H. H., D. C. West, and W. R. Emanuel. 1981. Patterns and dynamics of forests: an application of simulation models. Pages 74–94 *in* D. C. West, H. H. Shugart, and D. B. Botkin, eds. Forest succession. Springer-Verlag, New York.

Shulman, M. J., J. C. Ogden, J. P. Ebersole, W. N. McFarland, S. L. Miller, and N. G. Wolf. 1983. Priority effects in the recruitment of juvenile coral reef fishes. Ecology 64:1508–1513.

Siccama, T. G., M. Bless, and H. W. Vogelmann. 1982. Decline of red spruce in the Green Mountains of Vermont. Bull. Torrey Bot. Club 109:162–168.

Siegfried, W. R., and L. G. Underhill. 1975. Flocking as an anti-predator strategy in doves. Anim. Behav. 23:504–508.

Siever, R. 1983. The dynamic Earth. Sci. Am. September:46–55.

Silvertown, J. 1980. The dynamics of a grassland ecosystem: botanical equilibrium in the park grass experiment. J. Appl. Ecol. 17:491–504.

Simberloff, D. 1973. Population interactions and change in biotic communities. (Review) Stability and complexity in model ecosystems. R. M. May, Princeton University Press, Princeton, N.J. In Science 181:1157–1158.

———. 1986. Are we on the verge of a mass extinction in tropical rain forests? Pages 165–180 *in* D. K. Elliot, ed. Dynamics of extinction. Wiley, New York.

Simberloff, D., and W. Boecklen. 1981. Santa Rosalia reconsidered: size ratios and competition. Evolution 35:1206–1228.

Simberloff, D., and E. O. Wilson. 1969. Experimental zoogeography of islands: the colonization of empty islands. Ecology 50:278–296.

———. 1970. Experimental zoogeography of islands: A two-year record of colonization. Ecology 51: 934–937.

Simpson, G. G. 1964. Species density of North American recent mammals. Syst. Zool. 13:57–73.

———. 1976. Penguins, past and present, here and there. Yale University Press, New Haven, Conn.

Sims, P. L., J. S. Singh, and W. K. Lauenroth. 1978. The structure and function of ten western North American grasslands. Part I–IV. J. Ecol. 66:251–285, 547–572, 573–597, 983–1009.

Sinclair, A. R. E. 1974. The natural regulation of buffalo populations in East Africa. IV. The food supply as a regulating factor, and competition. E. Afr. Wildl. J. 12:291–311.

———. 1975. The resource limitation of trophic levels in tropical grassland ecosystems. J. Anim. Ecol. 44:497–520.

———. 1979a. Dynamics of the Serengeti ecosystem: process and pattern. Pages 1–30 *in* A. R. E. Sinclair and M. Norton-Griffiths, eds. Serengeti: dynamics of an ecosystem. University of Chicago Press, Chicago.

———. 1979b. The Serengeti environment. Pages 31–45 *in* A. R. E. Sinclair and M. Norton-Griffiths, eds. Serengeti: dynamics of an ecosystem. University of Chicago Press, Chicago.

Sinclair, A. R. E., and M. Norton-Griffiths, eds. 1979. Serengeti: dynamics of an ecosystem. University of Chicago Press, Chicago.

———. 1982. Does competition or facilitation regulate migrant ungulate populations in the Serengeti? A test of hypotheses. Oecologia (Berl.) 53:364–369.

Singer, M. C. 1972. Complex components of habitat suitability within a butterfly colony. Science 176:75–77.

———. 1981. Evolution of food-plant preference in the butterfly *Euphydryas editha*. Evolution 25:383–389.

Singer, M. C., and P. R. Ehrlich. 1979. Population dynamics of the checkerspot butterfly, *Euphydryas editha*. Popul. Ecol., Fortschr. Zool., no. 25, 2/3, pp. 53–60.

Skellam, J. G. 1951. Random dispersal in theoretical populations. Biometrica 38:196–218.

Slatyer, R. O., ed. n. d. Dynamic changes in terrestrial ecosystems. MAB Technical Note. 57pp.

Slobodkin, L. B., and L. Fishelson. 1974. The effect of the cleaner fish *Labroides dimidiatus* on the point diversity of fishes on the reef front at Eliat. Am. Nat. 108:369–376.

Slobodkin, L. B., F. E. Smith, and N. G. Hairston. 1967. Regulation in terrestrial ecosystems and the implied balance of nature. Am. Nat. 101:109–124.

Smigel, B. W., and M. L. Rosenzweig. 1974. Seed selection in *Dipodomys merriami* and *Perognathus penicillatus*. Ecology 55:329–339.

Smiley, J. 1978. Plant chemistry and the evolution of host specificity: new evidence from *Heliconius* and *Passiflora*. Science 201:745–747.

Smit, J. 1980. Terminal Cretaceous catastrophe. Nature 287:760.

Smit, J., and J. Hertogen. 1980. An extraterrestrial event at the Cretaceous–Tertiary boundary. Nature 285:198–200.

Smit, J., and S. van der Kaars. 1984. Terminal Cretaceous extinctions in the Hell Creek area, Montana: compatible with catastrophic extinction. Science 223:1177–1179.

Smith, A. C. 1973. Angiosperm evolution and the relationship of the floras of Africa and America. Pages 49–61 *in* B. J. Meggers, J. E. S. Ayensu, and W. D. Duckworth, eds. Tropical forest ecosystems in Africa and South America: a comparative review. Smithsonian Institution Press, Washington, D.C.

Smith, C. L., and J. C. Tyler. 1975. Succession and stability in fish communities of dome-shaped patch reefs in the West Indies. Am. Mus. Novitates 2572:1–18.

Smith, D., and S. Cooper. 1982. Competition among cladocera. Ecology 63:1004–1015.

Smith, F. E. 1961. Density dependence in the Australian thrips. Ecology 42:403–407.

Smith, J. M. 1977. Parental investment: a prospective analysis. Anim. Behav. 25:1–9.

Smith, N. G. 1968. The advantage of being parasitized. Nature 219:690–694.

———. 1970. On change in biological communities. Science 170:312–313.

Snell, T., and C. King. 1977. Lifespan and fecundity patterns in rotifers: the cost of reproduction. Evolution 31:882–890.

Snow, D. W. 1965. A possible selective factor in the evolution of fruiting seasons in tropical forest. Oikos. 15(2):274–281.

———. 1971. Evolutionary aspects of fruit-eating by birds. Ibis 113:194–202.

Snyder, R. L. 1962. Reproductive performance of a

population of woodchucks after a change in sex ratio. Ecology 43(3):506–515.

Snyder, R. L., and J. J. Christian. 1960. Reproductive cycle and litter size of the woodchuck. Ecology 41(4):647–656.

Snyder, R. L., D. E. Davis, and J. J. Christian. 1961. Seasonal changes in the heights of woodchucks. J. Mammal. 42:297–312.

Söderlund, R., and B. H. Svensson. 1976. The global nitrogen cycle. Pages 23–73 *in* B. H. Svensson and R. Söderlund, eds. Nitrogen, phosphorus and sulphur—global cycles. Scope Report No. 7. Ecol. Bull. 22:23–73.

Sokal, R. R. 1974. Senescence and genetic load: evidence from *Tribolium*. Science 167:1733–1734.

Sokal, R. R., and T. J. Crovello. 1970. The biological species concept: a critical evaluation. Am. Nat. 104:127–153.

Sokal, R. R., and F. J. Rohlf. 1981. Biometry. 2d ed. Freeman, San Francisco.

Solbrig, O. 1971. The population biology of dandelions. Am. Sci. 59:686–694.

———, ed. 1980. Demography and evolution in plant populations. University of California Press, Berkeley.

Sollins, P., C. C. Grier, F. M. McCorison, K. Cromack, Jr., R. Fogel, and R. L. Fredriksen. 1980. The internal element cycles of an old-growth douglas-fir ecosystem in Western Oregon. Ecol. Monogr. 50:261–285.

Solomon, A. M., H. R. Delcourt, D. C. West, and J. T. Blasing. 1980. Testing a simulation model for reconstruction of prehistoric forest-stand dynamics. Quat. Res. 14:275–293.

Soule, M. E., and B. A. Wilcox, eds. 1980. Conservation biology: an evolutionary-ecological perspective. Sinauer, Sunderland, Mass.

Sousa, W. 1979. Experimental investigations of disturbance and ecological succession in a rocky intertidal community. Ecol. Monogr. 49:227–254.

Southward, A. J. 1980. The western English Channel—an inconstant ecosystem? Nature 285:361–366.

Southwood, T. R. E. 1961. The number of species associated with various trees. J. Aniṃ. Ecol. 30: 1–8.

———. 1967. The interpretation of population change. J. Anim. Ecol. 36:519–529.

Southwood, T. R. E., V. K. Brown, and P. M. Reader. 1979. The relationships of plant and insect diversities in succession. Biol. J. Linn. Soc. 12:327–348.

Southwood, T. R. E., and H. N. Comins. 1976. A synoptic population model. J. Anim. Ecol. 45:949–965.

Spotila, J. R., O. H. Soule, and D. M. Gates. 1972. The biophysical ecology of the alligator: heat energy budgets and climate spaces. Ecology 53: 1094–1102.

Spradbery, J. P. 1969. The biology of *Pseudorhyssa sternata* Merrill (Hym., Ichneumonidae), a cleptoparasite of siricid woodwasps. Bull. Ent. Res. 59:291–297.

———. 1970. Host finding by *Rhyssa persuasoria* (L.), an ichneumonid parasite of siricid woodwasps. Anim. Behav. 18:103–114.

Springett, J. A. 1976. The effect of planting *Pinus pinaster* Ait. on populations of soil microarthropods and on litter decomposition at Gnangara, Western Australia. Aust. J. Ecol. 1:83–87.

Sprugel, D. G. 1984. Density, biomass, productivity, and nutrient-cycling changes during stand development in wave-regenerated balsam fir forests. Ecol. Monogr. 54:165–186.

Stamps, J. 1977. Rainfall, moisture and dry season growth rates in *Anolis aeneus*. Copeia 1977:415–419.

Stanley, S. M. 1975. A theory of evolution above the species level. Proc. Natl. Acad. Sci. USA 72:646–650.

———. 1979. Macroevolution: pattern and process. Freeman, San Francisco.

Stark, N., and C. Jordan. 1978. Nutrient retention by the root mat of an Amazonian rain forest. Ecology 59:434–437.

Staton, M. A. 1978. "Distress calls" of crocodilians—whom do they benefit? Am. Nat. 112(984): 327–332.

Stearns, S. 1976. Life history tactics: A review of ideas. Q. Rev. Biol. 51:3–47.

Stebbins, G. L. 1950. Variation and evolution in plants. Columbia University Press, New York.

———. 1974. Flowering plants: evolution above the species level. Harvard Univ. Press, Cambridge.

Steele, J. H. 1974. The structure of marine ecosystems. Harvard University Press, Cambridge, Mass.

Steenbergh, W. F., and C. H. Lowe. 1969. Critical factors during the first years of life of the saguaro (*Cereus giganteus*) at Saguaro National Monument, Arizona. Ecology 5:825–834.

Stephens, M. L. 1982. Mate takeover and possible infanticide by a female northern jacana (*Jacana spinosa*). Anim. Behav. 30:1253–1254.

Stephenson, T. A., and A. Stephenson. 1972. Life between tidemarks on rocky shores. Freeman, San Francisco.

Sternburg, J. G., G. P. Waldbauer, and M. R. Jeffords. 1977. Batesian mimicry: selective advantage of color pattern. Science 195:681–683.

Stiles, F. G. 1973. Food supply and the annual cycle of the Anna hummingbird. Univ. Calif. Publ. Zool. 97:1–109.

———. 1975. Ecology, flowering phenology and hummingbird pollination of some Costa Rican *Heliconia* species. Ecology 56:285–301.

———. 1977. Coadapted competitors: the flowering sensors of hummingbird pollinated plants in a tropical forest. Science 198:1177–1178.

Stiling, P. D., and D. R. Strong. 1983. Weak competition among *Spartina* stem borers, by means of murder. Ecology 64:770–778.

Stimson, J. 1970. Territorial behavior of the owl limpet, *Lottia gigantea*. Ecology 51:113–118.

Stott, P. 1981. Historical plant geography: an introduction. Allen and Unwin, London.

Strahler, A. N., and A. H. Strahler. 1973. Environmental geoscience: interaction between natural systems and man. Hamilton, Santa Barbara, Calif.

Streit, B. 1976. Energy flow in four different field populations of *Ancylus fluviatilis* (Gastropoda-Basommatophora). Oecologia (Berl.) 22:261–273.

———. 1978. Zur Strategie der Besiedlung, de Konsumption und des Wachstums bei benthiscen Primarkonsumentenpopulationen. Pages 223–232. Verhandlungen der Gesellschaft fur Okologie, Kiel 1977.

Strobeck, C. 1973. N species competition. Ecology 54:650–654.

Strong, D. R. 1983. Natural variability and the manifold mechanisms of ecological communities. Am. Nat. 122:636–660.

———. 1984. Banana's best friend. Nat. Hist. Dec.:51–57.

Strong, D. R., E. D. McCoy, and J. R. Rey. 1977. Time and the number of herbivore species: the pests of sugar cane. Ecology 58:167–175.

Strong, D. R., D. Simberloff, L. G. Abele, and A. B. Thistle. 1984. Ecological communities: conceptual issues and the evidence. Princeton University Press, Princeton, N.J.

Strong, D. R., L. Szyska, and D. Simberloff. 1979. Tests of community-wide character displacement against null hypotheses. Evolution 33:897–913.

Strong, D. R., Jr. 1974. Rapid asymptotic species accumulation in phyrophagous insect communities: the pests of Cacao. Science 185:1064–1066.

Suchanek, T. 1978. The ecology of *Mytilus edulis* L. in exposed rocky intertidal communities. J. Exp. Mar. Biol. Ecol. 31:105–120.

Sudzuki, F. 1969. Absorcion foliar de humedad atmosferica en tamarugo, *Prosopis tamarugo*. Phil. Universidad de Chile, Facultad de Agronomia, Boletin Tecnico 30:1–23.

Sugihara, A. 1982. Holes in niche space: A derived assembly rule and its relation to intervality. Pp. 25–37 in: D. L. DeAngelis, W. M. Post, and A. Sugihara, eds. Current trends in food web theory. Document ORNL-5983. Oak Ridge National Laboratory, Oak Ridge, Tennessee.

Sugihara, G. 1982. Niche Hierarchy: structure, organization, and assembly in natural communities. Ph.D. thesis. Princeton University, Princeton, New Jersey.

Sullivan, W. 1974. Continents in motion: the new Earth debate. McGraw-Hill, New York.

Sumich, J. L. 1980. An introduction to the biology of marine life. 2d ed. Brown, Dubuque, Iowa.

Sutherland, J. P. 1974. Multiple stable points in natural communities. Am. Nat. 108:859–873.

Sutherland, J. P., and R. H. Karlson. 1977. Development and stability of the fouling community at Beaufort, North Carolina. Ecol. Monogr. 47:425–446.

Sutton, S. L., T. C. Whitmore, and A. C. Chadwick, eds. 1983. Tropical rain forest: ecology and management. Blackwell, Oxford.

Swank, W. T., and J. E. Douglass. 1977. Nutrient budgets for undisturbed and manipulated hardwood forest ecosystems in the mountains of North Carolina. Pages 343–364 *in* D. L. Correll, ed. Watershed research in eastern North America. Vol. I. Smithsonian Institution, Edgewater, Md.

Swank, W. T., and J. B. Waide. 1980. Interpretation of nutrient cycling research in a management context: evaluating potential effects of alternative management strategies on site productivity. Pages 137–157 *in* R. H. Waring, ed. Forests: fresh perspectives from ecosystem analysis. Proc. 40th Annu. Biol. Coll. Oregon State University Press, Corvallis.

Symons, D. 1980. The evolution of human sexuality. Oxford, New York.

Tande, G. F. 1979. Fire history and vegetation pattern of coniferous forests in Jasper National Park, Alberta. Can. J. Bot. 1912–1931.

Tanner, J. T. 1966. Effects of population density on growth rates of animal populations. Ecology 47:733–745.

Tansley, A. G. 1914. Presidential address. The British Ecological Society. J. Ecol. 2:194–202.

———. 1917. On competition between *Galium saxatile* L. (*G. hercynicum* Weig.) and *Galium sylvestre* Poll. (*G. asperum* Schreb.) on different types of soil. J. Ecol. 5:173–179.

———. 1935. The use and abuse of vegetational concepts and terms. Ecology 16:284–307.

Tansley, A. G., and R. S. Adamson. 1925. Studies on the vegetation of the English chalk. III. The chalk grasslands of the Hampshire–Sussex border. J. Ecol. 13:177–223.

Teal, J. M. 1962. Energy flow in the salt marsh ecosystem of Georgia. Ecology 43:614–624.

Templeton, J. R. 1970. Reptiles. Pages 167–221 *in* G. C. Whittow, ed. Comparative physiology of thermoregulation. Academic Press, New York.

Tenaza, R. R. 1975. Territory and monogamy among Kloss' gibbons (*Hylobates klossi*) in Siberut Island, Indonesia. Folia Primatol. 24:60–80.

———. 1976. Songs, choruses, and countersinging of Kloss' gibbons (*Hylobates klossi*) in Siberut Island, Indonesia. Zeit. f. Tierpsych. 40:37–52.

Terborgh, J. 1971. Distribution on environmental gradients: theory and a preliminary interpretation of distributional patterns in the avifauna of the Cordillera Vilcabamba, Peru. Ecology 52:23–40.

Terry, K. L. 1982. Nitrate uptake and assimilation in *Thalassiosira weissflogii* and *Phaeodactylum tricornutum*: interactions with photosynthesis and with the uptake of other ions. Mar. Biol. 69:21–30.

Thomas, A. S. 1945. The vegetation of some hillsides in Uganda. J. Ecol. 33:10–43.

Thompson, J. H., and M. F. Willson. 1978. Disturbance and the dispersal of fleshy fruits. Science 200:1161–1163.

———. 1979. Evolution of temperate fruit/bird interactions: phenologyical strategies. Evolution 33(3):973–982.

Thompson, J. N. 1981. Reversed animal-plant interactions: the evolution of insectivorous and ant-fed plants. Biol. J. Linn. Soc. 16(2):147–155.

———. 1983. The use of ephemeral plant parts on small host plants: how *Depressaria leptotaeniae* (Lepidoptera: Oecophoridae) feeds on *Lomatium dissectum* (Umbelliferae). Anim. Ecol. 52:281–291.

Thorne, R. F. 1973. Floristic relationships between tropical Africa and tropical America. Pages 27–47 *in* B. J. Meggers, J. E. S. Ayensu, and W. D. Duckworth, eds. Tropical forest ecosystems in Africa and South America: a comparative review. Smithsonian Institution Press, Washington, D.C.

Thornhill, R., and J. Alcock. 1983. The evolution of insect mating systems. Harvard University Press, Cambridge, Mass.

Thorson, G. 1957. Bottom communities (sublittoral or shallow shelf). Pages 461–534 *in* J. Hedgepeth, ed. Treatise on marine ecology and paleoecology. Vol. I. Geological Society of America Memoir 67:461–534.

Thresher, R. 1983. Habitat effects on reproductive success in the coral reef fish, *Acanthochromis polyacanthus* (Pomacentridae). Ecology 64:1184–1199.

Thrower, N. J. W., and D. E. Bradbury, eds. 1977. Chile-California Mediterranean scrub atlas: a comparative analysis. Dowden, Hutchinson and Ross, Stroudsburg, Pa.

Tieszen, L. L., ed. 1978. Vegetation and production ecology of an Alaskan arctic tundra. Ecological Studies. Vol. 29. Springer-Verlag, New York.

Tilman, D. 1977. Resource competition between planktonic algae: an experimental and theoretical approach. Ecology 58:338–348.

———. 1982. Resource competition and community structure. Princeton University Press, Princeton, N.J.

———. 1986. Evolution and differentiation in terrestrial plant communities: the importance of the soil resource: light gradient. Pages 359–380 *in* J. Diamond and T. Case, eds. Community ecology. Harper & Row, New York.

Tinbergen, N. 1951. The study of instinct. Clarendon, London.

———. 1957. The functions of territory. Bird Study 4:14–27.

———. 1964. The evolution of signaling devices. Pages 206–230 *in* W. Etkin, ed. Social behavior and organization among vertebrates. University of Chicago Press, Chicago.

Tinbergen, N., and A. C. Perdeck. 1950. On the stimulus situation releasing the begging response in the newly hatched herring gull chick (*Larus argentatus* Pont.). Behaviour 3:1–39.

Tinkle, D. W., and R. E. Ballinger. 1972. *Sceloporus undulatus*: a study of the intraspecific comparative demography of a lizard. Ecology 53:570–584.

Toft, K., and T. Schoener. 1983. Abundance and diversity of orb spiders on 106 Bahamian islands: biogeography at an intermediate trophic level. Oikos 41:411–426.

Toksöz, M. N. 1975. The subduction of the lithosphere. Sci. Am. November:88–98.

Topp, R. W. 1969. Interoceanic sea-level canal: effects on the fish faunas. Science 165:1324–1327.

Toumey, J. W., and R. Keinholz. 1931. Trenched plots under canopies. Yale Univ. School Forest. Bull. 30.

Tracy, C. R. 1982. Biophysical modeling in reptilian physiology and ecology. Pages 275–321 *in* C. Gans and F. Pough, eds. Biology of the reptilia. Vol. 12.

Treat, A. E. 1954. A new gamasid inhabiting the tympanic organs of phalaenid moths. J. Parasit. 40:619–631.

———. 1957. Unilaterality in infestations of the moth ear mite. J. New York Entomol. Soc. 65: 411–450.

———. 1960. Experimental control of ear choice in the moth ear mite. Proc. 11th Int. Cong. Entomol. 1:619–621.

Treherne, J. E., and W. A. Foster. 1982. Group size and antipredator strategies in a marine insect. Anim. Behav. 32:536–542.

Trivers, R. L. 1972. Parental investment and sexual selection. Pages 136–179 *in* B. Campbell, ed. Sexual selection and the descent of man. Aldine, Chicago.

Trivers, R. L., and H. Hare. 1976. Haplodiploidy and the evolution of social insects. Science 191:249–263.

Trivers, R. L., and D. E. Willard. 1973. Natural selection of parental ability to vary the sex ratio of offspring. Science 179:90–92.

Tsumori, A. 1967. Newly acquired behavior and social interactions of Japanese monkeys. Pages 207–219 *in* S. A. Altmann, ed. Social communication among primates. Univ. of Chicago Press, Chicago.

Tuljapukar, S. D., and S. H. Orzack. 1980. Population dynamics in variable environments. I. Long-run growth rates and extinction. Theor. Popul. Biol. 18:314–342.

Turner, F. B. 1970. The ecological efficiency of consumer populations. Ecology 51:741–742.

Turner, J. 1977. Effect of nitrogen availability on nitrogen cycling in a Douglas-fir stand. Forest Sci. 23:307–316.

Turner, J. R. G. 1977. Butterfly mimicry: The genetical evolution of an adaptation. Evol. Biol. 10:163–206.

———. 1981. Adaptation and evolution in *Heliconius*: a defense of neo-Darwinism. Annu. Rev. Ecol. Syst. 12:99–121.

Turner, R. M., S. M. Alcorn, and G. Olin. 1969. Mortality of transplanted saguaro seedlings. Ecology 5:835–844.

Twitty, V., D. Grant, and O. Anderson. 1964. Long distance homing in the newt, *Taricha rivularis*. Proc. Natl. Acad. Sci. USA 51:51–58.

Udovic, D., and C. Aker. 1981. Fruit abortion and regulation of fruit number in Yucca whipplei. Oecologia 49:245–248.

Ulbrich, J. 1930. Die Bisamratte: Lebensweise, Gang ihrer Ausbreitung in Europa, wirschaftliche Bedeutung und Bekampfung. Dresden.

Ulrich, B., R. Mayer, and P. K. Khanna. 1980. Chemical changes due to acid precipitation in a loess-derived soil in central Europe. Soil Sci. 130:193–199.

Underwood, A., E. Denley, and M. Moran. 1983. Experimental analyses of the structure and dynamics of mid-shore rocky intertidal communities in New South Wales. Oecologia 56:202–219.

United Nations. 1977. Desertification: its causes and consequences. Pergamon Press, Oxford.

United States Department of Agriculture. 1975. Soil taxonomy. Handbook 436. U.S. Department of Agriculture, Soil Conservation Service. Washington, D.C.

Uyeda, S. 1978. The new view of the Earth: moving continents and moving oceans. Freeman, San Francisco.

Uyenoyama, M. 1984a. On the evolution of parthenogenesis: a genetic representation of the "cost of meiosis." Evolution 38:87–102.

———. 1984b. Inbreeding and the evolution of altruism under kin selection: effects on relatedness and group structure. Evolution 38:778–795.

Vance, R. R. 1973. On reproductive strategies in marine benthic invertebrates. Amer. Natur. 107:339–352.

Vandermeer, J. 1969. The competitive structure of communities: an experimental approach with protozoa. Ecology 50:362–371.

Vander Meer, R. K., and D. P. Wojcik. 1982. Chemical mimicry in the myrmecophilous beetle *Myrmecaphodius excavaticollis*. Science 218:806–808.

van der Pijl, L. 1972. Principles of dispersal in higher plants. Springer-Verlag. Berlin.

van der Pijl, L., and C. H. Dodson. 1966. Orchid flowers: their pollination and evolution. University of Miami Press, Coral Gables, Fla.

van der Valk, A. G. 1981. Succession in wetlands: a gleasonian approach. Ecology 62:688–696.

van der Valk, A. G., and C. B. Davis. 1978. The role of seed banks in the vegetation dynamics of prairie glacial marshes. Ecology 59:322–335.

Van Lawick-Goodall, J., and H. van Lawick. 1966. Use of tools by the Egyptian vulture, *Neophron peronopterus*. Nature 212:1468–1469.

van Noordwijk, A. J., J. H. van Balen, and W. Scharloo. 1980. Heritability of ecologically important traits in the great tit. Ardea 68:193–203.

Van Valen, L. 1973. A new evolutionary law. Evol. Theor. 1:1–30.

van Voris, P., R. V. O'Neill, W. R. Emanuel, and H. H. Shugart, Jr. 1980. Functional complexity and ecosystem stability. Ecology 61:352–360.

Varley, G. C. 1947. The natural control of population balance in the knap-weed gall-fly (*Urophora jaceana*). J. Anim. Ecol. 16:139–187.

Vaurie, C. 1951. Adaptive differences between two sympatric species of nuthatches (*Sitta*). Proc. 10th Int. Ornithol. Cong., Uppsala 1950:163–166.

Vehrencamp, S. L. 1976. The evolution of communal nesting in groove-billed anis. Ph.D. diss. Cornell University, Ithaca, N.Y.

———. 1977. Relative fecundity and parental effort in communally nesting anis *Crotophaga sulcirostris*. Science 197:403–405.

———. 1979. The roles of individual, kin and group selection in the evolution of sociality. Pages 351–394 *in* P. Marler and J. Vandenberg, eds. Social behavior and communication. Plenum, New York.

Veit, P. G. 1982. Gorilla society. Nat. Hist. March:48–59.

Vermeij, G. J. 1974. Marine faunal dominance and molluscan shell form. Evolution 28:656–664.

———. 1978. Biogeography and adaptation: patterns of marine life. Harvard University Press, Cambridge, Mass.

Verner, J. 1964. Evolution of polygamy in the long-billed marsh wren. Evolution 18:252–261.

Verner, J., and M. F. Willson. 1966. The influence of habitats on mating systems of North American passerine birds. Ecology 47:143–147.

Verrell, P. A. 1982. Male newts prefer large females as mates. Anim. Behav. 30:1254–1255.

Verwey, D. J. 1930. Coral reef studies. I. The symbiosis between damsel fishes and sea anemones in Batavia Bay. Treubia 12:305–366.

Victor, B. 1983. Recruitment and population dynamics of a coral reef fish. Science 219:419–420.

Vine, F. J., and D. H. Matthews. 1963. Magnetic anomalies over oceanic ridges. Nature 199:947–949.

Vines, G. 1982. War-games and sex-games in Britain's toads. New Scientist July 15:145–147.

Vitousek, P. M. 1977. The regulation of element concentrations in mountain streams in the northeastern United States. Ecol. Monogr. 47:65–87.

———. 1980. Nitrogen losses from disturbed ecosystems—ecological considerations. Pages 39–53 *in* T. Rosswall, ed. Nitrogen cycling in West African ecosystems. Royal Swedish Academy of Sciences, Stockholm, Sweden.

———. 1981. Clear-cutting and the nitrogen cycle. Ecol. Bull. 33:631–642.

———. 1983. Mechanisms of ion leaching in natural and managed ecosystems. Pages 129–144 *in* H. A. Mooney and M. Godron, eds. Disturbance and ecosystems: components of response. Springer-Verlag, Berlin.

Vitousek, P. M., P. R. Ehrlich, A. H. Ehrlich, and P. A. Matson. 1986. Human appropriation of the products of photosynthesis. BioSci. 36:368–373.

Vitousek, P. M., J. R. Gosz, C. C. Grier, J. M. Melillo, and W. A. Reiners. 1982. A comparative analysis of potential nitrification and nitrate mobility in forest ecosystems. Ecol. Monogr. 52:155–177.

Vitousek, P. M., and P. A. Matson. 1984. Mechanisms of nitrogen retention: a field experiment. Science 225:51–52.

———. 1985. Disturbance, nitrogen availability, and nitrogen losses in an intensively managed loblolly pine plantation. Ecology 66:1360–1376.

Vitousek, P. M., and W. A. Reiners. 1975. Ecosystem succession and nutrient retention: a hypothesis. BioSci. 25:376–331.

Vitousek, P. M., W. A. Reiners, J. M. Melillo, C. C. Grier, and J. R. Gosz. 1981. Nitrogen cycling and loss following forest perturbation: the components of response. Pages 115–127 *in* G. W. Barrett and R. Rosenberg, eds. Stress effects on natural ecosystems. Wiley-Interscience, New York.

Vitousek, P. M., and P. S. White. 1981. Process studies in succession. Pages 267–276 *in* D. C. West, H. H. Shugart, and D. B. Botkin, eds. Forest succession: concepts and application. Springer-Verlag, New York.

Vogelmann, H. W. 1982. Catastrophe on Camel's Hump. Nat. Hist. November:8–14.

Volterra, V. 1926. Variazioni e fluttuazioni del numero d'individui in specie animali conviventi. Mem. R. Accad. Naz. dei Lincei. Ser. VI, Vol. 2.

von Caemmerer, S., and G. D. Farquhar. 1981. Some relationships between the biochemistry of photosynthesis and the gas exchange of leaves. Planta 153:376–387.

Vuilleumier, F. 1967. Mixed species flocks in Patagonian forests, with remarks on interspecies flock formation. Condor 69(4):400–404.

Waage, J. K. 1979a. Dual function of the damselfly penis: sperm removal and transfer. Science 203:916–918.

———. 1979b. Adaptive significance of post-copulatory guarding of mates and nonmates by male *Calopteryx maculata* (Odonata). Behav. Ecol. Sociobiol. 6:147–154.

———. 1984. Sperm competition and the evolution of odonate mating systems. Pages 257–290 *in* R. L. Smith, ed. Sperm competition and the evolution of animal mating systems. Academic Press, Orlando, Fla.

Wade, M. J. 1976. Group selection among laboratory populations of *Tribolium*. Proc. Natl. Acad. Sci. USA 73:4604–4607.

———. 1977. An experimental study of group selection. Evolution 31:134–153.

Wade, M. J., and D. E. McCauley. 1984. Group selection: the interaction of local deme size and migration in the differentiation of small populations. Evolution 38:1047–1058.

Wainwright, C. M. 1977. Sun-tracking and related leaf movements in a desert lupine (*Lupinus arizonicus*). Am. J. Bot. 64:1032–1041.

Wainwright, S. A., W. D. Biggs, J. D. Currey, and J. M. Gosline. 1976. Mechanical design in organisms. Wiley, New York.

Walker, B. H. 1981. Is succession a viable concept in African savanna ecosystems? Pages 431–447 *in* D. C. West, H. H. Shugart, and D. B. Botkin, eds. Forest succession: concepts and application. Springer-Verlag, New York.

Walker, I. 1967. Effects of population density on the viability and fecundity in *Nasonia vitripennis* Walker (Hymenoptera, Pteromalidae). Ecology 48:294–301.

Walker, J., C. H. Thompson, I. F. Fergus, and B. R. Tunstall. 1981. Plant succession and soil development in coastal sand dunes of subtropical eastern Australia. Pages 107–131 *in* D. C. West, H. H. Shugart, and D. B. Botkin, eds. Forest succession: concepts and application. Springer-Verlag, New York.

Walker, T. W., and J. K. Syers. 1976. The fate of phosphorus during pedogenesis. Geoderma 15:1–9.

Wallace, A. R. 1869. The Malay archipelago: the land of the orangutan and the bird of paradise. Harper & Row, New York.

———. 1876. The geographical distribution of animals. Vols. I, II. Macmillan, London.

———. 1880. Island life, or the phenomena and causes of insular floras and faunas. Macmillan, London.

Wallace, B. 1963. The elimination of an autosomal lethal from an experimental population of *Drosophila melanogaster*. Am. Nat. 97:65–66.

———. 1981. Basic population genetics. Columbia University Press, New York.

Waloff, N., and O. W. Richards. 1977. The effect of insect fauna on growth, mortality and natality of broom, *Sarothamnus scoparius*. J. Appl. Ecol. 14:787–798.

Walter, H. 1979. Eleonora's falcon: adaptations to prey and habitat in a social raptor. University of Chicago Press, Chicago.

Wang, Chi-Wu. 1961. The forests of China, with a survey of grassland and desert vegetation. Maria Moors Cabot Foundation, Harvard University, Publ. 5:1–313.

Ward, P., and A. Zahavi. 1973. The importance of certain assemblages of birds as "information-centres" for food-finding. Ibis 115:517–534.

Waring, R. H. 1980. Vital signs of forest ecosystems. Pages 131–135 *in* R. H. Waring, ed. Forests: fresh perspectives from ecosystem analysis. Proc. 40th Annu. Bio. Colloq. Oregon State University Press, Corvallis.

Warner, R. R. 1978. Evolution of hermaphroditism and unisexuality in aquatic and terrestrial vertebrates. Pages 77–101 *in* E. S. Reese and F. J. Lighter, eds. Contrasts in behavior: adaptations in the aquatic and terrestrial environments. Wiley, New York.

———. 1982. Metamorphosis. Science 82 December:43–46.

Warner, R. R., and S. G. Hoffman. 1980. Population density and the economics of territorial defense in a coral reef fish. Ecology 61(4):772–780.

Waser, N. M. 1978. Competition for hummingbird pollination and sequential flowering in two Colorado wildflowers. Ecology 59:934–944.

———. 1983. Competition for pollination and floral character differences among sympatric plant species: a review of evidence. Pages 277–293 *in* C. D. Jones and R. J. Little, eds. Handbook of experimental pollination biology. Van Nostrand Reinhold, New York.

Waser, N. M., and M. V. Price. 1983. Optimal and actual outcrossing in plants and the nature of the plant–pollinator interaction. Pages 341–359 *in* C. D. Jones and R. J. Little, eds. Handbook of experimental pollination biology. Van Nostrand-Reinhold, New York.

Waser, N. M., and L. A. Real. 1979. Effective mutualism between sequentially flowering plant species. Nature 281:670–672.

Watt, A. S. 1933. Tree roots and the field layer. J. Ecol. 21:404–414.

———. 1947. Patterns and process in the plant community. J. Ecol. 35:1–22.

———. 1955. Bracken versus heather, a study in plant sociology. J. Ecol. 43:490–506.

Watt, K. E. F. 1975. Critique and comparison of biome ecosystem modeling. Pages 139–152 *in* B. C. Patten, ed. Systems analysis and simulation in ecology. Vol. 3. Academic Press, New York.

Watt, W. B. 1968. Adaptive significance of pigment polymorphisms in *Colias* butterflies. I. Variation of melanin pigment in relation to thermoregulation. Evolution 22:437–458.

Watt, W. B., P. A. Carter, and S. M. Blower. 1985. Adaptation at specific loci. IV. Differential mating success among glycolytic allozyme genotypes of *Colias* butterflies. Genetics 109:157–175.

Watt, W. B., R. Cassin, and M. Swan. 1983. Adaptation at specific loci. III. Field behavior and survivorship differences among Colias PGI genotypes are predictable from in vitro biochemistry. Genetics 103:725–739.

Watt, W. B., P. C. Hoch, and S. G. Mills. 1974. Nectar resource use by *Colias* butterflies: chemical and visual aspects. Oecologia 14:353–374.

Watt, W. D., and F. R. Hayes. 1963. Tracer study of the phosphorus cycle in sea water. Limnol. Oceanogr. 8:276–285.

Weatherhead, P. J., and R. J. Robertson. 1977. Harem size, territory quality, and reproductive success in the redwinged blackbird (*Agelaius phoeniceus*). Can. J. Zool. 55:1261–1267.

———. 1979. Offspring quality and the polygyny threshold: 'the sexy son hypothesis.' Am. Nat. 113:201–208.

———. 1981. In defense of the "sexy son" hypothesis. Am. Nat. 117:349–356.

Webb, S. D. 1977–1978. A history of savanna vertebrates in the New World. Parts I, II. Annu. Rev. Eccl. Syst. 8:355–380; 9:393–426.

Webb, T., III. 1981. The past 11,000 years of vegetational change in eastern North America. BioSci. 31:501–506.

Wells, K. D. 1977. The social behavior of anuran amphibians. Anim. Behav. 25:666–693.

Welty, J. C. 1979. The life of birds. 2d ed. Saunders, Philadelphia.

Went, F. W. 1975. Water vapor absorption in *Prosopis*. Pages 67–75 *in* F. J. Vernberg, ed. Physiological adaptation to the environment. Intext, New York.

Werner, E. 1977. Species packing and niche complementarity in three sunfishes. Am. Nat. 111:553–578.

Werner, E., and D. Hall. 1974. Optimal foraging and the size selection of prey by the bluegill sunfish (*Lepomis macrochirus*). Ecology 55:1042–1052.

———. 1976. Niche shifts in sunfishes: experimental evidence and significance. Science 191:404–406.

———. 1977. Competition and habitat shift in two sunfishes (*Centrarchidae*). Ecology 58:869–876.

Werner, E., J. Gilliam, D. Hall, and G. Mittelbach. 1983. An experimental test of the effects of predation risk on habitat use in fish. Ecology 64:1540–1548.

Werner, E., D. Hall, D. Laughlin, D. Wagner, L. Wilsmann, and F. Funk. 1977. Habitat partitioning in a freshwater fish community. J. Fish. Res. Bd. Canad. 34:360–370.

West Eberhard, M. 1969. The social biology of Polistine wasps. Univ. Mich. Mus. Zool. Misc. Publ. 140:1–101.

West, D.C., H. H. Shugart, and D. B. Botkin, eds. 1981. Forest succession: concepts and application. Springer-Verlag, New York.

West-Eberhard, M. J. 1975. The evolution of social behavior by kin selection. Q. Rev. Biol. 50:1–33.

———. 1981. Intragroup selection and the evolution of insect societies. Pages 3–17 *in* R. D. Alexander and D. W. Tinkle, eds. Natural selection and social behavior: recent research and new theory. Chiron, New York.

Westman, W. E. 1978. Measuring the inertia and resilience of ecosystems. BioSci. 28:705–710.

Wheye, D., and P. Ehrlich. 1985. The use of fluorescent pigments to study insect behavior: investigating mating patterns in butterfly populations. Ecol. Entomol. 10:231–234.

White, J. 1980. Demographic factors in populations of plants. Pages 21–48 *in* O. Solbrig, ed. Demography and evolution in plant populations. University of California Press, Berkeley.

White, M. J. D. 1978. Modes of speciation. Freeman, San Francisco.

Whitham, T. G. 1977. Coevolution of foraging in *Bombus*-nectar dispensing *Chilopsis*: a last dreg theory. Science 197:593–596.

———. 1983. Host manipulation of parasites: within-plant variation as a defense against rapidly evolving pests. Pages 15–41 *in* R. F. Denno and M. S. McClure, eds. Variable plants and herbivores in natural and managed systems. Academic Press, New York.

Whitmire, D. P., and A. A. Jackson IV. 1984. Are periodic mass extinctions driven by a distant solar companion? Nature 308:713–715.

Whitmore, T. C. 1974. Change with time and the role of cyclones in tropical rain forests on Kolombangara, Solomon Islands. Commonw. For. Inst. Pop. no. 46, Holywell, Oxford.

Whittaker, R. H. 1956. Vegetation in the Great Smoky Mountains. Ecol. Monogr. 26:1–80.

———. 1961. Experiments with radiophosphorus tracer in aquarium microcosms. Ecol. Monogr. 31:157–188.

———. 1967. Gradient analysis of vegetation. Biol. Rev. 42:207–264.

———. 1970. Communities and ecosystems. Macmillan, New York.

———. 1975. Communities and ecosystems. 2d ed. Macmillan, New York.

Whittaker, R. H., and G. E. Likens. 1973. Introduction to the primary production of the biosphere. Human Ecol. 1:301–369.

Whittaker, R. H., and W. A. Niering. 1965. Vegetation of the Santa Catalina Mountains, Arizona: a gradient analysis of the south slope. Ecology 46:429–452.

Whitton, B. A., ed. 1975. River ecology. Blackwell, Oxford.

Wickler, W. 1968. Mimicry in plants and animals. McGraw-Hill, New York.

Wiebes, J. J. 1979. Co-evolution of figs and their insect pollinators. Annu. Rev. Ecol. Syst. 10:1–12.

Wiens, D. 1978. Mimicry in plants. Evol. Biol. 111:365–403.

Wiens, J. 1982. On size ratios and sequences in ecological communities: are there no rules? Annales Zoologici Fennici 19:297–308.

Wiklund, C. 1974. The concept of oligophagy and the natural habitats and host plants of *Papilio machaon.* Entomol. Scanl. 5:151–160.

———. 1977. Oviposition, feeding and spatial separation of breeding and foraging habitats in a population of *Leptidea sinapis* (Lepidoptera). Oikos 28:56–68.

———. 1982a. Behavioural shift from courtship solicitation to mate avoidance in female ringlet butterflies (*Aphantopus hyperanthus*) after copulation. Anim. Behav. 30:790–793.

———. 1982b. Generalist versus specialist utilization of host plants among butterflies. Proc. Int. Symp. Insect–Plant Relationships 5:181–191.

Wiklund, C., and C. Ahrberg. 1978. Host plants, nectar source plants, and habitat selection of males and females of *Anthocharis cardamines* (Lepidoptera). Oikos 31:169–183.

Wilbur, H. 1972. Competition, predation and the structure of the *Ambystoma–Rana sylvatica* community. Ecology 53:3–20.

Wilbur, H. M., D. W. Tinkle, and J. P. Collins. 1974. Environmental certainty, trophic level and resource availability in life history evolution. Am. Nat. 108:805–817.

Wilcove, D. S. 1985. Nest predation in forest tracts and the decline of migratory songbirds. Ecology 66:1211–1212.

Wiley, R. H. 1973. Territoriality and non-random mating in sage grouse, *Centrocerous urophasianus.* Anim. Behav. Monogr. 6:89–169.

———. 1974. Evolution of social organization and life-history patterns among grouse. Q. Rev. Biol. 49:201–227.

———. 1978. The lek mating system of the sage grouse. Sci. Am. May:114–116.

Wille, A. 1963. Behavioral adaptations of bees for pollen collection from *Cassia* flowers. Rev. Biol. Trop. 11(2):205–210.

Williams, D. 1980. Dynamics of the pomacentrid community on small patch reefs in One Tree Island lagoon (Great Barrier Reef). Bull. Mar. Sci. 30:159–170.

Williams, E. E. 1969. The ecology of colonization as seen in the zoogeography of anoline lizards on small islands. Q. Rev. Biol. 44:345–389.

———. 1983. Ecomorphs, faunas, island size and diverse end points. Pages 326–370 *in* R. Huey, E. Pianka, and T. Schoener, eds. Lizard ecology: studies on a model organism. Harvard University Press, Cambridge, Mass.

Williams, G. C. 1957. Pleiotropy, natural selection and the evolution of senescence. Evolution 11:398–411.

———. 1964. Patterns in the balance of nature and related problems in quantitative ecology. Academic Press, New York.

———. 1975. Sex and evolution. Princeton University Press, Princeton, N.J.

———. 1978. Mysteries of sex and recombination. Q. Rev. Biol. 53:287–289.

Williams, G. C., and J. Mitton. 1973. Why reproduce sexually? J. Theor. Biol. 39:545–554.

Williams, K. S., and L. E. Gilbert. 1981. Insects as selective agents on plant vegetative morphology: egg mimicry reduces egg laying by butterflies. Science 212:467–469.

Williams, K. S., D. E. Lincoln, and P. R. Ehrlich. 1983a. The coevolution of *Euphydryas chalcedona* butterflies and their larval host plants. I. Larval feeding behavior and host plant chemistry. Oecologia 56:323–329.

———. 1983b. The coevolution of *Euphydryas chalcedona* butterflies and their larval host plants. II. Maternal and host plant effects on larval growth, development and food-use efficiency. Oecologia 56:330–335.

Williams, N. H. 1983. Floral fragrances as cues in animal behavior. Pages 50–72 *in* C. E. Jones and R. J. Little, eds. Handbook of experimental pollination biology. Van Nostrand Reinhold, New York.

Williams, W., and J. Lambert. 1959. Multivariate methods in plant ecology. I. Association-analysis in plant communities. J. Ecol. 47:83–101.

———. 1961. Multivariate methods in plant ecology. III. Inverse association analysis. J. Ecol. 49:717–729.

Williams, W., G. Lance, L. Webb, J. Tracey, and M. Dale. 1969. Studies in the numerical analysis of complex rain-forest communities. III. The analysis of successional data. J. Ecol. 57:515–535.

Williamson, M. 1983. The land-bird community of Skokholm: ordination and turnover. Oikos 41:378–384.

Willis, M. A., and M. C. Birch. 1982. Male lek formation and female calling in a population of the arctiid moth *Estigmene acrea.* Science 218:168–170.

Willson, M. F., and J. H. Thompson. 1982. Phenology and ecology of color in bird-dispersed fruits, or why some fruits are red when they are "green." Can. J. Bot. 60(5):701–713.

Wilson, D. S. 1980. The natural selection of populations and communities. Benjamin/Cummings, Menlo Park, Calif.

Wilson, E. O. 1961. The nature of the taxon cycle in the Melanesian ant fauna. Am. Nat. 95:169–193.

———. 1971. The insect societies. Belknap, Cambridge, Mass.

———. 1975. Sociobiology: the new synthesis. Belknap, Cambridge, Mass.

Wilson, E. O., and E. O. Willis. 1975. Applied biogeography. Pages 522–534 *in* M. L. Cody and J. Diamond, eds. Ecology and evolution of communities. Harvard University Press, Cambridge, Mass.

Wilson, R. E., and E. L. Rice. 1968. Allelopathy as expressed by *Helianthus annuus* and its role in old-field succession. B. Torrey Bot. Club. 95:432–448.

Wilson, W. 1983. The role of density dependence in a marine infaunal community. Ecology 64:295–306.

Wingate, D. 1965. Terrestrial herpetofauna of Bermuda. Herpetologica 21:202–218.

Wise, D. H. 1981. A removal experiment with darkling beetles: lack of evidence for interspecific competition. Ecology 62:727–738.

Wittenberger, J. F. 1981a. Male quality and polygyny: the 'sexy son' hypothesis revisited. Am. Nat. 117:329–342.

———. 1981b. Time: A hidden dimension in the polygyny threshold model. Am. Nat. 118:803–822.

Wolda, H., and R. Foster. 1978. *Zunacetha annulata* (Lepidoptera: Dioptidae), an outbreak insect in a tropical forest. Geo. Eco. Trop. 2:443–454.

Wolf, L. L., and F. G. Stiles. 1970. Evolution of pair cooperation in a tropical hummingbird. Evolution 24:759–773.

Wood, T., F. H. Bormann, and G. K. Voigt. 1984. Phosphorus cycling in a northern hardwood forest: biological and chemical control. Science 223:391–393.

Woodin, S. 1976. Adult-larval interactions in dense infaunal assemblages: patterns of abundance. J. Mar. Res. 34:25–41.

Woodmansee, R. G. 1978. Additions and losses of nitrogen in grassland ecosystems. BioSci. 28:448–453.

Woodmansee, R. G., and L. S. Wallach. 1981. Pages 379–400 *in* H. A. Mooney, T. M. Bonnicksen, N. L. Christensen, J. E. Lotan, and W. A. Reiners, eds. Fire regimes and ecosystem properties. U.S. Forest Service General Technical Report WO-26.

Woodwell, G. M. 1967. Radiation and the patterns of nature. Science 156:461–470.

———. 1970. Effects of pollution on the structure and physiology of ecosystems. Science 168:429–433.

———. 1974. Success, succession, and Adam Smith. BioSci. 24:81–87.

———. 1979. Leaky ecosystems: nutrient fluxes and succession in the pine barren vegetation. Pages 333–343 *in* R. T. T. Forman, ed. Pine barrens, ecosystem and landscape. Academic Press, New York.

———. 1983. The blue planet: of wholes and parts and man. Pages 2–10 *in* H. A. Mooney and M. Godron, eds. Disturbance and ecosystems: components of response. Springer-Verlag, Berlin.

Woodwell, G. M., and A. L. Rebuck, 1967. Effects of chronic gamma radiation on the structure and diversity of an oak-pine forest. Ecol. Monogr. 37:53–69.

Woolfenden, G. E. 1973. Nesting and survival in a population of Florida scrub jays. Living Bird 12:25–49.

———. 1975. Florida scrub jay helpers at the nest. Auk 92:1–15.

Woolfenden, G. E., and J. W. Fitzpatrick. 1984. The Florida scrub jay: demography of a cooperative-breeding bird. Princeton University Press, Princeton, N.J.

Wright, H. E. Jr. 1974. Landscape development, forest fires and wilderness management. Science 186:487–495.

Wrigley, G. 1982. Tropical agriculture: the development of production. 4th ed. Longman, London.

Wynne-Edwards, V. C. 1962. Animal dispersion in relation to social behaviour. Oliver and Boyd, Edinburgh.

Yarie, J. 1980. The role of understory vegetation in the nutrient cycle of forested ecosystems in the mountain hemlock biogeoclimatic zone. Ecology 61:1498–1514.

Yodzis, P. 1980. The connectance of real ecosystems. Nature 284:544–545.

Yonge, C. M. 1949. The sea shore. (Reprinted 1963.) Atheneum, New York.

Zaret, T. 1980. Predation and freshwater communities. Yale University Press, New Haven, Conn.

Zimmerman, M. 1979. Optimal foraging: a case for random movement. Oecologia 43:261–267.

Zimmerman, P., J. Greenberg, S. Wandiga, and P. Crutzen. 1982. Termites: a potentially large source of atmospheric methane, carbon dioxide, and molecular hydrogen. Science 218:563–565.

INDEX

Bold face numbers represent illustrations.
Asterisk (*) indicates terms listed in glossary.